PROGRESS IN

Nucleic Acid Research and Molecular Biology

Volume 71

PROGRESS IN
Nucleic Acid Research and Molecular Biology

edited by

KIVIE MOLDAVE

Department of Molecular Biology and Biochemistry
University of California, Irvine
Irvine, California

Volume 71

ACADEMIC PRESS
An imprint of Elsevier Science

Amsterdam Boston London New York Oxford Paris
San Diego San Francisco Singapore Sydney Tokyo

This book is printed on acid-free paper.

Copyright © 2002, Elsevier Science (USA).

All Rights Reserved.
No part of this publication may be reproduced or transmitted in any form or by any means, electronic or mechanical, including photocopy, recording, or any information storage and retrieval system, without permission in writing from the Publisher.

The appearance of the code at the bottom of the first page of a chapter in this book indicates the Publisher's consent that copies of the chapter may be made for personal or internal use of specific clients. This consent is given on the condition, however, that the copier pay the stated per copy fee through the Copyright Clearance Center, Inc. (222 Rosewood Drive, Danvers, Massachusetts 01923), for copying beyond that permitted by Sections 107 or 108 of the U.S. Copyright Law. This consent does not extend to other kinds of copying, such as copying for general distribution, for advertising or promotional purposes, for creating new collective works, or for resale. Copy fees for pre-2002 chapters are as shown on the title pages. If no fee code appears on the title page, the copy fee is the same as for current chapters.
0079-6603/2002 $35.00

Explicit permission from Academic Press is not required to reproduce a maximum of two figures or tables from an Academic Press chapter in another scientific or research publication provided that the material has not been credited to another source and that full credit to the Academic Press chapter is given.

Academic Press
An imprint of Elsevier Science
525 B Street, Suite 1900, San Diego, California 92101-4495, USA
http://www.academicpress.com

Academic Press
84 Theobalds Road, London WC1X 8RR, UK
http://www.academicpress.com

International Standard Book Number: 0-12-540071-3

PRINTED IN THE UNITED STATES OF AMERICA
02   03   04   05   06   07   MM   9   8   7   6   5   4   3   2   1

Contents

SOME ARTICLES PLANNED FOR FUTURE VOLUMES xi

DNA Modifications by Antitumor Platinum and Ruthenium Compounds: Their Recognition and Repair 1
Viktor Brabec

I. Introduction	2
II. Current State of Knowledge on DNA Interactions of "Classical" Antitumor Cisplatin and Its Clinically Ineffective *trans* Isomer	3
III. DNA Interactions of Cisplatin Analogs	25
IV. Activation of *trans* Geometry	38
V. Polynuclear Platinum Antitumor Drugs	42
VI. Antitumor Ruthenium Compounds	49
VII. Concluding Remarks	54
References	54

AMP- and Stress-Activated Protein Kinases: Key Regulators of Glucose-Dependent Gene Transcription in Mammalian Cells? 69
Isabelle Leclerc, Gabriela da Silva Xavier, and Guy A. Rutter

I. AMP-Activated Protein Kinase	70
II. SNF1 and Glucose Repression in Yeast	71
III. AMPK and Regulation of Gene Transcription in Mammals	71
IV. Downstream Targets of AMPK and Gene Transcription	77
V. Mitogen- and Stress-Activated Protein Kinases	78
VI. Conclusions	82
References	82

Molecular Basis of Fidelity of DNA Synthesis and Nucleotide Specificity of Retroviral Reverse Transcriptases . 91
Luis Menéndez-Arias

| I. Introduction .... | 92 |
| II. The Role of Reverse Transcriptase in Retroviral Mutagenesis .... | 93 |

III. Retroviral Reverse Transcriptases 94
IV. Fidelity of Retroviral Reverse Transcriptases 97
V. Control of Fidelity at Initiation of Reverse Transcription 108
VI. Fidelity of Strand Transfer: Implications for Retroviral Recombination ... 109
VII. Contribution of Accessory Proteins to Fidelity of Reverse Transcription .. 110
VIII. Mutational Analysis of HIV-1 Reverse Transcriptase: The Effects
of Mutations on Fidelity of DNA Synthesis 112
IX. Biological Consequences of Increasing or Decreasing Fidelity 129
X. Conclusions and Future Perspectives 131
References 132

Muc4/Sialomucin Complex, the Intramembrane ErbB2 Ligand, in Cancer and Epithelia: To Protect and To Survive   149

Kermit L. Carraway, Aymee Perez, Nebila Idris, Scott Jepson, Maria Arango, Masanobu Komatsu, Bushra Haq, Shari A. Price-Schiavi, Jin Zhang, and Coralie A. Carothers Carraway

I. Membrane Mucins 150
II. Muc4/SMC Structure and Functions 153
III. Muc4/SMC Contributions to Tumor Progression 160
IV. Muc4/SMC in Simple Epithelia 163
V. Muc4/SMC in Glandular Secretory Epithelia 170
VI. Muc4/SMC in Stratified Epithelia 177
VII. Conclusions and Future Directions 179
References 180

Functions of Alphavirus Nonstructural Proteins in RNA Replication 187

Leevi Kääriäinen and Tero Ahola

I. Introduction 187
II. Replication Cycle of Alphaviruses 188
III. Alphavirus-Like Superfamily 190
IV. Replication of Alphavirus RNAs 192
V. Processing of Alphavirus Nonstructural Polyprotein P1234 197
VI. nsP1: A Unique RNA-Capping Enzyme and Membrane Anchor 198
VII. nsP2: A Multifunctional Enzyme and Regulatory Protein 204
VIII. nsP3: An Ancient Conserved Protein and Phosphoprotein 208
IX. nsP4: A Catalytic RNA Polymerase Subunit 210
X. The Replication Complex 211
References 214

The Unique Biochemistry of Methanogenesis 223
Uwe Deppenmeier

I. Introduction	224
II. Methanogens: A Unique Group of Microorganisms	225
III. Biochemistry of Methanogenesis	228
IV. Mechanism of ATP Synthesis in Methanogenic Archaea	240
V. Energy-Conserving Systems in *Methanosarcina* Strains	242
VI. Energy Conservation in Obligate Hydrogenotrophic Methanogens	270
References	274

A History of Poly A Sequences: From Formation to Factors to Function 285
Mary Edmonds

I. Introduction	287
II. From Polymerases to Poly A(+) mRNA	290
III. Sequences Required for Polyadenylation	291
IV. The Biochemistry of Polyadenylation	296
V. Cleavage/Polyadenylation Proteins	306
VI. The Core Components of Cleavage/Polyadenylation	313
VII. Cloning, Sequencing, and Expressing the Core Proteins	320
VIII. Regulation of Polyadenylation	335
IX. Polyadenylation in Yeast	351
X. Polyadenylation in *E. coli*	364
XI. Polyadenylation in *Vaccinia Virus*	375
References	381

A Growing Family of Guanine Nucleotide Exchange Factors Is Responsible for Activation of Ras-Family GTPases 391
Lawrence A. Quilliam, John F. Rebhun, and Ariel F. Castro

I. Introduction	392
II. Regulation of *in Vivo* Ras-GTP Levels by Inhibition of GTPase-Activating Proteins	394
III. Early Identification of Ras-Family GEFs	395
IV. GEF Structure and the Nucleotide Exchange Reaction	398
V. Dominant Inhibitory Ras Proteins Target GEFs	404
VI. Biological Assays for GEF Activity	406
VII. Ras-Family GEFs	407
VIII. GEFs and Disease	427

IX. Are There More GEFs in Our Future?	428
References	428

Practical Approaches to Long Oligonucleotide-Based DNA Microarray: Lessons from Herpesviruses 445

Edward K. Wagner, J. J. Garcia Ramirez, S. W. Stingley, S. A. Aguilar, L. Buehler, G. B. Devi-Rao, and Peter Ghazal

I. A Rationale for Developing DNA Microarrays for Herpesviruses	446
II. Herpes Simplex and Cytomegaloviruses—Two Herpesviruses That Share Features of Productive Infection but Differ Markedly in Patterns of Latency and Reactivation	447
III. Design Criteria for Herpesvirus DNA Microarrays	451
IV. The Construction and Validation of an Oligonucleotide-Based Hsv-1 DNA Microarray on Glass Slides	455
V. Exemplary Applications	472
VI. Conclusions	486
References	487

Sphingosine Kinases: A Novel Family of Lipid Kinases ... 493

Hong Liu, Debyani Chakravarty, Michael Maceyka, Sheldon Milstien, and Sarah Spiegel

I. Pleiotropic Functions of Sphingosine-1-Phosphate	494
II. Sphingosine Kinase and Sphingosine-1-Phosphate in Yeast and Plants	495
III. Cellular Functions of Sphingosine Kinase in Mammalian Cells	497
IV. How Is Sphingosine Kinase Activated?	498
V. Cloning of Mammalian Sphingosine Kinases	500
VI. Sphingosine Kinase Family	504
VII. Five Conserved Domains of the SPHK Superfamily	505
VIII. Phylogenetic Analysis of Sphingosine Kinases	508
IX. Concluding Remarks	508
References	509

Mechanisms of EF-Tu, a Pioneer GTPase 513

Ivo M. Krab and Andrea Parmeggiani

I. Introduction	514
II. Structure–Function Relationships	524
III. EF-Ts as a Steric Chaperone for EF-Tu Folding	529

IV.	EF-Tu as Target of Antibiotics	531
V.	Specific Aspects of EF-Tu GTPase Activity	538
VI.	Conclusions and Perspectives	542
	References	543

INDEX 553

Some Articles Planned for Future Volumes

Tandem CCCH Zinc Finger Proteins in the Regulation of mRNA Turnover
  PERRY BLACKSHEAR

Initiation and Recombination: Early and Late Events in the Replication of Herpes Simplex Virus
  PAUL E. BOEHMER

CTD Phosphatase, Role in RNA Polymerase II Cycling and the Regulation of Transcript Elongation
  MICHAEL E. DAHMUS, NICK MARSHALL, AND PATRICK LIN

The Ubiquitous Nature of RNA Chaperone Proteins
  JEAN-LUC DARLIX AND GAEL CRISTOFARI

Deoxyribonucleotide Synthesis in Anaerobic Microorganisms: The Class II Ribonucleotide Reductase
  MARC FONTECAVE AND ETIENNE MULLIEZ

Hexameric RNA To Drive Viral DNA Translocating Motor
  PEIXUAN GUO

Dynamic O-Glycosylation of Nuclear and Cytosolic Proteins: A New Paradigm for Metabolic Control of Signal Transduction and Transcription
  GERALD W. HART AND KAZUO KAMEMURA

Viral Strategies of Translation Initiation: Ribosomal Shunt and Reinitiation
  THOMAS HOHN, L. A. RYABOVA, AND MIKHAIL M. POOGGIN

DNA–Protein Interactions Involved in the Initiation and Termination of Plasmid Rolling Circle Replication
  SALEEM A. KAHN, T.-L. CHANG, M. G. KRAMER, AND M. ESPINOSA

FGF3: A Gene with a Finely Tuned Spatiotemporal Pattern of Expression during Development
  CHRISTIAN LAVIALLE

Specificity and Diversity in DNA Recognition by *E. coli* Cyclic AMP Receptor Protein
  JAMES C. LEE

EIF4A, the Godfather of DEAD-Box Helicases
  WILLIAM C. MERRICK, GEORGE ROGERS, JR., AND ANTON A. KOMAR

Transcriptional Control of Multidrug Resistance in the Yeast Saccharomyces
SCOTT MOYE-ROWLEY

Initiation of Eucaryotic DNA Replication—Replication and Mechanisms
HEINZ-PETER NASHEUER, R. SMITH, C. BAUERSCHMIDT, F. GROSSE, AND K. WEISSHART

Translational Control of Gene Expression: Role of IRESes and Consequences in Cell Transformation and Angiogenesis
ANNE-CATHERINE PRATS AND HERVE PRATS

Protein Kinase CK2-Linked Gene Expression Control
WALTER PYERIN AND KARIN ACKERMANN

Mechanisms of Basal and Kinase-Inducible Transcription Activation by CREB
PATRICK G. QUINN

Steroid Hormone Regulation of mRNA Stability
DAVID J. SHAPIRO AND ROBIN E. DODSON

HIV Transcriptional Regulation in the Context of Chromatin
ERIC VERDIN

Jasmonates and Octadecanoids—Signals in Plant Stress Responses and Development
CLAUS WASTERNACK AND BETTINA HAUSE

DNA Modifications by Antitumor Platinum and Ruthenium Compounds: Their Recognition and Repair

VIKTOR BRABEC

Institute of Biophysics
Academy of Sciences of the Czech Republic
612 65 Brno, Czech Republic

I. Introduction	2
II. Current State of Knowledge on DNA Interactions of "Classical" Antitumor Cisplatin and Its Clinically Ineffective *trans* Isomer	3
A. DNA Adducts	3
B. Conformational Alterations Induced in DNA by the Adducts	7
C. Effects of the Adducts on DNA Replication and Transcription	9
D. Cellular Resistance to Cisplatin: Repair of the Adducts	11
E. Recognition of the Adducts by Cellular Proteins	14
F. Effects of the Adducts on Telomerase and Topoisomerases	23
G. Hypotheses for Mechanism of Antitumor Activity of Cisplatin	24
III. DNA Interactions of Cisplatin Analogs	25
A. Carboplatin	25
B. Oxaliplatin	26
C. Targeted Analogs	29
D. Tetravalent Analogs	32
E. Monodentate Platinum(II) Compounds	34
IV. Activation of *trans* Geometry	38
A. Transplatin Analogs Containing Planar Amine Ligand	38
B. Transplatin Analogs Containing Imino Ether Groups	40
C. Transplatin Analogs with Asymmetric Aliphatic Amines or Cyclohexylamine Ligands	42
V. Polynuclear Platinum Antitumor Drugs	42
A. Dinuclear Compounds	44
B. Trinuclear Compounds	47
VI. Antitumor Ruthenium Compounds	49
A. Dimethyl Sulfoxide Complexes	49
B. Heterocyclic Complexes	52
C. Chloropolypyridyl Compounds	53
D. Heterodinuclear (Ruthenium, Platinum) Compounds	53
VII. Concluding Remarks	54
References	54

The development of metal-based antitumor drugs has been stimulated by the clinical success of cis-diamminedichloroplatinum(II) (cisplatin) and its analogs and by the clinical trials of other platinum and ruthenium complexes with activity against resistant tumors and reduced toxicity including orally available platinum drugs. Broadening the spectrum of antitumor drugs depends on understanding existing agents with a view toward developing new modes of attack. It is therefore of great interest to understand the details of molecular and biochemical mechanisms underlying the biological efficacy of platinum and other transition-metal compounds. There is a large body of experimental evidence that the success of platinum complexes in killing tumor cells results from their ability to form various types of covalent adducts on DNA; thus, the research of DNA interactions of metal-based antitumor drugs has predominated. The present review summarizes current knowledge on DNA modifications by platinum and ruthenium complexes, their recognition by specific proteins, and repair. It also provides strong support for the view that either platinum or ruthenium drugs, which bind to DNA in a fundamentally different manner from that of 'classical' cisplatin, have altered pharmacological properties. The present article also demonstrates that this concept has already led to the synthesis of several new unconventional platinum or ruthenium antitumor compounds that violate the original structure–activity relationships. © 2002, Elsevier Science (USA).

I. Introduction

More than 30 years have elapsed since the accidental discovery of anticancer activity of cisplatin (cis-diamminedichloroplatinum(II)) (1). Cisplatin was the first purely inorganic antitumor drug introduced in the clinic. It is a very simple molecule consisting of only 11 atoms, 6 of which are hydrogen (Fig. 1) and, in spite of its simplicity, is one of the most potent drugs available for anticancer chemotherapy (2). Cisplatin has been successfully applied in the treatment of a variety of human cancers, especially germ cell tumors, lung cancer, head and neck cancer, ovarian cancer, and bladder cancer (3, 4). Despite its success, cisplatin has several disadvantages. Its applicability is still limited to a relatively narrow range of tumors—some tumors have natural resistance to cisplatin, while others develop resistance after the initial treatment. Cisplatin also has limited solubility in aqueous solution and is administered intravenously. There are also significant problems in terms of inducing severe side effects (especially kidney damage and vomiting/nausea). Because of the drawbacks coupled with cisplatin toxicity there has been an impetus toward the development of improved platinum drugs. Broadening the chemotherapeutic arsenal depends on understanding existing agents with a view toward developing new modes of attack. It is therefore of great interest to understand the details of molecular and biochemical mechanisms underlying the biological efficacy of the platinum compounds.

The first step in the studies of molecular mechanisms of the biological activity of chemotherapeutics is to determine their primary target in cells. The pharmacological target of platinum cytostatics may be defined as a site within the cell which is altered by the drug and whose modification leads to cell death. There is a large body of experimental evidence that DNA is the critical target for the cytostatic activity of platinum compounds (5, 6) and that the success of platinum complexes in killing tumor cells results from their ability to form various types of covalent adducts on DNA. There is also some evidence suggesting that sites other than DNA within the cell may be important in the mechanism of antitumor efficiency of platinum compounds. Nevertheless, it is generally accepted that the antitumor properties of platinum-containing drugs are attributable in large measure to their coordination to DNA; thus, the research of DNA interactions of platinum drugs has predominated. This topic has recently been frequently reviewed (see, for instance Refs. 6–10). The major part of this article reviews a recent development in the area of DNA interactions of novel classes of platinum and ruthenium anticancer drugs. This development is, however, often based on comparisons to the effects of "classical" cisplatin. Hence, this review first addresses the key aspects of the current knowledge on DNA modifications by cisplatin and their processing in cells. In the following chapters, DNA interactions of cisplatin analogs and other platinum and ruthenium compounds—potential antitumor drugs—are also reviewed.

II. Current State of Knowledge on DNA Interactions of "Classical" Antitumor Cisplatin and Its Clinically Ineffective *trans* Isomer

A. DNA Adducts

Cisplatin reacts with DNA in the cell nucleus, where the concentration of chlorides is markedly lower than that in extracellular fluids. The drug loses its chloride ligands in media containing low concentrations of chloride to form positively charged monoaqua and diaqua species. It has been shown (11) that only these aquated forms bind to DNA.

1. Adducts Formed *in Vitro*

Cisplatin forms monofunctional adducts on DNA, preferentially at the guanine residues which subsequently close to intrastrand cross-links (CLs) either between neighboring guanine residues or between adenine and guanine residues (~90%) (1,2-GG- or 1,2-AG-intrastrand CLs) (12, 13). Other minor adducts are intrastrand CLs between two purine nucleotides separated by one

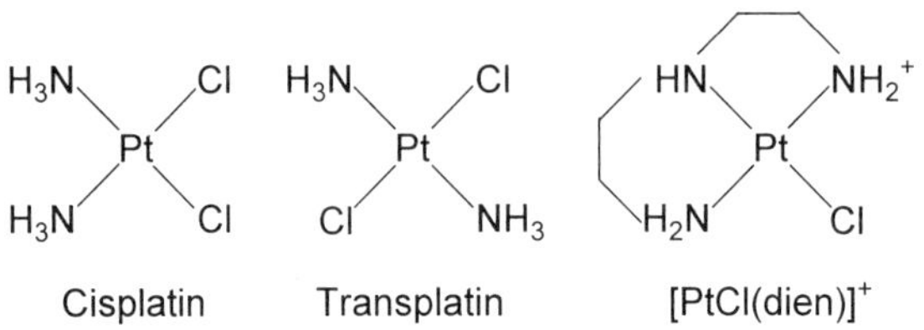

FIG. 1. Structures of "classical" platinum(II) compounds.

(1,3-GXG-intrastrand CLs, X = A, C, T) or more nucleotides, and few adducts remain monofunctional. Cisplatin also forms interstrand CLs (~6% in linear DNA), preferentially between guanine residues in the 5'-GC/5'-GC sequence (*14*). Interestingly, the frequency of interstrand CLs is noticeably higher in negatively supercoiled DNA (*15*). In all adducts, cisplatin is coordinated to the N7 atom of purine residues.

Structure–activity studies often employ inactive compounds (Fig. 1) such as *trans*-diamminedichloroplatinum(II) (*trans* isomer of cisplatin, transplatin) and chlorodiethylenetriamineplatinum(II) chloride ([PtCl(dien)]Cl, the monodentate complex, used to afford DNA monofunctional adducts to simulate the first step of the bifunctional reaction of cisplatin or transplatin) (Fig. 1) (*16*). These compounds have been widely used to investigate the mechanism of action of cisplatin. In this approach, the differences between active and inactive compounds that may be responsible for the pharmacological effect are explored. Hence, the clinical inactivity of the *trans* isomer of cisplatin is considered a paradigm for the classical structure–activity relationships of platinum drugs.

Transplatin–DNA adducts are mainly interstrand CLs (~12% in linear DNA), preferentially formed between guanine and complementary cytosine residues, and a relatively large portion (~50%) of adducts remain monofunctional (*17, 18*). Intrastrand CLs between nonadjacent guanine residues or between guanine and either adenine or cytosine residues have also been deduced from the results of the analyses of isolated DNA incubated with transplatin at a relatively high extent (1 or more Pt atoms per 100 nucleotides). However, it has been suggested (*19, 20*) that in cells transplatin forms only a small amount of interstrand CLs because of the slow closure of the monofunctional adducts coupled to their trapping by intracellular sulfur nucleophiles.

The adducts formed between double-helical DNA and [PtCl(dien)]Cl comprise mainly monofunctional adducts at N7 atoms of guanine residues (*21*).

Because the percentage of the 1,2-intrastrand adducts formed by cisplatin is larger than statistically expected, this CL has generally been assumed to be the important adduct for anticancer activity. Further support for this conclusion comes from transplatin's inability to form this type of intrastrand CL because of geometric constraints and its low antitumor activity. Both isomers can form interstrand CLs, although of distinctly different natures (14), and there is still support for interstrand CLs as the key lesions.

2. STABILITY OF THE ADDUCTS

A local distortion of the canonical conformation of DNA due to the formation of CLs by cisplatin can occur because the DNA conformation is dynamic. Double-helical DNA exists in solution in various transient and distorted conformations, which differ in the extent of, for instance, the base pair (bp) opening, the duplex unwinding, and the bending of the duplex axis (22, 23). In addition, accessibility of the binding sites, conformation of the duplex (its geometry), nucleotide sequence, electrostatic potential, flexibility, and the formation of transient reactive species can affect the DNA-binding mode of cisplatin, but each to a different extent.

In the past, the stability of the cisplatin-intrastrand and interstrand CLs was not systematically studied. It is well known that the Pt–N(7)(guanine) bond can be reverted by stronger nucleophilic ligands such as CN^- (24, 25), but it is still not entirely clear how stable this bond is in individual types of cisplatin adducts. Several examples of metastability of CLs of cisplatin have been described in recent reviews (14, 26), although further studies are required to understand the detailed mechanisms involved in the processes by means of which the Pt–N(7)(guanine) bond is ruptured.

The intrastrand CLs of transplatin are also unstable and this observation has led to an interesting application in the design of transplatin-modified oligonucleotides suitable for selective modulation of gene expression. The stable 1,3-GXG-intrastrand CLs of transplatin are readily formed in single-stranded DNA. There is only one exception as regards the stability of this CL in single-stranded DNA. If the base flanking this CL on its 5′ side is cytosine, then the 1,4-CGXG-intrastrand CL between cytosine and guanine residues is formed (27, 28). The 1,3-GXG-intrastrand CL does not prevent the formation of the duplex if the complementary strand is added, but the formation of such a duplex triggers the irreversible rearrangement of the 1,3-intrastrand CL into interstrand CL (29) (Fig. 2). The rate of this rearrangement is dependent on both the bases flanking the intrastrand CL and the nature (purine or pyrimidine) of the base complementary to the intervening base between the intrastrand cross-linked guanines. The bases involved in the interstrand CL are always the 5′ guanine of the initial 1,3-GXG-intrastrand CL and its complementary cytosine. Numerous attempts have been made to increase the rate of this rearrangement. While the replacement of the targeted DNA by RNA considerably reduces the rate of the rearrangement, the replacement of the triplet CYC (Y = A, G, T) within targeted RNA by the doublet 5′-UA or 5′-CA results in a large enhancement of the rate. In the latter case, the rearrangement is complete within a few minutes and the interstrand CLs are formed between 5′ guanine and the opposite adenine.

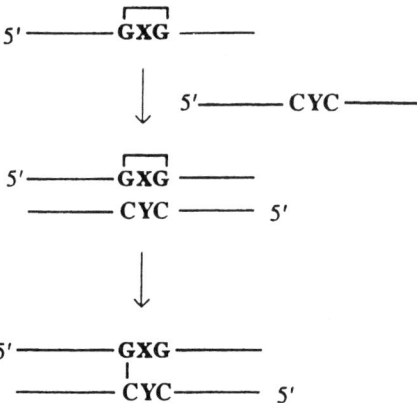

FIG. 2. Scheme of the rearrangement of 1,3-GXG-intrastrand CL of transplatin into interstrand CL. For other details, see the text. Reproduced from M. Boudvillain, R. Dalbies, and M. Leng, "Metal Ions in Biological Systems," pp. 87–104, Marcel Dekker, Inc., New York, 1996.

Importantly, the specificity of this isomerization reaction has been successfully tested in several *in vitro* and *ex vivo* systems (30–32).

3. ADDUCTS FORMED *IN VIVO*

Cisplatin adducts have also been identified on DNA, isolated from bacterial or cultured cells treated with cisplatin, from tumors and organs of cisplatin-treated, tumor-bearing experimental animals and from white blood cells of a number of cancer patients treated with the drug. Basically, similar spectrum and frequency of cisplatin adducts are found when the adducts are formed *in vitro* on isolated DNA (33). Recently, the effect of chromatin structure on cisplatin damage in intact human cells was investigated (34). Cisplatin preferentially binds to runs of consecutive guanines in intact cells; this is similar to the sequence specificity of cisplatin observed for purified DNA. Also recently, cisplatin and transplatin have been found to form interstrand CLs with nucleosomal DNA that were similar to those in free DNA (35). Interestingly, both agents react with histones, which negatively impacted nucleosome formation, but platinated DNA could be incorporated into unplatinated core particles. In addition, nucleosome appears to be a relatively minor inhibitor of DNA-interstrand cross-linking reactions. Nevertheless, while cisplatin exhibits some differences in site specificity of interstrand CLs in both the nucleosomal samples and the free DNA, transplatin does not.

Cisplatin is administered to human patients intravenously so that, on its way to the ultimate destination, platinum drugs also interact with many other biomolecules, especially those containing methionine and cysteine residues. In the blood and tissues several S-donor ligands are available for kinetic and thermodynamic competition. Inside the tumor cells, molecules like methionine (contained in peptides or proteins) and glutathione compete with bases in DNA for cisplatin.

Therefore, the effects of monomeric methionine and glutathione on reactions of cisplatin or transplatin with DNA or its monomeric nucleotides have also been analyzed. Although the kinetic reactivity of sulfur is high, the Pt–thioether bond is labile in the presence of other nucleophiles (36). Several studies revealed that 5′-GMP selectively displaces S-bound L-methionine and that the reaction of 5′-GMP with cisplatin is faster in the presence of L-methionine than in its absence (37). It is notable that thioethers such as L-methionine react with Pt(II) amines faster than thiols such as glutathione. Interestingly, intramolecular migration of [PtCl(dien)]Cl from sulfur to guanosine–N7 in S-guanosyl-L-homocysteine has also been observed (38). It has been suggested (36, 37, 39) that these results are consistent with novel routes to DNA modification by platinum drugs. However, our recent unpublished results indicate that the reaction of double-helical high-molecular-mass DNA with cisplatin is inhibited in the presence of L-methionine.

B. Conformational Alterations Induced in DNA by the Adducts

Much work has also been performed to reveal conformational distortions induced by the various DNA adducts of cisplatin, transplatin, and [PtCl(dien)]Cl. Most of the structural information currently available pertains to the bidentate adducts. The survey of the conformational changes induced by the various adducts of cisplatin, transplatin, and [PtCl(dien)]Cl described below reveals their predisposition to affect physical properties of DNA in distinctly different ways.

1. MONOFUNCTIONAL ADDUCTS

Monofunctional adducts have been studied less thoroughly, although a knowledge of alterations induced in DNA by monofunctional binding of platinum complexes may also be important for better understanding how antitumor cisplatin forms the bifunctional genotoxic lesions. Before the more systematic study of monofunctional DNA adducts of [PtCl(dien)]Cl and cisplatin was initiated, it was almost axiomatic that the monofunctional DNA adducts including those of [PtCl(dien)]Cl and cisplatin do not affect DNA conformation. However, it has been shown that the monofunctional adducts of [PtCl(dien)]Cl and cisplatin at the guanine sites distort DNA and reduce its thermal stability in a sequence-dependent manner (40–42). The most pronounced effects are observed if the platinated guanine residue is flanked by single pyrimidines on both 3′ and 5′ sides. Also importantly, the conformational distortion was always more pronounced in the flanking base pairs containing the base on the 5′ site of the monofunctional adduct. These distortions disturb stacking interactions in double-helical DNA and make bases around the adduct more accessible. In addition, the monofunctional adduct also unwinds the duplex (43)—the [PtCl(dien)]Cl–DNA structure exhibits an unwinding angle of 6°. On the other hand, no intrinsic bending is induced in DNA by monofunctional adducts of [PtCl(dien)]Cl (41).

2. CISPLATIN CROSS-LINKS

Numerous studies revealed marked conformational alterations induced in DNA by the major intrastrand CLs formed by cisplatin between adjacent purines [at d(GG) or at 5'-d(AG)] (6, 8). These adducts induce a roll of 26–50° between the platinated purine residues; a displacement of platinum atoms from the planes of the purine rings; a directional, rigid bend of the helix axis toward the major groove; a local unwinding of 9–11°; a severe perturbation of hydrogen bonding within the 5'-coordinated GC base pair; a widening and flattening of the minor groove opposite the cisplatin adduct; a creation of a hydrophobic notch; and a global distortion extending over 4–5 bp (Fig. 3A). Additional helical parameters characteristic of the A-form of DNA have also been reported, although modification of B-DNA by cisplatin inhibits its transition to the A-form (44).

The minor 1,3-intrastrand CL formed by cisplatin bends the helix axis toward a major groove by ∼30° and locally unwinds DNA (by ∼19°) (45, 46). In addition, DNA is locally denatured and flexible at the site of the adduct.

The important although less frequent interstrand CL, which is preferentially formed by cisplatin between opposite guanine residues in the 5'-GC/5'-GC sequence (25), also induces several irregularities in DNA (14, 47, 48) (Fig. 3B). The cross-linked guanine residues are not paired with hydrogen bonds to the complementary cytosines, which are located outside the duplex and not stacked with other aromatic rings. All other base residues are paired, but distortion extends over at least 4 bp at the site of the CL. In addition, the *cis*-diammineplatinum(II) bridge resides in the minor groove and the double helix is locally reversed to a left-handed, Z-DNA-like form. This adduct induces not only the helix unwinding by 76–80° relative to B-DNA but also the bending of 20–40° of the helix axis at the cross-linked site toward the minor groove. More detailed descriptions of structures of CLs of cisplatin may be found in other recent reviews (6, 8, 14, 20).

3. TRANSPLATIN CROSS-LINKS

Clinically ineffective transplatin also forms various types of adducts on DNA which affect DNA conformation less severely than the CLs of cisplatin. Monofunctional lesions, which are the major lesions, affect DNA conformation very slightly, similar to the adducts of [PtCl(dien)]Cl (see Section II,B,1). The conformational alterations induced by the interstrand CLs are, however, much less

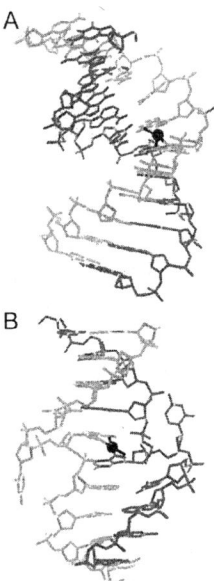

FIG. 3. (A) Crystal structure of a DNA duplex containing a 1,2-GG-intrastrand CL of cisplatin (241). Image of the duplex d(CCTCTG*G*TCTCC)·d(GGAGACCAGAGG), where the asterisks represent a cisplatin CL. The two strands are drawn in gray (the adducted strand) and black. (B) NMR structure of a DNA duplex containing an interstrand CL of cisplatin (47). Image of the duplex d(CATAG*CTATG)·d(CATAG*CTATG), where the asterisks represent a cisplatin CL. The two strands are drawn in gray and black. Adapted from *Journal of Biological Inorganic Chemistry*, Cisplatin adducts in DNA: Distortion and Recognition, D. M. J. Lilley, Vol. I, pp. 189–191, Copyright 1996, Springer Verlag.

severe than those induced by the CLs of cisplatin. The duplex is only slightly distorted on both sides of the CL, but all bases are still paired and hydrogen bonded. The interstrand CL of transplatin unwinds the double helix by $\sim 12°$ and induces a slight, flexible bending of $\sim 20°$ of its axis toward the minor groove (31, 49, 50). The 1,3-intrastrand CL of transplatin, which is less likely lesion in cells (19), unwinds the double helix by 45°, bends the DNA by 26°, and locally denatures, imparting a flexibility to the DNA which acts like a hinge joint without producing a directed bend (45, 51).

C. Effects of the Adducts on DNA Replication and Transcription

DNA replication and transcription are essential processes in rapidly proliferating tumor cells, so that their inhibition should result in cytostatic effects. As various adducts formed on DNA are capable of inhibiting DNA replication or transcription by DNA-dependent DNA or RNA polymerases, the effects of platinum adducts on these processes have been thoroughly investigated.

1. REPLICATION

Cisplatin inhibits DNA synthesis by prokaryotic and eukaryotic DNA polymerases both *in vitro* and *in vivo* (52–54). That inhibition of DNA synthesis occurs mainly at GG sites is consistent with the fact that these sites are preferential binding sites of cisplatin. On DNA, transplatin produces the arrest sites less selectively, and replication of platinated DNA is considerably less affected by the adducts of transplatin than by the lesions of cisplatin (52, 54). The monofunctional adducts of [PtCl(dien)]Cl, cisplatin, or transplatin affect replication only negligibly. Importantly, the methodology based on the use of DNA polymerases and DNA templates globally modified by platinum compounds has been used in the "replication" mapping experiments to identify the arrest sites of DNA replication and, consequently, the preferential DNA-binding sites of platinum compounds (55–58).

As demonstrated in the studies of replication of platinated DNA (54), the inhibition of DNA elongation is not complete; the polymerases are able to by-pass the adducts. The frequency of replication by-pass varies for the different polymerases. The capability of DNA polymerases to by-pass the lesions of platinum compounds is consistent with the view that these compounds are mutagenic through such replication by-pass. The mutagenic properties of drugs used in the clinic are important factors to consider, since mutagenicity may lead to the development of secondary tumors. Interestingly, the 1,2-GG-intrastrand CLs of cisplatin are considerably less mutagenic than the 1,2-AG CLs in *Escherichia coli* (59), whereas controversial mutagenicity has been reported for the 1,3-GXG-intrastrand CL (59, 60). Importantly, inhibition effects of cisplatin on DNA replication have also been observed in the *in vivo* studies (reviewed in Ref. 61). Mutagenesis induced by cisplatin *in vitro* and *in vivo* is similar and shows a correlation with DNA-binding sites of this drug. As concluded in several early studies, it would be reasonable to assume that inhibition of DNA replication by unrepaired cisplatin adducts is sufficient to explain cytostatic efficiency of cisplatin. However, it becomes evident that inhibition of DNA replication cannot fully explain the antitumor efficiency of this drug (62) and that this mechanism is more complex (63, 64). Nevertheless, there is no doubt that inhibition of DNA replication is an important part of the mechanism underlying antitumor effects of cisplatin.

2. TRANSCRIPTION

Another essential function of the cellular metabolism affected by lesions formed on DNA by platinum compounds is DNA transcription. Transcription by eukaryotic and prokaryotic RNA polymerases (RNA polymerase II and SP6 or T7 RNA polymerases, respectively) of DNA modified by cisplatin, transplatin, and [PtCl(dien)]Cl *in vitro* has been investigated in detail (*18, 25, 65–67*). The RNA polymerases react differently at the various platinum adducts. Bifunctional adducts of cisplatin strongly inhibit transcription of platinated DNA by RNA polymerases. In contrast, intrastrand CLs of transplatin do not provide an absolute block to this process so that RNA polymerase II by-passes transplatin adducts with a much higher efficiency than the adducts of cisplatin (*68*). The RNA polymerases entirely by-pass the monofunctional adducts of [PtCl(dien)]Cl (*18, 25, 68*). It has been suggested (*68*) that the platinum adducts which block the RNA polymerases not only constitute a physical barrier to the progress of the enzymes on the template but also specifically alter the properties of transcription complexes as a consequence of the specific conformational changes that they induce in template DNA. Similarly, as in the case of DNA polymerases (see Section II,C,1), the phenomenon that RNA synthesis by DNA-dependent RNA polymerases is prematurely terminated by bidentate adducts of platinum compounds has been used in the "transcription" mapping experiments to identify the arrest sites of DNA transcription and consequently also preferential DNA-binding sites of platinum compounds (*18, 25*).

The effects of cisplatin and transplatin adducts on mammalian transcription *in vivo* have also been investigated (*69–71*). For instance, a significant decrease in transcription level is observed when the β-galactosidase reporter gene is transfected into human or hamster cells (*69*). Hence, the data on inhibition of transcription by DNA adducts of platinum compounds support the view that this inhibition may also play a role in the mechanism of antitumor activity of platinum compounds.

D. Cellular Resistance to Cisplatin: Repair of the Adducts

1. MECHANISMS OF RESISTANCE

The drawbacks coupled with cisplatin toxicity are also associated with resistance of tumor cells to this drug; i.e., some tumors have natural resistance to cisplatin, while others develop resistance after the initial treatment. Resistance to cisplatin is multifactorial and in general it may consist of mechanisms either limiting the formation of DNA adducts and/or operating downstream of the interaction of cisplatin with DNA to promote cell survival. The formation of DNA adducts by cisplatin can be limited by reduced accumulation of the drug, enhanced drug efflux, and cisplatin inactivation by coordination to glutathione and thiol-containing proteins including metallothioneins, whose production may be increased as a consequence of cisplatin treatment. The second group of mechanisms includes not only enhanced repair of DNA adducts of cisplatin but also increased tolerance of the resulting DNA damage.

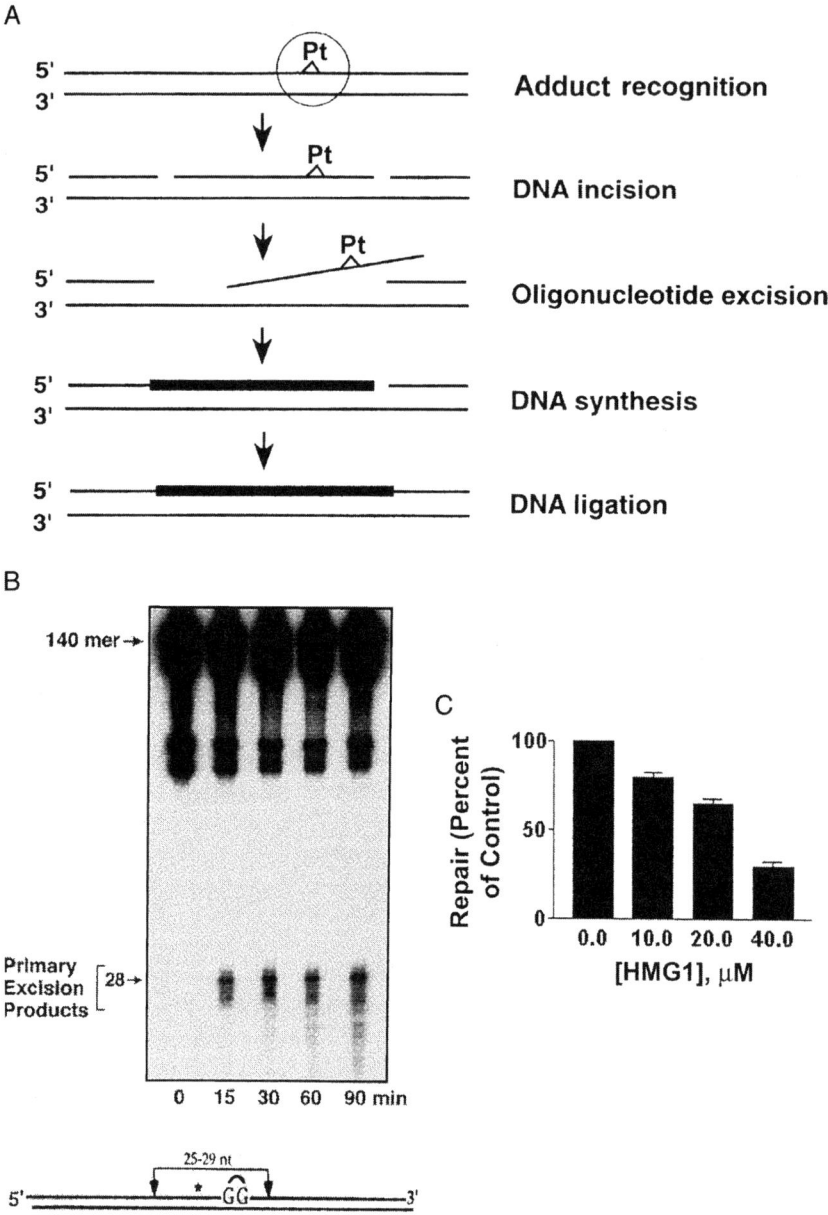

FIG. 4. (A) Scheme illustrating the primary biochemical steps of nucleotide excision repair. (B) Kinetics of removal of single 1,2-GG-intrastrand CL of cisplatin centrally located in the linear 140 bp duplex by HeLa excinuclease; after the excision reaction, DNA was resolved in 10% denaturing polyacrylamide gel to separate excision products from substrate DNA and visualized

2. NUCLEOTIDE EXCISION REPAIR

DNA repair pathways have a particularly intriguing role in the mechanism of the biological activity of agents damaging DNA. Early studies (72–75) indicate that the cells deficient in DNA repair are more sensitive to cisplatin than the cells proficient in repair, the effect not observed with transplatin.

In human cells, cisplatin-intrastrand adducts are removed from DNA mainly by the nucleotide excision repair (NER) (Fig. 4A), which is one of the many cellular defense mechanisms involved in the elimination of the toxic effects of cisplatin. This type of repair includes removal of the damaged base by hydrolyzing phosphodiester bonds on both sides of the lesion. It has been indicated that by using human and rodent cell-free extracts or a reconstituted system containing highly purified NER factors, the 1,2- and 1,3-intrastrand CLs of cisplatin are efficiently repaired (76) (Fig. 4B). Importantly, this repair of 1,2-, but not 1,3-intrastrand CL, is blocked upon addition of an HMG-domain protein (HMG = high-mobility group) (Fig. 4C) (see Section II,G).

An *in vitro* excision repair of a site-specific cisplatin-interstrand CL has also been studied using mammalian cell-free extracts containing HMG-domain proteins at levels not sufficient to block excision repair of the 1,2-intrastrand adducts (76). Repair of the interstrand CL formed by cisplatin has not been detected. Similarly, NER of cisplatin-interstrand CLs is not observed in cell strains derived from patients with Fanconi's anemia, although NER can readily occur in these cells (77, 78). Fanconi's anemia cells have been described as being extremely sensitive to cross-linking agents, so that their noticeably high sensitivity to cisplatin is explained by the incapability of these cells to repair cisplatin-interstrand CLs (79). On the other hand, repair of these lesions has been detected with the aid of a repair synthesis assay, which measures the amount of new DNA synthesized after the damage removal in whole cell extracts (80). In this way, however, the repair could also result from a mechanism different from that of nucleotide excision. The pathways for the repair of DNA-interstrand CLs of cisplatin and for other genotoxic agents in mammalian cells are poorly defined. DNA-interstrand CLs pose a special challenge to repair enzymes because they involve both strands of DNA and therefore cannot be repaired using the information in the complementary strand for resynthesis.

---

by autoradiography; the bands corresponding to 25–29 nucleotides located around the marker are excision products, as indicated on the scheme at the bottom, which marks the incision sites of the excinuclease. Adapted with permission from J. T. Reardon, A. Vaisman, S. G. Chaney, and A. Sancar, *Cancer Research*, Vol. 59, pp. 3968–3971, 1999. (C) Calf thymus HMGB1 inhibition of excision of the 1,2-GG intrastrand CL of cisplatin; the substrate (see Fig. 4A) was incubated with HMGB1 at the concentrations indicated for 10 min at 30°C followed by incubation for additional 45 min in the HeLa cell-free extract. The level of repair without any HMGB1 was designated 100%.

3. MISMATCH REPAIR

There is also evidence that other cellular repair mechanisms, such as recombination or mismatch repair (MMR), can affect antitumor efficiency of cisplatin. Recent observations support the view that MMR mediates the cytotoxicity of cisplatin in tumor cells (81, 82) and that dysfunction of this type of DNA repair may result either in the resistance of tumor cells to cisplatin or in the drug tolerance (83). The functions of MMR are to scan newly synthesized DNA and to remove mismatches that result from nucleotide incorporation errors made by the DNA polymerases.

To explain cisplatin tolerance, it is assumed that replication by-pass of DNA adducts of cisplatin (see Section II,C,1) leads to mutations. During MMR, the strand to be corrected is nicked, a short fragment containing the mismatch is excised, and the new DNA fragment is synthesized. The MMR system always replaces the incorrect sequence in the daughter strand, which would leave the cisplatin adduct unrepaired. This activity initiates a futile cycle. During DNA synthesis to replace the excised short fragment, the DNA polymerases would again incorporate mutations followed by attempts to remove them. The repeated nicks in DNA formed at each ineffective cycle of repair could trigger a cell death response. Thus, MMR recognition of cisplatin adducts on DNA may trigger a programmed cell death pathway rendering MMR-proficient cells more sensitive to DNA modification by cisplatin than MMR-deficient cells. In other words, loss of MMR would then increase the cell's ability to tolerate lesions formed on DNA by cisplatin. Also importantly, the MMR is likely to act mainly in the postreplicative phase, when the highest concentration of mispairs is expected in the newly synthesized DNA. This would be in contrast to NER, which probably acts independently of the cell cycle. Thus, rapidly proliferating tumor cells would be more prone to have futile repair attempts capable of triggering cell death. Although the level of resistance to cisplatin, which accompanies loss of MMR, is relatively small, loss of MMR has been shown to be sufficient in accounting for the failure of treatment with cisplatin in several model systems (84–86).

E. Recognition of the Adducts by Cellular Proteins

An important step in unraveling the mechanism of antitumor activity of cisplatin was also an identification of proteins that specifically bind to DNA modified by cisplatin. The search for these proteins has been initiated by results suggesting that the DNA adducts formed by cisplatin and transplatin in cells might be differentially processed (87).

To date, several classes of these proteins have been identified and mechanisms have been proposed to mediate antitumor effects of cisplatin. Extensive reviews addressing these questions have recently been published (6, 10, 88). This section describes several selected aspects of recognition of DNA adducts formed by "classical" platinum compounds, such as cisplatin and its ineffective *trans* isomer, by HMGB1 protein and its domains, i.e., by the proteins whose binding affinity to the adducts of these platinum compounds, including the structure of the complex between platinated DNA and the protein, has been most extensively investigated. In addition, the binding affinity of several DNA-binding proteins lacking an HMG domain to cisplatin adducts, such as histone H1, p53, and repair proteins, is also briefly described for comparative purposes, since recognition by these and HMG-domain proteins of DNA adducts of novel ("nonclassical") antitumor platinum and ruthenium compounds has also been studied and is discussed below.

1. HMG-Domain Proteins

HMGB1 and HMGB2 proteins belong to architectural chromatin proteins that play some kind of structural role in either the formation of functional higher order protein–DNA or protein–protein complexes or as signaling molecules in genetically regulated repair pathways. HMG-domain proteins bind selectively to the 1,2-GG or AG adducts of cisplatin but not to its 1,3-intrastrand CLs and monofunctional adducts or to the adducts of transplatin.

The full-length HMGB1 protein contains two tandem HMG-box domains, A and B, and an acidic C tail (*89, 90*). The domains A and B in HMGB1 protein (HMGB1a and HMGB1b) are linked by a short lysine-rich sequence (containing seven amino acids, A/B-linker region) (Fig. 5). Each "minimal" domain, A or B alone, specifically recognizes 1,2-GG-intrastrand CL of cisplatin (*91*). The affinity of HMGB1a for DNA containing 1,2-GG-intrastrand CL is generally higher than that of HMGB1b. It has been proposed that HMGB1a is the dominating domain in the full-length HMGB1 that binds to the site of the intrastrand CL, while HMGB1b facilitates binding providing additional protein–DNA interactions. Interestingly, the binding affinity of HMGB1a and HMGB1b with the duplex containing the site-specific 1,2-GG-intrastrand CL is modulated by the nature of the bases that flank the platinum lesion by more than two orders of magnitude (*91, 92*). The impact of flanking-base sequences on HMG-domain protein recognition correlates with the modulatory effect of the flanking bases on CL-induced alterations of duplex stability (*93, 94*).

The HMG domain binds to the minor groove of the DNA double helix opposite the platinum 1,2-intrastrand CL located in the major groove (*95*). Distortions such as prebending, unwinding at the site of platination, and preformation of a hydrophobic notch as a consequence of the 1,2-intrastrand CL formation are important for the recognition and affinity of the platinum-damaged DNA-binding proteins.

Moreover, the A/B-linker region in the full-length HMGB1 protein attached to the N terminus of the B domain markedly enhances binding of the B domain to DNA containing site-specific 1,2-GG-intrastrand CL (*96*). It has also been proposed that binding of the A/B-linker region within the major groove of DNA

A

HMGB1 domain A (HMGB1a): [1]MGKGDPKKPRGKMSSYAFFVQTCREEHKKKHPDASVNFSEFSKKCSERWKTMSAKEKGKFEDMAKADKARYEREMKTYIPPKGE[84]

HMGB1 A/B linker region: [85]TKKKFKD[91]

HMGB1 domain B (HMGB1b): [92]PNAPKRPPSAFFLFCSEYRPKIKGEHPGLSIGDVAKKLGEMWNNTAADDKQPYEKKAAKLKEKYEKDIAAY[162]

HMGB1 B/C linker region: [163]RAKGKPDAAKKGVVKAEKSKKK[185]

HMGB1 acidic C tail: [186]EEEDDEEDEEDEEEEEEEEDEEEDDDDE[215]

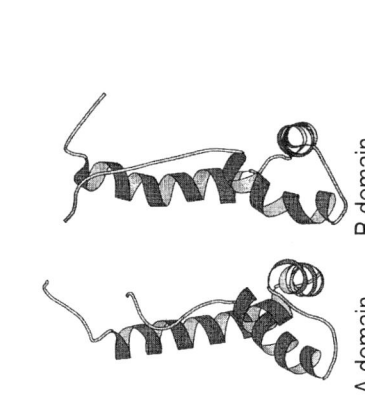

FIG. 5. (A) The amino acid sequence of rat HMGB1 protein (GenBank accession No. Y00463). (B) Schematic structure of HMGB1 protein. (C) Solution structure of the A and B boxes of HMGB1. Reprinted from *Trends in Biochemical Sciences*, Vol. 26, No. 3, J. O. Thomas and A. A. Travers, HMG1 and 2, and related 'architectural' DNA-binding proteins, Vol. 26, pp. 167–174, Copyright 2001, with permission from Elsevier Science.

helped the two HMGB1 domains to anchor to the minor DNA groove facilitating their DNA binding.

These studies have been extended to DNA-interstrand CLs produced by cisplatin or transplatin at the sites where these adducts are formed preferentially (97). Mammalian HMGB1 protein binds to the interstrand CL of cisplatin with an (slightly lower) affinity similar to that to the 1,2-GG-intrastrand CL. On the other hand, no binding of the HMGB1 protein to transplatin-interstrand CL has been noticed. Our most recent analysis revealed that isolated HMGB1a has no affinity to the duplex interstrand cross-linked by cisplatin, whereas the isolated HMG1b containing the A/B-linker region attached to its N terminus binds with a noticeable affinity, although lower than that of the "minimal" HMGB1a to 1,2-GG-intrastrand CL (unpublished results).

Thus far, available data show no clear correlation between the magnitude of bending and/or of unwinding induced in DNA by the individual types of the platinum adducts and by the resulting fixation of HMGB1 (14). In addition, the HMG-domain protein binding to the duplex containing 1,2-GG-intrastrand CL does not seem to considerably change the structure of this adduct (98). In addition, the structures of the free and protein-bound DNA are not very sequence dependent, suggesting that the sequence-dependent HMG-domain protein binding arises mainly from sequence-dependent differences in protein–DNA contacts (91, 92, 98). On the other hand, the structure of HMG-domain protein bound to the intrastrand CL is somewhat changed (95).

1,3-GXG-intrastrand CL of cisplatin is not recognized by the HMGB1, in spite of the fact that bending and unwinding induced in DNA by this adduct are rather similar to those induced by the 1,2-GG-intrastrand adduct. This suggests that there are other factors which control the recognition of platinum adducts by the DNA-binding proteins. At present, the factors which hinder the binding of these proteins to the platinum adducts capable of bending and unwinding of DNA are unknown. Some data on the local conformation of DNA around the individual types of platinum adducts are consistent with the hypothesis that the recognition of DNA bending and/or unwinding induced by these lesions might be obscured, if the formation of the platinum-induced DNA lesion is accompanied by a local denaturation and/or by a lowered rigidity of the duplex around the adduct (14, 97).

The mechanism of full-length HMGB1 protein binding to cisplatin-modified DNA has not been completely resolved. Thus far, the studies performed suggest that the binding of HMGB1-domain proteins to DNA modified by cisplatin can be modulated by conformational alterations induced by the CLs, changes in the stability and flexibility of the duplex, base–protein contacts, or by some combination of these factors. Further studies are therefore warranted to reveal all the details involved in the recognition of DNA adducts of platinum compounds by HMGB1 protein.

There are a variety of proteins, other than HMGB1 or HMGB2, that contain HMG boxes, e.g., a conserved 80-amino-acid domain found in a variety of eukaryotic DNA-binding proteins, including a number of transcription factors. Many of these proteins (for instance, HMG-D, human and *Drosophila* SSRP1, Cmb1, hUBF, Ixr1, mtTFA, LEF1, SRY, T160, and tsHMG) have been shown to recognize and bind to DNA modified by cisplatin, and several recent and excellent reviews describing these binding studies are available (*6, 10, 88*). In addition, very recently, interactions of a novel factor, human FACT (facilitates chromatin transcription), with DNA modified by cisplatin were investigated (*99*). This factor is composed of Spt16 and SSRP1 protein subunits and enables transcription elongation past nucleosomes. Interestingly, FACT exhibits both affinity and specificity for DNA modified globally by cisplatin compared to nonmodified or transplatin-modified DNA. While FACT binds the major 1,2-GG-intrastrand CL of cisplatin, its isolated SSRP1 subunit does not, suggesting that the Spt16 subunit primes SSRP1 for cisplatin-modified DNA recognition by unveiling its HMG domain.

2. Proteins without an HMG Domain

The *TATA-binding protein* (TBP) is essential for transcription initiation in eukaryotes. TBP recognizes and binds to the minor groove of a consensus sequence, TATAAA, known as the TATA box (*100*). The TBP binds selectively to and is sequestered by cisplatin-damaged DNA (*101, 102*). Interestingly, the structure of the TATA box in the human TBP–TATA complex is strikingly similar to the structure of the 1,2-GG-intrastrand CL. While the structure of the TATA box in the TBP–TATA complex is an induced fit, TBP binds to the 1,2-GG-intrastrand CL to a preformed, bent DNA in a lock-and-key fashion. In this way, cisplatin turns the GG sequence into a potential site for TBP. In addition, it was also recently demonstrated (*103*) that the 1,2-GG-intrastrand CLs flanking the TATA-box positions markedly enhance the binding affinity of TBP to the TATA box. Importantly, the binding affinity of TBP to 1,3-GXG-intrastrand CLs of cisplatin and DNA modified globally by either transplatin or [PtCl(dien)]Cl is noticeably weaker than that to DNA modified globally by cisplatin (*102*), supporting the view that TBP binds selectively to a specific three-dimensional structure.

Another protein lacking the HMG domain that has been shown to bind to cisplatin-modified DNA is the *Y-box binding protein 1* (*104*). It has been suggested that this transcription factor may also function as a recognition protein for cisplatin-damaged DNA either in DNA repair or in directing the cellular response to damage to DNA by cisplatin (*104, 105*).

A very abundant chromatin protein, the linker *histone H1*, also binds much more strongly to DNA modified by cisplatin than to transplatin-modified or nonmodified DNA (*106, 107*). Interestingly, linker histones share several important characteristics with HMGB1/2. Both bind to linker DNA in chromatin, specifically recognize four-way junction DNA, and unwind DNA upon binding, etc. (for review, see Refs. *108–110*). The adduct responsible for the strong binding of histone H1 to cisplatin-modified DNA has not been determined; nevertheless the fact that this binding is stronger than that of the HMGB1/2 makes histone H1 a good candidate for participating in the mechanism of antitumor activity of cisplatin.

A relatively large number of proteins that specifically bind to cisplatin adducts belong to DNA-repair proteins. The repair proteins that recognize cisplatin-modified DNA are mainly those involved in the first step in repair pathways, i.e., in damage recognition. Most of these proteins recognize not only cisplatin adducts but also other types of DNA lesions such as, for instance, those created by exposure to either other chemical agents or UV light. The repair proteins that have probably attracted the most attention in the studies of recognition of DNA modified by platinum drugs are those that are absent in patients suffering from the NER deficiency characteristic of the disease *Xeroderma pigmentosum* (XP). The minimal factors necessary for removal of damaged nucleotides also include *XPA protein* and *replication protein A* (RPA) which are among the major damage-recognition proteins involved in the early stage of NER. XPA (32-kDa protein) and RPA (composed of 70- and 34-kDa subunits) are able to bind damaged DNA independently, although RPA interaction stimulates XPA binding to damaged DNA. It has been suggested recently (*111*) that XPA in conjunction with RPA constitutes a regulatory factor that monitors DNA bending and unwinding to verify the damage-specific localization of NER complexes or control their three-dimensional assembly. These NER proteins have been tested for their capability to recognize and bind to cisplatin-modified DNA. Several studies have indicated that XPA binds to DNA globally modified by cisplatin and is able to selectively recognize the rigid bend induced by the 1,2-GG-intrastrand CL of cisplatin (*112–116*) (Fig. 6A). The low affinity of XPA to 1,3-GXG-intrastrand CL (threefold preference for platinated versus undamaged DNA) has also been reported (*117*). Further analyses of the binding affinity of RPA have revealed differential recognition of site-specific DNA adducts of cisplatin by this protein (*115, 118–120*). The adduct bound most efficiently is the 1,3-GXG-intrastrand CL, with the 1,2-GG-intrastrand CL also being bound preferentially compared to the undamaged control, but roughly 30% less compared to the 1,3-CL (Fig. 6B). The binding of a DNA substrate containing a single, site-specific interstrand CL is even less than that observed with the undamaged control DNA. The RPA affinity to individual cisplatin CLs has been correlated with thermal destabilization induced by these adducts (*115, 120*), and it has been proposed that RPA may bind to cisplatin-modified DNA via the generation of transient single-stranded regions. Importantly, the rate of association of RPA with cisplatin-modified DNA is smaller than that of HMGB1 for the same substrate (*115*).

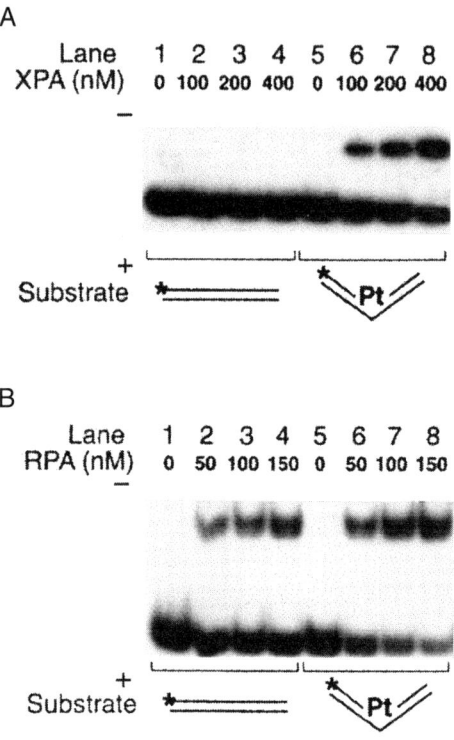

FIG. 6. Electrophoretic mobility shift assay demonstrating affinity of XPA (A) or RPA (B) protein to 20-bp duplex nonmodified (lanes 1–4) or containing a centrally located single, site specific 1,2-GG intrastrand CL of cisplatin (lanes 5–8). Adapted from M. Missura, T. Buterin, R. Hindges, U. Hübscher, J. Kasparkova, V. Brabec, and H. Naegeli, Initiation of mammalian nucleotide excision repair: Double-check probing of DNA helix conformation by repair protein XPA and replication factor RPA, *EMBO Journal*, 2001, Volume 20, pp. 3554–3564, by permission of The American Society for Biochemistry & Molecular Biology.

The proteins involved in the MMR pathway also display affinity for cisplatin-modified DNA. For instance, human MutSα binds to DNA containing 1,2-GG-intrastrand CL but fails to bind to the 1,3-GXG-intrastrand CL of transplatin (*121*). In addition, MSH2 (a component of MutSα) also binds specifically to DNA containing cisplatin adducts and displays selectivity for the DNA adducts of therapeutically active platinum complexes (*122, 123*).

DNA photolyase, involved in the repair of cyclobutane pyrimidine dimmers (*124*), and Ku autoantigen, taking part in recombination and double-strand break repair, are further examples of proteins lacking a HMG domain that bind to cisplatin-modified DNA. Interestingly, the latter protein binds to cisplatin-modified DNA but fails to stimulate DNA-activated protein kinase (*125*). Its activation requires translocation of its Ku subunits away from the DNA terminus, and the adducts formed on DNA by cisplatin inhibit this translocation (*126*).

The repair protein that has also been thoroughly tested, whether it specifically recognizes DNA adducts of cisplatin and transplatin, is an enzyme with deoxyribonuclease activity, T4 endonuclease VII (*97, 127*). This bacteriophage T4-encoded protein cleaves branched DNA structures, most notably four-way junctions, and is regarded as a repair enzyme. T4 endonuclease VII recognizes 1,2-intrastrand CLs of cisplatin, whereas 1,3-intrastrand CLs of both isomers are recognized much less efficiently (*127*). Later studies (*97*) demonstrated that the DNA duplex containing a single interstrand CL of cisplatin is also precisely cleaved in both strands by this DNA-debranching enzyme with an efficiency similar to that observed in DNA containing a single 1,2-GG-intrastrand CL or a four-way junction. In contrast, the duplex containing the interstrand CL of clinically ineffective transplatin is cleaved considerably less efficiently.

It has been hypothesized that sensitivity of tumor cells might be also associated with the processes involving p53 protein (*128, 129*). The p53 gene encodes a nuclear phosphoprotein consisting of 393 amino acids that is biologically activated in response to genotoxic stresses including treatment with cisplatin. Also interestingly, on average, cells with mutant p53 are more resistant to the effect of cisplatin (*130*). Hence, it seems reasonable to conclude that p53 can control the processing of DNA adducts of cisplatin depending on the cell type. The DNA-binding activity of p53 protein is crucial for its tumor suppressor function. DNA interactions of active wild-type human p53 protein with DNA modified by antitumor cisplatin and its clinically ineffective *trans* isomer (transplatin) have been investigated (*131*). DNA adducts of cisplatin reduce the binding affinity of the consensus DNA sequence to p53, whereas transplatin adducts do not (Fig. 7A). This result has been interpreted as indicating that the precise steric fit required for formation and stability of the tetrameric complex of p53 with the consensus sequence cannot be attained as a consequence of severe conformational perturbances induced in DNA by cisplatin adducts. The results also demonstrate an increase of the binding affinity of p53 to DNA lacking the consensus sequence and modified by cisplatin but not by transplatin (Fig. 7B). In addition, major 1,2-GG-intrastrand CLs of cisplatin are only responsible for this enhanced binding affinity of p53, which is, however, markedly lower than the affinity of p53 to unplatinated DNA containing the consensus sequence. The distinctive structural features of 1,2-intrastrand CLs of cisplatin have been suggested to play a unique role for this adduct in the binding of p53 to DNA lacking the consensus sequence.

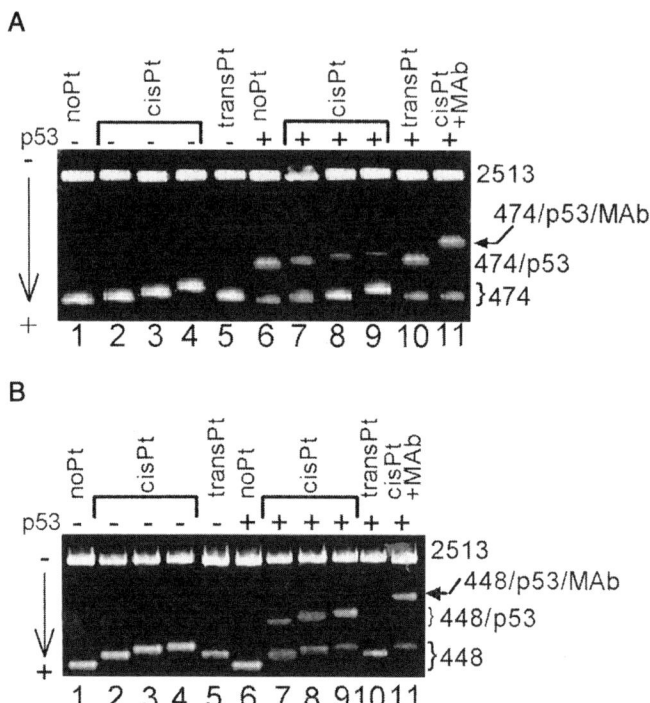

FIG. 7. (A) Binding of p53 to the 474-bp-long *Pvu*II fragment of pPMG1 DNA containing a 20-bp consensus DNA response element. The fragment was unplatinated (lanes 1, 6) and globally modified by cisplatin (lanes 2–4, 7–9) or transplatin (lanes 5, 10). Gel-mobility retardation assay was performed in the presence of the unplatinated 2513-bp nonspecific competitor (*Pvu*II fragment of pPMG1 lacking the consensus sequence) in 1% agarose gel. Concentrations of the 474- and 2513-bp fragments were 10 and 57 μg/ml (3.3 and 3.5 × 10^{-8} M), respectively, and concentration of the p53 protein was 0 (lanes 1–5) or 3.14 × 10^{-8} M (lanes 6–11). r_b values: 0 (lanes 1, 6, 11), 0.02 (lanes 2, 7), 0.04 (lanes 3, 8), 0.06 (lanes 4, 5, 9, 10). Lane 11: the same as in lane 6, but MAb DO-1 was added at the molar ratio MAb/p53 = 3. (B) Binding of p53 to the 448-bp-long *Pvu*II fragment of pBluescript SK II$^+$ DNA lacking the consensus DNA response element. The fragment was unplatinated (lanes 1, 6), globally modified by cisplatin (lanes 2–4, 7–9, 11) or transplatin (lanes 5, 10). Gel-mobility retardation assay was performed in the presence of the unplatinated 2513-bp nonspecific competitor (*Pvu*II fragment of pPMG1 lacking the consensus sequence) in 1% agarose gel; concentrations of the 448- and 2513-bp fragments were 10 and 57 μg/ml (3.1 and 3.5 × 10^{-8} M), respectively, and concentration of p53 was 0 (lanes 1–5) or 1.88 × $10^{-7} M$ (lanes 6–11). r_b values: 0 (lanes 1, 6), 0.01 (lanes 2, 7), 0.05 (lanes 3, 8), 0.08 (lanes 4, 5, 9–11). Lane 11: the same as in lane 9, but MAb DO-1 was added at the molar ratio MAb/p53 = 3. Reproduced from J. Kasparkova, S. Pospisilova, and V. Brabec, Different recognition of DNA modified by antitumor cisplatin and its clinically ineffective *trans* isomer by tumor suppressor protein p53, *J. Biol. Chem.*, Vol. 276, No. 19, pp. 16,064–16,069, 2001, by permission of The American Society for Biochemistry & Molecular Biology.

The cisplatin-damaged DNA-binding proteins apparently occur in nature for purposes other than specific recognition of platinum adducts in DNA, since platinum compounds do not belong to natural components of our environment. The capability of cisplatin-damaged DNA to bind DNA-binding proteins, which may have a fundamental relevance to the antitumor activity of cisplatin and its simple antitumor analogs, is probably a coincidence when the formation of some platinum adducts in double-helical DNA adopts a structure that mimics the recognition signal for these proteins.

F. Effects of the Adducts on Telomerase and Topoisomerases

Telomerase is a ribonucleoprotein, which elongates and/or maintains telomeres by adding TTAGGG tandem-repeat sequences using the RNA component of the enzyme as a template. Enzyme activity appears to be associated with cell immortalization and malignant progression as telomerase activity has been found in the majority of human tumors. Since telomeric tandem repeats as well as the gene of the human telomerase RNA component are guanine-rich and because cisplatin preferentially binds to guanine-rich sequences in nucleic acids, telomerase inhibition has, therefore, been proposed as a novel and potentially selective target for therapeutic intervention. However, several reports demonstrated (*132*) that cisplatin does not directly inhibit telomerase activity in cell-free assays. On the other hand, cisplatin reduces telomerase activity in a specific and concentration-dependent manner in human testicular tumor cells, while transplatin (and several other nonplatinum drugs) has no effect (*133*). Several later studies also supported the view that telomerase inhibition might be a component of the efficacy of cisplatin in the treatment of cancer (*134–138*). Much more work is required, however, to evaluate this possibility. Interestingly, cisplatin, transplatin, and monodentate $[Pt(H_2O)(NH_3)_3]^+$ form monofunctional adducts in the telomeric sequence $(T_4G_4)_4$ folded into an intramolecular G-quadruplex structure (*139*). The adducts are formed at the guanine residues not involved in G-quartets. In addition, cisplatin and transplatin also form a small amount of CLs in these quadruplexes, with transplatin being more effective. The platinated sites in these CLs are one of the guanine residues not involved in G-quartets and the guanine residue at the close end of the quadruplex structure. Restrained molecular dynamics simulations have suggested that these CLs do not perturb the overall structure and quartet stacking. The studies of telomerase interactions with platinated G-quadruplex structures are needed to provide more support for the view that the platination of these quadruplexes is relevant to biological effects of platinum compounds.

Because of their role in the control of the topological state of DNA, topoisomerases are ubiquitous and vital enzymes. As they alter the topology of DNA, topoisomerases participate in nearly all events related to DNA metabolism including replication, transcription, and recombination. The topoisomerases are now viewed as important therapeutic targets for cancer therapy; in particular, topoisomerase inhibitors are considered promising anticancer agents. More specifically, since the frequency of the individual adducts formed on DNA by platinum compounds (which may differ in their biological effects) is dependent on DNA topology (15, 140), the mechanism of the biological activity of platinum drugs could also be affected by the catalytic efficiency of topoisomerases. The information on direct effects of cisplatin on topoisomerase catalytic activity is relatively rare (141). Cisplatin inhibits topoisomerase II leading to a transient alteration of DNA supercoiling. *In vitro* studies of the effect of cisplatin on the supercoiling activity of purified topoisomerase II revealed that cisplatin is an efficient inhibitor of this enzyme as a consequence of a direct interaction of cisplatin with the enzyme. On the other hand, the results of several reports (142, 143) are consistent with the hypothesis that topoisomerase catalytic activity is also affected by DNA adducts of cisplatin. For instance, a synergism between cisplatin and several topoisomerase I inhibitors in inhibiting its catalytic activity has been demonstrated in human small-cell lung cancer cells. Severe conformational alterations induced in DNA by cisplatin adducts might modulate the stabilization of topoisomerase I–drug–DNA cleavable complexes, and/or topoisomerase inhibitors may inhibit repair of cisplatin adducts and enhance the antitumor activity of the platinum drug. Interestingly, the reaction of closed circular DNA containing cisplatin adducts with topoisomerase I yields unique topoisomers (144). Further investigation of the effect of DNA adducts of platinum compounds on the catalytic activity of topoisomerases is warranted to define their role in the mechanism of their cytotoxicity.

G. Hypotheses for Mechanism of Antitumor Activity of Cisplatin

More than 30 years after the discovery of antitumor activity of cisplatin, basic research has provided extensive information on how the drug exerts its antitumor effects and on how some tumors are, or become, resistant to this drug. It is generally believed that the key intracellular target for cisplatin is DNA on which cisplatin forms various types of adducts. These adducts distort DNA conformation, but it still remains uncertain which of these adducts is the most important in terms of producing anticancer effects. As briefly summarized in the preceding sections, DNA adducts of cisplatin inhibit replication and transcription, but they are also by-passed by DNA or RNA polymerases. In addition, cisplatin adducts are removed from DNA mainly by NER. They are, however, also recognized by a number of proteins which could block cisplatin DNA adducts from damage recognition needed for repair. The other hypothesis based on the observation that a number of various proteins recognize cisplatin-modified DNA is that cisplatin–DNA adducts hijack proteins away from their normal binding sites, thereby disrupting fundamental cellular processes. Experimental support of these hypothetical aspects of the mechanism underlying antitumor activity of cisplatin or resistance of some tumors to this drug has been recently thoroughly reviewed (6, 8–10, 83, 129).

Initially, inhibition of DNA replication was considered to be a process very likely relevant to antitumor efficiency of cisplatin (52). However, later studies indicated that cisplatin inhibits tumor cell growth at doses which are considerably lower than those needed to inhibit DNA synthesis (145). Subsequent observations revealed that cisplatin can trigger G2 cell-cycle arrest and programmed cell death (apoptosis) (62, 64), exposing a mechanism of cytotoxicity of this drug. However, since apoptosis is a very complex process, a number of possible pathways must still be explored for a complete understanding of how cisplatin triggers apoptosis (146).

III. DNA Interactions of Cisplatin Analogs

A. Carboplatin

In general, the toxicity and pharmacokinetic properties of platinum compounds are determined by the structure and characteristics of the leaving groups. The search for an agent less toxic than cisplatin led to the development of carboplatin [cis-diamminecyclobutanedicarboxylatoplatinum(II)] (Fig. 8), which has achieved routine clinical use. The leaving group of carboplatin is a cyclobutanodicarboxylato ligand which undergoes a slower rate of aquation than the

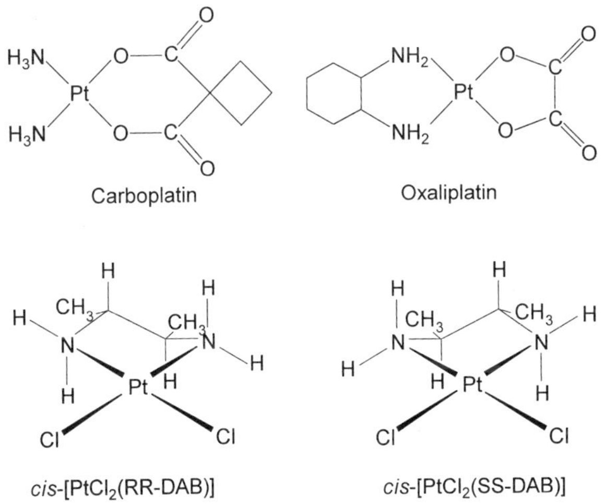

FIG. 8. Structures of direct analogs of cisplatin.

chlorides in cisplatin. Carboplatin forms a spectrum of DNA adducts similar to that of cisplatin with slightly different sequence preferences *in vitro* and in cells (*147*). The concentrations of carboplatin at approximately two orders of magnitude higher are needed to obtain DNA platination levels equivalent to cisplatin. Interestingly, increased DNA-binding affinity of carboplatin has been observed in the presence of nucleophiles, such as those present in human breast cancer cell cytoplasmatic extracts, thiourea, or glutathione (*148*). Global conformational alterations induced in DNA by carboplatin in cell-free media have been characterized by circular dichroism and differential pulse polarography (*149*), and nondenaturational changes similar to those observed for cisplatin at the same level of platination have been found. Although less toxic than cisplatin, carboplatin is still only active in the same range of tumors as cisplatin (*150*).

B. Oxaliplatin

Oxaliplatin [DACH-(oxalato)platinum(II) compound, DACH = 1,2-diaminocyclohexane] (Fig. 8) is another cisplatin analog which has received limited approval for use in some countries. In order for its reaction with DNA to occur, the parent compound must become aquated. The hydrolysis of oxaliplatin to form reactive species, 1,2-DACH diaqa platinum(II) (*cis*-[Pt(RR-DACH)(H$_2$O)$_2$]$^{2+}$, is a slower process than the hydrolysis of cisplatin, but it is facilitated by HCO$_3^-$ and H$_2$PO$_4^-$ ions (*151*). The site and region specificities of lesions induced by oxaliplatin in naked and intracellular DNA have been determined (*152*). The sites of oxaliplatin adducts and their spectrum and frequency (*153, 154*) are nearly identical to the situation when DNA is modified by cisplatin.

Thus far, the conformation of the major DNA adduct formed by oxaliplatin (1,2-GG-intrastrand CL) has only been studied by using molecular modeling (*155*). These studies suggest that the overall conformational alterations induced in DNA by the 1,2-GG-intrastrand CL of cisplatin and oxaliplatin are similar. The bulky DACH ring of the oxaliplatin adduct fills much of the DNA major groove, making it narrower and less polar at the site of the CL. These subtle differences in overall conformation have been suggested (*156*) to influence further processing of the oxaliplatin CL in the cell. Thus, full-length HMGB1 protein and mismatch repair proteins are found to have a slightly lower affinity to the CL of oxaliplatin than to that of cisplatin (*81, 156*). In addition, DNA lesions generated by oxaliplatin are also repaired *in vitro* by the mammalian nucleotide excision repair pathway with similar kinetics to those of the lesions of cisplatin (*157*). Another interesting finding is that eukaryotic DNA polymerases β, γ, ξ, and η by-pass 1,2-GG CLs of cisplatin less readily than the same CLs of oxaliplatin (*156, 158–160*). In addition, the misincorporation frequency of DNA polymerase β is slightly greater with 1,2-GG CLs of oxaliplatin than with those of cisplatin on primed single-stranded DNA. Thus, it may be possible that differences in the conformation of 1,2-intrastrand CLs of cisplatin and oxaliplatin can be responsible for differences exhibited by these compounds in both the fidelity and the efficiency of translesion synthesis by DNA polymerases.

Also interestingly, the MMR protein complex binds to DNA globally modified by cisplatin but not to that by oxaliplatin (*161*). Perhaps steric hindrance by the DACH ring may prevent the MMR complex from recognizing the lesion. The latter finding may also be related to the observation that MMR-deficient cells are slightly more resistant to cisplatin but not to oxaliplatin (*81, 162*).

Besides *cis*-[Pt(RR-DACH)]$^{2+}$ (directly derived from oxaliplatin), other enatiomeric forms of this complex exist, so that the biological activity of platinum complexes with enantiomeric amine ligands such as *cis*-[Pt(RR-DACH)]$^{2+}$ and *cis*-[Pt(SS-DACH)]$^{2+}$ and other enantiomeric pairs has been intensively investigated (*163–166*). For instance, the DACH carrier ligand has been shown to significantly affect the ability of platinum–DNA adducts to block essential processes such as replication and transcription (*167*). Also importantly, *cis*-[PtCl$_2$(N-N)] complexes with N–N = DACH, 2,3-diaminobutane (DAB) (Fig. 8), or 1,2-diaminopropane, having an S configuration at the asymmetric carbon atoms, are markedly more mutagenic toward several strains in *Salmonella typhimurium* than their R isomers (*168*). Hence, although the asymmetry in the amine ligand in these platinum complexes does not involve the coordinated nitrogen atom but rather an adjacent carbon atom, a dependence of the biological activity on the configuration of the amine is observed.

Recently, modifications of natural DNA and synthetic oligodeoxyribonucleotide duplexes in a cell-free medium by analogs of antitumor cisplatin containing enantiomeric amine ligands, such as *cis*-[PtCl$_2$(RR-DAB)] and *cis*-[PtCl$_2$(SS-DAB)] (Fig. 8), were also studied by using various methods of molecular biophysics and biophysical chemistry (*169*). The major differences resulting from the modification of DNA by the two enantiomers consist of the thermodynamical destabilization and conformational distortions induced in DNA by the 1,2-GG intrastrand CL. The CLs formed by *cis*-[PtCl$_2$(SS-DAB)] are more effective at inducing overall destabilization of the duplex and global conformational alterations than those formed by *cis*-[PtCl$_2$(RR-DAB)]. Bending and unwinding angles due to *cis*-[PtCl$_2$(SS-DAB)]-intrastrand CL (24 and 15°, respectively) are smaller than those due to the CL of *cis*-[PtCl$_2$(RR-DAB)] (35 and 20°, respectively). However, the overall destabilization of the duplex due to the CL of *cis*-[PtCl$_2$(SS-DAB)] is greater than that due to the CL of *cis*-[PtCl$_2$(RR-DAB)]. In addition, the duplex containing the CL of *cis*-[PtCl$_2$(RR-DAB)] shows considerably more pronounced distortion of the base pair adjacent to the CL on its 3' side (Fig. 9). In contrast, the duplex containing the CL of *cis*-[PtCl$_2$(SS-DAB)] shows a larger distortion of the base pair on the 5' side of the CL.

Interestingly, the intrastrand CLs of the *cis*-[PtCl$_2$(RR-DAB)] bind to HMGB1a and HMGB1b with an affinity similar to that to the same CL of

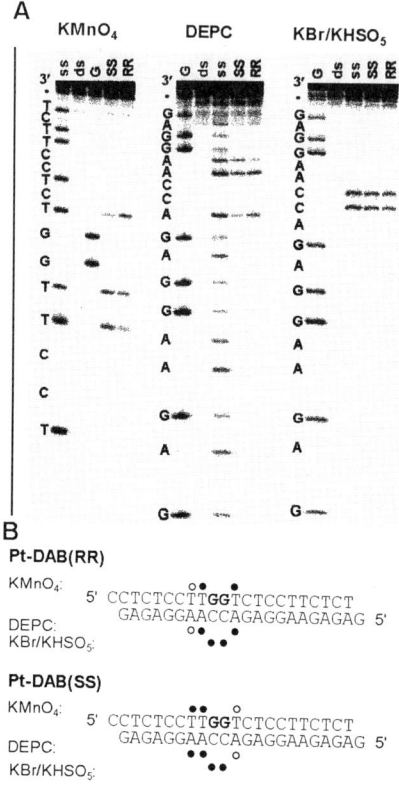

FIG. 9. Piperidine-induced specific strand cleavage at KMnO$_4$-modified (left), diethylpyrocarbonate-modified (center), and KBr/KHSO$_5$-modified (right) bases in unplatinated and platinated 20-bp DNA duplex containing central sequence TGGT.ACCA. The oligomers were 5′-end-labeled at either their top strands in the case of the modification by KMnO$_4$ or their bottom strands in the case of the modification by DEPC or KBr/KHSO$_5$. KMnO$_4$ (left). Lane ss is relative to the unplatinated top strand. Lane ds is relative to the unplatinated duplex. Lane G is a Maxam–Gilbert specific reaction for the unplatinated duplex that had only the top strand end-labeled. Lanes RR and SS are relative to the duplex containing 1,2-d(GpG)-intrastrand CL of either cis-[PtCl$_2$(RR-DAB)] or cis-[PtCl$_2$(SS-DAB)], respectively. DEPC (center): Lane G is a Maxam–Gilbert specific reaction for the unplatinated duplex that had only the bottom strand end-labeled. Lane ds is relative to the unplatinated duplex. Lane ss is relative to the unplatinated bottom strand. Lanes SS and RR are relative to the duplex containing cis-[PtCl$_2$(SS-DAB)] or cis-[PtCl$_2$(RR-DAB)] 1,2-d(GpG)-intrastrand cross-link, respectively. KBr/KHSO$_5$ (right). Lane G is a Maxam–Gilbert specific reaction for the unplatinated duplex. Lane ds is relative to the unplatinated duplex. Lane ss is relative to the unplatinated bottom strand that had only the bottom strand end-labeled. Lanes SS and RR are relative to the duplex containing cis-[PtCl$_2$(SS-DAB)] or cis-[PtCl$_2$(RR-DAB)] 1,2-d(GpG)-intrastrand cross-link, respectively, and had only the bottom strand end-labeled. (B) Summary of the reactivity of chemical probes. ● and ○ designate strong or weak reactivity, respectively. Reproduced from J. Malina, C. Hofr, L. Maresca, G. Natile, and V. Brabec, Biophysical Journal, 2000, Vol. 78, pp. 2008–2021, by permission of the Biophysical Society.

cisplatin (unpublished results). In contrast, the CL of cis-[PtCl$_2$(SS-DAB)] binds to the HMGB1a protein with a noticeably lower affinity, whereas the affinity of HMGB1b to the CL of cis-[PtCl$_2$(SS-DAB)] is only slightly lower than that to the CLs of cis-[PtCl$_2$(RR-DAB)] or cisplatin. As the intrastrand CL of the cis-[PtCl$_2$(SS-DAB)] thermodynamically destabilizes DNA more than the CL of the cis-[PtCl$_2$(RR-DAB)] (169), this result supports the previously established correlation (93, 94) that the increasing thermodynamic destabilization due to the 1,2-GG-intrastrand CL of cisplatin lowers its affinity to HMG-domain proteins.

The intrastrand CLs of both enantiomers are efficiently removed from DNA by NER in the *in vitro* assay using mammalian cell-free extracts capable of removing the 1,2-intrastrand CLs of cisplatin (unpublished results). Consistent with the different affinity of the intrastrand CLs of both enantiomers to HMG-domain proteins, the addition of HMGB1a only blocks the repair of the CLs of cis-[PtCl$_2$(RR-DAB)], whereas the repair of the CLs of cis-[PtCl$_2$(SS-DAB)] remains unaffected (unpublished results). Similar results have also been obtained with DNA modified by cis-[PtCl$_2$(DACH)] enantiomers. Thus, these results demonstrate that very fine structural modifications, such as those promoted by enantiomeric ligands in bifunctional platinum(II) compounds, can modulate the "downstream effects" such as specific protein recognition by DNA-processing enzymes and other cellular components. It may also be suggested that these differences are associated with a different biological activity of the two enantiomers observed previously (168).

In spite of some interesting and promising results describing DNA modifications by carboplatin, oxaliplatin and its derivatives, as of yet, have not demonstrated any substantial advantages over cisplatin. From a mechanistic DNA-binding point of view, this is not too surprising, since the adducts produced on DNA by cisplatin, carboplatin, and oxaliplatin are similar, though they differ in their relative rates of formation.

C. Targeted Analogs

One way to affect the biological activity of cisplatin is by targeting this platinum compound to DNA differently by the attachment of the platinum moiety to a suitable carrier (33, 170). Compared to untargeted analogs, such compounds may exhibit a different DNA-binding mode including sequence selectivity (171, 172). Importantly, this change of the DNA-binding selectivity has already been shown to result in good *in vitro* and *in vivo* activity in tumors resistant to conventional agents. Thus, the objective of the research focused on this type of cisplatin analogs is to design multifunctional molecules that bind DNA in predictable ways and that may have novel pharmacological activities.

One class of platinum complexes targeted to DNA involves cisplatin linked to an intercalator; for example, (1,2-diaminoethane)dichloroplatinum(II) linked to acridine orange by either trimethylene or hexamethylene chains (173) (Fig. 10A).

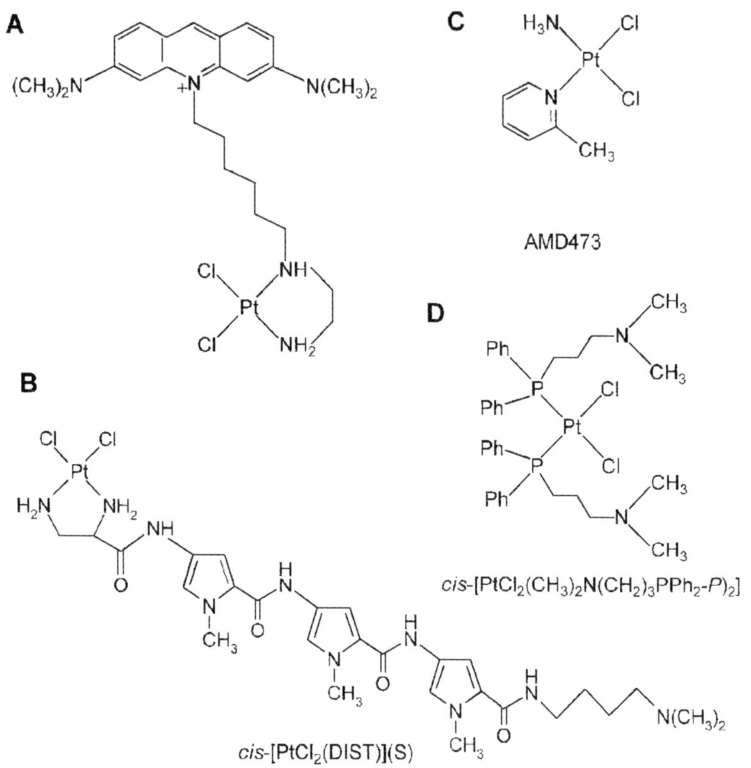

FIG. 10. Structures of targeted analogs of cisplatin. (A) (1,2-diaminoethane)dichloroplatinum(II) linked to acridine orange by hexamethylene chain. (B) S form of cisplatin–distamycin conjugate. (C) cis-amminedichloro(2-methylpyridine)platinum(II) (AMD473). (D) The bifunctional aminophosphine complex.

In the reaction with DNA, the platinum residues cross-link two base residues, while acridine orange is intercalated between the base pairs at a distance of 1 to 2 bp from the platinum-binding site.

DNA-targeted platinum compounds that show antitumor activity include analogs of cisplatin containing as a carrier DNA minor groove binders, netropsin and distamycin. These two oligopeptides are potent antibacterial, antiviral, and antineoplastic agents whose pharmacological activity has been correlated to their abilities to bind to DNA (*174, 175*). When free in solutions, these nonintercalative DNA-binding molecules for noncovalent complexes with double-helical DNA in the minor groove and exhibit a preference for AT-rich domains. The initial studies (*176*) performed with cisplatin conjugates of netropsin or distamycin revealed no radical improvement in cytotoxicity of these compounds in few tumor cells in culture, although a systematic evaluation of cytotoxicity and antitumor activity of these novel platinum(II) derivatives has yet to be performed.

Modifications of natural DNA in a cell-free medium by cisplatin tethered to AT-specific, minor groove-binder distamycin (Fig. 10B) were studied by various methods including biochemical analysis and molecular biophysics (*177, 178*). The attachment of distamycin to cisplatin changes several features of the DNA-binding mode of the parent platinum drug. The major differences consist of different conformational alterations in DNA and a considerably higher efficiency of the conjugated drug to form in DNA interstrand CLs. Cisplatin tethered to distamycin, however, coordinates to DNA with base-sequence preferences similar to those of the untargeted platinum drug. The attachment of distamycin to cisplatin affects the sites involved in the interstrand CLs, so that these adducts are preferentially formed between complementary guanine and cytosine residues and not between guanine residues in the 5'-GC/5'-GC sequences, as is the case of untargeted cisplatin. The interstrand CL of platinum–distamycin conjugates bends the helix axis by ~35° toward the minor groove, unwinds DNA by approximately 95°, and distorts DNA symmetrically around the adduct. In addition, the distamycin moiety in the interstrand CLs of these compounds interacts with DNA, which facilitates the formation of these adducts. It has been suggested that the platinum atom in the interstrand CL of platinum–distamycin conjugates is moved during the closure of this adduct into the minor groove and that this process is facilitated by the affinity of the distamycin moiety to bind preferentially in the minor groove of DNA (*174, 175*).

Cisplatin and all the "second-generation" platinum drugs are administered by intravenous infusion. The ability to deliver the drug orally would allow much greater flexibility in dosing and increases the potential for the use of platinum drugs. The poor oral bioavailability properties of cisplatin and carboplatin were the impetus for a search for oral activity in platinum drugs, which has led to the identification of new platinum compounds with suitable properties for oral administration and high antitumor activity. One compound selected for clinical evaluation from this class is also a bifunctional analog of cisplatin, AMD473 [*cis*-amminedichloro(2-methylpyridine)platinum(II)] (Fig. 10C). AMD473 is less reactive toward the sulfur-containing molecules such as methionine and thiourea (*179*). The DNA-binding properties of AMD473 differ from those of cisplatin. On naked DNA, several adducts unique to AMD473 have been observed (*179*). Within cells, AMD473 forms interstrand CLs much more slowly than cisplatin. An interesting finding is that antibodies elicited against DNA modified by AMD473 do not recognize DNA modified by cisplatin. Interestingly, the treatment of p53 wild-type cell lines with AMD473 induces p53, but at slower rate than that with cisplatin.

Mechanistic studies were also performed with a novel class of aminophosphine platinum(II) compounds which demonstrated activity against murine and human tumor cells including those resistant to cisplatin (*180*). An interesting feature of these complexes is that they can exist in chelate ring-opened and ring-closed forms in aqueous solution. The equilibrium can be controlled under conditions of biological relevance by a variation of pH and chloride concentration. Initial studies (*181*) indicated that the ring-opened forms, which contain chloride ligands, can bind to the nucleic acid bases including thymine or uracil contained in either the monomeric nucleotides or the short, single-stranded oligonucleotides, under physiologically relevant conditions. Later the modification of natural DNA and synthetic polydeoxyribonucleotides in cell-free media was studied by using various methods of molecular biophysics (*182*). For instance, the DNA-binding mode of bifunctional *cis*-[PtCl$_2$(Me$_2$N(CH$_2$)$_3$PPh$_2$-*P*)$_2$] (Fig. 10D) is noticeably different from that of cisplatin. In comparison to cisplatin, the aminophosphine complex also forms the adducts at purine residues but more frequently at adenine residues. In addition, the CLs of the aminophosphine complex terminate DNA synthesis *in vitro* more efficiently than the CLs of cisplatin, and conformational alterations induced in DNA by the aminophosphine complex have several characteristics distinctly different from those induced by cisplatin. A specific role of aminophosphine ligands in the modification of DNA by *cis*-[PtCl$_2$(Me$_2$N(CH$_2$)$_3$PPh$_2$-*P*)$_2$] is also evident from the observation that DNA platinated by this compound is not recognized by the antibodies elicited against DNA modified by cisplatin, despite an equally good specificity of these antibodies for DNA modified by several analogs of cisplatin having varied nonleaving amine groups. It has been suggested (*182*) that the different cellular processing of DNA modified by *cis*-dichloroplatinum(II) complexes containing nonleaving amine groups on the one hand and aminophosphine ligands on the other is relevant to the different biological activity of these two classes of platinum drugs.

D. Tetravalent Analogs

The idea of developing an orally available platinum drug has led to the design of antitumor platinum(IV) analogs (*183, 184*). In addition, being inert to substitution, platinum(IV) complexes theoretically have the advantage of demonstrating fewer side effects than their platinum(II) counterparts. The oxidation state of the platinum atom in platinum coordination compounds determines the steric configuration of the molecule. Platinum(II) structures are planar molecules, while platinum(IV) derivatives assume an octahedral shape.

Octahedral platinum(IV) complexes undergo ligand substitution reactions that are slow relative to those of their platinum(II) analogs and have been considered as the compounds which are unable to react directly with DNA. The antitumor activity of platinum(IV) compounds has been suggested to require *in vivo* reduction to the kinetically more labile, and therefore reactive, platinum(II) derivatives (*185–187*). Thus, platinum(IV) complexes are frequently designated as pro-drugs that have to be first, after their administration, activated by a reaction with reducing agents present in body liquids. On the other hand, in addition to the reduction of active platinum(II) species, other mechanisms may be important for the cytostatic efficiency of platinum(IV) drugs (*188*).

There is evidence that platinum(IV) species can enter cells (*189*) and may react with DNA (*190*). Several platinum(IV) analogs of cisplatin covalently bind simple nucleic acid bases, or their monomeric derivatives (*191, 192*), or an isolated DNA (*149*) directly without addition of any reducing agent. In contrast to cisplatin, however, the rate of the reaction of platinum(IV) complexes with DNA is markedly lower (*149*). In addition, at the same level of the modification in cell-free media, thermal stability, renaturation, and conformational alterations in DNA modified by the platinum(II) and platinum(IV) analogs are different (*149*).

It is likely that after an intravenous administration of a platinum(IV) complex a significant part of these molecules are reduced before they reach their biological target. Nevertheless, several platinum(IV) molecules could reach intracellular DNA without their reduction and modify it in the way which produces an antitumor effect. To address the direct binding mode of the platinum(IV) drugs, modifications of natural DNA and synthetic double-stranded oligodeoxyribonucleotides by *cis*-diamminedichloro–*trans*-dihydroxyplatinum(IV) (oxoplatin) (Fig. 11) have been studied in cell-free media (*193*). Oxoplatin could bind to DNA directly without the addition of a reducing agent. In addition, the antibodies elicited against DNA modified by cisplatin are not competitively inhibited by DNA modified by oxoplatin. However, DNA containing the adducts of oxoplatin becomes a strong inhibitor of these antibodies, if it is subsequently treated with a reducing agent, such as ascorbic acid (Fig. 12). These results have been interpreted to as indicating that oxoplatin can form DNA adducts containing the platinum moiety in the quadrivalent state. The DNA adducts containing the platinum(II) or platinum(IV) analogs differ in both the number of ligands and the formal charge on their platinum center. These differences are probably responsible for the distinct conformational features and the stability of DNA modified by platinum(II) or platinum(IV) complexes.

One of the tetravalent analogs of cisplatin is also JM216 [*bis*-acetatoamminedichloro(cyclohexylamine)platinum(IV)] (Fig. 11), the first platinum complex

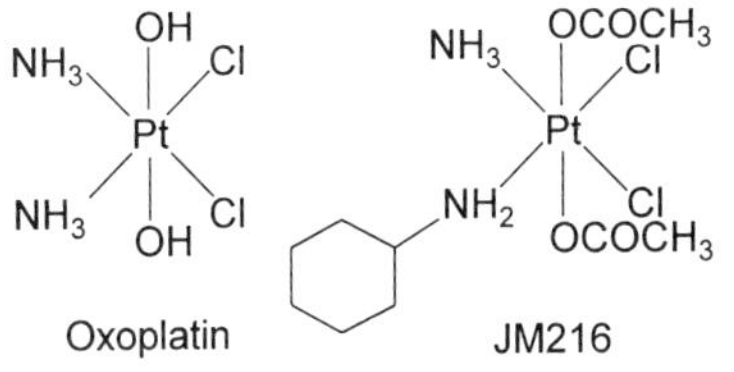

FIG. 11. Structures of tetravalent analogs of cisplatin.

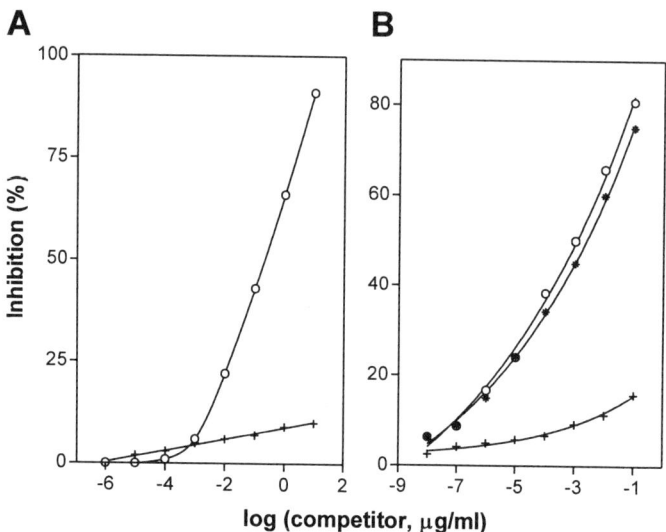

FIG. 12. Competitive inhibition in an ELISA of the cisplatin–DNA antibody binding to DNA modified by cisplatin (○) or oxoplatin (+). (A) The number of molecules of the platinum compound bound per nucleotide residue was 0.004. (B) The number of molecules of the platinum compound bound per nucleotide residue was 0.1; (●) the sample giving the curve passing through crosses in (B) but subsequently treated with ascorbic acid. The antibody binding is reported as the percentage of binding in the absence of any competitor. Competitor concentrations are given as μg of DNA/ml. Reproduced from (193) with permission.

orally administered in the clinic. The DNA-binding properties of JM216 on either naked DNA or within tumor cells are similar to those of cisplatin, although there are some differences in the nature of the adducts (194). JM216 forms interstrand CLs. As regards major intrastrand CLs, some differences between the adducts formed by JM216 and cisplatin may be deduced from the analysis by using antibodies elicited against DNA modified by cisplatin. DNA extracted from CH1 cells treated with JM216 is recognized by these antibodies around twofold less effectively than adducts formed by cisplatin.

E. Monodentate Platinum(II) Compounds

The exploration of new structural classes of platinum antitumor drugs has also resulted in the discovery of new platinum(II) complexes including those of formula cis-$[PtCl(NH_3)_2(A)]^+$ (where A is a heterocyclic amine ligand) (Fig. 13). Thus, these formally monofunctional complexes are analogs of cisplatin containing only one leaving chloride group similar to closely related and simpler, but inactive, platinum–triamine complexes such as either $[PtCl(dien)]Cl$ (Fig. 1) or $[PtCl(NH_3)_3]Cl$. The DNA-binding mode of these new "nonclassical"

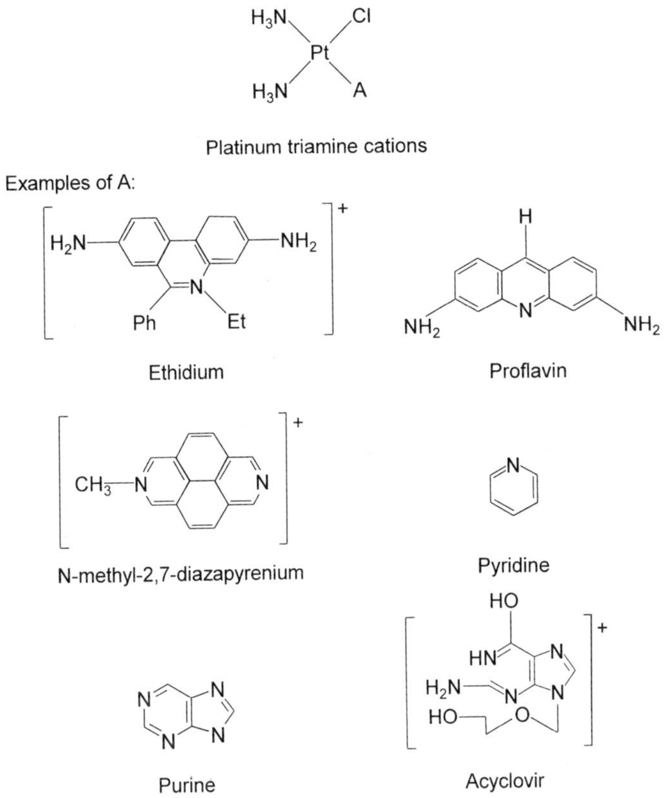

FIG. 13. Structures of platinum(II) triamine cations.

monodentate platinum compounds has also been intensively investigated. An interesting example of this class of platinum compound is the analog cis-[PtCl(NH$_3$)$_2$(N8-Etd)]$^+$, where Etd = ethidium (Fig. 13). This model compound coordinates to DNA forming the monofunctional adducts. The formation of this adduct results in DNA unwinding—the unwinding angle is 19° (43) and is much larger than those observed for the other monofunctional compounds (for instance, only 6° was found for the adducts of both [PtCl(dien)]Cl and [PtCl(NH$_3$)$_3$]Cl). The ethidium in the monofunctional adduct of cis-[PtCl(NH$_3$)$_2$(N8-Etd)]$^+$ (located cis to the DNA binding site) is geometrically well positioned to intercalate between the base pairs of the helix at the adjacent site. Thus, the large unwinding angle due to the binding of cis-[PtCl(NH$_3$)$_2$(N8-Etd)]$^+$ to DNA has been explained by a combination of intercalative and coordination binding modes. Consistent with this explanation, the DNA unwinding angle produced by trans-[PtCl(NH$_3$)$_2$(N8-Etd)]$^+$ is only 8° (43), i.e., very close to the value observed for the monofunctional adducts of both [PtCl(dien)]Cl and [PtCl(NH$_3$)$_3$]Cl. In the monofunctional adduct of trans-[PtCl(NH$_3$)$_2$(N8-Etd)]$^+$, ethidium is located trans to the DNA binding site, so that it projects away from the double helix. In this orientation, the intercalating moiety has little or no opportunity to interact with the double helix, thus providing no contribution to the unwinding induced by this compound.

The analysis of DNA reactions of the conjugated compound containing cisplatin and intercalator has revealed another interesting phenomenon. If cisplatin is allowed to react with double-stranded DNA in the presence of intercalators such as proflavine, ethidium bromide, or N-methyl-2,7-diazapyrenium (Fig. 13), the major DNA adduct is that in which the platinum atom is bound to a guanine residue and to the intercalator. Importantly, this adduct is not formed if double-stranded DNA is replaced by single-stranded DNA. Also importantly, under the same experimental conditions, but without DNA, cisplatin reacts poorly with the intercalator. Thus, double-stranded DNA promotes the binding of cisplatin to the intercalator by acting as a matrix which enables the reactants to adopt a favorable orientation according to

$$A + B \xrightarrow{\text{ds DNA}} AB,$$

where A is cisplatin, B is intercalator such as, for instance, Etd or proflavine, and ds DNA is double-stranded DNA. In other words, the DNA double helix behaves as a catalyst in this reaction (*195, 196*).

Antitumor agents belonging to monofunctional cisplatin analogs are also trisubstituted platinum(II) compounds in which A = pyridine, pyrimidine, purine, piperidine, or aniline (*197, 198*) (Fig. 13). These compounds demonstrated activity against a number of murine tumors and human tumor cell lines. These trisubstituted platinum(II) compounds form monofunctional adducts on DNA which have been suggested to be responsible for the antitumor activity of these agents. The monofunctional adducts of these compounds are capable of blocking DNA replication *in vitro* almost as efficiently as the major adducts of cisplatin, but are not recognized by cellular damaged DNA-recognition proteins. The principal sites at which the replication is blocked are at single guanines. Interestingly, the compound in which A = 4-methyl pyridine distorts double-helical DNA in the base pair flanking the adduct on its 5′ side due to significant contacts of the 4-methyl pyridine ring with the backbone of the DNA on the 5′ side of the adduct, but there is no indication for the intercalation of the pyridine ring (*199, 200*). Taken together, the platinum-triamines have been characterized as a new class of platinum anticancer agents which modify DNA different from cisplatin. These differences have been proposed to be associated with different biological effects of these monofunctional compounds in comparison to cisplatin.

In the search for new, therapeutically more effective platinum drugs, platinum(II) compounds containing, as a part of the coordination sphere, certain selected antiviral nucleosides have been synthesized as well (201, 202). Several compounds of this type exhibit similar or enhanced antiviral activities *in vitro* and in many instances are less toxic to normal cells than either component. *cis*-[PtCl(NH$_3$)$_2$(N7-ACV)]$^+$, a novel compound which has been synthesized (202), contains antiviral acyclovir (ACV) in the coordination sphere of cisplatin (Fig. 13), exhibits activity against various herpes viruses, is found to be as effective as cisplatin when equitoxic doses are administered *in vivo* to P388 leukemia-bearing mice, and is also found to be active against a cisplatin-resistant subline of the P388 leukemia. *cis*-[PtCl(NH$_3$)$_2$(N7-ACV)]$^+$ binds to DNA, forming monofunctional adducts preferentially at guanine sites. These adducts are capable of both terminating DNA and RNA synthesis *in vitro* (203) and affecting some structural and other physical properties of DNA in a manner similar to that produced by the major adduct of cisplatin (Fig. 14). It has been suggested that

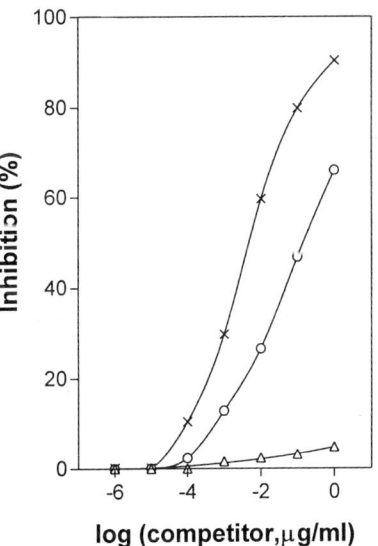

FIG. 14. Competitive inhibition in an ELISA of the cisplatin–DNA antibody binding to the competitor which was calf thymus DNA modified so that the number of molecules of the platinum compound bound per nucleotide residue was 0.05 by *cis*-[PtCl(NH$_3$)$_2$(N7-ACV)]$^+$ (○), cisplatin (x), or [PtCl(dien)]Cl (△). The inhibition expresses the antibody binding as percentage of binding in the absence of any competitor. Reproduced from Z. Balcarova, J. Kasparkova, A. Zakovska, O. Novakova, M. F. Sivo, and V. Brabec, DNA interactions of a novel platinum drug, *cis*-[PtCl(NH$_3$)$_2$ (N7-acyclovir)]$^+$. *Mol. Pharmacol.* **53**, 846–855 (1998), with permission by the American Society for Pharmacology and Experimental Therapeutics.

the ACV ligand itself interacts with DNA in a noncovalent manner producing a certain structural feature similar to the major adduct of cisplatin that could be recognized and further processed by some components of tumor cells in a manner fundamentally different from that of other monofunctional platinum adducts such as those of "classical" [PtCl(dien)]Cl.

Whatever the detailed mechanism of biological activity of this new class of monodentate antitumor platinum(II) agents, these compounds are interesting from a mechanistic point of view and, therefore, worthy of additional testing.

IV. Activation of *trans* Geometry

Since the discovery of the antitumor activity of cisplatin, none of the structural analogs of cisplatin which advanced to clinical trials are likely to display novel clinical properties in comparison to the parent drug. Therefore, the search continues for an improved platinum antitumor agent and is motivated by the desire to design a less toxic compound that is non-cross-resistant to cisplatin or carboplatin and oxaliplatin. In this search, the hypothesis that platinum drugs, which bind to DNA in a manner fundamentally different from that of cisplatin, alter pharmacological properties has been tested. This concept has already led to the synthesis of several new unconventional platinum antitumor compounds that violate the original structure–activity relationships. The clinical inactivity of transplatin is considered a paradigm for the classical structure–activity relationships of platinum drugs (7), but to this end several new analogs of transplatin, which exhibit a different spectrum of cytostatic activity including activity in tumor cells resistant to cisplatin, have been identified.

The antitumor *trans*-platinum complexes, whose DNA-binding mode will be reviewed in this section, include three distinct series (Fig. 15): (i) analogs containing planar amine ligand, general structure *trans*-[$PtCl_2(NH_3)(L)$], where L = planar amine such as quinoline or thiazole; (ii) analogs containing iminoether groups, general formula *trans*-[$PtCl_2(E$-iminoether$)_2$] (*trans-EE*); and (iii) analogs with asymmetric aliphatic ligands and *trans*-[$PtCl_2(NH_3)(L)$], where L = cyclohexylamine (CHA).

A. Transplatin Analogs Containing Planar Amine Ligand

In order to contribute to the understanding of mechanisms underlying the antitumor activity of this new class of platinum drugs (204), various biochemical and biophysical methods as well as molecular modeling techniques have been employed to study the modifications of DNA by antitumor *trans*-[$PtCl_2(NH_3)$(quinoline)] or *trans*-[$PtCl_2(NH_3)$(thiazole)] (205) (Fig. 15). These

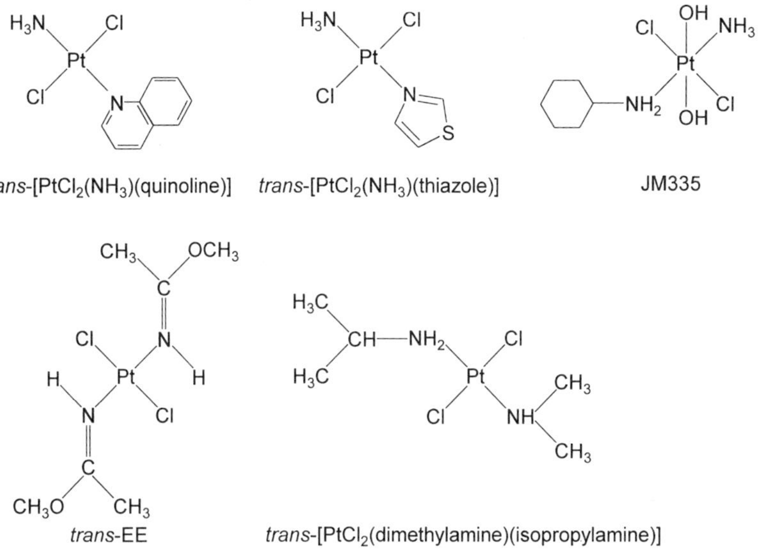

FIG. 15. Structures of antitumor transplatinum compounds.

compounds bind monofunctionally to DNA at a rate similar to that of transplatin. The overall rate of the rearrangement to bifunctional adducts is also similar to that observed in the case of DNA modification by transplatin; i.e., it is relatively slow (after 48 h ca. 34% adducts remain monofunctional). In contrast to transplatin, however, its analogs containing the planar ligand form considerably more interstrand CLs after 48 h ($\sim$30%) with a much shorter half-time ($\sim$5 h) ($\sim$12% for transplatin, $t_{1/2} > 11$ h). In addition, the planar ligand in all or in a significant fraction of DNA adducts of these transplatin analogs, in which platinum is coordinated by base residues, is well positioned to interact with the duplex.

The adducts of the transplatin analogs containing the planar ligand terminate RNA synthesis *in vitro*, preferentially at guanine residues. Interestingly, DNA modified by the *trans*-platinum compounds containing the planar ligand is recognized by cisplatin-specific antibodies, which suggests that these transplatin analogs behave in some respects like cisplatin. Models for both monofunctional adducts and bifunctional interstrand CLs have been proposed. Computer-generated models show that the combination of monofunctional covalent binding and a stacking interaction between the planar ligand and the DNA bases can produce a kink in the duplex, which is strongly suggestive of the directed bend produced by the major cisplatin–DNA adduct (1,2-GG-intrastrand CL).

Further investigations have been focused on the analysis of short duplexes containing the single, site-specific adduct of $trans$-[PtCl$_2$(NH$_3$)(thiazole)] (Fig. 15) (unpublished results). The monofunctional adduct creates a local conformational distortion, which is localized to 5 bp around the adduct and includes a stable curvature (34° toward the major groove) and unwinding (13°). Hence, this distortion is very similar to that produced in DNA by major 1,2-GG-intrastrand CLs of antitumor cisplatin (unpublished results). In addition, these monofunctional adducts are recognized by HMGB1-domain proteins with an affinity similar to that of the 1,2-GG-intrastrand CL of cisplatin. The effective removal of these adducts from DNA by NER has also been observed, but it is inhibited if the NER system is supplemented by either HMGB1a or full-length HMGB1 proteins.

Using Maxam–Gilbert footprinting additional work has shown (206) that $trans$-[PtCl$_2$(NH$_3$)(quinoline)] and $trans$-[PtCl$_2$(NH$_3$)(thiazole)] preferentially form DNA-interstrand CLs between guanine residues at the 5′-GC/5′-GC sites. Thus, DNA-interstrand cross-linking by transplatin analogs containing the planar ligand is formally equivalent to that by antitumor cisplatin but different from clinically ineffective transplatin which preferentially forms these adducts between complementary guanine and cytosine residues (see Section II,A,1).

These results show for the first time that the simple chemical modification of the structure of an inactive platinum compound alters its DNA-binding mode into that of an active drug and that the processing of the monofunctional DNA adducts of these $trans$-platinum analogs in tumor cells may be similar to that of the major bifunctional adducts of "classical" cisplatin.

B. Transplatin Analogs Containing Imino Ether Groups

The $trans$-[PtCl$_2$(E-imino ether)$_2$] complex ($trans$-EE) (Fig. 15) not only is more cytotoxic than its cis congener but also is endowed with significant antitumor activity (57, 207) (imino ether = HN=C(OCH$_3$)CH$_3$; it can have either E or Z configuration depending on the relative position of OCH$_3$ and N-bonded Pt with respect to the C=N double bond, with cis in the Z isomer and $trans$ in the E isomer). These results strongly imply a new mechanism of action for $trans$-EE. Analogous to the diamminedichloroplatinum(II) isomers, the inhibition of DNA synthesis by $trans$-EE implies a role for DNA binding in the mechanism of action. The presence of the imino ether group may result in altered hydrogen-bonding and steric effects affecting the kinetics of DNA binding, the structures and/or stability of the adducts formed, and the resulting local conformational alterations in DNA.

Bifunctional $trans$-EE preferentially forms stable monofunctional adducts at guanine residues in double-helical DNA even when DNA is incubated with the platinum complex for a relatively long time (48 h at 37°C in 10 mM NaClO$_4$) (208). The random modification of natural DNA in cell-free media results in the nondenaturational alterations in the conformation of DNA similar to the modification by cisplatin but different from the modification by transplatin which produces denaturational distortions in DNA. The most striking feature of the lesions of *trans-EE* is that they prematurely terminate RNA synthesis at similar sites and with an efficiency similar to that of the major DNA adducts of cisplatin (*209*). This is a very interesting finding, since the prevalent lesions formed on DNA by *trans-EE* are monofunctional adducts at guanine residues and the monofunctional DNA adducts of other simpler platinum(II) complexes (such as those of [PtCl(dien)]Cl, cisplatin, transplatin) do not terminate RNA synthesis. In addition, it is generally accepted that monofunctional DNA adducts of either cisplatin or transplatin are not relevant to the cytostatic effects of these metal complexes.

Recently, the short duplex containing the single, central monofunctional adduct of *trans-EE* at the guanine residue was analyzed by NMR spectroscopy (*210*). This analysis yielded the model from which it is possible to roughly estimate the bending induced by the monofunctional adduct of *trans-EE*. The bending angle estimated in this way was ~45° toward the minor groove.

We confirmed this result by studying the bending induced by the single, site-specific monofunctional adduct of *trans-EE* in the oligodeoxyribonucleotide duplexes (19–22 bp) using electrophoretic retardation as a quantitative measure of the extent of planar curvature (unpublished results). The bending angle estimated by this phasing assay was ~20° toward the minor groove. Hence, the distortion induced in DNA by the monofunctional adduct of *trans-EE* is apparently distinctly different from that produced in DNA by the major 1,2-GG-intrastrand CL of antitumor cisplatin. Then, it is also not surprising that the HMG-domain proteins have no affinity to the monofunctional adduct of *trans-EE* (unpublished results). On the other hand, this adduct is removed from DNA by NER with an efficiency similar to that of the cisplatin 1,2-GG-intrastrand CL of cisplatin. However, on the basis of the analogy with the mechanism of antitumor activity of cisplatin, the NER of *trans-EE* monofunctional adducts should be blocked allowing the damage to persist. As HMG-domain proteins do not bind to the *trans-EE* lesion, it is reasonable to expect that a pathway different from that including noncovalent binding of damaged DNA-binding proteins is effective in blocking NER of the *trans-EE* lesions. Importantly, monofunctional adducts formed by *trans-EE* on DNA can cross-link various types of proteins (including histone H1), and this cross-linking markedly enhances the efficiency of monofunctional adducts of *trans-EE* to terminate DNA and RNA synthesis *in vitro* (unpublished results). In addition, the cross-linking of proteins to DNA monofunctional adducts of *trans-EE* blocks NER of these lesions, suggesting a different mechanism for "repair shielding" of genotoxic *trans-EE* adducts. Unique properties of monofunctional DNA adducts of *trans-EE* and mainly their capability to cross-link proteins might be of a fundamental importance in explaining the anticancer activity of this class of *trans*-platinum(II) complexes.

C. Transplatin Analogs with Asymmetric Aliphatic Amines or Cyclohexylamine Ligands

Recently, the DNA-binding mode of a new cytotoxic *trans*-platinum compound containing different aliphatic amines (*trans*-[PtCl$_2$(dimethylamine)(isopropylamine)]) (Fig. 15) was evaluated in cell-free media (*211*). *trans*-[PtCl$_2$-(dimethylamine)(isopropylamine)] readily forms DNA-interstrand CLs. The amount of these CLs is considerably higher than the amount of these lesions formed under the same conditions by transplatin or cisplatin. In addition, the compound exhibits a preferential binding affinity to alternating purine–pyrimidine sequences in DNA, forms the adduct that inhibits the B-Z transition in DNA, and blocks DNA synthesis *in vitro* more efficiently than the adducts of cisplatin. These particular DNA-binding properties have been proposed to be related to the efficiency of *trans*-[PtCl$_2$(dimethylamine)(isopropylamine)] to induce apoptosis and some selective killing in a *H-ras* overexpressing cell line (*211–213*).

The search for a clinically useful orally available antitumor platinum drug also led to the design of several platinum(IV) compounds with *trans* geometry of leaving ligands (*183*), for example, JM335 [*trans*-amine(cyclohexylamine)dichlorodihydroxoplatinum(IV)] (Fig. 15). Mechanistic studies with JM335 in carcinoma cell lines (*194*) revealed DNA-binding properties different from those of cisplatin. JM335 induces in DNA single-strand breaks (this DNA lesion formation is cell-line dependent) and interstrand CLs. Interestingly, DNA extracted from cells treated with JM335 is not recognized by an antibody elicited against DNA modified by cisplatin. In addition, the kinetics of apoptosis is more rapid for JM335 than for its *cis* isomer. However, since JM335 is generally less cytotoxic than cisplatin probably due to its inactivation by thiols, it has not been selected for clinical trials; nevertheless, it continues to be one of the most interesting and active *trans*-platinum complexes.

V. Polynuclear Platinum Antitumor Drugs

Polynuclear platinum compounds represent a further class of antitumor drugs with chemical and biological properties different from those of cisplatin. The first drug of this class which already entered Phase II clinical trials is the trinuclear compound [{*trans*-PtCl(NH$_3$)$_2$}$_2\mu$-*trans*-Pt(NH$_3$)$_2${H$_2$N(CH$_2$)$_6$NH$_2$}$_2$]$^{4+}$ (Fig. 16). The compound, designated as BBR3464, is the lead representative of the entirely new structural class of DNA-modifying anticancer agents based on the poly(di,tri)nuclear platinum structural motif. Initial clinical trials revealed activity in pancreatic, lung, and melanoma cancers suggesting the complementary clinical anticancer activity of BBR3464 in comparison to cisplatin.

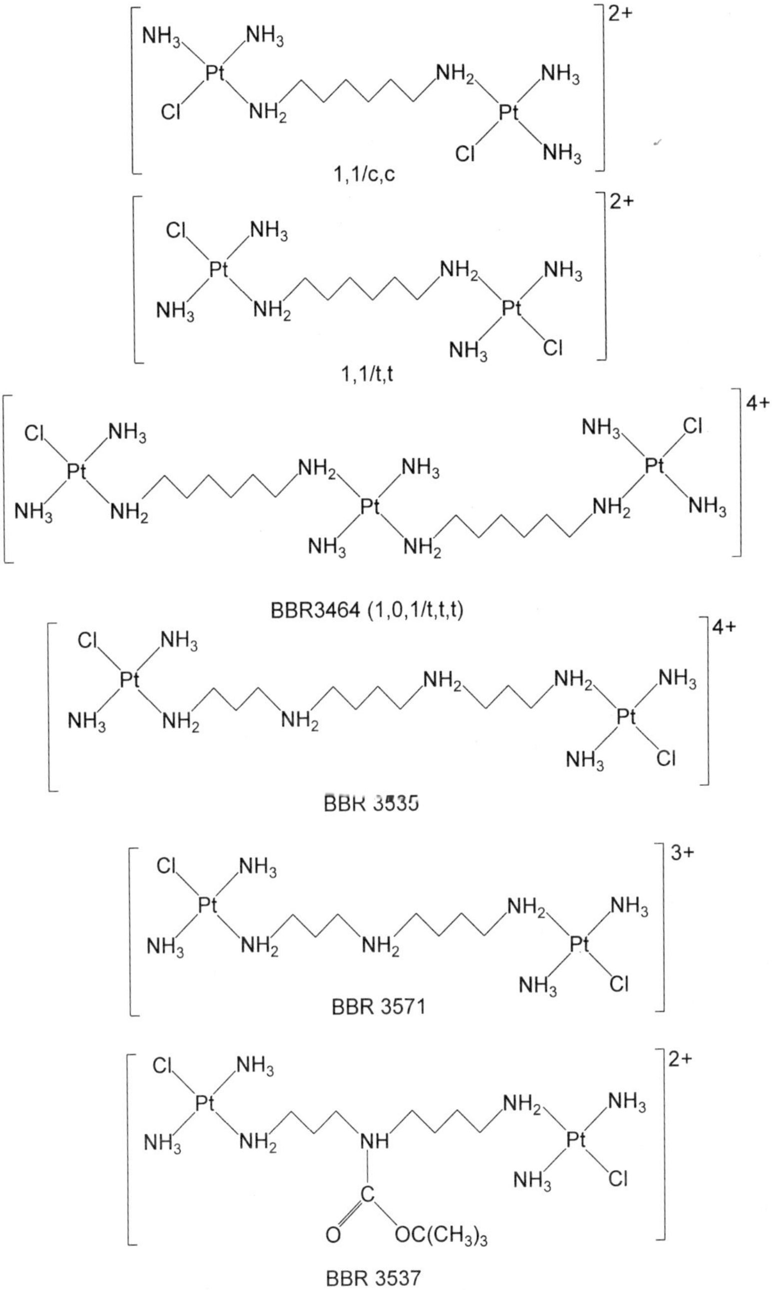

FIG. 16. Structures of antitumor dinuclear and trinuclear platinum(II) compounds.

The choice of BBR3464 as a clinical candidate also comes from preclinical studies showing cytotoxicity at a 10-fold lower concentration than that in cisplatin and collateral sensitivity in cisplatin-resistant cell lines. Importantly, BBR3464 also consistently displays high antitumor activity in human tumor xenografts characterized as mutant p53. This important feature suggests that the new agent may find utility in the over 60% of cancer cases where mutant p53 status has been indicated. DNA damage by chemotherapeutic agents is in many cases mediated through the p53 pathway. Consistently, cytotoxicity displayed in mutant cell lines would suggest an ability to by-pass this pathway.

BBR3464 has been designed on the basis of systematic studies on various dinuclear compounds. The developmental history of this class of platinum drugs has recently been thoroughly reviewed (*214, 215*). In the first part of this chapter, DNA interactions in cell-free media of the bifunctional dinuclear compounds, which are the agents most closely related to trinuclear BBR3464, were briefly summarized. The following section focuses on the studies of DNA modifications by BBR3464.

A. Dinuclear Compounds

The exploration of new structural classes of platinum antitumor drugs resulted in the discovery of dinuclear bis(platinum) complexes with equivalent coordination spheres with the single-chloride-leaving group on each platinum linked by a variable-length diamine chain, represented by the general formula $[\{PtCl(NH_3)_2\}_2(H_2N(CH_2)_nNH_2)]^{2+}$. The chloride-leaving ligands are either *cis* (1,1/*c,c*) or *trans* (1,1/*t,t*) (Fig. 16). These dinuclear platinum complexes exhibit antitumor activity *in vitro* and *in vivo* comparable to that of cisplatin, but importantly they retain activity in acquired cisplatin-resistant cell lines (*215*). However, the 1,1/*c,c* complexes have been shown to be less efficient in overcoming cisplatin resistance than their 1,1/*t,t* counterparts. This situation represents a fundamental difference between mononuclear and dinuclear platinum chemistry and biology—in mononuclear platinum chemistry cisplatin is active, while its direct isomer transplatin is antitumor inactive. These differences in biological activity between dinuclear and mononuclear platinum complexes on one hand and the differences between the dinuclear compounds themselves on the other provided the impetus for the studies of molecular mechanisms underlying these differences.

The dinuclear platinum complexes, similar to cisplatin or carboplatin, bind to DNA and inhibit DNA replication and transcription, which indicates that DNA modification by dinuclear platinum complexes plays an important role in the mechanism of their biological action. Initial studies have already shown some significant differences in DNA modification in cell-free media by individual dinuclear platinum complexes and cisplatin. 1,1/*t,t* isomers bind to DNA more readily than 1,1/*c,c* complexes. Both isomers unwind globally modified DNA by 10–12°, i.e., similar to the case of cisplatin. In contrast to cisplatin, both dinuclear isomers preferentially form interstrand CLs in DNA, 1,1/c,c isomer being more efficient. 1,1/t,t complexes have also been shown to form minor 1,2-GG-intrastrand CLs producing a flexible, nondirectional bend in DNA (216). Intrastrand CLs were not observed in double-helical DNA if it was modified by 1,1/c,c. A more recent study (217) investigated the reasons underlying the observed inability of 1,1/c,c complexes to form 1,2-GG-intrastrand CL with double-helical DNA. ^{1}H NMR spectroscopy of samples of very short, single-stranded di- or tetranucleotides containing the GG sequence modified by 1,1/c,c provided evidence for restricted rotation around the 3′ G in single-stranded 1,2-GG-intrastrand CLs of 1,1/c,c. This steric hindrance, not present in 1,2-GG-intrastrand CLs of 1,1/t,t complexes, has been suggested to be responsible for the inability of 1,1/c,c complexes to form 1,2-GG-intrastrand CLs with sterically more demanding double-helical DNA.

More recent studies have mainly been focused on the structural details of major interstrand CLs formed by 1,1/c,c and 1,1/t,t in DNA (218, 219). The interstrand CLs of 1,1/c,c are preferentially formed between guanine residues. Besides 1,2-interstrand CLs (between guanine residues in neighboring bps), 1,3- or 1,4-interstrand CLs are also possible. In the latter two long-range adducts, the sites involved in the cross-links are separated by 1 or 2 bp. 1,2-, 1,3-, and 1,4-interstrand CLs are formed with a similar rate and are preferentially oriented in the 5′ → 5′ direction. In addition, the DNA adducts of these complexes inhibit DNA transcription *in vitro*.

The properties of the interstrand CLs formed in DNA by 1,1/t,t complexes have been investigated more thoroughly. Preferential DNA-binding sites in these lesions are G residues in the base pairs separated by at least one other base pair. The CLs between G residues in neighboring base pairs (1,2-interstrand CLs) are formed at a pronouncedly slower rate. Importantly, the length of the diamine bridge linking the two platinum units does not appear to be a substantial factor affecting DNA-interstrand cross-linking by the bifunctional dinuclear platinum compounds. Similarly, as in the case of the interstrand CLs of 1,1/c,c, these lesions of 1,1/t,t complexes are also preferentially formed in the 5′ → 5′ direction, so that this feature of the interstrand cross-linking might be common for this class of platinum compounds. The reasons for this preference in the orientation of DNA-interstrand CLs of dinuclear compounds are unknown.

The conformational distortions induced in DNA by the major 1,3- or 1,4-interstrand CLs of 1,1/t,t have been further evaluated in more detail. Interestingly, this lesion results only in a very small directional bending of helix axis (∼10°) and duplex unwinding (9°). Despite these small bending and unwinding effects, the major interstrand CLs formed by 1,1/t,t compounds create a local conformational distortion extending over ∼5 bp at the site of the CL.

HMG-domain proteins play a role in sensitizing cells to cisplatin (see Sections II,E,1 and II,G). An important structural motif recognized by HMG-domain proteins on DNA modified by cisplatin is a stable, directional bend of the helix axis. The major interstrand CLs and even the minor 1,2-d(GpG) intrastrand cross-link of 1,1/t,t compounds bend the helix axis much less efficiently than the CLs of cisplatin. Therefore, it is not surprising that either very weak or no recognition of DNA adducts of 1,1/t,t complexes by HMGB1 protein has been observed (219). This is consistent with the assumption that an important structural motif recognized by HMG-domain proteins is a bent or kinked duplex axis. The affinity of HMG-domain proteins to the duplex containing a 1,2-GG-intrastrand CL of cisplatin is sequence dependent and is reduced with increasing destabilization of the duplex due to the CL. Thus, no affinity of the minor 1,2-GG-intrastrand CL of 1,1/t,t to HMG-domain proteins is consistent with the observation that this lesion reduces the thermal and thermodynamical stability of DNA markedly more than the same lesion of cisplatin (94). Thus, it is clear that the major DNA adducts of antitumor 1,1/t,t compounds may present a block to DNA or RNA polymerase (220, 221) but are not a substrate for recognition by HMG-domain proteins. From these considerations, it can be concluded that the mechanism of antitumor activity of bifunctional dinuclear platinum complexes does not involve recognition by HMG-domain proteins as a crucial step, in contrast to the proposals for cisplatin and its direct analogs.

One possible role for binding HMG-domain proteins to DNA modified by cisplatin is that these proteins shield damaged DNA from intracellular NER (see Section II,G). No recognition of the 1,2-intrastrand CL of 1,1/t,t by HMGB1 proteins has been noticed (219), but effective removal of these adducts from DNA by NER has been observed (111). These results suggest that the processing of the intrastrand CLs of 1,1/t,t in tumor cells sensitive to this drug may not be relevant to its antitumor effect. Major adducts formed in DNA by 1,1/t,t compounds are, however, interstrand CLs. In general, DNA-interstrand CLs pose a special challenge to repairing enzymes because they involve both strands of DNA and therefore cannot be repaired using the information in the complementary strand for resynthesis. The fact that interstrand CLs cannot be removed so readily by NER as intrastrand lesions is also corroborated by the observation that excision repair of the interstrand CL formed by cisplatin has not been detected under conditions when intrastrand adducts of this drug are readily removed by a reconstituted system containing highly purified NER factors (76). Similarly, 1,3- or 1,4-interstrand CLs of 1,1/t,t are not removed in the *in vitro* assay using mammalian cell-free extracts capable of removing the 1,2-intrastrand CLs of cisplatin (unpublished results). Hence, the major DNA adducts of bifunctional dinuclear platinum compounds would not have to be shielded by damaged DNA-recognition proteins, such as those containing HMG domains, to prevent their repair.

As mentioned earlier, the addition of the central tetraamine(platinum) unit, as in BBR3464, greatly enhances the cytotoxicty and antitumor activity in comparison to the prototype dinuclear compound linked by a simple diamine, as in 1,1/t,t or 1,1/c,c. In examining structure–activity relationships within this class of compounds, a logical step was to substitute the central tetraamine platinum unit with other H-bonding groups and retain the antitumor activity. Designed synthesis of linear polyamine-linked dinuclear platinum complexes display the same biological profile as BBR3464 and represent a further subclass of dinuclear compounds with a promising preclinical activity (214).

The DNA-binding profiles of three bifunctional dinuclear platinum(II) polyamine-linked compounds, [{$trans$-PtCl(NH$_3$)$_2$}$_2${μ-spermine-N^1,N^{12}}]$^{4+}$, [{$trans$-PtCl(NH$_3$)$_2$}$_2${μ-spermidine-N^1,N^8}]$^{3+}$, and [{$trans$-PtCl(NH$_3$)$_2$}$_2$ {μ-BOC-spermidine}]$^{2+}$ (BBR3535, BBR3571, and BBR3537, respectively, in Fig. 16), have been evaluated (222). All of the compounds bind preferentially in a bifunctional manner, according to the unwinding of the supercoiled DNA substrate. The kinetics of binding of these compounds correspond to their relative high charge (2+ to 4+). The preference for the formation of interstrand CLs, however, does not follow a charge-based pattern. The preferred binding sites of the dinuclear polyamine-linked compounds are similar to those of their trinuclear BBR3464 counterpart (see Section V,B), and charge differences do not contribute solely to the variances between the compounds.

B. Trinuclear Compounds

The identification of bifunctional trinuclear BBR3464 (Fig. 16) as eventually the first polynuclear platinum compound used in the clinic was also the impetus for its DNA-binding studies (223). The high charge on BBR3464 facilitates rapid binding to DNA with a $t_{1/2}$ of ~40 min, significantly faster than that for the neutral cisplatin. The melting temperature of DNA adducted by BBR3464 increases at low ionic strength but decreases in high salt for the same level of the modification. This unusual behavior is in contrast to that of cisplatin. BBR3464 produces an unwinding angle of 14° in negatively supercoiled pSP73 plasmid DNA, indicative of bifunctional DNA binding. Quantitation of interstrand DNA cross-linking in linearized plasmid DNA has indicated approximately 20% of the DNA to be interstrand cross-linked. While this is significantly higher than the value for cisplatin, it is, interestingly, lower than that for dinuclear platinum compounds such as 1,1/t,t or 1,1/c,c. Either the presence of charge in the linker backbone or the increased distance between platinating moieties may contribute to this relatively decreased ability of BBR3464 to induce DNA-interstrand cross-linking. Moreover, BBR3464 rapidly forms long-range delocalized lesions on DNA with sequence selectivity and strong sequence preference for single dG or d(GG) sites. By choosing an appropriate sequence on the basis of sequence specificity studies, molecular modeling on 1,4-GG-interstrand and 1,5-GG-intrastrand CLs has further confirmed the similarity in energy between the two forms of CL (Fig. 17). Finally, immunochemical analysis, which has shown that antibodies raised to cisplatin-adducted DNA do not recognize DNA modified by BBR3464, has confirmed the unique nature of the DNA adducts

5'- T G A A T T C G$_i$ A G C T C G G T A -3'

3'- A C T T A A G$_{iii}$ C T C G$_{ii}$ A G C C A T -5'

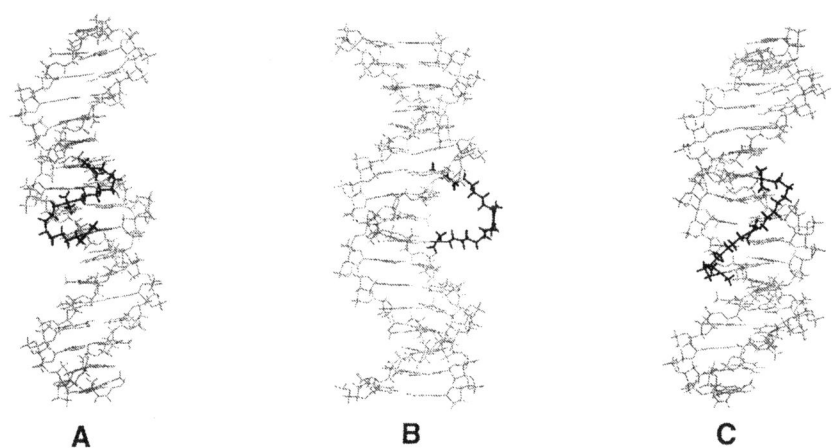

FIG. 17. Models for long-range intrastrand and interstrand CLs by trinuclear BBR3464 on DNA. (A, B) two conformations of a 1,4-interstrand CL formed between G$_i$ and G$_{ii}$, (C) A 1,5-intrastrand CL formed between G$_{ii}$ and G$_{iii}$. Reprinted with permission from V. Brabec, J. Kasparkova, O. Vrana, O. Novakova, J. W. Cox, Y. Qu, and N. Farrell, DNA Modifications by a novel bifunctional trinuclear platinum Phase I anticancer agent. *Biochemistry*, **38**, 6781–6790, Copyright 1999, American Chemical Society.

formed by BBR3464. In contrast, DNA modified by BBR3464 inhibits the binding of antibodies raised to transplatin-adducted DNA. Thus, the bifunctional binding of BBR3464 contains few similarities to that of cisplatin but may have a subset of adducts recognized as similar to those of the transplatinum species.

We examined how the structures of the various types of the CLs of BBR3464 affect conformational properties of DNA and how these adducts are recognized by HMGB1 protein and removed from DNA during *in vitro* NER reactions. The first analyses (*224*) focused on the 1,2-GG- and other long-range intrastrand CLs between guanine residues and revealed that these lesions create a local conformational distortion; however, none of these CLs result in a stable curvature. In addition, we observed no recognition of these CLs by HMGB1 protein, but we observed effective removal of these adducts from DNA by NER. Thus, similar to the case of the intrastrand CLs of dinuclear platinum compounds (see Section V,A), the processing of the intrastrand CLs of BBR3464 in tumor cells sensitive to this new drug may not be relevant to its antitumor effect. The analysis of interstrand CLs formed by BBR3464 in DNA is in progress, but it can be anticipated that, similar to the case of the dinuclear complex 1,1/t,t, these lesions could be more likely candidates for the genotoxic lesion relevant to the antitumor effects of BBR3464. On the other hand, the cytotoxic effects of BBR3464 may also be due to a cumulative effect of the structurally heterogeneous adducts produced by this polynuclear platinum drug.

In summary, BBR3464 exhibits a unique profile of DNA binding, strengthening the original hypothesis that modification of DNA binding in manners distinct from that of cisplatin would also lead to a distinct and unique profile of antitumor activity.

VI. Antitumor Ruthenium Compounds

One approach in the search for the new, metal-based anticancer agent which would exhibit antitumor activity markedly different from that of cisplatin and its direct analogs is to examine complexes that would contain another transition metal. The possible advantages in using transition-metal ions other than platinum may involve additional coordination sites, alterations in ligand affinity and substitution kinetics, changes in oxidation state, and photodynamic approaches to therapy (225). In the design of these new drugs, ruthenium complexes have raised great interest (225–227). Although the pharmacological target for antitumor ruthenium compounds has not been univocally identified, several ruthenium compounds have been found to inhibit DNA replication, exhibit mutagenic activity, induce SOS repair mechanism, bind to nuclear DNA, and reduce RNA synthesis, which is consistent with DNA binding of these compounds *in vivo*. Thus, analogous to platinum antitumor drugs, DNA interactions of antitumor ruthenium agents are of great interest as well. Since several review articles appeared recently (225–228) in which pharmacological properties of antitumor ruthenium compounds including their DNA interactions were examined, the main focus in this section is on the current discoveries from our laboratory in the field of DNA modifications by antitumor ruthenium agents in cell-free media.

A. Dimethyl Sulfoxide Complexes

Dimethyl sulfoxide complexes of both ruthenium(II) and ruthenium(III) constitute a relatively new group of anticancer compounds (225, 227). For instance, these complexes exhibit antiblastic activity comparable to cisplatin at equitoxic dosage in animal models of metastasizing tumors, but with less severe side effects and prolonged host survival times. A small series of complexes whose parent compounds are *cis*- and *trans*-[Ru(II)((CH_3)$_2$SO)$_4$Cl$_2$] (Fig. 18) constitute one class of dimethyl sulfoxide ruthenium compounds (227). The examination

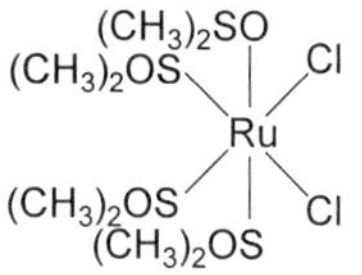

cis-$[Ru(II)((CH_3)_2SO)_4Cl_2]$

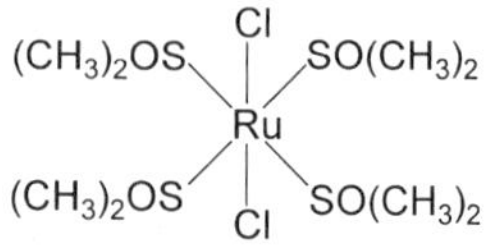

trans-$[Ru(II)((CH_3)_2SO)_4Cl_2]$

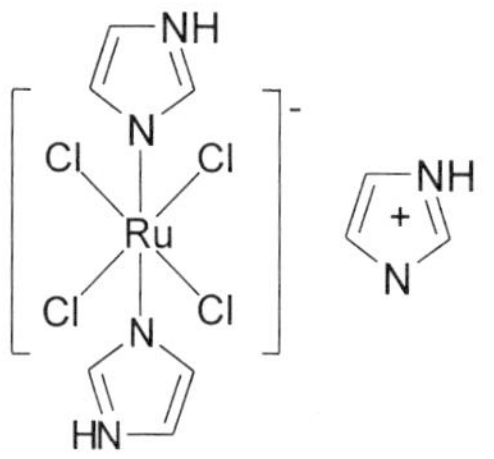

HIm[*trans*-$Ru(III)Cl_4Im_2$]

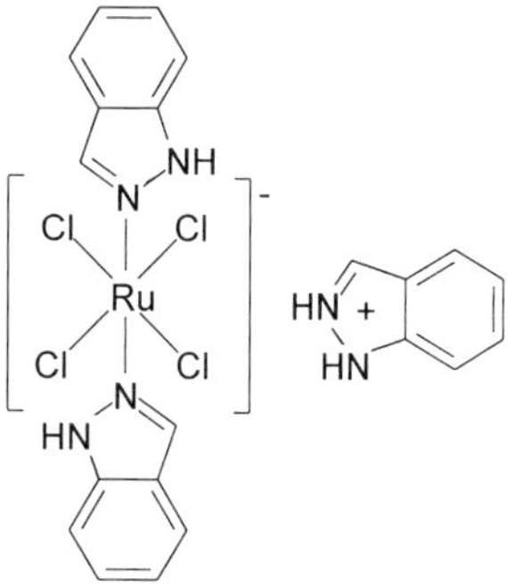

HInd[*trans*-$Ru(III)Cl_4Ind_2$]

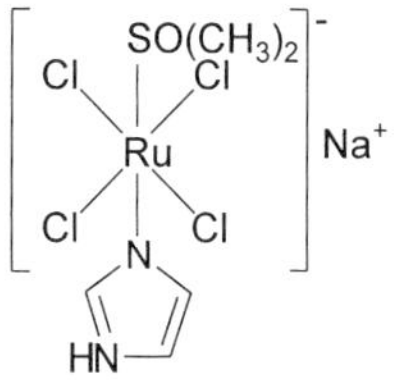

Na[*trans*-$Ru(III)((CH_3)_2SO)Cl_4Im$]

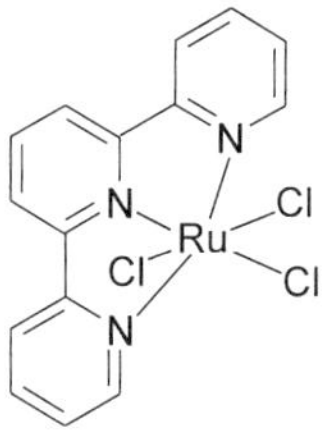

mer-$[Ru(II)Cl_2(terpy)]$

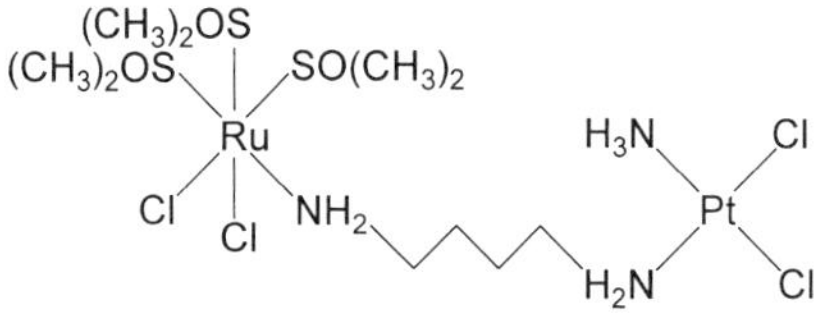

[{*cis*-$Ru(II)((CH_3)_2SO)_3Cl_2$}$NH_2(CH_2)_4NH_2${*cis*-$PtCl_2(NH_3)$}]

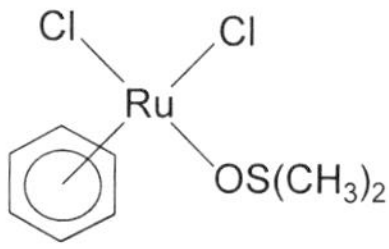

$Ru(II)(C_6H_6)((CH_3)_2SO)Cl_2$

FIG. 18. Structures of antitumor ruthenium compounds.

of their effect on both primary tumor and metastasis development revealed antimetastatic activities superior to the effects on primary tumor growth. The initial studies were performed with cis-[Ru(II)((CH$_3$)$_2$SO)$_4$Cl$_2$] because of its similarity to cisplatin. However, comparisons between the antitumor effects of cis- and trans-[Ru(II)((CH$_3$)$_2$SO)$_4$Cl$_2$] revealed the superiority of the latter.

cis- and trans-[Ru(II)((CH$_3$)$_2$SO)$_4$Cl$_2$] contain two chlorides in the octahedral structure (228). In cis-[Ru(II)((CH$_3$)$_2$SO)$_4$Cl$_2$], the three (CH$_3$)$_2$SO molecules are S-bound in a facial configuration and the fourth is O-bonded. In trans-[Ru(II)((CH$_3$)$_2$SO)$_4$Cl$_2$], all the (CH$_3$)$_2$SO's are S-bound. When dissolved in water the cis isomer immediately undergoes loss of the O-bonded dimethyl sulfoxide ligand, whereas the trans compound rapidly loses two S-bonded dimethyl sulfoxide ligands yielding cis-diaqua species. Both hydrolyzed isomers then undergo slow reversible chloride dissociation forming cationic compounds. After this step, the trans compound contains three reactive groups, while the cis isomer contains only two (228). In addition, the three remaining (CH$_3$)$_2$SO ligands in the cis isomer represent a considerable steric hindrance, which makes the cis-aqua species relatively inert in contrast to the trans isomer. This difference correlates with a higher potency of the trans isomer to act as the antitumor agent (229).

Both cis- and trans-[Ru(II)((CH$_3$)$_2$SO)$_4$Cl$_2$] bind to DNA in cell-free media (228–230). Some early studies based on the analysis of circular dichroism spectra of DNA suggest that coordination of the cis isomer to DNA does not significantly alter the conformation of B-DNA (228). The trans isomer binds to DNA more rapidly with some changes in the CD spectra indicating conformational alterations (228). Both isomers have a limited preference for bifunctional binding to neighboring guanine residues at their N7 atoms, with the trans isomer being more effective (230). The DNA-binding mode of trans-[Ru(II)((CH$_3$)$_2$SO)$_4$Cl$_2$] includes formation of bifunctional adducts such as intrastrand CLs between neighboring purine residues and a small amount (~1%) of interstrand CLs. cis-[Ru(II)((CH$_3$)$_2$SO)$_4$Cl$_2$] forms mainly monofunctional lesions on natural DNA. Both ruthenium isomers induce conformational alterations of nondenaturational character in DNA, with the trans compound being more effective. In addition, DNA adducts of trans-[Ru(II)((CH$_3$)$_2$SO)$_4$Cl$_2$] are capable of inhibiting RNA synthesis by DNA-dependent RNA polymerases, while the adducts of the cis isomer are not. Thus, several features of the DNA-binding mode of trans-[Ru(II)((CH$_3$)$_2$SO)$_4$Cl$_2$] are similar to those of antitumor cisplatin (230), which may be relevant to biological effects of this antitumor ruthenium drug. On the other hand, the different DNA-binding mode of cis-[Ru(II)((CH$_3$)$_2$SO)$_4$Cl$_2$] is consistent with its less pronounced biological effects.

Interesting results were also obtained in the studies of the mechanism of antitumor activity of Ru(II) (C$_6$H$_6$)((CH$_3$)$_2$SO)Cl$_2$ complex (Fig. 18) (231). This compound exhibits a strong DNA-binding affinity, but this binding does not substantially affect the conformation of DNA. On the other hand, it could completely inhibit the DNA-relaxation activity of topoisomerase II by forming a ternary complex containing DNA and the enzyme. A model has been proposed for this ternary complex, in which the ruthenium atom is coordinated by DNA, and the ligands of the ruthenium atom cross-link topoisomerase II.

B. Heterocyclic Complexes

Heterocyclic complexes of ruthenium (III) constitute a relatively new group of anticancer compounds (225, 226). The general formula of this class of compounds is HB[trans-Ru(III)B$_2$Cl$_4$], where B stands for a heterocyclic base, such as imidazole (Im) or indazole (Ind) (Fig. 18). These complexes exhibit activity in various tumor models and are particularly effective against cisplatin-resistant colorectal tumors (226). Later an analog of these complexes was developed, namely, Na[trans-Ru(III)((CH$_3$)$_2$SO)Cl$_4$Im], which exhibits encouraging antitumor and antimetastatic properties (227). Thus, the spectrum of the antitumor effects of these ruthenium compounds differs significantly from that of cisplatin while showing a lower systemic toxicity than that of platinum(II) compounds (232).

HB[trans-Ru(III)B$_2$Cl$_4$] and Na[trans-Ru(III)((CH$_3$)$_2$SO)Cl$_4$Im] are pseudo-octahedral with four equatorial chloride ligands, with the heterocyclic bases and (CH$_3$)$_2$SO as axial ligands. The complexes lose their chloride ligands and transform into the corresponding, more reactive, aquated species (225–227). The complexes HIm[trans-Ru(III)Cl$_4$Im$_2$], HInd[trans-Ru(III)Cl$_4$Ind$_2$], and Na[trans-Ru(III)((CH$_3$)$_2$SO)Cl$_4$Im] bind irreversibly to DNA (233). Their DNA-binding mode is, however, different from that of cisplatin. Interestingly, Na[trans-Ru(III)((CH$_3$)$_2$SO)Cl$_4$Im] binds to DNA considerably faster than the other two ruthenium compounds and cisplatin. In addition, when Na[trans-Ru(III)((CH$_3$)$_2$SO)Cl$_4$Im] binds to DNA, it exhibits an enhanced base-sequence specificity in comparison to the other two ruthenium complexes. Na[trans-Ru(III)((CH$_3$)$_2$SO)Cl$_4$Im] also forms bifunctional intrastrand adducts on double-helical DNA which are capable of terminating RNA synthesis *in vitro*, while the capability of the other two ruthenium compounds to form such adducts is markedly lower. This observation has been interpreted to indicate that the bifunctional adducts of HInd[trans-Ru(III)Cl$_4$Ind$_2$] and Na[trans-Ru(III)((CH$_3$)$_2$SO)Cl$_4$Im] formed on rigid double-helical DNA are sterically more crowded by their octahedral geometry than those of Na[trans-Ru(III)((CH$_3$)$_2$SO)Cl$_4$Im]. In addition, the adducts of all three heterocyclic ruthenium compounds affect DNA conformation, with Na[trans-Ru(III)((CH$_3$)$_2$SO)Cl$_4$Im] being the most effective. It has been suggested that, similar to the case of dimethyl sulfoxide ruthenium complexes, the altered DNA-binding mode of ruthenium compounds in comparison to that of cisplatin is an important factor responsible for the altered cytostatic activity of this class of ruthenium compounds in tumor cells.

C. Chloropolypyridyl Compounds

A third group of antitumor ruthenium compounds examined in this article includes the ruthenium complexes of polypyridyl ligands. Some of these complexes exist as chiral molecules capable of enantioselective recognition of DNA. Thus, DNA-binding and cleavage properties of various polypyridyl ruthenium compounds have been intensively investigated since they have also been proposed as possibly being useful probes of either DNA conformation (234) or DNA cleavage agents (235, 236). The analogs of these ruthenium complexes containing—besides polypyridyl ligands—aqua or chloro groups, were also synthesized and were found to bind DNA covalently in cell-free media (235, 237). The aqua or chloro ligands in these complexes represent leaving ligands in contrast to the kinetically more stable pyridyl groups.

The cytotoxicity of chloropolypyridyl ruthenium complexes of structural formulas [Ru(II)Cl(bpy)(terpy)]Cl, cis-[Ru(II)(bpy)$_2$Cl$_2$], and mer-[Ru(II)(terpy)Cl$_3$] (bpy = 2,2'-bipyridyl, terpy = 2,2':6',2''-terpyridine) has been demonstrated in murine and human tumor cell lines (238). mer-[Ru(II)Cl$_3$(terpy)] (Fig. 18) exhibits a remarkably higher cytotoxicity than the other complexes. Moreover, investigations of antitumor activity in a standard tumor screen, revealed the highest efficiency for mer-[Ru(II)Cl$_3$(terpy)]. In a cell-free medium, the ruthenium complexes coordinate to DNA, preferentially at guanine residues (238, 239). The resulting adducts terminate DNA synthesis in vitro. The reactivity of the complexes to DNA, their efficiency to unwind closed, negatively supercoiled DNA and a sequence preference of their DNA adducts do not show a correlation to biological activity. On the other hand, the cytotoxic mer-[Ru(II)Cl$_3$(terpy)] exhibits a significant DNA-interstrand cross-linking in contrast to the significantly less active complexes which exhibit no such efficacy. Thus, this potential new class of metal-based antitumor compounds may act by a mechanism involving DNA-interstrand cross-linking.

D. Heterodinuclear (Ruthenium, Platinum) Compounds

In general, ruthenium compounds are less reactive than platinum compounds, but ruthenium and platinum combined in one molecule may be used to both impart reactivity to particular DNA sequences and facilitate cross-linking of unique proteins with DNA. Therefore, a heteronuclear compound containing both platinum and ruthenium [{cis-Ru(II)((CH$_3$)$_2$SO)$_3$Cl$_2$}NH$_2$(CH$_2$)$_4$NH$_2${cis-PtCl$_2$(NH$_3$)}] (Fig. 18) has been synthesized (240). The ternary coordination complexes between DNA treated with the heterodinuclear compound and the E. coli UvrA and UvrB repair proteins have been identified. The DNA lesion responsible for efficient DNA–protein cross-linking is interstrand CL in which each metal atom is coordinated with one strand of the DNA double helix. The detailed structure of these ternary complexes awaits further study, but the formation of DNA-repair protein-associated DNA CLs suggests a novel action mechanism for these anticancer compounds.

Further studies should reveal the extent to which ruthenium complexes hold promise as clinically useful antitumor compounds.

VII. Concluding Remarks

Platinum and ruthenium antitumor drugs represent important agents for the treatment of several different tumors. It appears that changing the chemical structure of platinum and ruthenium compounds may substantially modulate their DNA-binding mode, subsequent processing of DNA damage, and consequently the mechanism of biological efficacy of these metal-based compounds. Thus, these structural alterations may also influence not only their spectrum of activity and the development of drug resistance but also the toxicity profile of these drugs. A further understanding of how new platinum and ruthenium compounds modify DNA and of how these modifications are further processed in cells should provide a rational basis for designing new chemotherapeutic strategies and metal-based antitumor drugs rather than only searching for cisplatin analogs.

Acknowledgments

Supported by Grant A5004101 from the Grant Agency of the Academy of Sciences of the Czech Republic and by the EC COST Chemistry Action D20. The author thanks Drs. J. Kasparkova and J. Kozelka for critical reading of the manuscript. Note that the reference list is not an exhaustive review of all literature associated with the subject of this article. In many cases, the most recent papers or reviews are cited to enable the reader to trace back to earlier contributions.

References

1. B. Rosenberg, L. Van Camp, J. E. Trosko, and V. H. Mansour, Platinum compounds: A new class of potent antitumor agents. *Nature* **222**, 385–386 (1969).
2. E. Wong and C. M. Giandomenico, Current status of platinum-based antitumor drugs. *Chem. Rev.* **99**, 2451–2466 (1999).
3. P. J. O'Dwyer, J. P. Stevenson, and S. W. Johnson, Clinical status of cisplatin, carboplatin, and other platinum-based antitumor drugs. *in* "Cisplatin. Chemistry and Biochemistry of a Leading Anticancer Drug" (B. Lippert, ed.), pp. 31–72. VHCA, Wiley-VCH, Zürich/Weinheim, 1999.
4. G. Giaccone, Clinical perspectives on platinum resistance. *Drugs* **59**, 9–17 (2000).
5. N. P. Johnson, J.-L. Butour, G. Villani, F. L. Wimmer, M. Defais, V. Pierson, and V. Brabec, Metal antitumor compounds: The mechanism of action of platinum complexes. *Prog. Clin. Biochem. Med.* **10**, 1–24 (1989).

6. E. R. Jamieson and S. J. Lippard, Structure, recognition, and processing of cisplatin-DNA adducts. *Chem. Rev.* **99**, 2467–2498 (1999).
7. J. Reedijk, Improved understanding in platinum antitumour chemistry. *Chem. Commun.* 801–806 (1996).
8. A. Gelasco and S. J. Lippard, Anticancer activity of cisplatin and related complexes. *in* "Metallopharmaceuticals I. DNA Interactions" (M. J. Clarke and P. J. Sadler, eds.), pp. 1–43. Springer, Berlin, 1999.
9. Z. J. Guo and P. J. Sadler, Metals in medicine. *Angew. Chem. Int. Ed.* **38**, 1513–1531 (1999).
10. D. B. Zamble and S. J. Lippard, The response of cellular proteins to cisplatin-damaged DNA. *in* "Cisplatin. Chemistry and Biochemistry of a Leading Anticancer Drug" (B. Lippert, ed.), pp. 73–110. VHCA, Wiley-VCH, Zürich/Weinheim, 1999.
11. D. P. Bancroft, C. A. Lepre, and S. J. Lippard, ^{195}Pt NMR kinetic and mechanistic studies of *cis*-diamminedichloroplatinum and *trans*-diamminedichloroplatinum(II) binding to DNA. *J. Am. Chem. Soc.* **112**, 6860–6871 (1990).
12. A. M. J. Fichtinger-Schepman, J. L. Van der Veer, J. H. J. Den Hartog, P. H. M. Lohman, and J. Reedijk, Adducts of the antitumor drug *cis*-diamminedichloroplatinum(II) with DNA: Formation, identification, and quantitation. *Biochemistry* **24**, 707–713 (1985).
13. A. Eastman, The formation, isolation and characterization of DNA adducts produced by anticancer platinum complexes. *Pharmacol. Ther.* **34**, 155–166 (1987).
14. V. Brabec, Chemistry and structural biology of 1,2-interstrand adducts of cisplatin. *in* "Platinum-Based Drugs in Cancer Therapy" (L. R. Kelland and N. P. Farrell, eds.), pp. 37–61. Humana Press Inc., Totowa, NJ, 2000.
15. O. Vrana, V. Boudny, and V. Brabec, Superhelical torsion controls DNA interstrand cross-linking by antitumor *cis*-diamminedichloroplatinum(II). *Nucleic Acids Res.* **24**, 3918–3925 (1996).
16. V. Brabec, V. Kleinwächter, J. L. Butour, and N. P. Johnson, Biophysical studies of the modification of DNA by antitumour platinum coordination complexes. *Biophys. Chem.* **35**, 129–141 (1990).
17. A. Eastman, M. M. Jennerwein, and D. L. Nagel, Characterization of bifunctional adducts produced in DNA by *trans*-diamminedichloroplatinum(II). *Chem.–Biol. Interact.* **67**, 71–80 (1988).
18. V. Brabec and M. Leng, DNA interstrand cross-links of *trans*-diamminedichloroplatinum(II) are preferentially formed between guanine and complementary cytosine residues. *Proc. Natl. Acad. Sci. U.S.A.* **90**, 5345–5349 (1993).
19. M. Boudvillain, R. Dalbies. C. Aussourd, and M. Leng, Intrastrand cross-links are not formed in the reaction between transplatin and native DNA: Relation with the clinical inefficiency of transplatin. *Nucleic Acids Res.* **23**, 2381–2388 (1995).
20. M. Leng, A. Schwartz, and M. J. Giraud-Panis, Transplatin-modified oligonucleotides as potential antitumor drugs. *in* "Platinum-Based Drugs in Cancer Therapy" (L. R. Kelland and N. P. Farrell, eds.), pp. 63–85. Humana Press, Clifton, NJ, 2000.
21. N. P. Johnson, J.-P. Macquet, J. L. Wiebers, and B. Monsarrat, Structures of adducts formed between [Pt(dien)Cl]Cl and DNA in vitro. *Nucleic Acids Res.* **10**, 5255–5271 (1982).
22. J. L. Leroy, E. Charretier, M. Kochyian, and M. Gueron, Evidence from base-pair kinetics for two types of adenine tract structures in solution: their reaction to DNA curvature. *Biochemistry* **27**, 8894–8898 (1988).
23. J. Ramstein and R. Lavery, Base pair opening pathways in B-DNA. *J. Biomol. Struct. Dyn.* **7**, 915–933 (1990).
24. A. Schwartz, M. Sip, and M. Leng, Sodium cyanide—A chemical probe of the conformation of DNA modified by the antitumor *cis*-diamminedichloroplatinum(II). *J. Am. Chem. Soc.* **112**, 3673–3674 (1990).

25. M. A. Lemaire, A. Schwartz, A. R. Rahmouni, and M. Leng, Interstrand cross-links are preferentially formed at the d(GC) sites in the reaction between cis-diamminedichloroplatinum(II) and DNA. *Proc. Natl. Acad. Sci. U.S.A.* **88**, 1982–1985 (1991).
26. D. Z. Yang and A. H. J. Wang, Structural studies of interactions between anticancer platinum drugs and DNA. *Prog. Biophys. Mol. Biol.* **66**, 81–111 (1996).
27. K. M. Comess, C. E. Costello, and S. J. Lippard, Identification and characterization of a novel linkage isomerization in the reaction of trans-diamminedichloroplatinum(II) with 5′-d(TCTACGCGTTCT). *Biochemistry* **29**, 2102–2110 (1990).
28. R. Dalbies, M. Boudvillain, and M. Leng, Linkage isomerization reaction of intrastrand cross-links in trans-diamminedichloroplatinum(II)-modified single-stranded oligonucleotides. *Nucleic Acids Res.* **23**, 949–953 (1995).
29. R. Dalbies, D. Payet, and M. Leng, DNA double helix promotes a linkage isomerization reaction in trans-diamminedichloroplatinum(II)-modified DNA. *Proc. Natl. Acad. Sci. U.S.A.* **91**, 8147–8151 (1994).
30. M. Boudvillain, M. Guerin, R. Dalbies, T. Saison-Behmoaras, and M. Leng, Transplatin-modified oligo(2′-O-methyl ribonucleotide)s: A new tool for selective modulation of gene expression. *Biochemistry* **36**, 2925–2931 (1997).
31. J.-M. Malinge and M. Leng, Interstrand cross-links in cisplatin or transplatin-modified DNA. in "Cisplatin. Chemistry and Biochemistry of a Leading Anticancer Drug" (B. Lippert, ed.), pp. 159–180. Verlag Helvetica Chimica Acta, Wiley-VCH, Zürich/Weinheim, 1999.
32. K. Aupeix-Scheidler, S. Chabas, L. Bidou, J.-P. Rousset, M. Leng, and J.-J. Toulme, Inhibition of in vitro and ex vivo translation by a transplatin-modified oligo(2′-O-methylribonucleotide) directed against the HIV-1 gag-pol frameshift signal. *Nucleic Acids Res.* **28**, 438–445 (2000).
33. M. Leng and V. Brabec, DNA adducts of cisplatin, transplatin and platinum-intercalating drugs. in "DNA Adducts: Identification and Biological Significance" (K. Hemminki, A. Dipple, D. E. G. Shuker, F. F. Kadlinbar, D. Segerbäck, H. Bartoch, eds.), pp. 339–348. International Agency for Research on Cancer, Lyon, 1994.
34. N. P. Davies, L. C. Hardman, and V. Murray, The effect of chromatin structure on cisplatin damage in intact human cells. *Nucleic Acids Res.* **28**, 2954–2958 (2000).
35. J. T. Millard and E. E. Wilkes, cis- and trans-Diamminedichloroplatinum(II) interstrand cross-linking of a defined sequence nucleosomal core particle. *Biochemistry* **39**, 16,046–16,055 (2000).
36. J. Reedijk, Why does cisplatin reach guanine-N7 with competing S-donor ligands available in the cell? *Chem. Rev.* 2499–2510 (1999).
37. K. J. Barnham, M. I. Djuran, P. d. S. Murdoch, J. D. Ranford, and P. J. Sadler, L.-Methionine increases the rate of reaction of 5′-guanosine monophosphate with anticancer drug cisplatin: mixed-ligand adducts and reversible methionine binding. *J. Chem. Soc., Dalton Trans.* 3721–3726 (1995).
38. S. S. G. E. van Boom and J. Reedijk, Unprecedented migration of [Pt(dien)]$^{2+}$ (dien = 1,5-diamino-3-azapentane) from sulfur to guanosine-N^7 in S-guanosyl-L-homocysteine (sgh). *J. Chem. Soc. Chem. Commun.* 1397–1398 (1993).
39. S. S. G. E. van Boom, B. W. Chen, J. M. Teuben, and J. Reedijk, Platinum-thioether bonds can be reverted by guanine-N7 bonds in Pt(dien)$^{2+}$ model. *Inorg. Chem.* **38**, 1450–1455 (1999).
40. C. J. Van Garderen, H. Van den Elst, J. H. Van Boom, J. Reedijk, and L. P. A. Van Houte, A double-stranded DNA fragment shows a significant decrease in double-helix stability after binding of monofunctional platinum amine compounds. *J. Am. Chem. Soc.* **111**, 4123–4125 (1989).
41. V. Brabec, J. Reedijk, and M. Leng, Sequence-dependent distortions induced in DNA by monofunctional platinum(II) binding. *Biochemistry* **31**, 12,397–12,402 (1992).
42. V. Brabec, V. Boudny, and Z. Balcarova, Monofunctional adducts of platinum(II) produce in DNA a sequence-dependent local denaturation. *Biochemistry* **32**, 1316–1322 (1994).

43. M. V. Keck and S. J. Lippard, Unwinding of supercoiled DNA by platinum ethidium and related complexes. *J. Am. Chem. Soc.* **114**, 3386–3390 (1992).
44. Z. Balcarova and V. Brabec, DNA modified by platinum derivatives cannot adopt the A-form. *Biochim. Biophys. Acta* **867**, 31–35 (1986).
45. S. F. Bellon and S. J. Lippard, Bending studies of DNA site-specifically modified by cisplatin, *trans*-diamminedichloroplatinum(II) and *cis*-Pt(NH$_3$)$_2$ (N3-cytosine)Cl$^+$. *Biophys. Chem.* **35**, 179–188 (1990).
46. J. M. Teuben, C. Bauer, A. H. J. Wang, and J. Reedijk, Solution structure of a DNA duplex containing a *cis*-diammineplatinum(II) 1,3-d(GTG) intrastrand cross-link, a major adduct in cells treated with the anticancer drug carboplatin. *Biochemistry* **38**, 12,305–12,312 (1999).
47. H. F. Huang, L. M. Zhu, B. R. Reid, G. P. Drobny, and P. B. Hopkins, Solution structure of a cisplatin-induced DNA interstrand cross-link. *Science* **270**, 1842–1845 (1995).
48. F. Coste, J.-M. Malinge, L. Serre, W. Shepard, M. Roth, M. Leng, and C. Zelwer, Crystal structure of a double-stranded DNA containing a cisplatin interstrand cross-link at 1.63 A resolution: Hydration at the platinated site. *Nucleic Acids Res.* **27**, 1837–1846 (1999).
49. V. Brabec, M. Sip, and M. Leng, DNA conformational distortion produced by site-specific interstrand cross-link of *trans*-diamminedichloroplatinum(II). *Biochemistry* **32**, 11,676–11,681 (1993).
50. F. Paquet, M. Boudvillain, G. Lancelot, and M. Leng, NMR solution structure of a DNA dodecamer containing a transplatin interstrand GN7-CN3 cross-link. *Nucleic Acids Res.* **27**, 4261–4268 (1999).
51. M. F. Anin and M. Leng, Distortions induced in double-stranded oligonucleotides by the binding of *cis*-diamminedichloroplatinum(II) or *trans*-diamminedichloroplatinum(II) to the d(GTG) sequence. *Nucleic Acids Res.* **18**, 4395–4400 (1990).
52. A. L. Pinto and S. J. Lippard, Sequence-dependent termination of in vitro DNA synthesis by *cis*- and *trans*-diamminedichloroplatinum(II). *Proc. Natl. Acad. Sci. U.S.A.* **82**, 4616–4620 (1985).
53. G. Villani, U. Hubscher, and J.-L. Butour, Sites of termination of *in vitro* DNA synthesis on *cis*-diamminedichloroplatinum(II) treated single-stranded DNA: A comparison between *E. coli* DNA polymerase I and eucaryotic DNA polymerases alpha. *Nucleic Acids Res.* **16**, 4407–4418 (1988).
54. K. M. Comess, J. N. Burstyn, J. M. Essigmann, and S. J. Lippard, Replication inhibition and translesion synthesis on templates containing site-specifically placed *cis*-diamminedichloroplatinum(II) DNA adducts. *Biochemistry* **31**, 3975–3990 (1992).
55. V. Murray, H. Motyka, P. R. England, G. Wickham, H. H. Lee, W. A. Denny, and W. D. McFadyen, The use of Taq DNA polymerase to determine the sequence specificity of DNA damage caused by *cis*-diamminedichloroplatinum(II), acridine-tethered platinum(II) diammine complexes or 2 analogues. *J. Biol. Chem.* **267**, 18,805–18,809 (1992).
56. V. Murray, H. Motyka, P. R. England, G. Wickham, H. H. Lee, W. A. Denny, and W. D. McFadyen, An investigation of the sequence-specific interaction of *cis*-diamminedichloroplatinum(II) and four analogues, including two acridine-tethered complexes, with DNA inside human cells. *Biochemistry* **31**, 11,812–11,817 (1992).
57. M. Coluccia, F. Nassii, F. Loseto, A. Boccarelli, M. A. Marigio, D. Giordano, F. P. Intini, P. Caputo, and G. Natile, A *trans*-platinum complex showing higher antitumor activity than the *cis* congeners. *J. Med. Chem.* **36**, 510–512 (1993).
58. V. Murray, J. Whittaker, M. D. Temple, and W. D. McFadyen, Interaction of 11 cisplatin analogues with DNA: Characteristic pattern of damage with monofunctional analogues. *Biochim. Biophys. Acta* **1354**, 261–271 (1997).
59. D. Burnouf, M. Daune, and R. P. Fuchs, Spectrum of cisplatin-induced mutations in *Escherichia coli*. *Proc. Natl. Acad. Sci. U.S.A.* **84**, 3758–3762 (1987).

60. K. J. Yarema, S. J. Lippard, and J. M. Essigmann, Mutagenic and genotoxic effects of DNA adducts formed by the anticancer drug cis-diamminedichloroplatinum(II). *Nucleic Acids Res.* **23**, 4066–4072 (1995).
61. G. Villani, N. T. Le Gac, and J.-S. Hoffmann, Replication of platinated DNA and its mutagenic consequences. In "Cisplatin. Chemistry and biochemistry of a leading anticancer drug" (B. Lippert, ed.), pp. 135–157. VHCA, Wiley-CH, Zürich/Weinheim, 1999.
62. C. M. Sorenson and A. Eastman, Mechanism of cis-diamminedichloroplatinum(II)-induced cytotoxicity—Role of G2 arrest and DNA double-strand breaks. *Cancer Res.* **48**, 4484–4488 (1988).
63. G. Chu, Cellular responses to cisplatin. The roles of DNA-binding proteins and DNA repair. *J. Biol. Chem.* **269**, 787–790 (1994).
64. M. J. Allday, G. J. Inman, D. H. Crawford, and P. J. Farrell, DNA damage in human B cells can induce apoptosis, proceeding from G1/S when p53 is transactivation competent and G2/M when it is transactivation defective. *EMBO J.* **14**, 4994–5005 (1995).
65. Y. Corda, C. Job, M. F. Anin, M. Leng, and D. Job, Transcription by eucaryotic and procaryotic RNA polymerases of DNA modified at a d(GG) or a d(AG) site by the antitumor drug cis-diamminedichloroplatinum(II). *Biochemistry* **30**, 222–230 (1991).
66. Y. Corda, M. F. Anin, M. Leng, and D. Job, RNA polymerases react differently at d(ApG) and d(GpG) adducts in DNA modified by cis-diamminedichloroplatinum(II). *Biochemistry* **31**, 1904–1908 (1992).
67. C. Cullinane, S. J. Mazur, J. M. Essigmann, D. R. Phillips, and V. A. Bohr, Inhibition of RNA polymerase II transcription in human cell extracts by cisplatin DNA damage. *Biochemistry* **38**, 6204–6212 (1999).
68. Y. Corda, C. Job, M.-F. Anin, M. Leng, and D. Job, Spectrum of DNA-platinum adduct recognition by prokaryotic and eukaryotic DNA-dependent RNA polymerases. *Biochemistry* **32**, 8582–8588 (1993).
69. J. A. Mello, S. J. Lippard, and J. M. Essigmann, DNA adducts of cis-diamminedichloroplatinum(II) and its *trans* isomer inhibit RNA polymerase II differentially *in vivo*. *Biochemistry* **34**, 14,783–14,791 (1995).
70. J. Mymryk, E. Zaniewski, and T. Archer, Cisplatin inhibits chromatin remodeling, transcription factor binding, and transcription from the mouse mammary tumor virus promoter *in vivo*. *Proc. Natl. Acad. Sci. U.S.A.* **92**, 2076–2080 (1995).
71. P. Jordan and M. Carmo-Fonseca, Cisplatin inhibits synthesis of ribosomal RNA *in vivo*. *Nucleic Acids Res.* **26**, 2831–2836 (1998).
72. D. J. Beck and R. R. Brubaker, Effect of cis-platinum(II)diamminodichloride on wild type and deoxyribonucleic acid repair-deficient mutants of *Escherichia coli*. *J. Bacteriol.* **116**, 1247–1252 (1973).
73. J. Brouwer, P. Van de Putte, A. M. J. Fichtinger-Schepman, and J. Reedijk, Base-pair substitution hotspots in GAG and GCG nucleotide sequences in E. coli K-12 induced by cis-diamminedichloroplatinum(II). *Proc. Natl. Acad. Sci. U.S.A.* **78**, 7010–7014 (1981).
74. D. J. Beck, S. Popoff, A. Sancar, and W. D. Rupp, Reactions of the UVRABC excision nuclease with DNA damaged by diamminedichloroplatinum(II). *Nucleic Acids Res.* **13**, 7395–7412 (1985).
75. E. Janovska and V. Kleinwächter, Inactivation effect of platinum(II) compounds on strains of *Escherichia coli*. *Stud. Biophys.* **114**, 187–192 (1986).
76. D. B. Zamble, D. Mu, J. T. Reardon, A. Sancar, and S. J. Lippard, Repair of cisplatin-DNA adducts by the mammalian excision nuclease. *Biochemistry* **35**, 10,004–10,013 (1996).
77. A. C. M. Plooy, M. Van Dijk, F. Berends, and P. H. M. Lohman, Formation and repair of DNA interstrand cross-links in relation to cytotoxicity and unscheduled DNA synthesis induced in control and mutant human cells treated with cis-diamminedichloroplatinum(II). *Cancer Res.* **45**, 4178–4184 (1985).

78. F. J. Dijt, A. M. J. Fichtinger-Schepman, F. Berends, and J. Reedijk, Formation and repair of cisplatin-induced adducts to DNA in cultured normal and repair-deficient human fibroblasts. *Cancer Res.* **48**, 6058–6062 (1988).
79. Y. Fujiwara, M. Tatsumi, and M. S. Sasaki, Cross-link repair in human cells and its possible defect in Fanconi's anemia cells. *J. Mol. Biol.* **113**, 635–649 (1977).
80. P. Calsou, P. Frit, and B. Salles, Repair synthesis by human cell extracts in cisplatin-damaged DNA is prefentially determined by minor adducts. *Nucleic Acids Res.* **20**, 6363–6368 (1992).
81. D. Fink, S. Nebel, S. Aebi, H. Zheng, B. Cenni, A. Nehme, R. D. Christen, and S. B. Howell, The role of DNA mismatch repair in platinum drug resistance. *Cancer Res.* **56**, 4881–4886 (1996).
82. S. Aebi, D. Fink, R. Gordon, H. K. Kim, H. Zheng, J. L. Fink, and S. B. Howell, Resistance to cytotoxic drugs in DNA mismatch repair-deficient cells. *Clin. Cancer Res.* **3**, 1763–1767 (1997).
83. S. W. Johnson, K. V. Ferry, and T. C. Hamilton, Recent insights into platinum drug resistance in cancer. *Drug Resistance Updates* **1**, 243–254 (1998).
84. J. T. Drummond, A. Anthoney, R. Brown, and P. Modrich, Cisplatin and adriamycin resistance are associated with MutL alpha and mismatch repair deficiency in an ovarian tumor cell line. *J. Biol. Chem.* **271**, 19,645–19,648 (1996).
85. D. Fink, S. Nebel, S. Aebi, A. Nehme, and S. B. Howell, Loss of DNA mismatch repair due to knockout of MSH2 or PMS2 results in resistance to cisplatin and carboplatin. *Int. J. Oncol.* **11**, 539–542 (1997).
86. K. V. Ferry, D. Fink, S. W. Johnson, S. Nebel, T. C. Hamilton, and S. B. Howell, Decreased cisplatin damage-dependent DNA synthesis in cellular extracts of mismatch repair deficient cells. *Biochem. Pharmacol.* **57**, 861–867 (1999).
87. R. B. Ciccarelli, M. J. Solomon, A. Varshavsky, and S. J. Lippard, In vivo effects of cis- and trans-diamminedichloroplatinum(II) on SV 40 chromosomes: Differential repair, DNA-protein cross-linking, and inhibition of replication. *Biochemistry* **24**, 7533–7540 (1985).
88. J. Zlatanova, J. Yaneva, and S. H. Leuba, Proteins that specifically recognize cisplatin-damaged DNA: A clue to anticancer activity of cisplatin. *FASEB J.* **12**, 791–799 (1998).
89. R. Reeves, Structure and function of the HMGI(Y) family of architectural transcription factors. *Environ. Health Persp.* **108**, 803–809 (2000).
90. J. O. Thomas and A. A. Travers, HMG1 and 2, and related 'architectural' DNA-binding proteins. *Trends Biochem. Sci.* **26**, 167–174 (2001).
91. S. U. Dunham and S. J. Lippard, DNA sequence context and protein composition modulate HMG-domain protein recognition of ciplatin-modified DNA. *Biochemistry* **36**, 11428–11436 (1997).
92. S. M. Cohen, Y. Mikata, Q. He, and S. J. Lippard, HMG-Domain protein recognition of cisplatin 1,2-intrastrand d(GpG) cross-links in purine-rich sequence contexts. *Biochemistry* **39**, 11,771–11,776 (2000).
93. D. S. Pilch, S. U. Dunham, E. R. Jamieson, S. J. Lippard, and K. J. Breslauer, DNA sequence context modulates the impact of a cisplatin 1,2-d(GpG) intrastrand cross-link an the conformational and thermodynamic properties of duplex DNA. *J. Mol. Biol.* **296**, 803–812 (2000).
94. C. Hofr, N. Farrell, and V. Brabee, Thermodynamic properties of duplex DNA containing a site-specific d(GpG) intrastrand crosslink formed by an antitumor dinuclear platinum complex. *Nucleic Acids Res.* **29**, 2034–2040 (2001).
95. U. M. Ohndorf, M. A. Rould, Q. He, C. O. Pabo, and S. J. Lippard, Basis for recognition of cisplatin-modified DNA by high-mobility-group proteins. *Nature* **399**, 708–712 (1999).
96. M. Stros, Two mutations of basic residues within the N-terminus of HMG-1 B domain with different effects on DNA supercoiling and binding to bent DNA. *Biochemistry* **40**, 4769–4779 (2001).

97. J. Kasparkova and V. Brabec, Recognition of DNA interstrand cross-links of *cis*-diamminedichloroplatinum(II) and its trans isomer by DNA-binding proteins. *Biochemistry* **34,** 12,379–12,387 (1995).
98. L. G. Marzilli, J. S. Saad, Z. Kuklenyik, K. A. Keating, and Y. H. Xu, Relationship of solution and protein-bound structures of DNA duplexes with the major intrastrand cross-link lesions formed on cisplatin binding to DNA. *J. Am. Chem. Soc.* **123,** 2764–2770 (2001).
99. A. T. Yarnell, S. Oh, D. Reinberg, and S. J. Lippard, Interaction of FACT, SSRP1, and the high mobility group (HMG) domain of SSRP1 with DNA damaged by the anticancer drug cisplatin. *J. Biol. Chem.* **276,** 25,736–25,741 (2001).
100. G. Orphanides, T. Lagrange, and D. Reinberg, The general transcription factors of RNA polymerase II. *Genes Dev.* **10,** 2657–2683 (1996).
101. P. Vichi, F. Coin, J. P. Renaud, W. Vermeulen, J. H. J. Hoeijmakers, D. Moras, and J. M. Egly, Cisplatin- and UV-damaged DNA lure the basal transcription factor TFIID/TBP. *EMBO J.* **16,** 7444–7456 (1997).
102. F. Coin, P. Frit, B. Viollet, B. Salles, and J. M. Egly, TATA binding protein discriminates between different lesions on DNA, resulting in a transcription decrease. *Mol. Cell. Biol.* **18,** 3907–3914 (1998).
103. S. M. Cohen, E. R. Jamieson, and S. J. Lippard, Enhanced binding of the TATA-binding protein to TATA boxes containing flanking cisplatin 1,2-cross-links. *Biochemistry* **39,** 8259–8265 (2000).
104. T. Ise, G. Nagatani, T. Imamura, K. Kato, H. Takano, M. Nomoto, H. Izumi, H. Ohmori, T. Okamoto, T. Ohga, T. Uchiumi, M. Kuwano, and K. Kohno, Transcription factor Y-box binding protein 1 binds preferentially to cisplatin-modified DNA and interacts with proliferating cell nuclear antigen. *Cancer Res.* **59,** 342–346 (1999).
105. T. Ohga, K. Koike, M. Ono, Y. Makino, Y. Itagaki, M. Tanimoto, M. Kuwano, and K. Kohno, Role of the human Y box-binding protein YB-1 in cellular sensitivity to the DNA-damaging agents cisplatin, mitomycin C, and ultraviolet light. *Cancer Res.* **56,** 4224–4228 (1996).
106. J. Yaneva, S. H. Leuba, K. van Holde, and J. Zlatanova, The major chromatin protein histone H1 binds preferentially to *cis*-platinum-damaged DNA. *Proc. Natl. Acad. Sci. U.S.A.* **94,** 13,448–13,451 (1997).
107. E. G. Paneva, N. C. Spassovska, K. C. Grancharov, J. S. Zlatanova, and J. N. Yaneva, Interaction of histone H1 with *cis*-platinum modified DNA. *Z. Naturforsch.* **53c,** 135–138 (1998).
108. M. Bustin and R. Reeves, High-mobility-group chromosomal proteins: Architectural components that facilitate chromatin function. *Prog. Nucleic Acid Res. Mol. Biol.* **54,** 35–100 (1996).
109. J. Zlatanova and K. van Holde, The linker histones and chromatin structure: New twists. *Prog. Nucleic Acid Res. Mol. Biol.* **52,** 217–259 (1996).
110. J. Zlatanova and K. Van Holde, Binding to four-way junction DNA: A common property of architectural proteins? *FASEB J.* **12,** 421–431 (1998).
111. M. Missura, T. Buterin, R. Hindges, U. Hübscher, J. Kasparkova, V. Brabec, and H. Naegeli, Initiation of mammalian nucleotide excision repair: Double-check probing of DNA helix conformation by repair protein XPA and replication factor RPA. *EMBO J.* **20,** 3554–3564 (2001).
112. C. J. Jones and R. D. Wood, Preferential binding of the Xeroderma Pigmentosum group A complementing protein to damaged DNA. *Biochemistry* **32,** 12,096–12,104 (1993).
113. H. Asahina, I. Kuraoka, M. Shirakawa, E. Morita, N. Miura, I. Miyamoto, E. Ohtsuka, Y. Okada, and K. Tanaka, The XPA protein is a zinc metalloprotein with an abilito to recognize various kinds of DNA-damage. *Mutat. Res.* **315,** 229–237 (1994).
114. I. Kuraoka, E. H. Morita, M. Saijo, T. Matsuda, K. Morikawa, M. Shirakawa, and K. Tanaka, Identification of a damaged-DNA binding domain of the XPA protein. *Mutat. Res.* **362,** 87–95 (1996).

115. J. J. Turchi, K. M. Henkels, I. L. Hermanson, and S. M. Patrick, Interactions of mammalian proteins with cisplatin-damaged DNA. *J. Inorg. Biochem.* **77,** 83–87 (1999).
116. N. X. Xu, L. Pasa-Tolic, R. D. Smith, S. S. Ni, and B. D. Thrall, Electrospray ionization-mass spectrometry study of the interaction of cisplatin-adducted oligonucleotides with human XPA minimal binding domain protein. *Anal. Biochem.* **272,** 26–33 (1999).
117. T. Hey, G. Lipps, and G. Krauss, Binding of XPA and RPA to damaged DNA investigated by fluorescence anisotrophy. *Biochemistry* **40,** 2901–2910 (2001).
118. C. K. Clugston, K. McLaughlin, M. K. Kenny, and R. Brown, Binding of human single-stranded-DNA binding-protein to DNA damaged by the anticancer drug cis-diamminedichloroplatinum(II). *Cancer Res.* **52,** 6375–6379 (1992).
119. S. M. Patrick and J. J. Turchi, Human replication protein a preferentially binds cisplatin-damaged duplex DNA in vitro. *Biochemistry* **37,** 8808–8815 (1998).
120. S. M. Patrick and J. J. Turchi, Replication protein A (RPA) binding to duplex cisplatin-damaged DNA is mediated through the generation of single-stranded DNA. *J. Biol. Chem.* **274,** 14,972–14,978 (1999).
121. D. R. Duckett, J. T. Drummond, A. I. H. Murchie, J. T. Reardon, A. Sancar, D. M. Lilley, and P. Modrich, Human MutSα recognizes damaged DNA base pairs containing O^6-methylguanine, O^4-methylthymine, or the cisplatin-d(GpG) adduct. *Proc. Natl. Acad. Sci. U.S.A.* **93,** 6443–6447 (1996).
122. J. A. Mello, S. Acharya, R. Fishel, and J. M. Essigmann, The mismatch-repair protein hMSH2 binds selectively to DNA adducts of the anticancer drug cisplatin. *Chem. Biol.* **3,** 579–589 (1996).
123. M. Yamada, E. O'Regan, R. Brown, and P. Karran, Selective recognition of a cisplatin-DNA adduct by human mismatch repair proteins. *Nucleic Acids Res.* **25,** 491–495 (1997).
124. Z. Özer, J. T. Reardon, D. S. Hsu, K. Malhotra, and A. Sancar, The other function of DNA photolyase: Stimulation of excision repair of chemical damage to DNA. *Biochemistry* **34,** 15,886–15,889 (1995).
125. J. J. Turchi and K. Henkels, Human Ku autoantigen binds cisplatin-damaged DNA but fails to stimulate human DNA-activated protein kinase. *J. Biol. Chem.* **271,** 13,861–13,867 (1996).
126. J. J. Turchi, K. M. Henkels, and Y. Zhou, Cisplatin-DNA adducts inhibit translocation of the Ku subunits of DNA-PK. *Nucleic Acids Res.* **28,** 4634–4641 (2000).
127. A. I. H. Murchie and D. M. J. Lilley, T4 endonuclease VII cleaves DNA containing a cisplatin adduct. *J. Mol. Biol.* **233,** 77–82 (1993).
128. C. M. Riva, Restoration of wild-type p53 activity enhances the sensitivity of pleural metastasis to cisplatin through an apoptotic mechanism. *Anticancer Res.* **20,** 4463–4471 (2000).
129. P. Jordan and M. Carmo-Fonseca, Molecular mechanisms involved in cisplatin cytotoxicity. *Cell. Mol. Life Sci.* **57,** 1229–1235 (2000).
130. P. M. O'Connor, J. Jackman, Bae, I., T. G. Myers, S. Fan, M. Mutoh, D. A. Scudiero, A. Monks, E. A. Sausville, J. N. Weinstein, S. Friend, A. J. Fornace, and K. W. Kohn, Characterization of the p53 tumor suppressor pathway in cell lines of the National Cancer Institute anticancer drug screen and correlations with the growth-inhibitory potency of 123 anticancer agents. *Cancer Res.* **57,** 4285–4300 (1997).
131. J. Kasparkova, S. Pospisilova, and V. Brabec, Different recognition of DNA modified by antitumor cisplatin and its clinically ineffective trans isomer by tumor suppressor protein p53. *J. Biol. Chem.* **276,** 16,064–16,069 (2001).
132. P. J. Perry and L. R. Kelland, Telomeres and telomerase: targets for cancer chemotherapy? *Expert Opinion on Therapeutic Patents* **8,** 1567–1586 (1998).
133. A. M. Burger, J. A. Double, and D. R. Newell, Inhibition of telomerase activity by cisplatin in human testicular cancer cells. *Eur. J. Cancer* **33,** 638–644 (1997).

134. Y. Kondo, S. Kondo, Y. Tanaka, T. Haqqi, B. P. Barna, and J. K. Cowell, Inhibition of telomerase increases the susceptibility of human malignant glioblastoma cells to cisplatin-induced apoptosis. *Oncogene* **16,** 2243–2248 (1998).
135. X. H. Wang, S. C. H. Wong, J. Pan, S. W. Tsao, K. H. Y. Fung, D. L. W. Kwong, J. S. T. Sham, and J. M. Nicholls, Evidence of cisplatin-induced senescent-like growth arrest in nasopharyngeal carcinoma cells. *Cancer Res.* **58,** 5019–5022 (1998).
136. R. Villa, M. Folini, P. Perego, R. Supino, E. Setti, M. G. Daidone, F. Zunino, and N. Zaffaroni, Telomerase activity and telomere length in human ovarian cancer and melanoma cell lines: Correlation with sensitivity to DNA damaging agents. *Int. J. Oncol.* **16,** 995–1002 (2000).
137. Y. Kiyozuka, D. Yamamoto, J. H. Yang, Y. Uemura, H. Senzaki, S. Adachi, and A. Tsubura, Correlation of chemosensitivity to anticancer drugs and telomere length, telomerase activity and telomerase RNA expression in human ovarian cancer cells. *Anticancer Res.* **20,** 203–212 (2000).
138. H. Mese, Y. Ueyama, A. Suzuki, S. Nakayama, A. Sasaki, H. Hamakawa, and T. Matsumura, Inhibition of telomerase activity as a measure of tumor cell killing by cisplatin in squamous cell carcinoma cell line. *Chemotherapy* **47,** 136–142 (2001).
139. S. Redon, S. Bombard, M.-A. Elionzo-Riojas, and J.-C. Chottard, Platination of the $(T_2G_4)_4$ telomeric sequence: A structural and cross-linking study. *Biochemistry* **40,** 8463–8470 (2001).
140. K. Bouayadi, P. Calsou, A. M. Pedrini, and B. Salles, *In vitro* evolution of cisplatin DNA monoadducts into diadducts is dependent upon superhelical density. *Biochem. Biophys. Res. Commun.* **189,** 111–118 (1992).
141. S. Neumann, H. Simon, C. Zimmer, and A. Quinones, The antitumor agent cisplatin inhibits DNA gyrase and preferentially induces gyrB gene expression in *Escherichia coli. Biol. Chem.* **377,** 731–739 (1996).
142. M. Fukuda, K. Nishio, F. Kanzawa, H. Ogasawara, T. Ishida, H. Arioka, K. Bojanowski, M. Oka, and N. Saijo, Synergism between cisplatin and topoisomerase I inhibitors, NB-506 and SN-38, in human small cell lung cancer cells. *Cancer Res.* **56,** 789–793 (1996).
143. N. Masumoto, S. Nakano, T. Esaki, H. Fujishima, T. Tatsumoto, and Y. Niho, Inhibition of *cis*-diamminedichloroplatinum(II)-induced DNA interstrand cross-link removal by 7-ethyl-10-hydroxy-camptothecin in HST-1 human squamous-carcinoma cells. *Int. J. Cancer* **62,** 70–75 (1995).
144. S. Kobayashi, M. Furukawa, C. Dohi, H. Hamashima, T. Arai, and A. Tanaka, Topology effect for DNA structure of cisplatin: Topological transformation of cisplatin-closed circular DNA adducts by DNA topoisomerase I. *Chem. Pharm. Bull.* **47,** 783–790 (1999).
145. B. Salles, J. L. Butour, C. Lesca, and J. P. Macquet, *Cis*-Pt(NH$_3$)$_2$Cl$_2$ and *trans*-Pt(NH$_3$)$_2$Cl$_2$ inhibit DNA synthesis in cultured L1210 leukemia cells. *Biochem. Biophys. Res. Commun.* **112,** 555–563 (1983).
146. V. M. Gonzalez, M. A. Fuertes, C. Alonso, and J. M. Perez, Is cisplatin-induced cell death always produced by apoptosis? *Mol. Pharmacol.* **59,** 657–663 (2001).
147. F. A. Blommaert, H. C. M. Van Dijk-Knijnenburg, F. J. Dijt, L. Denengelse, R. A. Baan, F. Berends, and A. M. J. Fichtinger-Schepman, Formation of DNA adducts by the anticancer drug carboplatin: Different nucleotide sequence preferences in vitro and in cells. *Biochemistry* **34,** 8474–8480 (1995).
148. G. Natarajan, R. Malathi, and E. Holler, Increased DNA binding activity of *cis*-1,1-cyclobutanedicarboxylatodiammineplatinum(II) (Carboplatin) in the presence of nucleophiles and human breast cancer MCF-7 cell cytoplasmic extracts: Activation theory revisited. *Biochem. Pharmacol.* **58,** 1625–1629 (1999).
149. O. Vrana, V. Brabec, and V. Kleinwächter, Polarographic studies on the conformation of some platinum complexes: relations to anti-tumour activity. *Anti-Cancer Drug Design* **1,** 95–109 (1986).

150. M. S. Highley and A. H. Calvert, Clinical experience with cisplatin and carboplatin. in "Platinum-Based Drugs in Cancer Therapy" (L. R. Kelland and N. P. Farrell, eds.), pp. 171–194. Humana Press, Totowa, NJ, 2000.
151. S. K. Mauldin, M. Plescia, F. A. Richard, S. D. Wyrick, R. D. Voyksner, and S. G. Chaney, Displacement of the bidentate malonate ligand from (d,1-*trans*-1,2-diaminecyclohexane)malonatoplatinum(II) by physiologically important compounds *in vitro*. *Biochem. Pharmacol.* **37,** 3321–3333 (1998).
152. J. M. Woynarowski, W. G. Chapman, C. Napier, M. C. S. Herzig, and P. Juniewicz, Sequence- and region-specificity of oxaliplatin adducts in naked and cellular DNA. *Mol. Pharmacol.* **54,** 770–777 (1998).
153. M. M. Jennerwein, A. Eastman, and A. Khokhar, Characterization of adducts produced in DNA by isomeric 1,2-diaminocyclohexaneplatinum(II) complexes. *Chem.–Biol. Interact.* **70,** 39–50 (1989).
154. C. P. Saris, P. J. M. van de Vaart, R. C. Rietbroek, and F. A. Blommaert, *In vitro* formation of DNA adducts by cisplatin, lobaplatin and oxaliplatin in calf thymus DNA in solution and in cultured human cells. *Carcinogenesis* **17,** 2763–2769 (1996).
155. E. D. Scheeff, J. M. Briggs, and S. B. Howell, Molecular modeling of the intrastrand guanine-guanine DNA adducts produced by cisplatin and oxaliplatin. *Mol. Pharmacol.* **56,** 633–643 (1999).
156. A. Vaisman, S. E. Lim, S. M. Patrick, W. C. Copeland, D. C. Hinkle, J. J. Turchi, and S. G. Chaney, Effect of DNA polymerases and high mobility group protein 1 on the carrier ligand specificity for translesion synthesis past platinum-DNA adducts. *Biochemistry* **38,** 11,026–11,039 (1999).
157. J. T. Reardon, A. Vaisman, S. G. Chaney, and A. Sancar, Efficient nucleotide excision repair of cisplatin, oxaliplatin, and bis-aceto-ammine-dichloro-cyclohexylamine-platinum(IV) (JM216) platinum intrastrand DNA diadducts. *Cancer Res.* **59,** 3968–3971 (1999).
158. A. Vaisman and S. G. Chaney, The efficiency and fidelity of translesion synthesis past cisplatin and oxaliplatin GpG adducts by human DNA polymerase beta. *J. Biol. Chem.* **275,** 13,017–13,025 (2000).
159. A. Vaisman, C. Masutani, F. Hanaoka, and S. G. Chaney, Efficient translesion replication past oxaliplatin and cisplatin GpG adducts by human DNA polymerase eta. *Biochemistry* **39,** 4575–4580 (2000).
160. A. Vaisman, M. W. Warren, and S. G. Chaney, The effect of DNA structure on the catalytic efficiency and fidelity of human DNA polymerase β on templates with platinum-DNA adducts. *J. Biol. Chem.* **276,** 18,999–19,005 (2001).
161. S. Nebel, D. Fink, S. Aebi, A. Nehme, R. D. Christen, and S. B. Howell, Role of the DNA mismatch repair proteins in the recognition of platinum DNA adducts. *Proc. Am. Assoc. Cancer. Res.* **38,** A2402 (1997).
162. D. Fink, H. Zheng, S. Nebel, P. S. Norris, S. Aebi, T. P. Lin, A. Nehme, R. D. Christen, M. Haas, C. L. MacLeod, and S. B. Howell, *In vitro* and *in vivo* resistance to cisplatin in cells that have lost DNA mismatch repair. *Cancer Res.* **57,** 1841–1845 (1997).
163. Y. Kidani, K. Inagaki, M. Iigo, A. Hoshi, and K. Kuretani, Antitumor activity of 1,2-diamminocyclohexane-platinum complexes against Sarcoma 180 ascites form. *J. Med. Chem.* **21,** 1315–1318 (1978).
164. M. Coluccia, M. Correale, D. Giordano, M. A. Mariggio, S. Moscelli, F. P. Fanizzi, G. Natile, and L. Maresca, Mutagenic activity of some platinum complexes with monodentate and bidentate amines. *Inorg. Chim. Acta* **123,** 225–229 (1986).
165. A. Pasini and F. Zunino, New cisplatin analogues—On the way to better antitumor agents. *Angew. Chem. Int. Ed.* **26,** 615–624 (1987).
166. K. Vickery, A. M. Bonin, R. R. Fenton, S. O'Mara, P. J. Russell, L. K. Webster, and T. W. J. Hambley, Preparation, characterization, cytotoxicity, and mutagenicity of a pair of enantiomeric

platinum(II) complexes with the potential to bind enantioselectively to DNA. *J. Med. Chem.* **36,** 3663–3668 (1993).
167. J. D. Page, I. Husain, A. Sancar, and S. G. Chaney, Effect of the diaminocyclohexane carrier ligand on platinum adduct formation, repair, and lethality. *Biochemistry* **29,** 1016–1024 (1990).
168. F. P. Fanizzi, F. P. Intini, L. Maresca, G. Natile, R. Quaranata, M. Coluccia, L. Di Bari, D. Giordano, and M. A. Mariggio, Biological activity of platinum complexes containing chiral centers on the nitrogen or carbon atoms of a chelate diamine ring. *Inorg. Chim. Acta* **137,** 45–51 (1987).
169. J. Malina, C. Hofr, L. Maresca, G. Natile, and V. Brabec, DNA interactions of antitumor cisplatin analogs containing enantiomeric amine ligands. *Biophys. J.* **78,** 2008–2021 (2000).
170. W. I. Sundquist and S. J. Lippard, The coordination chemistry of platinum anticancer drugs and related compounds with DNA. *Coord. Chem. Rev.* **100,** 293–322 (1990).
171. M. Broggini, E. Erba, M. Ponti, D. Ballinari, C. Geroni, F. Spreafico, and M. Dincalci, Selective DNA Interaction of the Novel Distamycin Derivative FCE 24517. *Cancer Res.* **51,** 199–204 (1991).
172. K. M. Church, R. L. Wurdeman, Y. Zhang, F. X. Chen, and B. Gold, N-(2-chloroethyl)-N-ntrosoureas covalently bound to nonionic and monocationic lexitropsin dipeptides. Synthesis, DNA affinity binding characteristics, and reactions with P-32-end-labeled DNA. *Biochemistry* **29,** 6827–6838 (1990).
173. B. E. Bowler, L. S. Hollis, and S. J. Lippard, Synthesis and DNA binding and photonicking properties of acridine orange linked by a polymethylene tether to (1,2-diaminoethane)dichloroplatinum(II). *J. Am. Chem. Soc.* **106,** 6102–6104 (1984).
174. C. Zimmer, Effects of the antibiotics netropsin and distamycin A on the structure and function of nucleic acids. *Prog. Nucleic Acid Res. Mol. Biol.* **15,** 285–318 (1975).
175. C. Zimmer and U. Wahnert, Nonintercalating DNA-binding ligands: Specificity of the interaction and their use as tools in biophysical, biochemical and biological investigations of the genetic material. *Prog. Biophys. Mol. Biol.* **47,** 31–112 (1986).
176. M. Lee, J. E. Simpson, A. J. Burns, S. Kupchinsky, N. Brooks, J. A. Hartley, and L. R. Kelland, Novel platinum(II) derivatives of analogues of netropsin and distamycin: Synthesis, DNA binding and cytotoxic properties. *Med. Chem. Res.* **6,** 365–371 (1996).
177. H. Loskotova and V. Brabec, DNA interactions of cisplatin tethered to the DNA minor groove binder distamycin. *Eur. J. Biochem.* **266,** 392–402 (1999).
178. H. Kostrhunova and V. Brabec, Conformational analysis of site-specific DNA cross-links of cisplatin-distamycin conjugates. *Biochemistry* **39,** 12,639–12,649 (2000).
179. J. Holford, F. Raynaud, B. A. Murrer, K. Grimaldi, J. A. Hartley, M. Abrams, and L. R. Kelland, Chemical, biochemical and pharmacological activity of the novel sterically hindered platinum co-ordination complex, *cis*-[amminedichloro(2-methylpyridine)]platinum(II) (AMD473). *Anti-Cancer Drug Design* **13,** 1–18 (1998).
180. A. Habtemariam and P. Sadler, Design of chelate ring-opening platinum anticancer complexes: Reversible binding to guanine. *Chem. Commun.* 1785–1786 (1996).
181. N. Margiotta, A. Habtemariam, and P. J. Sadler, Strong, rapid binding of a platinum complex to thymine and uracil under physiological conditions. *Angew. Chem. Int. Ed.* **36,** 1185–1187 (1997).
182. K. Neplechova, J. Kasparkova, O. Vrana, O. Novakova, A. Habtemariam, B. Watchman, P. J. Sadler, and V. Brabec, DNA interactions of new antitumor aminophosphine platinum(II) complexes. *Mol. Pharmacol.* **56,** 20–30 (1999).
183. L. R. Kelland, New platinum drugs. The pathway to oral therapy. *in* "Platinum-Based Drugs in Cancer Therapy" (L. R. Kelland and N. P. Farrell, eds.), pp. 299–319. Humana Press, Totowa, NJ, 2000.
184. V. Juraskova and V. Brabec, Evaluation of cytotoxic and antitumour effects of tetravalent analog of carboplatin. *Neoplasma* **36,** 297–303 (1989).

185. E. Rotondo, V. Fimiani, A. Cavallaro, and T. Ainis, Does the antitumoral activity of platinum(IV) derivatives result from their in vivo reduction. *Tumori* **69**, 31–36 (1983).
186. E. E. Blatter, J. F. Vollano, B. S. Krishnan, and J. C. Dabrowiak, Interaction of the antitumor agents cis,cis,trans-Pt^{IV}(NH^3)$_2$Cl$_2$(OH)$_2$ and cis,cis,trans-Pt^{IV}[(CH$_3$)$_2$CHNH$_2$]$_2$Cl$_2$(OH)$_2$ and their reduction products with PM2 DNA. *Biochemistry* **23**, 4817–4820 (1984).
187. L. Pendyala, A. V. Arakali, P. Sansone, J. W. Cowens, and P. J. Creaven, DNA binding of iproplatin and its divalent metabolite cis-dichloro-bis-isopropylamine platinum(II). *Cancer Chemother. Pharmacol.* **27**, 248–250 (1990).
188. L. R. Kelland, B. A. Murrer, G. Abel, C. M. Giandomenico, P. Mistry, and K. R. Harrap, Ammine/Amine platinum(IV) dicarboxylates: A novel class of platinum complex exhibiting selective cytotoxicity to intrinsically cisplatin-resistant human ovarian carcinoma cell lines. *Cancer Res.* **52**, 822–828 (1992).
189. S. G. Chaney, S. Wyrick, and G. K. Till, In vitro biotransformations of tetrachloro (d,l-*trans*)-1, 2-diaminocyclohexaneplatinum(IV) (tetraplatin) in rat plasma. *Cancer Res.* **50**, 4539–4545 (1990).
190. M. Defais, M. Germanier, and N. P. Johnson, Detection of DNA strand breaks in *Escherichia Coli* treated with platinum(IV) antitumor compounds. *Chem.–Biol. Interact.* **74**, 343–352 (1990).
191. H. K. Choi, S. K.-S. Huang, and R. Bau, Octahedral complex of anticancer Pt(IV)(cyclohexyldiamine) agents with 9-methylguanine. *Biochem. Biophys. Res. Commun.* **156**, 1125–1129 (1988).
192. R. M. Roat and J. Reedijk, Reaction of *mer*-trichloro(diethylenetriamine)-platinum(IV) chloride, (*mer*-[Pt(dien)Cl$_3$]Cl), with purine nucleosides and nucleotides results in formation of platinum(II) as well as platinum(IV) complexes. *J. Inorg. Biochem.* **52**, 263–274 (1993).
193. O. Novakova, O. Vrana, V. I. Kiseleva, and V. Brabec, DNA interactions of antitumor platinum(IV) complexes. *Eur. J. Biochem.* **228**, 616–624 (1995).
194. K. J. Mellish, C. F. J. Barnard, B. A. Murrer, and L. R. Kelland, DNA-binding properties of novel cis- and trans platinum-based anticancer agents in 2 human ovarian carcinoma cell lines. *Int. J. Cancer* **62**, 717–723 (1995).
195. W. I. Sundquist, L. Bancroft, L. Chassot, and S. J. Lippard, DNA promotes the reaction of cis-diamminedichloroplatinum(II) with the exocyclic amino groups of ethidium bromide. *J. Am. Chem. Soc.* **110**, 8559–8561 (1988).
196. M. Boudvillain, R. Dalbies, and M. Leng, Evidences for a catalytic activity of the DNA double helix in the reaction between DNA, platinum(II), and intercalators. in "Metal Ions in Biological Systems" (A. Sigel and H. Sigel, eds.), pp. 87–104. Dekker, New York, 1996.
197. L. S. Hollis, A. R. Amundsen, and E. W. Stern, Chemical and biological properties of a new series of cis-diammineplatinum(II) antitumor agents containing three nitrogen donors: cis-[Pt(NH$_3$)$_2$(N-donor)Cl]$^+$. *J. Med. Chem.* **32**, 128–136 (1989).
198. L. S. Hollis, W. I. Sundquist, J. N. Burstyn, W. J. Heiger-Bernays, S. F. Bellon, K. J. Ahmed, A. R. Amundsen, E. W. Stern, and S. J. Lippard, Mechanistic studies of a novel class of trisubstituted platinum(II) antitumor agents. *Cancer Res.* **51**, 1866–1875 (1991).
199. C. Bauer, T. Peleg-Shulman, D. Gibson, and A. H.-J. Wang, Monofunctional platinum amine complexes destabilize DNA significantly. *Eur. J. Biochem.* **256**, 253–260 (1998).
200. T. Peleg-Shulman, J. Katzhendler, and D. Gibson, Effects of monofunctional platinum binding on the thermal stability and conformation of a self-complementary 22-mer. *J. Inorg. Biochem.* **81**, 313–323 (2000).
201. R. C. Taylor and S. G. Ward, in "Lectures in Bioinorganic Chemistry" (M. Nicolini and L. Sindellari, eds.), pp. 63–90. Raven Press, New York, 1991.
202. M. Coluccia, A. Boccarelli, C. Cermelli, M. Portolani, and G. Natile, Platinum(II)-acyclovir complexes: Synthesis, antiviral and antitumour activity. *Metal-Based Drugs* **2**, 249–256 (1995).

203. Z. Balcarova, J. Kasparkova, A. Zakovska, O. Novakova, M. F. Sivo, G. Natile, and V. Brabec, DNA interactions of a novel platinum drug, cis-[PtCl(NH$_3$)$_2$(N7-acyclovir)]$^+$. Mol. Pharmacol. **53**, 846–855 (1998).
204. N. Farrell, Current status of structure-activity relationships of platinum anticancer drugs: Activation of the trans geometry. in "Metal Ions in Biological Systems" (A. Sigel and H. Sigel, eds.), pp. 603–639. Dekker, New York/Basel/Hong Kong, 1996.
205. A. Zakovska, O. Novakova, Z. Balcarova, U. Bierbach, N. Farrell, and V. Brabec, DNA interactions of antitumor trans-[PtCl$_2$(NH$_3$)(quinoline)]. Eur. J. Biochem. **254**, 547–557 (1998).
206. V. Brabec, K. Neplechova, J. Kasparkova, and N. Farrell, Steric control of DNA interstrand cross-link sites of trans platinum complexes: Specificity can be dictated by planar nonleaving groups. J. Biol. Inorg. Chem. **5**, 364–368 (2000).
207. M. Coluccia, A. Boccarelli, M. A. Mariggio, N. Cardellicchio, P. Caputo, F. P. Intini, and G. Natile, Platinum(II) complexes containing iminoethers: A trans platinum antitumour agent. Chem.–Biol. Interact. **98**, 251–266 (1995).
208. V. Brabec, O. Vrana, O. Novakova, V. Kleinwächter, F. P. Intini, M. Coluccia, and G. Natile, DNA adducts of antitumor trans-[PtCl$_2$(E-imino ether)$_2$]. Nucleic Acids Res. **24**, 336–341 (1996).
209. R. Zaludova, A. Zakovska, J. Kasparkova, Z. Balcarova, O. Vrana, M. Coluccia, G. Natile, and V. Brabec, DNA modifications by antitumor trans-[PtCl$_2$(E-iminoether)$_2$]. Mol. Pharmacol. **52**, 354–361 (1997).
210. B. Andersen, N. Margiotta, M. Coluccia, G. Natile, and E. Sletten, Antitumor trans platinum DNA adducts: NMR and HPLC study of the interaction between a trans-Pt iminoether complex and the deoxy decamer d(CCTCGCTCTC)·d(GAGAGCGAGG). Metal-Based Drugs **7**, 23–32 (2000).
211. J. M. Perez, E. I. Montero, A. M. Gonzalez, X. Solans, M. Font-Bardia, M. A. Fuertes, C. Alonso, and C. Navarro-Ranninger, X-ray structure of cytotoxic trans-[PtCl$_2$(dimethylamine)(isopropylamine)]: Interstrand cross-link efficiency, DNA sequence specificity, and inhibition of the B-Z transition. J. Med. Chem. **43**, 2411–2418 (2000).
212. J. M. Perez, E. I. Montero, A. M. Gonzalez, A. Alvarez-Valdes, C. Alonso, and C. Navarro-Ranninger, Apoptosis induction and inhibition of H-ras overexpression by novel trans-[PtCl$_2$(isopropylamine)(amine)] complexes. J. Inorg. Biochem. **77**, 37–42 (1999).
213. E. I. Montero, S. Diaz, A. M. Gonzalez-Vadillo, J. M. Perez, C. Alonso, and C. Navarro-Ranninger, Preparation and characterization of novel trans-[PtCl$_2$(amine)(isopropylamine)] compounds: Cytotoxic activity and apoptosis induction in ras-transformed cells. J. Med. Chem. **42**, 4264–4268 (1999).
214. N. Farrell, Polynuclear charged platinum compounds as a new class of anticancer agents: Toward a new paradigm. in "Platinum-Based Drugs in Cancer Therapy" (L. R. Kelland and N. P. Farrell, eds.), pp. 321–338. Humana Press, Totowa, NJ, 2000.
215. N. Farrell, Y. Qu, U. Bierbach, M. Valsecchi, and E. Menta, Structure-activity relationship within di- and trinuclear platinum phase I clinical agents. in "Cisplatin. Chemistry and Biochemistry of a Leading Anticancer Drug" (B. Lippert, ed.), pp. 479–496. VHCA, Wiley-VCH, Zürich/Weinheim, 1999.
216. J. Kasparkova, K. J. Mellish, Y. Qu, V. Brabec, and N. Farrell, Site-specific d(GpG) intrastrand cross-links formed by dinuclear platinum complexes. Bending and NMR studies. Biochemistry **35**, 16,705–16,713 (1996).
217. K. J. Mellish, Y. Qu, N. Scarsdale, and N. Farrell, Effect of geometric isomerism in dinuclear platinum antitumour complexes on the rate of formation and structure of intrastrand adducts with oligonucleotides. Nucleic Acids Res. **25**, 1265–1271 (1997).
218. J. Kasparkova, O. Novakova, O. Vrana, N. Farrell, and V. Brabec, Effect of geometric isomerism in dinuclear platinum antitumor complexes on DNA interstrand cross-linking. Biochemistry **38**, 10,997–11,005 (1999).

219. J. Kasparkova, N. Farrell, and V. Brabec, Sequence specificity, conformation, and recognition by HMG1 protein of major DNA interstrand cross-links of antitumor dinuclear platinum complexes. *J. Biol. Chem.* **275,** 15,789–15,798 (2000).
220. R. Zaludova, A. Zakovska, J. Kasparkova, Z. Balcarova, V. Kleinwächter, O. Vrana, N. Farrell, and V. Brabec, DNA interactions of bifunctional dinuclear platinum(II) antitumor agents. *Eur. J. Biochem.* **246,** 508–517 (1997).
221. Y. Zou, B. Vanhouten, and N. Farrell, Sequence specificity of DNA-DNA interstrand cross-link formation by cisplatin and dinuclear platinum complexes. *Biochemistry* **33,** 5404–5410 (1994).
222. T. D. McGregor, J. Kasparkova, K. Neplechova, O. Novakova, H. Penazova, O. Vrana, V. Brabec, and N. Farrell, A. Comparison of DNA binding profiles of dinuclear platinum compounds with polyamine linkers and the trinuclear platinum phase II clinical agent BBR3464. *J. Biol. Inorg. Chem.* Published online, D01 10.1007s 00775-001-0312-4.
223. V. Brabec, J. Kasparkova, O. Vrana, O. Novakova, J. W. Cox, Y. Qu, and N. Farrell, DNA modifications by a novel bifunctional trinuclear platinum Phase I anticancer agent. *Biochemistry* **38,** 6781–6790 (1999).
224. J. Zehnulova, J. Kasparkova, N. Farrell, and V. Brabec, Conformation, recognition by high mobility group domain proteins, and nucleotide excision repair of DNA intrastrand cross-links of novel antitumor trinuclear platinum complex BBR3464. *J. Biol. Chem.* **276,** 22,191–22,199 (2001).
225. M. J. Clarke, F. Zhu, and D. R. Frasca, Non-platinum chemotherapeutic metallopharmaceuticals. *Chem. Rev.* **99,** 2511–2533 (1999).
226. B. K. Keppler, K.-G. Lipponer, B. Stenzel, and F. Kratz, New tumor-inhibiting ruthenium complexes. *In* "Metal complexes in cancer chemotherapy" (B. Keppler, ed.), pp. 187–220. VCH Verlagsgesellschaft, VCH Publishers, Weinheim/New York, 1993.
227. G. Sava, E. Alessio, A. Bergano, and G. Mestroni, Sulfoxide ruthenium complexes: Nontoxic tools for the selective treatment of solid tumour metastases. *in* "Topics in Biological Inorganic Chemistry. Metallopharmaceuticals" (M. J. Clarke and P. J. Sadler, eds.), pp. 143–169. Springer, Berlin, 1999.
228. G. Mestroni, E. Alessio, M. Calligaris, W. M. Attia, F. Quadrifoglio, S. Cauci, G. Sava, S. Zorzet, S. Pacor, C. Monti-Bragadin, M. Tamaro, and L. Dolzani, Chemical, biological and antitumor properties of ruthenium(II) complexes with dimethylsulfoxide. *Prog. Clin. Biochem. Med.* **10,** 71–87 (1989).
229. F. Loseto, F. Alessio, G. Mestroni, G. Lacidogna, A. Nassi, D. Giordano, and M. Coluccia, Interaction of $RuCl_2(dimethylsulphoxide)_4$ isomers with DNA. *Anticancer Res.* **11,** 1549–1553 (1991).
230. O. Novakova, C. Hofr, and V. Brabec, Modification of natural, double-helical DNA by antitumor *cis*- and *trans*-$[Cl_2(Me_2SO)_4Ru]$ in cell-free media. *Biochem. Pharmacol.* **60,** 1761–1771 (2000).
231. Y. N. V. Gopal, D. Jayaraju, and A. K. Kondapi, Inhibition of topoisomerase II catalytic activity by two ruthenium compounds: A ligand-dependent mode of action. *Biochemistry* **38,** 4382–4388 (1999).
232. B. K. Keppler, "Metal Complexes in Cancer Chemotherapy" VCH Verlagsgesellschaft, VCH Publishers, Weinheim/New York, 1993.
233. J. Malina, O. Novakova, B. K. Keppler, E. Alessio, and V. Brabec, Biophysical analysis of natural, double-helical DNA modified by anticancer heterocyclic complexes of ruthenium(III) in cell-free media. *J. Biol. Inorg. Chem.* **6,** 435–445 (2001).
234. J. K. Barton, Metals and DNA: Molecular left-handed complements. *Science* **233,** 727–734 (1986).
235. N. Grover, T. W. Welch, T. A. Fairley, M. Cory, and H. H. Thorp, Covalent binding of aquaruthenium complexes to DNA. *Inorg. Chem.* **33,** 3544–3548 (1994).

236. G. A. Neyhart, C. C. Cheng, and H. H. Thorp, Kinetics and mechanism of the oxidation of sugars and nucleotides by oxoruthenium(IV): Model studies for predicting cleavage patterns in polymeric DNA and RNA. *J. Am. Chem. Soc.* **117,** 1463–1471 (1995).
237. J. K. Barton and E. Lolis, Chiral discrimination in the covalent binding of bis(phenanthroline) dichlororuthenium(II) to B-DNA. *J. Am. Chem. Soc.* **107,** 708–709 (1985).
238. O. Novakova, J. Kasparkova, O. Vrana, P. M. van Vliet, J. Reedijk, and V. Brabec, Correlation between cytotoxicity and DNA binding of polypyridyl ruthenium complexes. *Biochemistry* **34,** 12,369–12,378 (1995).
239. P. M. van Vliet, S. M. S. Toekimin, J. G. Haasnoot, J. Reedijk, O. Novakova, O. Vrana, and V. Brabec, *mer*-[Ru(terpy)Cl$_3$] (terpy = 2,2′:6′,2″-terpyridine) shows biological activity, forms interstrand cross-links in DNA and binds two guanine derivatives in a *trans* configuration. *Inorg. Chim. Acta* **231,** 57–65 (1995).
240. B. Van Houten, S. Illenye, Y. Qu, and N. Farrell, Homodinuclear (Pt,Pt) and heterodinuclear (Ru,Pt) metal compounds as DNA-protein cross-linking agents: Potential suicide DNA lesions. *Biochemistry* **32,** 11,794–11,801 (1993).
241. P. M. Takahara, A. C. Rosenzweig, C. A. Frederick, and S. J. Lippard, Crystal structure of double-stranded DNA containing the major adduct of the anticancer drug cisplatin. *Nature* **377,** 649–652 (1995).

AMP- and Stress-Activated Protein Kinases: Key Regulators of Glucose-Dependent Gene Transcription in Mammalian Cells?

Isabelle Leclerc,
Gabriela da Silva Xavier,
and Guy A. Rutter

Department of Biochemistry
School of Medical Sciences
University of Bristol
Bristol, BS8 1TD, United Kingdom

I. AMP-Activated Protein Kinase.. 70
II. SNF1 and Glucose Repression in Yeast.. 71
III. AMPK and Regulation of Gene Transcription in Mammals............... 71
    A. Liver Gene Expression... 71
    B. Pancreatic β-Cell Gene Expression... 74
    C. Gene Expression in Muscle Tissue... 77
IV. Downstream Targets of AMPK and Gene Transcription................... 77
V. Mitogen- and Stress-Activated Protein Kinases............................. 78
    A. Glucose-Regulated Preproinsulin Gene Expression in Pancreatic β-Cell. 81
VI. Conclusions.. 82
    References.. 82

This article will discuss the role of two classes of serine/threonine protein kinases in the regulation of gene transcription in mammals. The first is AMP-activated protein kinase (AMPK), which is responsive to changes in the intracellular energy status. The second is the "stress-activated" family of protein kinases, members of the mitogen-activated protein (MAP) kinase superfamily, whose regulation by a number of extracellular agents (including osmotic stresses, cytokines, and heat) is less well understood. Interest in these enzymes has grown in the past

Abbreviations: AICAR, 5-aminoimidazole-4-carboxamide riboside; ACC, acetyl-CoA carboxylase; AMPK, AMP-activated protein kinase; FAS, fatty acid synthase; GLUT, glucose transporter; HNF, hepatocyte nuclear factor; L-PK, L-type pyruvate kinase; MAPK, mitogen-activated protein kinase; PEPCK, phospho*enol*pyruvate carboxykinase; PI3-kinase, phosphatidylinositol-3 kinase; PPI, preproinsulin; SAPK, stress-activated protein kinase; SNF, sucrose nonfermenting; SREBP, sterol responsive element binding protein.

few years due to mounting evidence (both pharmacological and genetic) which has implicated them in the regulation of a number genes important in mammalian metabolism [G. A. Rutter, J. M. Tavaré and D. G. Palmer, *News Physiol. Sci.* 15, 149–154 (2000)]. © 2002, Elsevier Science (USA).

I. AMP-Activated Protein Kinase

AMPK is a multisubstrate heterotrimeric serine/threonine protein kinase consisting of a catalytic α-subunit and noncatalytic β- and γ-subunits, all of which are necessary for full kinase activity (2–5). Two isoforms exist for the α- ($\alpha 1$ and $\alpha 2$) (3, 6) and the β-subunits ($\beta 1$ and $\beta 2$) (7); and three for the γ-subunit ($\gamma 1$, $\gamma 2$, and $\gamma 3$) (8). The $\alpha 1$-subunit appears to be ubiquitous and located mainly in the cytosol, whereas $\alpha 2$-subunit expression seems more restricted to heart, muscle, and liver, and is present in both the cytosol and the nucleus of the cells (9–11). AMPK was described in 1975 by Brown and Goldstein as the protein kinase responsible for the phosphorylation and inactivation of 3-hydroxy-3-methyl-glutaryl-coenzyme A (HMG-CoA) reductase, the key enzyme of cholesterol synthesis (12). In 1980, a protein kinase distinct from cyclic AMP-dependent protein kinase, whose activity was dependent on AMP and capable of phosphorylating and thus inhibiting acetyl-CoA carboxylase (ACC), the key enzyme of fatty acid synthesis, was characterized (13). It was in 1987 that David Carling and Grahame Hardie showed that the HMG-CoA kinase and the ACC kinase were the same enzyme (14). They also defined its energy-sensitive status and described the mechanism of activation of AMPK (15). AMPK is activated allosterically by both AMP and reversible phosphorylation at residue threonine 172 of the α-subunit by an upstream kinase (16), AMPK kinase, which is also a metabolite-sensing protein kinase and activated by AMP (15, 17, 18). Stresses such as heat shock, hypoxia, and arsenic that cause a rise in AMP/ATP ratio activate AMPK. Starvation and exercise have also been reported to activate AMPK in the liver (19–21). AMPK is also activated in muscle during contraction by a decrease in the phosphocreatine/creatine ratio (22). By phosphorylating and inactivating HMG-CoA reductase and ACC, AMPK shuts down cholesterol and fatty acid synthesis, two energy-consuming pathways; and by inhibition of ACC replenishes the ATP stores by promoting fatty acid oxidation via a decrease in malonyl-CoA levels (14, 23). It is now established that AMPK phosphorylates many other target proteins, namely, hormone-sensitive lipase, glycogen synthase, phosphorylase kinase (24), creatine kinase (22), endothelial nitric oxide synthase (25), and phosphofructokinase-2 (26), reinforcing its important role in energy metabolism.

TABLE I
SUBUNITS HOMOLOGY BETWEEN AMPK AND SNF1

Subunit function	SNF1	AMPK	Gene map locus in human
Catalytic	Snf1p	$\alpha 1$	5p12
		$\alpha 2$	1p31
Adaptator/cellular localization	Sip1/Sip2/Gal83	$\beta 1$	12q24.1
		$\beta 2$	?
Regulatory	Snf4p	$\gamma 1$	12q13.1
		$\gamma 2$	7q35–q36
		$\gamma 3$	?

II. SNF1 and Glucose Repression in Yeast

Cloning of AMPK has revealed its striking homology with the yeast sucrose nonfermenting-1 complex (SNF1) (27–30). SNF1 complex in yeast is formed by the association of Snf1p, Snf4p, and Sip1/Sip2/Gal83 proteins and is the key regulator of glucose-dependent gene repression (31, 32). All three subunits of AMPK are related to the proteins forming the SNF1 complex in yeast (Table I). When glucose is removed from the growth medium of yeast *Saccharomyces cerevisiae*, a phosphorylation cascade is activated, involving Snf1p, signaling glucose absence to the nucleus and switching on the transcription of genes whose products are necessary for the metabolism of alternate carbon sources, such as sucrose, maltose, and galactose. Mutants for the *SNF1* gene are unable to grow in the absence of glucose (15). This homology between AMPK and Snf1p has led to the hypothesis that AMPK, in addition to its role in regulating fuel partitioning by phosphorylation of key enzymes, may also be involved in glucose-dependent gene transcription in mammalian cells. Beginning with studies in 1998 (33, 34), considerable evidence has now accumulated demonstrating that AMPK is indeed involved in the regulation of gene expression in mammals, at least in liver, pancreatic islet β- and muscle cells (11, 33–38).

III. AMPK and Regulation of Gene Transcription in Mammals

A. Liver Gene Expression

Glucose is an important regulator in gene expression in the liver. According to glucose availability, different metabolic pathways might be turned on and off to ensure a correct handling of energy supply and the maintenance of normal blood glucose level. Glucose positively regulates the expression of many glycolytic and lipogenic enzymes (*39*) and negatively regulates the expression of the key gluconeogenic gene, phospho*enol*pyruvate carboxykinase (PEPCK) (*40*). So, as in yeast, glucose promotes the expression of genes encoding proteins involved in its own metabolism.

In vivo, it is difficult to distinguish the exact part devoted to glucose, since elevation of glucose concentration provokes insulin secretion and inhibits glucagon secretion. These two hormones were long known as important regulators of gene expression, but recently carbohydrates have been shown to play a key role in gene transcription by themselves (*39, 41*). Analysis in cultured rat or mouse hepatocytes made it possible to distinguish between purely glucose-dependent genes, purely insulin-dependent genes, and genes with mixed dependence (*1*). The glucose-signaling pathway in hepatocytes begins by the entry of glucose by the glucose transporter GLUT2, which is bidirectional and allows rapid equilibrium between intra- and extracellular glucose concentrations (*42*). The glucose must then be phosphorylated into glucose 6-phosphate (G6P) by glucokinase (GK) in order to transmit its signal to the transcriptional machinery. The hepatic *GK* gene is now considered as the prototype of a purely "insulin-dependent" gene. Insulin induction of *GK* gene expression is thought to be mediated by the transcription-factor sterol-response element-binding protein 1c (SREBP1c) (*43*). The signaling steps subsequent to the formation of G6P are, however, still disputed. Some experimental evidence has suggested that the glucose signal could be generated by an intermediate of the nonoxidative branch of the pentose–phosphate pathway (*44–46*), but others have suggested that the signal is the metabolite G6P itself (*47*). In any case, whether the metabolite signal is a pentose- or a hexose-phosphate, a phosphorylation/dephosphorylation cascade is next apparently activated to transmit the glucose signal to the nucleus (*34, 48*).

It is now widely accepted that AMPK activation, whether by the use of the pharmacological activator 5-aminoimidazole-4-carboxamide riboside (AICAR) (*49*) or by the adenoviral gene transfer of a constitutively active form of the kinase (*16, 50*), inhibits the glucose-stimulated transcription of the L-type pyruvate kinase (*L-PK*), spot 14, fatty acid synthase (*FAS*), and acetyl-CoA carboxylase (*ACC*) genes (*33, 34, 36*), and conversely, AICAR stimulates the expression of the glucose-inhibited PEPCK gene (*35*). However, it must be emphasized that conflicting results have been published for the *PEPCK* gene. As well as a report describing stimulatory effect of AICAR on *PEPCK* gene transcription in rat hepatocytes in primary culture (*35*), one publication reported no effect of AMPK activation on *PEPCK* gene expression (*33*), and another one (*38*) reported that AMPK activation by AICAR mimicked insulin action and inhibited *PEPCK* gene transcription. The experimental conditions used by Leclerc *et al.* (*33*) to visualize *PEPCK* mRNA by northern blotting included the addition of cyclic

AMP in the culture medium of primary hepatocytes. This would have already maximally stimulated *PEPCK* gene expression and explains why no additional effect of AICAR was detected (35). On the other hand, it is more difficult to interpret the data presented by Lochhead et al. (38), suggesting that AMPK activation would mimic the effect of insulin on gluconeogenic gene expression. First, this is opposite to the notion that AMPK is a stress-activated protein kinase (51) and is therefore expected to mimic the effect of stress-stimulated hormones like cortisol or glucagon, which are potent stimulators of *PEPCK* gene expression (52). One possible explanation for the results of Lochhead et al. is that these workers exclusively used a rat hepatoma-derived cell line (H4IIE) in their study. Extreme caution should be taken when working with cell lines, since they very often lose the expression of proteins necessary for the correct regulation of gene expression in response to glucose or insulin (53, 54).

Attempts were made to inhibit AMPK activity in hepatocytes in primary culture by the use of an adenovirus encoding a dominant negative form of the α1-subunit of AMPK (Ad.α1DN), and to study the resulting effect on the expression of glucose-dependent genes at low glucose concentration (36). Hepatocytes infected with this Ad.α1DN did not show any increase in *L-PK, spot 14, FAS,* or *ACC* mRNA abundance at low glucose concentration, as one would have expected if, in hepatocytes, AMPK inhibition was the mechanism by which glucose induces the expression of these genes. However, AMPK inhibition achieved by the use of Ad.α1DN was partial, being 75% of the AICAR-stimulated AMPK activity and 60% in the basal state. Unfortunately, no positive control of AMPK inhibition, such as stimulation of ACC activity (13, 14, 55), was presented. It is also possible that α2 activity in the nucleus was not efficiently inhibited by the use of Ad.α1DN, explaining the absence of effect of this dominant negative form on gene expression.

It is now clear that in hepatocytes in primary culture, AMPK activation leads to inhibition of the effect of glucose on gene expression (33–36). By contrast, definitive proof of a role for AMPK in the glucose-signaling pathway to gene expression in liver cells is still lacking. In fact, the presence or absence of glucose in the medium of hepatocytes in primary culture, with or without insulin, does not modify AMPK activity measured in cell homogenates (34, 36). Interestingly, in the hepatoma cell line, mhAT3F (56), glucose inhibits AMPK activity (Isabelle Leclerc and Axel Kahn, unpublished data, see Fig. 1), and insulin has a similar effect in the hepatoma Fao cell line (55). In the whole animal, starvation stimulates AMPK activity in the liver, whereas a carbohydrate-rich diet inhibits it (19, 20). The precise nutritional modulator of AMPK activity has yet to be defined.

Conclusive answers concerning the role of AMPK isoforms inhibition on glucose-dependent gene expression in the liver should come from the study of the hepatocytes of knockout mice for α1- and α2-subunits of AMPK (57).

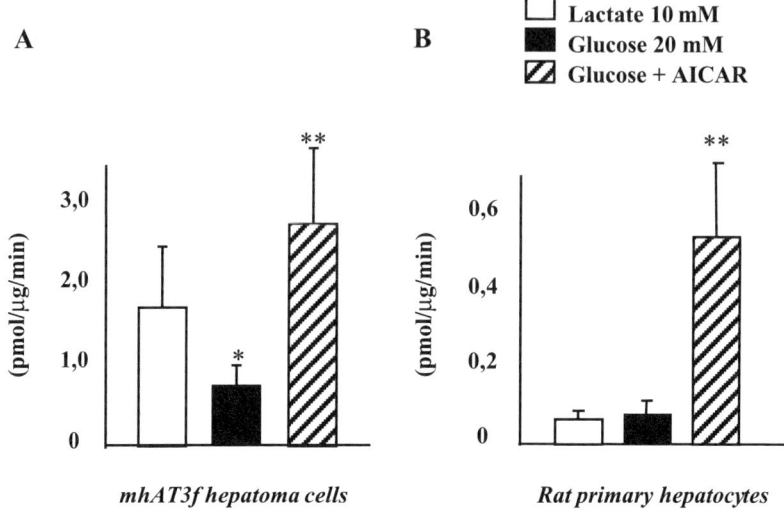

FIG. 1. Effect of glucose on AMPK activity in liver cells. A.mhAT3f hepatoma cells were cultured in medium 199 without glucose, in the presence of fructose 20 mM, 5% FCS, insulin 20 nM, triiodothyronine 1 μM, dexamethasone 1 μM, and penicillin/streptomycin until 60–70% confluence, then in medium DMEM/HamF12, glucose 17.5 mM, no serum but hormones for 6 h, and then for 60 min in either lactate 10 mM (white bar) or glucose 20 mM (black bar) +/− AICAR 500 μM (hatched bar). (B) Rat primary hepatocytes were isolated and cultured as described in Ref. 33 and then in the presence of either lactate 10 mM or glucose 20 mM +/− AICAR 200 μM for 1 h before protein extraction in the presence of phosphatase inhibitors. AMPK activity was measured against the SAMS peptide as described in Ref. 55. (∗) $P < 0.05$ for the effect of glucose in mhAT3f cells. (∗∗) $P < 0.005$ for the effect of AICAR.

B. Pancreatic β-Cell Gene Expression

In the pancreatic β-cell, glucose positively regulates the expression of glycolytic enzymes (58, 59) and preproinsulin (*PPI*) (60–63) genes and negatively regulates the expression of the peroxisome proliferator-activated receptor-α gene (64), a transcription factor responsible for the entire program of fatty acid oxidation (65). In MIN6 pancreatic islet β-cells, we have shown that AMPK activation by AICAR inhibited the activities of the L-PK and human preproinsulin (PPI) promoters at high glucose concentration, whereas microinjection of antibodies directed against the α2-subunit of AMPK into the cytosol and the nucleus of single living MIN6 cells stimulated the activity of these promoters at low glucose concentration (11). It is significant that microinjection of antibodies directed against the α1-subunit was without effect on promoter activities, while microinjection of antibodies directed against the β2-subunit, which co-immunoprecipitate both the α1- and α2-subunits (I. Leclerc, unpublished data) was effective, indicating that the α2-containing complexes are those involved in the regulation of gene transcription.

Several factors could explain the discrepancy between the results obtained with Ad.α1DN in primary cultured hepatocytes and the microinjection of antibodies against the α2-subunit of AMPK in MIN6 β-cells. First, the β-cells used (MIN6) were derived from a mouse insulinoma and could theoretically respond differently from primary β-cells. It has long been believed that the glucose-signaling pathway through gene expression was identical in both hepatocytes and β-cells (58), but this idea was recently challenged, and it is now accepted that the glucose-signaling pathway differs between the hepatocyte and the β-cell. In β-cells, glucose can act by inducing insulin secretion. Thus insulin, and not glucose, is the positive regulator for the *PPI* and *L-PK* genes expression (66–68). Intriguingly, however, at least one glucose-regulated β-cell gene, *SREBP1c*, appears to be regulated by the sugar through a mechanism not involving insulin release (C. Andreolas, G. A. Rutter *et al.*, submitted), in contrast to the regulation of this gene in the liver (43). Again, this observation emphasizes the striking difference between the regulation of the same gene in different cellular contexts.

In pancreatic MIN6, INS1, or HIT-T15 β-cell lines, stimulatory glucose concentrations (>8 mM) inhibit AMPK activity (11, 69). Whether this a specific feature of pancreatic β-cells or a more generalized feature of cell lines is not yet clearly defined. It is noteworthy that in pancreatic β-cells, glucose metabolism causes an increase in ATP concentration (70–72) and so would be expected to have an inhibitory effect on AMPK activity (73). By contrast, increases in glucose concentration have near detectable effects on ATP levels in the liver (74).

Data from Hardie's group (69) indicated that incubation of primary rodent pancreatic islets with AICAR leads to a weak activation of insulin secretion at low glucose concentrations but a marked inhibition of release at stimulatory concentrations of glucose (16 mM). Moreover, AICAR inhibited insulin secretion in MIN6 cells (da Silva Xavier *et al.*, in preparation) and in INS-1 cells (75) at high glucose concentrations, although others (H. Mulder and C. Newgard, personal communication) have reported that insulin secretion from clonal INS-1 cells (clone 832/13) (76) was unaffected by this agent. Given the uncertain specificity of this pharmacological agent, we have explored the effects of overexpressing constitutively active or dominant-negative forms of AMPKα1 on insulin release (Fig. 2). Supporting the earlier work on MIN6 cells and islets, forced overexpression or inactivation, respectively, of AMPKα1, caused marked stimulation or inhibition of insulin secretion (measured using cotransfected human growth hormone as a surrogate molecule). Given the central role of insulin release in the regulation of several glucose-responsive genes in MIN6 β-cells (68), these data argue that the downstream effects of modulating AMPK activity on gene expression may be modulated in part through alterations in insulin release. However, the molecular targets of AMPK, whose phosphorylation apparently

A

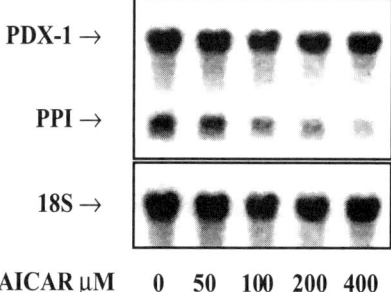

B

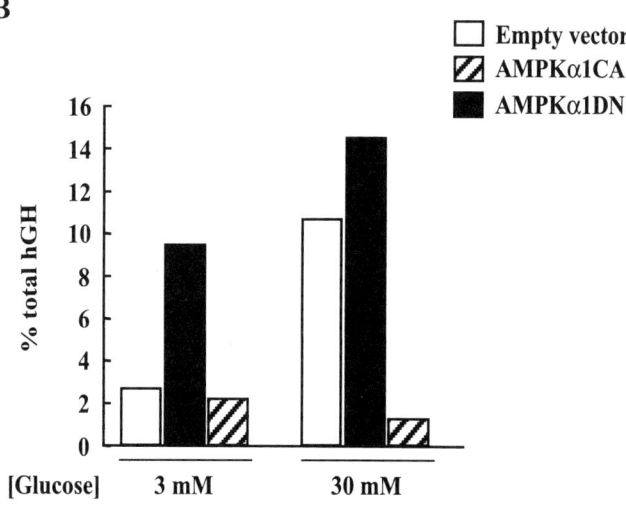

FIG. 2. Effect of AMPK activation on insulin gene expression and insulin secretion in MIN6 pancreatic β-cells. (A) Northern blot analysis was performed as described in Ref. 33 using 15 µg of total RNA per lane. The blot was hybridized simultaneously with PDX-1 and PPI cDNA probes, then stripped and rehybridized with a ribosomal 18S cDNA probe for normalization of RNA content. MIN6 cells were cultured as described in Ref. 11 and left overnight in the presence of the indicated AICAR concentration before RNA extraction. (B) MIN6 cells were cotransfected with either an empty vector (white bars) or plasmids encoding for a constitutively active (AMPKα1CA) or dominant-negative (AMPKα1DN) forms of AMPK and a vector encoding human growth hormone (hGH). The cells were cultured 24 h in 3 mM glucose and then incubated in KREBS medium containing 3 or 30 mM glucose for 20 min. The medium was collected, and the cells were lysed in 1% triton X-100 buffer. Secreted and total hGH content was quantitated with the aid of a commercially available hGH assay kit. The amount of secreted hGH is expressed as a percentage of the total amount of hGH present in each sample.

leads to the inhibition of insulin release, remain obscure. A good candidate would be the ATP-sensitive K^+ channel, whose closure by glucose is central to triggering insulin release (77). Alternatively, phosphorylation and inactivation of acetyl-CoA carboxylase (75) and inhibition of malonyl-CoA formation may also inhibit insulin secretion (78), though some evidence argues against a role for malonyl-CoA formation in regulated insulin secretion (76). Alternatively, AMPK could act on the insulin-signaling pathway involving phosphatidylinositol 3′-kinase (PI3K) through gene expression (68). Interestingly, PI3K inhibition by either the LY 294002 compound or the use of a negative dominant of the p85 subunit prevents *L-PK* gene stimulation by glucose and insulin, but also by the microinjection of anti-AMPKα2 antibody at low glucose concentration (68). This latter experiment suggests that AMPK inhibition may stimulate PI3K activity in β-cells and, more strikingly, that AMPK inhibition is a prerequisite for PI3K activation by glucose and insulin in MIN6 pancreatic β-cells. On the contrary, it seems unlikely that PI3K could lie upstream of AMPK, since in ES PDK1-/-(3-phosphoinositide-dependent protein kinase 1) cells, AMPK is regulated normally (79).

C. Gene Expression in Muscle Tissue

AMPK is activated in skeletal muscle following exercise (22, 80) and promotes contraction-dependent, insulin-independent glucose transport via GLUT4 translocation to the plasma membrane (81). The molecular confirmation of AMPK's role in exercise-induced glucose transport has been shown in transgenic mice overexpressing dominant-negative form of the kinase (82). Exercise is also known to modulate gene expression in muscle tissue, and it has been proposed that this transcriptional effect of exercise could be mediated by AMPK (37).

IV. Downstream Targets of AMPK and Gene Transcription

In order to determine which transcription factor would be involved in the AMPK-signaling pathway to gene expression in hepatocytes, electrophoretic mobility gel shift assays were performed using elements of the L-PK promoter as DNA probes and nuclear extracts of hepatocytes in primary culture incubated in the presence of AICAR (83). This resulted in the disappearance of the hepatocyte nuclear factor-4α (HNF4) retarded band. The expression of all the HNF4-dependent genes tested so far was down-regulated when hepatocytes were incubated in the presence of AICAR, whether they were glucose-responsive genes like *GLUT2*, *L-PK*, and *Aldolase B,* or glucose-unresponsive genes like *HNF1, ApoB,* an *ApoC*III (*83*). The mechanism by which AMPK inhibits HNF4 function is not completely resolved at the present time but seems to occur at the post-transcriptional level. HNF4 in the adult liver is mainly involved in the control of lipid export (*84*). By decreasing HNF4 protein levels, AMPK would decrease the apolipoprotein synthesis and stop lipid export by the liver, therefore preserving this carbon source for ATP repletion at the cellular level.

It is likely that AMPK regulates gene expression by interfering with other transcription factors, since the glucose-regulated genes *spot 14, FAS, ACC,* and *PEPCK,* in which transcription is controlled by AMPK (*33–36*), are not strictly HNF4-dependent genes. Other transcription factors involved in the expression of these genes that could be serious candidates for AMPK's inhibitory effect are the insulin-responsive factor *SREBP1c* (*43, 85, 86*) and the newly described carbohydrate-responsive factor (*87–89*). Another exciting possibility would be that AMPK acts as a histone-kinase and has a more profound effect on transcription, as has recently been shown for its yeast homolog Snf1p (*90*).

V. Mitogen- and Stress-Activated Protein Kinases

Stress-activated protein kinases (SAPKs) are a subgroup of the mitogen-activated protein kinase (MAPKs) family (*91*). The MAPKs belong in the "C-M-G-C" (Cyclin-dependent kinases, MAPKs, Glycogen synthase kinases 3, and Cell-division cycle-dependent-like kinases) group of protein kinase families in the eukaryotic protein kinase superfamily (*92*). This family of protein kinases have received much attention, because many of these enzymes translocate to the nucleus to regulate the activity of transcription factors (*93*). Full-length sequences of more than 100 MAPKs have been reported in the major databases since the first member was identified (*94*). Distinct MAPK pathways, made up of kinase cascades which lead to the activation of discrete MKKs and MAPKs, have been identified in both yeast and higher eukaryotes (*95–99*). Four subgroups of MAPKs have been identified in the latter: (i) extracellular signal-regulated kinase (ERKs) (*100, 101*), (ii) c-Jun amino-terminal kinases or stress-activated protein kinase (JNK or SAPK) (*102, 103*), (iii) p38 MAPK (*104*), and (iv) BMK/Erk5 (*105, 106*). However, there is increasing evidence that the MAPK signal transduction pathways in mammalian cells are much more complex than homologous systems in yeast; i.e., there is cross-talk between the pathways in higher cell types (*96, 97, 107, 108*).

MAPKs are proline-directed serine/threonine kinases that phosphorylate only substrates with a proline residue in the P+1 site of the substrate recognition consensus motif ψX[ST]P, where ψ represents P or an aliphatic amino acid (*109*). Their target serine and threonine residues are closely followed by one or more prolines in a motif recognized preferentially by a particular MAPK.

The "classical" MAPKs, also termed extracellular regulated kinases (ERKs) 1 and 2, or p42/p44, are involved in the control of growth and differentiation in response to ligands which act to stimulate receptor tyrosine kinases and activate the GTP-binding protein, Ras. This then results in the activation of a cascade of protein kinase activation, ultimately leading to phosphorylation of Erks (*110*).

ERK2 was the first family member to be identified as a 42-kDa protein serine/threonine kinase that phosphorylates microtubule-associated protein (*94*). ERK is related in sequence to the *FUS3* and *KSS1* gene products in *S. cereviseae* and to the *SPK1* gene product in *S. pombe* (*100*), while its upstream kinase, MEK, is homologous to *STE7* of *S. cereviseae* and *byr1* of *S. pombe* (*99*). ERK2 is phosphorylated on Thr185 and Tyr187 (*111*) *in vivo*.

In contrast to ERKs, JNK, and p38/MAPK are activated in cells exposed to stresses such as heat shock and osmotic changes, respectively. Elevated glucose concentrations and a consequent increase in the intracellular concentration of glucose metabolites, are expected to impose an osmotic stress in tissues (e.g., the liver, islet β- and adipose tissue cells) in which a number of glycolytic and lipogenic genes are regulated by glucose. Given the involvement of the p38 homolog, HOG1, in osmo-sensing in yeast, this has led to a number of groups exploring the potential role of stress-activated protein kinases in the transcriptional response to nutrients, notably in the islet β-cell (*112, 113*). We will first explore the properties and regulation of these enzymes in other cell types before considering the latter.

JNK/SAPK1 activity responds to a variety of other extracellular stimuli such as proinflammatory cytokines (interleukin IL-1, TNF-α) (*114*) and ultraviolet light (UV) (*103*) as well as hyperosmolarity (*115*). There are two well-characterized isoforms of JNK: the 46-kDa JNK1 and the 55-kDa JNK2 (*102, 103, 115*). The conserved phosphorylation sequence in domain VIII for JNK is TPY (*103*). It is activated by a MEK-related kinase called SAPK/ERK kinase (SEK) *in vitro* and *in vivo* (*96–98*). SEK itself is activated by phosphorylation at a Ser and Thr in the activation domain. JNK may also be involved in cellular functions other than those induced by stress and has been linked to thymocyte activation (*116*) and cell proliferation (*117*).

p38/MAPK2 (SAPK2) was identified as the mammalian homolog of the yeast HOG-1 protein, which is involved in the response to osmotic stress (*104, 118*). Mouse p38 (human homologs: CSBP1 and 2) has a unique Thr–Gly–Tyr motif as the site of Thr/Tyr phosphorylation and is phosphoryated by MAPK kinase 3/ reactivating kinase kinase (MKK3/RKK) dual-specificity kinase (*118, 119*). Two mammalian MEK homologs, SEK1 and MKK3, can phosphorylate and activate SAPK2 (*96, 97*), although it seems that the more likely *in vivo* upstream kinase is MKK3 rather than SEK1 (*96*). SAPK2 was found to be identical to one of the two protein targets of cytokine-suppressive anti-inflammatory drugs (CSAIDs), CSAID-binding protein 2 (CSBP2) (*120*). SAPK2 can, therefore, activate the

JNK pathway if it is involved in the regulation of inflammatory cytokine biosynthesis. p38/MAPK is potently inhibited by the pharmacological agent SB203580. This interaction is mediated between T106 in human CSBP2 (SAPK2α) and the 4 phenyl ring of SB203580.

p38/MAPK phosphorylates the transcription factor ATF2 at Thr69 and Thr71 in its N-terminal activation domain, and phosphorylation at these two sites leads to increased transcriptional activity. The activity of CSBP1 but not 2 was increased under hyperosmolar conditions when the two were expressed in yeast (*121*).

IL-1 and hyperosmolar conditions also induce p38 phosphorylation. MKK3 and MKK6 preferentially phosphorylate p38 *in vitro* and have been proposed as potential physiological p38 activators (*97, 98, 108, 122, 123*). In contrast to ERKs, p38 is not significantly activated by Ras-controlled pathways but is efficiently activated by MKK3 and MKK6 and can activate ATF-2 and Elk-1 (*124*).

One known downstream target of p38/MAPK is MAPK-activated protein kinase 2 (MAPKAPK-2), a 50-kDa protein kinase that can be activated in KB, HeLa, PC12 cells, monocytes, and macrophages by chemical or physical stress such as sodium arsenite, heat shock, and osmotic stress, and by certain cytokines and inflammatory mediators, e.g. IL-1 and LPS (*104*). It phosphorylates Hsp25/Hsp27, where phosphorylation of the latter facilitates actin polymerization, thereby contributing to the repair of stress-damaged actin filaments (*125*). p38 activates MAPKAPK-2 by phosphorylating any two of the three residues Thr222, Ser272, and Thr334, and is itself activated by RKK (*104, 118, 119, 126*). MAPKAPK-2 can be phosphorylated by the ERKs *in vitro* but not *in vivo*. However, LPS and hyperosmolarity can activate both ERK and p38 (*104*). CSBP p38 can phosphorylate MAPKAPK-3 when activated by stress, LPS, IL-1, or TNFα (*127*). Biochemical studies indicate that the p38-MAPK signaling module activates CREB and ATF1 (*128, 129*), ATF2 (*123, 124*), CHOP (*130*), and MEF-2C (*131*), which are all transcription factors, and other protein kinases such as MAPKAPK2 (*118, 119, 132*), MAPKAPK3 (*127*), and Mnk1/2 (*133, 134*).

Various p38 isoforms have been identified and three of these (p38α, p38β2, p38γ) phosphorylate ATF2, Elk1, and MBP, although p38β2 showed less activation than the other two isoforms (*135*). However, the target sites of ATF2 phosphorylation of the three kinases are slightly different and only p38α phosphorylated MAPKAPK2; i.e., the different isoforms of the protein kinase have different substrate specificities. There is also evidence for the existence of p38δ/SAPK4 (*136, 137*) and alternative splice forms of p38α (*120, 138*). p38β is not actually a functional protein kinase *in vivo* or *in vitro,* but p38β2, isolated from a human brain cDNA library, encodes for a functional protein which is not activable by MKK3 but inhibitable by SB203580 (*135*). p38β differs from p38β2 by an 8-amino acid insert close to what corresponds to the substrate-binding site of JNK.

Specific activating kinases lie upstream of the MAPKs. The protein kinase upstream of p38-MAPK, termed MKK3/6, exists in a variety of splice forms with distinct patterns of tissue distribution (122). In the study of Han et al. (122), two human splice isoforms of MKK6 were identified: one of 278 and the other of 334 amino acids. One murine protein of 237 amino acids has also been found. High levels of MKK6 mRNA of 1.7 kb were found in skeletal muscle. Four different splice forms of MKK6b were found, but the longest of these only was found in skeletal muscle and the pancreas. This particular splice form is enriched in heart, skeletal muscle, pancreas, and liver, and detectable in the brain, placenta, and lung. Ser151 and Thr155 of MKK6 may be phosphorylation sites required for enzymatic activity. Lys25 may be part of the ATP-binding site.

A. Glucose-Regulated Preproinsulin Gene Expression in Pancreatic β-Cell

ERK 1 and 2 were first shown to be activated by glucose in INS-1 cells by Frodin et al. (139). Classical MAPKs were shown to be activated up to 15-fold; and p38/MAPK, by 1.5-fold, by physiological elevations in glucose concentration (3–12 mM) in INS-1 cells (140). The activated-ERK content in the nucleus increased following glucose treatment. ERK activation was found to be dependent on glucose metabolism and subsequent calcium influx and presumably reflected an increase in Ras activity. JNK, on the other hand, was not activated by glucose, arguing against a role for ERKs1/2 activation in the regulation of insulin secretion, and thus enhanced gene expression. Insulin secretion was not inhibited by a blockade of MEK activities by PD 98059 (141). Moreover NGF strongly activated ERKs1/2 without affecting insulin release.

Later data from MacFarlane et al. (112) suggested that p38/MAPK may have a role in the transcriptional regulation of the preproinsulin gene through its effects on insulin upstream factor 1 (IUF1), now termed pancreatic duodenum homeobox 1 (PDX-1) (112). PDX-1, a homeodomain factor, binds to A boxes on the human insulin gene promoter: A1, 2, 3, and 5. Binding of PDX-1 to the A3 site has been shown to be triggered by exposure of rat islets of Langerhans to high glucose or insulin (67, 142) and abolished by low glucose. This binding was reportedly prevented by the p38/MAPK specific inhibitor, SB203580 (112). High glucose led to activation of MAPKAPK2 (112), which was, again, prevented by SB203580. Sodium arsenite and heat shock triggered the activation of PDX1 DNA binding, and glucose stimulated the transcription of a reporter gene that included nucleotides −250 to −50 of the human insulin promoter, bearing the PDX1-binding sites.

However, studies from other laboratories, including our own, have been unable to obtain data to support these findings (66, 68, 113). In particular, we observed no effect of SB203580 on stimulation by glucose of the preproinsulin promoter in MIN6 cells, and activation of p38/SAPK2, by expression of a constitutively-active form of MKK6, had no effect on the PPI promoter (while strongly transactivating an ATF2/Gal4 fusion construct). Finally, we were unable to detect any increase in p38 activity in response to glucose (*113*). However, MIN6 cell preculture conditions were subtly different in these studies, a variable which could explain the discrepancies observed in the activation of p38/MAPK in response to high glucose.

VI. Conclusions

We have discussed here the role of both AMPK and SAPKs in the regulation of gene expression by glucose and other nutrients in mammalian cells. While a growing body of evidence implicates the former, the role of the latter is more controversial.

Future work will probably include examining the effects of conditional inactivation or super-activation of the genes encoding these kinases in the liver and β-cells of transgenic animals. However, it should be stressed that such manipulations may lead to complex phenotypes, including developmental changes, given the likely involvement of MAPK family members in a number of key-signaling pathways. Nevertheless, the study of each of these enzymes is likely to be important, given their potential as new therapeutic targets for the treatment of diabetes mellitus and other metabolic disorders.

References

1. G. A. Rutter, J. M. Tavaré, and D. G. Palmer, Regulation of mammalian gene expression by glucose. *News Physiol. Sci.* **15,** 149–154 (2000).
2. G. Gao, C. S. Fernandez, D. Stapleton, A. S. Auster, J. Widmer, J. R. Dyck, B. E. Kemp, and L. A. Witters, Non-catalytic beta- and gamma-subunit isoforms of the 5′-AMP-activated protein kinase. *J. Biol. Chem.* **271,** 8675–8681 (1996).
3. D. Stapleton, K. I. Mitchelhill, G. Gao, J. Widmer, B. J. Michell, C. M. House, C. S. Fernandez, T. Cox, L. A. Witters, and B. E. Kemp, Mammalian AMP-activated protein kinase subfamily. *J. Biol. Chem.* **271,** 611–614 (1996).
4. J. R. B. Dyck, G. Gao, J. Widmer, D. Stapleton, C. S. Fernandez, B. E. Kemp, and L. A. Witters, Regulation of 5′-AMP-activated protein kinase activity by the noncatalytic beta and gamma subunits. *J. Biol. Chem.* **271,** 17,798–17,803 (1996).
5. C. Thornton, M. A. Snowden, and D. Carling, Identification of a novel AMP-activated protein kinase beta subunit isoform that is highly expressed in skeletal muscle. *J. Biol. Chem.* **273,** 12,443–12,450 (1998).
6. A. Woods, I. Salt, J. Scott, D. G. Hardie, and D. Carling, The alpha1 and alpha2 isoforms of the AMP-activated protein kinase have similar activities in rat liver but exhibit differences in substrate specificity in vitro. *FEBS Lett.* **397,** 347–351 (1996).
7. A. Woods, P. C. Cheung, F. C. Smith, M. D. Davison, J. Scott, R. K. Beri, and D. Carling, Characterization of AMP-activated protein kinase beta and gamma subunits. Assembly of the heterotrimeric complex in vitro. *J. Biol. Chem.* **271,** 10,282–10,290 (1996).

8. P. C. F. Cheung, I. P. Salt, S. P. Davies, D. G. Hardie, and D. Carling, Characterization of AMP-activated protein kinase gamma-subunit isoforms and their role in AMP binding. *Biochem. J.* **346,** 659–669 (2000).
9. I. Salt, J. W. Celler, S. A. Hawley, A. Prescott, A. Woods, D. Carling, and D. G. Hardie, AMP-activated protein kinase: greater AMP dependence, and preferential nuclear localization, of complexes containing the alpha2 isoform. *Biochem. J.* **334,** 177–187 (1998).
10. A. M. Turnley, D. Stapleton, R. J. Mann, L. A. Witters, B. E. Kemp, and P. F. Bartlett, Cellular distribution and developmental expression of AMP-activated protein kinase isoforms in mouse central nervous system. *J. Neurochem.* **72,** 1707–1716 (1999).
11. G. da Silva Xavier, I. Leclerc, I. P. Salt, B. Doiron, D. G. Hardie, A. Kahn, and G. A. Rutter, Role of AMP-activated protein kinase in the regulation by glucose of islet beta cell gene expression. *Proc. Natl. Acad. Sci. U.S.A.* **97,** 4023–4028 (2000).
12. M. S. Brown, G. Y. Brunschede, and J. L. Goldstein, Inactivation of 3-hydroxy-3-methylglutaryl coenzyme A reductase in vitro. An adenine nucleotide-dependent reaction catalyzed by a factor in human fibroblasts. *J. Biol. Chem.* **250,** 2502–2509 (1975).
13. L. A. Yeh, K. H. Lee, and K. H. Kim, Regulation of rat liver acetyl-CoA carboxylase. Regulation of phosphorylation and inactivation of acetyl-CoA carboxylase by the adenylate energy charge. *J. Biol. Chem.* **255,** 2308–2314 (1980).
14. D. Carling, V. A. Zammit, and D. G. Hardie, A common bicyclic protein kinase cascade inactivates the regulatory enzymes of fatty acid and cholesterol biosynthesis. *FEBS Lett.* **223,** 217–222 (1987).
15. D. G. Hardie, D. Carling, and M. Carlson, The AMP-activated/SNF1 protein kinase subfamily: metabolic sensors of the eukaryotic cell? *Annu. Rev. Biochem.* **67,** 821–855 (1998).
16. S. C. Stein, A. Woods, N. A. Jones, M. D. Davison, and D. Carling, The regulation of AMP-activated protein kinase by phosphorylation. *Biochem. J.* **345,** 437–443 (2000).
17. D. G. Hardie and D. Carling, The AMP-activated protein kinase–fuel gauge of the mammalian cell? *Eur. J. Biochem.* **246,** 259–273 (1997).
18. B. E. Kemp, K. I. Mitchelhill, D. Stapleton, B. J. Michell, Z. P. Chen, and L. A. Witters, Dealing with energy demand: the AMP-activated protein kinase. *Trends Biochem. Sci.* **24,** 22–25 (1999).
19. M. R. Munday, M. R. Milic, S. Takhar, M. J. Holness, and M. C. Sugden, The short-term regulation of hepatic acetyl-CoA carboxylase during starvation and re-feeding in the rat. *Biochem. J.* **280,** 733–737 (1991).
20. L. A. Witters, G. Gao, B. E. Kemp, and B. Quistorff, Hepatic 5′-AMP-activated protein kinase: zonal distribution and relationship to acetyl-CoA carboxylase activity in varying nutritional states. *Arch. Biochem. Biophys.* **308,** 413–419 (1994).
21. C. L. Carlson and W. W. Winder, Liver AMP-activated protein kinase and acetyl-CoA carboxylase during and after exercise. *J. Appl. Physiol.* **86,** 669–674 (1999).
22. M. Ponticos, Q. L. Lu, J. E. Morgan, D. G. Hardie, T. A. Partridge, and D. Carling, Dual regulation of the AMP-activated protein kinase provides a novel mechanism for the control of creatine kinase in skeletal muscle. *EMBO J.* **17,** 1688–1699 (1998).
23. D. G. Hardie, D. Carling, and A. T. R. Sim, The AMP-activated protein kinase: a multisubstrate regulator of lipid metabolism. *Trends Biochem. Sci.* **14,** 20–23 (1989).
24. D. Carling and D. G. Hardie, The substrate and sequence specificity of the AMP-activated protein kinase. Phosphorylation of glycogen synthase and phosphorylase kinase. *Biochim. Biophys. Acta* **1012,** 81–86 (1989).
25. Z. P. Chen, K. I. Mitchelhill, B. J. Michell, D. Stapleton, I. Rodriguez-Crespo, L. A. Witters, D. A. Power, P. R. Ortiz de Montellano, and B. E. Kemp, AMP-activated protein kinase phosphorylation of endothelial NO synthase. *FEBS Lett.* **443,** 285–289 (1999).

26. A. S. Marsin, L. Bertrand, M. H. Rider, J. Deprez, C. Beauloye, M. F. Vincent, G. Van den Berghe, D. Carling, and L. Hue, Phosphorylation and activation of heart PFK-2 by AMPK has a role in the stimulation of glycolysis during ischaemia. *Curr. Biol.* **10,** 1247–1255 (2000).
27. D. Carling, K. Aguan, A. Woods, A. J. Verhoeven, R. K. Beri, C. H. Brennan, C. Sidebottom, M. D. Davison, and J. Scott, Mammalian AMP-activated protein kinase is homologous to yeast and plant protein kinases involved in the regulation of carbon metabolism. *J. Biol. Chem.* **269,** 11,442–11,448 (1994).
28. D. Stapleton, G. Gao, B. J. Michell, J. Widmer, K. Mitchelhill, T. Teh, C. M. House, L. A. Witters, and B. E. Kemp, Mammalian 5′-AMP-activated protein kinase non-catalytic subunits are homologs of proteins that interact with yeast Snf1 protein kinase. *J. Biol. Chem.* **269,** 29,343–29,346 (1994).
29. K. I. Mitchelhill, D. Stapleton, G. Gao, C. House, B. Michell, F. Katsis, L. A. Witters, and B. E. Kemp, Mammalian AMP-activated protein kinase shares structural and functional homology with the catalytic domain of yeast Snf1 protein kinase. *J. Biol. Chem.* **269,** 2361–2364 (1994).
30. G. Gao, J. Widmer, D. Stapleton, T. Teh, T. Cox, B. E. Kemp, and L. A. Witters, Catalytic subunits of the porcine and rat 5′-AMP-activated protein kinase are members of the SNF1 protein kinase family. *Biochim. Biophys. Acta* **1266,** 73–82 (1995).
31. M. Carlson, Glucose repression in yeast. *Curr. Opin. Microbiol.* **2,** 202–207 (1999).
32. M. Johnston, Feasting, fasting and fermenting. Glucose sensing in yeast and other cells. *Trends Genet.* **15,** 29–33 (1999).
33. I. Leclerc, A. Kahn, and B. Doiron, The 5′AMP-activated protein kinase inhibits the transcriptional stimulation by glucose in liver cells, acting through the glucose response complex. *FEBS Lett.* **431,** 180–184 (1998).
34. M. Foretz, D. Carling, C. Guichard, P. Ferre, and F. Foufelle, AMP-activated protein kinase inhibits the glucose-activated expression of fatty acid synthase gene in rat hepatocytes. *J. Biol. Chem.* **273,** 14,767–14,771 (1998).
35. A. Hubert, A. Husson, A. Chedeville, and A. Lavoinne, AMP-activated protein kinase counteracted the inhibitory effect of glucose on the phosphoenolpyruvate carboxykinase gene expression in rat hepatocytes. *FEBS Lett.* **481,** 209–212 (2000).
36. A. Woods, D. Azzout-Marniche, M. Foretz, S. C. Stein, P. Lemarchand, P. Ferre, F. Foufelle, and D. Carling, Characterization of the role of AMP-activated protein kinase in the regulation of glucose-activated gene expression using constitutively active and dominant negative forms of the kinase. *Mol. Cell. Biol.* **20,** 6704–6711 (2000).
37. M. Zhou, B. Z. Lin, S. Coughlin, G. Vallega, and P. F. Pilch, UCP-3 expression in skeletal muscle: effects of exercise, hypoxia, and AMP-activated protein kinase. *Am. J. Physiol. Endocrinol. Metab.* **279,** E622–629 (2000).
38. P. A. Lochhead, I. P. Salt, K. S. Walker, D. G. Hardie, and C. Sutherland, 5-aminoimidazole-4-carboxamide riboside mimics the effects of insulin on the expression of the 2 key gluconeogenic genes PEPCK and glucose-6-phosphatase. *Diabetes* **49,** 896–903 (2000).
39. S. Vaulont, M. Vasseur-Cognet, and A. Kahn, Glucose regulation of gene transcription. *J. Biol. Chem.* **275,** 31,555–31,558 (2000).
40. D. K. Scott, R. M. O'Doherty, J. M. Stafford, C. B. Newgard, and D. K. Granner, The repression of hormone-activated PEPCK gene expression by glucose is insulin-independent but requires glucose metabolism. *J. Biol. Chem.* **273,** 24,145–24,151 (1998).
41. H. C. Towle, E. N. Kaytor, and H. M. Shih, Regulation of the expression of lipogenic enzyme genes by carbohydrate. *Annu. Rev. Nutr.* **17,** 405–433 (1997).
42. B. Thorens, H. K. Sarkar, H. R. Kaback, and H. F. Lodish, Cloning and functional expression in bacteria of a novel glucose transporter present in liver, intestine, kidney, and beta-pancreatic islet cells. *Cell* **55,** 281–290 (1988).

43. M. Foretz, C. Guichard, P. Ferre, and F. Foufelle, Sterol regulatory element binding protein-1c is a major mediator of insulin action on the hepatic expression of glucokinase and lipogenesis-related genes. *Proc. Natl. Acad. Sci. U.S.A.* **96**, 12,737–12,742 (1999).
44. B. Doiron, M. H. Cuif, R. Chen, and A. Kahn, Transcriptional glucose signaling through the glucose response element is mediated by the pentose phosphate pathway. *J. Biol. Chem.* **271**, 5321–5324 (1996).
45. D. Massillon, W. Chen, N. Barzilai, D. Prus-Wertheimer, M. Hawkins, R. Liu, R. Taub, and L. Rossetti, Carbon flux via the pentose phosphate pathway regulates the hepatic expression of the glucose-6-phosphatase and phosphoenolpyruvate carboxykinase genes in conscious rats. *J. Biol. Chem.* **273**, 228–234 (1998).
46. D. Massillon, Regulation of the glucose-6-phosphatase gene by glucose occurs by transcriptional and post-transcriptional mechanisms. Differential effet of glucose and xylitol. *J. Biol. Chem.* **276**, 4055–4062 (2001).
47. F. Mourrieras, F. Foufelle, M. Foretz, J. Morin, S. Bouche, and P. Ferre, Induction of fatty acid synthase and S14 gene expression by glucose, xylitol and dihydroxyacetone in cultured rat hepatocytes is closely correlated with glucose 6-phosphate concentrations. *Biochem. J.* **326**, 345–349 (1997).
48. Y. Sudo and C. N. Mariash, Two glucose-signaling pathways in S14 gene transcription in primary hepatocytes: a common role of protein phosphorylation. *Endocrinology* **134**, 2532–2540 (1994).
49. J. M. Corton, J. G. Gillespie, S. A. Hawley, and D. G. Hardie, 5-aminoimidazole-4-carboxamide ribonucleoside. A specific method for activating AMP-activated protein kinase in intact cells? *Eur. J. Biochem.* **229**, 558–565 (1995).
50. B. E. Crute, K. Seefeld, J. Gamble, B. E. Kemp, and L. A. Witters, Functional domains of the alpha1 catalytic subunit of the AMP-activated protein kinase. *J. Biol. Chem.* **273**, 35,347–35,354 (1998).
51. D. G. Hardie, Molecular physiology. Ways of coping with stress. *Nature* **370**, 599–600 (1994).
52. R. W. Hanson and L. Reshef, Regulation of phosphoenolpyruvate carboxykinase (GTP) gene expression. *Annu. Rev. Biochem.* **66**, 581–611 (1997).
53. M. C. Meienhofer, E. De Medicis, M. Cognet, and A. Kahn, Regulation of genes for glycolytic enzymes in cultured rat hepatoma cell lines. *Eur. J. Biochem.* **169**, 237–243 (1987).
54. A. M. Lefrancois-Martinez, M. J. Diaz-Guerra, V. Vallet, A. Kahn, and B. Antoine, Glucose-dependent regulation of the L-pyruvate kinase gene in a hepatoma cell line is independent of insulin and cyclic AMP. *Faseb J.* **8**, 89–96 (1994).
55. L. A. Witters and B. E. Kemp, Insulin activation of acetyl-CoA carboxylase accompanied by inhibition of the 5′-AMP-activated protein kinase. *J. Biol. Chem.* **267**, 2864–2867 (1992).
56. B. Antoine, F. Levrat, V. Vallet, T. Berbar, N. Cartier, N. Dubois, P. Briand, and A. Kahn, Gene expression in hepatocyte-like lines established by targeted carcinogenesis in transgenic mice. *Exp. Cell. Res.* **200**, 175–185 (1992).
57. M. Holzenberger, C. Lenzner, P. Leneuve, R. Zaoui, G. Hamard, S. Vaulont, and Y. L. Bouc, Cre-mediated germline mosaicism: a method allowing rapid generation of several alleles of a target gene. *Nucleic Acids. Res.* **28**, E92 (2000).
58. S. Marie, M. J. Diaz-Guerra, L. Miquerol, A. Kahn, and P. B. Iynedjian, The pyruvate kinase gene as a model for studies of glucose-dependent regulation of gene expression in the endocrine pancreatic beta-cell type. *J. Biol. Chem.* **268**, 23,881–23,890 (1993).
59. E. Roche, F. Assimacopoulos-Jeannet, L. A. Witters, B. Perruchoud, G. Yaney, B. Corkey, M. Asfari, and M. Prentki, Induction by glucose of genes coding for glycolytic enzymes in a pancreatic beta-cell line (INS-1). *J. Biol. Chem.* **272**, 3091–3098 (1997).
60. M. A. Permutt and D. M. Kipnis, Insulin biosynthesis. I. On the mechanism of glucose stimulation. *J. Biol. Chem.* **247**, 1194–1199 (1972).

61. D. A. Nielsen, M. Welsh, M. J. Casadaban, and D. F. Steiner, Control of insulin gene expression in pancreatic beta-cells and in an insulin-producing cell line, RIN-5F cells. I. Effects of glucose and cyclic AMP on the transcription of insulin mRNA. *J. Biol. Chem.* **260,** 13,585–13,589 (1985).
62. S. Efrat, M. Surana, and N. Fleischer, Glucose induces insulin gene transcription in a murine pancreatic beta-cell line. *J. Biol. Chem.* **266,** 11,141–11,143 (1991).
63. B. Leibiger, T. Moede, T. Schwarz, G. R. Brown, M. Kohler, I. B. Leibiger, and P. O. Berggren, Short-term regulation of insulin gene transcription by glucose. *Proc. Natl. Acad. Sci. U.S.A.* **95,** 9307–9312 (1998).
64. R. Roduit, J. Morin, F. Masse, L. Segall, E. Roche, C. B. Newgard, F. Assimacopoulos-Jeannet, and M. Prentki, Glucose down-regulates the expression of the peroxisome proliferator-activated receptor-alpha gene in the pancreatic beta-cell. *J. Biol. Chem.* **275,** 35,799–35,806 (2000).
65. B. Desvergne and W. Wahli, Peroxisome proliferator-activated receptors: Nuclear control of metabolism. *Endocrinol. Rev.* **20,** 649–688 (1999).
66. I. B. Leibiger, B. Leibiger, T. Moede, and P. O. Berggren, Exocytosis of insulin promotes insulin gene transcription via the insulin receptor/PI-3 kinase/p70 s6 kinase and CaM kinase pathways. *Mol. Cell* **1,** 933–938 (1998).
67. H. Wu, W. M. MacFarlane, M. Tadayyon, J. R. Arch, R. F. James, and K. Docherty, Insulin stimulates pancreatic-duodenal homoeobox factor-1 (PDX1) DNA-binding activity and insulin promoter activity in pancreatic beta cells. *Biochem. J.* **344,** 813–818 (1999).
68. G. da Silva Xavier, A. Varadi, E. K. Ainscow, and G. A. Rutter, Regulation of gene expression by glucose in pancreatic beta-cells (MIN6) via insulin secretion and activation of phosphatidylinositol 3′-kinase. *J. Biol. Chem.* **275,** 36,269–36,277 (2000).
69. I. P. Salt, G. Johnson, S. J. Ashcroft, and D. G. Hardie, AMP-activated protein kinase is activated by low glucose in cell lines derived from pancreatic beta cells, and may regulate insulin release. *Biochem. J.* **335,** 533–539 (1998).
70. W. J. Malaisse and A. Sener, Glucose-induced changes in cytosolic ATP content in pancreatic islets. *Biochim. Biophys. Acta* **927,** 190–195 (1987).
71. H. J. Kennedy, A. E. Pouli, E. K. Ainscow, L. S. Jouaville, R. Rizzuto, and G. A. Rutter, Glucose generates sub-plasma membrane ATP microdomains in single islet beta-cells. Potential role for strategically located mitochondria. *J. Biol. Chem.* **274,** 13,281–13,291 (1999).
72. E. K. Ainscow and G. A. Rutter, MItochondrial priming modifies Ca^{2+} oscillations and insulin secretion in pancreatic islets. *Biochem. J.* **353,** 175–180 (2001).
73. D. G. Hardie, I. P. Salt, S. A. Hawley, and S. P. Davies, AMP-activated protein kinase: An ultrasensitive system for monitoring cellular energy charge. *Biochem. J.* **338,** 717–722 (1999).
74. E. K. Ainscow and M. D. Brand, Internal regulation of ATP turnover, glycolysis and oxidative phosphorylation in rat hepatocytes. *Eur. J. Biochem.* **266,** 737–749 (1999).
75. A. Kowluru, H.-Q. Chen, L. M. Modrick, and C. Stefanell, Activation of acetyl-CoA carboxylase by a glutamate- and magnesium-sensitive protein phosphatase in the islet beta-cell. *Diabetes* **50,** 1580–1587 (2001).
76. H. Mulder, D. Lu, J. 4th. Finley, J. An, J. Cohen, P. A. Antinozzi, J. D. McGarry, and C. B. Newgard, Overexpression of a modified human malonyl-CoA decarboxylase blocks the glucose-induced increase in malonyl-CoA level but has no impact on insulin secretion in INS-1 derived (832/13) beta cells. *J. Biol. Chem.* **276,** 6479–6484 (2001).
77. L. Aguilar-Bryan and J. Bryan, Molecular biology of adenosine triphosphate-sensitive potassium channels. *Endocrinol. Rev.* **20,** 101–135 (1999).
78. M. Prentki, S. Vischer, M. C. Glennon, R. Regazzi, J. T. Deeney, and B. E. Corkey, Malonyl-CoA and long chain acyl-CoA esters as metabolic coupling factors in nutrient-induced insulin secretion. *J. Biol. Chem.* **267,** 5802–5810 (1992).

79. M. R. Williams, J. S. Arthur, A. Balendran, J. van der Kaay, V. Poli, P. Cohen, and D. R. Alessi, The role of 3-phosphoinositide-dependent protein kinase 1 in activating AGC kinases defined in embryonic stem cells. *Curr. Biol.* **10**, 439–448 (2000).
80. W. W. Winder and D. G. Hardie, Inactivation of acetyl-CoA carboxylase and activation of AMP-activated protein kinase in muscle during exercise. *Am. J. Physiol.* **270**, E299–E304 (1996).
81. E. J. Kurth-Kraczek, M. F. Hirshman, L. J. Goodyear, and W. W. Winder, 5′ AMP-activated protein kinase activation causes GLUT4 translocation in skeletal muscle. *Diabetes* **48**, 1667–1671 (1999).
82. J. Mu, J. Brozinick, J. T., O. Valladares, M. Bucan, and M. J. Birnbaum, A role for AMP-activated protein kinase in contraction- and hypoxia-regulated glucose transport in skeletal muscle. *Mol. Cell.* **7**, 1085–1094 (2001).
83. I. Leclerc, C. Lenzner, L. Gourdon, S. Vaulont, A. Kahn, and B. Viollet, Hepatocyte nuclear factor 4alpha involved in maturity-onset diabetes of the young (MODY1) is a novel target of AMP-activated protein kinase. *Diabetes* **50**, 1515–1521 (2001).
84. G. P. Hayhurst, Y. H. Lee, G. Lambert, J. M. Ward, and F. J. Gonzalez, Hepatocyte nuclear factor 4alpha (Nuclear Receptor 2A1) is essential for maintenance of hepatic gene expression and lipid homeostasis. *Mol. Biol. Cell.* **21**, 1393–1403 (2001).
85. M. Foretz, C. Pacot, I. Dugail, P. Lemarchand, C. Guichard, X. Le Liepvre, C. Berthelier-Lubrano, B. Spiegelman, J. B. Kim, P. Ferre, and F. Foufelle, ADD1/SREBP-1c is required in the activation of hepatic lipogenic gene expression by glucose. *Mol. Cell. Biol.* **19**, 3760–3768 (1999).
86. K. Chakravarty, P. Leahy, D. Becard, P. Hakimi, M. Foretz, P. Ferre, F. Foufelle, and R. W. Hanson, Sterol regulatory element-binding protein 1c mimics the negative effect of insulin on phosphoenolpyruvate carboxykinase (GTP) gene transcription. *J. Biol. Chem.* **276**, 34,816–34,823 (2001).
87. S. H. Koo and H. C. Towle, Glucose regulation of mouse S(14) gene expression in hepatocytes. Involvement of a novel transcription factor complex. *J. Biol. Chem.* **275**, 5200–5207 (2000).
88. S. H. Koo, A. K. Dutcher, and H. C. Towle, Glucose and insulin function through two distinct transcription factors to stimulate expression of lipogenic enzyme genes in liver. *J. Biol. Chem.* **276**, 9437–9445 (2001).
89. H. Yamashita, M. Takenoshita, M. Sakurai, R. K. Bruick, W. J. Henzel, W. Shillinglaw, D. Arnot, and K. Uyeda, A glucose-responsive transcription factor that regulates carbohydrate metabolism in the liver. *Proc. Natl. Acad. Sci. U.S.A.* Early Edition, July (2001).
90. W. S. Lo, L. Duggan, N. C. Tolga, Emre, R. Belotserkovskya, W. S. Lane, R. Shiekhattar, and S. L. Berger, Snf1—A histone kinase that works in concert with the histone acetyltransferase gcn5 to regulate transcription. *Science* **293**, 1142–1146 (2001).
91. D. T. Denhardt, Signal-transducing protein phosphorylation cascades mediated by Ras/Rho proteins in the mammalian cell: the potential for multiplex signalling. *Biochem. J.* **318**, 729–747 (1996).
92. S. K. Hanks and T. Hunter, Protein kinases 6. The eukaryotic protein kinase superfamily: kinase (catalytic) domain structure and classification. *FASEB J.* **9**, 576–596 (1995).
93. R. Treisman, Regulation of transcription by MAP kinase cascades. *Curr. Opin. Cell Biol.* **8**, 205–215 (1996).
94. L. B. Ray and T. W. Sturgill, Rapid stimulation by insulin of a serine/threonine kinase in 3T3-L1 adipocytes that phosphorylates microtubule-associated protein 2 in vitro. *Proc. Natl. Acad. Sci. U.S.A.* **84**, 1502–1506 (1987).
95. R. J. Davis, MAPKs: New JNK expands the group. *Trends Biochem. Sci.* **19**, 470–473 (1994).
96. A. Lin, A. Minden, H. Martinetto, F. X. Claret, C. Lange-Carter, F. Mercurio, G. L. Johnson, and M. Karin, Identification of a dual specificity kinase that activates the Jun kinases. *Science* **268**, 286–290 (1995).

97. B. Derijard, J. Raingeaud, T. Barrett, I. H. Wu, J. Han, R. J. Ulevitch, and R. J. Davis, Independent human MAP-kinase signal transduction pathways defined by MEK and MKK isoforms. *Science* **267**, 682–685 (1995).
98. I. Sanchez, R. T. Hughes, B. J. Mayer, K. Yee, J. R. Woodgett, J. Avruch, J. M. Kyriakis, and L. I. Zon, Role of SAPK/ERK kinase-1 in the stress-activated pathway regulating. *Nature* **372**, 794–798 (1994).
99. C. M. Crews, A. Alessandrini, and R. L. Erikson, The primary structure of MEK, a protein kinase that phosphorylates the ERK gene product. *Science* **258**, 478–480 (1992).
100. T. C. Boulton, S. H. Nye, D. J. Robbins, N. Y. Ip, E. Radziejewska, S. D. Morgenbesser, R. A. DePinho, N. Panayotatos, M. H. Cobb, and G. D. Yancopoulos, ERKs: A family of protein-serine/threonine kinases that are activated and tyrosine phosphorylated in response to insulin and NGF. *Cell* **65**, 663–675 (1991).
101. R. Seger, N. G. Ahn, T. G. Boulton, G. D. Yancopoulos, N. Panayotatos, E. Radziejewska, L. Ericsson, R. L. Bratlien, M. H. Cobb, and E. G. Krebs, Microtubule-associated protein 2 kinases, ERK1 and ERK2, undergo autophosphorylation on both tyrosine and threonine residues: implication for their mechanism of activation. *Proc. Natl. Acad. Sci. U.S.A.* **88**, 6142–6146 (1991).
102. J. M. Kyriakis, P. Banerjee, E. Nikolakaki, T. Dai, E. A. Rubie, M. F. Ahmad, J. Avruch, and J. R. Woodgett, The stress-activated protein kinase subfamily of c-Jun kinases. *Nature* **369**, 156–160 (1994).
103. B. Derijard, M. Hibi, I. H. Wum, T. Barrett, B. Su, T. Deng, M. Karin, and R. J. Davis, JNK1: a protein kinase stimulated by UV light and Ha-Ras that binds and phosphorylates the c-Jun activation domain. *Cell* **76**, 1025–1037 (1994).
104. J. Han, J. D. Lee, L. Bibbs, and R. J. Ulevitch, A MAP kinase targeted by endotoxin and hyperosmolarity in mammalian cells. *Science* **265**, 808–811 (1994).
105. J. D. Lee, R. J. Ulevitch, and J. Han, Primary structure of BMK1: a new mammalian map kinase. *Biochem. Biophys. Res. Commun.* **213**, 715–724 (1995).
106. G. Zhou, Z. Q. Bao, and J. E. Dixon, Components of a new human protein kinase signal transduction pathway. *J. Biol. Chem.* **270**, 12,665–12,669 (1995).
107. C. A. Lange-Carter, C. M. Pleiman, A. M. Gardner, K. J. Blumer, and G. L. Johnson, A divergence in the MAP kinase regulatory network defined by MEK kinase. *Science* **260**, 315–319 (1993).
108. M. Yan, T. Dai, J. C. Deak, J. M. Kyriakis, L. I. Zon, J. R. Woodgett, and D. J. Templeton, Activation of stress-activated protein kinase by MEKK1 phosphorylation of its activator SEK1. *Nature* **372**, 798–800 (1994).
109. I. Clark-Lewis, J. S. Sanghera, and S. L. Pelech, Definition of a consensus sequence for peptide substrate recognition by p44mpk, the meiosis-activated myelin basic protein kinase. *J. Biol. Chem.* **266**, 15,180–15,184 (1991).
110. R. J. Davis, The mitogen-activated protein kinase signal transduction pathway. *J. Biol. Chem.* **268**, 14,553–14,556 (1993).
111. D. M. Payne, A. J. Rossomando, P. Martino, A. K. Erickson, J. H. Her, J. Shabanowitz, D. F. Hunt, M. J. Weber, and T. W. Sturgill, Identification of the regulatory phosphorylation sites in in pp42/mitogen-activated protein kinase (MAP kinase). *EMBO J.* **10**, 885–892 (1991).
112. W. M. MacFarlane, S. B. Smith, R. F. James, A. D. Clifton, Y. N. Doza, P. Cohen, and K. Docherty, The p38/reactivating kinase mitogen-activated protein kinase cascade mediates the activation of the transcription factor insulin upstream factor 1 and insulin gene transcription by high glucose in pancreatic beta-cells. *J. Biol. Chem.* **272**, 20,936–20,944 (1997).
113. I. Rafiq, G. da Silva Xavier, S. Hooper, and G. A. Rutter, Glucose-stimulated preproinsulin gene expression and nuclear trans-location of pancreatic duodenum homeobox-1 require activation

of phosphatidylinositol 3-kinase but not p38 MAPK/SAPK2. *J. Biol. Chem.* **275,** 15,977–15,984 (2000).
114. H. K. Sluss, T. Barrett, B. Derijard, and R. J. Davis, Signal transduction by tumor necrosis factor mediated by JNK protein kinases. *Mol. Cell. Biol.* **14,** 8376–8384 (1994).
115. Z. Galcheva-Gargova, B. Derijard, I. H. Wu, and R. J. Davis, An osmosensing signal transduction pathway in mammalian cells. *Science* **265,** 806–808 (1994).
116. B. Su, E. Jacinto, M. Hibi, T. Kallunki, M. Karin, and Y. Ben-Neriah, JNK is involved in signal integration during costimulation of T lymphocytes. *Cell* **77,** 727–736 (1994).
117. D. Bohmann, T. J. Bos, A. Admon, T. Nishimura, P. K. Vogt, and R. Tjian, Human proto-oncogene c-jun encodes a DNA binding protein with structural and functional properties of transcription factor AP-1. *Science* **238,** 1386–1392 (1987).
118. J. Rouse, P. Cohen, S. Trigon, M. Morange, A. Alonso-Llamazares, D. Zamanillo, T. Hunt, and A. R. Nebreda, A novel kinase cascade triggered by stress and heat shock that stimulates MAPKAP kinase-2 and phosphorylation of the small heat shock proteins. *Cell* **78,** 1027–1037 (1994).
119. N. W. Freshney, L. Rawlinson, F. Guesdon, E. Jones, S. Cowley, J. Hsuan, and J. Saklatvala, Interleukin-1 activates a novel protein kinase cascade that results in the phosphorylation of Hsp27. *Cell* **78,** 1039–1049 (1994).
120. J. C. Lee, J. T. Laydon, P. C. McDonnell, T. F. Gallagher, S. Kumar, D. Green, D. McNulty, M. J. Blumenthal, J. R. Heys, S. W. Landvatter, J. E. Strickler, M. M. McLaughlin, I. R. Siemens, S. M. Fisher, G. P. Livi, J. R. White, J. L. Adams, and P. R. Young, A protein kinase involved in the regulation of inflammatory cytokine biosynthesis. *Nature* **372,** 739–746 (1994).
121. S. Kumar, M. M. McLaughlin, P. C. McDonnell, J. C. Lee, G. P. Livi, and P. R. Young, Human mitogen-activated protein kinase CSBP1, but not CSBP2, complements a hog1 deletion in yeast. *J. Biol. Chem.* **270,** 29,043–29,046 (1995).
122. J. Han, J. D. Lee, Y. Jiang, Z. Li, L. Feng, and R. J. Ulevitch, Characterization of the structure and function of a novel MAP kinase. *J. Biol. Chem.* **271,** 2886–2891 (1996).
123. J. Raingeaud, S. Gupta, J. S. Rogers, M. Dickens, J. Han, R. J. Ulevitch, and R. J. Davis, Pro-inflammatory cytokines and environmental stress cause p38 mitogen-activated protein kinase activation by dual phosphorylation on tyrosine and threonine. *J. Biol. Chem.* **270,** 7420–7426 (1995).
124. J. Raingeaud, A. J. Whitmarsh, T. Barrett, B. Derijard, and R. J. Davis, MKK3- and MKK6-regulated gene expression is mediated by the p38 mitogen-activated protein kinase signal transduction pathway. *Mol. Cell. Biol.* **16,** 1247–1255 (1996).
125. J. N. Lavoie, H. Lambert, E. Hickey, L. A. Weber, and J. Landry, Modulation of cellular thermoresistance and actin filament stability accompanies phosphorylation-induced changes in the oligomeric structure of heat shock protein 27. *Mol. Cell. Biol.* **15,** 505–516 (1995).
126. R. Ben-Levy, I. A. Leighton, Y. N. Doza, P. Attwood, N. Morrice, C. J. Marshall, and P. Cohen, Identification of novel phosphorylation sites required for activation of MAPKAP kinase-2. *EMBO J.* **14,** 5920–5930 (1995).
127. M. M. McLaughlin, S. Kumar, P. C. McDonnell, S. Van Horn, J. C. Lee, G. P. Livi, and P. R. Young, Identification of mitogen-activated protein (MAP) kinase-activated protein kinase-3, a novel substrate of CSBP p38 MAP kinase. *J. Biol. Chem.* **271,** 8488–8492 (1996).
128. Y. Tan, J. Rouse, A. Zhang, S. Cariati, P. Cohen, and M. J. Comb, FGF and stress regulate CREB and ATF-1 via a pathway involving p38 MAP kinase and MAPKAP kinase-2. *EMBO J.* **15,** 4629–4642 (1996).
129. M. Iordanov, K. Bender, T. Ade, W. Schmid, C. Sachsenmaier, K. Engel, M. Gaestel, H. J. Rahmsdorf, and P. Herrlich, CREB is activated by UVC through a p38/HOG-1-dependent protein kinase. *EMBO J.* **16,** (1997).

130. X. Z. Wang and D. Ron, Stress-induced phosphorylation and activation of the transcription factor CHOP (GADD153) by p38 MAP kinase. *Science* **271**, 1347–1349 (1996).
131. J. Han, Y. Jiang, Z. Li, V. V. Kravchenko, and R. J. Ulevitch, Activation of the transcription factor MEF2C by the MAP kinase p38 in inflammation. *Nature* **386**, 296–299 (1997).
132. D. Stokoe, D. G. Campbell, S. Nakielny, H. Hidaka, S. J. Leevers, C. Marshall, and P. Cohen, MAPKAP kinase-2; a novel protein kinase activated by mitogen-activated protein kinase. *EMBO J.* **11**, 3985–3994 (1992).
133. R. Fukunaga and T. Hunter, MNK1, a new MAP kinase-activated protein kinase, isolated by a novel expression screening method for identifying protein kinase substrates. *EMBO J.* **16**, 1921–1933 (1997).
134. A. J. Waskiewicz, A. Flynn, C. G. Proud, and J. A. Cooper, Mitogen-activated protein kinases activate the serine/threonine kinases Mnk1 and Mnk2. *EMBO J.* **16**, 1909–1920 (1997).
135. H. Enslen, J. Raingeaud, and R. J. Davis, Selective activation of p38 mitogen-activated protein (MAP) kinase isoforms by the MAP kinase kinases MKK3 and MKK6. *J. Biol. Chem.* **273**, 1741–1748 (1998).
136. X. S. Wang, K. Diener, C. L. Manthey, S. Wang, B. Rosenzweig, J. Bray, J. Delaney, C. N. Cole, P. Y. Chan-Hui, N. Mantlo, H. S. Lichenstein, M. Zukowski, and Z. Yao, Molecular cloning and characterization of a novel p38 mitogen-activated. *J. Biol. Chem.* **272**, 23,668–23,674 (1997).
137. M. Goedert, A. Cuenda, M. Craxton, R. Jakes, and P. Cohen, Activation of the novel stress-activated protein kinase SAPK4 by cytokines. *EMBO J.* **16**, 3563–3571 (1997).
138. A. S. Zervos, L. Faccio, J. P. Gatto, J. M. Kyriakis, and R. Brent, Mxi2, a mitogen-activated protein kinase that recognizes and phosphorylates Max protein. *Proc. Natl. Acad. Sci. U.S.A.* **92**, 10,531–10,534 (1995).
139. M. Frodin, N. Sekine, E. Roche, C. Filloux, M. Prentki, C. B. Wollheim, and E. Van Obberghen, Glucose, other secretagogues, and nerve growth factor stimulate stimulate mitogen-activated protein kinase in the insulin-secreting beta-cell line, INS-1. *J. Biol. Chem.* **270**, 7882–7889 (1995).
140. S. Khoo and M. H. Cobb, Activation of mitogen-activating protein kinase by glucose is not required for insulin secretion. *Proc. Natl. Acad. Sci. U.S.A.* **94**, 5599–5604 (1997).
141. C. J. Burns, S. L. Howell, P. M. Jones, and S. J. Persaud, Glucose-stimulated insulin secretion from rat islets of Langerhans is independent of mitogen-activated protein kinase activation. *Biochem. Biophys. Res. Commun.* **239**, 447–450 (1997).
142. W. M. MacFarlane, M. L. Read, M. Gilligan, I. Bujalska, and K. Docherty, Glucose modulates the binding activity of the beta-cell transcription factor IUF1 in a phosphorylation-dependent manner. *Biochem. J.* **303**, 625–631 (1994).

Molecular Basis of Fidelity of DNA Synthesis and Nucleotide Specificity of Retroviral Reverse Transcriptases

LUIS MENÉNDEZ-ARIAS

Centro de Biología Molecular "Severo Ochoa," Consejo Superior de Investigaciones Científicas Universidad Autónoma de Madrid Cantoblanco, 28049 Madrid, Spain

I. Introduction 92
II. The Role of Reverse Transcriptase in Retroviral Mutagenesis 93
III. Retroviral Reverse Transcriptases 94
   A. Structure 94
   B. Synthesis of Proviral DNA 97
IV. Fidelity of Retroviral Reverse Transcriptases 97
   A. Determination of Viral Mutation Rates in a Single Cycle of Infection . 98
   B. *In Vitro* Assays of Polymerase Fidelity 99
V. Control of Fidelity at Initiation of Reverse Transcription 108
VI. Fidelity of Strand Transfer: Implications for Retroviral Recombination ... 109
VII. Contribution of Accessory Proteins to Fidelity of Reverse Transcription .. 110
VIII. Mutational Analysis of HIV-1 Reverse Transcriptase: The Effects of Mutations on Fidelity of DNA Synthesis 112
   A. Amino Acid Residues of the dNTP-Binding Site 114
   B. Residues Involved in Interactions with the Template Strand 121
   C. Residues Involved in Interactions with the Primer Strand 123
   D. Role of Other Amino Acids in the Fidelity of DNA Synthesis 127
IX. Biological Consequences of Increasing or Decreasing Fidelity 129
   A. Enhanced Polymerase Fidelity in Drug-Resistant Viruses 129
   B. Mutator Reverse Transcriptases and Error Catastrophe 129
   C. Extrinsic Factors Leading to Mutation Rate Variations 130
X. Conclusions and Future Perspectives 131
   References 132

Abbreviations: AMV, avian myeloblastosis virus; BLV, bovine leukemia virus; CAEV, caprine arthritis encephalitis virus; dNTP, deoxynucleoside triphosphate; EIAV, equine infectious anemia virus; FIV, feline immunodeficiency virus; HIV-1, human immunodeficiency virus type 1; HIV-2, human immunodeficiency virus type 2; HTLV-I, human T-cell leukemia virus type I; LTR, long terminal repeat; MLV, murine leukemia virus; MMTV, mouse mammary tumor virus; Mo-MLV, Moloney murine leukemia virus; MPMV, Mason–Pfizer monkey virus; PBS, primer binding site; PPT, polypurine tracts; R, repeat; RSV, Rous sarcoma virus; RT, reverse transcriptase; SIV, simian immunodeficiency virus; SNV, spleen necrosis virus; U3′, unique 3′; U5′, unique 5′.

Reverse transcription involves the conversion of viral genomic RNA into proviral double-stranded DNA that integrates into the host cell genome. Cellular DNA polymerases replicate the integrated viral DNA and RNA polymerase II transcribes the proviral DNA into RNA genomes that are packaged into virions. Although mutations can be introduced at any of these replication steps, reverse transcriptase (RT) errors play a major role in retroviral mutation. This review summarizes our current knowledge on fidelity of reverse transcriptases. Estimates of retroviral mutation rates or fidelity of retroviral RTs are discussed in the context of the different techniques used for this purpose (i.e., retroviral vectors replicated in culture, misinsertion and mispair extension fidelity assay, etc.). *In vitro* fidelity assays provide information on the RT's accuracy during the elongation reaction of DNA synthesis. In addition, other steps such as initiation of reverse transcription, or strand transfer, and factors including viral proteins such as Vpr [in the case of the human immunodeficiency virus type 1 (HIV-1)] have been shown to influence fidelity. A comprehensive description of the effect of amino acid substitutions on the fidelity of HIV-1 RT is presented. Published data point to certain dNTP-binding residues, as well as to various amino acids involved in interactions with the template or the primer strand, and to residues in the minor groove-binding track as major components of the fidelity center of retroviral RTs. Implications of these studies include the design of novel therapeutic strategies leading to virus extinction, by increasing the viral mutation rate beyond a tolerable threshold. © 2002, Elsevier Science (USA).

I. Introduction

Retrovirus variation has important implications not only in virus diversity and evolution but also in virulence, pathogenesis, and in the ability to develop effective antiviral drugs and vaccines. Howard Temin was one of the first investigators to demonstrate variation in retroviruses, in his early studies of avian leukosis-sarcoma viruses (1). In these experiments, it was shown that clonal stocks of Rous sarcoma virus (RSV) could lead to different morphological changes in infected cells: virus-producing foci of round refractile cells, or virus-producing foci of long fusiform cells. It was also observed that focus morphology was an inheritable trait of the virus. These characteristics placed retroviruses together with other RNA viruses (yellow fever, influenza, and tobacco mosaic viruses), whose phenotypic variation and quick adaptability in cell culture were earlier demonstrated. It is now clear that all RNA viruses are highly mutable and exist as mixtures of closely related but divergent virions, designated as "quasispecies" (2).

Genetic diversity in retroviruses has been widely documented, particularly in the case of human immunodeficiency virus type 1 (HIV-1) and simian immunodeficiency virus (SIV) (3–6). Estimates of the rates of variation of HIV-1 and SIV have been determined from nucleotide sequence analysis of genes from sequential isolates taken from individuals infected with a known source of virus. The extent of variation (measured as nucleotide substitutions per site per year) was estimated to be at least 10^{-3} for the *env* gene and 10^{-4} for the *gag* gene. Genetic variation also occurs on continuous passage of retroviruses in tissue culture. Retroviral mutation rates have been estimated in wild-type viruses and in retroviral vectors replicated in culture (for a review, see Ref. 7). Reported values range from 10^{-6} to 10^{-4} mutations per nucleotide per cycle (7–9), well above the values reported for DNA-based microbes which range from 10^{-7} to 10^{-11} (10).

II. The Role of Reverse Transcriptase in Retroviral Mutagenesis

Retroviruses are RNA viruses which replicate through a DNA intermediate. Upon infection, the single-stranded viral RNA serves as template for minus-strand DNA synthesis by the viral reverse transcriptase (RT; deoxynucleoside triphosphate:DNA deoxynucleotyltransferase, RNA-directed E.C. 2.7.7.49), followed by plus-strand DNA synthesis by RT using the minus-strand DNA as template. The double-stranded DNA resulting from reverse transcription is then integrated into the genome of the host cell where it is stably maintained as a provirus. Transcription of the provirus by cellular RNA polymerase II and packaging of the newly synthesized RNA into progeny virions complete the retroviral replication cycle. Along this process, there are four polymerization steps that can affect the retroviral mutation rate: (i) RNA-templated minus-strand DNA synthesis by the viral RT, (ii) DNA-templated plus-strand DNA synthesis by the viral RT, (iii) proviral DNA replication that involves the participation of cellular DNA polymerases, and (iv) DNA-templated plus-strand RNA synthesis catalyzed by cellular RNA polymerase II.

Unlike cellular DNA polymerases, viral reverse transcriptases are devoid of $3' \rightarrow 5'$ exonucleolytic proofreading activity. Average base substitution error rates for proofreading-proficient DNA polymerases (i.e., DNA polymerases δ, γ, and/or ε) are about 10^{-6} to 10^{-7} (11, 12) and, therefore, 10 to 100 times lower than those for retroviral RTs (13–18). Based on those studies, it has been proposed that the high mutation rate during retroviral replication is, at least in part, a consequence of errors made by RT.

The potential contribution of cellular RNA polymerase II is real but largely unknown. RNA polymerase II transcribes retroviral genomic RNA from proviral DNA and is absolutely essential for virus replication. The fidelity of wheat-germ RNA polymerase II *in vitro* ranges from 1 error every 250 to 1 error in every 200,000 nucleotides polymerized, depending on the mutation type and on the sequence and structure of the template (19). Assuming similar error rates in other organisms, these data suggest that RNA transcription is relatively error prone and contributes to retroviral mutagenesis. However, accurate measurements of mutation rates by RNA polymerase II are not available, and there is evidence that indicates that cellular factors may enhance RNA polymerase II transcription fidelity. For example, the eukaryotic elongation factor SII (or TFIIS) enhances fidelity (20) in a manner analogous to the role played by the bacterial Gre A protein in maintaining the fidelity of *Escherichia coli* RNA polymerase-mediated transcription (21). The protein SII stimulates the intrinsic 3' → 5' nuclease activity of RNA polymerase II and is able to remove misincorporated nucleotides from the nascent transcript during rapid chain extension (22). These observations suggest that the fidelity of RNA polymerase II is likely to be much higher than originally thought. Although errors at the level of cellular RNA polymerase II may occur, the low fidelity of RT emerges as a major determinant of the high mutation rate in HIV-1 and other retroviruses.

III. Retroviral Reverse Transcriptases

A. Structure

The RT of all retroviruses is encoded by the *pol* gene, a large open reading frame located near the center of the viral genome. In most retroviruses, RT is expressed as part of a Gag–Pol precursor polyprotein, that contains in addition to Gag sequences, the viral enzymes protease and integrase. The synthesis of Gag–Pol is controlled at the transcriptional level either by reading through the *gag* termination codon (nonsense suppression) or by frameshifting (23). Proteolytic cleavage of the Gag–Pol polyprotein mediated by the retroviral protease generates the mature RT. The cleavage sites in Gag–Pol vary from retrovirus to retrovirus and result in RTs with different subunit compositions (for reviews, see Refs. 24 and 25). Thus, in murine leukemia virus (MLV), the Pol region of Gag–Pol is fully cleaved to render three nonoverlapping proteins: the protease, the 78-kDa monomeric RT, and the integrase. In the avian sarcoma–leukemia viruses [i.e., avian myeloblastosis virus (AMV) or Rous sarcoma virus (RSV)], the situation is different since the protease-coding region is not part of the *pol* gene and the protease is synthesized together with Gag. The Pol region of Gag–Pol is only partially processed (cleavage at the RT–IN boundary is incomplete). The mature RT is an α–β heterodimer, with a larger β-subunit (95 kDa) containing the full RT sequence plus the C-terminal integrase domain, and an α-subunit (68 kDa) bearing the RT domain.

In HIV-1 as well as in other lentiviruses [i.e., equine anemia infectious virus (EIAV), feline immunodeficiency virus (FIV), SIV, etc.], the RT domain is completely cleaved from the protease and integrase regions, as in MLV. However, the RT is partially cut an additional time, removing a small sequence corresponding to an RNase H domain located at the C terminus (26, 27). The mature form of HIV-1 RT is a heterodimer composed of two subunits of 560 and 440 amino acids, termed as p66 and p51, respectively. Biochemical studies have shown that both p66/p51 heterodimers and p66/p66 homodimers display significant polymerase activity, while p51/p51 homodimers and p66 and p51 monomers show little or no activity (28, 29).

The crystal structure of HIV-1 RT has been determined in the absence of ligands (30) and bound to nonnucleoside RT inhibitors (31), double-stranded DNA (32, 33), or a polypurine tract RNA/DNA complex (34). The structure of a catalytic complex formed by the RT, a DNA/DNA template primer, and a deoxynucleoside triphosphate (dNTP), has also been determined (35). Structural analysis revealed that both RT subunits (p66 and p51) contain four common subdomains termed as "fingers," "palm," "thumb," and "connection," while the RNase H domain is absent from p51 (Fig. 1). The overall folding of the RT subunits is similar in p66 and p51, but spatial arrangements of the subdomains

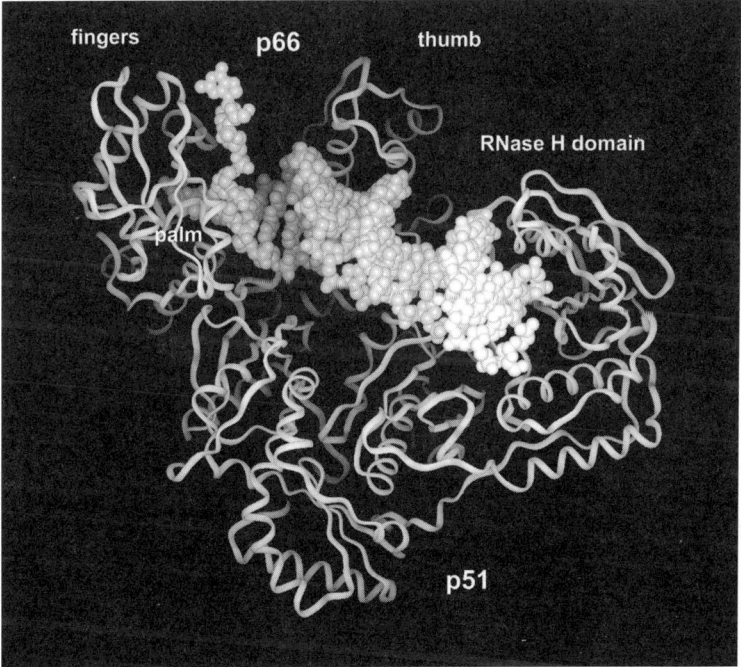

FIG. 1. Crystal structure of a ternary complex of HIV-1 RT, double-stranded DNA, and dTTP (35). The p66 and p51 subunits are represented with a ribbon diagram. Template and primer strands are represented with a CPK model.

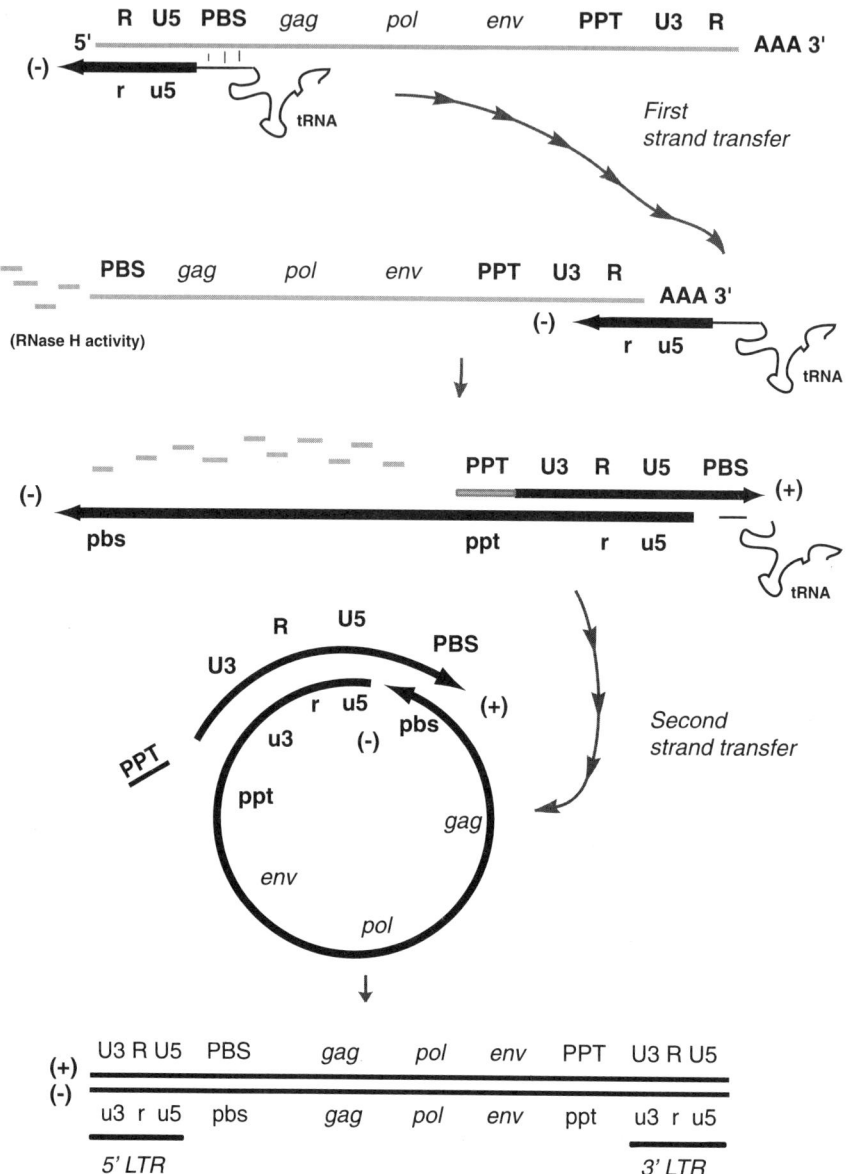

FIG. 2. Process of reverse transcription. R, repeat; U5, unique 5'; PBS, primer-binding site; PPT, polypurine tract; U3, unique 3'; LTR, long terminal repeat. Adapted from Ref. 36.

differ markedly. The p66 subunit adopts an "open" conformation, where highly conserved amino acid residues in the "fingers" and "palm" subdomains together with two α-helices of the "thumb" subdomain act as a clamp to position the template/primer relative to the polymerase active site. The catalytic site of the enzyme includes three conserved aspartic acid residues at positions 110, 185, and 186 in the 66-kDa subunit, which are exposed in the "open" conformation of p66. On the other hand, the p51 subunit is closely folded, with its catalytic residues occupying an internal position in the molecule.

B. Synthesis of Proviral DNA

Retroviral RTs are multifunctional enzymes, possessing RNA- and DNA-dependent DNA polymerase, RNase H, and strand transfer and strand displacement activities. All these activities are essential for reverse transcription (for recent reviews, see Refs. 25, 36, and 37). A graphic description of the major steps involved in this process is depicted in Fig. 2.

RT initiates DNA synthesis by adding nucleotides at the 3' end of a tRNA primer; it then copies the short distance toward the 5' end of the viral RNA genome, making the so-called "minus-strand DNA." The resulting RNA–DNA hybrid is hydrolyzed via a synthesis-dependent RNase H activity. When the 5' end is reached, the exposed copy of the repeat (R) can base pair with the 3' R sequence, through a strand transfer event. Thereafter, minus-strand DNA synthesis continues along the RNA genome. Concomitant with minus-strand DNA synthesis, RNase H activity hydrolyzes RNA of the replicative intermediate, with the exception of polypurine tracts (PPT), which resist hydrolysis to serve as primers for plus-strand DNA synthesis. Plus-strand DNA synthesis proceeds along the minus-strand DNA template and over the first 18 nucleotides of the tRNA primer. Homology between primer binding site (PBS) regions of minus- and plus-strand DNAs allows a second-strand-transfer event, relocating nascent plus-strand DNA to the 3' end of the fully elongated minus strand. Following the second-strand transfer, DNA-dependent DNA polymerase activity completes synthesis of minus and plus strands to yield a double-stranded preintegrative intermediate that, once translocated into the nucleus, will be integrated into the host cell genome.

IV. Fidelity of Retroviral Reverse Transcriptases

The number of mutations per nucleotide and replication cycle as determined using retroviral vectors genetically restricted to a single cycle of replication reflects the contribution of the retroviral RT, cellular DNA polymerases, and RNA polymerase II, as well as other cellular factors that may influence retroviral mutation. Several *in vitro* fidelity assays have been used to measure the accuracy of retroviral RTs. Gel-based assays include misinsertion and mispair extension fidelity assays, while genetic assays are represented by codon reversion and forward mutation assays.

A. Determination of Viral Mutation Rates in a Single Cycle of Infection

To understand more about the mutational processes involved in retroviral replication, protocols have been developed to determine retroviral mutation rates (i.e., the mutation frequency in a single cycle of retroviral replication) (7, 38). These methods imply the use of a vector virus genome which contains a target gene for scoring mutations (i.e., neomycin, *lacZα* or thymidine kinase), a marker gene to quantify total virus and all the *cis*-acting elements required for genomic transcription, encapsidation, and proviral DNA synthesis [long terminal repeat (LTR), primer binding site (PBS), encapsidation sequence (ψ), and the polypurine tract (PPT)]. One or more essential genes (*gag*, *pol*, *env*, and/or accessory genes) are deleted or disrupted in the vector virus, and provided *in trans* by packaging (or helper) cells that express those genes.

Packaging cells containing a single copy of integrated vector proviral DNA transcribe the vector genome using RNA polymerase II, directed by the viral LTR promoter and enhancer sequences present in the vector. The transcribed RNA is packaged together with a full complement of normal viral proteins supplied by the packaging cell and shed into the medium. These vector-derived viral particles are used to infect target cells lacking the *gag–pol–env* viral genes. Vector RNA is then integrated into the host genome as proviral DNA by the sequential actions of the RT and the integrase. At this point, an appropriate selection of cultured cells can determine the wild-type or mutant phenotype of the target gene, and mutant frequencies can be obtained.

Mutation frequencies for spleen necrosis virus (SNV), MLV, bovine leukemia virus (BLV), HIV-1, and human T-cell leukemia virus I (HTLV-I) have been measured by using this approach, and obtained mutation rates range from 2×10^{-5} to 6×10^{-6} per nucleotide per replication cycle (9, 39–46). However, these rates were determined using different vectors and different packaging and target cells, and sometimes the obtained values were not comparable. For example, reported estimates of the mutation rate of MLV ranged from 2×10^{-6} (44) to 2×10^{-5} (47), depending on the experimental system used. In addition, it should be noted that mutation rates have been obtained by observing the phenotypic changes of marker genes (i.e., *neo*, *lacZα*, or *tk*), and some mutations may not cause phenotypic changes. This has been considered for error rates obtained using *lacZ*, since the number of mutational targets in this gene is known (i.e., 113 out of 280 bases for substitutions, 150 for frameshifts, and 280 for deletions).

However, in some cases (i.e., in assays using the *tk* gene), the mutational targets are not known, and therefore the obtained values represent minimal mutation rates.

Reliable comparisons of mutation rates have been reported for SNV, BLV, HIV-1, and HTLV-I, using the *lacZα* gene as a mutational target. Mutation rates for the related retroviruses BLV and HTLV-I were similar and were estimated to be around 4×10^{-6} and 7×10^{-6}, respectively (*40, 46*). On the other hand, the mutation rate of HIV-1 was 3 to 4×10^{-5} (*41, 43*), and the mutation rate of SNV was estimated to be around 1×10^{-5} (*9*). Base-pair substitutions and frameshifts were the most commonly detected errors during a single round of HIV-1 replication. Base-pair substitution mutations were predominantly G-to-A and C-to-T transition mutations, while −1 frameshifts occurred mainly in runs of T's or A's (*41, 43*). Error distribution frequencies were not affected by the target cells used in these experiments (i.e., CEM-A or HeLa cells in HIV-1) (*43*). Similar types of errors were observed for HTLV-I, with the majority of the G-to-A transitions located at GpA dinucleotides (*46*). Most of these mutations occurred at sites adjacent to runs of nucleotides, suggesting that dislocation mutagenesis plays an important role in the generation of errors. On the other hand, the types of mutations observed for both SNV and BLV included a relatively higher proportion of deletions and deletions with insertions, in comparison to the number of base-pair substitutions and frameshifts, perhaps revealing a higher propensity of these viruses to produce recombinants (*9, 40*). Recently, single-cycle replication experiments were used to determine the replication fidelity of the yeast retrotransposon Ty1 (*48*). The observed base substitution rate of 2.5×10^{-5} per nucleotide and replication cycle suggests that error-prone replication is not exclusive to infectious retroviruses.

B. *In Vitro* Assays of Polymerase Fidelity

The fidelity of purified RTs has been analyzed *in vitro* by using enzymological and genetic assays. Classical biochemical assays included measurements of the incorporation of dCMP in place of dTMP in cell-free DNA synthesis with polyadenylic acid as template (*49*). These assays revealed that HIV-1 RT was around threefold less accurate than the RTs of AMV, Moloney-MLV (Mo-MLV), and Rous-associated virus 2. Further evidence of the lower fidelity of HIV-1 RT was obtained from qualitative assessments of misincorporations opposite multiple template sites by using the so-called "minus"-sequencing gel assay. In the presence of only three dNTPs, DNA polymerases or viral RTs are able to extend a primer up to, but not opposite to, the template positions that normally pair with the absent dNTP. Extensions beyond this position arise by nucleotide misincorporation. Comparison of AMV, Mo-MLV, and HIV-1 RTs using this approach revealed that the HIV-1 RT was highly error-prone in comparison to the other retroviral RTs (13). AMV RT was slightly less accurate than Mo-MLV RT in these assays. Although these methods provide a quick estimate of fidelity, reactions must be done in the presence of high concentrations of dNTPs. Therefore, the fidelity estimates obtained with these assays are more sensitive to differences in the incorporation rate (V_{max} or k_{cat}) of correct or incorrect dNTPs than to differences in their corresponding nucleotide-binding affinities (i.e., apparent K_M values).

1. MISINSERTION FIDELITY ASSAYS

Steady-state kinetic assays based on the determination of catalytic constants (k_{cat} and K_M) for the incorporation of nucleotides at the 3′ end of a primer provide a valuable assessment of the nucleotide selectivity of the RT (50, reviewed in Ref. 51). In these assays, the template–primer complex used as substrate is preincubated with the analyzed RT to produce a binary complex prior to the addition of dNTPs. Saturating concentrations of template–primer should be used, and therefore template–primers showing low K_D values (i.e., less than 10 nM) are preferred (52, 53).

The determination of kinetic parameters is performed after labeling the primer at its 5′ end with [γ-^{32}P]ATP, by measuring nucleotide incorporation at the 3′ end of the primer in the presence of different concentrations of dNTP, once the binary complex RT/template–primer is formed (Fig. 3). The extension products resulting from the incorporation reaction are quantitated after electrophoresis in polyacrylamide–urea gels. Elongation measurements are then fitted to the Michaelis–Menten equation and the kinetic parameters k_{cat} and K_M are determined. These procedures can be applied to either correct or incorrect nucleotides. Obviously, the incorporation rate (k_{cat} value, or alternatively V_{max}) is expected to be higher for a correct nucleotide than for an incorrect one. On the other hand, incorrect nucleotides bind the RT/template–primer complex with less affinity than correct nucleotides and show a higher apparent K_M value. Misinsertion efficiency (f_{ins}) or nucleotide selectivity is here defined as the ratio between the catalytic efficiencies of the polymerase when an incorrect dNTP is used as substrate relative to a correct dNTP:

$$f_{ins} = \frac{k_{cat}/K_M \text{ (incorrect)}}{k_{cat}/K_M \text{ (correct)}}.$$

Therefore, higher polymerase fidelity implies lower misinsertion efficiency. The objective of the gel assay is to measure polymerase fidelity at any target site along a DNA or RNA template strand. The method is also adequate to compare the incorporation of dNTPs versus rNTPs, dNTPs versus ddNTPs, etc.

Reported misinsertion efficiencies of retroviral RTs range from 10^{-3} to 10^{-6} depending on the type of mispair and the template–primer used in the experiments (13, 17, 53–58). RT forms transversions with lower efficiencies than transitions. Misinsertion ratios are especially low for A:G, G:G, and C:C (54), while

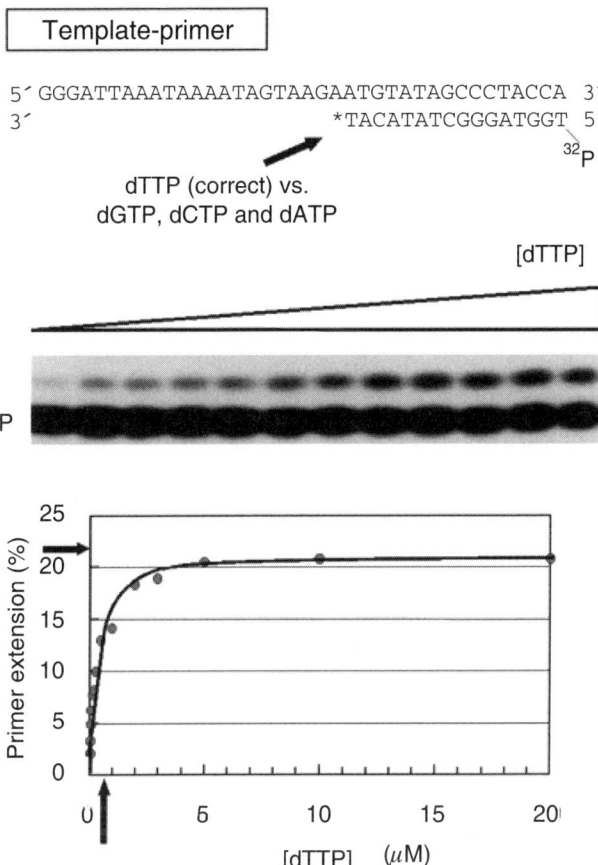

FIG. 3. Determination of kinetic constants for the incorporation of correct or incorrect dNTPs at the 3′ end of a ^{32}P-labeled primer. A duplex containing a 5′-^{32}P-labeled primer is used as substrate (in the example shown, this is a 38/16-mer DNA/DNA template–primer). Incorporation of dTTP (or any other nucleotide) is determined at different concentrations of nucleotide (the incorporation site is indicated with an asterisk). Then, primer elongation measurements are obtained after resolving the extended and unextended primers in a polyacrylamide–urea gel. Primer extension values are represented versus the dNTP concentrations used in these assays, and the V_{max} and K_M values were obtained after fitting the data to the Michaelis–Menten equation. The catalytic constant, k_{cat}, can be obtained from the equation $k_{cat} = V_{max}/[E]$, if the concentration of active enzyme [E] used in the assays is known. A similar procedure is followed to obtain kinetic parameters for incorrect nucleotides, and misinsertion efficiencies are then obtained from the ratios of catalytic efficiencies for incorrect and correct nucleotides (see text for details).

the highest misinsertion ratios have been reported for G:T and G:U mismatches (56, 59). Elongation reactions are usually performed at room temperature or 37°C. In a recent report, it has been shown that increasing the reaction temperature up to 55°C produces a significant enhancement of misinsertion fidelity, as determined with MLV RT (60).

In general, discrimination between a correct and an incorrect dNTP is determined both by the K_M and the k_{cat} values. For example, the K_M values for the incorporation of a correct nucleotide in heteropolymeric template–primers by the HIV-1 RT ranged from 10 nM to 5 μM, while the corresponding K_M values for incorrect dNTPs are usually at least 500 times higher. In contrast, reported k_{cat} values for correct dNTPs are less than 25 times higher than those for incorrect nucleotides (13, 17, 53–56). In an extensive analysis using a large collection of template–primers made of single-strand M13 DNA and short oligonucleotide primers, it was concluded that the 5′ nearest neighbor nucleotide of the primer influences the fidelity of AMV RT, since the presence of purine or pyrimidine at this position affects the kinetic parameters for the incorporation of correct or incorrect nucleotides (54).

Discrimination between a correct and an incorrect nucleotide operates at the level of nucleotide recognition. Recent work with other DNA polymerases indicates that hydrogen bonds in a base pair are not absolutely required for efficient nucleotide insertion (61, 62), supporting the proposal of geometric selection as the dominant mechanism determining insertion specificity (63, 64). According to this view, geometrical and electrostatic properties of the polymerase active site are likely to have a profound influence on nucleotide-insertion specificities.

a. Misinsertion Fidelity of Retroviral RTs: Comparative Studies. Misinsertion efficiency is a function of many parameters, such as the type of polymerase, the nature of the template (DNA or RNA), the analyzed mispair, the concentration of nucleoside triphosphates, and the buffer conditions of the reaction. Misinsertion efficiencies of HIV-1 and AMV RTs have been compared using different template–primers. For example, misinsertion of C opposite A was sixfold more efficient with AMV RT than with HIV-1 RT, with a single-stranded ϕX174 am3 DNA template and a short 15-nucleotide primer (13). The lower fidelity of HIV-1 RT relative to the AMV polymerase was further confirmed using DNA/DNA template–primers of different sizes (17, 56). In an extensive analysis, Ricchetti and Buc (17) measured misinsertion efficiencies with different complexes, prepared with templates of 29 to 57 nucleotides and primers of 12 to 20 bases. Their results revealed higher misinsertion efficiencies for transitions (i.e., misinsertion of C opposite A, T opposite G, etc.) than for transversions, for all enzymes. Misinsertion efficiency values were usually higher for HIV-1 RT than for AMV RT. However, this general trend was not observed in 2 out of 10 template–primers analyzed (17). Mo-MLV RT was also more accurate than HIV-1 RT in misinsertion fidelity assays (17). However, it was not possible to produce a reliable comparison of the fidelity of Mo-MLV and AMV RTs due to experimental limitations of the assays.

Other RTs whose misinsertion fidelity has been analyzed are EIAV RT, mouse mammary tumor virus (MMTV) RT, and yeast retrotransposon Ty1 RT (57, 58, 65). Misinsertion ratios for C, G, or A opposite A were similar for EIAV and HIV-1 RTs and ranged from 1.3×10^{-5} to 3.3×10^{-4} depending on type of misinsertion analyzed. MMTV RT was somewhat more error-prone than AMV RT, with 2- to >4-fold higher misinsertion efficiencies. Relative to lentiviral RTs (HIV-1, HIV-2, or EIAV), the yeast retrotransposon Ty1 RT was around 10-fold more accurate, showing misinsertion efficiencies similar to those reported for AMV RT (65).

b. Comparison of Misinserion Fidelity Using RNA Versus DNA Templates. Reverse transcription involves the synthesis of DNA using both RNA and DNA templates. It has been reported that HIV-1 and AMV RTs display similar kinetics and fidelity of nucleotide insertion on RNA and DNA templates (56). However, it should be noted that in these assays the size and nucleotide sequences of the compared RNA and DNA templates were different (i.e., a 592-base-long RNA transcript versus a single-stranded M13 DNA, or a DNA oligomer template of 40 bases), although the sequence context at the incorporation site (i.e., neighboring nucleotides) was identical for RNA and DNA templates. In addition, the RNase H activity of both RTs was not inhibited, and the influence of potential endonucleolytic cleavage on the template RNA was not taken into account in these experiments. Studies with HIV-1 RT using template–primers of the same size and nucleotide sequence gave similar results (53). Misinsertion ratios for C, G, or A opposite A were 1.7×10^{-5}, 2.3×10^{-6}, and 1.9×10^{-6}, respectively, when using RNA/DNA template–primers. These values were similar for DNA/DNA complexes, although misinsertion of G opposite A was around 10-fold more efficient in this case. These results were almost identical to those obtained with a mutant RT devoid of RNase H activity, and lacking the last 20 residues of the C terminus of p66 (53).

2. Mispair Extension Fidelity Assays

Efficient extension of mismatched 3' termini of DNA was recognized as a major determinant of the infidelity of HIV-1 RT, in assays carried out with template–primers made of bacteriophage ϕX174 DNA as template and an oligonucleotide DNA primer of 15 bases, carrying one mispaired nucleotide at its 3' terminus (66). Mispair extension frequencies were around 50-fold higher by HIV-1 RT than by the mammalian replicative enzyme DNA polymerase α (66). Further studies have shown that in cell-free assays, HIV-1 RT can extend mismatches of up to 3 bases (56), an observation which was confirmed with SNV-based retroviral vectors in a single cycle of replication experiments (67).

Mispair extension assays involve the determination of kinetic constants for the incorporation of a correct nucleotide, using template–primers with matched or mismatched 3' termini. In these assays, mispair extension ratios (f_{ext}) can be obtained by using

$$f_{ext} = \frac{k_{cat}/K_M \text{ (mismatched template–primer)}}{k_{cat}/K_M \text{ (matched template–primer)}}.$$

This equation is valid only if the K_D values of the polymerase for both matched and mismatched template–primers are roughly similar, as demonstrated in a number of cases (51, 68, 69). Retroviral RTs show mispair extension ratios that are usually one to three orders of magnitude higher than misinsertion ratios. Reported values for AMV RT using all 16 possible mismatched 3' termini ranged from 6.9×10^{-7} to 0.18 (70). The highest mispair extension efficiencies were observed with purine:pyrimidine or pyrimidine:purine mismatches. HIV-1 and AMV RTs were found to extend U:G mispairs with high efficiencies (2.7×10^{-2} and 4.5×10^{-2}, respectively) (56). Differences in mispair extension ratios between both enzymes were not significant and were largely dependent on the type of mispair analyzed and the template–primer used in the experiments.

A systematic analysis of mispair extension fidelity, using retroviral RTs from different sources and different template–primers, has revealed that lentiviral RTs (i.e., HIV-1, HIV-2, and EIAV RTs) show the highest ability to extend mismatched template–primers, ahead of the MMTV RT, AMV RT, and the less accurate Mo-MLV RT (55, 57, 58, 71). HIV-1 and HIV-2 RTs showed around 5- to 10-fold higher efficiency of mispair extension than the Mo-MLV RT (71). However, it should be noted that the buffer conditions used in these experiments were similar for all enzymes tested, except for Mo-MLV RT which was tested in the presence of 0.5 mM MnCl$_2$, instead of 10 mM MgCl$_2$, as reported for the other enzymes. The presence of magnesium (Mg^{2+}) or manganese cations (Mn^{2+}) as cofactors has a remarkable influence in both misinsertion and mispair extension fidelity (72). Mispair extension ratios obtained with HIV-1 RT and template–primers with mismatched A:C, A:G, or A:A 3' termini were around two orders of magnitude higher in assays carried out in the presence of Mn^{2+} instead of Mg^{2+} (72).

All tested retroviral RTs showed equal or higher mispair extension ratios when using an RNA template instead of DNA, and a less than 5- to 10-fold difference was observed in all cases (53, 55–57, 71).

3. Nucleotide Selectivity Determinations Using Pre-Steady-State Kinetic Assays

A pre-steady-state kinetic analysis of nucleotide incorporation by several DNA polymerases has provided insight into the mechanisms of polymerase fidelity (for reviews see Refs. 63 and 73). Rapid transient kinetic methods allow the determination of kinetics of nucleotide incorporation after time intervals ranging from 3 ms to several seconds and therefore the measurement of the equilibrium dissociation constant (K_D) for the interaction of dNTP and the binary complex RT/template–primer, as well as the reaction turnover (k_{pol}). Studies with HIV-1 RT suggest that nucleotide binding is a two-step process (74). The initial binding complex serves to bring the dNTP into the polymerase active site and involves base pairing with the template strand. The second step serves as an additional check for proper base pairing that, if correct, leads to rapid isomerization of the ternary complex and catalysis of the nucleoside phosphoryl transfer reaction. Nucleotide selectivity can be estimated from the ratio between the k_{pol}/K_D values for incorporation of incorrect and correct base pairs. The data reported for HIV-1 RT show that fidelity of DNA-directed DNA synthesis can be attributed to a one- to two-order-of-magnitude reduction in affinity for noncomplementary dNTPs, and a one- to four-order-of-magnitude reduction in the rate of conformational change that limits the rate of nucleotide addition (69, 74).

Reported nucleotide selectivity values were in the range of 10^{-3} to 10^{-6} and were lower for purine:purine and pyrimidine:pyrimidine base pairs, as revealed using steady-state kinetic analysis. In addition, HIV-1 RT showed a very low misinsertion efficiency of rUTP or rGTP incorporation instead of the correct dTTP or dGTP ($4-9 \times 10^{-6}$) but similar to the values reported using conventional steady-state misinsertion fidelity assays (69, 72).

Pre-steady-state kinetic assays have shown that extension of mispaired primer terminus proceeds at a rate of one to three orders of magnitude faster than the dissociation of HIV-1 RT from DNA, providing a mechanism for fixation of misincorporated nucleotides (69). Studies with oligonucleotide RNA/DNA and DNA/DNA template–primers, bearing the same nucleotide sequences (except for the presence of U instead of T in the RNA template), showed that copying fidelity was around 10–25 times more accurate with the RNA template (75). Although data were limited to the analysis of misinsertion of C, A, or G instead of T, at the 3′ end of the primer, the higher fidelity of RNA-templated nucleotide incorporations relates to a better discrimination at the catalytic step of the reaction, with relatively small differences in nucleotide-binding discrimination (K_D values for the incorporation of correct or incorrect dNTPs were similar to both RNA/DNA and DNA/DNA template–primers).

4. GENETIC ASSAYS

Although gel-based fidelity assays provide accurate measurements of the nucleotide selectivity of the polymerase at a given position, these assays are time-consuming, and the analysis is usually limited to a relatively small number of incorporation sites. Genetic assays provide an alternative assessment of fidelity that can facilitate information on specific types of errors that RTs may introduce while copying the RNA or DNA template. In addition, some of these assays allow the exploration of a wide variety of sequence contexts that may help the identification of relevant hot spots related to reverse transcription errors. Genetic assays used to estimate the fidelity of retroviral RTs can be classified into two major groups: codon reversion and forward reversion assays. Codon reversion assays are based on the use of templates having an in-frame termination codon, typically an amber (TAG) mutant of the single-stranded circular DNA phage ϕX174 (13, 16) or an opal (TGA) mutant at position 89 of the lacZ gene of bacteriophage M13mp2 (14, 76). The reversion of the amber or opal codons to a coding triplet can be visualized as a revertant plaque upon transformation of bacteria in the first case or in the second case, as a dark blue M13 plaque using X-gal indicator plates. These assays have not revealed large differences in fidelity between HIV-1, SIV_{agm}, and AMV RTs, with error rates in the range of 10^{-4}–10^{-5} (Table I).

The most widely used forward mutation assays utilize a gapped double-stranded M13mp2 DNA duplex, where the lacZ sequence of one of the DNA strands has been deleted. Gap-filling reactions are carried out in the presence of RT and relatively high concentrations of all four dNTPs. Upon transformation of bacteria with the reaction products, mutants are identified by the white/blue color of M13 plaques revealed using X-gal indicator plates (see a detailed description of the method in Ref. 92). Forward mutation assays have revealed significant differences in fidelity between the retroviral RTs (Table I). Despite the differences in error rates reported by different authors (Table I), HIV-1 RT was around 10-fold less accurate than the AMV RT and 20-fold less accurate than the Mo-MLV RT, in assays carried out under the same conditions (14, 15, 77).

The higher inaccuracy of HIV-1 RT has been attributed to its strong tendency to produce -1 frameshifts which represent more than 60% of the total number of errors found in these assays, while these errors represent 30% or less than 15%, in reactions carried out with Mo-MLV RT and AMV RT, respectively (15, 77). The mechanisms leading to -1 frameshifts, which predominate in runs of T's or A's, have been extensively studied. Several models have been proposed to explain errors involving misaligned template–primers (for a recent review, see Ref. 93). Template–primer slippage can result in deletion or addition errors that can produce frameshifts. Frameshift errors can also be initiated by nucleotide misincorporations producing a misaligned intermediate containing a correct terminal base pair that can be extended by the polymerase (94). This model is supported by studies of base substitution hot spots generated by HIV-1 RT (87, 95). Interestingly, increases and decreases in frameshift fidelity are frequently correlated with increases or decreases in the probability that HIV-1 RT terminated processive synthesis within the run (95).

TABLE I
In Vitro Fidelity Measurements with Retroviral RTs, Using Genetic Assays.

Enzyme	Method[a]	Template	Estimated error rate	References
HIV-1 RT	Codon reversion assay (ϕX174 am16)	DNA	1.4–2.0×10^{-4}	16
	Codon reversion assay (ϕX174 am3)	DNA	2.5×10^{-4}	13
	Codon reversion assay (M13mp2 A89)	DNA	5.6×10^{-5}	14
	Forward mutation assay (M13mp2, lacZ)	DNA	0.6–6.7×10^{-4}[b]	14, 77–81
	Forward mutation assay (pBluescript, lacZ)	DNA	1.7×10^{-4}	82
	Forward mutation assay (pBluescript, lacZ)	RNA	1.4×10^{-4}	82
	Forward mutation assay (M13, env V-1)	DNA	1.9×10^{-4}	83
	Forward mutation assay (M13, env V-1)	RNA	2.0×10^{-4}	83
	U-DNA method (Litmus 29, lacZ)	DNA	1.6×10^{-4}	84
SIV$_{agm}$ RT	Codon reversion assay (ϕX174 am16)	DNA	5.3×10^{-5}	18
AMV RT	Codon reversion assay (M13mp2 A89)	DNA	4.2–5.6×10^{-5}	14, 15
	Codon reversion assay (ϕX174 am3)	DNA	0.6–1.7×10^{-4}	13, 15
	Forward mutation assay (M13mp2, lacZ)	DNA	5.9×10^{-5}	14, 15
Mo-MLV RT	Forward mutation assay (M13mp2, lacZ)	DNA	3.3×10^{-5}	14, 15
	Forward mutation assay (pBluescript, lacZ)	DNA	3.4×10^{-5}	82
	Forward mutation assay (pBluescript, lacZ)	RNA	2.7×10^{-5}	82

[a] Plasmids containing sequences used as templates in the polymerization reaction are indicated within parentheses, followed by the target gene.

[b] Reported error rate estimates based on this method show large variability, which appears to originate in the gap-filling reaction, since differences are already detected while counting the number of mutant plaques. For example, the number of reported mutant frequencies (ratio of mutant to total number of plaques) range from 65–100×10^{-4} (80, 81, 85), to 450–540×10^{-4} (14, 77, 78), with intermediate values of 150–250×10^{-4} (86–91). These estimates have been obtained with RTs from different strains of HIV-1 (i.e., NL4-3, HXB2, BH10, and NY5). The highest mutant frequencies have been observed with the HIV-1$_{NY5}$ RT, although the molecular basis of this different behavior is still uncertain. In addition, increasing the gap-filling reaction pH from 6.2 to 8.0 can result in an eightfold increase of the mutant frequency obtained in these assays (79).

In a recent report, the relevance of frameshifts in HIV-1 variation has been questioned, since genetic assays based on the use of a deoxyuracil-containing DNA template reveal a preponderance of transitions and tranversions in contrast to frameshifts that represented less than 15% of the total number of mutations found in these assays (*84*). A clear explanation for the observed reduction in the number of frameshift errors is missing, although many factors including the *Escherichia coli* mismatch repair machinery could have an influence on the obtained results. Variations in the nucleic acid secondary structure may also affect the frameshift error rate, and the presence of deoxyuracil instead of thymine in the template strand can affect the overall DNA folding. In agreement with this proposal, it has been shown that frameshift mutations (particularly, one-nucleotide deletions) were less frequently observed with RNA templates than with DNA templates, using an M13 *lacZ* forward mutation assay (*78, 82, 83*).

$G \to A$ and $C \to T$ transitions were the most frequently found base substitutions in assays carried out with a template sequence derived from the hypervariable region 1 (V-1) of the HIV-1 *env* gene (*83*). However, their frequency was not really high and did not suggest that the preponderance of $G \to A$ substitutions observed *in vivo* could be attributed exclusively to the HIV-1 RT (*9, 96*). Aberrant RNA editing (*39, 97, 98*), and fluctuations of nucleotide pools (*99*) are additional factors that may contribute to the appearance of clustered $A \to G$ or $G \to A$ hypermutations which are occasionally observed in HIV, SIV, and other retroviruses (*5, 9, 100, 101*).

It has been reported that fidelity of retroviral RTs, including HIV-1 RT, was several-fold higher with RNA than with DNA templates in genetic assays (*78, 102*). However, differences reported by other authors were not significant (*82, 83*). These results are consistent with *in vivo* mutation rates determined for RNA- and DNA-dependent DNA synthesis, which were within twofold of each other (*97*). Genetic assays have revealed different error patterns when copying RNA or DNA, but these assays failed to provide a consistent answer to the question of which mutations are predominantly associated with each template.

V. Control of Fidelity at Initiation of Reverse Transcription

Genetic and gel-based fidelity assays have shown that error-generating mechanisms relevant to retroviral RTs include poor discrimination between correct and incorrect dNTPs (*17, 69, 70*), lack of proofreading activity resulting in high mismatch extension efficiencies (*56, 66, 70*), and template–primer slippage that frequently leads to frameshifts (*77, 95*). These mechanisms operate during the elongation reactions of reverse transcription. However, there are other steps in reverse transcription where the control of fidelity plays an important role. Thus, it has been shown that incorporation of the first two nucleotides during initiation of HIV-1 plus-strand DNA synthesis is severely diminished in the presence of low concentrations of the correct dNTPs. Furthermore, formation of mismatches was almost absent at this stage of initiation, and elongation of mispaired primers was blocked (103). This process appears to be controlled by the RT-associated RNase H activity, which mediates premature RNA primer removal once the polymerization process is aggravated. Although fidelity of nucleotide incorporation in tRNA/RNA duplexes during initiation of minus-strand DNA synthesis has not been extensively analyzed, experiments with two or three of the four dNTPs showed that misinsertions were rarely observed (104, 105), thereby revealing the higher fidelity of both RNA-primed reactions.

VI. Fidelity of Strand Transfer: Implications for Retroviral Recombination

Errors generated during strand transfer can also influence the fidelity of reverse transcription. Strand transfer occurs when DNA synthesized on one template is translocated to another region on the same or a different template. Retroviral reverse transcription involves two DNA strand transfer reactions. In vitro assays have shown that retroviral RTs catalyze the addition of nontemplated bases at the end of both DNA and RNA template ends (106–108). Blunt-end additions are more efficiently catalyzed by HIV-1 and Mo-MLV RTs than by the AMV RT, and preferred substrates were purine nucleotides, particularly dATP (107).

This process may be a source of errors during strand transfer, since after transfer to the second template the RT is able to extend the DNA containing a nontemplated base at its 3′ end (106, 107). The frequency of blunt-end additions in vitro can be as high as 30–50% in the case of Mo-MLV RT (109). Both HIV-1 and Mo-MLV RTs are extremely unfaithful at the site of first strand transfer, where error rates are around 1000-fold higher than reported average RT error rates for template-internal positions (107, 109).

Another source of errors arising during strand transfer result from nucleotide misincorporation during RNA-directed DNA synthesis (110). Nucleotide misincorporation appears to be higher at pause sites due to an increased RT dissociation from the template–primer. Pause sites in DNA and RNA templates occur frequently in homopolymeric nucleotide runs and at regions of predicted secondary structure (111). Thus, elimination of pause sites with a minimal change in sequence decreased the frequency of strand transfer in the same area (112). Consequently, the observed enhancement of the strand transfer activity of HIV-1 RT, which is triggered by mismatched nucleotides at the 3′ primer terminus, appears to be mediated by pausing (110, 112, 113).

Despite those observations, genetic assays have consistently shown that the fidelity of strand transfer catalyzed by HIV-1 RT is relatively high. Reported error rates were similar or somewhat lower in comparison to values obtained for RNA-dependent DNA synthesis (*114, 115*). These results are in good agreement with the analysis of DNA sequences at recombination junctions generated during retroviral replication. No mutations were found in any of the 30 recombinant MLV-based vectors in the region where recombination occurred (*116*). In any case, these data do not exclude the possibility that recombination may be mutagenic in certain circumstances, for example, in specific sequence contexts, in the presence of altered intracellular dNTP pools (*117*), or when stimulated by the presence of broken RNA templates, usually promoting the occurrence of mismatch mutations (*118*).

VII. Contribution of Accessory Proteins to Fidelity of Reverse Transcription

Retroviral proteins that influence RT function include the nucleocapsid protein (NC), dUTP pyrophosphatase (dUTPase), and the HIV accessory proteins Vif and Vpr. NC is a small basic protein that coats the genomic RNA inside the virion core. It binds single-stranded nucleic acids nonspecifically and provides chaperone-like functions that enhance other nucleic acid-dependent steps in the life cycle, including reverse transcription (for reviews, see Refs. *119* and *120*). NC promotes annealing of the tRNA primer, melting of RNA secondary structures, and strand transfer reactions during reverse transcription. Biochemical studies have shown differences in the ability of the Mo-MLV RT to extend different mutated primers using RNA or DNA in the presence of the NC protein (*121*). These results led authors to propose that NC could be implicated in the genetic variability of Mo-MLV. In contrast, the presence of NC did not have an influence on fidelity of strand transfer in assays carried out with HIV-1 RT (*115*).

The *vif* gene of HIV-1 is essential for viral replication, but it is absent from several retroviruses including MLV or AMV. Its functional target remains elusive, but a recent report suggests that Vif could have a role in the regulation of efficient reverse transcription (*122*). The influence of this gene product in fidelity is unknown.

Retroviruses must avoid misincorporation of dUTP during reverse transcription. Intracellular concentrations of dUTP can influence mutation rate by affecting the level of dUTP incorporation into DNA which results in T → C transitions mediated by G:U base pairing. Cellular mechanisms that control nucleotide specificity at this level involve two key enzymes: dUTPase and uracil-N-glycosylase (*123*). The presence of an active dUTPase encoded within the viral genome has been demonstrated for several retroviruses, including FIV, EIAV, caprine arthritis encephalitis virus (CAEV), visna virus, Mason–Pfizer monkey virus (MPMV), and MMTV (*124, 125*). The presence of dUTPase in lentivirus appears to be an adaptation to facilitate replication within macrophages which contain high intracellular levels of dUTP (*126*). HIV-1, HIV-2, or SIV do not encode a dUTPase. However, purified cellular uracil-N-glycosylase has been shown to interact with HIV-1 and SIV Vpr, while retaining its enzymatic activity (*127, 128*).

Vpr is a protein of 96 amino acids which plays an important role in regulating nuclear import of HIV-1 preintegration complex and is required for virus replication in nondividing cells. Vpr can also induce G2 cell cycle arrest prior to nuclear envelope breakdown and chromosome condensation (for a recent review on Vpr functions, see Ref. *129*). Vpr may also have a role on initiation of reverse transcription, based on its interaction with lysine–tRNA synthetase (*130*). Binding of Vpr to lysine–tRNA synthetase inhibited aminoacylation of tRNA^{Lys}, which needs a free 3' end to function as a primer for reverse transcription. Incorporation of Vpr into viral particles requires a direct interaction with the viral *gag*-encoded protein p6 (*131, 132*). In addition to uracil-N-glycosylase, lysine–tRNA synthetase, and p6, Vpr has been shown to interact with the DNA repair protein HHR23A (*133, 134*). Both proteins localize primarily in the nucleus (*135*). However, the Vpr–HHR23A interaction did not influence the HIV-1 *in vivo* mutation rate (*135*). In addition, experiments showing a Vpr-mediated reduction in the frequency of deletions in a shuttle vector system showed no correlation to changes in nucleotide excision repair or double-strand break repair on cellular DNA (*136*).

Vpr function could help to explain the lower mutation rate found in a single cycle of replication experiments (estimated around $3-4 \times 10^{-5}$) (*41, 13*) compared to the higher error rates usually reported for purified HIV-1 RT in cell-free systems (Table I). Experimental evidence on the role of Vpr in controlling the *in vivo* mutation rate has been obtained from assays using a variant of HIV-1 deficient in Vpr function. Vpr-deficient viruses showed a four-fold increase in the mutation rate, an effect which was eliminated when Vpr was provided *in trans* (*137*). A correlation between Vpr binding to uracil-N-glycosylase and its influence on the HIV-1 *in vivo* mutation rate has been shown (*138*). Replacement of Trp-54 by Arg is sufficient to disrupt Vpr binding to uracil-N-glycosylase. Unlike the wild-type Vpr, the mutant W54R did not reduce the mutation rate when used to *trans*-complement the *vpr* null mutant HIV-1. In contrast, this reduction was observed with another Vpr variant that contained the R90K substitution. This amino acid change impairs the G2 arrest function of Vpr but allows efficient binding to uracil-N-glycosylase. It was also shown that uracil-N-glycosylase is recruited into HIV-1 particles through Vpr incorporation (*138*). The encoding of dUTPase by uracil-N-glycosylase nonprimate lentiviruses and the incorporation of uracil-*N*-glycosylase by primate lentiviruses such as HIV-1 through its interaction with Vpr support the view of these different mechanisms as an adaptation to improve the fidelity of DNA synthesis by reducing the uracil content in the proviral DNA, a strategy that may be necessary under certain conditions as, for example, while replicating in nondividing cells.

VIII. Mutational Analysis of HIV-1 Reverse Transcriptase: The Effects of Mutations on Fidelity of DNA Synthesis

HIV-1 RT is an important target for antiretroviral therapy. Many studies on the effects of mutations in the RT-associated activities (i.e., DNA polymerase, RNase H, strand transfer, etc.), as well as on the role of different amino acids in the acquisition of resistance, have been published during the last 12 years. These efforts have led to a better understanding of RT biochemistry and helped to design RT inhibitors which are currently being used for treatment of HIV infection (for reviews, see Refs. *139–141*).

Amino acid sequence alignments of the polymerase domain of retroviral RTs (Fig. 4) reveal a series of homology regions which are conserved in RT-related DNA polymerases as well as in RNA-dependent RNA polymerases (*142, 143*). Motif 1–2 is located in the "fingers" subdomain and contains highly conserved basic residues such as Lys-65 and Arg-72 which participate in triphosphate binding. The catalytic residues are found in motifs A (Asp-110) and C (Asp-185 and Asp-186). Motif B′ participates in dNTP binding through Gln-151, which is part of the highly conserved sequence Leu–Pro–Gln–Gly; motif E is part of the "primer grip," formed by residues involved in maintaining the primer terminus in an appropriate orientation for nucleophilic attack of an incoming dNTP. Motif D is less conserved among retroviral RTs. This motif is located in the "palm" subdomain and forms a small loop spanning residues 215 to 233. Residues of motif D have been implicated in drug resistance (i.e., Thr-215 and Lys-219 of HIV-1 RT), and the conserved lysine at position 220 has been shown to participate in nucleotide selection and catalysis (*144*).

The role of different residues in polymerase function has been assessed through systematic evaluation of RNA- and DNA-dependent DNA polymerase activity displayed by mutant RTs obtained by random mutagenesis (*145–151*). More than 400 different mutants have been obtained, most of them involving amino acid residues within positions 65 and 205. In most of these studies, RT variants were not purified and the available enzymological data were limited to rough estimates of polymerase activity. These experiments provided further support to the essential role of Asp-110, Asp-185, and Asp-186 in DNA polymerization (*28, 29, 152, 153*). Several residues whose replacement was usually detrimental

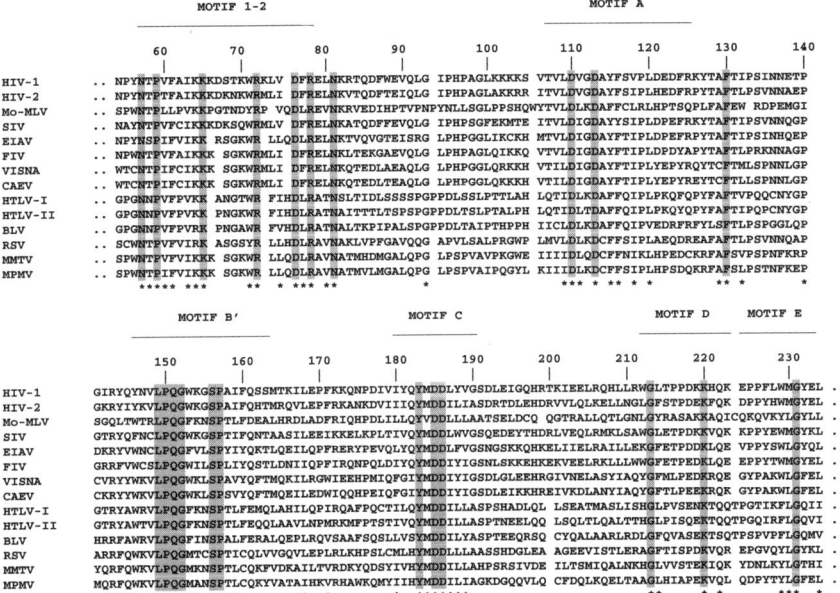

FIG. 4. Amino acid sequence alignments of the polymerase domain of retroviral RTs, showing the conserved motifs 1–2, A, B', C, D, and E. Numbering corresponds to the HIV-1 RT sequence. Conserved residues are indicated with an asterisk below the alignment. Strictly conserved amino acids are indicated with a grey background. Abbreviations used: HIV-1, human immunodeficiency virus type 1; HIV-2, human immunodeficiency virus type 2; Mo-MLV, Moloney murine leukemia virus; SIV, simian immunodeficiency virus; EIAV, equine infectious anemia virus; FIV, feline immunodeficiency virus; CAEV, caprine arthritis encephalitis virus; HTLV-I, human T-cell leukemia virus I; HTLV-II, human T-cell leukemia virus II; BLV, bovine leukemia virus; RSV, Rous sarcoma virus; MMTV, mouse mammary tumor virus; and MPMV, Mason–Pfizer monkey virus.

for polymerase function were identified. Examples are Trp-71, Arg-72, Arg-78, Val-111, Leu-120, Pro-150, Gly-152, Gly-155, Gln-182, Tyr-183, and Leu-187, among others. An extensive biochemical characterization is now available for a number of purified recombinant mutant RTs, and a significant number of studies have been devoted to the analysis of fidelity of DNA synthesis by HIV-1 RT and, in some cases, by Mo-MLV RT. In the following sections, the most relevant data in the context of the HIV-1 RT structure are discussed. It should be noted that studies have usually been carried out with either heterodimeric p66/p51 RT or homodimeric p66/p66 RT. However, in many cases, mutations have been introduced in both subunits, and often there are no data on which subunit mediates the observed effects. Despite these limitations, all the available evidence on the effect of mutations in the fidelity of HIV-1 RT is consistent with a p66-mediated effect.

A. Amino Acid Residues of the dNTP-Binding Site

The crystal structure of a covalently trapped catalytic complex of HIV-1 RT containing a DNA template–primer and a dNTP is available (35). According to this structure, the triphosphate moiety of the dNTP is coordinated with the side chains of Lys-65 and Arg-72, the main chains of Asp-113 and Ala-114, and two Mg^{2+} ions. The side chains of Arg-72 and Gln-151 pack against the outer surface of the incoming dNTP, and the ribose moiety of the incoming dNTP binds in a pocket defined by the side chains of Tyr-115, Phe-116, and Gln-151.

1. TYR-115: FIDELITY, DISCRIMINATION AGAINST rNTPs AND INTERACTION WITH PHE-160

Tyr-115 is located below the sugar ring of the incoming dNTP (Fig. 5). The substitution of Tyr-115 by Phe renders a fully active RT (52, 154, 155). However, nonconservative substitutions at this position led to reduced polymerase activity that resulted from increased K_M values for the incorporation of dNTP (52, 156). This effect was observed not only with natural dNTPs but also with dNTPs with substitutions at the pyrimidine base (157).

The substitution of Val, Ala, or Ser for Tyr-115 rendered RTs with reduced misinsertion fidelity (52, 72). Although different values can be obtained depending on the template–primer and the type of mispair analyzed, our most recent estimates indicate that Y115V is around 2 to 5 times less accurate than the wild-type RT. On the other hand, Y115A is 5- to 10-fold less accurate than the wild-type RT (72). This effect was due to the lower catalytic efficiency (k_{cat}/K_M) for the

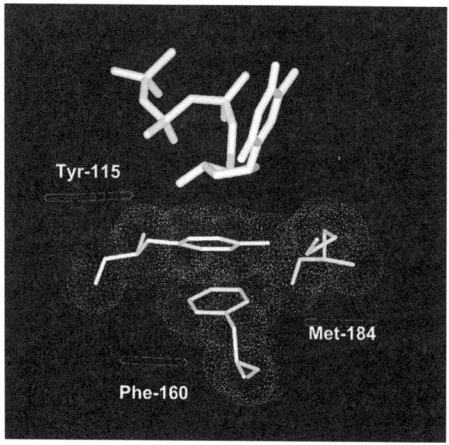

FIG. 5. Structural localization of the side-chains of Tyr-115, Phe-160, and Met-184 in the crystal structure of HIV-1 RT.

incorporation of a correct nucleotide shown by the mutant RTs relative to the wild-type enzyme, while the catalytic efficiency of incorporation of incorrect nucleotides was not largely affected by mutations at position 115. Interestingly, all these data were obtained in the presence of Mg^{2+}, the preferred cation for HIV-1 RT polymerase activity. Nevertheless, HIV-1 RT retains polymerase activity in the presence of Mn^{2+}; albeit fidelity is reduced (72) and aberrant polymerization products such as repetitive additions of nucleotides could be generated (158). The mutagenic effect of Mn^{2+} on HIV replication has been shown in cell culture, where it may produce a more than 10-fold increase in viral mutation frequency (159). Misinsertion fidelity assays carried out in the presence of Mn^{2+} showed that when Tyr-115 was replaced by Val or Ala, the accuracy of HIV-1 RT was not reduced (72).

The data obtained from mispair extension fidelity assays also supported the role of Tyr-115 in fidelity of DNA synthesis (72, 156). Twelve RT variants with substitutions at position 115 were analyzed. The largest effects were observed with mutant Y115G that showed a 33- to 50-fold higher A:C mispair extension ratio (f_{ext}) than the wild-type RT. In contrast, mutants Y115F, Y115I, Y115V, Y115W, Y115M, and Y115A showed A:C mispair extension ratios that were only 2.5–8 times higher than those for wild-type HIV-1 RT. Other mutants (i.e., Y115L, Y115H, Y115C, Y115N, and Y115S) showed intermediate f_{ext} values ranging from 8- to 25-fold higher than those for the wild-type RT (156). As in the case of misinsertion fidelity assays, the largest effects of substituting Tyr-115 were observed in the presence of Mg^{2+} (72). In Mn^{2+}-catalyzed reactions, wild-type RT was 100-fold less accurate than in the presence of Mg^{2+}, but the mispair extension fidelity of mutants Y115V and Y115G was higher than that for the wild-type RT with most of the tested mismatches.

The role of Tyr-115 in fidelity of DNA synthesis was further supported by data provided by lacZ-based forward mutation fidelity assays using a gapped M13mp2 DNA template (160). Mutant Y115A showed 3.5-fold lower fidelity than the wild-type RT in these assays. However, Y115F and Y115V showed less than 2-fold differences in their error rates, when assayed with a forward mutation assay using a deoxyuracil-containing DNA template (84). These relatively small differences in the reported error rates for Y115V and Y115A compared to those for the wild-type RT are surprising. However, it should be noted that RT-catalyzed reactions in these assays are carried out in the presence of relatively high concentrations of dNTPs. Under these conditions, the contribution of the apparent K_M, which is a critical factor leading to inaccuracy of Tyr-115 mutants in steady-state fidelity assays, is minimized in relation to the contribution of the catalytic constant, k_{cat}.

The effect of substitutions at position 115 has also been tested in a virus grown in cell culture. Nonconservative substitutions (i.e., Y115W, Y115L, Y115A, and Y115D) usually lead to the loss of replication capacity of HIV-1 (161). Phe-155 of

Mo-MLV RT occupies the equivalent position of Tyr-115 in HIV-1 RT (Fig. 4). As suggested for HIV-1 RT (52, 155), substituting Tyr for Phe-155 had no effect on the mutation rate of the virus. However, the substitution of Trp for Phe-155 of MLV RT produces a 2.8-fold increase in the mutation rate, as determined by using an *in vivo* fidelity assay based on the inactivation of *lacZ* (162).

Crystallographic analysis of the ternary complex of HIV-1 RT bound to template–primer and dNTP showed that modifications at the 2' and 3' positions of the incoming nucleotide could interfere with the side chain of Tyr-115 of HIV-1 RT (or Phe-155 of MLV RT) (Fig. 5). Studies with Mo-MLV RT showed that the wild-type RT exhibited a 15,000-fold preference for dNTPs compared to that for rNTPs in the presence of Mn^{2+}, which is the usual cofactor in MLV RT-catalyzed polymerization reactions. However, this selectivity value was reduced to about 25-fold for mutant Y115V (163). These results were confirmed for the wild-type HIV-1 RT that incorporates dNTPs with a $>10^5$-fold higher efficiency than rNTPs or 3'-dNTPs (i.e., 3'-dATP) and with a 100-fold higher efficiency than ddNTPs (72).

Mutational analysis of HIV-1 RT has shown that the substitution of Tyr-115 by Val, Ala, or Gly leads to a 10^3-, 10^4-, or 10^5-fold increase in the enzyme's ability to incorporate rNTPs versus dNTPs. In contrast, the substitution of Trp for Tyr-115 did not affect rNTP/dNTP discrimination (72). These effects were observed with both Mg^{2+} and Mn^{2+}. Interestingly, discrimination between dNTPs and ddNTPs was not significantly affected by substitutions at position 115 in Mg^{2+}-catalyzed reactions. On the other hand, these amino acid changes have a significant effect on the discrimination between dNTPs and nucleotides bearing a 2'-hydroxyl group (i.e., 3'-dATP), although the magnitude of the effect is smaller than that on rNTP/dNTP discrimination (72). In addition to its role in fidelity of DNA synthesis, the side chain of Tyr-115 of HIV-1 RT (or the equivalent residue in other retroviral RTs) acts as a "steric gate," preventing the incorporation of nucleotides with a 2'-hydroxyl group.

The effects on fidelity of DNA synthesis and rNTP/dNTP discrimination reported for Tyr-115 mutants of HIV-1 RT and Phe-155 mutants of Mo-MLV RT were also observed with variants of other DNA polymerases with substitutions at the equivalent amino acid residue. Tyr-115 of HIV-1 RT is conserved in DNA polymerases α (164, 165). Mutations at this position in human DNA polymerase α (Tyr-865) rendered enzymes with lower misinsertion fidelity (166). In addition, the substitution of Val for Tyr in the equivalent position of *Thermococcus litoralis* (Vent^{TM}) DNA polymerase (167), and bacteriophage φ29 DNA polymerase (168) led to polymerases with reduced ability to discriminate between rNTPs and dNTPs.

The analysis of the crystal structure of HIV-1 RT shows that the side chain of Tyr-115 interacts with the side chain of Phe-160 (Fig. 5). Tyr-115 is also located in the vicinity of Met-184, a residue that may also influence the positioning of the

3'-primer terminus. These three residues appear to be critical to facilitating the correct positioning of the deoxyribose moiety in the dNTP-binding pocket. The role of Phe-160 has been demonstrated by introducing conservative and nonconservative substitutions at this position (169). An aromatic side chain at position 160 (i.e., Phe, Tyr, or Trp) was required to maintain a functional polymerase and a viable HIV. Misinsertion and mispair extension fidelity assays revealed that the fidelity of mutant F160Y was similar to that of the wild-type RT. The substitution of Trp for Phe rendered an RT which was only slightly more accurate than the wild-type enzyme (169). These studies showed that interactions between the aromatic side chains of residues at positions 115 and 160 are important for the structural stabilization of the dNTP-binding pocket and influence the correct positioning of the incoming nucleotide in the RT polymerase active site.

2. MET-184

Met-184 of HIV-1 RT (Fig. 5) is an important residue for the acquisition of resistance to the antiretroviral drug lamivudine (2',3'-dideoxy-3'-thiacytidine, 3TC). The substitution of Ile or Val for Met-184, mediated by a single nucleotide change, appears quickly upon lamivudine treatment both *in vivo* and *in vitro*. The M184I variant emerged first and then it was outcompeted by M184V (170, 171). Both mutant enzymes showed decreased processivity in comparison to the wild-type RT (170).

Crystal structures of the lamivudine-resistant mutant M184I, obtained in the presence or absence of a DNA/DNA template–primer, suggested that steric hindrance between the oxathiolane ring of lamivudine triphosphate (3TCTP) and the side chain of β-branched amino acids (Val, Ile, Thr) at position 184 perturbs inhibitor binding, leading to reduced incorporation of the drug (172). It is not clear whether steric hindrance blocks 3TCTP binding or, alternatively, if binding takes place, but in a configuration that interferes with incorporation. The available kinetic data are controversial, since authors have reported that binding of 3TCTP to M184V is minimally affected (173), as well as the opposite results (174, 175). Recent work monitoring changes in RT conformation upon 3TCTP binding indicates that drug-resistant RTs bind 3TCTP less efficiently than the wild-type enzyme, and in a strained configuration that is not catalytically competent (176). The equivalent position of Met-184 of HIV-1 RT is occupied by Val-223 in Mo-MLV RT (Fig. 4). Surprisingly, the proposed steric hindrance model is not valid for MLV RT, since both the wild-type enzyme and the mutant V223M were highly resistant to lamivudine (177).

The drug-resistant HIV-1 RT mutant M184V displayed a 1.4- to 17-fold enhanced misinsertion fidelity in gel-based assays, using DNA/DNA oligonucleotide template–primers (178). The largest differences between the mutant and the wild-type enzyme were observed for the incorporation of T opposite C (17-fold) and T opposite T (12.6-fold), while the misinsertion of G opposite T,

A, or G was very inefficient with both enzymes (differences less than 2.2-fold). These observations led authors to propose that the increased fidelity of the drug-resistant mutant could help to delay the emergence of virus resistant to other antiretroviral drugs subsequently administered to the patient. This proposal has been questioned on theoretical grounds, since viral fitness, population size, and the number of replication cycles emerged as important factors that could shadow the observed effects on fidelity (*179*). Further studies in the clinical setting have shown that increased polymerase fidelity of lamivudine-resistant variants neither delays development of resistance to other inhibitors nor limits their evolutionary potential (*180–183*).

The fidelity of drug-resistant mutant M184V was further analyzed in comparison to the wild-type RT using mispair extension assays. Both enzymes showed similar mispair extension efficiencies for most mismatches in DNA-dependent DNA polymerization reactions (*184, 185*). In some cases (i.e., for G:T mismatch), it has been reported that mutant M184V is 3.6 times more efficient in mispair extension reactions (*185*). In contrast, M184V showed a much higher fidelity of RNA-dependent DNA mispair extension (*186, 187*). Reported f_{ext} values were 3.4 to 48.6 times lower for mutant M184V than for the wild-type RT (*186*). These observations were all in good agreement with qualitative assessments of misincorporations opposite multiple template sites, by using a "minus" sequencing gel assay. In the presence of only three dNTPs, the RT mutant M184V was less efficient than the wild-type, in extending a primer using either an RNA or a DNA template (*154, 187, 188*). In comparison to M184V, the mutant M184I shows reduced processivity (*170*). Gel-based fidelity assays with this mutant revealed its higher accuracy, especially in RNA-dependent DNA polymerase reactions (*186, 187*).

The higher fidelity of mutants M184I and M184V, determined by using gel-based fidelity assays, was confirmed by genetic assays only in the case of mutant M184I (*81*). M184I showed an overall error rate 4 times lower than the wild-type HIV-1 (strain HXB2) RT and 2.5 times lower than the M184V variant (*81*). Interestingly, M184I generated fewer mutations at known wild-type RT hot spots. In addition, the number of frameshift mutations found outside homopolymeric runs was much higher with M184I than with either the mutant M184V or the wild-type HIV-1 RT (*81*). *In vitro* genetic assays and mutation rate estimates determined on a single cycle of replication did not detect significant differences in fidelity between wild-type RT and mutant M184V, and the mutational spectra were similar for both enzymes (*80, 81, 160, 189*).

Other Met-184 mutants analyzed for fidelity of DNA synthesis were M184L and M184A. M184L displayed higher misinsertion and mispair extension fidelity of DNA-dependent DNA synthesis compared to the wild-type enzyme, as determined for A:C, A:G, and A:A mispairs (*190*). The effect of substituting Ala for Met-184 on the RT's accuracy is not clear. Qualitative assessments based on primer extensions in the presence of three dNTPs suggest that M184A is more accurate than the wild-type RT during RNA-dependent DNA polymerization reactions, although the differences were not significant when copying a DNA template (*160, 184*). In contrast, M184A showed more than a 10-fold higher A:A and A:G mispair extension efficiency compared to wild-type HIV-1 RT in assays carried out with a 47/20-mer DNA/DNA template–primer (*184*).

The substitution of Ala or Met for Val-223 of Mo-MLV RT renders an active polymerase. The misinsertion efficiency on an RNA template has been compared for all three enzymes. Interestingly, wild-type MLV RT was 9-fold more accurate than the mutant RTs for most misinsertions in misincorporation fidelity assays (*191*). *In vivo* fidelity assays based on the inactivation of *lacZ* showed that mutation V223I had no effect on the accuracy of the polymerase. However, other mutations (i.e., V223A, V223M, and V223S) exhibited frequencies of *lacZ* inactivation 1.7- to 2.3-fold higher than the wild-type MLV (*192*). These results suggest that Val-223 is a relevant component of the fidelity center of MLV RT.

3. GLN-151 AND THE INFLUENCE OF RESIDUES OF THE $\beta 8$-αE LOOP

The side chain of Gln-151 interacts with the incoming nucleotide, with its amido group protuding into the space left between the nitrogen base and the sugar ring (Fig. 6). Mutants of RT and *Thermus aquaticus* polymerase (whose equivalent position is Phe-667) have altered sensitivities to ddNTPs (*193–196*). Substitution of Ala or Lys for Gln-151 leads to significant reduction of polymerase activity (*197–199*). However, Met instead of Gln-151 of HIV-1 RT appears *in vivo* during antiretroviral treatment with a combination of nucleoside analogue RT inhibitors (*200*). Q151M has reduced ddNTP-binding affinity compared to the wild-type RT and shows very efficient discrimination between ddNTPs and

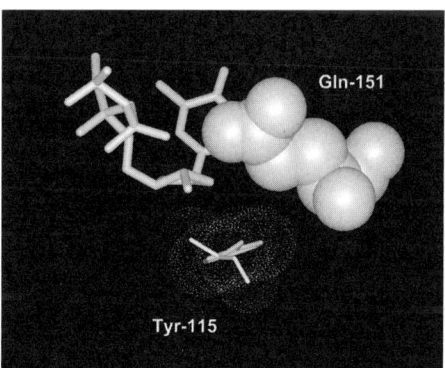

FIG. 6. Structural positioning of the side chain of Gln-151 protuding into the incoming dNTP. The localization of Tyr-115 under the deoxyribose ring of the incoming dNTP is also shown.

dNTPs (*193, 195*). In addition, the HIV-1 RT mutant Q151N and its equivalent of Mo-MLV (Q190N) were partially resistant to ddNTPs (*195, 201*). These studies suggest that the side chain at this position influences the interaction with the 3′-hydroxyl group of the incoming dNTP.

The fidelity of DNA synthesis has been analyzed in reactions catalyzed by Q151N and Q151M. Qualitative assessments based on primer extensions in the presence of three dNTPs indicated that mutants Q151N of HIV-1 RT and Q190N of Mo-MLV RT were both more accurate than the wild-type RT when copying RNA or DNA (*199, 201*). Moreover, these mutants also displayed reduced capability to extend a mismatched primer. These observations were in good agreement with determinations of misincorporation ratios, that were 8- to 26.5-fold lower for mutant Q151N than for the wild-type HIV-1 RT (*195*). In addition, further confirmation *in vitro* of the higher accuracy of this mutant was obtained using the M13 *lacZ* forward mutation assay. In these experiments, the mutant frequency of Q151N was around 13-fold reduced in comparison to the wild-type RT (*199*).

The substitution of Met for Gln-151 had a relatively minor effect on fidelity of DNA synthesis. Misinsertion ratios obtained with this mutant were only two- to six-fold higher than those obtained with wild-type HIV-1 RT (*195*), and forward mutation assays did not reveal significant differences in the overall error rates calculated for wild-type HIV-1 RT, Q151M, and the multidrug resistant mutant A62V/V75I/F77L/F116Y/Q151M (*85*). Nevertheless, genetic assays revealed at least two hot spots in the *lacZ* gene that were specific for reactions catalyzed by Q151M. Available data for the equivalent mutation of MLV RT (Q190M) are also consistent with previous results, indicating that this amino acid change has no effect on fidelity *in vivo* (*162*).

In addition to Gln-151, the $\beta 8$-αE loop includes residues Gly-152, Trp-153, and Lys-154. Gly-152 and Trp-153 are highly conserved, and mutations at these positions often lead to inactive polymerases. Mutant K154A shows a slightly increased fidelity relative to the wild-type RT, as demonstrated using qualitative assays as well as a forward mutation assay (*199*).

4. Residues Interacting with dNTP Phosphates: Lys-65 and Arg-72

Lys-65 and Arg-72 of HIV-1 RT are located in a highly flexible loop, and their side chains stabilize dNTP binding through interactions with the γ- and α-phosphates of the incoming dNTP (Fig. 7). The substitution of Arg for Lys-65 is commonly found during antiretroviral treatment, conferring viral resistance to didanosine (ddI), zalcitabine (ddC), and lamivudine, among other inhibitors (*202–205*). The HIV-1 RT bearing Arg at position 65 was equally or more processive than the wild-type enzyme (*203, 206*) and less prone to incorporate ddNTPs (*203, 206–208*). M13 *lacZ* forward mutation assays showed that the K65R variant

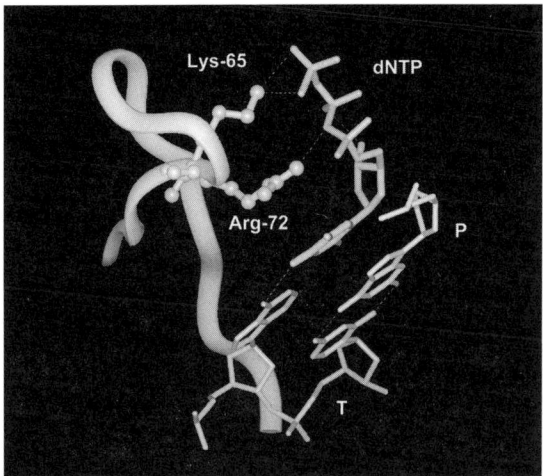

FIG. 7. Ribbon representation of the β3–β4 hairpin in the fingers subdomain of p66 of HIV-1 RT, showing the interactions of Lys-65 and Arg-72 with the phosphate moiety of the incoming dNTP.

displayed an eight-fold higher accuracy compared to the wild-type enzyme (209). Although the mutant RT showed a general decrease in the frequency of most types of errors, there was a larger reduction in the number of frameshift mutations at homopolymeric runs. These results were not consistent with data obtained for the equivalent mutation in MLV RT (K103R), whose effects on fidelity *in vivo* were almost negligible (162).

The substitution of Ala for Arg-72 impairs polymerase activity by interfering with the pyrophosphate removal function of the RT (210, 211), and viruses carrying mutations at this position may not be viable (149, 162). However, mutant R72A retains some polymerase activity *in vitro*. Although the variant RT displays higher fidelity at most sites as revealed using a *lac*Z forward mutation assay, it was highly error-prone for misincorporations opposite template T in the sequence context 5'-CTGG (212). This error is a consequence of the loss of discrimination resulting from a 1200-fold increase in the K_M for correct dAMP insertion for R72A as compared to the wild-type HIV-1 RT (212). This is a good example for showing that RTs yielding higher than normal average replication fidelity may place specific sequences at very high risk of mutation.

B. Residues Involved in Interactions with the Template Strand

The role on fidelity of DNA synthesis of residues that interact with the template strand around the nucleotide incorporation site has been studied for HIV-1 RT variants having substitutions at positions 74, 76, 78, and 89 (Fig. 8).

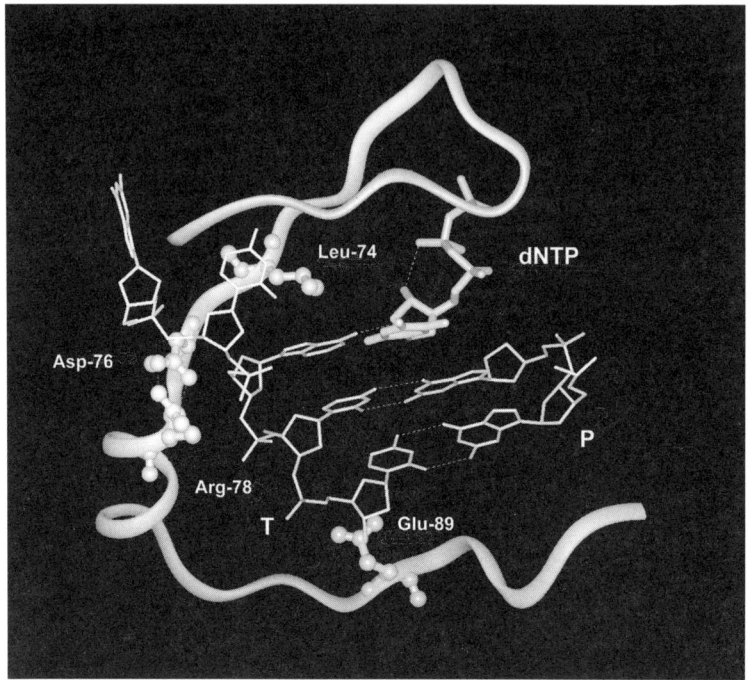

FIG. 8. Ribbon representation of the polypeptide chain at the fingers subdomain of p66 of HIV-1 RT, showing the interactions of Leu-74, Asp-76, Arg-78, and Glu-89 with the template strand of the DNA/DNA duplex. Template–primer base pairs at positions −1 and −2 as well as the incoming dNTP and three nucleotides of the overhanging template strand are shown.

Leu-74 interacts with the templating base and the incoming dNTP and with the side chains of Arg-72 and Gln-151 (35) and helps to lock the templating base tightly in place. The L74V mutation confers partial resistance to didanosine (ddI) and zalcitabine (ddC) and reduces RT processivity (205, 213, 214). This amino acid substitution produced a significant increase in the misinsertion and mispair extension fidelity of HIV-1 RT (215) that resulted from an approximately 5- to 10-fold reduction of the k_{cat} with the incorrect dNTP or template–primer mismatch. The effects on fidelity observed upon introduction of this mutation in HIV-1 RT were not observed with HIV-2 RT, where this mutation did not affect fidelity of DNA synthesis in gel-based kinetic assays (216). Results obtained from genetic assays showed either no significant difference between the mutant L74V and the wild-type HIV-1 RT (209) or a slightly higher fidelity (3.5-fold difference) in the case of mutant L74V (160).

Asp-76 and Arg-78 interact with the sugar–phosphate backbone of the templating base, apart from making potential hydrogen bonds between their side chains (33, 35) (Fig. 8). Interaction with the single-stranded template influences the accuracy of DNA synthesis. Forward mutation assays have shown that mutants D67V and R78A display nearly 9-fold higher accuracy than the wild-type RT (90, 91). Although a detailed analysis on preferred types of errors or hot spots is missing, kinetic data showed that mutant D67V displayed 1.5 to 14 times higher misinsertion ratios than the wild-type RT for all mispairs except T:T. The highest differences were observed for misinsertion of T opposite G (14-fold difference) (90). Qualitative measurements of fidelity based on primer extension assays using a 63-mer DNA template and biased dNTP pools were consistent with the higher accuracy of D76V compared to the wild-type RT (90). A similar effect was observed in these assays with mutants D76R and D76I, while other substitutions (i.e., D67E, D67S, and D67C) had little effect on fidelity.

Glu-89 of HIV-1 RT interacts with the sugar–phosphate backbone of the template strand around the template position −2 (33–35) (Fig. 8). The substitution of Gly for Glu-89 confers resistance to ddGTP (217) as well as increased polymerase processivity and diminished pausing on DNA templates (218, 219). Misinsertion and mispair extension fidelity of mutant E89G was significantly higher compared to that of the wild-type RT (185, 215, 220), and these properties were not affected by the presence of additional mutations such as M184V (80, 185). In contrast, E89G had a minor effect on misinsertion and mispair extension fidelity when introduced in HIV-2 RT (216).

The fidelity of mutant E89G was less than twofold higher than the fidelity of the wild-type RT, as determined using an M13 lacZ forward mutation assay (80, 160). Although the forward mutation assays revealed significant differences in the mutational spectra between mutant E89G and the wild-type RT (80), no correlation was found between the preferred misinsertions or mispair extensions in gel-based fidelity assays and the most frequently observed types of errors. The E89G mutation had an enhancing effect on frameshift mutagenesis (both run and non-run-associated) (80). This could be explained by the smaller size of Gly that could result in a looser template grip and therefore facilitate the formation of misaligned template–primers.

C. Residues Involved in Interactions with the Primer Strand

1. TYR-183

Tyr-183 (as well as Met-184) is involved in interactions with the 3′ terminus of the primer strand (33). The side chain of Tyr-183 interacts with the second nucleotide base pair of the duplex region in the minor groove (Fig. 9), where it can form a hydrogen bond with the corresponding nitrogen base (33). The side chain of Tyr-183 may also influence positioning of the side chain of Met-184 (Fig. 9). It has been shown that the replacement of Met-184 by Val can compensate for the reduction of polymerase activity produced by the substitution of Phe for Tyr-183 (188). The role of Tyr-183 of HIV-1 RT and the equivalent residue in

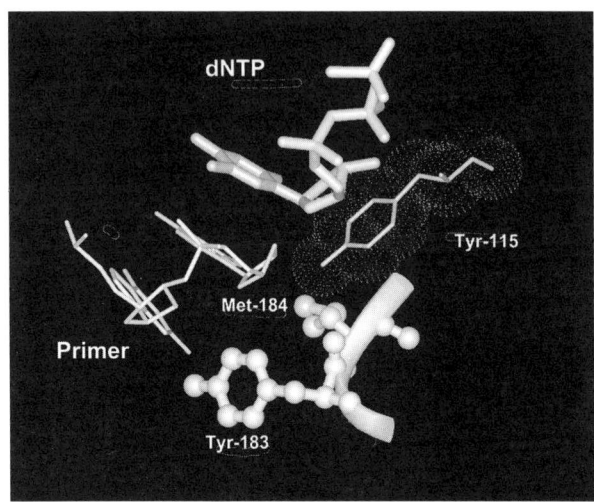

FIG. 9. Amino acid residues interacting with the primer terminal nucleotide. The side chains of Tyr-183 and Met-184 are shown. Met-184 makes interactions with Tyr-115, which contacts the sugar ring of the incoming dNTP.

Mo-MLV RT (Tyr-222) have been investigated by using misinsertion and mispair extension fidelity assays (*190, 221*) as well as primer extension experiments carried out in the presence of three dNTPs (*154, 188*). Qualitative assessments indicate that mutants Y183F and Y183A have reduced processivity and increased fidelity of DNA synthesis on both RNA and DNA templates (*154, 188*). Similar results were obtained with the equivalent mutants of Mo-MLV RT, although the effects were most significant with Y222F in RNA-dependent DNA polymerization (*221*). Qualitative estimates also showed that all mutants (Y183F and Y183A of HIV-1 RT and their equivalents in MLV RT) had improved discrimination efficiency against rNTPs (*154, 221*).

Accurate determinations of misinsertion and mispair extension efficencies have been reported only for mutant Y183F of HIV-1 RT (*190*). This mutation had no significant effect on misinsertion fidelity, although it produced a three- to five-fold decrease in A:C, A:A, or A:G mispair extension efficiency compared to the wild-type enzyme (*190*). In agreement with these observations, error rates of mutant Y183F were similar to those of the wild-type RT, as determined using a forward mutation assay (*160*).

2. Primer Grip Residues

The $\beta 12$–$\beta 13$ hairpin of the p66 subunit of HIV-1 RT (corresponding to residues 227–235 of the palm subdomain) has been designated as the *primer grip*. It is involved in maintaining the primer terminus in an orientation appropriate for nucleophilic attack on an incoming dNTP. Mutational analysis of the primer grip has shown its influence on many different RT functions including polypurine tract removal, RNase H activity, and template–primer utilization, among others (222–227). Published data indicate that mutations in the primer grip produce a low to moderate increase of fidelity of DNA-dependent DNA synthesis. Thus, RTs having Ala instead of Phe-227, Trp-229, Met-230, Gly-231, or Tyr-232 were 40 to 76% less efficient than the wild-type enzyme in extending a mismatch in primer extension assays performed in the presence of all four dNTPs (228). The higher accuracy of mutants F227A and W229A compared to the wild-type was further confirmed in misinsertion fidelity assays measuring the incorporation of A, G, C, or T opposite T (228).

Met-230 is located at the tip of the $\beta12$–$\beta13$ hairpin (Fig. 10). The side chain of Met-230 interacts with the deoxyribose ring at position -2 of the primer and together with Phe-183 lies in the minor groove of the template–primer. The substitution of Ala for Met-230 renders a polymerase with reduced affinity for the incoming dNTP (227) and leads to a noninfectious virus (229). However, unlike mutant M230A, the substitution of Ile for Met-230 does not impair

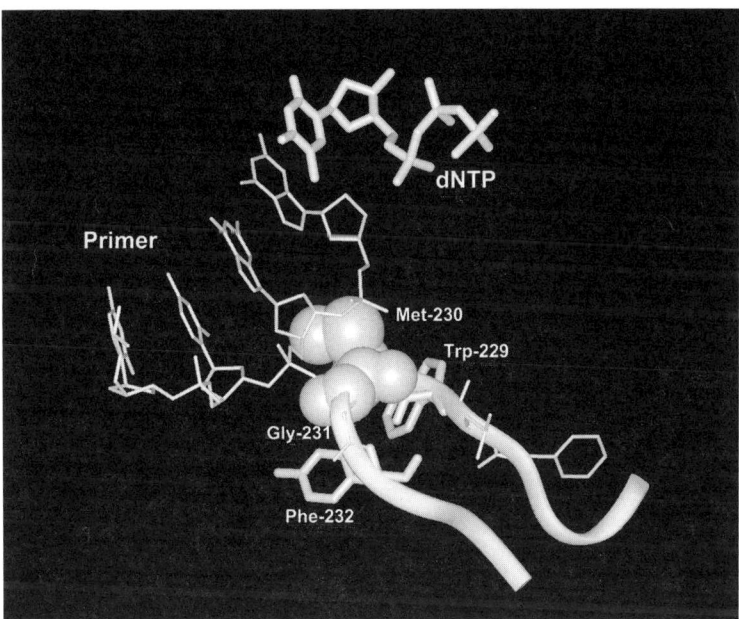

FIG. 10. Amino acid residues of the primer grip and their interaction with the template–primer. The position of the side chains of Trp-229, Met-230, Gly-231, and Phe-232 are shown. Met-230 is represented with a CPK model. This residue interacts with the deoxyribose ring at position -2 of the primer.

polymerase function or virus viability. It has been observed that M230I compensates for the dNTP-binding defect shown by an HIV-1 RT-bearing Trp at position 115 (*161*). In contrast to other primer grip mutants, studies on the fidelity of M230I revealed that this RT variant showed reduced fidelity in misincorporation fidelity assays, measuring insertion of A, G, C, and T opposite T (*229a*). Interestingly, fidelity of this variant RT was not influenced by nucleotide substitutions at the template–primer around the incorporation site (i.e., at base pairs flanking primer position −2). On the other hand, its accuracy was influenced by the structure of the 5′ overhang of the template strand.

3. MINOR GROOVE BINDING TRACK RESIDUES

The minor groove binding track interactions occur in the DNA minor groove from the second to the sixth base pair from the primer 3′ terminus where the DNA undergoes a structural transition from A-like to B-form DNA (*33, 35*). Five residues are considered to be of potential importance for binding to the duplex template–primer in this region: Ile-94, Gln-258, Gly-262, Trp-266, and Gln-269 (*230*). Gln-258, Gly-262, and Trp-266 are part of α-helix H. This α-helix together with α-helix I are located in the thumb subdomain of p66 of HIV-1 RT, where they constitute the so-called "helix clamp" (Fig. 11). Alanine-scanning mutagenesis of residues 253–271 (including α-helix H) and residues 277–287 (including α-helix I) have shown that these structures have a role in template–primer binding and processivity and influence the RNase H-catalyzed removal of the polypurine primer tract (*86, 88, 231*).

M13 *lac*Z forward mutation assays have been carried out with 17 alanine mutants of these region (D256A, Q258A, K259A, L260A, G262A, K263A, W266A, Q269A, R277A, Q278A, L279A, C280A, K281A, L282A, R284A, G285A, and K287A). For most of them, the estimated mutation frequencies were less than twofold different than those for the wild-type enzyme (*86, 88*). The largest differences were observed with G262A and W266A that showed 3- to 4-fold higher error rates than the wild-type RT (*86, 87*). Interestingly, the lower accuracy of mutants G262A and W266A correlated with their large dissociation rate constants for template–primer, relative to the wild-type enzyme (*86*). Both enzymes showed reduced processivity and increased strand slippage error rates, particularly dislocation errors and −1 frameshifts in runs of up to four bases (*87*). A relatively high frequency of −1 frameshift mutations has also been observed with the minor groove mutants Q258A and Q269A in a template TTTT run (*230*). In addition, mutants having Glu, Ala, Leu, Ile, Val, or Arg instead of Trp-266 also displayed significantly reduced frameshift fidelity (3- to 4.5-fold higher error rate), while the accuracy of mutant W266Y was similar to that shown by the wild-type RT (*89*). These effects correlate to binding free energies for the side chain at position 266 and suggest that hydrophobic interactions at the binding cleft are important to maintain wild-type frameshift fidelity.

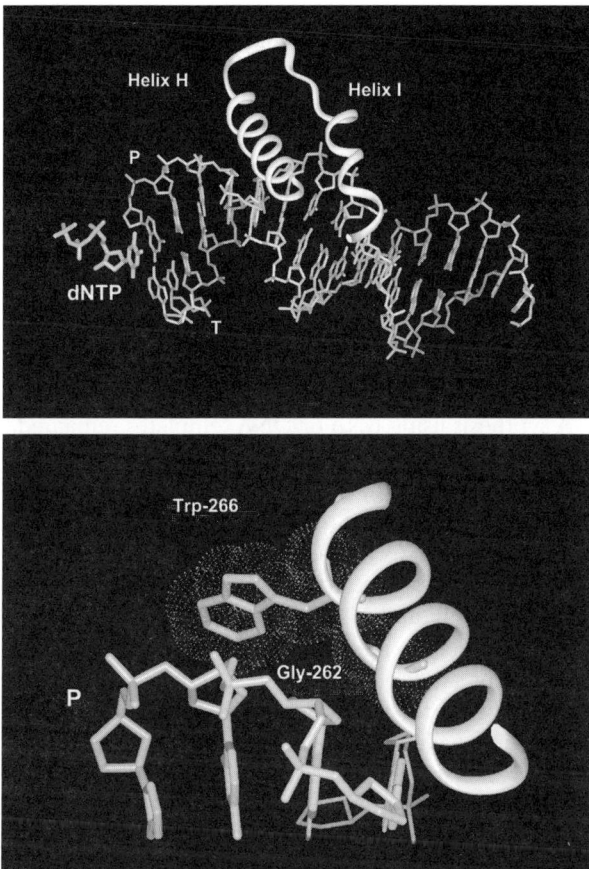

FIG. 11. Minor groove interactions and structural localization of Gly-262 and Trp-266. The interaction of α-helix H with the minor groove of the template–primer substrate is illustrated above. The α-helices (H and I) of the thumb subdomain of HIV-1 RT are shown with a ribbon representation, while the DNA duplex is shown using a stick diagram. Below, a detail of the interaction between residues of α-helix H (Gly-262 and Trp-266) and primer positions −3 and −4 is shown.

D. Role of Other Amino Acids in the Fidelity of DNA Synthesis

In addition to minor groove mutants and other relevant amino acid substitutions whose influence in fidelity of DNA synthesis has been discussed in previous sections, there are a number of residues whose contribution to fidelity has been assessed less extensively, most of them showing a relatively minor influence on the RT's accuracy. Thus, Cys-38 and Cys-280 of HIV-1 RT had a minor effect on mispair extension fidelity, as determined with the double mutant C38S/C280S (*232*). A similar conclusion was obtained for mutants Y181L and Y188L, although in this case kinetic parameters were not determined (*233*). Large insertions between codons 66 and 67 in the $\beta3-\beta4$ hairpin loop conferred increased DNA polymerase processivity (*234*) but had no effect on fidelity as determined with an M13 *lacZ* forward mutation assay (*235*).

An extensive analysis has been carried out for mutants involving C-terminal deletions in the RNase H domain of HIV-1 RT. The α-helix E' (residues 544–555) contains the conserved Asp-549, and deletions affecting α-helix E' uncouple the endonuclease and directional processing activities of RT, interfering with strand transfer (*236*). Although partial or total removal of α-helix E' weakens the interaction of RT with the template–primer, misinsertion and mispair extension fidelity of DNA synthesis using RNA or DNA templates remained largely unaffected by deletions of up to 20 residues at the C-terminal region of p66, which includes α-helix E' (*53*). In addition to these data, the fidelity of three additional mutants of the RNase H domain of MLV RT has been studied using a sensitive *in vivo* assay based on the inactivation of *lacZ*. Results showed that two mutations (S526A and R657S) exhibited somewhat higher frequencies of *lacZ* inactivation than the wild-type RT (1.6- and 1.4-fold increase, respectively), while the third one (Y598V) had no effect on fidelity (*192*).

The substitution of Tyr for Thr-215 appears during treatment with zidovudine (AZT) and is critical for the acquisition of resistance to this drug (for a review, see Ref. *237*). However, Thr-215 is located away from the polymerase active site in the crystal structure of HIV-1 RT, and AZT resistance mediated by T215Y does not appear to be a result of enhanced discrimination between dNTPs and AZT triphosphate but is more likely achieved through an alternative mechanism (i.e., ATP-dependent phosphorolysis) (*238–240* and references therein). The effect of mutation T215Y in misinsertion and mispair extension fidelity has been studied in the context of HIV-2 RT (*216*). This mutation alone, or in combination with L74V or E89G, did not affect misinsertion fidelity, although it exhibited a moderately higher accuracy in mispair extension assays. The error rate obtained for the AZT-resistant HIV-1 RT containing mutations D67N, K70R, T215F, and K219Q was roughly similar to that of the wild-type enzyme, as determined using the M13 *lacZ* forward mutation assay (*241*). However, estimates obtained from a single cycle of HIV-1 shuttle vector replication showed that mutants M41L/T215Y and M41L/D67N/K70R/T215Y had a three- to fourfold increased mutation rate relative to the wild-type and the single-mutant T215Y (*189*). These data are consistent with the higher HIV mutation rate observed during treatment with zidovudine (AZT) (*242*), although other factors may also contribute to the observed effects.

IX. Biological Consequences of Increasing or Decreasing Fidelity

A. Enhanced Polymerase Fidelity in Drug-Resistant Viruses

The high level of genetic diversity in HIV-infected individuals profoundly influences the outcome of clinical strategies for disease therapy and prevention. The low fidelity of the retroviral RT is thought to be a major driving force behind the generation of new viral variants, including drug-resistant escape mutants. Controlling genetic variation could be very helpful in improving current treatment strategies. An enhanced fidelity could theoretically limit viral variation and delay the appearance of drug-resistant viruses. This proposal was tested with lamivudine-resistant HIV-1, bearing the mutation M184V in its RT-coding region and displaying higher accuracy than the wild-type RT (*178*). However, the overall fidelity increase was relatively small (see Section VIII,A,2), and clinical studies revealed that the drug-resistant mutation had a minor impact in the evolutionary potential of the virus (*182, 183*).

A theoretical model considering the viral replication rate, fitness, and fidelity of the virally encoded RT suggested that, in order for increases in fidelity to make an impact on the rate at which a subsequent variant appears, increases in polymerase fidelity should be very large (*179, 243*). The largest increases in fidelity found for HIV-1 RT variants have been observed with mutants K65R, D76V, R78A, and Q151N (*90, 91, 199, 209*), although their relevance *in vivo* has not been addressed. Several drug-resistant variants showed slightly or moderately increased fidelity, while displaying reduced processivity (i.e., E89G, M184V, and M184I). However, no correlation has been found between fidelity and processivity (*218, 219*). In some cases, reduced fidelity has been associated with drug-resistant mutations, suggesting that viral evolution during the course of infection could correlate with increased viral replication accuracy. This hypothesis is consistent with the higher misinsertion accuracy observed for a SIV RT variant obtained from virus isolated during the late symptomatic phase of infection, relative to the RT of the parental viral strain (*244*), although further studies will be necessary to confirm this proposal.

B. Mutator Reverse Transcriptases and Error Catastrophe

Retroviruses have high mutation rates and are particularly vulnerable to the loss of genome replication accuracy. Theoretically, an increase in the average error rate above a critical threshold during DNA synthesis should result in the loss of genetic information in a process termed "error catastrophe" (*2*). This possibility has been speculated as being a rational approach for the antiviral treatment of RNA virus infections. Several mutations in HIV-1 RT, including the substitution of Ala for Arg-72 (*212*), nonconservative substitutions of Tyr-115 (*52, 72, 156*), and amino acid replacements at positions 262 and 266 of α-helix H (*86, 87, 89*) produced significant reductions in the fidelity of DNA synthesis. However, most of these mutations impair DNA polymerase activity and are not functional *in vivo*. In this scenario, the relationship between the RT mutator phenotype and viral extinction due to violation of the polymerase's error threshold and the dynamics of the process leading to error catastrophe constitute a challenge for future research.

C. Extrinsic Factors Leading to Mutation Rate Variations

Apart from the polymerase itself, extrinsic factors can also influence fidelity of DNA synthesis *in vivo*. Examples are mutagenic cations such as Mn^{2+} (*159*) or alterations in the deoxynucleotide pools. Intracellular dNTP pool imbalances affect virus titers, as well as the rate and spectrum of retroviral mutations (*245*). Deoxynucleotide pools have an important effect on the rate of HIV-1 reverse transcription, which proceeds more slowly in mononuclear phagocytes than in T cells (*170, 246*). Limiting dNTP concentrations can also enhance the processivity defect of HIV-1 variants, including those bearing the lamivudine-resistance mutation M184V (*247*). The fidelity of retroviral reverse transcription is sensitive to biased dNTP concentrations. Genetic assays designed to estimate the frequency of G $\rightarrow$ A mutations showed that the RNA-dependent DNA polymerase activity of HIV-1 RT was very sensitive to the depletion of intracellular dCTP concentrations (*248, 249*). HIV-1 RT displayed greater sensitivity to biased [dCTP]/[dTTP] ratios in comparison to the AMV and Mo-MLV RTs (*250*).

Certain drugs may alter dNTP pools and influence the viral mutation rate and its mutational pattern. Thus, increasing the dCTP pools in HIV-1-infected cell cultures through the administration of 2'-deoxycytidine and tetrahydrouridine can reverse the characteristic G $\rightarrow$ A mutational bias found in HIV-1 (*251*). The ribonucleotide reductase inhibitor hydroxyurea is known to deplete dATP pools more severely than other dNTP pools and, in combination with didanosine (ddI), has a synergistic inhibitory effect without increasing toxicity in activated peripheral blood mononuclear cells (for reviews see Refs. *252* and *253* and references therein). Treatment of SNV- or MLV-infected cells with hydroxyurea produced a two- to three-fold increase in the mutation rate, an effect that could be mimicked by supplementing the cell culture growth medium with 0.5 m*M* thymidine (*245*). In addition, alterations of the intracellular dNTP pools observed in hydroxyurea-treated and nontreated cells were found to affect the frequency of template switching, thereby increasing the frequency of deletion errors in MLV replication (*117, 254*).

Zidovudine has been shown to increase the *in vivo* mutation rate of various retroviruses, including SNV, MLV, FIV, and HIV-1 (*189, 242, 255*). The largest effects were those reported for SNV that showed a 7- to 10-fold increase in the mutant frequency after treatment of target-infected cells with 0.1 and 0.5 μM AZT (242). This effect does not appear to be a consequence of altered dNTP pools, at least in the case of SNV (245). The molecular mechanisms contributing to the AZT-mediated increase of the mutation rate are still not clear.

The high mutation rate of HIV suggests that the viral population exists near the threshold for viral viability. Mutagenic compounds altering the fidelity of the retroviral RT could drive the virus into "error catastrophe." Potential candidates include the minor groove binding peptide netropsin, which enhances pausing of HIV-1 RT at the polydenosine tract and increases the frequency of frameshift mutations as determined in a forward mutation assay (256). Nucleoside analogs have also shown a mutagenic effect in retroviral replication. For example, 5-azacytidine, which is incorporated into RNA and inhibits protein synthesis, was found to increase the *in vivo* SNV mutation rate by a factor of 13 (257). In addition, there are other mutagenic dNTPs whose incorporation by retroviral RTs has been documented (258–260). It has been reported that sequential passage of HIV-1 in cells grown in the presence of 5-hydroxydeoxycytidine resulted in the loss of viral replication, in a process that has been termed lethal mutagenesis of HIV-1 (261). Viral extinction induced by 5-hydroxydeoxycytidine correlated to an accumulation of G → A mutations that resulted from its strong tendency to base pair with a template A or G during DNA synthesis catalyzed by HIV-1 RT.

X. Conclusions and Future Perspectives

The mutation rate has a significant influence on retrovirus genetic variation. At the same time, RTs play a pivotal role in retroviral replication. The accuracy of this process has to be tightly controlled, since increased fidelity may decrease virus adaptability, while decreased fidelity puts the virus at risk of extinction through violation of the error threshold. Significant progress has been made toward understanding the molecular basis of fidelity of DNA polymerases, including retroviral RTs. As shown in this review, the fidelity properties of many mutant RTs have been studied in detail. However, despite a better knowledge on which residues participate in nucleotide selectivity, our basic understanding is still incomplete. Important issues such as accurate predictions of mutational hot and cold spots remain unavailable, since the molecular determinants of fidelity are not well understood.

Control of retroviral replication accuracy may not reside exclusively in the viral RT. The role of RNA polymerase II is largely unknown, although its contribution cannot be overlooked. In fact, authors have proposed using mutagenic ribonucleosides as a strategy to increase the viral mutation rate (262). In addition, cell DNA repair systems and viral proteins such as Vpr may also contribute to the mutation rate either in an independent way or through complex interactions with other cell or viral components.

One goal of the application of our knowledge in this field is to generate mutator and antimutator conditions that attenuate viral replication and disease progression. For example, the possibility of treating viral infections with mutagens is provocative, but success will probably depend on a delicate balance between an increased mutation rate that would facilitate the emergence of mutagen-resistant virions and the deleterious (and hopefully lethal) effect of the mutant. Clearly, a great deal remains to be investigated, and we can look forward to an exciting time in this area of research.

ACKNOWLEDGMENTS

I express my gratitude to A. M. Martín-Hernández, M. Gutiérrez-Rivas, C. E. Cases-González, A. Mas, and A. L. Zakharenko for their commitment to the RT fidelity projects carried out in our laboratory; to E. Domingo for his support; and to our collaborators in other institutions for their valuable contributions. I am also grateful to B. Canard, M. Modak, and V. Pathak for critical reading of the manuscript. Work in our laboratory is supported by Fondo de Investigación Sanitaria Grant 01/0067-01. An institutional grant of Fundación Ramón Areces to the C.B.M.S.O. is also acknowledged.

REFERENCES

1. H. M. Temin, The control of cellular morphology in embryonic cells infected with Rous sarcoma virus in vitro. *Virology* **10**, 182–197 (1960).
2. E. Domingo, C. K. Biebricher, M. Eigen, and J. J. Holland, "Quasispecies and RNA Virus Evolution: Principles and Consequences." Landes Bioscience, Georgetown, TX, (2001).
3. B. H. Hahn, G. M. Shaw, M. E. Taylor, R. R. Redfield, P. D. Markham, S. Z. Salahuddin, F. Wong-Staal, R. C. Gallo, E. S. Parks, and W. P. Parks, Genetic variation in HTLV-III/LAV over time in patients with AIDS or at risk for AIDS. *Science* **232**, 1548–1553 (1986).
4. D. P. W. Burns and R. C. Desrosiers, Selection of genetic variants of simian immunodeficiency virus in persistently infected rhesus monkeys. *J. Virol.* **65**, 1843–1854 (1991).
5. P. R. Johnson, T. E. Hamm, S. Goldstein, S. Kitov, and V. M. Hirsch, The genetic fate of molecularly cloned simian immunodeficiency virus in experimentally infected macaques. *Virology* **185**, 217–228 (1991).
6. J. Overbaugh, L. M. Rudensey, M. D. Papenhausen, R. E. Benveniste, and W. R. Morton, Variation in simian immunodeficiency virus *env* is confined to V1 and V4 during progression to simian AIDS. *J. Virol.* **65**, 7025–7031 (1991).
7. B. D. Preston and J. P. Dougherty, Mechanisms of retroviral mutation. *Trends in Microbiol.* **4**, 16–21 (1996).
8. J. M. Leider, P. Palese, and F. I. Smith, Determination of the mutation rate of a retrovirus. *J. Virol.* **62**, 3084–3091 (1988).
9. V. K. Pathak and H. M. Temin, Broad spectrum of *in vivo* forward mutations, hypermutations, and mutational hotspots in a retroviral shuttle vector after a single replication cycle: Deletions and deletions with insertions. *Proc. Natl. Acad. Sci. U.S.A.* **87**, 6024–6028 (1990).

10. J. W. Drake, A constant rate of spontaneous mutation in DNA-based microbes. *Proc. Natl. Acad. Sci. U.S.A.* **88,** 7160–7164 (1991).
11. T. A. Kunkel and P. S. Alexander, The base substitution fidelity of eucaryotic DNA polymerases. *J. Biol. Chem.* **261,** 160–166 (1986).
12. T. Matsuda, K. Bebenek, C. Masutani, F. Hanaoka, and T. A. Kunkel, Low fidelity DNA synthesis by human DNA polymerase-η. *Nature* **404,** 1011–1013 (2000).
13. B. D. Preston, B. J. Poiesz, and L. A. Loeb, Fidelity of HIV-1 reverse transcriptase. *Science* **242,** 1168–1171 (1988).
14. J. D. Roberts, K. Bebenek, and T. A. Kunkel, The accuracy of reverse transcriptase from HIV-1. *Science* **242,** 1171–1173 (1988).
15. J. D. Roberts, B. D. Preston, L. A. Johnston, A. Soni, L. A. Loeb, and T. A. Kunkel, Fidelity of two retroviral reverse transcriptases during DNA-dependent DNA synthesis in vitro. *Mol. Cell. Biol.* **9,** 469–476 (1989).
16. J. Weber and F. Grosse, Fidelity of human immunodeficiency virus type 1 reverse transcriptase in copying natural DNA. *Nucleic Acids Res.* **17,** 1379–1393 (1989).
17. M. Ricchetti and H. Buc, Reverse transcriptases and genomic variability: the accuracy of DNA replication is enzyme specific and sequence dependent. *EMBO J.* **9,** 1583–1593 (1990).
18. A. Manns, H. König, M. Baier, R. Kurth, and F. Grosse, Fidelity of reverse transcriptase of the simian immunodeficiency virus from African green monkey. *Nucleic Acids Res.* **19,** 533–537 (1991).
19. L. de Mercoryol, Y. Corda, C. Job, and D. Job, Accuracy of wheat-germ RNA polymerase II: General enzymatic properties and effect of template conformational transition from right-handed B-DNA to left-handed Z-DNA. *Eur. J. Biochem.* **206,** 49–58 (1992).
20. C. J. Jeon and K. Agarwal, Fidelity of RNA polymerase II transcription controlled by elongation factor TFIIS. *Proc. Natl. Acad. Sci. U.S.A.* **93,** 13,677–13,682 (1996).
21. D. A. Erie, O. Hajiseyedjavadi, M. C. Young, and P. H. von Hippel, Multiple RNA polymerase conformations and GreA: Control of the fidelity of transcription. *Science* **262,** 867–873 (1993).
22. M. J. Thomas, A. A. Platas, and D. K. Hawley, Transcriptional fidelity and proofreading by RNA polymerase II. *Cell* **93,** 627–637 (1998).
23. J. G. Levin, D. L. Hatfield, S. Oroszlan, and A. Rein, Mechanisms of translational suppression used in biosynthesis of reverse transcriptase. in "Reverse Transcriptase" (A. M. Skalka and S. P. Goff, eds.), pp. 5–31. Cold Spring Harbor Laboratory Press, Plainview, New York, 1993.
24. S. P. Goff, Retroviral reverse transcriptase: Synthesis, structure and function. *J. Acquir. Immune Def. Syndr.* **3,** 817–831 (1990).
25. A. Telesnitsky and S. P. Goff, Reverse transcription and the generation of retroviral DNA. in "Retroviruses" (J. Coffin, S. H. Hughes, and H. Varmus, eds.), pp. 121–160. Cold Spring Harbor Laboratory Press, Plainview, New York (1997).
26. F. Di Marzo Veronese, T. D. Copeland, A. L. DeVico, R. Rahman, S. Oroszlan, R. C. Gallo, and M. G. Sarngadharan, Characterization of highly immunogenic p66/p51 as the reverse transcriptase of HTLV-III/LAV. *Science,* **231,** 1289–1291 (1986).
27. M. M. Lightfoote, J. E. Coligan, T. M. Folks, A. S. Fauci, M. A. Martin, and S. Venkatesan, Structural characterization of reverse transcriptase and endonuclease polypeptides of the acquired immunodeficiency syndrome retrovirus. *J. Virol.* **60,** 771–775 (1986).
28. S. F. J. Le Grice, T. Naas, B. Wohlgensinger, and O. Schatz, Subunit-selective mutagenesis indicates minimal polymerase activity in heterodimer-associated p51 HIV-1 reverse transcriptase. *EMBO J.* **10,** 3905–3911 (1991).
29. Z. Hostomsky, Z. Hostomska, T.-B. Fu, and J. Taylor, Reverse transcriptase of human immunodeficiency virus type 1: Functionality of subunits of the heterodimer in DNA synthesis. *J. Virol.* **66,** 3179–3182 (1992).

30. D. W. Rodgers, S. J. Gamblin, B. A. Harris, S. Ray, J. S. Culp, B. Hellmig, D. J. Woolf, C. Debouck, and S. C. Harrison, The structure of unliganded reverse transcriptase from the human immunodeficiency virus type 1. *Proc. Natl. Acad. Sci. U.S.A.* **92**, 1222–1226 (1995).
31. L. A. Kohlstaedt, J. Wang, J. M. Friedman, P. A. Rice, and T. A. Steitz, Crystal structure at 3.5 Å resolution of HIV-1 reverse transcriptase complexed with an inhibitor. *Science* **256**, 1783–1790 (1992).
32. A. Jacobo-Molina, J. Ding, R. G. Nanni, A. D. Clark, Jr., X. Lu, C. Tantillo, R. L. Williams, G. Kamer, A. L. Ferris, P. Clark, A. Hizi, S. H. Hughes, and E. Arnold, Crystal structure of human immunodeficiency virus type 1 reverse transcriptase complexed with double-stranded DNA at 3.0 Å resolution shows bent DNA. *Proc. Natl. Acad. Sci. U.S.A.* **90**, 6320–6324 (1993).
33. J. Ding, K. Das, Y. Hsiou, S. G. Sarafianos, A. D. Clark, Jr., A. Jacobo-Molina, C. Tantillo, S. H. Hughes, and E. Arnold, Structure and functional implications of the polymerase active site region in a complex of HIV-1 RT with a double-stranded DNA template-primer and an antibody Fab fragment at 2.8 Å resolution. *J. Mol. Biol.* **284**, 1095–1111 (1998).
34. S. G. Sarafianos, K. Das, C. Tantillo, A. D. Clark, Jr., J. Ding, J. M. Whitcomb, P. L. Boyer, S. H. Hughes, and E. Arnold, Crystal structure of HIV-1 reverse transcriptase in complex with a polypurine tract RNA:DNA. *EMBO J.* **20**, 1449–1461 (2001).
35. H. Huang, R. Chopra, G. L. Verdine, and S. C. Harrison, Structure of a covalently trapped catalytic complex of HIV-1 reverse transcriptase: Implications for drug resistance. *Science* **282**, 1669–1675 (1998).
36. E. J. Arts and S. F. J. Le Grice, Interaction of retroviral reverse transcriptase with template-primer duplexes during replication. *Progr. Nucleic Acid Res. Mol. Biol.* **58**, 339–393 (1998).
37. M. Götte, X. Li, and M. A. Wainberg, HIV-1 reverse transcription: A brief overview focused on structure-function relationship between molecules involved in initiation of the reaction. *Archs. Biochem. Biophys.* **365**, 199–210 (1999).
38. H. Yu, A. E. Jetzt, and J. P. Dougherty, Use of single-cycle analysis to study rates and mechanisms of retroviral mutation. *Methods: A Comp. Methods Enzymol.* **12**, 325–336 (1997).
39. J. P. Dougherty and H. M. Temin, High mutation rate of a spleen necrosis virus-based retrovirus vector. *Molec. Cell. Biol.* **6**, 4387–4395 (1986).
40. L. M. Mansky and H. M. Temin, Lower mutation rate of bovine leukemia virus relative to that of spleen necrosis virus. *J. Virol.* **68**, 494–499 (1994).
41. L. M. Mansky and H. M. Temin, Lower in vivo mutation rate of human immunodeficiency virus type 1 than that predicted from the fidelity of purified reverse transcriptase. *J. Virol.* **69**, 5087–5094 (1995).
42. S. Parthasarathi, A. Varela-Echevarría, Y. Ron, B. D. Preston, and J. P. Dougherty, Genetic rearrangements occurring during a single cycle of murine leukemia virus vector replication: characterization and implications. *J. Virol.* **69**, 7991–8000 (1995).
43. L. Mansky, Forward mutation rate of human immunodeficiency virus type 1 in a T lymphoid cell line. *AIDS Res. Human Retrovir.* **12**, 307–314 (1996).
44. A. Varela-Echavarría, N. Garvey, B. D. Preston, and J. P. Dougherty, Comparison of Moloney murine leukemia virus mutation rate with the fidelity of its reverse transcriptase *in vitro*. *J. Biol. Chem.* **267**, 24,681–24,688 (1992).
45. A. Varela-Echavarría, C. M. Prorock, Y. Ron, and J. P. Dougherty, High rate of genetic rearrangement during replication of a Moloney murine leukemia virus-based vector. *J. Virol.* **67**, 6357–6364 (1993).
46. L. M. Mansky, In vivo analysis of human T-cell leukemia virus type 1 reverse transcription accuracy. *J. Virol.* **74**, 9525–9531 (2000).
47. R. J. Monk, F. G. Malik, D. Stokesberry, and L. H. Evans, Direct determination of the point mutation rate of a murine retrovirus. *J. Virol.* **66**, 3683–3689 (1992).

48. A. Gabriel, M. Willems, E. H. Mules, and J. D. Boeke, Replication infidelity during a single cycle of Ty1 retrotransposition. *Proc. Natl. Acad. Sci. U.S.A.* **93,** 7767–7771 (1996).
49. Y. Takeuchi, T. Nagumo, and H. Hoshino, Low fidelity of cell-free DNA synthesis by reverse transcriptase of human immunodeficiency virus. *J. Virol.* **62,** 3900–3902 (1988).
50. M. S. Boosalis, J. Petruska, and M. F. Goodman, DNA polymerase insertion fidelity-Gel assay for site-specific kinetics. *J. Biol. Chem.* **262,** 14,689–14,696 (1987).
51. M. F. Goodman, S. Creighton, L. B. Bloom, and J. Petruska, Biochemical basis of DNA replication fidelity. *Crit. Rev. Biochem. Mol. Biol.* **28,** 83–126 (1993).
52. A. M. Martín-Hernández, E. Domingo, and L. Menéndez-Arias, Human immunodeficiency virus type 1 reverse transcriptase: role of Tyr115 in deoxynucleotide binding and misinsertion fidelity of DNA synthesis. *EMBO J.* **15,** 4434–4442 (1996).
53. L. Menéndez-Arias, Studies on the effects of truncating α-helix E' of p66 human immunodeficiency virus type 1 reverse transcriptase on template-primer binding and fidelity of DNA synthesis. *Biochemistry* **37,** 16,636–16,644 (1998).
54. L. V. Mendelman, M. S. Boosalis, J. Petruska, and M. F. Goodman, Nearest neighbor influences on DNA polymerase insertion fidelity. *J. Biol. Chem.* **264,** 14,415–14,423 (1989).
55. M. Bakhanashvili and A. Hizi, Fidelity of the reverse transcriptase of human immunodeficiency virus type 2. *FEBS Lett.* **306,** 151–156 (1992).
56. H. Yu and M. F. Goodman, Comparison of HIV-1 and avian myeloblastosis virus reverse transcriptase fidelity on RNA and DNA templates. *J. Biol. Chem.* **267,** 10,888–10,896 (1992).
57. M. Bakhanashvili and A. Hizi, Fidelity of DNA synthesis exhibited in vitro by the reverse transcriptase of the lentivirus equine infectious anemia virus. *Biochemistry* **32,** 7559–7567 (1993).
58. R. Taube, O. Avidan, M. Bakhanashvili, and A. Hizi, DNA synthesis exhibited by the reverse transcriptase of mouse mammary tumor virus: Processivity and fidelity of misinsertion and mispair extension. *Eur. J. Biochem.* **258,** 1032–1039 (1998).
59. M. Sala, S. Wain-Hobson, and F. Schaeffer, Human immunodeficiency virus type 1 reverse transcriptase $_t$G:T mispair formation on RNA and DNA templates with mismatched primers: a kinetic and thermodynamic study. *EMBO J.* **14,** 4622–4627 (1995).
60. C. M. Malboeuf, S. J. Isaacs, N. H. Tran, and B. Kim, Thermal effects on reverse transcription: Improvement of accuracy and processivity in cDNA synthesis. *Biotechniques* **30,** 1074–1084 (2001).
61. S. Moran, R. X.-F. Ren, and E. T. Kool, A thymidine triphosphate shape analog lacking Watson-Crick pairing ability is replicated with high sequence selectivity. *Proc. Natl. Acad. Sci. U.S.A.* **94,** 10,506–10,511 (1997).
62. J. C. Morales and E. T. Kool, Efficient replication between non-hydrogen bonded nucleoside shape analogs. *Nature Struct. Biol.* **5,** 950–954 (1998).
63. H. Echols and M. F. Goodman, Fidelity mechanisms in DNA replication. *Annu. Rev. Biochem.* **60,** 477–511 (1991).
64. M. F. Goodman, Hydrogen bonding revisited: Geometric selection as a principal determinant of DNA replication fidelity. *Proc. Natl. Acad. Sci. U.S.A.* **94,** 10,493–10,495 (1997).
65. M. Boutabout, M. Wilhelm, and F.-X. Wilhelm, DNA synthesis fidelity by the reverse transcriptase of the yeast retrotransposon Ty1. *Nucleic Acids Res.* **29,** 2217–2222 (2001).
66. F. W. Perrino, B. D. Preston, L. L. Sandell, and L. A. Loeb, Extension of mismatched 3' termini of DNA is a major determinant of the infidelity of human immunodeficiency virus type 1 reverse transcriptase. *Proc. Natl. Acad. Sci. U.S.A.* **86,** 8343–8347 (1989).
67. G. A. Pulsinelli and H. M. Temin, High rate of mismatch extension during reverse transcription in a single round of retrovirus replication. *Proc. Natl. Acad. Sci. U.S.A.* **91,** 9490–9494 (1994).

68. S. Creighton, M.-H. Huang, H. Cai, N. Arnheim, and M. F. Goodman, Base mispair extension kinetics—Binding of avian myeloblastosis reverse transcriptase to matched and mismatched base pair termini. *J. Biol. Chem.* **267**, 2633–2639 (1992).
69. S. Zinnen, J.-C. Hsieh, and P. Modrich, Misincorporation and mispaired primer extension by human immunodeficiency virus reverse transcriptase. *J. Biol. Chem.* **269**, 24,195–24,202 (1994).
70. L. V. Mendelman, J. Petruska, and M. F. Goodman, Base mispair extension kinetics- Comparison of DNA polymerase α and reverse transcriptase. *J. Biol. Chem.* **265**, 2338–2346 (1990).
71. M. Bakhanashvili and A. Hizi, Fidelity of the RNA-dependent DNA synthesis exhibited by the reverse transcriptases of human immunodeficiency virus types 1 and 2 and of murine leukemia virus: Mispair extension frequencies. *Biochemistry* **31**, 9393–9398 (1992).
72. C. E. Cases-González, M. Gutiérrez-Rivas, and L. Menéndez-Arias, Coupling ribose selection to fidelity of DNA synthesis: The role of Tyr-115 of human immunodeficiency virus type 1 reverse transcriptase. *J. Biol. Chem.* **275**, 19,759–19,767 (2000).
73. K. A. Johnson, Conformational coupling in DNA polymerase fidelity. *Annu. Rev. Biochem.* **62**, 685–713 (1993).
74. W. M. Kati, K. A. Johnson, L. F. Jerva, and K. S. Anderson, Mechanism and fidelity of HIV reverse transcriptase. *J. Biol. Chem.* **267**, 25,988–25,997 (1992).
75. S. G. Kerr and K. S. Anderson, RNA dependent DNA replication fidelity of HIV-1 reverse transcriptase: Evidence of discrimination between DNA and RNA substrates. *Biochemistry* **36**, 14,056–14,063 (1997).
76. T. A. Kunkel and A. Soni, Mutagenesis by transient misalignment. *J. Biol. Chem.* **263**, 14,784–14,789.
77. K. Bebenek, J. Abbotts, J. D. Roberts, S. H. Wilson, and T. A. Kunkel, Specificity and mechanism of error-prone replication by human immunodeficiency virus-1 reverse transcriptase. *J. Biol. Chem.* **264**, 16,948–16,956 (1989).
78. J. C. Boyer, K. Bebenek, and T. A. Kunkel, Unequal human immunodeficiency virus type 1 reverse transcriptase error rates with RNA and DNA templates. *Proc. Natl. Acad. Sci. U.S.A.* **89**, 6919–6923 (1992).
79. K. A. Eckert and T. A. Kunkel, Fidelity of DNA synthesis catalyzed by human DNA polymerase α and HIV-1 reverse transcriptase: Effect of reaction pH. *Nucleic Acids Res.* **21**, 5212–5220 (1993).
80. W. C. Drosopoulos and V. R. Prasad, Increased misincorporation fidelity observed for nucleoside analog resistance mutations M184V and E89G in human immunodeficiency virus type 1 reverse transcriptase does not correlate with the overall error rate measured in vitro. *J. Virol.* **72**, 4224–4230 (1998).
81. L. F. Rezende, W. C. Drosopoulos, and V. R. Prasad, The influence of 3TC resistance mutation M184I on the fidelity and error specificity of human immunodeficiency virus type 1 reverse transcriptase. *Nucleic Acids Res.* **26**, 3066–3072 (1998).
82. J. Ji and L. A. Loeb, Fidelity of HIV-1 reverse transcriptase copying RNA in vitro. *Biochemistry* **31**, 954–958 (1992).
83. J. Ji and L. A. Loeb, Fidelity of HIV-1 reverse transcriptase copying a hypervariable region of the HIV-1 env gene. *Virology* **199**, 323–330 (1994).
84. P. L. Boyer and S. H. Hughes, Effects of amino acid substitutions at position 115 on the fidelity of human immunodeficiency virus type 1 reverse transcriptase. *J. Virol.* **74**, 6494–6500 (2000).
85. L. F. Rezende, K. Curr, T. Ueno, H. Mitsuya, and V. R. Prasad, The impact of multidideoxynucleoside resistance-conferring mutations in human immunodeficiency virus type 1 reverse transcriptase on polymerase fidelity and error specificity. *J. Virol.* **72**, 2890–2895 (1998).

86. W. A. Beard, S. J. Stahl, H.-R. Kim, K. Bebenek, A. Kumar, M.-P. Strub, S. P. Becerra, T. A. Kunkel, and S. H. Wilson, Structure/function studies of human immunodeficiency virus type 1 reverse transcriptase—Alanine scanning mutagenesis of an α-helix in the thumb subdomain. *J. Biol. Chem.* **269**, 28,091–28,097 (1994).
87. K. Bebenek, W. A. Beard, J. R. Casas-Finet, H.-R. Kim, T. A. Darden, S. H. Wilson, and T. A. Kunkel, Reduced frameshift fidelity and processivity of HIV-1 reverse transcriptase mutants containing alanine substitutions in helix H of the thumb subdomain. *J. Biol. Chem.* **270**, 19,516–19,523 (1995).
88. W. A. Beard, D. T. Minnick, C. L. Wade, R. Prasad, R. L. Won, A. Kumar, T. A. Kunkel, and S. H. Wilson, Role of the "helix clamp" in HIV-1 reverse transcriptase catalytic cycling as revealed by alanine-scanning mutagenesis. *J. Biol. Chem.* **271**, 12,213–12,220 (1996).
89. W. A. Beard, K. Bebenek, T. A. Darden, L. Li, R. Prasad, T. A. Kunkel, and S. H. Wilson, Vertical-scanning mutagenesis of a critical tryptophan in the minor groove binding track of HIV-1 reverse transcriptase—Molecular nature of polymerase-nucleic acid interactions. *J. Biol. Chem.* **273**, 30,435–30,442 (1998).
90. B. Kim, T. R. Hathaway, and L. A. Loeb, Fidelity of mutant HIV-1 reverse transcriptases: Interaction with the single-stranded template influences the accuracy of DNA synthesis. *Biochemistry* **37**, 5831–5839 (1998).
91. B. Kim, J. C. Ayran, S. G. Sagar, E. T. Adman, S. M. Fuller, N. H. Tran, and J. Horrigan, New human immunodeficiency virus type 1 reverse transcriptase (HIV-1 RT) mutants with increased fidelity of DNA synthesis—Accuracy, template binding, and processivity. *J. Biol. Chem.* **274**, 27,666–27,673 (1999).
92. K. Bebenek and T. A. Kunkel, Analyzing fidelity of DNA polymerases. *Methods Enzymol.* **262**, 217–232 (1995).
93. T. A. Kunkel and K. Bebenek, DNA replication fidelity. *Annu. Rev. Biochem.* **69**, 497–529 (2000).
94. K. Bebenek and T. A. Kunkel, Frameshift errors initiated by nucleotide misincorporation. *Proc. Natl. Acad. Sci. U.S.A.* **87**, 4946–4950 (1990).
95. K. Bebenek, J. Abbotts, S. H. Wilson, and T. A. Kunkel, Error-prone polymerization by HIV-1 reverse transcriptase—Contribution of template-primer misalignment, miscoding, and termination probability to mutational hot spots. *J. Biol. Chem.* **268**, 10,324–10,334 (1993).
96. J.-P. Vartanian, A. Meyerhans, B. Asjo, and S. Wain-Hobson, Selection, recombination, and G → A hypermutation of human immunodeficiency virus type 1 genomes. *J. Virol.* **65**, 1779–1788 (1991).
97. T. Kim, R. A. Mudry Jr., C. A. Rexrode II, and V. K. Pathak, Retroviral mutation rates and A-to-G hypermutations during different stages of retroviral replication. *J. Virol.* **70**, 7594–7602 (1996).
98. K. Bourara, S. Litvak and A. Araya, Generation of G-to-A and C-to-U changes in HIV-1 transcripts by RNA editing. *Science* **289**, 1564–1566 (2000).
99. J.-P. Vartanian, A. Meyerhans, M. Sala, and S. Wain-Hobson, G → A hypermutation of the human immunodeficiency virus type 1 genome: Evidence for dCTP pool imbalance during reverse transcription. *Proc. Natl. Acad. Sci. U.S.A.* **91**, 3092–3096 (1994).
100. J. E. Fitzgibbon, S. Mazar, and D. T. Dubin, A new type of G → A hypermutation affecting human immunodeficiency virus. *AIDS Res. Human Retrovir.* **9**, 833–838 (1993).
101. S. Wain-Hobson, P. Sonigo, M. Guyader, A. Gazit, and M. Henry, Erratic G → A hypermutation within a complete caprine arthritis-encephalitis virus (CAEV) provirus. *Virology* **209**, 297–303 (1995).
102. A. Hübner, M. Kruhoffer, F. Grosse, and G. Krauss, Fidelity of human immunodeficiency virus type I reverse transcriptase in copying natural RNA. *J. Mol. Biol.* **223**, 595–600 (1992).

103. M. Götte, M. Kameoka, N. McLellan, L. Cellai, and M. A. Wainberg, Analysis of efficiency and fidelity of HIV-1 (+)-strand DNA synthesis reveals a novel rate-limiting step during retroviral reverse transcription. *J. Biol. Chem.* **276**, 6711–6719 (2001).
104. J. M. Lanchy, C. Ehresmann, S. F. J. Le Grice, B. Ehresmann, and R. Marquet, Binding and kinetic properties of HIV-1 reverse transcriptase markedly differ during initiation and elongation of reverse transcription. *EMBO J.* **15**, 7178–7187 (1996).
105. B. B. Oude Essink and B. Berkhout, The fidelity of reverse transcription differs in reactions primed with RNA versus DNA primers. *J. Biomed. Sci.* **6**, 121–132 (1999).
106. J. A. Peliska and S. J. Benkovic, Mechanism of DNA strand transfer reactions catalyzed by HIV-1 reverse transcriptase. *Science* **258**, 1112–1118 (1992).
107. P. H. Patel and B. D. Preston, Marked infidelity of human immunodeficiency virus type 1 reverse transcriptase at RNA and DNA template ends. *Proc. Natl. Acad. Sci. U.S.A.* **91**, 549–553 (1994).
108. J. A. Peliska and S. J. Benkovic, Fidelity of *in vitro* DNA strand transfer reactions catalyzed by HIV-1 reverse transcriptase. *Biochemistry* **33**, 3890–3895 (1994).
109. D. Kulpa, R. Topping, and A. Telesnitsky, Determination of the site of first strand transfer during Moloney murine leukemia virus reverse transcription and identification of strand transfer-associated reverse transcriptase errors. *EMBO J.* **16**, 856–865 (1997).
110. C. Palaniappan, M. Wisniewski, W. Wu, P. J. Fay, and R. A. Bambara, Misincorporation of HIV-1 reverse transcriptase promotes recombination via strand transfer synthesis. *J. Biol. Chem.* **271**, 22,331–22,338 (1996).
111. G. J. Klarmann, C. A. Schauber, and B. D. Preston, Template-directed pausing of DNA synthesis by HIV-1 reverse transcriptase during polymerization of HIV-1 sequences *in vitro*. *J. Biol. Chem.* **268**, 9793–9802 (1993).
112. W. Wu, B. M. Blumberg, P. J. Fay, and R. A. Bambara, Strand transfer mediated by human immunodeficiency virus reverse transcriptase *in vitro* is promoted by pausing and results in misincorporation. *J. Biol. Chem.* **270**, 325–332 (1995).
113. L. Diaz and J. J. DeStefano, Strand transfer is enhanced by mismatched nucleotides at the 3' primer terminus: a possible link between HIV reverse transcriptase fidelity and recombination. *Nucleic Acids Res.* **24**, 3086–3092 (1996).
114. W. Wu, C. Palaniappan, R. A. Bambara, and P. J. Fay, Differences in mutagenesis during minus strand, plus strand and strand transfer (recombination) synthesis of the HIV-1 *nef* gene *in vitro*. *Nucleic Acids Res.* **24**, 1710–1718 (1996).
115. J. J. DeStefano, J. Ghosh, B. Prasad, and A. Raja, High fidelity of internal strand transfer catalyzed by human immunodeficiency virus reverse transcriptase. *J. Biol. Chem.* **273**, 1483–1489 (1998).
116. J. Zhang and H. M. Temin, Retrovirus recombination depends on the length of sequence identity and is not error prone. *J. Virol.* **68**, 2409–2414 (1994).
117. J. K. Pfeiffer, R. S. Topping, N.-H. Shin, and A. Telesnitsky, Altering the intracellular environment increases the frequency of tandem repeat deletion during Moloney murine leukemia virus reverse transcription. *J. Virol.* **73**, 8441–8447 (1999).
118. J. J. DeStefano, A. Raja, and J. V. Cristofaro, In vitro strand transfer from broken RNAs results in mismatch but not frameshift mutations. *Virology* **276**, 7–15 (2000).
119. J.-L. Darlix, M. Lapadat-Tapolsky, H. de Rocquigny, and B. P. Roques, First glimpses of structure-function relationships of the nucleocapsid protein of retroviruses. *J. Mol. Biol.* **254**, 523–537 (1995).
120. A. Rein, L. E. Henderson, and J. G. Levin, Nucleic-acid-chaperone activity of retroviral nucleocapsid proteins: significance for viral replication. *Trends Biochem. Sci.* **23**, 297–301 (1998).
121. J.-B. Rascle, D. Ficheux, and J.-L. Darlix, Possible roles of nucleocapsid protein of MoMuLV

in the specificity of proviral DNA synthesis and in the genetic variability of the virus. *J. Mol. Biol.* **280**, 215–225 (1998).
122. M. Dettenhofer, S. Cen, B. A. Carlson, L. Kleiman, and X.-F. Fu, Association of human immunodeficiency virus type 1 Vif with RNA and its role in reverse transcription. *J. Virol.* **74**, 8938–8945 (2000).
123. D. G. Vassylyev and K. Morikawa, Precluding uracil from DNA. *Structure* **4**, 1381–1385 (1996).
124. J. H. Elder, D. L. Lerner, C. S. Hasselkus-Light, D. J. Fontenot, E. Hunter, P. A. Luciw, R. C. Montelaro, and T. R. Phillips, Distinct subsets of retroviruses encode dUTPase. *J. Virol.* **66**, 1791–1794 (1992).
125. B. Köppe, L. Menéndez-Arias and S. Oroszlan, Expression and purification of the mouse mammary tumor virus *gag-pro* transframe protein p30 and characterization of its dUTPase activity. *J. Virol.* **68**, 2313–2319 (1994).
126. D. L. Lerner, P. C. Wagaman, T. R. Phillips, O. Prospero-García, S. J. Henriksen, H. S. Fox, F. E. Bloom, and J. H. Elder, Increased mutation frequency of feline immunodeficiency virus lacking a functional deoxyuridine-triphosphatase. *Proc. Natl. Acad. Sci. U.S.A.* **92**, 7480–7484 (1995).
127. M. Bouhamdan, S. Benichou, F. Rey, J.-M. Navarro, I. Agostini, B. Spire, J. Camonis, G. Slupphaug, R. Vigne, R. Benarous, and J. Sire, Human immunodeficiency virus type 1 Vpr protein binds to the uracil DNA glycosylase DNA repair enzyme. *J. Virol.* **70**, 697–704 (1996).
128. L. Selig, S. Benichou, M. E. Rogel, L. I. Wu, M. A. Wodicka, J. Sire, R. Benarous, and M. Emerman, Uracil DNA glycosylase specifically interacts with Vpr of both human immunodeficiency virus type 1 and simian immunodeficiency virus of sooty mangabeys, but binding does not correlate with cell cycle arrest. *J. Virol.* **71**, 4842–4846 (1997).
129. M. Bukrinsky and A. Adzhubei, Viral protein R of HIV-1. *Rev. Med. Virol.* **9**, 39–49 (1999).
130. L. A. Stark and R. T. Hay, Human immunodeficiency virus type 1 (HIV-1) viral protein R (Vpr) interacts with Lys-tRNA synthetase: Implications for priming of HIV-1 reverse transcription. *J. Virol.* **72**, 3037–3044 (1998).
131. F. Bachand, X.-J. Yao, M. Hrimech, N. Rougeau, and E. A. Cohen, Incorporation of Vpr into human immunodeficiency virus type 1 requires a direct interaction with the p6 domain of the p55 gag precursor. *J. Biol. Chem.* **274**, 9083–9091 (1999).
132. L. Selig, J.-C. Pages, V. Tanchou, S. Préveral, C. Berlioz-Torrent, L. X. Liu, L. Erdtman, J.-L. Darlix, R. Benarous, and S. Benichou, Interaction with the p6 domain of the Gag precursor mediates incorporation into virions of Vpr and Vpx proteins from primate lentiviruses. *J. Virol.* **73**, 592–600 (1999).
133. E. S. Withers-Ward, J. B. M. Jowett, S. A. Stewart, Y.-M. Xie, A. Garfinkel, Y. Shibagaki, S. A. Chow, N. Shah, F. Hanaoka, D. G. Sawitz, R. W. Armstrong, L. M. Souza, and I. S. Y. Chen, Human immunodeficiency virus type 1 Vpr interacts with HHR23A, a cellular protein implicated in nucleotide excision DNA repair. *J. Virol.* **71**, 9732–9742 (1997).
134. T. Dieckmann, E. S. Withers-Ward, M. A. Jarosinski, C.-F. Liu, I. S. Y. Chen, and J. Feigon, Structure of a human DNA repair protein UBA domain that interacts with HIV-1 Vpr. *Nature Struct. Biol.* **5**, 1042–1047 (1998).
135. L. M. Mansky, S. Preveral, E. Le Rouzic, L. C. Bernard, L. Selig, C. Depienne, R. Benarous, and S. Benichou, Interaction of human immunodeficiency virus type 1 Vpr with the HHR23A DNA repair protein does not correlate with multiple biological functions of Vpr. *Virology* **282**, 176–185 (2001).
136. J. B. Jowett, Y.-M. Xie, and I. S. Y. Chen, The presence of human immunodeficiency virus type 1 Vpr correlates with a decrease in the frequency of mutations in the plasmid shuttle vector. *J. Virol.* **73**, 7132–7137 (1999).

137. L. Mansky, The mutation rate of human immunodeficiency virus type 1 is influenced by the *vpr* gene. *Virology* **222**, 391–400 (1996).
138. L. M. Mansky, S. Preveral, L. Selig, R. Benarous, and S. Benichou, The interaction of Vpr with uracil DNA glycosylase modulates the human immunodeficiency virus type 1 in vivo mutation rate. *J. Virol.* **74**, 7039–7047 (2000).
139. J. Balzarini, Suppression of resistance to drugs targeted to human immunodeficiency virus reverse transcriptase by combination therapy. *Biochem. Pharmacol.* **58**, 1–27 (1999).
140. H. Jonckheere, J. Anné, and E. De Clercq, The HIV-1 reverse transcription (RT) process as target for RT inhibitors. *Med. Res. Rev.* **20**, 129–154 (2000).
141. E. Domingo, A. Mas, E. Yuste, N. Pariente, S. Sierra, M. Gutiérrez-Rivas, and L. Menéndez-Arias, Virus population dynamics, fitness variations and the control of viral disease: an update. *Prog. Drug Res.* **57**, 77–115 (2001).
142. Y. Xiong and T. H. Eickbush, Origin and evolution of retroelements based upon their reverse transcriptase sequences. *EMBO J.* **9**, 3353–3362 (1990).
143. T. M. Nakamura, G. B. Morin, K. B. Chapman, S. L. Weinrich, W. H. Andrews, J. Lingner, C. B. Harley, and T. R. Cech, Telomerase catalytic subunit homologs from fission yeast and human. *Science* **277**, 955–959 (1997).
144. B. Canard, K. Chowdhury, R. Sarfati, S. Doublié, and C. C. Richardson, The motif D loop of human immunodeficiency virus type 1 reverse transcriptase is critical for nucleoside 5′-triphosphate selectivity. *J. Biol. Chem.* **274**, 35,768–35,776 (1999).
145. P. L. Boyer, A. L. Ferris, and S. H. Hughes, Cassette mutagenesis of the reverse transcriptase of human immunodeficiency virus type 1. *J. Virol.* **66**, 1031–1039 (1992).
146. P. L. Boyer, A. L. Ferris, and S. H. Hughes, Mutational analysis of the fingers domain of human immunodeficiency virus type 1 reverse transcriptase. *J. Virol.* **66**, 7533–7537 (1992).
147. P. L. Boyer, A. L. Ferris, P. Clark, J. Whitmer, P. Frank, C. Tantillo, E. Arnold, and S. H. Hughes, Mutational analysis of the fingers and palm subdomains of human immunodeficiency virus type-1 (HIV-1) reverse transcriptase. *J. Mol. Biol.* **243**, 472–483 (1994).
148. S.-F. Chao, V. L. Chan, P. Juranka, A. H. Kaplan, R. Swanstrom, and C. A. Hutchison III, Mutational sensitivity patterns define critical residues in the palm subdomain of the reverse transcriptase of human immunodeficiency virus type 1. *Nucleic Acids Res.* **23**, 803–810 (1995).
149. B. Kim, T. R. Hathaway, and L. A. Loeb, Human immunodeficiency virus reverse transcriptase—Functional mutants obtained by random mutagenesis coupled with genetic selection in *Escherichia coli*. *J. Biol. Chem.* **271**, 4872–4878 (1996).
150. H.-Q. Gao, P. L. Boyer, E. Arnold, and S. H. Hughes, Effects of mutations in the polymerase domain on the polymerase, RNase H and strand transfer activities of human immunodeficiency virus type 1 reverse transcriptase. *J. Mol. Biol.* **277**, 559–572 (1998).
151. J. A. Wrobel, S.-F. Chao, M. J. Conrad, J. D. Merker, R. Swanstrom, G. J. Pielak, and C. A. Hutchison III, A genetic approach for identifying critical residues in the fingers and palm subdomains of HIV-1 reverse transcriptase. *Proc. Natl. Acad. Sci. U.S.A.* **95**, 638–645 (1998).
152. B. A. Larder, S. D. Kemp, and D. J. M. Purifoy, Infectious potential of human immunodeficiency virus type 1 reverse transcriptase mutants with altered inhibitor sensitivity. *Proc. Natl. Acad. Sci. U.S.A.* **86**, 4803–4807 (1989).
153. N. Kaushik, N. Rege, P. N. S. Yadav, S. G. Sarafianos, M. J. Modak, and V. N. Pandey, Biochemical analysis of catalytically crucial aspartate mutants of human immunodeficiency virus type 1 reverse transcriptase. *Biochemistry* **35**, 11,536–11,546 (1996).
154. D. Harris, N. Kaushik, P. K. Pandey, P. N. S. Yadav, and V. N. Pandey, Functional analysis of amino acid residues constituting the dNTP binding pocket of HIV-1 reverse transcriptase. *J. Biol. Chem.* **273**, 33,624–33,634 (1998).

155. P. L. Boyer, S. G. Sarafianos, E. Arnold, and S. H. Hughes, Analysis of mutations at positions 115 and 116 in the dNTP binding site of HIV-1 reverse transcriptase. *Proc. Natl. Acad. Sci. U.S.A.* **97**, 3056–3061 (2000).
156. A. M. Martín-Hernández, M. Gutiérrez-Rivas, E. Domingo, and L. Menéndez-Arias, Mispair extension fidelity of human immunodeficiency virus type 1 reverse transcriptases with amino acid substitutions affecting Tyr115. *Nucleic Acids Res.* **25**, 1383–1389 (1997).
157. A. L. Zakharenko, D. M. Kolpashchikov, S. N. Khodyreva, O. I. Lavrik, and L. Menéndez-Arias, Investigation of the dNTP-binding site of HIV-1 reverse transcriptase using photoreactive analogs of dNTP. *Biochemistry (Moscow)* **66**, 999–1007 (2001).
158. M. Ricchetti and H. Buc, A reiterative mode of DNA synthesis adopted by HIV-1 reverse transcriptase after a misincorporation, *Biochemistry* **35**, 14,970–14,983 (1996).
159. J.-P. Vartanian, M. Sala, M. Henry, S. Wain-Hobson, and A. Meyerhans, Manganese cations increase the mutation rate of human immunodeficiency virus type 1 ex vivo. *J. Gen. Virol.* **80**, 1983–1986 (1999).
160. H. Jonckheere, E. De Clercq, and J. Anné, Fidelity analysis of HIV-1 reverse transcriptase mutants with an altered amino-acid sequence at residues Leu74, Glu89, Tyr115, Tyr183 and Met184. *Eur. J. Biochem.* **267**, 2658–2665 (2000).
161. I. Olivares, V. Sánchez-Merino, M. A. Martínez, E. Domingo, C. López-Galíndez, and L. Menéndez-Arias, Second-site reversion of a human immunodeficiency virus type 1 reverse transcriptase mutant that restores enzyme function and replication capacity. *J. Virol.* **73**, 6293–6298 (1999).
162. E. K. Halvas, E. S. Svarovskaia, and V. K. Pathak, Role of murine leukemia virus reverse transcriptase deoxyribonucleoside triphosphate-binding site in retroviral replication and in vivo fidelity. *J. Virol.* **74**, 10,349–10,358 (2000).
163. G. Gao, M. Orlova, M. M. Georgiadis, W. A. Hendrickson, and S. P. Goff, Conferring RNA polymerase activity to a DNA polymerase: A single residue in reverse transcriptase controls substrate selection. *Proc. Natl. Acad. Sci. U.S.A.* **94**, 407–411 (1997).
164. M. Delarue, O. Poch, N. Tordo, D. Moras, and P. Argos, An attempt to unify the structure of polymerases. *Protein Eng.* **3**, 461–467 (1990).
165. D. K. Braithwaite and J. Ito, Compilation, alignment, and phylogenetic relationships of DNA polymerases. *Nucleic Acids Res.* **21**, 787–802 (1993).
166. Q. Dong, W. C. Copeland, and T. S.-F. Wang, Mutational studies of human DNA polymerase α. Identification of residues critical for deoxynucleotide binding and misinsertion fidelity of DNA synthesis. *J. Biol. Chem.* **268**, 24,163–24,174 (1993).
167. A. F. Gardner and W. E. Jack, Determinants of nucleotide sugar recognition in an archaeon DNA polymerase. *Nucleic Acids Res.* **27**, 2545–2553 (1999).
168. A. Bonnin, J. M. Lázaro, L. Blanco, and M. Salas, A single tyrosine prevents insertion of ribonucleotides in the eukaryotic-type φ29 DNA polymerase. *J. Mol. Biol.* **290**, 241–251 (1999).
169. M. Gutiérrez-Rivas, A. Ibáñez, M. A. Martínez, E. Domingo, and L. Menéndez-Arias, Mutational analysis of Phe-160 within the 'palm' subdomain of human immunodeficiency virus type 1 reverse transcriptase. *J. Mol. Biol.* **290**, 615–625 (1999).
170. N. K. T. Back, M. Nijhuis, W. Keulen, C. A. B. Boucher, B. B. Oude Essink, A. B. P. van Kuilenburg, A. H. van Gennip, and B. Berkhout, Reduced replication of 3TC-resistant HIV-1 variants in primary cells due to a processivity defect of the reverse transcriptase enzyme. *EMBO J.* **15**, 4040–4049 (1996).
171. W. Keulen, N. K. T. Back, A. van Wijk, C. A. B. Boucher, and B. Berkhout, Initial appearance of the 184Ile variant in lamivudine-treated patients is caused by the mutational bias of human immunodeficiency virus type 1 reverse transcriptase. *J. Virol.* **71**, 3346–3350 (1997).

172. S. G. Sarafianos, K. Das, A. D. Clark, Jr., J. Ding, P. L. Boyer, S. H. Hughes, and E. Arnold, Lamivudine (3TC) resistance in HIV-1 reverse transcriptase involves steric hindrance with β-branched amino acids. *Proc. Natl. Acad. Sci. U.S.A.* **96**, 10,027–10,032 (1999).
173. R. Krebs, U. Immendörfer, S. H. Thrall, B. M. Wöhrl, and R. S. Goody, Single-step kinetics of HIV-1 reverse transcriptase mutants responsible for virus resistance to nucleoside inhibitors zidovudine and 3TC. *Biochemistry* **36**, 10,292–10,300 (1997).
174. J. E. Wilson, A. Aulabaugh, B. Caligan, S. McPherson, J. K. Wakefield, S. Jablonski, C. D. Morrow, J. E. Reardon, and P. A. Furman, Human immunodeficiency virus type-1 reverse transcriptase—Contribution of Met-184 to binding of nucleoside 5'-triphosphate. *J. Biol. Chem.* **271**, 13,656–13,662 (1996).
175. J. Y. Feng and K. S. Anderson, Mechanistic studies examining the efficiency and fidelity of DNA synthesis by the 3TC-resistant mutant (184V) of HIV-1 reverse transcriptase. *Biochemistry* **38**, 9440–9448 (1999).
176. H.-Q. Gao, P. L. Boyer, S. G. Sarafianos, E. Arnold, and S. H. Hughes, The role of steric hindrance in 3TC resistance of human immunodeficiency virus type-1 reverse transcriptase. *J. Mol. Biol.* **300**, 403–418 (2000).
177. E. K. Halvas, E. S. Svarovskaia, E. O. Freed, and V. K. Pathak, Wild-type and YMDD mutant murine leukemia virus reverse transcriptases are resistant to 2',3'-dideoxy-3'-thiacytidine. *J. Virol.* **74**, 6669–6674 (2000).
178. M. A. Wainberg, W. C. Drosopoulos, H. Salomon, M. Hsu, G. Borkow, M. A. Parniak, Z. Gu, Q. Song, J. Manne, S. Islam, G. Castriota, and V. R. Prasad, Enhanced fidelity of 3TC-selected mutant HIV-1 reverse transcriptase. *Science* **271**, 1282–1285 (1996).
179. B. D. Preston, Reverse transcriptase fidelity and HIV-1 variation. *Science* **275**, 228–229 (1997).
180. W. Keulen, M. Nijhuis, R. Schuurman, B. Berkhout, and C. Boucher, Reverse transcriptase fidelity and HIV-1 variation. *Science* **275**, 229 (1997).
181. J. Balzarini, H. Pelemans, E. De Clercq, A. Karlsson, and J.-P. Kleim, Reverse transcriptase fidelity and HIV-1 variation. *Science* **275**, 229–230 (1997).
182. H. Jonckheere, M. Witvrouw, E. De Clercq, and J. Anné, Lamivudine resistance of HIV type 1 does not delay development of resistance to nonnucleoside HIV type 1-specific reverse transcriptase inhibitors as compared with wild-type HIV type 1. *AIDS Res. Human Retrovir.* **14**, 249–253 (1998).
183. W. Keulen, A. van Wijk, R. Schuurman, B. Berkhout, and C. A. B. Boucher, Increased polymerase fidelity of lamivudine-resistant HIV-1 variants does not limit their evolutionary potential. *AIDS* **13**, 1343–1349 (1999).
184. V. N. Pandey, N. Kaushik, N. Rege, S. G. Sarafianos, P. N. S. Yadav, and M. J. Modak, Role of methionine 184 of human immunodeficiency virus type-1 reverse transcriptase in the polymerase function and fidelity of DNA synthesis. *Biochemistry* **35**, 2168–2179 (1996).
185. M. E. Hamburgh, W. C. Drosopoulos, and V. R. Prasad, The influence of 3TC-resistance mutations E89G and M184V in the human immunodeficiency virus reverse transcriptase on mispair extension efficiency. *Nucleic Acids Res.* **26**, 4389–4394 (1998).
186. M. Hsu, P. Inouye, L. Rezende, N. Richard, Z. Li, V. R. Prasad, and M. A. Wainberg, Higher fidelity of RNA-dependent DNA mispair extension by M184V drug-resistant than wild-type reverse transcriptase of human immunodeficiency virus type 1. *Nucleic Acids Res.* **25**, 4532–4536 (1997).
187. B. B. Oude Essink, N. K. T. Back, and B. Berkhout, Increased polymerase fidelity of the 3TC-resistant variants of HIV-1 reverse transcriptase. *Nucleic Acids Res.* **25**, 3212–3217 (1997).

FIDELITY OF RETROVIRAL REVERSE TRANSCRIPTASES 143

188. D. Harris, P. N. S. Yadav, and V. N. Pandey, Loss of polymerase activity due to Tyr to Phe substitution in the YMDD motif of human immunodeficiency virus type-1 reverse transcriptase is compensated by Met to Val substitution within the same motif. *Biochemistry* **37,** 9630–9640 (1998).
189. L. M. Mansky and L. C. Bernard, 3′-azido-3′-deoxythymidine (AZT) and AZT-resistant reverse transcriptase can increase the in vivo mutation rate of human immunodeficiency virus type 1. *J. Virol.* **74,** 9532–9539 (2000).
190. M. Bakhanashvili, O. Avidan, and A. Hizi, Mutational studies of human immunodeficiency virus type 1 reverse transcriptase: the involvement of residues 183 and 184 in the fidelity of DNA synthesis. *FEBS Lett.* **391,** 257–262 (1996).
191. N. Kaushik, K. Chowdhury, V. N. Pandey, and M. J. Modak, Valine of the YVDD motif of Moloney murine leukemia virus reverse transcriptase: Role in the fidelity of DNA synthesis. *Biochemistry* **39,** 5155–5165 (2000).
192. E. K. Halvas, E. S. Svarovskaia, and V. K. Pathak, Development of an in vivo assay to identify structural determinants in murine leukemia virus reverse transcriptase important for fidelity. *J. Virol.* **74,** 312–319 (2000).
193. T. Ueno, T. Shirasaka, and H. Mitsuya, Enzymatic characterization of human immunodeficiency virus type 1 reverse transcriptase resistant to multiple 2′,3′-dideoxynucleoside 5′-triphosphates. *J. Biol. Chem.* **270,** 23,605–23,611 (1995).
194. T. Ueno and H. Mitsuya, Comparative enzymatic study of HIV-1 reverse transcriptase resistant to 2′,3′-dideoxynucleotide analogs using the single-nucleotide incorporation assay. *Biochemistry* **36,** 1092–1099 (1997).
195. N. Kaushik, T. T. Talele, P. K. Pandey, D. Harris, P. N. S. Yadav, and V. N. Pandey, Role of glutamine 151 of human immunodeficiency virus type-1 reverse transcriptase in substrate selection as assessed by site-directed mutagenesis. *Biochemistry* **39,** 2912–2920 (2000).
196. S. Tabor and C. C. Richardson, A single residue in DNA polymerases of the *Escherichia coli* DNA polymerase I family is critical for distinguishing between deoxy- and dideoxyribonucleotides. *Proc. Natl. Acad. Sci. U.S.A.* **92,** 6339–6343 (1995).
197. S. G. Sarafianos, V. N. Pandey, N. Kaushik, and M. J. Modak, Glutamine 151 participates in the substrate dNTP binding function of HIV-1 reverse transcriptase. *Biochemistry* **34,** 7207–7216 (1995).
198. N. Kaushik, D. Harris, N. Rege, M. J. Modak, P. N. S. Yadav, and V. N Pandey, Role of glutamine-151 of human immunodeficiency virus type-1 reverse transcriptase in RNA-directed DNA synthesis. *Biochemistry* **36,** 14,430–14,438 (1997).
199. K. K. Weiss, S. J. Isaacs, N. H. Tran, E. T. Adman, and B. Kim, Molecular architecture of the mutagenic active site of human immunodeficiency virus type 1 reverse transcriptase: Roles of the $\beta 8$-αE loop in fidelity, processivity and substrate interactions. *Biochemistry* **39,** 10,684–10,694 (2000).
200. T. Shirasaka, M. F. Kavlick, T. Ueno, W.-Y. Gao, E. Kojima, M. L. Alcaide, S. Chokekijchai, B. M. Roy, E. Arnold, R. Yarchoan, and H. Mitsuya, Emergence of human immunodeficiency virus type 1 variants with resistance to multiple dideoxynucleosides in patients receiving therapy with dideoxynucleosides. *Proc. Natl. Acad. Sci. U.S.A.* **92,** 2398–2402 (1995).
201. K. Singh, N. Kaushik, J. Jin, M. Madhusudanan, and M. J. Modak, Role of Q190 of MuLV RT in ddNTP resistance and fidelity of DNA synthesis: a molecular model of interactions with substrates. *Protein Eng.* **13,** 635–643 (2000).
202. Z. Gu, Q. Gao, H. Fang, H. Salomon, M. A. Parniak, E. Goldberg, J. Cameron, and M. A. Wainberg, Identification of a mutation at codon 65 in the IKKK motif of reverse transcriptase that encodes human immunodeficiency virus resistance to 2′,3′-dideoxycytidine and 2′,3′-dideoxy-3′-thiacytidine. *Antimicrob. Agents Chemother.* **38,** 275–281 (1994).

203. Z. Gu, R. S. Fletcher, E. J. Arts, M. A. Wainberg, and M. A. Parniak, The K65R mutant reverse transcriptase of HIV-1 cross-resistant to 2′,3′-dideoxycytidine, 2′,3′-dideoxy-3′-thiacytidine, and 2′,3′-dideoxyinosine shows reduced sensitivity to specific dideoxynucleoside triphosphate inhibitors in vitro. *J. Biol. Chem.* **269**, 28,118–28,122 (1994).

204. D. Zhang, A. M. Caliendo, J. J. Eron, K. M. DeVore, J. C. Kaplan, M. S. Hirsch, and R. T. D'Aquila, Resistance to 2′,3′-dideoxycytidine conferred by a mutation in codon 65 of the human immunodeficiency virus type 1 reverse transcriptase. *Antimicrob. Agents Chemother.* **38**, 282–287 (1994).

205. M. A. Winters, R. W. Shafer, R. A. Jellinger, G. Mamtora, T. Gingeras, and T. C. Merigan, Human immunodeficiency virus type 1 reverse transcriptase genotype and drug susceptibility changes in infected individuals receiving dideoxyinosine monotherapy for 1 to 2 years. *Antimicrob. Agents Chemother.* **41**, 757–762 (1997).

206. Z. Gu, E. J. Arts, M. A. Parniak, and M. A. Wainberg, Mutated K65R recombinant reverse transcriptase of human immunodeficiency virus type 1 shows diminished chain termination in the presence of 2′,3′-dideoxycytidine 5′-triphosphate and other drugs. *Proc. Natl. Acad. Sci. U.S.A.* **92**, 2760–2764 (1995).

207. D. Arion, G. Borkow, Z. Gu, M. A. Wainberg, and M. A. Parniak, The K65R mutation confers increased DNA polymerase processivity to HIV-1 reverse transcriptase. *J. Biol. Chem.* **271**, 19,860–19,864 (1996).

208. N. Sluis-Cremer, D. Arion, N. Kaushik, and H. Lim, Mutational analysis of Lys65 of HIV-1 reverse transcriptase. *Biochem. J.* **348**, 77–82 (2000).

209. F. S. Shah, K. A. Curr, M. E. Hamburgh, M. Parniak, H. Mitsuya, J. G. Arnez, and V. R. Prasad, Differential influence of nucleoside analog-resistance mutations K65R and L74V on the overall mutation rate and error specificity of human immunodeficiency virus type 1 reverse transcriptase. *J. Biol. Chem.* **275**, 27,037–27,044 (2000).

210. S. G. Sarafianos, V. N. Pandey, N. Kaushik, and M. J. Modak, Site-directed mutagenesis of arginine 72 of HIV-1 reverse transcriptase—Catalytic role and inhibitor sensitivity. *J. Biol. Chem.* **270**, 19,729–19,735 (1995).

211. K. Chowdhury, N. Kaushik, V. N. Pandey, and M. J. Modak, Elucidation of the role of Arg 110 of murine leukemia virus reverse transcriptase in the catalytic mechanism: Biochemical characterization of its mutant enzymes. *Biochemistry* **35**, 16,610–16,620 (1996).

212. D. A. Lewis, K. Bebenek, W. A. Beard, S. H. Wilson, and T. A. Kunkel, Uniquely altered DNA replication fidelity conferred by an amino acid change in the nucleotide binding pocket of human immunodeficiency virus type 1 reverse transcriptase. *J. Biol. Chem.* **274**, 32,924–32,930 (1999).

213. M. H. St. Clair, J. L. Martin, G. Tudor-Williams, M. C. Bach, C. L. Vavro, D. M. King, P. Kellam, S. D. Kemp, and B. A. Larder, Resistance to ddI and sensitivity to AZT induced by a mutation in HIV-1 reverse transcriptase. *Science* **253**, 1557–1559 (1991).

214. P. L. Sharma and C. S. Crumpacker, Decreased processivity of human immunodeficiency virus type 1 reverse transcriptase (RT) containing didanosine-selected mutation Leu74Val: A comparative analysis of RT variants Leu74Val and lamivudine-selected Met184Val. *J. Virol.* **73**, 8448–8456 (1999).

215. T. Rubinek, M. Bakhanashvili, R. Taube, O. Avidan, and A. Hizi, The fidelity of 3′ misinsertion and mispair extension during RNA synthesis exhibited by two drug-resistant mutants of the reverse transcriptase of human immunodeficiency virus type 1 with Leu74 → Val and Glu89 → Gly. *Eur. J. Biochem.* **247**, 238–247 (1997).

216. R. Taube, O. Avidan, and A. Hizi, The fidelity of misinsertion and mispair extension throughout DNA synthesis exhibited by mutants of the reverse transcriptase of human immunodeficiency virus type 2 resistant to nucleoside analogs. *Eur. J. Biochem.* **250**, 106–114 (1997).

217. V. R. Prasad, I. Lowy, T. de los Santos, L. Chiang, and S. P. Goff, Isolation and characterization of a dideoxyguanosine triphosphate-resistant HIV-1 reverse transcriptase expressed in bacteria. *Proc. Natl. Acad. Sci. U.S.A.* **88**, 11,363–11,367 (1991).
218. O. Avidan and A. Hizi, The processivity of DNA synthesis exhibited by drug-resistant variants of human immunodeficiency virus type-1 reverse transcriptase. *Nucleic Acids Res.* **26**, 1713–1717 (1998).
219. Y. Quan, P. Inouye, C. Liang, L. Rong, M. Götte, and M. A. Wainberg, Dominance of the E89G substitution in HIV-1 reverse transcriptase in regard to increased polymerase processivity and patterns of pausing. *J. Biol. Chem.* **273**, 21,918–21,925 (1998).
220. W. C. Drosopoulos and V. R. Prasad, Increased polymerase fidelity of E89G, a nucleoside analog-resistant variant of human immunodeficiency virus type 1 reverse transcriptase. *J. Virol.* **70**, 4834–4838 (1996).
221. N. Kaushik, K. Singh, I. Alluru, and M. J. Modak, Tyrosine 222, a member of the YXDD motif of MuLV RT, is catalytically essential and is a major component of the fidelity center. *Biochemistry* **38**, 2617–2627 (1999).
222. P. S. Jacques, B. M. Wöhrl, M. Ottmann, J.-L. Darlix, and S. F. J. Le Grice, Mutating the "primer grip" of p66 HIV-1 reverse transcriptase implicates tryptophan-229 in template-primer utilization. *J. Biol. Chem.* **269**, 26,472–26,478 (1994).
223. M. Ghosh, P. S. Jacques, D. W. Rodgers, M. Ottman, J.-L. Darlix, and S. F. J. Le Grice, Alterations to the primer grip of p66 HIV-1 reverse transcriptase and their consequences for template-primer utilization. *Biochemistry* **35**, 8553–8562 (1996).
224. M. Ghosh, J. Williams, M. D. Powell, J. G. Levin, and S. F. J. Le Grice, Mutating a conserved motif of the HIV-1 reverse transcriptase palm subdomain alters primer utilization. *Biochemistry* **36**, 5758–5768 (1997).
225. C. Palaniappan, M. Wisniewski, P. S. Jacques, S. F. J. Le Grice, P. J. Fay, and R. A. Bambara, Mutations within the primer grip region of HIV-1 reverse transcriptase result in loss of RNase H function. *J. Biol. Chem.* **272**, 11,157–11,164 (1997).
226. M. D. Powell, M. Ghosh, P. S. Jacques, K. J. Howard, S. F. J. Le Grice, and J. G. Levin, Alanine-scanning mutations in the "primer grip" of p66 HIV-1 reverse transcriptase result in selective loss of RNA priming activity. *J. Biol. Chem.* **272**, 13,262–13,269 (1997).
227. B. M. Wöhrl, R. Krebs, S. H. Thrall, S. F. J. Le Grice, A. J. Scheidig, and R. S. Goody, Kinetic analysis of four HIV-1 reverse transcriptase enzymes mutated in the primer grip region of p66—Implications for DNA synthesis and dimerization. *J. Biol. Chem.* **272**, 17,581–17,587 (1997).
228. M. Wisniewski, C. Palaniappan, Z. Fu, S. F. J. Le Grice, P. Fay, and R. A. Bambara, Mutations in the primer grip region of HIV reverse transcriptase can increase replication fidelity. *J. Biol. Chem.* **274**, 28,175–28,184 (1999).
229. Q. Yu, M. Ottmann, C. Pechoux, S. Le Grice, and J.-L. Darlix, Mutations in the primer grip of human immunodeficiency virus type 1 reverse transcriptase impair proviral DNA synthesis and virion maturation. *J. Virol.* **72**, 7676–7680 (1998).
229a. M. Gutiérrez-Rivas and L. Menéndez-Arias, A mutation in the primer grip region of HIV-1 reverse transcriptase that confers reduced fidelity of DNA synthesis. *Nucleic Acids Res.* **29**, 4963–4972 (2001).
230. K. Bebenek, W. A. Beard, T. A. Darden, L. Li, R. Prasad, B. A. Luxon, D. G. Gorenstein, S. H. Wilson, and T. A. Kunkel, A minor groove binding track in reverse transcriptase. *Nature Struct. Biol.* **4**, 194–197 (1997).
231. M. D. Powell, W. A. Beard, K. Bebenek, K. J. Howard, S. F. J. Le Grice, T. A. Darden, T. A. Kunkel, S. H. Wilson, and J. G. Levin, Residues in the αH and αI helices of the HIV-1 reverse transcriptase thumb subdomain required for the specificity of RNase H-catalyzed removal of the polypurine tract primer. *J. Biol. Chem.* **274**, 19,885–19,893 (1999).

232. M. Bakhanashvili and A. Hizi, A possible role for cysteine residues in the fidelity of DNA synthesis exhibited by the reverse transcriptases of human immunodeficiency viruses type 1 and type 2. *FEBS Lett.* **304,** 289–293 (1992).
233. S. Loya, M. Bakhanashvili, R. Tal, S. H. Hughes, P. L. Boyer, and A. Hizi, Enzymatic properties of two mutants of reverse transcriptase of human immunodeficiency virus type 1 (Tyrosine 181→Isoleucine and Tyrosine 188→Leucine) resistant to nonnucleoside inhibitors. *AIDS Res. Human Retrovir.* **10,** 939–946 (1994).
234. Y. Kew, L. R. Olsen, A. J. Japour, and V. R. Prasad, Insertions into the $\beta 3$-$\beta 4$ hairpin loop of HIV-1 reverse transcriptase reveal a role for fingers subdomain in processive polymerization. *J. Biol. Chem.* **273,** 7529–7537 (1998).
235. L. F. Rezende, Y. Kew, and V. R. Prasad, The effect of increased processivity on overall fidelity of human immunodeficiency virus type 1 reverse transcriptase. *J. Biomed. Sci.* **8,** 197–205 (2001).
236. M. Ghosh, K. J. Howard, C. E. Cameron, S. J. Benkovic, S. H. Hughes, and S. F. J. Le Grice, Truncating α-helix E' of p66 human immunodeficiency virus reverse transcriptase modulates RNase H function and impairs DNA strand transfer. *J. Biol. Chem.* **270,** 7068–7076 (1995).
237. B. A. Larder, Interactions between drug resistance mutations in human immunodeficiency virus type 1 reverse transcriptase. *J. Gen. Virol.* **75,** 951–957 (1994).
238. P. R. Meyer, S. E. Matsuura, A. M. Mian, A. G. So, and W. A. Scott, A mechanism of AZT resistance: an increase in nucleotide-dependent primer unblocking by mutant HIV-1 reverse transcriptase. *Mol. Cell* **4,** 35–43 (1999).
239. N. Sluis-Cremer, D. Arion, and M. A. Parniak, Molecular mechanisms of HIV-1 resistance to nucleoside reverse transcriptase inhibitors (NRTIs). *CMLS, Cell. Mol. Life Sci.* **57,** 1408–1422 (2000).
240. A. Mas, M. Parera, C. Briones, V. Soriano, M. A. Martínez, E. Domingo, and L. Menéndez-Arias, Role of a dipeptide insertion between codons 69 and 70 of HIV-1 reverse transcriptase in the mechanism of AZT resistance. *EMBO J.* **19,** 5752–5761 (2000).
241. S. F. Lacey, J. E. Reardon, E. S. Furfine, T. A. Kunkel, K. Bebenek, K. A. Eckert, S. D. Kemp, and B. A. Larder, Biochemical studies on the reverse transcriptase and RNase H activities from human immunodeficiency virus strains resistant to 3'-azido-3'-deoxythymidine. *J. Biol. Chem.* **267,** 15,789–15,794 (1992).
242. J. G. Julias, T. Kim, G. Arnold, and V. K. Pathak, The antiretrovirus drug 3'-azido-3'-deoxythymidine increases the retrovirus mutation rate. *J. Virol.* **71,** 4254–4263 (1997).
243. J. M. Coffin, HIV population dynamics in vivo: implications for genetic variation, pathogenesis and therapy. *Science* **267,** 483–489 (1995).
244. T. L. Diamond, J. Kimata, and B. Kim, Identification of a simian immunodeficiency virus reverse transcriptase variant with enhanced replicational fidelity in the late stage of viral infection. *J. Biol. Chem.* **276,** 23,624–23,631 (2001).
245. J. G. Julias and V. K. Pathak, Deoxyribonucleoside triphosphate pool imbalances in vivo are associated with an increased retroviral mutation rate. *J. Virol.* **72,** 7941–7949 (1998).
246. W. A. O'Brien, A. Namazi, H. Kalhor, S.-H. Mao, J. A. Zack, and I. S. Y. Chen, Kinetics of human immunodeficiency virus type 1 reverse transcriptase in blood mononuclear phagocytes are slowed by limitations of nucleotide precursors. *J. Virol.* **68,** 1258–1263 (1994).
247. N. K. T. Back and B. Berkhout, Limiting deoxynucleoside triphosphate concentrations emphasize the processivity defect of lamivudine-resistant variants of human immunodeficiency virus type 1 reverse transcriptase. *Antimicrob. Agents Chemother.* **41,** 2484–2491 (1997).
248. M. A. Martínez, J.-P. Vartanian, and S. Wain-Hobson, Hypermutagenesis of RNA using human immunodeficiency virus type 1 reverse transcriptase and biased dNTP concentrations. *Proc. Natl. Acad. Sci. U.S.A.* **91,** 11,787–11,791 (1994).

249. J.-P. Vartanian, U. Plikat, M. Henry, R. Mahieux, L. Guillemot, A. Meyerhans, and S. Wain-Hobson, HIV genetic variation is directed and restricted by DNA precursor availability. *J. Mol. Biol.* **270,** 139–151 (1997).
250. M. A. Martínez, M. Sala, J.-P. Vartanian, and S. Wain-Hobson, Reverse transcriptase and substrate dependence of the RNA hypermutagenesis reaction. *Nucleic Acids Res.* **23,** 2573–2578 (1995).
251. J. Balzarini, M.-J. Camarasa, M.-J. Pérez-Pérez, A. San-Félix, S. Velázquez, C.-F. Perno, E. De Clercq, J. N. Anderson, and A. Karlsson, Exploitation of the low fidelity of human immunodeficiency virus type 1 (HIV-1) reverse transcriptase and the nucleotide composition bias in the HIV-1 genome to alter the drug resistance development of HIV. *J. Virol.* **75,** 5772–5777 (2001).
252. D. G. Johns and W.-Y. Gao, Selective depletion of DNA precursors: An evolving strategy for potentiation of dideoxynucleoside activity against human immunodeficiency virus. *Biochem. Pharmacol.* **55,** 1551–1556 (1998).
253. J. Balzarini, Effect of antimetabolite drugs of nucleotide metabolism on the anti-human immunodeficiency virus activity of nucleoside reverse transcriptase inhibitors. *Pharmacol. Therapeut.* **87,** 175–187 (2000).
254. J. K. Pfeiffer, M. M. Georgiadis, and A. Telestnitsky, Structure-based Moloney murine leukemia virus reverse transcriptase mutants with altered intracellular direct-repeat deletion frequencies. *J. Virol.* **74,** 9629–9636 (2000).
255. R. A. LaCasse, K. M. Remington, and T. W. North, The mutation frequency of feline immunodeficiency virus enhanced by 3′-azido-3′-deoxythymidine. *J. Acquir. Immune Def. Syndr. Humar Retrovir.* **12,** 26–32 (1996).
256. J. Ji, J.-S. Hoffmann, and L. A. Loeb, Mutagenicity and pausing of HIV reverse transcriptase during HIV plus-strand DNA synthesis. *Nucleic Acids Res.* **22,** 47–52 (1994).
257. V. K. Pathak and H. M. Temin, 5-Azacytidine and RNA secondary structure increase the retrovirus mutation rate. *J. Virol.* **66,** 3093–3100 (1992).
258. J. E. Mott, J. Van Arsdell, and T. Platt, Targeted mutagenesis *in vitro*: *lac* repressor mutations generated using AMV reverse transcriptase and dBrUTP. *Nucleic Acids Res.* **12,** 4139–4152 (1984).
259. A. S. Kamath-Loeb, A. Hizi, H. Kasai, and L. A. Loeb, Incorporation of the guanosine triphosphate analogs 8-oxo-dGTP and 8-NH$_2$-dGTP by reverse transcriptases and mammalian DNA polymerases. *J. Biol. Chem.* **272,** 5892–5898 (1997).
260. A. Hizi, A. S. Kamath-Loeb, K. D. Rose, and L. A. Loeb, Mutagenesis by human immunodeficiency virus reverse transcriptase: Incorporation of O^6-methyldeoxyguanosine triphosphate. *Mutation Res.* **374,** 41–50 (1997).
261. L. A. Loeb, J. M. Essigmann, F. Kazazi, J. Zhang, K. D. Rose, and J. I. Mullins, Lethal mutagenesis of HIV with mutagenic nucleoside analogs. *Proc. Natl. Acad. Sci. U.S.A.* **96,** 1492–1497 (1999).
262. L. A. Loeb and J. I. Mullins, Lethal mutagenesis of HIV by mutagenic ribonucleoside analogs. *AIDS Res. Human Retrovir.* **16,** 1–3 (2000).

Muc4/Sialomucin Complex, the Intramembrane ErbB2 Ligand, in Cancer and Epithelia: To Protect and To Survive

KERMIT L. CARRAWAY,*
AYMEE PEREZ,* NEBILA IDRIS,*
SCOTT JEPSON,*
MARIA ARANGO,*
MASANOBU KOMATSU,*
BUSHRA HAQ,*
SHARI A. PRICE-SCHIAVI,*
JIN ZHANG,* AND CORALIE A.
CAROTHERS CARRAWAY[†]

*Department of Cell Biology and Anatomy
University of Miami School of Medicine
Miami, Florida 33101
[†]Department of Biochemistry and
  Molecular Biology
University of Miami School of Medicine
Miami, Florida 33101

I. Membrane Mucins	150
II. Muc4/SMC Structure and Functions	153
III. Muc4/SMC Contributions to Tumor Progression	160
A. Metastasis	160
B. Primary Tumor Growth and Repression of Apoptosis	161
IV. Muc4/SMC in Simple Epithelia	163
A. Airway: A Proposed Role in Mucociliary Transport	163
B. Uterus and Blastocyst Implantation	164
C. Oviduct and ErbB2 Localization	167
V. Muc4/SMC in Glandular Secretory Epithelia	170
A. Mammary Acinar Cells, Milk, and Breast Cancer	170
B. Salivary: Cell-Type-Specific Expression	175
C. Lacrimal: Acinar-Specific Expression	176
VI. Muc4/SMC in Stratified Epithelia	177
A. Vagina/Cervix: Differentiation, Localization, and Cell Survival	177
B. Cornea and Conjunctiva: Ocular Protection	177
VII. Conclusions and Future Directions	179
References	180

The membrane mucin Muc4, also called sialomucin complex (SMC), is a heterodimeric complex of two subunits, ASGP-1 and ASGP-2, derived from a single gene. It is produced by multiple epithelia in both membrane and soluble forms and serves as a protective agent for the epithelia. The membrane form of Muc4 acts as a steric barrier to the apical cell surface of epithelial or tumor cells. An important example is the uterus of the rat, in which Muc4 expression is downregulated for blastocyst implantation. The soluble form facilitates the protection and lubrication of epithelia by mucous gels composed of gel-forming mucins, as in the airway, where Muc4 is proposed to participate in mucociliary transport as a constituent of the periciliary fluid. The soluble form is also found in body fluids, such as milk, tears, and saliva. The transmembrane subunit ASGP-2 acts as an intramembrane ligand and activator for the receptor tyrosine kinase ErbB2. Formation of this ligand–receptor complex is proposed to repress apopotosis in epithelial and cancer cells in which the ligand–receptor complex is formed, providing a second type of cell protective mechanism. Muc4 expression is regulated in epithelial tissues in a cell- and tissue-specific manner during epithelial differentiation. In stratified epithelia, it is predominantly in the most superficial, differentiated layers, often coincident with ErbB2. Dysregulation of Muc4 expression may contribute to cell and tissue dysfunction, such as the proposed contribution of Muc4 to mammary tumor progression. These observations clearly show that Muc4 has multiple roles in epithelia, which may provide insights into aberrant behaviors of these tissues and their derivative carcinomas. © 2002, Elsevier Science (USA).

I. Membrane Mucins

Mucins are large, highly glycosylated proteins found in association with most epithelia (1). They are characterized by two features: a high content of serine and threonine residues, many of which are present in tandem repeat domains of the protein, and O-linked oligosaccharides attached to those serine and threonine residues (2). The attachment of the first sugar of the oligosaccharides results in a restriction of rotation of the polypeptide chain, thus causing the proteins to have extended and rigid structures. Three classes of mucins can be defined, gel-forming, membrane, and soluble, based on their sequence characteristics. Gel-forming mucins (MUCs 2, 5AC, 5B, and 6) are large and contain von Willebrand factor domains which are disulfide-crosslinked to form gels (3). Membrane mucins (MUCs 1, 3, 4, 12, and 13) have transmembrane domains which bind them tightly to cell surfaces (4). The soluble mucins (MUC7) are small compared to the others. The term mucin originally referred to the glycoproteins of the mucus overlying mucosal surfaces, primarily gel-forming mucins (5). However, studies on rodent mammary ascites tumor cells demonstrated the presence of mucin-like molecules associated with the cell surfaces of the tumor cells (6, 7), characterized by a strong association with the plasma membrane (8). Related mucin-like components were also found in physiological fluids, such as milk (9), in both membrane and soluble forms (4), thus lending legitimacy to their description as mucins. The advent of molecular cloning resulted in the definition

TABLE I
Human Membrane Mucins

Mucin	Class	Cloned from	Expressed in	Special domains	Chromosome
MUC1	Membrane	Breast tumor	Most epithelia	cytoplasmic	1q21-24
MUC3	Membrane	Intestine	Intestine, airway	EGF[a]	7q22
MUC4	Membrane	Airway	Most epithelia	EGF	3q29
MUC11	Membrane	Intestine	Intestine	EGF	7q22
MUC12	Membrane	Intestine	Intestine	EGF	7q22
MUC13	Membrane	Database	Intestine, trachea	EGF	3q13.3

[a]EGF, epidermal growth factor.

of a mucin gene family (2), which encompassed both the membrane mucins and the mucus gel-derived mucins. At this writing, there are 16 designated members of the human mucin gene family (*MUC*), though the characterization of some of these is still incomplete. The fact that mucin repeat domains are highly variable and are not highly conserved across species makes the identification of mucin gene family members somewhat difficult. By convention, human mucin proteins are designated as MUCs; those of other species are Mucs.

Membrane mucins are defined by the presence of a transmembrane domain which associates them with membranes (4). Several membrane mucins have been described by molecular cloning (Table I), but only two of these have been well characterized biochemically, MUC1 and rat Muc4 (also known as sialomucin complex). These two mucins exhibit both similar and disparate properties. Both of these membrane mucins are present at the cell surface as heterodimers (Fig. 1). The heterodimers are formed from high M_r precursors, which are cleaved proteolytically early in the transit of the mucins to the cell surface (10, 11). The resulting two subunits can be defined as mucin and transmembrane. The role of this heterodimeric structure is unclear, but it may be necessary to permit glycosylation of the polypeptide by glycosyltransferases in intracellular membranes. Functionally, both MUC1 and Muc4/SMC have been proposed to provide steric protection of cell surfaces by the rigid and extended mucin subunit (12, 13). MUC1 and Muc4/sialomucin complex (SMC) have also been implicated in cellular signaling (14, 15), though by different mechanisms. MUC1 cytoplasmic domain can be phosphorylated on tyrosine (14, 16); thus, it is potentially capable of acting as a docking site at the membrane for adaptor proteins which initiate signaling pathways (17). In contrast, Muc4/SMC has an EGF-like domain in its extracellular region (18) which has been shown to act as a novel intramembrane ligand for the receptor tyrosine kinase ErbB2 (19), triggering its specific phosphorylation (Jepson *et al.*, submitted for publication).

Both MUC1 and Muc4/SMC are found in soluble as well as in membrane forms (20, 21). Thus, the term membrane mucin is somewhat of a misnomer (4).

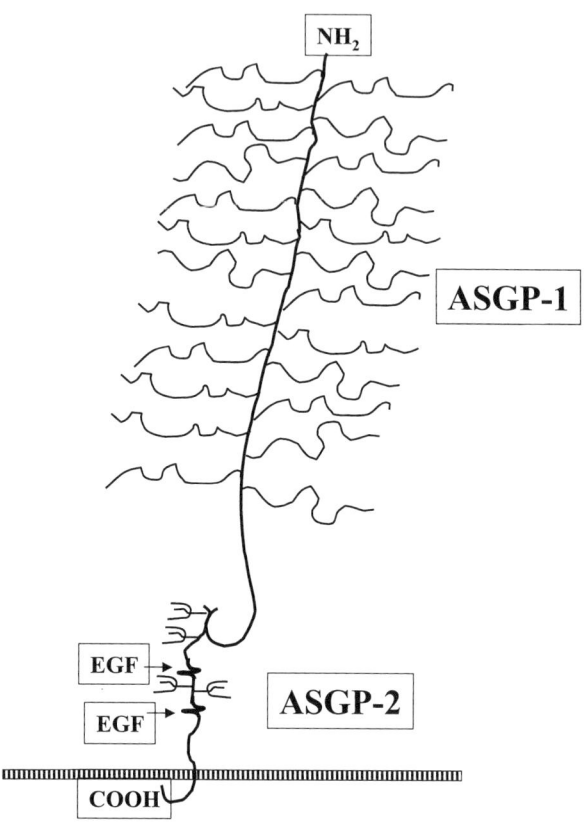

FIG. 1. Model for Muc4/SMC exhibiting the heterodimeric structure observed for both MUC1 and MUC4. Reprinted from K. L. Carraway, S. A. Price-Schiavi, M. Komatsu, S. Jepson, A. Perez, and C. A. C. Carraway, *J. Mammary Gland Biol. Neoplasia*, **6**, 323–337, with permission by Kluwer Academic/Plenum Publisher.

The soluble forms can be produced by either alternative splicing (22, 23) or proteolysis (20). Depending on the site of splicing or proteolysis, the soluble form may contain only mucin subunit or mucin subunit plus part of the transmembrane subunit. The function of the soluble forms is unknown, but their presence in physiological fluids, such as milk (21), saliva (24), and tears (25, 26), has suggested that they may play a role in protection from microbes (27). Consistent with that role, some mucins have been shown to bind bacteria and viruses (28).

Though less well characterized, three other membrane mucins appear to belong to the subfamily including MUC4: MUCs 3, 12, and 13 (Table I). Notably,

```
hEGF          -NSDSECPLSHDG-YCLHDGVCMYIEALDKY-ACNCVVGYIG---ERC
hMUC4-EGFI    -CQNQSCPVN----YCYNQGHCYISQTLGCQPMCTCPPAHTD--SRC
hMUC3-EGFI    ----CACLPGFSGDRCQLQTRCQNGGQWDGLK-CQCPSTHYG---SC
hMUC12-EGFI   --SFTSTIVSTESLETLAPGLCQEGQIWNGKQ-CVCPQSYVGY--QC
hMUC13-EGFI   ------CQDD----PCADNSLCVKLHNTSF---CLCLECHYYNSSTC

hEGF          -NSDSECPLSHDG--YC-LHDGVCMY-IEALDKYACNCVVSYI-----GERC
hMUC4-EGFII   GFTCVSPCSRG-----YC-DHGGQCQHLPSCPR---CSCVSFSIYTAW-GEHC
hMUC3-EGFII   RLRCVTKCTSGVDNAIDC--HQGCVLETSCPT---CRCYSTDTHWES-GPRC
hMUC12-EGFII  KLACVNKCTKGTKSQMNC--NLGTCQLQRSGPR---CLCPNTNTHWHW-GETC
hMUC13-EGFII  NYDLTLRCDYYG-----CNQTADDCIN---GLA---CDCKSDLQRPNPQSPFC

hEGF          -NSDSECPLSHDGYC-LHDGVCMY-IEALDKYACNCVVGTIGE---RCQ
hMUC13-EGFIII ------C-----DACNAQHKQCLIKKSG-GAPACACVPGYQEDANGNCQ
```

FIG. 2. Sequence comparisons of EGF-like domains of membrane mucins.

all of these contain EGF-like domains, as indicated by sequence alignments, particularly those of cysteine residues (Fig. 2). Since EGF1 of Muc4/SMC has been shown to act as a ligand for ErbB2, it will be interesting to discover if any of the other MUCs also exhibit receptor associations.

II. Muc4/SMC Structure and Functions

Muc4/SMC was originally isolated from membranes of highly malignant, metastatic rat 13762 mammary adenocarcinoma ascites sublines as a stable complex of two subunits, termed ASGP-1 (ascites sialoglycoprotein) (29) and ASGP-2 (8). The stability of the complex, designated sialomucin complex (SMC), was indicated by the fact that it survives fractionation in CsCl, a strong chaotrope, and is resistant to dissociation except by protein-denaturing agents (8). However, the association is noncovalent, and the two subunits can readily be separated either by sodium dodecyl sulfate electrophoresis (SDS PAGE) or by fractionation in guanidine or urea. The mucin subunit ASGP-1 was characterized biochemically as a high M_r glycoprotein, with about 70% carbohydrate as O-linked oligosaccharides (29). ASGP-2 behaved as an intrinsic membrane glycoprotein with about 50% carbohydrate, primarily as N-linked oligosaccharides (30). Biosynthesis studies showed that both subunits were produced from a single ≈300-kDa precursor, which was cleaved to its two subunits early during transit to the cell surface (11).

Molecular cloning verified that the two subunits are derived from a single gene (18). The full-length clone of SMC showed that ASGP-1 consisted of three serine- and threonine-rich regions (31). The middle region contained 11 full and 2 partial repeats of a sequence of about 124 amino acids, with considerable variability between repeats (31) compared to MUC1 (2). Surprisingly,

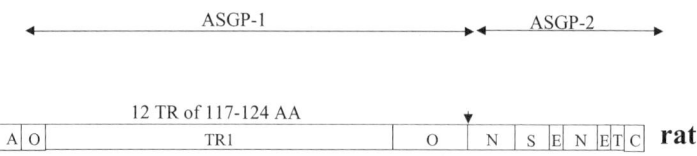

FIG. 3. Comparisons of domain structures of human MUC4 and rat Muc4/SMC. Note absence of 16-amino acid repeat in rat protein. Thus, the human protein is substantially larger. Arrows indicate cleavage site GDPH between two subunits. A, signal sequence; O,O-glycosylated unique (nonrepeat) domains of ASGP-1; TR, tandem repeat domains of ASGP-1; N,N-glycosylated domains of ASGP-2; S, cysteine-rich domain; EGF-like domains; T, transmembrane domain; C, cytoplasmic domain.

SMC was established to be the rat homolog of human MUC4 (32) when a full-length sequence of MUC4 became available. MUC4 had previously been partially cloned and sequenced, yielding a 16-amino acid repeat domain (33). However, this repeat domain is not present in the rat Muc4/SMC (Fig. 3). Thus, the human MUC4 has two types of tandem repeats (3 repeats of 126 amino acids and 146–400 repeats of 16 amino acids), while the rat has only the first one (11 repeats of 124–126 amino acids). Because of the presence of large numbers of the 16 amino acid repeat, the human glycoprotein is much larger than the rat glycoprotein (32). The transmembrane subunit ASGP-2 (called MUC4β in the human) is more conserved, exhibiting 60–70% amino acid identities between the rat and the human (32). It contains two N-glycosylated regions, two EGF-like domains, a separate cysteine-rich region, a transmembrane domain, and a cytoplasmic domain (Fig. 3) (18, 32).

The availability of purified ASGP-1 and ASGP-2 permitted the production of antibodies (Table II). Several different types of antibodies have been produced (11, 21, 34). Sequential immunoprecipitation and immunoblot analyses have shown that ASGP-1 and ASGP-2 are present as a complex in all tissues and tumors examined (35). Thus, we have primarily used antibodies against ASGP-2 for our analyses, because the size of ASGP-2 is more conducive to SDS PAGE and immunoblot analyses. The most important antibodies recognizing rat ASGP-2 are a polyclonal antibody made against intact purified ASGP-2 (anti-ASGP-2 pAb), a polyclonal made against a peptide from the cytoplasmic domain of ASGP-2 (anti-C-pep), and a monoclonal made against purified intact ASGP-2 (mAb 4F12) (Table II) (21). The last of these recognizes an epitope in the N-terminal 53 amino acids of ASGP-2 and has been particularly useful for

TABLE II
ANTIBODIES RECOGNIZING Muc4 TRANSMEMBRANE SUBUNIT ASGP-2

Specificity	Type	Examples	Comments
ASGP-2	Polyclonal	Anti-ASGP-2	Made against purified rat ascites ASGP-2
ASGP-2	Monoclonal	4F12, 13C4	Made against rat ascites Muc4/SMC
ASGP-2	Polyclonal (anti-peptide)	Anti-C-pep	Made against ASGP-2 cytoplasmic domain peptide
ASGP-2	Monoclonal	1G8	Made against rat tumor ASGP-2, but selected against recombinant human MUC4β

immunoblot and immunocytochemical studies. This antibody is commercially available from the Developmental Studies Hybridoma Bank at the University of Iowa (http://www.uiowa.edu/~dshbwww). Recently, we have produced a new monoclonal antibody against rat ASGP-2. The hybridomas were screened against recombinant human ASGP-2 expressed in two different cell lines (Cos7 and HC11 mouse mammary cells) and selected for the strongest reactivity to human MUC4β compared to rat ASGP-2. This antibody (mAb 1G8) appears to be particularly promising for analyses of human tissue and tumor samples.

The diverse nature of the two subunits suggests the possibility of multiple functions. Since the early studies on the role of the membrane mucin epiglycanin in allotransplantation of mammary tumors (6), these molecules have been postulated to act as antirecognition and antiadhesion agents. These functions have been validated for both MUC1 and Muc4/SMC by transfection experiments (12, 36). In the case of Muc4/SMC, constructs containing different numbers of mucin repeats under the control of a tetracycline-regulated promoter were transfected into A375 melanoma cells (36). Both cell–matrix and cell–cell adhesion were shown to be abrogated by expression of the Muc4/SMC (36). The fact that repression of these functions was dependent on the number of mucin repeats in the constructs indicated that the effects were due to steric hindrance from the mucin at the cell surface. Similarly, expression of Muc4/SMC in the A375 cells blocked killing of the cells by lymphokine-activated killer cells (37). These observations describe a mechanism by which Muc4/SMC can contribute to tumor progression and metastasis, as in the case of the 13762 mammary ascites cells from which it was isolated. They also suggest that expression of Muc4/SMC will have to be tightly controlled in any normal cell or tissue in which it is found to prevent these properties from disrupting the tissue.

In addition to the cell surface form which provides the steric protection mechanism, Muc4/SMC is also found in a soluble form in many epithelial tissues

TABLE III
Muc4/SMC Protein Expression in Selected Rat Epithelial Tissues

Tissue	Expression level	Membrane/ soluble	Cell/tissue type	Regulation
Mammary gland milk	Low to high	M/S	Luminal/secretory	Post-transcriptional
13762 Mammary ascites tumors	Overexpressed	M	Ascites	Loss of transcriptional and post-transcriptional regulation
Colon	High	S	Goblet	Unknown
Small intestine	Moderate	M/S	Crypt	Unknown
Airway	Moderate	M/S	Luminal	Constitutive?
Eye/tear film	Moderate	M/S	Sratified epithelial	Hormonal?
Lacrimal gland	Moderate	M/S	Luminal/secretory	Hormonal?
Uterus	Low to high	M/S	Luminal. glandular	Transcript level, hormonal
Cervix, vagina	High	M/S	Stratified epithelial	Constitutive?
Oviduct	Low to high	M/S	Luminal	Constitutive?

(Table III) (15) and in fluids such as milk (21), saliva (24), and tears (25). This form is most readily analyzed by sequential immunoprecipitations with anti-C-pep, which recognizes and precipitates the membrane form but not the soluble form, and anti-ASGP-2 pAb, which precipitates the remaining soluble form (21). Potential functions of this soluble form in specific tissues are discussed below. These functions raise the question of how soluble form is produced and regulated. Soluble human MUC4 has been shown to be produced by alternative splicing (23), but no evidence for splicing was observed in rat tissues which are known to produce soluble Muc4/SMC (21).

To address the question of the production of soluble Muc4/SMC, we analyzed Muc4/SMC transfectants of several cell lines (Komatsu et al., manuscript in preparation). Both sequential immunoprecipitations with anti-C-pep and anti-ASGP-2 and direct immunoblots of culture fluid with mAb 4F12 demonstrated the production of soluble form by all of the cell lines tested, including Cos7, MCF-7 breast cancer cell line, and HBL-100 epithelial cells (Komatsu et al., manuscript in preparation). Alternative splicing cannot contribute to production of soluble Muc4/SMC in these transfectants. A cleavage product of ≈20 kDa was detected in cell lysates but not in culture fluids by immunoblotting with anti-C-pep. A similar cleavage product was demonstrated in lysates of rat colon, which produces primarily soluble Muc4/SMC (21). Surprisingly, the size of the released extracellular domain of ASGP-2 (≈120 kDa) in the culture fluid does not differ discernibly from the uncleaved membrane form (21), in spite of the loss of the 20-kDa fragment. One possible explanation is that the cleaved and membrane forms are differentially glycosylated, thus compensating for the change in polypeptide size. Biosynthesis studies suggest that cleavage to yield the soluble form occurs within the cell (Komatsu et al., manuscript in preparation). This cleavage step was kinetically indistinguishable from the cleavage to produce the two subunits ASGP-1 and ASGP-2. Thus, the release can best be described as a secretion rather than as a shedding of the Muc4/SMC. This mechanism is consistent with the observation of Muc4/SMC in granular structures in the cytoplasm of cells of the colon (21) and lacrimal gland (38) and with the observation that Muc4/SMC can be released from the colon by secretagogue (21).

The most intriguing aspect of Muc4/SMC is its ability to act as a ligand for the receptor tyrosine kinase ErbB2/HER2/Neu (19). The ErbB family of growth factor receptor tyrosine kinases (also known as the EGF receptor family) contains four members (39, 40) and plays an important role in development and neoplasia (41). Only two of those, ErbB1 (EGF receptor) and ErbB4, appear to signal via the "classical" scheme, in which ligand binding by the receptor triggers homodimerization and phosphorylation (42). No soluble ligand has been discovered for ErbB2 despite extensive effort, and the ErbB3 kinase domain appears to be inactive (43). By those criteria, neither of these two receptors should be able to participate in signaling by ligand-induced homodimerization. In fact, the primary mechanism of signaling through the ErbB receptors in most physiological contexts appears to involve heterodimerization (39, 40, 44), with ErbB2 frequently acting as a "coreceptor" with one of the other receptors (45). The initial step in ligand–induced signaling through an ErbB receptor is the binding of ligand to the receptor. The specificity of ligand–receptor binding is complicated, since some ligands bind more than one receptor, and some receptors can bind more than one ligand (46). This step also determines the type of receptor dimer or multimer which forms and the pattern of phosphorylation of receptor tyrosine residues which results (47). Even the relationship between the dimer (or multimer) formed and the phosphorylation pattern is complex. For example, neuregulins 1 and 2 each induce formation of ErbB2–ErbB3 heterodimers, but the tyrosine residues phosphorylated in the receptors are different for the two ligands (48). Since downstream signaling pathways are dictated by the complement of phosphorylated tyrosine residues in the receptor cytoplasmic tails, these results indicate that the ligands play a subtle and complex role in determining cellular behaviors.

"Classical" growth factors usually work by endocrine or paracrine mechanisms, in which the factor is secreted by one cell and acts at the surface of a different cell (49). In some instances, stimulation may occur by an autocrine mechanism, in which the growth factor acts on the same cell population by which it is secreted (49). The demonstration of membrane-associated growth factors has led to the suggestion of alternative stimulation mechanisms. Massague (50) proposed a "juxtacrine" mechanism, a version of paracrine stimulation in which cell-surface growth factor on one cell stimulates growth by interaction with a receptor on the cell surface of a second cell. An intracellular (intracrine), autostimulatory mechanism has been suggested for effects of the *v-sis* oncogene (51). We have recently discovered a unique mechanism for formation of ligand–receptor complexes, an association of the two in the same membrane (Fig. 4C) (19), which we call the intramembrane interaction. Membrane-bound ligands

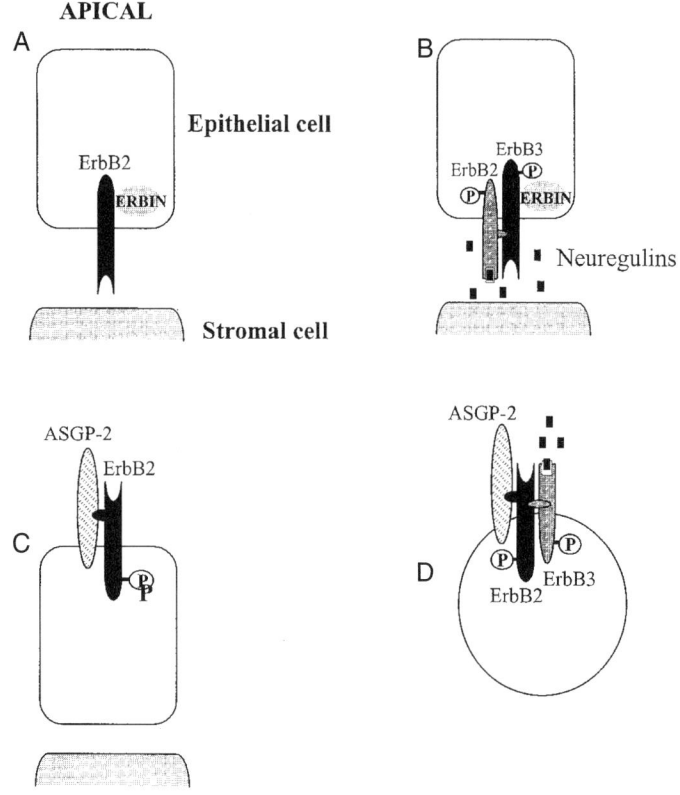

FIG. 4. Models for the intramembrane complexes of Muc4/SMC with ErbB2, ErbB3, and neuregulin, and their localization and activation in polarized epithelial cells. In the absence of Muc4/SMC expression, ErbB2 is restricted to the basolateral domain of the cells by its interaction with the PDZ domain-containing protein ERBIN (A). At this surface the ErbB2 is accessible to neuregulin, when it is produced by stromal cells. Neuregulin interaction with ErbB3 at the basolateral surface can trigger heterodimerization of the receptors, phosphorylation of both receptors and downstream signaling (B). Expression of Muc4/SMC leads to intramembrane complex formation, phosphorylation of the ErbB2 and localization at the apical surface (C), where it is inaccessible to neuregulin. Neoplastic transformation or some types of epithelial damage leads to a loss of cell–cell interactions, changes in morphology, and relocation of membrane proteins. Under these conditions the Muc4–SMC–ErbB2 complex will be accessible to neuregulin and able to form the "quad" complexes with potentiated phosphorylation of both ErbB2 and ErbB3 (D). Each of these different scenarios (B, C, and D) can potentially lead to different downstream signaling cascades.

may act on signaling pathways in two important ways. (i) They can activate the receptor to initiate a pathway. (ii) They may also act to localize the receptor in its cell (Fig. 4) (52), and thus influence the cellular response to signaling pathways. Such localization is an inherent feature of juxtacrine signaling but also appears to be important for many signaling pathways. We have shown that Muc4/SMC can act as an intramembrane ligand for the receptor ErbB2, providing a constitutive mechanism for activating the receptor in cells in which they are both expressed. This mechanism was originally suggested by two findings. (i) The EGF domains of Muc4/SMC contain the consensus residues found in ligands for the ErbB receptors (Fig. 2). (ii) ErbB2 of the ascites cells from which Muc4/SMC was originally isolated is constitutively activated in these cells (53) and present in a complex with Muc4/SMC (19).

The large size and complexity of the ascites cell complex (54, 55) did not permit interpretation of the nature of the interaction of the ErbB2 and Muc4/SMC in the ascites cells. Therefore, additional experiments were done to probe the nature of the association. The interaction of ASGP-2 and ErbB2 was demonstrated in insect cells, which have no ErbB receptors to complicate analyses and interpretation. Co-infection studies with each of the ErbB receptors and ASGP-2 demonstrated a specific association of ASGP-2 with ErbB2, but not ErbB1 (EGF receptor), ErbB3, or ErbB4 (19). Association of the ASGP-2 with ErbB2 induced phosphorylation of the ErbB2 but not of other co-expressed receptors. Based on these studies, we postulated that ASGP-2 (or Muc4/SMC) can form a novel intramembrane complex with ErbB2, which results in constitutive activation and phosphorylation of the receptor (Fig. 4C). To investigate complex formation further, insect cells were infected separately or co-infected with constructs for the extracellular domains of ErbB2 and ASGP-2. The stable ErbB2–ASGP-2 complex was secreted by the co-infected cells and shown by metabolic labeling to consist only of ASGP-2 and ErbB2 (19). Deletion mutants of ASGP-2 demonstrated a requirement for EGF1, the most N terminal of the two EGF domains (Figs. 1–3), for complex formation. In contrast, mixing the media from cells expressing ASGP-2 and ErbB2 extracellular domains separately failed to produce complex. Similarly, recombinant soluble ASGP-2 was unable to form complex with or induce phosphorylation of ErbB2 at cell surfaces. These results indicate that the ligand and receptor must be present in the same cell to form complex and suggest that the ligand and receptor form complex within the cell, though they do not necessarily have to be membrane bound. This observation may have significant consequences for signaling through this complex, as described later.

Complex formation between ErbB2 and Muc4/SMC could also be demonstrated by transfection studies with the tetracycline-regulatable Muc4/SMC construct in different cell types (36; Jepson et al., manuscript in preparation). Since Muc4/SMC contains both ASGP-1 and ASGP-2, whereas the insect cell experiments used only ASGP-2, these results indicate that ASGP-1 has little if any

TABLE IV
RELATIVE DEGREES OF ErbB2 AND ErbB3 PHOSPHORYLATION IN DIFFERENT
COMPLEXES WITH Muc4 AND NEUREGULIN

Complex	ErbB2 phosphorylation	ErbB3 phosphorylation
ErbB2 alone	—	—
Muc4–ErbB2	1	—
ErbB2–Neuregulin–ErbB3	2	2
Muc4–ErbB2–Neuregulin–ErbB3	4	4

influence on Muc4/SMC as a ligand for ErbB2. Most importantly, Muc4/SMC can induce specific tyrosine phosphorylation of ErbB2, demonstrated by tetracycline regulation of Muc4/SMC expression (19). Furthermore, Muc4/SMC can potentiate the phosphorylation of ErbB2 and ErbB3 in the presence of the ErbB3 ligand neuregulin, as indicated in Table IV. Upregulation of Muc4/SMC stimulates ErbB2 phosphorylation, though to a lesser extent than neuregulin, which stimulates phosphorylation of both ErbB2 and ErbB3. However, the combination of Muc4/SMC expression and neuregulin addition stimulates phosphorylation of both ErbB2 and ErbB3 to a greater exent than the sum of the two (Table IV) (19). One of the pathways stimulated specifically by the phosphorylation of ErbB3 results in the activation of phosphoinositol 3-kinase and its downstream effect Akt (protein kinase B). Interestingly, the potentiation of the phosphorylation of ErbB3 by Muc4/SMC is manifested in the activation of the downstream signaling kinase Akt (Jepson et al., manuscript in preparation). Muc4/SMC expression alone also does not activate Akt. However, Muc4/SMC expression potentiates the activation of Akt by neuregulin (Jepson et al., manuscript in preparation).

III. Muc4/SMC Contributions to Tumor Progression

A. Metastasis

The ability of Muc4/SMC to repress cell adhesions and block tumor cell killing suggests that it could play a role in tumor dissemination, tumor progression, and metastasis. To test this possibility, we have used A375 melanoma cells transfected with tetracycline-regulatable Muc4/SMC. Measurements of Muc4/SMC in tumors growing as xenotransplants in nude mice demonstrated that providing tetracycline in the drinking water could regulate Muc4/SMC over a 30- to 100-fold range *in vivo* (56), a level similar to that observed *in vitro* in the cell adhesion experiments (36). An experimental lung metastasis assay, in which the tumor cells were injected into the tail vein of the mice, was then used to determine whether Muc4/SMC could enhance metastasis. Three different conditions were used. (i) Cells were grown *in vitro* under conditions which induced expression of Muc4/SMC, and then injected into animals maintained under conditions which favored Muc4/SMC expression. (ii) Cells were grown *in vitro* under conditions which induced expression of Muc4/SMC, and then injected into animals maintained under conditions which repressed Muc4/SMC expression. (iii) Cells were grown *in vitro* under conditions which repressed expression of Muc4/SMC, and then injected into animals under conditions which repressed Muc4/SMC expression. Lung metastases were 10- to 20-fold more abundant under conditions (i) and (ii) than under condition (iii) (56). These observations suggest two conclusions. First, expression of Muc4/SMC does enhance metastasis. Second, formation of the metastatic foci occurs more rapidly than the turnover of Muc4/SMC resulting from the repression of its expression.

These results show that Muc4/SMC can contribute to the latter stages of the metastatic process. However, the question remained whether Muc4/SMC can also promote metastasis from a primary tumor. Since the Muc4/SMC-expressing A375 tumors are highly aggressive and invasive, we used a distant site for testing metastasis from the primary tumor (56). Cells expressing Muc4/SMC were injected into the foot pad of nude mice, which were maintained under conditions favoring Muc4/SMC expression. As a control, mice were injected with cells not expressing Muc4/SMC and maintained under conditions which repress Muc4/SMC expression. The tumor-bearing animals were amputated at the knee to remove the tumors and then maintained under the previous conditions until sacrificed to examine for lung metastases. Half (5/10) of the animals expressing Muc4/SMC were shown to have metastases compared to none of the control animals. These results again demonstrate the ability of Muc4/SMC to contribute to the metastatic process (56).

B. Primary Tumor Growth and Repression of Apoptosis

The experiments showing tetracycline regulation of Muc4/SMC *in vivo* and metastasis from the primary tumors also demonstrated that Muc4/SMC caused a significant increase (three- to five-fold) in primary tumor growth. The ability of Muc4/SMC to act as a ligand for ErbB2 suggested that the ErbB2 activation might stimulate tumor proliferation. However, immunocytochemical analyses of the tumors for proliferating cells failed to show any significant increase. Similarly, Muc4/SMC expression *in vitro* did not increase cell proliferation, progression through the cell cycle, or thymidine incorporation (19, 57). In contrast, a decrease in apoptosis could be demonstrated by TUNEL assay in the Muc4/SMC-expressing tumors. However, this effect could be due to an alteration in the host–tumor interaction induced by the antiadhesive Muc4/SMC. To verify the Muc4/SMC suppression of apoptosis, we used several assays on A375 cells expressing or not expressing Muc4/SMC *in vitro,* including Annexin V/propidium iodide binding, DNA fragmentation, and PARP cleavage (57). These results clearly demonstrate that Muc4/SMC can repress apoptosis. Moreover, the repression of PARP cleavage implies that Muc4/SMC is able to reduce caspase activity in the tumor cells.

The mechanism connecting apoptosis to Muc4/SMC is yet unclear. Since the Muc4/SMC abrogation of cell adhesion would be expected to increase apoptosis (58), we have proposed that the repression of apoptosis involves the activity of Muc4/SMC as an ErbB2 ligand. The first step of this pathway is the phosphorylation of ErbB2. To investigate this phosphorylation, we used a specific antibody for phospho-ErbB2 made against a phosphorylated peptide from the C-terminal domain of ErbB2 (59). As a control we used an antibody against the unphosphorylated peptide. Immunocytochemical and immunofluorescene analyses were performed of tumors grown *in vivo* and *in vitro,* respectively, with each antibody. These results demonstrated conclusively that Muc4/SMC upregulation leads to essentially complete phosphorylation of the ErbB2 at the site recognized by the anti-phospho-ErbB2 both *in vitro* and *in vivo*. The site recognized by the antibody is Tyr1248 of ErbB2, which has been implicated in neoplastic transformation (60). This phosphorylated Tyr is recognized by the adaptor protein SHC, which can initate the Ras–Erk signal transduction pathway (61). However, Muc4/SMC expression does not appear to activate Erk, thus suggesting an alternate pathway for downstream signaling by the Muc4/SMC–ErbB2 complex. Since Muc4/SMC plus neuregulin results in additional phosphorylation of ErbB2 (Table IV), additional sites must be phosphorylated in the presence of the ErbB3 ligand. Moreover, the combination of neuregulin and Muc4/SMC potentiates phosphorylation of sites on ErbB3 (Table IV). These sites are most likely responsible for the potentiated Akt activity induced by Muc4/SMC plus neuregulin compared to Muc4/SMC or neuregulin alone, as described above.

To address the question of the involvement of Muc4/SMC in apoptosis, investigations of signaling pathways commonly implicated in the regulation of apoptosis have been undertaken. Muc4/SMC expression did not activate phosphorylation of any of the common MAPKs (Erk1/2, p38, or JunK). Similarly, Muc4/SMC expression did not activate PI3K or Akt, nor did PI3K inhibitors block the effects of Muc4/SMC on apoptosis. As noted above, Muc4/SMC did potentiate the activation of Akt by neuregulin. This effect is not too surprising, since the PI3K–Akt pathway is stimulated via ErbB3 phosphorylation. Muc4/SMC alone does not trigger ErbB3 phosphorylation, but potentiates ErbB3 phosphorylation induced by neuregulin. These results are consistent with a model of at least two different types of Muc4/SMC signaling, one involving ErbB2 alone and a second involving ErbB2 plus ErbB3 and neuregulin through the PI3 kinase/Akt pathway. Interestingly, we have recently shown that Muc4/SMC expression in the A375 cells upregulates one signaling component which has been implicated in apoptosis, the cell cycle inhibitor $p27^{kip}$ (Jepson *et al.*, manuscript in preparation). Significantly, the effect of Muc4/SMC expression on $p27^{kip}$ can be overridden by neuregulin. Recent studies suggest that $p27^{kip}$ may play a key role as a regulator of cell fate. Expression of $p27^{kip}$ is best known to block cell cycle progression (62). However, in some contexts it can also repress apoptosis (63). These and other considerations have led us to propose that regulation of $p27^{kip}$ expression may provide a mechanism by which Muc4/SMC acts as an intrinsic survival factor in epithelia, described below.

IV. Muc4/SMC in Simple Epithelia

A. Airway: A Proposed Role in Mucociliary Transport

The airway is one of the most vulnerable tissues in the body. Thus, it is not surprising to find multiple mechanisms protecting the airway epithelium. One of the primary mechanisms involves synthesis and secretion of airway mucus, a gel composed of mucin glycoproteins and other associated factors (64). This mucus flows across the airway epithelium, trapping particulates and other materials and carrying them through the trachea for elimination from the airway. The protection of the airway epithelial surface by mucus is complex. Inhaled particulates are removed by mucociliary transport by the action of ciliated epithelial cells of the airway epithelial surface. The mucus protective barrier of the airway consists of two parts: a mucus gel at the level of the tips of the cilia and a more fluid layer, the periciliary fluid, between the mucin gel layer and the cell surfaces, interspersed among the cilia (65). The mucus layer is primarily composed of secreted gel-forming mucins from glandular and goblet cells which form a highly crosslinked viscous gel (66). The periciliary fluid is proposed to play an important role in the ciliary transport mechanism by providing the proper consistency for ciliary beating and for support of the mucus gel above the apical surface of the epithelial cells (65).

As a second protective mechanism involving mucins, the luminal epithelial cells of the airway are covered by membrane mucins. A major component of this membrane mucin layer is Muc4/SMC, which is highly expressed on the epithelium of rat conducting airways, with the highest levels occurring in the proximal trachea and progressively decreasing into the bronchioles and alveoli (67). Airway Muc4/SMC consists of two forms: a soluble form that lacks the C-terminal cytoplasmic and transmembrane domains and accounts for about 70% of the total and a membrane-associated form that has the C-terminal domains. Immunocytochemical analyses show that Muc4/SMC is predominantly present on the apical surfaces of the airway epithelium but not in goblet cells (67). The soluble form can be removed from the trachea by rinsing, consistent with a location in the periciliary fluid. To our knowledge, Muc4/SMC is the first mucin shown to be present in both soluble and membrane forms at the apical cell surface of an epithelium.

Based on these results, we propose a protective mechanism in which membrane and soluble forms of Muc4/SMC are produced by airway luminal epithelial cells to provide a cell-associated epithelial glycoprotein barrier which also serves as an interface with flowing mucus. In this model the membrane form of Muc4/SMC is directly associated with the cell surface, and soluble Muc4/SMC is an important constituent of the periciliary fluid between the plasma membrane and the mucus gel layer. Several observations are consistent with this model. (i) Muc4/SMC does not form crosslinked multimers, so it would not be expected to be incorporated into the mucin gel (67). (ii) By immunocytochemistry, Muc4/SMC is found at the luminal surface of the epithelium in a glycocalyx-like layer. (iii) The soluble form of Muc4/SMC is loosely associated with the luminal surface of the trachea and can be removed by gentle washing (67). (iv) Muc4/SMC was not observed in the submucosal glands or goblet cells of the trachea involved in the production of the gel-forming mucins (67). In contrast, Muc4/SMC is readily observed in secretory granules of goblet cells in the colon (21). Therefore, we suggest that both forms of Muc4/SMC are constitutively expressed by the luminal epithelial cells and transported to their apical cell surfaces where the transmembrane form is associated with the extracellular face of the plasma membrane and the soluble form is released into the pericellular fluid of the tracheal lumen. In support of this mechanism, we demonstrated secretion of soluble Muc4/SMC by primary cultures of isolated tracheal surface epithelial cells (67). This model suggests that Muc4/SMC is a critical element in the protective barrier of the airway epithelium.

B. Uterus and Blastocyst Implantation

The uterus is a critical organ for mammalian reproduction, which must accomodate two divergent functions. It must be protected from external insults, injury and invasion, but it must also be accessible for implantation of the blastocyst (68). The protection is provided in part by the expression of the membrane mucins Muc1 and Muc4/SMC at the luminal surface of the uterus (69, 70). However, these cell surface mucins create an unfavorable aspect for blastocyst implantation. Both the blastocyst and the uterus are epithelial surfaces, which are nonadhesive (71). This antiadhesivity of the uterine surface is created by the presence of the mucins. Thus, the expression of the mucins is tightly regulated during pregnancy and the estrous cycle. In rats and mice, Muc1 and Muc4/SMC disappear at the time of implantation to create a narrow window of receptivity established by changes in hormone levels (72).

To address the role of Muc4/SMC in the uterus, we performed immunoblot, sequential immunoprecipitation and immunocytochemical analyses. By sequential immunoprecipitation and immunoblot analyses with anti-C-pep and anti-ASGP-2 (Table II), both membrane and nonmembrane forms of Muc4/SMC were found in rat uterus as a complex of ASGP-1 and ASGP-2 (70). Immunocytochemical analyses indicate that the primary site of Muc4/SMC expression in the adult rat is at the luminal surface of the endometrium, though expression in endometrial glands was also observed. About 40% of the Muc4/SMC, corresponding to the nonmembrane (soluble) fraction, is removed by rinsing uterine preparations with saline, indicating that the soluble form is adsorbed loosely to the apical cell surfaces (70), as in the airway (67). Muc4/SMC is most highly expressed in the virgin animal. Moreover, its expression varies during the estrous cycle with the steady-state level of transcript, suggesting that expression is regulated at the transcript level. The complex is present in a location consistent with steric inhibition of microbe invasion. However, this location creates a problem for blastocyst implantation. To remove this problem, Muc4/SMC is downregulated at the beginning of the period of receptivity for implantation during day 5 of pregnancy in the rat (70). Moreover, the complex reappears immediately after the receptivity window on day 6. The key to the downregulation of Muc4/SMC is the rise in progesterone, as determined by the effects of estrogen and progesterone on the expression of Muc4/SMC in ovariectomized rats. Thus, the regulation of Muc4/SMC at the luminal uterine surface has the appropriate characteristics for a critical event in the implantation process.

Since implantation does not occur at other luminal surfaces in the rat female reproductive tract, we were interested to learn whether regulation of Muc4/SMC expression occurs similarly in the cervix, vagina, and oviduct. Both immunoblot and immunocytochemical analyses showed that Muc4/SMC expression changes quantitatively during the estrous cycle in the uterine luminal epithelium (70, 73). It is high during estrogen-dominated phases and low during progesterone-dominated phases, consistent with the observations during pregnancy. In contrast, no major quantitative changes are seen in the expression of Muc4/SMC in the cervix, vagina, oviduct, or uterine glandular epithelial cells during the phases of the estrous cycle (73). Thus, hormonal regulation of Muc4/SMC is specific to the uterine luminal epithelium, presumably as a mechanism to permit blastocyst implantation. However, significant changes are noted in the cellular localization of Muc4/SMC in the different regions of the cervix and vagina as they undergo differentiative morphological changes during the estrous cycle (73). Most importantly for its protective function, Muc4/SMC is always found at the apical surface of the most superficial layers of the stratified epithelia. The mucin is rarely found in the basal layers and is absent from the keratinized layer when it is present. Muc4/SMC expression in the medial layers varies with the stage of the cycle.

Muc4/SMC expression thus appears to be regulated differently in the uterus, where it must be lost from the endometrial surface before blastocyst implantation can occur (35, 70), than it is in other regions of the female reproductive tract. During the estrous cycle the amount of Muc4/SMC protein parallels transcript levels in the luminal epithelial cells. Uterine luminal Muc4/SMC is absent from ovariectomized rats, but is upregulated when they are treated with estrogen, an effect which is overridden by progesterone. In contrast to the uterine luminal epithelia, Muc4/SMC is present in the uterine glandular epithelia, cervix, and vagina in ovariectomized rats (73) and does not change substantially with hormone treatments. Behavior of Muc4/SMC in the rat resembles observations made on Muc1 in the mouse. Both implantation and loss of Muc4/SMC expression can be blocked with the anti-progestin RU486 (70). Interestingly, these results suggest that the antiprogestin RU486 may be able to prevent pregnancy by preventing progesterone downregulation of the luminal epithelial mucins, though it clearly has other effects as well. Regardless, these results clearly show that Muc4/SMC expression is differentially regulated by steroid hormones in different regions of the female reproductive tract. Moreover, these results support a model in which progesterone-dependent downregulation of Muc4/SMC and its loss from the apical surface of the rat uterine lining contribute to the generation of the receptive state for implantation.

The question of the regulation of Muc4/SMC expression in the uterus has been further investigated using primary cultures of rat uterine luminal epithelial cells. Hormone-sensitive rat uterine luminal epithelial cells (RULEC) were isolated and tested by the methods described previously by Glasser, Carson, and co-workers (74) and cultured on Matrigel. Although these cells are hormone responsive (74), no increase in Muc4/SMC expression in the cells was observed when the cells were cultured in the presence of estradiol. Similarly, progesterone did not decrease Muc4/SMC expression. These results, together with our observations on the constitutive expression of Muc4/SMC in the hormone-sensitive cervical luminal epithelium, suggested that the hormone effects on the luminal epithelial cells might be indirect. An analogous indirect mechanism has been proposed for the regulation of matrix metalloproteases in the human uterus (75). In that case hormones appear to act on uterine stromal cells to induce production of active $TGF\beta$, which acts on the epithelial cells to block expression of the protease. Similarly, when RULEC were continuously cultured in the presence of $TGF\beta$, no expression of Muc4/SMC was observed. Loss of Muc4/SMC paralleled loss of its transcript, indicating that regulation occurred at the transcript level rather than post-transcriptionally.

To test for an indirect mechanism of regulation, RULEC were co-cultured with uterine stromal fibroblasts. Under those conditions Muc4/SMC expression was repressed (Fig. 5). Likewise, Muc4/SMC expression was repressed when culture medium from stromal cells was added to the cultured luminal epithelial cells. Two additional experiments suggest that RULEC Muc4/SMC is indirectly

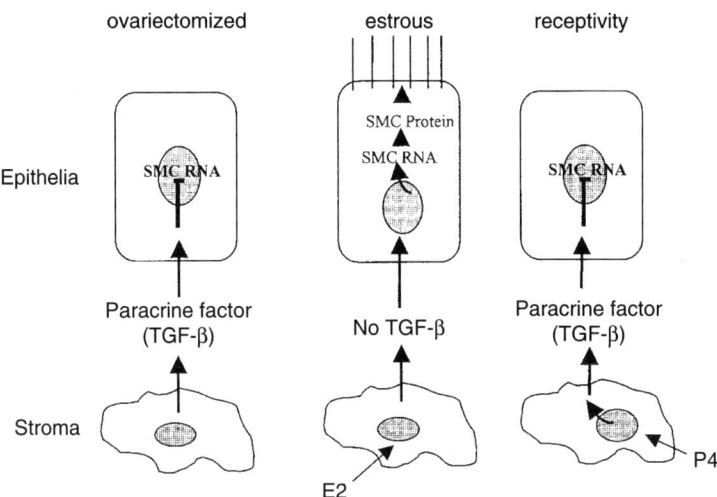

FIG. 5. Model for indirect hormone regulation of Muc4/SMC in rat uterine luminal epithelial cells. In this model, hormone acts on uterine stromal cells to regulate their production of active TGFβ. The action of TGFβ on the luminal epithelial cells represses transcript levels of Muc4/SMC, thus providing a hormone control of Muc4/SMC at the uterine luminal surface.

regulated by stromal TGFβ in response to steroids. First, anti-TGFβ substantially blocks the ability of culture medium from the stromal cells to downregulate Muc4/SMC. Second, treatment of the co-cultures with estradiol prevents the downregulation of Muc4/SMC (Fig. 5), providing additional evidence that the hormone acts through the stromal cells. These results clearly show that expression of Muc4/SMC responds to TGFβ produced by co-cultured fibroblasts but not directly to steroid hormones. However, estrogen can block the stromal cell effect on epithelial cell expression of Muc4/SMC. These results suggest that Muc4/SMC expression at the uterine cell surface is specifically regulated by hormones by an indirect effect via stromal cell production of TGFβ. A model for this effect is shown in Fig. 5.

C. Oviduct and ErbB2 Localization

The oviduct functions as a tube for the transport of gametes and is an active secretory organ, whose secretions maintain a suitable environment for continued maturation of male gametes, interaction between gametes, and early embryonic development (76). It consists of three different regions: isthmus, ampulla, and infundibulum. The isthmus is composed of highly branched folds whose epithelium consists of a single layer of ciliated and secretory columnar cells and stains intensely for Muc4/SMC at the apical surface. The infundibulum and ampulla are also composed of branched folds, whose epithelium is lined with a single layer of ciliated and secretory columnar cells, but they stain more weakly for Muc4/SMC, though also at the apical surface. Expression levels of Muc4/SMC in the oviduct are unchanged during the estrous cycle (73).

The ability of Muc4/SMC to act as a ligand for ErbB2 in tumor cells and transfected cells in culture (19) raises the question whether a similar phenomenon occurs in normal epithelia. One aspect of this question is whether the ligand and receptor are co-expressed and co-localized in epithelia. Therefore, immunocytochemical analyses for ErbB2 were performed on tissues of the female reproductive tract. The results were particularly striking in the isthmus of the oviduct, which exhibits a high level of apical membrane Muc4/SMC. ErbBs are usually considered to be basolateral proteins, though some tissues exhibit apical expression of ErbBs (77). We used a battery of ErbB2 antibodies which recognize epitopes in the cytoplasmic domains of ErbB2 for immunocytochemical analyses (78). Surprisingly, the ErbB2 localization differed depending on the antibody used, suggesting that the different antibodies recognize different forms of ErbB2. A similar phenomenon was observed in early studies of ErbB2 in breast cancer, because one of the commonly used antibodies did not recognize a specific phosphorylated form of ErbB2 (59). In our work, the DAKO anti-ErbB2 (Herceptest antibody) localized ErbB2 to the basolateral surface of oviduct luminal epithelial and mammary acinar cells. In contrast, two different Neomarkers antibodies (of about five tested) showed an apical localization of part of the ErbB2 in both of these tissues, co-localizing this fraction with Muc4/SMC (78). Both types of ErbB2 antibodies (DAKO and Neomarkers) showed circumferential staining of cells of the stratified epithelia of the cervix and vagina and partial co-localization with ASGP-2. Moreover, all of the antibodies, regardless of localization pattern, stained predominantly a 185-kDa band on immunoblots. Thus, the localization differences observed by the different antibodies are unlikely to be due only to differences in phosphorylation, which should be observed on the immunoblots as well.

Our explanation of these results is that the association of Muc4/SMC with ErbB2 at the apical plasma membrane in the oviduct results in the recruitment of cytoplasmic signaling components to the region of the ErbB2 epitope for the DAKO antibody. Therefore, the DAKO epitope in the apical ErbB2 is blocked from recognition (78). A similar phenomenon was observed in the studies of the interaction of ErbB2 with the PDZ domain-containing protein ERBIN, which was shown to direct ErbB2 to the basolateral surface (79). To pursue this issue further, we used an anti-phospho-ErbB2, which was made against a phosphorylated peptide containing phosphorylated Tyr1248. This antibody also stained predominantly the apical surfaces of oviduct epithelia, consistent with our observations that ErbB2 is phosphorylated at this position when it is in a complex with Muc4/SMC (Jepson et al., manuscript submitted for publication). We envision that the phosphorylated Tyr, which is near the site of the DAKO epitope, is involved in forming a signaling complex that blocks the DAKO antibody binding.

Our results indicate for the first time the presence of differentially localized ErbB2 forms in polarized epithelia, which are recognized by different anti-ErbB2 antibodies. These ErbB2 forms may be differentially associated with Muc4, which is strictly apically localized. This differential localization has important implications for cell signaling through ErbB2. We have therefore proposed a model for the modulation of ErbB2 signaling in polarized epithelial cells via its complexes with Muc4/SMC and other ErbB receptors. This model is described in Fig. 4. As previously shown, ErbB2 in polarized epithelial cells is localized to the basolateral surface in a complex with ERBIN, a PDZ-containing protein (79), in the absence of Muc4/SMC. In this location the ErbB2 is accessible to ligands produced by the stroma (Fig. 4), which can induce receptor heterodimerization (Fig. 4B). For example, neuregulin can bind ErbB3 and induce formation of the ErbB2–ErbB3 complex, with a resulting initiation of downstream signaling pathways. This scenario is changed by the expression of Muc4/SMC in epithelial cells. Formation of the Muc4/SMC-ErbB2 complex results in phosphorylation of the ErbB2, which causes a loss of its ability to bind ERBIN, as shown previously (79). We envision that the ErbB2 is then translocated to the apical surface as a complex with Muc4/SMC. Thus, ErbB2 can participate in two different kinds of signaling in the polarized epithelial cells, depending on the expression of Muc4/SMC. One type (Fig. 4C) involves formation of the Muc4/SMC–ErbB2 complex, results in specific phosphorylation of the ErbB2, occurs at the apical surface, and has been implicated in the repression of apoptosis (57). The second type (Fig. 4B) involves formation of heterodimeric ErbB2 complexes triggered by ligands, results in more extensive phosphorylation of ErbB2 and its partner receptor, and occurs at the basolateral surface.

Loss of epithelial cell polarity through damage or neoplastic transformation can create a new scenario. In this case, the Muc4/SMC-ErbB2 complex becomes accessible to ligands and can form heterodimeric receptor complexes, for example, Muc4/SMC–ErbB2–ErbB3–neuregulin, also known as the "quad complex" (Fig. 4D). As noted above, phosphorylation of both ErbB2 and ErbB3 are potentiated in this scenario, potentially initiating additional signaling responses not triggered by either Muc4/SMC or neuregulin alone. From these results we envision that Muc4/SMC might act as a sensor for epithelial injury through this set of mechanisms, with formation of the quad complex (Fig. 4) triggering a set of cellular responses, such as repression of apoptosis of nonadhered cells, cell motility, and cell proliferation, to respond to the damage. It is worth noting that these are the same kind of behaviors of epithelial cells which occur as they undergo neoplastic transformation and tumor progression.

V. Muc4/SMC in Glandular Secretory Epithelia

A. Mammary Acinar Cells, Milk, and Breast Cancer

The history of Muc4/SMC is intimately connected to mammary biology. As noted previously, the glycoprotein was first isolated from rat mammary adenocarcinoma ascites sublines (8, 29), and the rat form was first cloned from one of those sublines (18, 31). From a physiological standpoint, two mucins were described in human milk over a decade ago (80). One of these has been identified as MUC1 (9). Based on two of our observations, the other is almost certainly MUC4 (21). First, Muc4/SMC has been shown to be a prominent component of rat milk (21). Second, our new monoclonal antibody 1G8 demonstrates strong staining of the transmembrane subunit of MUC4 on immunoblots of samples of human milk (J. Zhang et al., manuscript in preparation). The function of these mucins in milk is still unclear. One possibility is that the mucins help to protect the newborn from bacteria and viruses by binding these microbes (27). Mucin binding to microbes has been demonstrated (28), but the physiological consequences of that binding remain to be demonstrated.

These observations have led us to investigate Muc4/SMC in the rat mammary gland. Immunolocalization studies showed that the mucin in the rat lactating gland is predominantly in acinar cells, which are responsible for the production of proteins secreted into milk, though some mucin can also be found on ductal structures (21, 81; Price-Schiavi et al., manuscript in preparation). Localization is primarily apical. In the lactating gland, both membrane and soluble forms of Muc4/SMC are produced (21). At least part of the membrane Muc4/SMC is present in the rat mammary gland as a complex with ErbB2 (19, Price-Schiavi et al., manuscript in preparation). As previously described in the oviduct, ErbB2 is present in rat mammary acinar cells at both the basolateral and apical surfaces, recognized by the DAKO and Neomarkers1 antibodies, respectively (Price-Schiavi et al., manuscript in preparation). Phospho-ErbB2 is also found at both of these surfaces. This observation is consistent with our model in Fig. 4. In this model, Muc4/SMC is at the apical surface in the lactating mammary gland and will form a complex with ErbB2, thus triggering ErbB2 phosphorylation on Tyr1248. Moreover, neuregulin should be present at the basolateral surface, since it is produced in lactating mammary gland (82), presumably by stromal cells. The neuregulin will associate with ErbB3 at the epithelial cell basolateral surface and trigger heterodimerization of ErbB2 and ErbB3 (see Fig. 4B), also leading to phosphorylation of ErbB2 Tyr1248 but inducing phosphorylation of other ErbB2 and ErbB3 tyrosines as well. The function of the activated ErbB2 in the lactating gland is unknown. Significantly, the tight junctions in the lactating mammary gland are sealed, restricting access of stromal factors such as neuregulin to the apical surface of the acinar cells and limiting formation of the "quad complex," described in Fig. 4D. Our model for the function of Muc4/SMC and ErbB2 in epithelial cells suggests that the acinar cell during lactation receives two signals: one from the Muc4/SMC-ErbB2 complex at the apical membrane and the second from the neuregulin-induced ErbB2-ErbB3 heterodimer at the basolateral surface. How these signals contribute to acinar cell function is uncertain, but one consequence of the apical signal could be a repression of apoptosis (57, Jepson *et al.*, manuscript in preparation), promoting survival of the cells.

This model is predicated in part on the timing of Muc4/SMC, ErbB2, and neuregulin expression in the mammary gland. In the rat, Muc4/SMC is present at low levels in the virgin animal at the apical surfaces of ductal cells (81). At about day 11 of pregnancy the level begins to increase dramatically and remains high through lactation. Loss of Muc4/SMC production occurs with involution (81, 83). This time course clearly indicates that Muc4/SMC is a product of mammary differentiation and is consistent with our hypothesis that Muc4/SMC may serve as an intrinsic epithelial cell survival factor. Our results indicate that ErbB2 is also expressed in the mammary gland in acinar cells during late pregnancy and lactation, though with a somewhat later time course than Muc4/SMC (Price-Schiavi *et al.*, manuscript in preparation). Results from studies on mouse mammary gland showed that neuregulin is produced during a relatively narrow window of time during pregnancy (82). Interpreting and understanding these observations will require more information, but they do provide support for our hypothesis that Muc4/SMC can play an important role in mammary function.

This hypothesis is supported further by the recent development of a Muc4/SMC transgenic mouse in collaboration with Bill Muller at McMaster University. A Muc4/SMC construct was placed under the regulation of a mouse mammary tumor virus promoter to direct the expression of the transgene primarily to the mammary gland. Preliminary results on the first two founder lines show two interesting phenotypic changes compared to normal mammary gland. First, the developing gland exhibits hyperplasia. Second, the extension of the gland into the fat pad is "bifurcated." These changes are particularly significant, because they have also been observed in a transgenic model for ErbB2, in which the activated mutant form of ErbB2 under its normal promoter was directed to the mammary gland using a Cre-lox targeting strategy (84). The hyperplasia is consistent with our hypothesis that Muc4/SMC can repress apoptosis, while the specific effect on mammary morphogenesis (bifurcation) supports our model that Muc4/SMC acts on mammary gland through its effects on ErbB2 signaling.

The results above suggest that Muc4/SMC must be tightly regulated in the mammary gland. To examine the mechanism of regulation, we measured Muc4/SMC transcripts in the virgin rat and during pregnancy. Although Muc4/SMC protein increased dramatically during mid-pregnancy (83), the transcript levels were as high in the virgin animals as they were during pregnancy (81, 83). These results indicate that Muc4/SMC is post-transcriptionally downregulated in the mammary gland in the virgin animal and during early pregnancy. The mechanism of this regulation was investigated using primary cultures of rat mammary epithelial cells taken from either virgin or pregnant animals. In either case expression of Muc4/SMC in cells grown on plastic was promoted by serum. Significantly, growth of the cells in Matrigel, which induces differentiation and casein synthesis, repressed expression of Muc4/SMC (83). Moreover, Matrigel reduced Muc4/SMC protein levels without affecting transcript levels, thus mimicking the behavior *in vivo*. This Matrigel effect may represent the repression of Muc4/SMC production by basement membrane in the normal gland, since overproduction of the glycoprotein could lead to changes in cell adhesion or to hyperplasia. Biosynthesis studies showed that Matrigel represses the synthesis of Muc4/SMC precursor, suggesting that the effect occurs at the translational level (83). The mechanism by which extracellular matrix regulates protein synthesis is unknown. However, translational control of proteins is often exerted through noncoding genetic sequences, either 5′ or 3′ of the coding sequence. In the case of Muc4/SMC both noncoding sequences may be important. Reporter assays of translation have suggested that 5′ and 3′ sequences can alter translational efficiency.

Since Matrigel is a complex mixture of extracellular matrix components, growth factors, and other proteins, we tested a number of individual components for their ability to block Muc4/SMC synthesis (85). None of the extracellular matrix components tested individually (collagen, laminin, fibronectin) was able to repress Muc4/SMC synthesis. Most hormones, cytokines, and growth factors, including estrogen, EGF, TGFα, heregulin, and hepatocyte growth factor, have no effect on Muc4/SMC expression in mammary epithelial cells (85). However, TGFβ strongly downregulates it, suggesting that TGFβ may play an important role in Muc4/SMC regulation in the normal mammary gland (83, 86). Furthermore, TGFβ affects only protein, not transcript, levels. These results indicate that Muc4/SMC is post-transcriptionally regulated by TGFβ in normal mammary gland. The mechanism by which TGFβ downregulates Muc4/SMC has been investigated by pulse–chase analyses in mammary epithelial cells (86). Synthesis of precursor was not affected by TGFβ, ruling out a translational mechanism. The primary effect was a reduction in the amount of ASGP-2 produced from precursor, suggesting that cleavage of the precursor is repressed in the presence of TGFβ. This mechanism is different from that observed with

Matrigel, indicating that TGFβ is not responsible for the effects of Matrigel on the mammary epithelial cells (86).

The mechanism by which TGFβ can block processing of the Muc4/SMC precursor is still unclear. However, we have observed that interferon-γ can prevent the TGFβ effect (Price-Schiavi et al., manuscript in preparation). Previous studies have shown that interferon-γ, acting through the Jak-STAT pathway, can induce synthesis of SMAD7, one of the inhibitory SMADs that can block transcriptional regulation by TGFβ (87). Thus, we envision that TGFβ acts by blocking transcription and synthesis of a factor involved in the processing of Muc4/SMC. A likely candidate would be the protease responsible for the cleavage to yield the two subunits ASGP-1 and ASGP-2.

Muc4/SMC is highly overexpressed in the rat mammary adenocarcinoma cells from which it was originally isolated (88). Similarly, highly aggressive breast cancer cells from patient effusions frequently exhibit high levels of MUC4 (37). These results suggest that the mechanisms regulating Muc4/SMC may be lost in the tumor cells. Examination of the 13762 rat mammary ascites cells showed that neither Matrigel nor TGFβ had an effect on the expression of Muc4/SMC (83, 86). These results are not too surprising, since neoplastic transformation results in loss of adhesion to basement membrane, and many cancers lose their responsiveness to TGFβ during tumor progression (89). These findings permit us to draw a model for the role of Muc4/SMC in breast cancer (Fig. 6). Neoplastic transformation of mammary epitheilial cells results in the loss of cell adhesions. Loss of adhesion to the basement membrane releases one mechanism for repressing Muc4/SMC synthesis, increasing the level of the glycoprotein in the cell membrane. Loss of cell–cell adhesion and polarity removes barriers which prevent growth factor access to apical Muc4/SMC and ErbB2 and allows mixing of receptors from the apical and basolateral surfaces. The resultant formation of quad complexes (Fig. 4D) increases cellular signaling that stimulates proliferation, repression of apoptosis, and cell motility. Loss of TGFβ responsiveness will further increase Muc4/SMC levels, exacerbating these behaviors. Importantly, all of these could arise from a single cellular event, overexpression of ErbB2, which has been shown to trigger neoplastic transformation and to cause a loss of TGFβ responsiveness.

Although post-transcriptional effects can explain the changes in Muc4/SMC expression during pregnancy and rationalize its overexpression in mammary tumors, Muc4/SMC transcript levels must increase in the mammary gland at some time before maturation of the rat. Although the timing has not been determined, two putative mechanisms contributing to the increase have been uncovered. First, studies on a potential rat mammary stem cell, called Rama 37, have shown that prolactin concomitantly induces differentiation to an alveolar cell type and increases Muc4/SMC expression (81), probably through a change in transcript

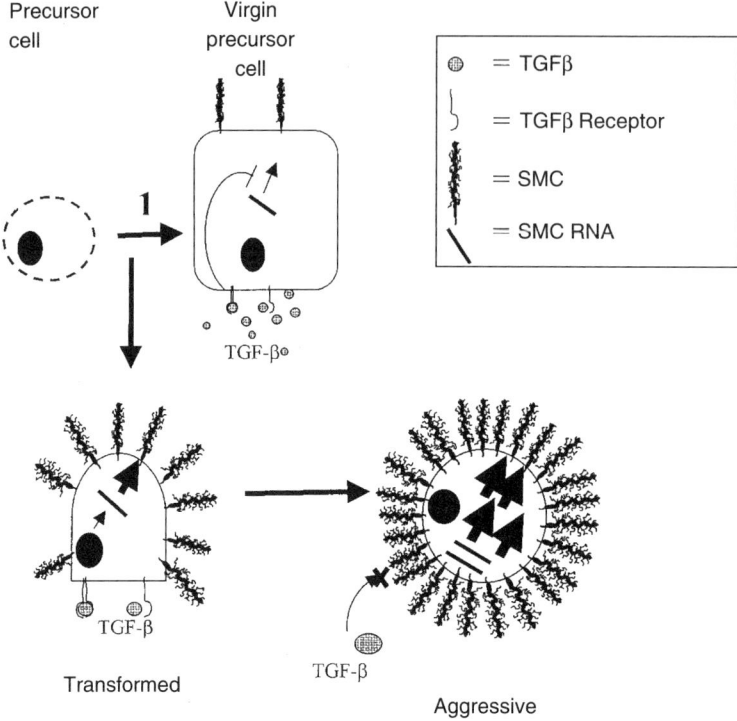

FIG. 6. Model for the dysregulation of Muc4/SMC in breast cancer. Reprinted from K. L. Carraway, S. A. Price-Schiavi, M. Komatsu, S. Jepson, A. Perez, and C. A. C. Carraway, *J. Mammary Gland Biol. Neoplasia*, **6**, 323–337, with permission by Kluwer Academic/Plenum Publisher.

levels. More specifically, Muc4/SMC transcript levels in primary mammary epithelial cells are enhanced by serum, insulin, or insulin-like growth factor (85). Inhibitor and transfection studies indicate that this effect requires a sustained increase in Erk phosphorylation, thus implicating the Ras–Erk pathway.

In previous studies we showed that overepression of antiadhesive mucin Muc4/SMC in A375 or MCF-7 cells blocked their killing by cells of the immune system (37). Moreover, overexpression of Muc4/SMC reduces the binding of antibodies against some cell surface components, such as the MHC complex (37). Since ErbB2 forms a complex with Muc4/SMC at the cell surface, these observations have potential implications for anti-ErbB2 reagents, such as Herceptin, used in treatments of cancer patients. To determine whether Muc4/SMC overexpression will reduce binding of anti-ErbB2s, we used flow cytometry to measure antibody binding of several extracellular domain anti-ErbB2s to A375 and MCF-7 cells with Muc4/SMC ON and OFF (36). In all examples, Muc4/SMC overexpression caused reductions in antibody binding, though the reduction could be overcome in some cases by increasing the antibody level (Price-Schiavi et al., manuscript in preparation). That the reduced antibody binding was due to Muc4/SMC–ErbB2 complex formation was shown by a simple experiment. When Muc4/SMC was aggregated in the cell membrane by anti-ASGP-2, anti-ErbB2 binding was further diminished. In contrast, when Muc4/SMC was aggregated in the cell membrane by anti-ASGP-2, binding of anti-MHC, which does not form a complex with ASGP-2, was enhanced. These results show that Muc4/SMC expression can act as an antagonist of anti-ErbB2 antibody binding and may be one of the factors which contributes to the relatively low efficacy of Herceptin as a therapeutic agent in breast cancer.

B. Salivary: Cell-Type-Specific Expression

The oral cavity plays complex roles in the physiology of respiration, mastication, digestion, and articulation of speech. Mucins cover all of the exposed surfaces of the oral cavity and are secreted by both mucous and serous cells of multiple glandular elements (90, 91). Salivary mucins are generally described by two classes: MG1 and MG2 (90, 91). MG1 is a high M_r mucin containing multiple disulfide-linked subunits capable of forming mucous gels which establish mobile protective barriers at the surfaces of epithelia (91). Recent studies indicate that MG1 is composed, at least in part, of MUC5B (92, 93), an important respiratory mucin (63). MG2 contains MUC7, the small soluble mucin (94). Salivary constituents are produced by paired parotid, submandibular, and sublingual glands. The parotid glands produce primarily serous secretions while the submandibular and sublingual glands excrete primarily mucous secretions.

Immunoblot and immunocytochemical analyses of rat parotid, submandibular, and sublingual glands showed expression of Muc4/SMC in all three. Staining was most abundant in parotid glands, with intermediate levels in submaxillary glands and lower levels in sublingual glands (24). Muc4/SMC was located in serous-type acini in the parotid glands. In contrast, Muc4/SMC in sublingual glands was localized to caps of cells around mucous acini, known as serous demilunes, which are also present in submaxillary glands. By immunoblot analyses two forms of each subunit were observed. The lower M_r form of ASGP-1 is probably a degradation product. However, the higher M_r form of the transmembrane subunit ASGP-2 cannot be so readily explained. One possibility is that this form results from altered glycosylation, possibly due to addition of polylactosamine oligosaccharides to the ASGP-2. As noted below, a similar higher M_r form of ASGP-2 has also been observed in the lacrimal gland. Immunocytochemical staining of MUC4 in human parotid glands was localized to epithelial cells of serous acini and ducts. However, the staining pattern of human parotid epithelial cells was heterogeneous, with MUC4 present in some acinar and ductal epithelial cells but not in others.

C. Lacrimal: Acinar-Specific Expression

The ocular surface is covered by a tear film whose primary function is to maintain a smooth and clear refractive optical surface for vision in a hostile external environment. The tear film is biochemically complex, composed of an outer layer of lipid and an inner mucous gel layer separated by a midlayer of aqueous fluid (95). This aqueous layer contains electrolytes, water and proteins, including mucins. Immunoblot analyses have demonstrated that Muc4/SMC is a prominent component of rat tear fluid (25). Since the components of the tear fluid are primarily secreted by the lacrimal gland, we have used immunoblot and immunocytochemical analyses to determine whether the lacrimal gland can be a source of the mucin of tears (38). The lacrimal gland consists of lobules of acini, similar to those of the mammary gland and salivary glands. Immunocytochemical staining of rat lacrimal glands for ASGP-2 showed two patterns of staining in different types of acini. In the major type the Muc4/SMC was localized in granular structures inside the acinar cells. In the other type the staining was predominantly on the membranes of the acinar cells and appeared to be located on basolateral as well as apical membranes. Muc4/SMC was also shown by immunocytochemical analyses with anti-C-pep (Table II) to be present on ductule cells sufaces of the gland.

The presence of membrane-localized Muc4/SMC raised the question whether it was present in a complex with ErbB2. Immunoblot and immunocytochemical analyses showed the presence of ErbB2 in the lacrimal gland, primarily localized to the membrane, and present in most, if not all, of the acini (38). Other ErbBs were also present in the lacrimal gland, but in more restricted distributions. Two color immunofluorescence analyses of merged images showed the co-localization of the Muc4/SMC and ErbB2 in the cells expressing membrane Muc4/SMC but not in the cells expressing Muc4/SMC in granules. Moreover, ErbB2 and Muc4/SMC could be co-immunoprecipitated from lacrimal gland lysates. These results suggest that the two types of acini in the lacrimal glands differ in the presence of Muc4/SMC–ErbB2 complexes.

The possible multifunctionality of Muc4/SMC in the lacrimal gland was supported by biochemical analyses. Sequential immunoprecipitation and immunoblotting demonstrated the presence of both soluble and membrane forms of Muc4/SMC (38). Presumably the soluble form is present in the granular structures in the acinar cells. As was previously shown in the salivary glands, ASGP-2 exhibited two forms in the lacrimal gland. The common 120- to 140-kDa form was present in both soluble and membrane Muc4/SMC. A larger (200-kDa) form was only found in the membrane Muc4/SMC. The origin of the larger form is unknown, though it may be produced by altered glycosylation. Whether this larger form plays any special role in the functions of Muc4/SMC is unknown.

VI. Muc4/SMC in Stratified Epithelia

A. Vagina/Cervix: Differentiation, Localization, and Cell Survival

Stratified epithelia are found at surfaces which are subject to mechanical forces. These multilayered structures are associated at their basal layers with basement membrane, while the surface of the most superficial layers provides a free interface interacting with the environment. Cells in the basal layers are proliferative and differentiate as they move toward the free surface, from which they are eventually shed. Thus, the most highly differentiated cells are those of the most superficial layer. In the rat cervix, vagina, oral cavity mucosa, cornea, and conjunctiva, Muc4/SMC is expressed only by cells of the most superficial half of the stratified layer (25, 73). In the cervix and vagina the localization of Muc4/SMC changes during the estrous cycle (73). The most superficial layer of vaginal epithelial cells is stained, but the cornefied layer produced during estrous stages is not stained. However, the expression of Muc4/SMC in the cervix and vagina does not change quantitatively during the estrous cycle and is unaffected by ovariectomy.

ErbB2 is also expressed in the cervix and vagina, though its distribution pattern differs somewhat from that of Muc4/SMC (78). Significantly, there is overlap in the distributions in the differentiated cell layers, and a Muc4/SMC-ErbB2 complex can be immunoprecipitated from cervix and vagina (78). These observations suggest that Muc4/SMC may participate in signaling in these differentiated cells. The purpose of this signaling is unknown, but one possibility is that Muc4/SMC acts as an intrinsic survival factor in the nonproliferative epithelial cells of the superficial layers as they wait to be desquamated from the epithelial surface.

B. Cornea and Conjunctiva: Ocular Protection

The ocular surface is one of the most accessible and vulnerable epithelia in the body. Moreover, it has to be maintained in an optically clear state to permit unobscured vision. Thus, the composition and organization of the ocular surface are absolutely critical to good vision. As noted above, the corneal surface on which light impinges is covered by a tripartite tear film consisting of a mucin layer next to the corneal epithelium, a liquid layer containing both gel-forming and soluble membrane mucin, and a lipid covering (95). The corneal epithelium is a stratified squamous epithelium and is continuous with the conjunctival epithelium covering the inner surface of the eyelid. Both epithelia are multilayered, though the conjunctival epithelium is somewhat thicker and contains goblet cells which secrete gel-forming mucin that becomes a constituent of the tear fluid.

A major question in ocular surface research has been the nature and source of the glycocalyx adjacent to the corneal epithelium in the tear film. Using immunoblot and immunoflurorescence analyses, we were able to show that Muc4/SMC is highly expressed in both corneal and conjunctival epithelia of the rat (25) and human (26). Both soluble and membrane forms of the mucin were present, and the former could be removed from the rat epithelial preparations by gentle rinsing, indicating that it is loosely bound. Interestingly, the Muc4/SMC was expressed only in the superficial half of each of the epithelia and was absent from the cytoplasmic granules of the goblet cells of the conjunctiva. These results suggest that a major component of the cell surface glycocalyx is being produced by the epithelial cells themselves. Thus, membrane Muc4/SMC is present at the superficial surface of the epithelia as a protective agent and provides continuity with the cell membrane. Furthermore, we envision that soluble Muc4/SMC is bound to the membrane mucin, but in equilibrium with mucin from the tear fluid. This equilibrium may serve to smooth the transition between the cell surface and tear fluid, eliminating a discontinuous boundary which might distort vision. In our model (Fig. 7) both membrane and soluble mucin forms contribute to the function of the ocular surface.

A second potential function of Muc4/SMC in these epithelia relates to its ability to act as an antiapoptotic agent when in complex with ErbB2. Both Muc4/SMC and ErbB2 are present in the corneal and conjunctival epithelia.

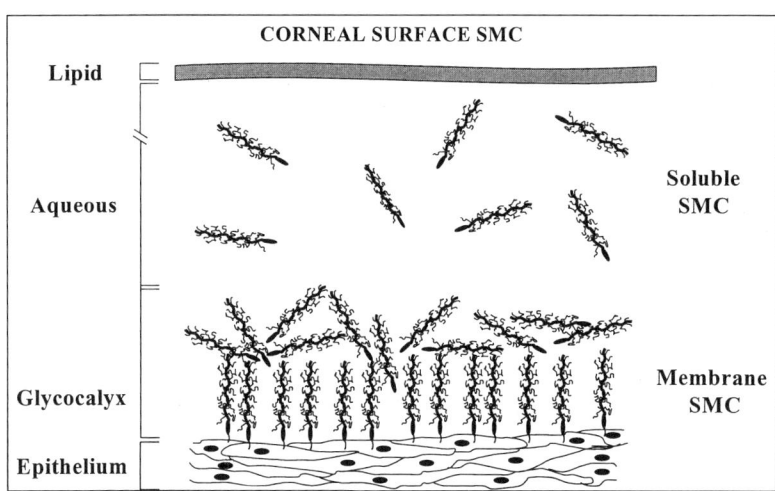

FIG. 7. Model for Muc4/SMC at the ocular surface. Membrane Muc4/SMC is produced by the epithelial cells and associated with the corneal surface glycocalyx. This glycocalyx also has loosely associated soluble Muc4/SMC, presumably produced by the epithelial cells and in equilibrium with soluble Muc4/SMC in the tear fluid. Soluble Muc4/SMC is additionally produced by the lacrimal gland for secretion into the tear fluid.

As noted above, the former is present in the most superficial half of the stratified layers. Thus, it is only present in the most differentiated cells, those which are not dividing. In contrast, ErbB2 appears to be expressed in all of the layers of these stratified epithelia. ErbB2 could therefore contribute to proliferation in the basal layers of cells by interactions with growth factors from the adjacent stroma. In contrast, we envision that ErbB2 complexes with Muc4/SMC in the superficial layers provide an antiapoptotic mechanism, by which Muc4/SMC can act as an intrinsic survival factor in the cornea and conjunctiva. As is the case with polarized epithelial cells, damage to the stratified epithelium would allow growth factors from the basal surface to have access to the superficial layers. This scenario could permit formation of quad complexes of Muc4/SMC–ErbB2–ErbBX–growth factor, with potentiation of the ErbB phosphorylation and signaling. This hypothetical scenario could act as a sensing mechanism for epithelial damage at the ocular surface.

VII. Conclusions and Future Directions

Muc4/SMC is a complicated molecule which is able to contribute to multiple functions in epithelia. The simplest function is its ability to act as a steric barrier at the plasma membrane to prevent the approach and association of noxious elements, such as microbes. This barrier function also can prevent the binding of the blastocyst to the uterine luminal epithelium in some species unless the mucin is downregulated. Similarly, the mucin can block killing of tumor cells by cellular mechanisms of the immune system when the Muc4/SMC is overexpressed in the tumors. A second function of the membrane form is to act as an intramembrane ligand and activator of the receptor tyrosine kinase ErbB2. This activation probably contributes to the antiapoptotic effect of Muc4/SMC to protect epithelial layers from extensive damage during injury. This protective mechanism may be particularly important in stratified epithelia which are subject to mechanical stresses. The production and roles of the soluble form have seen less attention. Presumably, this form also contributes to epithelial protection, but how it integrates with mucous gels is still uncertain in most cases. The proposal that Muc4/SMC contributes to mucociliary transport in the airway provides one example of how gel and soluble mucin may cooperate.

The last two decades have provided substantial information on properties, expression, and localization of Muc4/SMC in epithelia and tumors. The story is very incomplete, particularly with respect to understanding the functions of the molecule. Studies in cell culture have provided some insights into functions, but it is clear that animal models will be required for progress in this area. The mammary-targeted transgene provides one limited example. Studies are in progress to produce a Muc4/SMC knockout. Eventually, tissue-specific knockouts will be necessary to investigate specific roles of the glycoprotein. Such approaches should begin to produce more definitive answers about the multiple functions of this intriguing molecule.

ACKNOWLEDGMENTS

This work was supported by grants from the National Institutes of Health (CA52498, CA74072, EY12343, and HD35472), Department of Defense DAMD 17-00-1-0685, State of Florida Biomedical Research Program, and the Sylvester Cancer Center of the University of Miami.

REFERENCES

1. G. Strous and J. Dekker, Mucin-type glycoproteins. *Crit. Rev. Biochem. Mol. Biol.* **27**, 57–92 (1992).
2. S. J. Gendler and A. P. Spicer, Epitheial mucin genes. *Annu. Rev. Physiol.* **57**, 607–634 (1995).
3. J. Perez-Vilar and R. L. Hill, The structure and assembly of secreted mucins. *J. Biol. Chem.* **274**, 31,751–31,754 (1999).
4. K. L. Carraway, Preparation of membrane mucin. *Methods Mol. Biol.* **125**, 15–26 (2000).
5. A. Allen, G. Flemstrom, A. Garner, and E. Kivilaakso, Gastroduodenal mucosal protection. *Physiol. Rev.* **73**, 823–857 (1993).
6. J. F. Codington and S. Haavik, Epiglycanin—A carcinoma-specific mucin-type glycoprotein of the mouse TA3 tumour. *Glycobiology* **2**, 173–180 (1992).
7. K. L. Carraway and J. Spielman, Structural and functional aspects of tumor cell sialomucins. *Mol. Cell. Biochem.* **72**, 108–120 (1986).
8. A. P. Sherblom and K. L. Carraway, A complex of two cell surface glycoproteins from ascites mammary adenocarcinoma cells. *J. Biol. Chem.* **255**, 12,051–12,059 (1980).
9. S. Patton, S. J. Gendler, and A. P. Spicer, The epithelial mucin, MUC1, of milk, mammary gland and other tissues. *Biochim. Biophys. Acta* **1241**, 407–424 (1995).
10. M. J. Ligtenberg, L. Kruijshaar, F. Buijs, M. van Meijer, S. V. Litvinov, and J. Hilkens, Cell-associated episialin is a complex containing two proteins derived from a common precursor. *J. Biol. Chem.* **267**, 6171–6177 (1992).
11. Z. Sheng, S. R. Hull, and K. L. Carraway, Biosynthesis of the cell surface sialomucin complex of ascites 13762 rat mammary adenocarcinoma cells from a high Mr precursor. *J. Biol. Chem.* **265**, 8505–8510 (1990).
12. J. Hilkens, M. J. L. Ligtenberg, H. L. Vos, and S. Litvinov, Cell membrane-associated mucins and their adhesion-modulating property. *Trends Biochem. Sci.* **17**, 359–363 (1992).
13. K. L. Carraway, N. Fregien, C. A. C. Carraway, and K. L. Carraway III, Tumor sialomucin complexes as tumor antigens and modulators of cellular interactions and proliferation. *J. Cell Sci.* **103**, 299–307 (1992).
14. S. J. Gendler, MUC1, the renaissance molecule. *J. Mammary Gland Biol. Neoplasia* **6**, 339–353 (2001).
15. K. L. Carraway, S. A. Price-Schiavi, M. Komatsu, N. Idris, A. Perez, P. Li, S. Jepson, X. Zhu, M. E. Carvajal, and C. A. C. Carraway, Multiple facets of sialomucin complex/Muc4, a membrane mucin and Erbb2 ligand, in tumors and tissues. *Front. Biosci.* **5**, d95–d107 (2000).
16. S. Zrihan-Licht, A. Baruch, O. Elroy-Stein, I. Keydar, and D. H. Wreschner, Tyrosine phosphorylation of the MUC1 breast cancer membrane proteins. Cytokine receptor-like molecules. *FEBS Lett.* **356**, 130–136 (1994).

17. K. L. Carraway, C. A. C. Carraway, and K. L. Carraway III, "Signaling and the Cytoskeleton." Springer, Berlin, 1998.
18. Z. Sheng, K. Wu, K. L. Carraway, and N. Fregien, Molecular cloning of the transmembrane component of the 13762 mammary adenocarcinoma sialomucin complex: A new member of the epidermal growth factor superfamily. *J. Biol. Chem.* **267**, 16,341–16,346 (1992).
19. K. L. Carraway III, E. A. Rossi, M. Komatsu, S. A. Price-Schiavi, D. Huang, M. E. Carvajal, P. M. Guy, N. Fregien, C. A. C. Carraway, and K. L. Carraway, K. L., An intramembrane modulator of the ErbB2 receptor tyrosine kinase that potentiates neuregulin signaling. *J. Biol. Chem.* **274**, 5263–5266 (1999).
20. M. Boshell, E.-N. Lalani, L. Pemberton, J. Burchell, S. Gendler, and J. Taylor-Papadimitriou, The product of the human MUC1 gene when secreted by mouse cells transfected with the full-length cDNA lacks the cytoplasmic tail. *Biochem. Biophys. Res. Commun.* **185**, 1–8 (1992).
21. E. A. Rossi, R. McNeer, S. A. Price-Schiavi, M. Komatsu, J. M. H. Van den Brande, J. F. Thompson, C. A. C. Carraway, N. L. Fregien, and K. L. Carraway, K. L., Sialomucin complex, a heterodimeric glycoprotein complex: Expression as a soluble, secretable form in lactating mammary gland and colon. *J. Biol. Chem.* **271**, 33,476–33,485 (1996).
22. C. J. Williams, D. H. Wreschner, A. Tanaka, I. Tsarfaty, I. Keydar, and A. S. Dion, Multiple protein forms of the human breast tumor-associated epithelial membrane antigen (EMA) are generated by alternative splicing and induced by hormonal stimulation. *Biochem. Biophys. Res. Commun.* **170**, 1331–1338 (1990).
23. N. Moniaux, F. Escande, S. K. Batra, N. Porchet, A. Laine, and J.-P. Aubert, Alternative splicing generates a family of putative secreted and membrane-associated MUC4 mucins. *Eur. J. Biochem.* **267**, 4536–4544 (2000).
24. P. Li, M. E. Arango, R. E., Perez, C. A. Reis, E. L. Bonfante, D. Weed, and K. L. Carraway, Expression and localization of immunoreactive-sialomucin complex (Muc4) in salivary glands. *Tissue Cell* **33**, 111–118 (2001).
25. S. A. Price-Schiavi, D. Meller, X. Jing, J. Merritt, M. E. Carvajal, S. C. G. Tseng, and K. L. Carraway, Sialomucin complex at the rat ocular surface: a new model for ocular surface protection. *Biochem. J.* **335**, 457–463 (1998).
26. S. C. Pflugfelder, Z. Liu, D. Monroy, D.-Q. Li, M. E. Carvajal, S. A. Price-Schiavi, N. Idris, A. Solomon, A. Perez, and K. L. Carraway, Detection of Sialomucin Complex (MUC4) in human ocular surface epithelium and tear fluid. *Invest. Ophthalmol. Vis. Sci.* **41**, 1316–1326 (2000).
27. D. S. Newburg, Oligosaccharides and glycoconjugates in human milk: Their role in host defense. *J. Mammary Gland Biol. Neoplasia* **1**, 271–283 (1996).
28. N. A. McNamara, R. A. Sack, and S. M. Fleiszig, Mucin-bacterial binding assays. *Methods Mol. Biol.* **125**, 429–437 (2000).
29. A. P. Sherblom, R. L. Buck, and K. L. Carraway, Purification of the major sialoglycoproteins of 13762 MAT-B1 and MAT-C1 rat ascites mammary adenocarcinoma cells by density gradient centrifugation in cesium chloride and guanidine hydrochloride. *J. Biol. Chem.* **255**, 783–790 (1980).
30. S. R. Hull, Z. Sheng, O. Vanderpuye, C. David, and K. L. Carraway, Isolation and partial characterization of ascites sialoglycoprotein-2 (ASGP-2) of the cell surface sialomucin complex of 13762 rat mammary adenocarcinoma cells. *Biochem. J.* **265**, 121–129 (1990).
31. K. Wu, N. Fregien, and K. L. Carraway, Molecular cloning and sequencing of the mucin subunit of a heterodimeric, bifunctional cell surface glycoprotein complex of ascites rat mammary adenocarcinoma cells. *J. Biol. Chem.* **269**, 11,950–11,955 (1994).
32. N. Moniaux, S. Nollet, N. Porchet, P. Degand, A. Laine, and J.-P. Aubert, Complete sequence of the human mucin MUC4: a putative cell membrane-associated mucin. *Biochem. J.* **328**, 325–333 (1999).

33. N. Porchet, V. C. Nguyen, J. Dufosse, J. P. Audie, V. Guyonnet Duperat, M. S. Gross, C. Denis, P. Degand, A. Bernheim, and J.-P. Aubert, Molecular cloning and chromosomal localization of a novel human tracheo-bronchial mucin cDNA containing tandemly repeated sequences of 48 base pairs. *Biochem. Biophys. Res. Commun.* **175,** 414–422 (1991).
34. O. A. Vanderpuye, C. A. C. Carraway, and K. L. Carraway, Microfilament association of ASGP-2, the concanavalin A-binding glycoprotein of the cell surface sialomucin complex of 13762 rat mammary ascites tumor cells. *Exp. Cell Res.* **178,** 211–223 (1988).
35. R. R. McNeer, S. Price-Schiavi, M. Komatsu, N. Fregien, C. A. C. Carraway, and K. L. Carraway, K. L., Sialomucin complex in tumors and tissues. *Front. Biosci.* **2,** 449–459 (1997).
36. M. Komatsu, C. A. C. Carraway, N. L. Fregien, and K. L. Carraway, Reversible disruption of cell-matrix and cell-cell interactions by overexpression of sialomucin complex. *J. Biol. Chem.* **272,** 33,245–33,254 (1997).
37. M. Komatsu, L. Yee, and K. L. Carraway, Overexpression of sialomucin complex, a rat homolog of MUC4, inhibits tumor killing by lymphokine-activated killer cells. *Cancer Res.* **59,** 2229–2236 (1999).
38. M. E. Arango, P. Li, M. Komatsu, C. Montes, C. A. C. Carraway, and K. L. Carraway, Production and localization of Muc4/sialomucin complex and its receptor tyrosine kinase ErbB2 in the rat lacrimal gland. *Invest. Ophthalmol. Vis. Sci.* **42,** 2749–2756 (2001).
39. K. L. Carraway III and L. C. Cantley, A neu acquaintance for erbB3 and erbB4: A role for receptor heterodimerization in growth signaling. *Cell* **78,** 5–8 (1994).
40. L. N. Klapper, M. H. Kirschbaum, M. Sela, and Y. Yarden, Biochemical and clinical implications of the ErbB/HER signaling network of growth factor receptors. *Adv. Cancer Res.* **77,** 25–79 (2000).
41. I. Alroy and Y. Yarden, The ErbB signaling network in embryogenesis and oncogenesis: Signal diversification through combinatorial ligand-receptor interactions. *FEBS Lett.* **410,** 83–86 (1997).
42. A. Ullrich and J. Schlessinger, Signal transduction by receptors with tyrosine kinase activity. *Cell* **61,** 203–212 (1990).
43. P. M. Guy, J. V. Platko, L. C. Cantley, R. A. Cerione, and K. L. Carraway III, Insect cell-expressed p180erbB3 possesses an impaired tyrosine kinase activity. *Proc. Natl. Acad. Sci. U.S.A.* **91,** 8132–8136 (1994).
44. D. J. Riese II and D. F. Stern, Specificity within the EGF family/ErbB receptor family signaling network. *Bioessays* **20,** 41–48 (1998).
45. L. N. Klapper, S. Glathe, N. Vaisman, N. E. Hynes, G. C. Andrews, M. Sela, and Y. Yarden, The ErbB-2/HER2 oncoprotein of human carcinomas may function solely as a shared coreceptor for multiple stroma-derived growth factors. *Proc. Natl. Acad. Sci. U.S.A.* **96,** 4995–5000 (1999).
46. E. Tzahar and Y. Yarden, The ErbB-2/HER2 oncogenic receptor of adenocarcinomas: from orphanhood to multiple stromal ligands. *Biochim. Biophys. Acta* **1377,** M25–M37 (1998).
47. M. A. Olayioye, D. Graus-Porta, R. R. Beerli, J. Rohrer, B. Gay, and N. E. Hynes, ErbB-1 and ErbB-2 acquire distinct signaling properties dependent upon their dimerization partner. *Mol. Cell. Biol.* **18,** 5042–5051 (1998).
48. C. S. Crovello, C. Lai, L. C. Cantley, and K. L. Carraway III, Differential signaling by the epidermal growth factor-like growth factors neuregulin-1 and neuregulin-2. *J. Biol. Chem.* **273,** 26,954–26,961 (1998).
49. M. B. Sporn and G. J. Todaro, Autocrine secretion and malignant transformation of cells. *New Engl. J. Med.* **303,** 878–880 (1980).
50. J. Massague, Transforming growth factor-α—A model for membrane-anchored growth factors. *J. Biol. Chem.* **265,** 21,393–21,396 (1990).
51. B. E. Bejcek, D. Y. Li, and T. F. Deuel, Transformation by v-sis occurs by an internal autoactivation mechanism. *Science* **245,** 1496–1499 (1989).

52. K. L. Carraway III and C. Sweeney, Localization, and modulation of ErbB receptor tyrosine kinases. *Curr. Opin. Cell Biol.* **13,** 125–130 (2001).
53. S.-H. Juang, M. E. Carvajal, M. Whitney, Y. Liu, and C. A. C. Carraway, Tyrosine phosphorylation at the membrane-microfilament interface: a p185^{neu}-associated signal transduction particle containing Src, Abl and phosphorylated p58, a membrane- and microfilament-associated retroviral gag-like protein. *Oncogene* **12,** 1033–1042 (1996).
54. C. A. C. Carraway, M. E. Carvajal, and K. L. Carraway, Association of the Ras/MAP kinase signal transduction pathway with microfilaments. Evidence for a p185neu-containing cell surface signal transduction particle. *J. Biol. Chem.* **274,** 25,659–25,667 (1999).
55. C. A. C. Carraway, M. E. Carvajal, Y. Li, and K. L. Carraway, Association of p185^{neu} with microfilaments via a large glycoprotein complex in mammary carcinoma microvilli. Evidence for a microfilament-associated signal transduction particle. *J. Biol. Chem.* **268,** 5582–5587 (1993).
56. M. Komatsu, L. Tatum, N. H. Altman, C. A. C. Carraway, and K. L. Carraway, Potentiation of metastasis by cell surface sialomucin complex (rat Muc4), a multifunctional anti-adhesive glycoprotein. *Int. J. Cancer* **87,** 480–486 (2000).
57. M. Komatsu, S. Jepson, M. E. Arango, C. A. C. Carraway, and K. L. Carraway, Muc4/Sialomucin Complex, an intramembrane modulator of ErbB2/HER2/Neu, potentiates primary tumor growth and suppresses apoptosis in a xenotransplanted tumor. *Oncogene* **20,** 461–470 (2001).
58. S. M. Frisch and E. Ruoslahti, Integrins and anoikis. *Curr. Opin. Cell Biol.* **9,** 701–706 (1997).
59. M. P. DiGiovanna and D. F. Stern, Activation state-specific monoclonal antibody detects tyrosine phosphorylated p185neu/erbB-2 in a subset of human breast tumors overexpressing this receptor. *Cancer Res.* **55,** 1946–1955 (1995).
60. R. Ben-Levy, H. F. Paterson, C. J. Marshall, and Y. Yarden, A single autophosphorylation site confers oncogenicity to the Neu/ErbB-2 receptor and enables coupling to the MAP kinase pathway. *EMBO J.* **13,** 3302–3311 (1994).
61. K. L. Carraway and C. A. C. Carraway, Signaling, mitogenesis and the cytoskeleton: Where the action is. *BioEssays* **17,** 171–175 (1995).
62. A. Sgambato, A. Cittadini, A., B. Faraglia, and I. B. Weinstein, Multiple functions of p27^{kip} and its alterations in tumor cells: a review. *J. Cell. Physiol.* **183,** 18–27 (2000).
63. B. Eymin, M. Haugg, N. Droin, O. Sordet, M. T. Dimanche-Boitrel, and E. Solary, p27Kip1 induces drug resistance by preventing apoptosis upstream of cytochrome c release and procaspase-3 activation in leukemic cells. *Oncogene* **18,** 1411–1418 (1999).
64. M. C. Rose, Mucins: structure, function, and role in pulmonary diseases. *Am. J. Physiol.* **263,** L413–L429 (1992).
65. A. Wanner, M. Salathe, and T. G. O'Riordan, Mucociliary clearance in the airways. *Am. J. Respir. Crit. Care Med.* **154,** 1868–1902 (1996).
66. J. K. Sheehan, D. J. Thornton, M. Somerville, and I. Carlstedt, Mucin structure. The structure and heterogeneity of respiratory mucus glycoproteins. *Am. Rev. Respir. Dis.* **144,** S4–S9 (1991).
67. R. R. McNeer, D. Huang, N. L. Fregien, and K. L. Carraway, Sialomucin complex in the rat respiratory tract: a model for its role in epithelial protection. *Biochem. J.* **330,** 737–744 (1998).
68. K. L. Carraway and N. Idris, Regulation of sialomucin complex/Muc4 in the female rat reproductive tract. *Biochem. Soc. Trans.* **29,** 162–166 (2001).
69. E. Lagow, M. M. DeSouza, and D. D. Carson, Mammalian reproductive tract mucins. *Hum. Reprod. Update* **5,** 280–292 (1999).
70. R. R. McNeer, C. A. C. Carraway, N. L. Fregien, and K. L. Carraway, Characterization of the expression and steroid hormone control of sialomucin complex in the rat uterus: Implications for uterine receptivity. *J. Cell. Physiol.* **176,** 110–119 (1998).
71. H.-W. Denker, Implantation: A cell biological paradox. *J. Exp. Zool.* **266,** 541–558 (1993).
72. A. Psychoyos, Hormonal control of ovoimplantation. *Vitamins Horm.* **31,** 201–256 (1973).

73. N. Idris and K. L. Carraway, Sialomucin complex (Muc4) expression in the rat female reproductive tract. *Biol. Reprod.* **61,** 1431–1438 (1999).
74. S. R. Glasser, J. Julian, G. L. Decker, J.-P. Tang, and D. D. Carson, Development of morphological and functional polarity in primary cultures of immature rat uterine epithelial cells. *J. Cell Biol.* **107,** 2409–2423 (1988).
75. K. L. Bruner, W. H. Rodgers, L. I. Gold, M. Korc, J. T. Hargrove, L. M. Matrisian, and K. G. Osteen, Transforming growth factor β mediates the progesterone suppression of an epithelial metalloproteinase by adjacent stroma in the human endometrium. *Proc. Natl. Acad. Sci. U.S.A.* **92,** 7362–7366 (1995).
76. F. Gandolfi, T. A. L. Brevini, L. Richardson, C. R. Brown, and R. M. Moor, Characterization of proteins secreted by sheep oviduct epithelial cellls and their function in embryonic development. *Development* **106,** 303–312 (1989).
77. C. R. De Potter, J. Quatacker, G. Maertens, S. Van Daele, C. Pauwels, C. Verhofstede, W. Eechaute, and H. Roels, The subcellular localization of the neu protein in human normal and neoplastic cells. *Int. J. Cancer* **44,** 969–974 (1989).
78. N. Idris, C. A. C. Carraway, and K. L Carraway, Differential localization of ErbB2 in different tissues of the rat female reproductive tract: implications for the use of specific antibodies for erbb2 analysis. *J. Cell. Physiol.* **189,** 162–170 (2001).
79. J. P. Borg, S. Marchetto, A. Le Bivic, V. Ollendorff, F. Jaulin-Bastard, H. Saito, E. Fournier, J. Adelaide, B. Margolis, and D. Birnbaum, ERBIN: a basolateral PDZ protein that interacts with the mammalian ERBB2/HER2 receptor. *Nature Cell Biol.* **2,** 407–414 (2000).
80. M. Shimizu, K. Yamauchi, Y. Miyauchi, T. Sakurai, K. Tokugawa, and R. A. McIlhinney, High-Mr glycoprotein profiles in human milk serum and fat-globule membrane. *Biochem. J.* **233,** 725–730 (1986).
81. P. Li, S. A. Price-Schiavi, P. S. Rudland, and K. L. Carraway, Sialomucin Complex (rat Muc4) transmembrane subunit binds the differentiation marker peanut lectin in the normal rat mammary gland. *J. Cell. Physiol.* **186,** 397–405 (2001).
82. Y. Yang, E. Spitzer, D. Meyer, M. Sachs, C. Niemann, G. Hartmann, K. M. Weidner, C. Birchmeier, and W. Birchmeier, Sequential requirement of hepatocyte growth factor and neuregulin in the morphogenesis and differentiation of the mammary gland. *J. Cell Biol.* **131,** 215–226 (1995).
83. S. A. Price-Schiavi, C. A. C. Carraway, N. Fregien, and K. L. Carraway, Post-transcriptional regulation of a milk membrane protein, the sialomucin complex (ascites sialoglycoprotein (ASGP)-1/ASGP-2, rat Muc4) by TGFβ. *J. Biol. Chem.* **273,** 35,228–35,237 (1998).
84. E. R. Andrechek, W. R. Hardy, P. M. Siegel, M. A. Rudnicki, R. D. Cardiff, and W. J. Muller, Amplification of the neu/erbB-2 oncogene in a mouse model of mammary tumorigenesis. *Proc. Natl. Acad. Sci. U.S.A.* **97,** 3444–3449 (2000).
85. X. Zhu, S. A. Price-Schiavi, and K. L. Carraway, Extracellular regulated kinase (erk)-dependent regulation of sialomucin complex/MUC4, the intramembrane ErbB2 ligand, in mammary epithelial cells. *Oncogene* **19,** 4354–4361 (2000).
86. S. A. Price-Schiavi, X. Zhu, R. Aquinin, and K. L. Carraway, Sialomucin complex (rat muc4) is regulated by transforming growth factor in mammary gland by a novel post-translational mechanism. *J. Biol. Chem.* **275,** 17,800–17,807 (2000).
87. L. Ulloa, J. Doody, and J. Massague, Inhibition of transforming growth factor-beta/SMAD signalling by the interferon-gamma/STAT pathway. *Nature* **397,** 710 (1999).
88. A. P. Sherblom, J. W. Huggins, R. W. Chesnut, R. L. Buck, C. L. Ownby, G. B. Dermer, and K. L. Carraway, Cell surface properties of ascites sublines of the 13762 rat mammary adenocarcinoma. Relationship of the major sialoglycoprotein to xenotransplantability. *Exp. Cell Res.* **126,** 417–426 (1980).

89. C. L. Arteaga and H. L. Moses, TGF-beta in mammary development and neoplasia. *J. Mammary Gland Biol. Neoplasia* **1,** 327–329 (1996).
90. L. A. Tabak, In defense of the oral cavity: Structure, biosynthesis, and function of salivary mucins. *Annu. Rev. Physiol.* **57,** 547–564 (1995).
91. A. V. N. Amerongen, J. G. M. Bolscher, and E. C. I. Veerman, Salivary mucins: Protective functions in relation to their diversity. *Glycobiology* **5,** 733–740 (1995).
92. J.-L. Desseyn, J.-P. Aubert, I. van Seuningen, N. Porchet, and A. Laine, Genomic organization of the 3′ region of the human mucin gene MUC5B. *J. Biol. Chem.* **272,** 16,873–16,783 (1997).
93. P. A. Nielsen, E. P. Bennett, H. H. Wandall, M. H. Therkildsen, J. Hannibal, and H. Clausen, Identification of a major human high molecular weight salivary mucin (MG1) as tracheobronchial mucin MUC5B. *Glycobiology* **7,** 413–419 (1997).
94. L. A. Bobek, H. Tsai, A. R. Biesbrock, and M. J. Levine, Molecular cloning, sequence, and specificity of expression of the gene encoding the low molecular weight human salivary mucin (MUC7). *J. Biol. Chem.* **268,** 20,563–20,569 (1993).
95. S. C. Pflugfelder, Tear fluid influence on the ocular surface. *Adv. Exp. Med. Biol.* **438,** 611–617 (1998).
96. K. L. Carraway, S. A. Price-Schiavi, M. Komatsu, S. Jepson, A. Perez, and C. A. C., Carraway Muc/Sialomucin complex in the mammary gland and breast cancer. *J. Mammary Gland Biol. Neoplasia* **6,** 323–337 (2001).

Functions of Alphavirus Nonstructural Proteins in RNA Replication

LEEVI KÄÄRIÄINEN AND
TERO AHOLA

Institute of Biotechnology, Biocenter Viikki
FIN-00014 University of Helsinki, Finland

I. Introduction 187
II. Replication Cycle of Alphaviruses 188
III. Alphavirus-like Superfamily 190
IV. Replication of Alphavirus RNAs.................................. 192
   A. Synthesis of RNA Complementary to the Genome................ 192
   B. Plus-Strand RNA Synthesis 194
   C. Synthesis of 26S mRNA of the Structural Proteins............. 196
V. Processing of Alphavirus Nonstructural Polyprotein P1234............ 197
VI. nsP1: A Unique RNA-Capping Enzyme and Membrane Anchor 198
   A. Methyltransferase and Guanylyltransferase Activities................ 198
   B. Membrane Association 201
   C. Role in Minus-Strand RNA Synthesis 203
VII. nsP2: A Multifunctional Enzyme and Regulatory Protein 204
   A. Nucleosidetriphosphatase and RNA Helicase Activities............. 204
   B. RNA Triphosphatase Activity 205
   C. Protease Activity 206
   D. Nuclear Transport and Neuropathogenicity 206
VIII. nsP3: An Ancient Conserved Protein and Phosphoprotein 208
   A. Sequence Conservation 208
   B. Phosphorylation... 209
   C. Other Features 210
IX. nsP4: A Catalytic RNA Polymerase Subunit 210
X. The Replication Complex... 211
References... 214

I. Introduction

Alphaviruses are enveloped positive-strand RNA viruses transmitted to vertebrate hosts by mosquitoes. Several alphaviruses are pathogenic to humans or domestic animals, causing serious central nervous system infections or milder infections with, e.g., arthritis, rash, and fever (*1*). The structure and replication of *Semliki Forest virus* (SFV) and *Sindbis virus* (SIN) have been studied extensively during the last 30 years (*2*).

Alphaviruses have been important probes in cell biology to study translation, glycosylation, folding, and transport of membrane glycoproteins, as well as endocytosis and membrane fusion mechanisms (*3*). A new organelle, the intermediate compartment, operating between the endoplasmic reticulum and the Golgi complex was found by the aid of SFV (*4*). During the last 10 years, alphavirus replicons have been increasingly used as expression vectors for basic research, for generation of vaccines, and for production of recombinant proteins in industrial scale. The first attempts to use them as gene therapy vectors, and even in cancer therapy, have already been reported (*5, 6*).

Taken together with some well-studied alphavirus-like plant viruses, notably *Brome mosaic virus* (BMV) and *Tobacco mosaic virus* (TMV), alphaviruses form the most advanced model system to study eukaryotic positive-strand RNA virus replication. This review focuses on the functions and properties of alphavirus nonstructural or replicase proteins, which form the viral RNA-dependent RNA polymerase complex. A deeper understanding of RNA replication also provides insight to a fundamental stage in the history of life, as RNA replication, currently the exclusive property of viruses, is thought to antedate development of DNA-based genetic information. From a practical point of view, it is needed to combat viral diseases and to improve the properties of RNA virus expression vectors.

The main approaches of our laboratory in the recent years have been twofold. On one hand, we have discovered and characterized the enzymatic activities of the individual replicase proteins, and on the other hand, we have studied the localization, membrane association, and other cell biological aspects of the replication complex.

II. Replication Cycle of Alphaviruses

The alphavirus particles are icosahedral, with a diameter of 700 Å. The virus envelope consists of dimers of transmembrane glycoproteins E1 and E2, which form 80 projections, each consisting of three heterodimers following the symmetry $T = 4$. In SFV a third extrinsic glycoprotein, E3, is associated with the heterodimers. The envelope surrounds the icosahedral nucleocapsid consisting of 240 capsid proteins, arranged in pentamers and hexamers, and a single-stranded RNA molecule of about 11.5 kb (*7*).

After attachment to cellular receptors, the virus is internalized by adsorptive endocytosis (*8*). The acid milieu induces a conformational change in the virus envelope, resulting in fusogenic E1 homotrimers, which mediate the fusion of the virus envelope with the endosomal membranes (*3*). As a result, the nucleocapsid enters the cytoplasm, where uncoating of the genome RNA is carried out by ribosomes (*9*). The 5′ two-thirds of the SFV 42S RNA genome is translated into a large polyprotein of 2432 aa, designated P1234, which is then autocatalytically

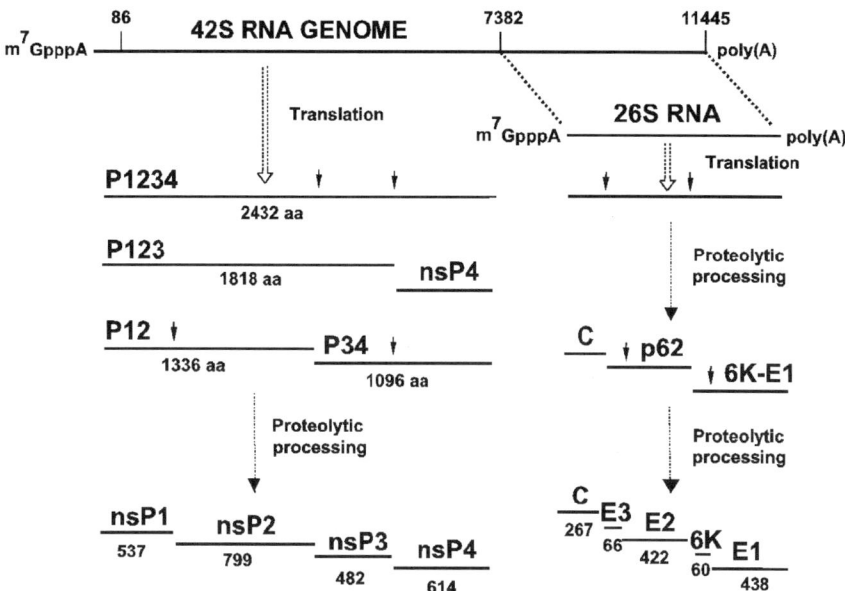

FIG. 1. Protein synthesis directed by SFV mRNA molecules. The primary translation products of the genomic 42S RNA and subgenomic 26S RNA and their proteolytical processing intermediates and individual nonstructural and structural proteins are shown.

cleaved to yield nonstructural protein 4 (nsP4) and P123 (Fig. 1). This is the "early RNA polymerase" responsible for the synthesis of full-size complementary RNA (42S RNA minus strand). After further cleavage of P123 into nsP1, nsP2, and nsP3, the minus strand in turn is used as template for the synthesis of new 42S RNA plus strands, as well as subgenomic 26S mRNA of the structural proteins.

The 26S RNA of SFV, which corresponds to the 3' third of 42S RNA, is translated to structural polyprotein of 1250 aa consisting of capsid protein, E3, E2, 6K, and E1 proteins (Fig. 1). Translation of 26S RNA starts on free ribosomes, but nascent autocatalytic cleavage of the capsid protein releases the signal peptide at the N terminus of E3, which guides the ribosomes to the endoplasmic reticulum membrane. The capsid protein associates transiently with the 60S subunit of the translating ribosome, and is then transferred to the 42S RNA genome during the assembly of the nucleocapsid (10). The assembly of the nucleocapsid is poorly understood, but RNA–protein interactions probably play an important role (11–13). E3 plus E2 are translocated as a precursor protein p62, which is co-translationally glycosylated. E1 is preceded by a hydrophobic 6K protein, the N terminus of which serves as a signal peptide. The cellular signal peptidase cleaves between p62 and 6K, and between 6K and E1. The E1 and p62, which acquire N-linked complex glycans (14, 15), are transported via the Golgi complex to the plasma membrane. p62 is cleaved into E3 and E2 by a furin-like protease during transport from the *trans*-Golgi network to the plasma membrane. In SIN-infected cells, E3 is secreted to the medium, whereas SFV E3 remains associated with the E1–E2 heterodimer and is incorporated into virions during their budding at the plasma membrane.

III. Alphavirus-like Superfamily

Positive-strand RNA viruses can be divided into large groups termed superfamilies, the members of which share common features in their encoded replicase proteins, genome organization, and replication strategies. Thus, the superfamily concept is biologically useful, although superfamilies have no officially recognized taxonomic status. A comprehensive analysis of RNA virus sequences suggests that the superfamilies might be as few as three in number, i.e., the picornavirus-like, alphavirus-like, and flavivirus-like viruses (16), although it could be argued that the coronavirus-like viruses and positive-strand RNA bacteriophages could also be placed in their own groups. A striking feature of the superfamilies is that although the replicase proteins share sequence similarity, and therefore a common descendent within the superfamily, the viruses may exhibit widely variable structures (nonenveloped and enveloped) and infect different kinds of hosts (plants and animals). Even within the superfamilies, the replicase proteins often show rearrangements and acquisition or deletion of domains.

Members of the alphavirus-like superfamily of positive-strand RNA viruses include the animal viruses of genus *Alphavirus*, and *Rubella virus* (comprising family *Togaviridae*) and *Hepatitis E virus*, as well as the insect viruses of *Tetraviridae* family, and numerous groups of plant viruses, including *Bromoviridae*, *Closteroviridae*, and the genera *Tobamo-*, *Tobra-*, *Hordei-*, *Furo-*, *Beny-*, *Capillo-*, *Tymo-*, *Carla-*, and *Potexvirus*, and other plant virus groups. The genome organization of three superfamily members is illustrated in Fig. 2 to highlight some of the similarities and differences. The replicase proteins always feature three conserved domains (16), an RNA-dependent RNA polymerase module, a conserved helicase-like domain of helicase superfamily SF1, and a module termed methyltransferase, or more recently capping enzyme. The presence of the methyltransferase/capping enzyme is the distinguishing hallmark of the alphavirus-like superfamily, as it is always present in superfamily members, and on the other hand, its relationship to other viral or cellular polypeptides is exceedingly distant (17, 18). The three conserved domains are always organized in the order methyltransferase–helicase–polymerase (in cases where all three

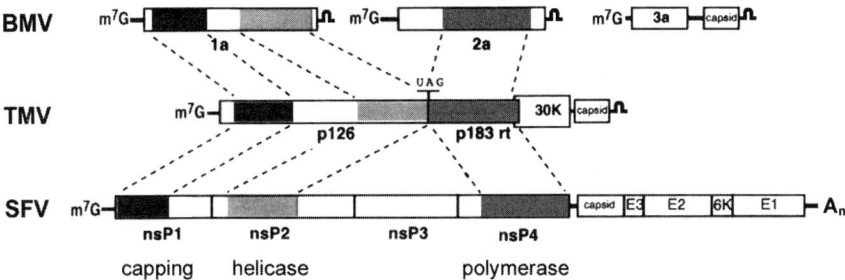

FIG. 2. Comparison of the genome structures of three alphavirus superfamily members: SFV, BMV (brome mosaic virus), and TMV (tobacco mosaic virus). Conserved replicase domains (capping enzyme, helicase, and polymerase) are indicated. Most of the alphavirus-like plant viruses do not encode a protease function, and none contains an nsP3-like domain. The tRNA-like structures at the 3′ end of TMV and BMV genome RNAs are indicated by a schematic structure, and the translational readthrough site of the TMV nonstructural protein (UAG) is marked.

are encoded by the same RNA), although they may be separated by other domains or be finally present in separate polypeptides as a result of proteolytic processing.

Other similarities shared by members of the alphavirus superfamily include (i) membrane-associated replication, (ii) presence of cap0 structure at the 5′ end of positive-strand mRNAs, (iii) an extra untemplated G residue at the 3′ end of minus strand, (iv) asymmetry of replication producing an excess of plus-strand RNA, (v) shut-off of minus-strand RNA synthesis late in infection, and (vi) regulated production of subgenomic mRNA molecules by internal initiation. In many other respects the viruses can be highly dissimilar. For instance, the genome can either consist of a single RNA or be segmented. It can end in a poly(A) or a tRNA-like structure, and the nonstructural proteins may or may not be subject to proteolytic processing (19).

The pairwise sequence identity of, for instance, SFV and BMV methyltransferase, helicase, and polymerase domains at the amino acid level is only 15–18%. As the remaining sequence similarity of even the conserved portions of the replicase is so low, it would be surprising if they had not adapted in many ways to different functions during the evolution of different lineages within the superfamily. Nevertheless, the basic replication complex with the conserved capping enzyme, helicase and polymerase domains is likely to be similar throughout the alphavirus-like superfamily, as all viruses have to perform a closely similar set of coordinated reactions. These similarities, when experimentally discovered and proven, will validate the virus superfamily concept. We hope that the information presented here on alphaviruses will prove useful and stimulating for investigators working on other virus groups.

IV. Replication of Alphavirus RNAs

Early work with SFV and SIN identified in infected cells two major single-stranded RNAs sedimenting in sucrose gradients at 42S (49S in SIN) and 26S (20–22). In addition, there is a heterogenous, partly double-stranded replicative intermediate (RI) RNA sedimenting between 20S and 29S. When this heterogenous material is isolated and subjected to RNase treatment, double-stranded RNAs arise. The replicative form RFI consists of a complete duplex of full-size plus and minus strands, whereas RFIII represents 26S RNA plus strand together with a respective portion of minus strand. RFII is a double-stranded form of the 5′ two-thirds of the genome. After short pulses with tritiated uridine, label is found in RFI and RFIII, whereas labeling of RFII takes a much longer time. These results lead to the deduction that the genome RNA is synthesized via a replicative intermediate RI_a, which after ribonuclease treatment yields RFI, whereas another structure RI_b is involved in the synthesis of the subgenomic 26S RNA. RNase treatment of RI_b yields both RFII and RFIII. The slow labeling of RFII was interpreted to mean a slow interconversion between RI_a and RI_b, i.e., between synthesis of genomic and subgenomic RNAs (23, 24) (Fig. 3).

Numerous studies have shown that virus-specific RNA polymerase activity is associated with cytoplasmic membranes together with heterogenous RNA (25–27). The role of 26S RNA as messenger for the structural proteins was finally proven by its *in vitro* translation in the presence of ER membranes, which yields all structural proteins (28). The role of the 42S RNA genome as the messenger for nonstructural proteins of SFV has been established (29–31) and confirmed by determination of the complete nucleotide sequence of SFV (32) and SIN (33).

A. Synthesis of RNA Complementary to the Genome

The genomes of alphaviruses have four conserved sequence motifs, two at the 5′ end within the first about 200 nucleotides, one of about an 20-nt-long motif at the 3′ end preceding the poly(A) sequence, and one of about 20 nt, which in minus-strand RNA serves as the promoter for the synthesis of 26S RNA (34). In addition, the genome has a recognition signal for its incorporation into nucleocapsid particles, the so-called encapsidation signal. In SIN RNA it is located in the coding region of nsP1 (35); and in SFV, in the region coding for nsP2 (36). The conserved 5′ and 3′ end sequences plus the encapsidation sequences have been regularly found in defective-interfering RNAs (DI-RNAs) of alphaviruses, although often in rearranged order (37–40). The extreme 5′ end of some DI-RNAs differs from that of the genome RNA (41, 42). This suggests that the conserved sequences downstream of the extreme 5′ terminus act as the RNA replication signals (43, 44) (Fig. 3).

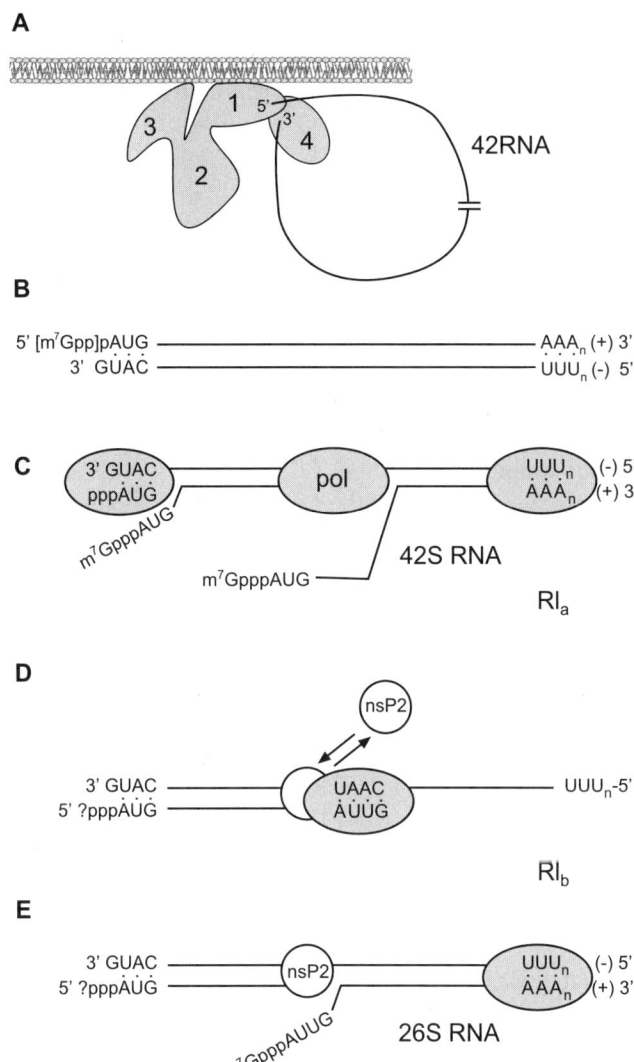

FIG. 3. Models for different stages in the replication of alphavirus RNA. (A) Initiation of minus-strand synthesis. Membrane-bound P123 + nsP4 recognize both ends of the template, which start to cycle through the polymerase. (B) A complete minus strand with a 3′ extra G, annealed with the plus strand, is shown. A puzzling observation indicates that double-stranded RNA in alphavirus-infected cells lacks a cap structure, shown in brackets (156), but it is not clear whether this form is an intermediate in replication or an aberrant dead-end product. (C) Classical model of the replicative intermediate RI_a, during the synthesis of genomic plus-strand RNAs with polymerase complexes (pol) consisting of nsP1–nsP4, engaged in initiation, elongation, and termination. (D and E) Initiation and termination of subgenomic RNA synthesis. A regulatory transcription factor (nsP2), interacting with the subgenomic promoter, is shown.

Synthesis of SFV and SIN 42S RNA minus strands takes place within the first hours of infection. At 37–40°C, minus-strand RNA synthesis is maximal around 2.5–3 h and is shut off approximately 4 h after infection. At 28°C, the shut-off takes place about 6 h after infection. Minus-strand RNA synthesis is accompanied by synthesis of plus strands at a ratio of about 1 to 5. The synthesis of minus-strand RNA is strictly dependent on protein synthesis and ceases within 15 min after addition of, e.g., cycloheximide, unlike the synthesis of RNA plus strands (45). Thus, continuous protein synthesis is required to produce the "minus-strand RNA polymerase." The virus specificity of this early polymerase has been confirmed by analyzing temperature-sensitive virus mutants. An RNA-negative SIN mutant, ts11, is defective only in minus-strand RNA synthesis, when infected cultures are shifted from the permissive (28°C) to the restrictive (39°C) temperature. The defect of ts11 is due to one amino acid replacement Ala348Thr in nsP1 (46), indicating that nsP1 is specifically involved in the regulation of minus-strand RNA synthesis (47, 48). Another SIN mutant, ts4, with an amino acid change Ala268Val in the nsP3 protein, is also defective in minus-strand synthesis (49).

An explanation for the short half-life of minus-strand RNA synthesis was generated by a series of studies on the processing intermediates of SIN nonstructural polyprotein (see Section V). Expression of recombinant vaccinia virus constructions, encoding various uncleavable polyproteins, showed that minus-strand RNA synthesis requires uncleaved P123, together with correctly cleaved nsP4 (50–54) (Fig. 3A). Processing of P123 into nsP1, nsP2, and nsP3 stops minus-strand RNA synthesis and enables their use as templates for synthesis of plus strands. Late in infection, new replication complexes cannot be assembled, as nsP2 protease rapidly cleaves P1234 through an alternative pathway first producing P12 and P34.

However, a number of alphavirus ts mutants have been characterized, which can reactivate minus-strand RNA synthesis after it has been shut off. The reactivation takes place in the presence of protein synthesis inhibitors (55–58). Many of the mutations have been mapped to the carboxy-terminal half of nsP2 (59); and one mutation, to nsP4 (57). Studies of the reactivation have suggested specific interactions between nsP1 and nsP4 (48) as well as between nsP2 and nsP4 (58). The mechanism of this puzzling phenomenon is still unknown, but it implies that in these mutants the late RNA polymerase, in which all proteolytic cleavages have taken place, can rearrange into the conformation of the early RNA polymerase after a temperature shift from 28 to 39°C. Thus, the structural differences between the two functions of the RNA polymerase must be small.

B. Plus-Strand RNA Synthesis

Quantitation by [^{32}P]orthophosphate equilibrium labeling has shown that about 200,000 molecules of both 42S and 26S RNAs have been synthesized at

8 h after infection in an SFV-infected BHK21 cell (60). As the number of minus strands has been estimated by dot-blotting to be approximately 5000 (48), RNA replication as a whole is highly asymmetric. The rate of 42S and 26S RNA synthesis is almost linear between 4 and 8 h after infection and is not affected by inhibition of protein synthesis (61), indicating that stable RNA replication complexes are continuously producing the plus-strand RNAs in replicative intermediates RI_a and RI_b. Interestingly, plus-strand RNA synthesis is not increased in mutant-infected cells, in which minus-strand RNA synthesis is not shut off (56). Neither is RNA synthesis affected by overproduction of viral nonstructural proteins (62).

The site of late RNA synthesis has been localized to alphavirus-specific structures, designated cytoplasmic vacuoles (CPVs) (63–66). Their diameter varies from 0.2 to 1 μm, and their surface consists of small invaginations, or spherules, with a diameter of about 50 nm (Fig. 4). The four nonstructural proteins as well as nascent RNA molecules are associated with CPVs and more closely associated with the spherules (66). The time of appearance of CPVs depends on the amount of infecting virus. With 10–20 plaque-forming units per cell, they first appear 3 h after infection, and their number starts to decline at 6 h. Addition of cycloheximide at 2 h after infection reduces the number of CPVIs, whereas addition after 4 h has no effect (67). This time scale coincides with the cessation of minus-strand RNA synthesis (45) and the cessation of the synthesis of nsPs

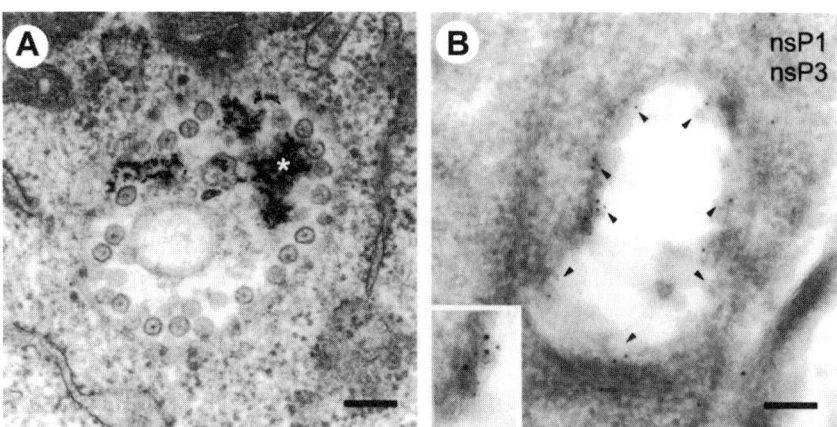

FIG. 4. Electron microscopic images of alphavirus-specific cytoplasmic vacuoles. (A) In an Epon section, invaginations or spherules line the membrane of the vacuole, whose endosomal nature is shown by endocytosed gold-labeled bovine serum albumin (asterisk). (B) Cryo-immunoelectron-microscopic image of spherules (arrowheads) double-labeled with antibodies against nsP1 (small gold particles) and nsP3 (large gold particles). Scale bars = 200 nm. Reproduced from Kujala et al., J. Virol., 75, 3873–3884 (2001), with permission by the American Society for Microbiology.

(68), suggesting that CPVs arise through the action of nsPs like the minus-strand RNA. Later in infection, the synthesis of plus strands continues in association with stable CPVs using the previously synthesized minus strands as templates. CPVs stain for endosomal and lysosomal markers, indicating that they are modified endosomes and lysosomes (64–66). Similar structures have been described in rubella virus-infected cells (69, 70).

Cytoplasmic membrane fractions, from SFV- and SIN-infected cells sedimenting at 15,000g contain essentially all virus-specific RNA polymerase activity of the cell (26, 71). The polymerase activity is associated with a smooth membrane fraction (27, 71, 72), which synthesises 42S and 26S RNAs and replicative intermediate RNAs. Attempts to isolate template-free RNA polymerase have so far failed. Expression of P123 plus nsP4 yields membrane fractions, which can support double-stranded RNA synthesis after addition of suitable positive-sense template RNAs containing the conserved 5' and 3' sequences of the genome RNA (54).

C. Synthesis of 26S mRNA of the Structural Proteins

At a time when the function of 26S RNA was not known, evidence accumulated that its synthesis was not coupled with the synthesis of 42S RNA. Puromycin, given early in SIN infection, inhibits only 26S RNA production (45, 73). RNA-negative ts mutants of SFV and SIN, which are unable to replicate at 39°C, have been isolated and characterized (74–77). With some of the mutants, the synthesis of only 26S RNA stops after shift to the restrictive temperature. One particular SFV mutant, ts4, turned out to be interesting in this respect. When ts4-infected cells were incubated at 28°C for 6 h to start the RNA synthesis, followed by a shift to 39°C in the presence of cycloheximide, the synthesis of 26S RNA was specifically shut off. However, it was resumed, once the cultures were shifted back to 28°C. The reversal was independent of protein synthesis (78). Careful analysis of the labeled RIs and double-stranded RNAs from ts4-infected cells revealed a reversible shift of RI_b to RI_a. This leads us to suggest that a virus-specific protein regulates the initiation of 26S RNA synthesis by binding reversibly to a promoter on the minus-strand RNA within the replicative intermediate RI_b (79) (Fig. 3D and 3E). SFV ts4 has a single-amino acid replacement Met781Thr in nsP2. Reversible cross-linking experiments suggested that an increased amount of nsP2 was associated with the other nsPs at 28°C as compared to 39°C (59). All SIN ts mutants with a defect in 26S RNA synthesis have been mapped to the C-terminal half on nsP2 (46, 59). However, direct proof that nsP2 interacts with the 26S RNA promoter is still lacking.

The subgenomic RNA promoter in the 42S minus strand has been carefully mapped for SIN RNA (80). The minimal region is 19 nucleotides upstream and 5 nucleotides downstream of the initiation site of 26S RNA transcription.

However, a larger region containing nucleotides −98 to +14 is required for optimal transcription (*81*). More detailed mapping revealed that optimal transcription could be obtained with sequences −40 to −20 and +6 to +14 together with the core promoter (*82*).

V. Processing of Alphavirus Nonstructural Polyprotein P1234

Short-lived processing intermediates P123, P12, and P34 have been identified as precursors for the four nonstructural proteins, later designated nsP1–nsP4 (*2, 83–85*) (Fig. 1). Proteins of similar size were also identified in SIN-infected cells (*76, 86*). Sequencing of the SIN (*33*) and SFV (*32*) genome RNAs revealed differences in the translation strategies of the nonstructural proteins. In SFV-infected cells, the entire nonstructural region coding for 2432 aa is translated as polyprotein P1234. In the SIN genome, there is an opal termination codon close to the carboxy terminus of nsP3. This is occasionally suppressed, giving rise to P1234 (2513 aa). Thus, in SIN-infected cells, an excess of P123 is produced, and nsP4 can only be produced by processing of P1234 (*34*). *In vitro* translation of SIN genome RNA has been used to study the processing of P123 and P1234, produced from an RNA where the opal codon was mutated to a codon for cysteine. Both P123 and P1234 are processed *in vitro* autocatalytically due to protease activity of nsP2 (*87–90*). This complex process was approached by using constructions in which the cleavage sites were mutated either alone or in combination with substrates in which the protease activity had been eliminated (*34, 89*). The experiments resulted in cleavage "rules": (i) P1234 can cleave autocatalytically to produce P123 and nsP4; (ii) P123 cannot cleave in *cis* but can be cleaved only in a bimolecular reaction; (iii) P12 can slowly cleave autoproteolytically to nsP1 and nsP2; (iv) all three sites can be cleaved *in trans* (*34, 91*). However, the most important result from this work was the possibility to express stable precursor proteins, which revealed the nature of the "early RNA polymerase" (P123 plus nsP4) (*52, 53*).

Many of these experiments have been repeated with SFV nonstructural polyproteins translated *in vitro*. The polyproteins were also expressed alone and in combinations in insect cells using baculovirus vectors. Coexpression of protease P23 with proteolytically defective P1234 or P123 yields P12 in addition to nsP3 from P123 and nsP3 and nsP4 from P1234. Expression of P12 or P23 in insect cells and *in vitro* resulted in their cleavage to nsP1 and nsP2 and to nsP2 and nsP3, respectively. All attempts to cleave proteolytically defective P12 or P123 *in trans*, using, e.g., P123 as the protease, have failed. P34 and P1234 yielded regularly nsP4 by P23 protease, indicating that cleavage at the 3/4 site takes place *in trans* (*92*; Merits *et al.*, unpublished results).

VI. nsP1: A Unique RNA-Capping Enzyme and Membrane Anchor

In recent years, nsP1 has been one of the main objects of our studies. We have discovered a novel kind of coupled methyltransferase and guanylyltransferase activity needed in the capping of virus-specific mRNA molecules and conserved in the alphavirus-like superfamily. On the other hand, we have studied the membrane interaction of nsP1, which is mediated by covalent palmitoylation and by direct interaction of the polypeptide chain with anionic membrane lipids (Fig. 5). Genetic evidence obtained by us and others indicates that nsP1 is involved in the synthesis of minus-strand RNAs.

A. Methyltransferase and Guanylyltransferase Activities

NsP1 is a guanine-7-methyltransferase transferring a methyl group from S-adenosyl-methionine (AdoMet) to GTP (93, 94), and it forms a covalent complex with 7-methyl-GMP (m^{7}GMP) in a guanylyltransferase-like reaction (95). These results indicate that nsP1 is an enzyme acting in the formation of m^{7}GpppA cap structures on virus-specific genomic and subgenomic mRNAs. The substrate specificity of nsP1 differs from cellular RNA-capping enzymes. In the methyltransferase reaction, nsP1 prefers GTP (and dGTP) as methyl-accepting substrates, whereas cellular enzymes methylate cap analogs and unmethylated guanosine-capped RNAs (94, 96) (Fig. 6). In the guanylyltransferase reaction, nsP1 forms a covalent complex exclusively with m^{7}GMP, whereas cellular enzymes form a GMP–enzyme complex. These specificities demonstrate that alphaviruses possess a novel RNA-capping pathway, where GTP is first methylated and only thereafter forms a covalent m^{7}GMP intermediate with the enzyme (95) (Fig. 6). In capping of eukaryotic nuclear RNA, guanylyltransferase first transfers GMP to RNA via a covalent intermediate, and 7-methylation is the final

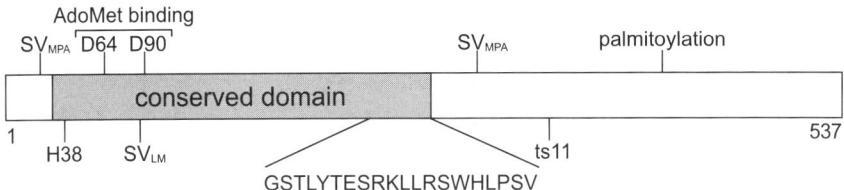

FIG. 5. Scheme of nsP1, showing the location of mutations and other features discussed in the text. SV$_{LM}$ is a SIN mutant resistant to low methionine, and SV$_{MPA}$ is a mutant resistant to mycophenolic acid. The sequence of the lipid-binding peptide, which is located at the C terminus of the conserved domain, is given in single-letter code. The numbering follows SFV sequence.

A

(1) pppN$_1$N$_2$N$_3$... $\longrightarrow$ ppN$_1$N$_2$N$_3$... + P$_i$

(2) GTP + AdoMet $\longrightarrow$ m^{7}GTP + AdoHcy

(3) m^{7}GTP + nsP1 $\longrightarrow$ m^{7}GMP-nsP1 + PP$_i$

(4) m^{7}GMP-nsP1 + ppN$_1$N$_2$N$_3$... $\longrightarrow$ m^{7}GpppN$_1$N$_2$N$_3$... + nsP1

B

(1) pppN$_1$N$_2$N$_3$... $\longrightarrow$ ppN$_1$N$_2$N$_3$... + P$_i$

(2) GTP + enzyme $\longrightarrow$ GMP-enzyme + PP$_i$

(3) GMP-enzyme + ppN$_1$N$_2$N$_3$... $\longrightarrow$ GpppN$_1$N$_2$N$_3$... + enzyme

(4) GpppN$_1$N$_2$N$_3$... + AdoMet $\longrightarrow$ m^{7}GpppN$_1$N$_2$N$_3$... + AdoHcy

FIG. 6. (A) RNA-capping reactions catalyzed by alphavirus proteins nsP2 (reaction 1, RNA triphosphatase) and nsP1 (reactions 2–4, guanine-7-methyltransferase and mRNA 7-methyl guanylyltransferase). Although nsP1 can release m^{7}GTP to solution (2), the synthesis of the covalent complex m^{7}GMP–nsP1 requires methylation and complex formation (reactions 2 and 3) to take place concomitantly, without the release of m^{7}GTP (18). The transfer of the bound nucleotide to RNA acceptor (step 4) has not been demonstrated. (B) RNA-capping reactions catalyzed by cellular RNA triphospahatase (1), mRNA guanylyltransferase (2–3), and mRNA guanine-7-methyltransferase (4) enzymes.

step taking place on the RNA molecule (for reviews, see Refs. 97 and 98). In both patways, these reactions are preceded by RNA 5' triphosphatase, which removes the 5' gamma phosphate of the nascent RNA. In alphaviruses, nsP2 has RNA triphosphatase activity (see Section VII,B).

NsP1-related proteins encoded by members of the alphavirus-like superfamily catalyze similar virus-specific RNA-capping reactions. Methyltransferase activity and covalent complex formation exclusively with m^{7}GMP have been demonstrated for TMV p126 (99), BMV 1a (100, 101), and the N-terminal fragments of bamboo mosaic potexvirus (102) and HEV replicase proteins (103). Thus, all members of the alphavirus-like superfamily are likely to share a similar conserved RNA-capping activity. Since it differs from cellular RNA capping, it offers a target for virus-specific inhibitors of RNA replication. So far, it has been shown that some cap analogs can inhibit SFV nsP1 and HEV-capping enzyme (96, 103). Importantly, in this respect, it has been demonstrated by mutational analysis that the capping enzyme activities are essential for alphavirus and BMV replication (104, 105). In a BMV replication system in yeast cells, it was possible to directly demonstrate the role of 1a protein in viral RNA capping *in vivo,* as mutation of the host exonuclease XRN1 permitted replication and thus the study of uncapped viral RNAs (105).

Some mechanistic details of the reactions catalyzed by nsP1 and related proteins have been studied. The guanylyltransferase reaction absolutely requires divalent cations (Mg^{2+} or Mn^{2+}), whereas the methyltransferase reaction of at least some family members can proceed in the presence of EDTA (94, 95, 100, 103). For BMV 1a, it has been directly demonstrated by NMR spectroscopy analysis of reaction products that methylation takes place only on the 7-position of guanylate (100), which is in keeping with the fact that alphavirus-like viral mRNAs are capped with cap0 structure m^7GpppN lacking further methyl groups. Although nsP1 exclusively forms a covalent complex with m^7GMP, it cannot accept m^7GTP directly as a substrate for complex formation. Instead, GTP and AdoMet are required, indicating that methylation and covalent complex formation are normally coupled (18). dGTP is a good substrate for methylation by both nsP1 and BMV 1a, but yet these enzymes preferentially form a covalent complex with m^7GMP, as compared to m^7dGMP, indicating that further selection takes place at the guanylyltransfer step (95, 100, 101). BMV 1a with a covalently bound guanylate appears to be inhibited in further methyltransferase reactions (100). SFV nsP1 has to be associated with anionic phospholipids in order to be active in the capping reactions (106; see below). This feature may not be shared among all members of the alphavirus-like superfamily (103).

As indicated above, the RNA-capping pathway is conserved within the alphavirus-like superfamily, and the methyltransferase and guanylyltransferase activities are associated with the conserved methyltransferase/capping enzyme domain, which is the hallmark of the alphavirus-like superfamily. It was initially thought that this domain would bear no resemblance to any cellular proteins (17). Only when the structures of several cellular methyltransferases became available, it was demonstrated that the predicted secondary structure of a portion of nsP1 and related viral proteins was similar to known methyltransferase structures (18). It was also shown by cross-linking that two conserved residues Asp64 and Asp90 in two adjacent loops of SFV nsP1, which correspond to the AdoMet-binding region of structurally studied methyltransferases, are needed for AdoMet binding and methyltransferase activity (18, 100). The same region is implicated in AdoMet binding by a double-mutation Arg87Leu and Ser88Cys in SIN nsP1, which enables virus growth in methionine-deprived cells and lowers the K_M of the methyltransferase for AdoMet (107, 108). Together, these results strongly suggest that nsP1-related proteins structurally resemble methyltransferases and pinpoint the AdoMet-binding site (Fig. 5).

In contrast, it is less clear where the methyl-accepting GTP substrate binds, as mutations in conserved residues do not abrogate GTP cross-linking to the protein (18). However, a SIN mutant resistant to mycophenolic acid and ribavirin, compounds which lower intracellular GTP concentration by inhibiting inosine monophosphate dehydrogenase, appears highly interesting in this respect (109). An nsP1 double-mutation Ser23Asn and Val302Met is required for resistance (110, 111), which may indicate that these two regions of the protein flanking the conserved methyltransferase domain may come together in the three-dimensional structure to form the GTP-binding site.

NsP1 does not seem to have any similarity to the cellular guanylyltransferase family of proteins (18). In cellular enzymes, a conserved lysine residue forms the covalent linkage with GMP. There are no conserved lysines in nsP1-like proteins, although the properties of the covalent linkage resemble that of a phosphoamide-type bond (95). Instead, an absolutely conserved histidine (His38 in SFV nsP1) is a good candidate for covalent binding of m^{7}GMP, in that mutations of this residue permit retention of methyltransferase activity but completely abolish covalent nucleotide binding (18, 100). Thus, joining of guanylyltransferase with the methyltransferase region and coupling of these two activities may represent an evolutionary innovation, which took place in a progenitor of the alphavirus superfamily. It should be noted that although the region showing high conservation within the alphavirus-like superfamily encompasses only 200 aa (17), at least alphavirus nsP1s require approximately 500 aa for enzymatic activities (18, 104). Considerable structural variation may therefore exist in the capping enzymes of alphavirus-like viruses.

Finally, the active site of nsP1-related proteins may have an additional role in viral RNA replication. Mutations of BMV 1a active site residues can either abolish or enhance the recruitment of viral RNA to a membrane-bound stable form prior to replication. This is a complex pathway requiring host proteins in addition to 1a, and even 1a appears to have several roles in the process. Based on the mutation data, one of these roles could either be influenced by binding of substrates involved in RNA capping or more directly involve a direct recognition of the RNA cap structure by 1a, as one step of recruiting RNA from translation to replication (105).

B. Membrane Association

NsP1 is tightly associated with membranes both in alphavirus-infected cells and when expressed alone in mammalian cells (112, 113). It can only be released to solution by detergents or highly alkaline solutions (pH 12) (112, 114). The tight membrane interaction of nsP1 is mediated by covalent palmitoylation of one or more of the cysteine residues 418–420 in SFV nsP1, and a single-cysteine 420 in SIN nsP1 (113, 114) (Fig. 5). Conserved cysteine(s) are found in this position in all alphavirus nsP1s, suggesting that palmitoylation is a conserved feature. However, if the palmitoylation sites of nsP1 are removed by mutagenesis, the protein still retains a weak peripheral affinity for membranes (114). Nonpalmitoylated nsP1 can also function in SFV or SIN replication in complexes that appear normal in their localization and morphology. The resultant viruses show only very slightly reduced growth rates in cell culture, but neuropathogenesis of at least SFV encoding nonpalmitoylated nsP1 is abolished in mice (113).

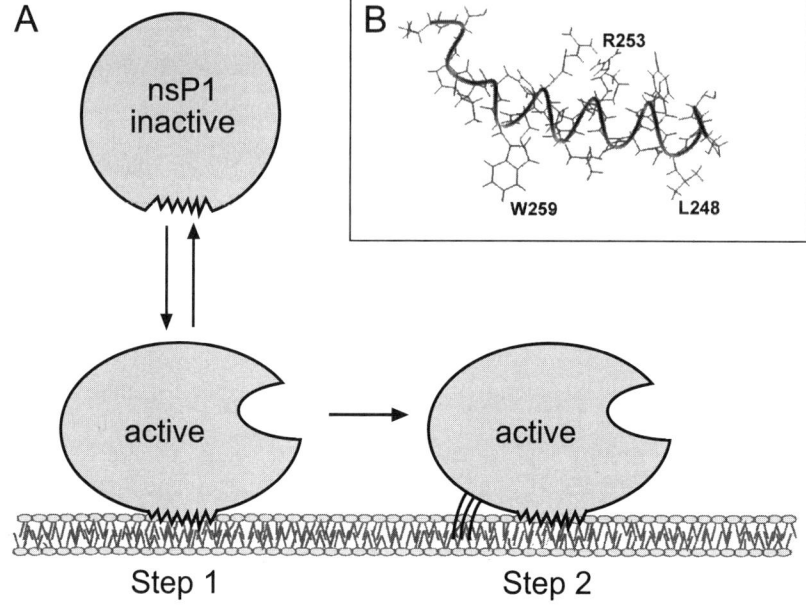

FIG. 7. (A) A schematic model for the membrane binding of nsP1. Association of the lipid-binding peptide with membranes containing anionic phopholipids changes the conformation of nsP1 and activates it as capping enzyme. Association via the binding peptide is weak (step 1) and reversible by high salt, whereas palmitoylated form is bound tightly (step 2) resembling integral membrane proteins. (B) NMR solution structure of the synthetic binding peptide (aa 245–264) of nsP1. W259, which points downwards in the image, intercalates deeply into the lipid bilayer. Figure 7B reproduced from Lampio et al., J. Biol. Chem. **275**, 37,853–37,859 (2000), with permission by the American Society for Biochemistry & Molecular Biology.

Nonpalmitoylated nsP1 binds to membranes through direct interaction of the polypeptide chain with anionic phospholipids (Fig. 7). This interaction can be observed for nsP1 expressed in *Escherichia coli* or by *in vitro* translation and is mainly mediated by amino acids 245–264, where both positively charged and hydrophobic residues are required for binding (*106*). As studied by NMR spectroscopy, the respective synthetic peptide forms an amphipathic alpha-helix, which can interact with liposomes containing acidic phospholipids. Trp259 is crucial for association and intercalates deep in the bilayer (*115*). These results are consistent with monotopic binding of nsP1 to membranes, mediated by an amphipathic alpha-helical peptide using both ionic and hydrophobic interactions. Such a mechanism appears to be common for several diverse groups of proteins (*116*). The nsP1 polypeptide segment 245–264 is among the most highly conserved in the alphavirus-like superfamily (*17*) and could play a role in the membrane association of other superfamily members. All nsP1-related proteins studied to date are associated with membranes (*100, 102, 103*), although the binding mechanism has not been studied in other cases. The palmitoylation site is not conserved in other superfamily members and, indeed, they do not appear to be palmitoylated.

Association with anionic membrane phospholipids is also required for the enzymatic activities of nsP1 (*106*) (Fig. 7). The methyltransferase and guanylyltransferase activities of nonpalmitoylated nsP1 are strongly inhibited by detergents and reactivated by added phospholipid vesicles or mixed micelles containing anionic phospholipids. Phospholipids appear to act by inducing a conformational change is nsP1, as cross-linking of the substrate AdoMet is also inhibited by detergents and reactivated by anionic lipids (*106*). The behavior of palmitoylated nsP1 is slightly more complex. It is also inactivated by several detergents, but activated by deoxycholate and octylglucoside (*94*), which may interact favorably with the palmitoylated form of the protein and permit maintenance of the active conformation. These results indicate that nsP1 is designed to function in a membraneous environment containing anionic phospholipids.

When expressed alone in mammalian cells, both palmitoylated and nonpalmitoylated forms of nsP1 localize predominantly to the cytoplasmic surface of the plasma membrane and also to that of endosomes and lysosomes (*112, 114*). The plasma membrane and the endosomal apparatus, which are connected by vesicle transport, are rich in anionic phospholipids, and therefore the hypothesis arises that the affinity of nsP1 to these lipids directs its intracellular localization (*66*). Palmitoylation could serve a secondary role by fixing the protein tightly to membranes after the initial interaction. However, only palmitoylated forms of nsP1 expressed in animal cells either by themselves by transfection or in the context of alphavirus infection are capable of inducing thin filopodia-like protrusions of the cell surface and selectively disrupting the actin cytoskeleton (*117*). The significance of these phenotypic changes is not clear, but it is intriguing to speculate that they would be connected with the increased pathogenicity of the wild-type virus, as compared to virus-encoding palmitoylation-deficient nsP1 (*113*).

C. Role in Minus-Strand RNA Synthesis

So far, only one of the alphavirus ts mutants, SIN ts11, where Ala348 is changed to Thr, has been mapped to nsP1 (*46*). The phenotype of ts11 has been highly informative, as it exhibits a rapid and selective cessation of minus-strand RNA synthesis but does not interfere with plus-strand synthesis catalyzed by stable replication complexes (*47*). Ts11 suppresses nsP4 mutant ts24 (Gln191Lys), which in otherwise wild-type background allows reactivation of minus-strand synthesis by the stable replication complexes. This is interpreted to mean that nsP1 always directly participates in minus-strand synthesis during either initiation or elongation (*48*). A mutation of the adjacent amino acid in SIN nsP1, Thr349Lys, is able to suppress the effect of a normally nonviable change in nsP4 of the N-terminal Tyr to Ala, Arg, or Leu. As an alteration of the nsP4 amino terminus appears to also cause a defect in minus-strand synthesis, it can be hypothesized that an interaction of these regions of nsP1 and nsP4 is required for minus-strand synthesis or promoter recognition (*118*) (Fig. 3A).

VII. nsP2: A Multifunctional Enzyme and Regulatory Protein

NsP2 is the largest replicase protein consisting of 794–807 aa residues among different alphaviruses. Several SIN RNA-negative ts mutants of group A have been mapped to various parts of nsP2 (*46, 59, 119*). They indicate that nsP2 functions in both the regulation of subgenomic RNA synthesis and the regulation of shut-off of minus-strand synthesis (see Section IV). According to sequence comparisons, the N-terminal half of nsP2 has sequence motifs typical for RNA and DNA helicases, whereas the C-terminal part resembles papain-like cysteine proteinases (Fig. 8) (*88, 120, 121*).

A. Nucleosidetriphosphatase and RNA Helicase Activities

Direct evidence of nucleosidetriphosphatase (NTPase) activity has been demonstrated for purified nsP2 preparations expressed in *E. coli* (*122*). The activity is associated also with the 470 N-terminal amino acids of the protein. NTPase activity is stimulated by poly(A), poly(U), oligo(A), and tRNA. Mutation of the putative NTP-binding site GVPGSG**K**$_{192}$S to GVPGSG**N**$_{192}$S inhibits the NTPase activity. Both nsP2 and nsP2-N bind to single-stranded RNA and can be cross-linked to oxidized ATP, GTP, CTP, and UTP. When the GNS mutation was introduced to the SFV genome, no virus replication could be detected. After longer incubation, a revertant virus appeared, in which GNS had been back-mutated to GKS (*122, 123*).

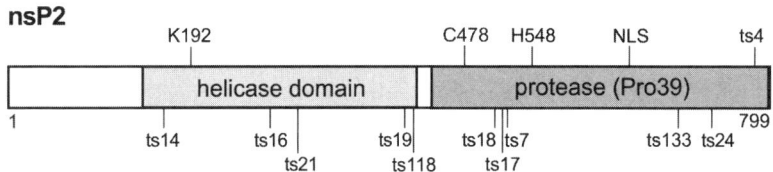

FIG. 8. Scheme of nsP2 showing the conserved helicase domain as well as the protease domain (Pro39). Some of the active site residues are indicated together with the numerous ts mutations mapped to nsP2. All mutants, except SFV ts4 are SIN mutants. NLS is the nuclear localization signal of nsP2.

NsP2 is also an RNA helicase, which can unwind partially double-stranded RNA preparations in the presence of NTPs and dNTPs (124). The reaction is inhibited by the GNS mutation suggesting that NTPase activity is necessary to drive the unwinding (125). The helicase activity is inhibited in NaCl concentrations higher than 100 mM. So far, helicase activity has not been demonstrated for the nsP2-N fragment.

After demonstration of RNA helicase activity for nsP2, it could be presumed that all homologous proteins encoded by members of the alphavirus-like superfamily have helicase activity. The role of helicase in the replication cycle of positive-strand RNA viruses remains mysterious, although it is often thought to either separate double-stranded RNAs or remove secondary structures from template RNAs. A ts mutant within the helicase domain of BMV 1a encodes a protein defective in all forms of RNA synthesis, consistent with this line of thinking (126). On the other hand, the phenotype of SIN RNA-negative ts mutants located in the helicase is not very clear-cut but has been interpreted to support a role in the conversion from negative-strand to positive-strand synthesis (119). Finally, mutations in the helicase motifs of BMV 1a completely prevent minus-strand synthesis but also disrupt severely the initial recruitment of viral RNAs to the replication complex (105). Thus, the helicase proteins are likely to have multiple functions at different stages of the replication cycle, which need to be untangled in further experiments.

B. RNA Triphosphatase Activity

As described above, the methyltransferase and guanylyltransferase activities needed in the capping of alphavirus RNAs are carried out by nsP1 (see Section VI,A). However, nsP1 lacks RNA triphosphatase activity essential for mRNA capping (97, 98). Using short γ-^{32}P-labeled RNA molecules for screening, nsP2 has been identified as an RNA triphosphatase for both SFV and SIN. This activity, which removes specifically only the outmost γ-phosphate, is also associated with the nsP2-N fragment, previously shown to have NTPase activity (122, 127). Unlike the helicase activity, the RNA triphosphatase activity tolerates relatively high-salt concentrations. There is an absolute requirement for divalent cations, indicating that the alphavirus enzyme belongs to the metal-dependent RNA triphosphatases (98). The GNS mutation in the NTP-binding site inactivated the triphosphatase activity, suggesting that NTPase and RNA triphosphatase reactions take place in the same reaction center. If so, NTPs should competitively inhibit the RNA triphosphatase reaction, as shown for, e.g., vaccinia virus-capping enzyme (128, 129). However, this is not the case, as addition of GTP rather enhanced the triphosphatase activity, suggesting that hydrolysis of γ-phosphate of NTPs and that of RNA take place in different reaction centers. This mode has been suggested for flavivirus NS3 protein, which is a 68-kDa multifunctional protein with NTPase/RNA triphosphatase/protease/RNA helicase activities (130). It must be remembered that triphosphatase activity is needed only once during the synthesis of an RNA molecule, while RNA helicase/NTPase activities may be required throughout the synthesis.

C. Protease Activity

Both mapping of active site residues (90) and assaying the sensitivity to inhibitors (131) strongly support the hypothesis that the C-terminal part of nsP2 is a papain-like cysteine protease. To better understand the complex enzymatic processing of the nonstructural polyprotein (see Section V), nsP2 and a soluble C-terminal fragment, Pro39, consisting of amino acid residues 459–799 of SFV nsP2, was recently expressed and purified (131). In vitro synthesized, labeled, proteolytically inactive polyproteins $P12^{CA}$, $P2^{CA}3$, $P12^{CA}$, P34, $P12^{CA}3$, $P12^{CA}34$, with the Cys478Ala mutation in the predicted active site of the enzyme, were used as substrates for Pro39 and nsP2. Cleavage was observed at all sites but with different efficiencies. Site 3/4 (between nsP3 and nsP4) was cleaved most readily, site 1/2 next, and site 2/3 poorly (Fig. 1).

In order to eliminate possible shielding effects within the polyprotein, short cleavage site sequences were joined to a thioredoxin carrier. By this means, it was possible to study the proteolysis in vitro with purified reagents. The cleavage products were isolated, and their masses and N-terminal sequences were determined. These results show that (i) the cleavage takes place at the same sites as those determined in virus-infected cells (132, 133); (ii) the cleavage at site 3/4 is most efficient; (iii) about 5000-fold more enzyme is required for complete cleavage at site 1/2, whereas cleavage at site 2/3 remains poor even with huge excess of Pro39-enzyme (131). These results are compatible with the previous observations that processing at site 3/4 takes readily place in trans. The poor cleavability of site 1/2 cannot be due to inaccessibility for the enzyme in this in vitro system. Rather, it suggests that some cofactor(s) is needed. This hypothetical "factor" is evidently present in polyproteins P12, P123, and P1234, in which the cleavage at site 1/2 takes place in cis. The very inefficient cleavage at site 2/3 is surprising, since P23 is cleaved rapidly after expression in insect cells and after in vitro translation (92; Merits et al., unpublished). In the polyprotein precursors P1234 and P123 the cleavage of 2/3 site must be preceded by cleavage at the site 1/2 yielding P23 or P234 (89, 92). Thus, cleavage at site 2/3 must also require a "cofactor," which is probably different from that needed in the cleavage at site 1/2. The specificity of the cleavage at all three sites is determined solely by the protease moiety of nsP2, but the efficiency of the cleavage requires unidentified cofactors.

D. Nuclear Transport and Neuropathogenicity

Immunofluorescence microscopy has shown that of the nsPs, only nsP2 is transported to the nuclei and nucleoli of SFV-infected cells (134). Cell fractionation and immunoprecipitation experiments showed that about 25% of labeled nsP2 was associated with cytoplasmic membranes sedimenting at 15,000g, 25%

remained in the supernatant fraction, and about 50% in the nuclear fraction. The transport took place already early in infection, and nsP2 was detectable in the nuclear matrix fraction within 5 min in pulse–chase experiments, reaching 50% in 20 min. NsP2 had to be cleaved from its precursors (P12 and P123) before transport took place. When nsP2 was expressed alone in BHK21 cells using an SV40-based vector, almost all of the protein was transported to the nucleus.

A deletion analysis was carried out to find the putative nuclear localization signal (NLS) of nsP2. We identified a pentapeptide region $PR_{648}RRV$ responsible for the nuclear transport of nsP2 (Fig. 8). Mutations changing arginines 648–649 to aspartic acids (RDR, DRR, and DDR) resulted in cytoplasmic nsP2. A fusion protein, consisting of β-galactosidase and 232 C-terminal amino acid residues of nsP2, was transported to the nucleus only if it had an intact PRRRV sequence. Further sequences, especially within region 470–490, were required for nucleolar targeting (135). The NLS of nsP2 was also functional in the yeast *Saccharomyces cerevisiae*, since expression of SFV P1234 resulted in the accumulation of nsP2 in the nucleus (136).

When the RDR mutation was introduced to infectious cDNA, transfection of cells with the 42S transcript resulted in virus production (SFV-RDR), showing that functional NLS of nsP2 is not vital for SFV replication. A closer analysis of the synthesis of nsP2 and its precursors showed that the synthesis of nsPs was delayed early in infection. Later in infection, there was a considerable delay in the processing of the polyproteins (P1234, P123, and P12). There was also a clear delay in the virus release from the SFV-RDR mutant-infected cells. However, the final yields were the same as those with the wild type. Normally, SFV inhibits cellular protein and RNA and DNA syntheses efficiently (137). There is a clear difference in the inhibition of host cell DNA synthesis at 6 h after infection between wild type (13% of uninfected control cells) and SFV-RDR (about 50%). No significant difference in the inhibition of host protein synthesis was found between the wild type and the mutant. Most interestingly, SFV-RDR has lost its pathogencity for adult mouse injected intraperitoneally (123).

Both wild-type and SFV-RDR are neuroinvasive after intraperitoneal inoculation. However, wild-type SFV spreads rapidly, infecting cells throughout the brain, while SFV-DRD infection is confined to small foci of cells. When mice deficient in type I interferon responses were infected with SFV-RDR, the virus was distributed widely in the brain causing death (137a). These results stress the importance of the interferon system as a host defence mechanism. They also show that a component of RNA polymerase complex can be vital for the neuropathogenicity of SFV. The slower processing of P1234 strongly suggests that the RDR mutation affects the protease moiety of nsP2, causing a delay in early replication events. This delay may be vital for the host defense system to be able to limit the infection to the primary foci of infected cells. Thus, the nuclear localization of nsP2 may reflect the structural requirements of the multifunctional nsP2 protein rather than a separate nuclear function.

Mutation of the nsP1 palmitoylation site also results in a viable but apathogenic SFV mutant, which causes viremia in mice, but no infectious virus is detectable in the brain (see Section VI,B). In HeLa and BHK cells, there is a delay in virus replication during infection (113). Furthermore, deletion of 50 amino acid residues from the variable region of SFV nsP3 results in viable but apathogenic virus mutant (see Section VIII,B). There is a clear reduction in the RNA synthesis rate between 2.5 and 6 h after infection, which results in a delay of virus release from the cells, although final virus yields are same as those with the wild type (138). One simple possibility is that, with all these mutants, the delay in virus replication is enough for the host to successfully limit the infection of brain cells, which usually causes the death of the animals in wild-type SFV infection. However, further studies may reveal that each of these proteins has specific functions in the pathogenesis of alphaviruses.

VIII. nsP3: An Ancient Conserved Protein and Phosphoprotein

Alphavirus nsP3 is an intriguing hybrid protein. Its N-terminal domain of unknown function exhibits a high degree of similarity to some cellular proteins, whereas its carboxy terminus is hypervariable in length and sequence even among alphaviruses. Approximately 320 N-terminal aa of nsP3 show conservation between different alphaviruses. These are followed by a "tail" region of 150–250 aa, rich in acidic residues serine and threonine and devoid of predicted secondary structure (Fig. 9A). A main line of investigation by us and others has been the characterization of nsP3 "tail" phosphorylation.

A. Sequence Conservation

The most N-terminal 160 aa of alphavirus nsP3 form a small domain, which shows an unusual pattern of conservation. Related sequences are found not only in the genomes of rubella virus and hepatitis E virus but also in otherwise unrelated coronaviruses (but not in alphavirus-like plant viruses or coronavirus-like arteriviruses) (120). Recently, it has become apparent that this domain is widely, although not universally, distributed in bacteria, archae, and eukaryotes (139). It usually exists on its own as a small protein, but the human genome also contains open reading frames in which this domain is repeated. A gene termed *BAL*, which contains a duplicated nsP3-like domain, is a highly expressed risk factor in certain aggressive lymphomas (140). The conservation between viral and cellular homologs is unusually high for an RNA virus protein, reaching up to 35–40% sequence identity (139). Furthermore, an unusual class of histone variants found in animal cells predominantly associated with the inactivated X chromosome, the macrohistones H2A, contains a more distantly related nsP3-like domain following the core histone domain (139). The biochemical function of the nsP3 domain remains unknown, but the high conservation points to a basic and essential role in catalysis or binding. A possible clue is provided by the yeast gene YBR022w, the product of which was shown to hydrolyze ADP-ribose-1″phosphate (*141*). However, the yeast homolog is much less related to human and bacterial proteins than the latter are to each other or to nsP3, pointing to a possible divergence of function. Furthermore, a more general role for such an unusual nucleotide compound may be difficult to envision. Nevertheless, nsP3-like proteins could potentially be phosphoesterases hydrolyzing an as yet unidentified substrate.

B. Phosphorylation

NsP3 is phosphorylated on threonine and more heavily on serine, but not on tyrosine (*142, 143*). Deletion mapping suggested that the phosphorylation sites of SIN and SFV nsP3 are mainly located in the variable C-terminal region (*138, 144*). This has been proven in the case of SFV nsP3, where the phosphorylation sites have been mapped by mass spectrometric methods (*145*), supported by mutation analysis (*138*). The phosphorylation pattern is complex, as a total of 16 possible sites (Ser320, 327, 332, 335, 347, 348, 349, 352, 356, 359, 362, 367, and Thr344, 345, 350, 354) can become phosphorylated. Differentially phosphorylated forms of the protein exist, since forms containing from 7 up to the 12 possible phosphates in the region Thr344–Ser367 have been detected. Based on point mutational analysis, Thr344/Thr345 appear to be the major threonine phosphorylation sites. Thus, phosphorylation is confined to a small subregion of the tail (Fig. 9A), the deletion of which abolishes phosphorylation both in the cells expressing nsP3 alone and in the context of virus infection (*138, 145*).

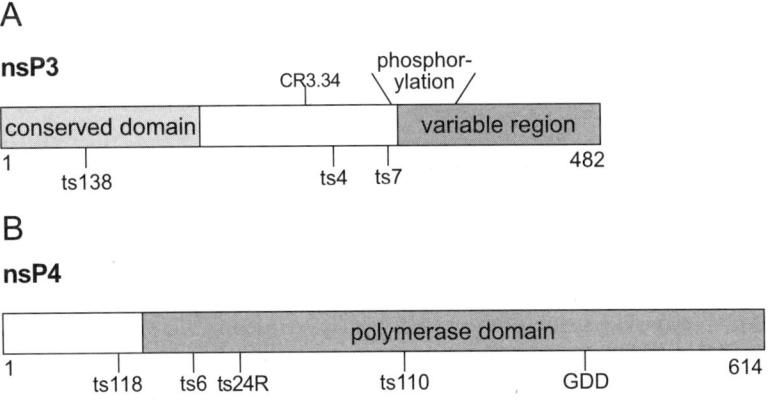

FIG. 9. Schemes of nsP3 (A) and nsP4 (B). Domains and mutations discussed in the text are indicated. CR3.34 is the linker insertion mutation after aa 252 of SIN nsP3, and GDD denotes the most conserved polymerase motif in nsP4.

The kinases responsible for phosphorylation of nsP3 remain unknown. Although a kinase resembling casein kinase II in its biochemical properties can phosphorylate SIN nsP3 (*143*), the presence of multiple sites with dissimilar sequence contexts suggests phosphorylation by multiple cellular kinases (*138*). Similarly, the role of phosphorylation in RNA replication is not clear. Although phosphorylation is not essential, as deletions reducing or abolishing phosphorylation are tolerated, the deletions have biological effects. Deletions in SIN nsP3 tail reduce plaque formation in mosquito cells (*144*), and deletions in SFV nsP3 tail reduce the level of RNA synthesis in BHK cells, although virus growth remains unchanged (*138*). Most interestingly, a SFV variant devoid of nsP3 phosphorylation also exhibits greatly reduced mouse pathogenicity (*138*). Another study has stressed the importance of the C-terminal variable region of nsP3 for neurovirulence of SFV in mice (*146*). Phosphorylation and the entire variable tail of nsP3, which has been subject to rapid alteration during alphavirus evolution, may be involved in the optimization and fine-tuning of replication in diverse host cell types.

C. Other Features

NsP3 has a weak peripheral affinity for membranes (*138*) and it may contribute to the membrane association of the replication complex. When expressed alone in mammalian cells, nsP3 associates with unidentified cytoplasmic vesicles, which do not contain endosomal or lysosomal markers (*65*). Membrane association is mediated by the region of nsP3 conserved within alphaviruses and it does not require phosphorylation (*138*). However, in SFV-infected cells, nsP3 found in the membrane-associated replication complex fraction is phosphorylated to a greater extent than soluble nsP3 (*142*).

Genetic experiments have also failed to define a precise role for nsP3, although it is essential for RNA synthesis (*46*). Both classical and linker insertion ts mutants indicate that nsP3 functions in the formation of early replication complexes synthesizing negative-strand RNA (*49, 147*), where it presumably is present in precursors P123 or P23. More surprisingly, an insertion at SIN nsP3 aa 252 leads to a specific defect in subgenomic RNA synthesis (*147*). No information is available concerning the functions of nsP3-like sequences of other viruses.

IX. nsP4: A Catalytic RNA Polymerase Subunit

NsP4 (607–614 aa in different alphaviruses) contains a large C-terminal domain of almost 500 aa related in sequence and predicted secondary structure to RNA-dependent RNA polymerases (*16, 148*) (Fig. 9B). Experiments with ts mutants strongly support the contention that nsP4 is the catalytic polymerase subunit. SIN mutations Gly153Glu (ts6) and Gly324Glu (ts110) map to nsP4 (*149*). Ts6 causes a rapid cessation of all RNA synthesis when cells are shifted to nonpermissive temperature (47, 76), and ts110 has a similar but milder phenotype (149). The ts6 defect is reproduced in polymerase extracts prepared from infected cells, which enabled an elegant demonstration that the defect affects elongation of RNA chains first initiated at the permissive temperature (71). SIN mutations causing resistance to pyrazofurin, a compound lowering cellular UTP and CTP levels, also map to nsP4 and may increase the affinity of the polymerase for pyrimidine nucleotides (150). Finally, a mutation Gln191 Lys in SIN nsP4 causes a reactivation of minus-strand synthesis at the restrictive temperature, a phenotype similar to many nsP2 mutants, indicating that nsP4 is also involved in the regulation of minus-strand synthesis shut-off (57) (see Section IV,B).

The small N-terminal extension of nsP4 has no conserved counterparts in other viruses and it has no known enzymatic activity (92). It may be involved in interactions with other components of the replication complex. A double-mutation Val425Ala in nsP2 and Gln93Arg in nsP4 is required for the phenotype of SIN ts118, which might among other possibilities suggest that the helicase domain of nsP2 interacts with the polymerase to form a functional complex (149). The N-terminal amino acid of mature nsP4 is normally a tyrosine, and only an aromatic amino acid or histidine is functional at the N terminus of SIN nsP4 (51, 151). Alteration of the nsP4 N terminus to normally nonviable residues appears to cause a defect in minus-strand RNA synthesis, which can be suppressed by a mutation in nsP1 (see Section VI,C). This is interpreted to support a role for nsP1–nsP4 interaction in minus-strand synthesis or promoter interaction (118).

The bulk of nsP4 is unstable in alphavirus-infected cells, although a stable fraction presumably associated with the polymerase complex also exists (85, 86, 152, 153). NsP4 is degraded by the N-end rule pathway, as it bears a destabilizing N-terminal Tyr residue (154), and it is accordingly stabilized by inhibitors of proteosome activity (92). The amount of nsP4 can thus be regulated at multiple levels. Translation of nsP4 requires a readthrough of a termination codon in many alphaviruses, production of active polymerase complexes requires cleavage of nsP4 from the nonstructural polyprotein, and nsP4 is susceptible for degradation by the proteosome system.

X. The Replication Complex

After alphavirus entry and uncoating, which may take place in the close vicinity of endosomes, the genome RNA is subjected to translation. As a single virus particle can cause infection (155), we have to think that one RNA molecule is first translated, followed by its recruitment as a template for minus-strand synthesis. These primary events cannot be detected by available techniques. Experiments described in Section IV,A indicate that P1234, cleaved at site 3/4, yields the minus-strand RNA polymerase, P123 plus nsP4, although it is not known whether one heterodimer constitutes an active complex. Our recent results with expression of uncleavable polyproteins P1234 and P123 in insect and mammalian cells have shown that they both bind to intracellular membranes and become palmitoylated (Salonen et al., unpublished). Thus, we can expect that the minus-strand RNA polymerase is membrane bound, probably by a mechanism similar to that demonstrated for nsP1 (see Section VI,B). We assume that several rounds of translation take place, resulting in a small pool of membrane-bound polyprotein.

A polymerase complex recognizes the conserved sequences at the 5' and 3' ends of the parental 42S RNA genome, possibly through joint activity of the nsP1 portion of P123, and nsP4 (Fig. 3A). A complete uncapped complementary RNA is synthesized, which has an unpaired extra G residue at the 3' end of the minus strand, the origin of which is not known (156, 157). A similar extra G residue at the 3' end of cucumber mosaic satellite virus minus strand RNA is essential for RNA replication (158). As there is no evidence of the existence of free 42S RNA minus-strands, a double-stranded RNA may be synthesized (23, 24, 45, 47, 55, 159). The polymerase may utilize the plus-strand template for only one round of minus-strand synthesis. This would explain why only about 20% of synthesized RNA is of negative polarity early in infection (45). The occurrence of homologous recombination and the ease of creation of defective-interfering RNAs indicate that template switching can take place, presumably during minus-strand synthesis (37, 160).

The shift from minus-strand to plus-strand RNA synthesis requires the proteolytic processing of P123 (see Section IV,B). The 42S RNA minus strand acts as a template, but whether it is part of either double- or single-stranded RNA is not definitively known. It can be deduced that the synthesis of the plus strand starts from the penultimate base. The presence of replicative intermediate RNA with nascent plus strands has suggested a model, according to which a single minus strand is copied by several polymerase molecules (23, 24) (Fig. 3C). According to this model, the double-stranded replicative forms RFI, II, and III arise during RNA isolation by hybridization of the excess plus strands to the minus-strand template. The same replication complex can alternate between synthesis of 42S and 26S RNA, regulated by an extra "soluble" nsP2 transcription factor (59, 79) (Fig. 3D and 3E). An important feature of the late replication complex is its stability, as it continues to make both 42S RNA and 26S RNA for several hours using the same minus strand as template.

We have recently shown that all four nsPs as well as nascent RNA are associated with small invaginations, or spherules, of alphavirus-specific large cytoplasmic vacuoles, CPVs (Fig. 4) (66). CPVs are modified endosomes and lysosomes (64–66, 112). We have suggested that each spherule represents one unit of replication, carrying a 42S RNA minus-strand template as well as a set of late RNA polymerases, each consisting of nsP1–nsP4 (66). This hypothesis is supported by electron microscopic studies. Autoradiography showed a clear association of nascent virus-specific RNA with spherules (63), while on the other hand protrusion of RNA-like structures from the spherules was described (64).

The membrane association of the replication complex has interesting consequences. Since the polymerase is fixed to the membrane, the template RNA must move. If a single spherule constitutes a unit of replication, the template RNA of about 11.5 kb must be packed within the spherule of a diameter of about 50 nm. If the template were double-stranded RNA, the packing would be even tighter. In order to achieve continuous synthesis of plus strands, the template should preferably be circular. The classical model of RI, with one template occupied by several polymerases (Fig. 3C), is difficult to reconcile with polymerases that are fixed to the membrane. The problem could be solved, assuming that the same template is successively passing from one membrane-associated polymerase to another in a circular manner (Fig. 10). As both 42S RNA and 26S RNA have a cap structure at their 5' end, the coordinated capping reactions catalyzed by

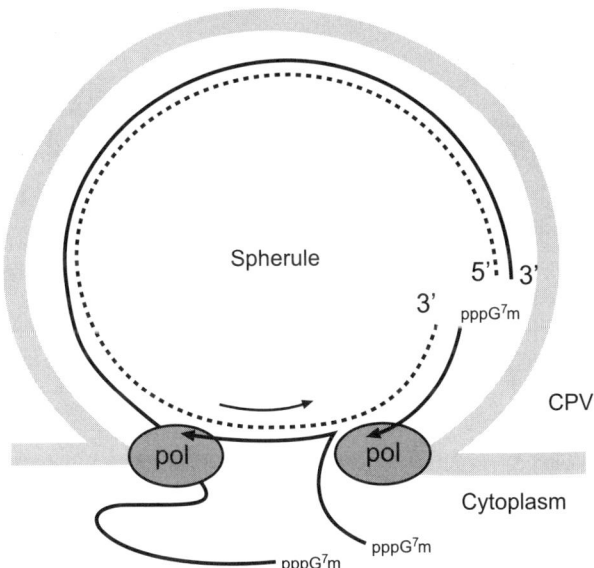

FIG. 10. A speculative model of the alphavirus spherule-associated unit of replication synthesizing positive 42S RNA strands. The minus-strand template (dotted line) is shown in association with a complementary plus strand (solid line). The late RNA polymerase molecules (pol), consisting of nsP1–nsP4, localized to the orfice of the spherule, are fixed to the membrane. The dsRNA template rotates through two different polymerases (in this image) in the direction shown by the long arrow. Nascent, capped, positive RNA strands, arise by semiconservative replication. Upon reaching the cytoplasm they are utilized for translation early in infection and later for the assembly of nucleocapsids (64).

nsP1 and nsP2 (Fig. 6) must presumably take place soon after initiation. Anionic membrane lipids are a vital cofactor for nsP1 at this stage. So far, there is no direct experimental data concerning the role of the RNA helicase activity of nsP2.

The presence of spherules also on the outer surface of the plasma membrane, in addition to CPVs (63, 64, 66), has led us to propose the following hypothesis. The primary assembly of the replication complex takes place in association with endosomes, or the plasma membrane, through the affinity of the lipid-binding peptide of nsP1 to phosphatidyserine-rich lipids (Fig. 7). Only a fraction of the replication complexes assemble correctly and manage to recruit a template. After proteolytic cleavages, the components of unsuccessful complexes dissociate from each other, and nsP1 associates permanently with the plasma membrane. NsP2 is transported to the nucleus, and nsP3 and nsP4 remain distributed in the cytoplasm. The correctly assembled complexes participate in the endosomal circulation, redistributing the complexes to new endosomes and finally to lysosomes (66).

We have aimed at emphasizing throughout this review how genetic, virological, biochemical, and cell biological approaches have complemented each other in providing a better understanding of alphavirus RNA replication and the replication complex itself. Even though progress has been made, our knowledge of the molecular mechanisms of RNA replication is far from complete. The three-dimensional structures of the nsPs are required to further define their roles in RNA replication. An even more demanding task is to isolate, or reconstitute, the unit of replication in functional form.

ACKNOWLEDGMENTS

We acknowledge the financial support we have received for the studies described from the Academy of Finland (grants no 8397, 62496, 78763) the Finnish Technology Development Center (Tekes), the Sigrid Juselius Foundation, the Helsinki University Foundation, and the Centre for International Mobility. We thank Dr. Marja Makarow for critical reading of the manuscript. LK has been a Biocentrum Helsinki fellow from 1995–2000.

REFERENCES

1. R. E. Johnston and C. J. Peters, Alphaviruses, in "Fields Virology," 3rd ed. (B. N. Fields, D. M. Knipe, and P. M. Howley, eds.), pp. 843–898. Lippincott-Raven, Philadelphia, 1996.
2. L. Kääriäinen and H. Söderlund, Structure and replication of alphaviruses. Curr. Top. Microbiol. Immunol. 82, 15–69 (1978).
3. A. Helenius, Alphavirus and flavivirus glycoproteins: Structures and functions. Cell 81, 651–653 (1995).

4. J. Saraste and E. Kuismanen, Pre- and post-Golgi vacuoles operate in the transport of Semliki Forest virus membrane glycoproteins to the cell surface. *Cell* **38,** 535–543 (1984).
5. P. Berglund, I. Tubulekas, and P. Liljeström, Alphaviruses as vectors for gene delivery. *Trends Biotechnol.* **14,** 130–134 (1996).
6. S. Schlesinger and T. W. Dubensky, Alphavirus vectors for gene expression and vaccines. *Curr. Opin. Biotechnol.* **10,** 434–439 (1999).
7. E. J. Mancini, M. Clarke, B. E. Gowen, T. Rutten, and S. D. Fuller, Cryo-electron microscopy reveals the functional organization of an enveloped virus, Semliki Forest virus. *Mol. Cell* **5,** 255–266 (2000).
8. M. Marsh and A. Helenius, Virus entry into animal cells. *Adv. Virus Res.* **36,** 107–151 (1989).
9. I. Singh and A. Helenius, Role of ribosomes in Semliki Forest virus nucleocapsid uncoating. *J. Virol.* **66,** 7049–7058 (1992).
10. I. Ulmanen, H. Söderlund, and L. Kääriäinen, Role of protein synthesis in the assembly of Semliki forest virus nucleocapsid. *Virology* **99,** 265–276 (1979).
11. H. Söderlund, L. Kääriäinen, and C. H. von Bonsdorff, Properties of Semliki Forest virus nucleocapsid. *Med. Biol.* **53,** 412–417 (1975).
12. I. Ulmanen, Assembly of Semliki Forest virus nucleocapsid: Detection of a precursor in infected cells. *J. Gen. Virol.* **41,** 353–365 (1978).
13. K. M. Coombs and D. T. Brown, Form-determining functions in Sindbis virus nucleocapsids: Nucleosomelike organization of the nucleocapsid. *J. Virol.* **63,** 883–891 (1989).
14. M. Pesonen and O. Renkonen, Serum glycoprotein-type sequence of monosaccharides in membrane glycoproteins of Semliki Forest virus. *Biochim. Biophys. Acta* **455,** 510–525 (1976).
15. M. Pesonen, Sequence analysis of lactosamine type glycans of individual membrane proteins of Semliki Forest virus. *J. Gen. Virol.* **45,** 479–487 (1979).
16. E. V. Koonin and V. V. Dolja, Evolution and taxonomy of positive-strand RNA viruses: Implications of comparative analysis of amino acid sequences. *Crit. Rev. Biochem. Mol. Biol.* **28,** 375–430 (1993).
17. M. N. Rozanov, E. V. Koonin, and A. E. Gorbalenya, Conservation of the putative methyltransferase domain: a hallmark of the "Sindbis-like" supergroup of positive-strand RNA viruses. *J. Gen. Virol.* **73,** 2129–2134 (1992).
18. T. Ahola, P. Laakkonen, H. Vihinen, and L. Kääriäinen, Critical residues of Semliki Forest virus RNA capping enzyme involved in methyltransferase and guanylyltransferase-like activities. *J. Virol.* **71,** 392–397 (1997).
19. K. W. Buck, Comparison of the replication of positive-stranded RNA viruses of plants and animals. *Adv. Virus Res.* **47,** 159–251 (1996).
20. R. M. Friedman, H. B. Levy, and W. B. Carter, Replication of Semliki Forest virus: Three forms of viral RNA produced during infection. *Proc. Natl. Acad. Sci. U.S.A.* **56,** 440–446 (1966).
21. J. A. Sonnabend, E. M. Martin, and E. Mecs, Viral specific RNAs in infected cells. *Nature* **213,** 365–367 (1967).
22. L. Kääriäinen and P. J. Gomatos, A kinetic analysis of the synthesis in BHK 21 cells of RNAs specific for Semliki Forest virus. *J. Gen. Virol.* **5,** 251–265 (1969).
23. D. T. Simmons and J. H. Strauss, Replication of Sindbis virus: I. Relative size and genetic content of 26S and 49S RNA. *J. Mol. Biol.* **71,** 599–613 (1972).
24. D. T. Simmons and J. H. Strauss, Replication of Sindbis virus: II. Multiple forms of double-stranded RNA isolated from infected cells. *J. Mol. Biol.* **71,** 615–631 (1972).
25. R. M. Friedman and I. K. Berezesky, Cytoplasmic fractions associated with Semliki Forest virus ribonucleic acid replication. *J. Virol.* **1,** 374–383 (1967).
26. M. Ranki and L. Kääriäinen, Solubilized RNA replication complex from Semliki Forest virus-infected cells. *Virology* **98,** 298–307 (1979).

27. P. J. Gomatos, L. Kääriäinen, S. Keränen, M. Ranki, and D. L. Sawicki, Semliki Forest virus replication complex capable of synthesizing 42S and 26S nascent RNA chains. *J. Gen. Virol.* **49,** 61–69 (1980).
28. H. Garoff, K. Simons, and B. Dobberstein, Assembly of the Semliki Forest virus membrane glycoproteins in the membrane of the endoplasmic reticulum in vitro. *J. Mol. Biol.* **124,** 587–600 (1978).
29. N. Glanville, M. Ranki, J. Morser, L. Kääriäinen, and A. E. Smith, Initiation of translation directed by 42S and 26S RNAs from Semliki Forest virus *in vitro. Proc. Natl. Acad. Sci. U.S.A.* **73,** 3059–3063 (1976).
30. N. Glanville, B.-E. Lachmi, A. E. Smith, and L. Kääriäinen, Tryptic peptide mapping of the nonstructural proteins of Semliki Forest virus and their precursors. *Biochim. Biophys. Acta* **518,** 497–506 (1978).
31. P. Lehtovaara, I. Ulmanen, L. Kääriäinen, S. Keränen, and L. Philipson, Synthesis and processing of Semliki Forest virus-specific nonstructural proteins *in vivo* and *in vitro. Eur. J. Biochem.* **112,** 461–468 (1980).
32. K. Takkinen, Complete nucleotide sequence of the nonstructural protein genes of Semliki Forest virus. *Nucleic Acids Res.* **14,** 5667–5682 (1986).
33. E. G. Strauss, C. M. Rice, and J. H. Strauss, Complete nucleotide sequence of the genomic RNA of Sindbis virus. *Virology* **133,** 92–110 (1984).
34. J. H. Strauss and E. G. Strauss, The alphaviruses: Gene expression, replication, and evolution. *Microbiol. Rev.* **58,** 491–562 (1994).
35. E. Frolova, I. Frolov, and S. Schlesinger, Packaging signals in alphaviruses. *J. Virol.* **71,** 248–258 (1997).
36. C. L. White, M. Thomson, and N. J. Dimmock, Deletion analysis of a defective interfering Semliki Forest virus RNA genome defines a region in the nsP2 sequence that is required for efficient packaging of the genome into virus particles. *J. Virol.* **72,** 4320–4326 (1998).
37. P. Lehtovaara, H. Söderlund, S. Keränen, R. F. Pettersson, and L. Kääriäinen, 18S defective interfering RNA of Semliki Forest virus contains a triplicated linear repeat. *Proc. Natl. Acad. Sci. U.S.A.* **78,** 5353–5357 (1981).
38. P. Lehtovaara, H. Söderlund, S. Keränen, R. F. Pettersson, and L. Kääriäinen, Extreme ends of the genome are conserved and rearranged in the defective interfering RNAs of Semliki Forest virus. *J. Mol. Biol.* **156,** 731–748 (1982).
39. S. S. Monroe and S. Schlesinger, Common and distinct regions of defective-interfering RNAs of Sindbis virus. *J. Virol.* **49,** 865–872 (1984).
40. M. Thomson and N. J. Dimmock, Common sequence elements in structurally unrelated genomes of defective interfering Semliki Forest virus. *Virology* **199,** 354–365 (1994).
41. R. F. Pettersson, 5′-Terminal nucleotide sequence of Semliki forest virus 18S defective interfering RNA is heterogeneous and different from the genomic 42S RNA. *Proc. Natl. Acad. Sci. U.S.A.* **78,** 115–119 (1981).
42. S. S. Monroe and S. Schlesinger, RNAs from two independently isolated defective interfering particles of Sindbis virus contain a cellular tRNA sequence at their 5′ ends. *Proc. Natl. Acad. Sci. U.S.A.* **80,** 3279–3283 (1983).
43. H. G. M. Niesters and J. H. Strauss, Defined mutations in the 5′ nontranslated sequence of Sindbis virus RNA. *J. Virol.* **64,** 4162–4168 (1990).
44. H. G. M. Niesters and J. H. Strauss, Mutagenesis of the conserved 51-nucleotide region of Sindbis virus. *J. Virol.* **64,** 1639–1647 (1990).
45. D. L. Sawicki and S. G. Sawicki, Short-lived minus-strand polymerase for Semliki Forest virus. *J. Virol.* **34,** 108–118 (1980).
46. Y. S. Hahn, E. G. Strauss, and J. H. Strauss, Mapping of RNA$^-$ temperature-sensitive mutants of Sindbis virus: Assignment of complementation groups A, B, and G to nonstructural proteins. *J. Virol.* **63,** 3142–3150 (1989).

47. D. L. Sawicki, S. G. Sawicki, S. Keränen, and L. Kääriäinen, Specific Sindbis virus-coded function for minus-strand RNA synthesis. *J. Virol.* **39**, 348–358 (1981).
48. Y.-F. Wang, S. G. Sawicki, and D. L. Sawicki, Sindbis virus nsP1 functions in negative-strand RNA synthesis. *J. Virol.* **65**, 985–988 (1991).
49. Y.-F. Wang, S. G. Sawicki, and D. L. Sawicki, Alphavirus nsP3 functions to form replication complexes transcribing negative-strand RNA. *J. Virol.* **68**, 6466–6475 (1994).
50. J. A. Lemm and C. M. Rice, Assembly of functional Sindbis virus RNA replication complexes: Requirement for coexpression of P123 and P34. *J. Virol.* **67**, 1905–1915 (1993).
51. J. A. Lemm and C. M. Rice, Roles of nonstructural polyproteins and cleavage products in regulating Sindbis virus RNA replication and transcription. *J. Virol.* **67**, 1916–1926 (1993).
52. J. A. Lemm, T. Rümenapf, E. G. Strauss, J. H. Strauss, and C. M. Rice, Polypeptide requirements for assembly of functional Sindbis virus replication complexes: A model for the temporal regulation of minus- and plus-strand RNA synthesis. *EMBO J.* **13**, 2925–2934 (1994).
53. Y. Shirako and J. H. Strauss, Regulation of Sindbis virus RNA replication: Uncleaved P123 and nsP4 function in minus-strand RNA synthesis, whereas cleaved products from P123 are required for efficient plus-strand RNA synthesis. *J. Virol.* **68**, 1874–1885 (1994).
54. J. A. Lemm, A. Bergqvist, C. M. Read, and C. M. Rice, Template-dependent initiation of Sindbis virus RNA replication *in vitro. J. Virol.* **72**, 6546–6553 (1998).
55. S. G. Sawicki, D. L. Sawicki, L. Kääriäinen, and S. Keränen, A Sindbis virus mutant temperature-sensitive in the regulation of minus-strand RNA synthesis. *Virology* **115**, 161–172 (1981).
56. S. G. Sawicki and D. L. Sawicki, The effect of loss regulation of minus-strand RNA synthesis on Sindbis virus replication. *Virology* **151**, 339–349 (1986).
57. D. L. Sawicki, D. B. Barkhimer, S. G. Sawicki, C. M. Rice, and S. Schlesinger, Temperature sensitive shut-off of alphavirus minus strand RNA synthesis maps to a nonstructural protein, nsP4. *Virology* **174**, 43–52 (1990).
58. D. L. Sawicki and S. G. Sawicki, A second nonstructural protein functions in the regulation of alphavirus negative-strand RNA synthesis. *J. Virol.* **67**, 3605–3610 (1993).
59. J. Suopanki, D. L. Sawicki, S. G. Sawicki, and L. Kääriäinen, Regulation of alphavirus 26S mRNA transcription by replicase component nsP2. *J. Gen. Virol.* **79**, 309–319 (1998).
60. K. Tuomi, L. Kääriäinen, and H. Söderlund, Quantitation of Semliki Forest virus RNAs in infected cells using ^{32}P equilibrium labelling. *Nucleic Acids Res.* **2**, 555–565 (1975).
61. G. Wengler and Gi. Wengler, Studies on the synthesis of viral RNA-polymerase-template complexes in BHK 21 cells infected with Semliki Forest virus. *Virology* **66**, 322–326 (1975).
62. S. G. Sawicki and D. L. Sawicki, The effect of overproduction of nonstructural proteins on alphavirus plus-strand and minus-strand RNA synthesis. *Virology* **152**, 507–512 (1986).
63. P. M. Grimley, I. K. Berezesky, and R. M. Friedman, Cytoplasmic structures associated with an arbovirus infection: Loci of viral ribonucleic acid synthesis. *J. Virol.* **2**, 1326–1338 (1968).
64. S. Froshauer, J. Kartenbeck, and A. Helenius, Alphavirus RNA replicase is located on the cytoplasmic surface of endosomes and lysosomes. *J. Cell Biol.* **107**, 2075–2086 (1988).
65. J. Peränen and L. Kääriäinen, Biogenesis of type I cytopathic vacuoles in Semliki Forest virus-infected BHK cells. *J. Virol.* **65**, 1623–1627 (1991).
66. P. Kujala, A. Ikäheimonen, N. Ehsani, H. Vihinen, P. Auvinen, and L. Kääriäinen, Biogenesis of the Semliki Forest virus RNA replication complex. *J. Virol.* **75**, 3873–3884 (2001).
67. P. M. Grimley, J. G. Levin, I. K. Berezesky, and R. M. Friedman, Specific membranous structures associated with the replication of group A arboviruses. *J. Virol.* **10**, 492–503 (1972).
68. B.-E. Lachmi and L. Kääriäinen, Control of protein synthesis in Semliki Forest virus infected cells. *J. Virol.* **22**, 142–149 (1977).
69. D. Magliano, J. A. Marshall, D. S. Bowden, N. Vardaxis, J. Meanger, and J.-Y. Lee, Rubella virus replication complexes are virus-modified lysosomes. *Virology* **240**, 57–63 (1998).

70. P. Kujala, T. Ahola, N. Ehsani, P. Auvinen, H. Vihinen, and L. Kääriäinen, Intracellular distribution of rubella virus nonstructural protein P150. *J. Virol.* **73**, 7805–7811 (1999).
71. D. J. Barton, S. G. Sawicki, and D. L. Sawicki, Demonstration *in vitro* of temperature-sensitive elongation of RNA in Sindbis virus mutant *ts6*. *J. Virol.* **62**, 3597–3602 (1988).
72. D. J. Barton, S. G. Sawicki, and D. L. Sawicki, Solubilization and immunoprecipitation of alphavirus replication complexes. *J. Virol.* **65**, 1496–1506 (1991).
73. C. M. Scheele and E. R. Pfefferkorn, Inhibition of interjacent ribonucleic acid (26S) synthesis in cells infected by Sindbis virus. *J. Virol.* **4**, 117–122 (1969).
74. S. Keränen and L. Kääriäinen, Isolation and basic characterization of temperature-sensitive mutants from Semliki Forest virus. *Acta Pathol. Microbiol. Scand.* B **82**, 810–820 (1974).
75. E. G. Strauss, E. M. Lenches, and J. H. Strauss, Mutants of Sindbis virus. I. Isolation and partial characterization of 89 new temperature-sensitive mutants. *Virology* **74**, 154–168 (1976).
76. S. Keränen and L. Kääriäinen, Functional defects of RNA-negative temperature-sensitive mutants of Sindbis and Semliki Forest viruses. *J. Virol.* **32**, 19–29 (1979).
77. D. L. Sawicki and S. G. Sawicki, Functional analysis of the A complementation group mutants of Sindbis HR virus. *Virology* **144**, 20–34 (1985).
78. J. Saraste, L. Kääriäinen, H. Söderlund, and S. Keränen, RNA synthesis directed by a temperature-sensitive mutant of Semliki Forest virus. *J. Gen. Virol.* **37**, 399–406 (1977).
79. D. L. Sawicki, L. Kääriäinen, C. Lambek, and P. J. Gomatos, Mechanism for control of synthesis of Semliki Forest virus 26S and 42S RNA. *J. Virol.* **25**, 19–27 (1978).
80. R. Levis, S. Schlesinger, and H. V. Huang, Promoter for Sindbis virus RNA-depedent subgenomic RNA transcription. *J. Virol.* **64**, 1726–1733 (1990).
81. R. Raju and H. V. Huang, Analysis of Sindbis virus promoter recognition *in vivo*, using novel vectors with two subgenomic mRNA promoters. *J. Virol.* **65**, 2501–2510 (1991).
82. M. M. Wielgosz, R. Raju, and H. V. Huang, Sequence requirements for Sindbis virus subgenomic mRNA promoter function in cultured cells. *J. Virol.* **75**, 3509–3519 (2001).
83. B.-E. Lachmi and L. Kääriäinen, Sequential translation of nonstructural proteins in cells infected with a Semliki Forest virus mutant. *Proc. Natl. Acad. Sci. U.S.A.* **73**, 1936–1940 (1976).
84. L. Kääriäinen, D. L. Sawicki, and P. J. Gomatos, Cleavage defect in the non-structural polyprotein of Semliki Forest virus has two separate effects on virus RNA synthesis. *J. Gen. Virol.* **39**, 463–473 (1978).
85. S. Keränen and L. Ruohonen, Nonstructural proteins of Semliki Forest virus: Synthesis, processing, and stability in infected cells. *J. Virol.* **47**, 505–551 (1983).
86. W. R. Hardy and J. H. Strauss, Processing the nonstructural polyproteins of Sindbis virus: Study of the kinetics *in vivo* by using monospecific antibodies. *J. Virol.* **62**, 998–1007 (1988).
87. M. Ding and M. J. Schlesinger, Evidence that Sindbis virus nsP2 is an autoprotease which processes the virus nonstructural polyprotein. *Virology* **171**, 280–284 (1989).
88. W. R. Hardy and J. H. Strauss, Processing the nonstructural polyproteins of Sindbis virus: Nonstructural proteinase is in the C-terminal half of nsP2 and functions both in cis and in trans. *J. Virol.* **63**, 4653–4664 (1989).
89. Y. Shirako and J. H. Strauss, Cleavage between nsP1 and nsP2 initiates the processing pathway of Sindbis virus nonstructural polyprotein P123. *Virology* **177**, 54–64 (1990).
90. E. G. Strauss, R. J. deGroot, R. Levinson, and J. H. Strauss, Identification of the active site residues in the nsP2 proteinase of Sindbis virus. *Virology* **191**, 932–940 (1992).
91. E. ten Dam, M. Flint, and M. D. Ryan, Virus-coded proteinases of the Togaviridae. *J. Gen. Virol.* **80**, 1879–1888 (1999).
92. A. Merits, L. Vasiljeva, T. Ahola, L. Kääriäinen, and P. Auvinen, Proteolytic processing of Semliki Forest virus-specific non-structural polyprotein by nsP2 protease. *J. Gen. Virol.* **82**, 765–773 (2001).

93. S. Mi and V. Stollar, Expression of Sindbis virus nsP1 and methyltransferase activity in *Escherichia coli*. *Virology* **184**, 423–427 (1991).
94. P. Laakkonen, M. Hyvönen, J. Peränen, and L. Kääriäinen, Expression of Semliki Forest virus nsP1-specific methyltransferase in insect cells and in *Escherichia coli*. *J. Virol.* **68**, 7418–7425 (1994).
95. T. Ahola and L. Kääriäinen, Reaction in alphavirus mRNA capping: Formation of a covalent complex of nonstructural protein nsP1 with 7-methyl-GMP. *Proc. Natl. Acad. Sci. U.S.A.* **92**, 507–511 (1995).
96. A. Lampio, T. Ahola, E. Darzynkiewicz, J. Stepinski, M. Jankowska-Anyszka, and L. Kääriäinen, Guanosine nucleotide analogs as inhibitors of alphavirus mRNA capping enzyme. *Antiviral Res.* **42**, 35–46 (1999).
97. Y. Furuichi and A. J. Shatkin, Viral and cellular mRNA capping: Past and prospects. *Adv. Virus Res.* **55**, 135–184 (2000).
98. S. Shuman, Structure, mechanism, and evolution of the mRNA capping apparatus. *Prog. Nucleic Acid Res. Mol. Biol.* **66**, 1–40 (2000).
99. A. Merits, R. Kettunen, K. Mäkinen, A. Lampio, P. Auvinen, L. Kääriäinen, and T. Ahola, Virus-specific capping of tobacco mosaic virus RNA: Methylation of GTP prior to formation of covalent complex p126-m7GMP. *FEBS Lett.* **455**, 45–48 (1999).
100. T. Ahola and P. Ahlquist, Putative RNA capping activities encoded by brome mosaic virus: Methylation and covalent binding of guanylate by replicase protein 1a. *J. Virol.* **73**, 10,061–10,069 (1999).
101. F. Kong, K. Sivakumaran, and C. Kao, The N-terminal half of the brome mosaic virus 1a protein has RNA capping-associated activities: Specificity for GTP and S-adenosylmethionine. *Virology* **259**, 200–210 (1999).
102. Y. I. Li, Y. J. Chen, Y. H. Hsu, and M. Meng, Characterization of the AdoMet-dependent guanylyltransferase activity that is associated with the N terminus of bamboo mosaic virus replicase. *J. Virol.* **75**, 782–788 (2001).
103. J. Magden, N. Takeda, T. Li, P. Auvinen, T. Ahola, T. Miyamura, A. Merits, and L. Kääriäinen, Virus-specific mRNA capping enzyme encoded by hepatitis E virus. *J. Virol.* **75**, 6249–6255 (2001).
104. H.-L. Wang, J. O'Rear, and V. Stollar, Mutagenesis of the Sindbis virus nsP1 protein: Effects on methyltransferase activity and viral infectivity. *Virology* **217**, 527–531 (1996).
105. T. Ahola, J. A. den Boon, and P. Ahlquist, Helicase and capping enzyme active site mutations in brome mosaic virus protein 1a cause defects in template recruitment, negative-strand RNA synthesis, and viral RNA capping. *J. Virol.* **74**, 8803–8811 (2000).
106. T. Ahola, A. Lampio, P. Auvinen, and L. Kääriäinen, Semliki Forest virus mRNA capping enzyme requires association with anionic membrane phospholipids for activity. *EMBO J.* **18**, 3164–3172 (1999).
107. S. Mi, R. K. Durbin, H. V. Huang, C. M. Rice, and V. Stollar, Association of the Sindbis virus RNA methyltransferase activity with the nonstructural protein nsP1. *Virology* **170**, 385–391 (1989).
108. L. M. Scheidel, R. K. Durbin, and V. Stollar, SV_{LM21}, a Sindbis virus mutant resistant to methionine deprivation, encodes an altered methyltransferase. *Virology* **173**, 408–414 (1989).
109. L. M. Scheidel, R. K. Durbin, and V. Stollar, Sindbis virus mutants resistant to mycophenolic acid and ribavirin. *Virology* **158**, 1–7 (1987).
110. L. M. Scheidel and V. Stollar, Mutations that confer resistance to mycophenolic acid and ribavirin on Sindbis virus map to the nonstructural protein nsP1. *Virology* **181**, 490–499 (1991).
111. C. I. Rosenblum, L. M. Scheidel, and V. Stollar, Mutations in the nsP1 coding sequence of Sindbis virus which restrict viral replication in secondary cultures of chick embryo fibroblasts prepared from aged primary cultures. *Virology* **198**, 100–108 (1994).

112. J. Peränen, P. Laakkonen, M. Hyvönen, and L. Kääriäinen, The alphavirus replicase protein nsP1 is membrane-associated and has affinity to endocytic organelles. *Virology* **208**, 610–620 (1995).
113. T. Ahola, P. Kujala, M. Tuittila, T. Blom, P. Laakkonen, A. Hinkkanen, and P. Auvinen, Effects of palmitoylation of replicase protein nsP1 on alphavirus infection. *J. Virol.* **74**, 6725–6733 (2000).
114. P. Laakkonen, T. Ahola, and L. Kääriäinen, The effects of palmitoylation on membrane association of Semliki Forest virus RNA capping enzyme. *J. Biol. Chem.* **271**, 28,567–28,571 (1996).
115. A. Lampio, I. Kilpeläinen, S. Pesonen, K. Karhi, P. Auvinen, P. Somerharju, and L. Kääriäinen, Membrane binding mechanism of an RNA virus-capping enzyme. *J. Biol. Chem.* **275**, 37,853–37,859 (2000).
116. J. E. Johnson and R. B. Cornell, Amphitropic proteins: Regulation by reversible membrane interactions. *Mol. Membr. Biol.* **16**, 217–235 (1999).
117. P. Laakkonen, P. Auvinen, P. Kujala, and L. Kääriäinen, Alphavirus replicase protein Nsp1 induces filopodia and rearrangement of actin filaments. *J. Virol.* **72**, 10,265–10,269 (1998).
118. Y. Shirako, E. G. Strauss, and J. H. Strauss, Suppressor mutations that allow Sindbis virus RNA polymerase to function with nonaromatic amino acids at the N-terminus: Evidence for interaction between nsP1 and nsP4 in minus-strand RNA synthesis. *Virology* **276**, 148–160 (2000).
119. I. Dé, S. G. Sawicki, and D. L. Sawicki, Sindbis virus RNA-negative mutants that fail to convert from minus-strand to plus-strand synthesis: Role of the nsP2 protein. *J. Virol.* **70**, 2706–2719 (1996).
120. A. E. Gorbalenya, E. V. Koonin, and M.-C. Lai, Putative papain-related thiol proteases of positive-strand RNA viruses. Identification of rubi- and aphtovirus proteases and delineation of a novel conserved domain associated with proteases of rubi-, alpha- and coronaviruses. *FEBS Lett.* **288**, 201–205 (1991).
121. A. E. Gorbalenya and E. V. Koonin, Helicases: Amino acid sequence comparisons and structure-function relationships. *Curr. Opin. Cell Biol.* **3**, 419–429 (1993).
122. M. Rikkonen, J. Peränen, and L. Kääriäinen, ATPase and GTPase activities associated with Semliki Forest virus nonstructural protein nsP2. *J. Virol.* **68**, 5804–5810 (1994).
123. M. Rikkonen, Functional significance of the nuclear-targeting and NTP-binding motifs of Semliki Forest virus nonstructural protein nsP2. *Virology* **218**, 352–361 (1996).
124. M. Gomez de Cedrón, N. Ehsani, M. L. Mikkola, J. A. García, and L. Kääriäinen, RNA helicase activity of Semliki Forest virus replicase protein NSP2. *FEBS Lett.* **448**, 19–22 (1999).
125. L. E. Bird, H. S. Subramanya, and D. B. Wigley, Helicases: A unifying structural theme?. *Curr. Opin. Struct. Biol.* **8**, 14–18 (1998).
126. P. A. Kroner, B. M. Young, and P. Ahlquist, Analysis of the role of brome mosaic virus 1a protein domains in RNA replication, using linker insertion mutagenesis. *J. Virol.* **64**, 6110–6120 (1990).
127. L. Vasiljeva, A. Merits, P. Auvinen, and L. Kääriäinen, Identification of a novel function of the *Alphavirus* capping apparatus. RNA 5′ triphosphatase activity of Nsp2. *J. Biol. Chem.* **275**, 17,281–17,287 (2000).
128. J. R. Myette and E. G. Niles, Characterization ot the vaccinia virus RNA 5′-triphosphatase and nucleoside triphosphate phosphohydrolase activities. Demonstration that both activities are carried out at the same active site. *J. Biol. Chem.* **271**, 11,945–11,952 (1996).
129. C. K. Ho, Y. Pei, and S. Shuman, Yeast and viral RNA 5′ triphosphatases comprise a new nucleoside triphosphatase family. *J. Biol. Chem.* **273**, 34,151–34,156 (1998).
130. G. Wengler and Gi. Wengler, The NS3 nonstructural protein of flaviviruses contains an RNA triphosphatase activity. *Virology* **197**, 265–273 (1993).

131. L. Vasiljeva, L. Valmu, L. Kääriäinen, and A. Merits, Site-specific protease activity of the carboxyl-terminal domain of Semliki Forest virus replicase protein nsP2. *J. Biol. Chem.* **276,** 30,786–30,793 (2001).
132. N. Kalkkinen, M. Laaksonen, H. Söderlund, and H. Jörnvall, Radio-sequence analysis of in vivo multilabeled nonstructural protein ns86 of Semliki Forest virus. *Virology* **113,** 188–195 (1981).
133. N. Kalkkinen, Radio-sequence analysis: An ultra-sensitive method to align protein and nucleotide sequences. in "Advanced Methods in Protein Microsequence Analysis" (B. Wittmann-Liebold, J. Salnikow, and V. A. Erdmann, Eds.), pp. 194–206. Springer-Verlag, Berlin, 1986.
134. J. Peränen, M. Rikkonen, P. Liljeström, and L. Kääriäinen, Nuclear localization of Semliki Forest virus-specific nonstructural protein nsP2. *J. Virol.* **64,** 1888–1896 (1990).
135. M. Rikkonen, J. Peränen, and L. Kääriäinen, Nuclear and nucleolar targeting signals of Semliki Forest virus nonstructural protein nsP2. *Virology* **189,** 462–473 (1992).
136. P. Russo, P. Laakkonen, T. Ahola, and L. Kääriäinen, Synthesis of Semliki Forest virus RNA polymerase components nsP1 through nsP4 in *Saccharomyces cerevisiae* by expression of cDNA encoding the nonstructural polyprotein. *J. Virol.* **70,** 4086–4089 (1996).
137. L. Kääriäinen and M. Ranki, Inhibition of cell functions by RNA-virus infections. *Annu. Rev. Microbiol.* **38,** 91–109 (1984).
137a. J. K. Fazakerley, A. Boyd, M. L. Mikkola, and L. Kääriäinen, A single amino acid change in the nuclear localization sequence of the nsP2 protein affects the neurovirulence of Semliki Forest virus. *J. Virol.* **76,** 392–396, (2002).
138. H. Vihinen, T. Ahola, M. Tuittila, A. Merits, and L. Kääriäinen, Elimination of phosphorylation sites of Semliki Forest virus replicase protein nsP3. *J. Biol. Chem.* **276,** 5745–5752 (2001).
139. J. R. Pehrson and R. N. Fuji, Evolutionary conservation of histone macroH2A subtypes and domains. *Nucleic Acids Res.* **26,** 2837–2842 (1998).
140. R. C. Aguiar, Y. Yakushijin, S. Kharbanda, R. Salgia, J. A. Fletcher, and M. A. Shipp, BAL is a novel risk-related gene in diffuse large B-cell lymphomas that enhances cellular migration. *Blood* **96,** 4328–4334 (2000).
141. M. R. Martzen, S. M. McCraith, S. L. Spinelli, F. M. Torres, S. Fields, E. J. Grayhack, and E. M. Phizicky, A biochemical genomics approach for identifying genes by the activity of their products. *Science* **286,** 1153–1155 (1999).
142. J. Peränen, K. Takkinen, N. Kalkkinen, and L. Kääriäinen, Semliki Forest virus-specific nonstructural protein nsP3 is a phosphoprotein. *J. Gen. Virol.* **69,** 2165–2178 (1988).
143. G. Li, M. W. LaStarza, W. R. Hardy, J. H. Strauss, and C. M. Rice, Phosphorylation of Sindbis virus nsP3 *in vivo* and *in vitro*. *Virology* **179,** 416–427 (1990).
144. M. W. LaStarza, A. Grakoui, and C. M. Rice, Deletion and duplication mutations in the C-terminal nonconserved region of Sindbis virus nsP3: Effects on phoshorylation and on virus replication in vertebrate and invertebrate cells. *Virology* **202,** 224–232 (1994).
145. H. Vihinen and J. Saarinen, Phosphorylation site analysis of Semliki Forest virus nonstructural protein 3. *J. Biol. Chem.* **275,** 27,775–27,783 (2000).
146. M. T. Tuittila, M. G. Santagati, M. Röytta, J. A. Määttä, and A. E. Hinkkanen, Replicase complex genes of Semliki Forest virus confer lethal neurovirulence. *J. Virol.* **74,** 4579–4589 (2000).
147. M. W. LaStarza, J. A. Lemm, and C. M. Rice, Genetic analysis of the nsP3 region of Sindbis virus: Evidence for roles in minus-strand and subgenomic RNA synthesis. *J. Virol.* **68,** 5781–5791 (1994).
148. E. K. O'Reilly and C. C. Kao, Analysis of RNA-dependent RNA polymerase structure and function as guided by known polymerase structures and computer predictions of secondary structure. *Virology* **252,** 287–303 (1998).

149. Y. S. Hahn, A. Grakoui, C. M. Rice, E. G. Strauss, and J. H. Strauss, Mapping of RNA⁻ temperature-sensitive mutants of Sindbis virus: Complementation group F mutants have lesions in nsP4. *J. Virol.* **63,** 1194–1202 (1989).
150. Y. H. Lin, P. Yadav, R. Ravatn, and V. Stollar, A mutant of Sindbis virus that is resistant to pyrazofurin encodes an altered RNA polymerase. *Virology* **272,** 61–71 (2000).
151. Y. Shirako and J. H. Strauss, Requirement for an aromatic amino acid or histidine at the N terminus of Sindbis virus RNA polymerase. *J. Virol.* **72,** 2310–2315 (1998).
152. G. Li and C. M. Rice, Mutagenesis of the in-frame opal termination codon preceding nsP4 of Sindbis virus: Studies of translational readthrough and its effect on virus replication. *J. Virol.* **63,** 1326–1337 (1989).
153. K. Takkinen, J. Peränen, and L. Kääriäinen, Proteolytic processing of Semliki Forest virus-specific non-structural polyprotein. *J. Gen. Virol.* **72,** 1627–1633 (1991).
154. R. J. deGroot, T. Rümenapf, R. J. Kuhn, E. G. Strauss, and J. H. Strauss, Sindbis virus RNA polymerase is degraded by the N-end rule pathway. *Proc. Natl. Acad. Sci. U.S.A.* **88,** 8967–8971 (1991).
155. P.-Y. Cheng, Purification, size, and morphology of a mosquito-borne animal virus, Semliki Forest virus. *Virology* **14,** 124–131 (1961).
156. G. Wengler, Gi. Wengler, and H. J. Gross, Replicative form of Semliki Forest virus RNA contains an unpaired guanosine. *Nature* **282,** 754–756 (1979).
157. G. Wengler, Gi. Wengler, and H. J. Gross, Terminal sequences of Sindbis virus-specific nucleic acids: Identity in molecules synthesized in vertebrate and insect cells and characteristic properties of the replicative form RNA. *Virology* **123,** 273–283 (1982).
158. G. Wu and J. M. Kaper, Requirement of 3′-terminal guanosine in (-)-stranded RNA for *in vitro* replication of cucumber mosaic virus satellite RNA by viral RNA-dependent RNA polymerase. *J. Mol. Biol.* **238,** 655–657 (1994).
159. D. L. Sawicki and P. J. Gomatos, Replication of Semliki Forest virus: Polyadenylate in plus-strand RNA and polyuridylate in minus-strand RNA. *J. Virol.* **20,** 446–464 (1976).
160. K. R. Hill, M. Hajjou, J. Y. Hu, and R. Raju, RNA-RNA recombination in Sindbis virus: Roles of the 3′ conserved motif, poly(A) tail, and nonviral sequences of template RNAs in polymerase recognition and template switching. *J. Virol.* **71,** 2693–2704 (1997).

The Unique Biochemistry of Methanogenesis

UWE DEPPENMEIER

Department of Microbiology and Genetics
Universität Göttingen
37077 Göttingen, Germany

I. Introduction... 224
II. Methanogens: A Unique Group of Microorganisms................... 225
   A. Ecological Importance 225
   B. Taxonomy and Cellular Characteristics of Methanogens 227
III. Biochemistry of Methanogenesis 228
   A. Unusual Cofactors... 228
   B. Methanogenesis from $H_2 + CO_2$ and Formate 232
   C. Methane Formations from Methylated C_1 Compounds.............. 235
   D. Conversion of Acetate to Carbon Dioxide and Methane.............. 238
   E. Pathways of Methanogenesis: A Summary........................ 238
IV. Mechanism of ATP Synthesis in Methanogenic Archaea 240
V. Energy-Conserving Systems in *Methanosarcina* Strains................. 242
   A. Redox-Driven Proton Translocation in *Methanosarcina mazei*......... 242
   B. Components of the H^+-Translocating Systems 245
   C. Further Analysis of the Membrane-Bound Electron Transport
      Systems of *Ms. mazei* 255
   D. Proton Translocating during Growth on $H_2 + CO_2$ or Methylated
      C_1 Compounds: A Summary 256
   E. Proton Translocating during Growth on Acetate 256
   F. Proton-Translocating Pyrophosphatases........................... 258
   G. The Membrane-Bound Methyltransferase in Methanogenic Archaea:
      A Primary Sodium Ion Pump.................................. 259
   H. The Formyl-Methanofuran Dehydrogenase System................. 261
   I. ATP Synthases... 265
VI. Energy Conservation in Obligate Hydrogenotrophic Methanogens........ 270
References.. 274

Methanogenic archaea have an unusual type of metabolism because they use $H_2 + CO_2$, formate, methylated C_1 compounds, or acetate as energy and carbon sources for growth. The methanogens produce methane as the major end product of their metabolism in a unique energy-generating process. The organisms received much attention because they catalyze the terminal step in the anaerobic breakdown of organic matter under sulfate-limiting conditions and are essential for both the recycling of carbon compounds and the maintenance of the global carbon flux on Earth. Furthermore, methane is an important greenhouse gas that directly contributes to climate changes and global warming. Hence, the understanding of the biochemical processes leading to methane formation are of major interest. This review focuses on the metabolic pathways of methanogenesis that are rather unique and involve a number of unusual enzymes and coenzymes. It will be shown how the previously mentioned substrates are converted to CH_4 via the CO_2-reducing, methylotrophic, or aceticlastic pathway. All catabolic processes finally lead to the formation of a mixed disulfide from coenzyme M and coenzyme B that functions as an electron acceptor of certain anaerobic respiratory chains. Molecular hydrogen, reduced coenzyme F_{420}, or reduced ferredoxin are used as electron donors. The redox reactions as catalyzed by the membrane-bound electron transport chains are coupled to proton translocation across the cytoplasmic membrane. The resulting electrochemical proton gradient is the driving force for ATP synthesis as catalyzed by an A_1A_o-type ATP synthase. Other energy-transducing enzymes involved in methanogenesis are the membrane-integral methyltransferase and the formylmethanofuran dehydrogenase complex. The former enzyme is a unique, reversible sodium ion pump that couples methyl-group transfer with the transport of Na^+ across the membrane. The formylmethanofuran dehydrogenase is a reversible ion pump that catalyzes formylation and deformylation of methanofuran. Furthermore, the review addresses questions related to the biochemical and genetic characteristics of the energy-transducing enzymes and to the mechanisms of ion translocation. © 2002, Elsevier Science (USA).

I. Introduction

The analysis of 16S rRNA molecules has shown that life on Earth consists of three primary lineages referred to as the domains of *Eukarya, Bacteria,* and *Archaea* (1–3). All three groups of organisms are believed to have shared common ancestors before they separated about 3 billion years ago (4). The archaeal domain consists of three phylogenetically distinct groups, the Crenarchaeota (extremely thermophilic and thermoacidophilic organisms), the Euryarchaeota (e.g., methanogenic and extremely halophilic archaea), and the Korarchaeota (unculturable microbes from terrestrial hot springs which have been classified based on rRNA sequences (5). *Archaea* lack eukaryotic cell nuclei and resemble eubacteria in morphology and genomic organization, but their molecular design reveals many features of eukaryotes (4). Their common ancestry with eukaryotes is evident from the similarity of replication, transcription, and translation proteins (6–9). Because of the apparent refinement of cellular information processing, it seems reasonable to argue that *Archaea* and *Eukaryota* share common evolutionary pathways independent of bacterial lineage (10, 11). On the other hand, *Archaea* and *Bacteria* are closely related with respect to their prokaryotic mode of cellular and genetic organization which the most recent bacterial ancestors might have possessed, in particular circular genomes and genes organized as operons (12, 13). *Archaea* are also related to *Bacteria* metabolically because both groups seem to have derived central biochemical pathways from a common ancestor (14, 15). However, the genomic sequences available from several *Archaea* now indicate that these organisms also contain many genes which are unique and are not found in other living beings (9, 16). The unshared genetic information establish archaea's separate identity. Some solely archaeal characteristics include distinctive cell envelopes and membrane lipids with respect to the presence of isoprenyl ether lipids, the absence of acyl ester lipids and fatty acid synthetase, as well as the stereochemistry of the glycerophosphate backbone of membrane lipids (17, 18). In summary, the cellular, biochemical, and genetic features of archaea might provide new information to the evolution of early life on the Earth (4).

I will focus on methanogenic archaea which have gained much attention due to their important ecological function and because of their unique biochemical features. In this context, I like to mention that several other reviews have been published in recent years which emphasize the taxonomy and ecology (19–22), the biochemistry (23–28), the energetics (29–32), and the molecular biology (33–37) of methanogenic archaea.

II. Methanogens: A Unique Group of Microorganisms

A. Ecological Importance

Methanogenic archaea are characterized by their ability to form methane as the major end product of metabolism. All methanogens are strict anaerobes and are widespread in anoxic environments which are limited in light, sulfate, and nitrate but contain complex organic compounds. Thus, typical habitats of methanogens are fresh-water sediments of lakes and rivers, swamps, tundra areas, rice fields and anaerobic digesters of sewage plants, as well as the intestinal tract of ruminants and termites.

Concerning major global bioelement fluxes, the formation of methane is one of the most important biological processes on Earth (20). Organic materials which enter anoxic environments are mineralized by anaerobic microorganisms in a four-step process which is called the anaerobic food chain (38). First, biopolymers (polysaccharides, proteins, nucleic acids, and lipids) are degraded to oligomers and finally to monomers (mainly sugars, amino acids, purines, pyrimidines, and fatty acids) by hydrolytic microorganisms. In a second step, fermentative bacteria convert these organic compounds to simple carbonic acids (e.g., propionate, butyrate, acetate, formate, succinate, and lactate), alcohols (e.g., ethanol propanol and butanol), and other compounds (e.g., H_2, CO_2, and ketones). Some of these products (e.g., fatty acids longer than two carbon atoms, alcohols longer than one carbon atom, and aromatic fatty acids) are used by acetogenic or syntrophic bacteria and are converted to acetate and C-1 compounds (39–41). These substances are used by methanogens as carbon and energy sources resulting in the formation of methane and CO_2 as end products of the anaerobic food chain (42). Because hydrogen is a critical intermediate in anaerobic fermentations, another important function of methanogens is the consumption of H_2 (43). In general, degradation of fatty acids and alcohols to acetate and hydrogen as performed by syntrophic bacteria is an endergonic process under standard conditions. Consequently, the hydrogen partial pressure has to be decreased substantially by hydrogen-scavenging organisms. This prerequisite for anaerobic breakdown of organic matter is achieved by methanogenic archaea. The hydrogen partial pressure in methanogenic environments containing an active hydrogen-utilizing community is maintained below 10 Pa. This low pressure allows electrons at the redox potential of NADH to be released as hydrogen. As a consequence, part of the intermediates of the fermentative pathways in synthrophic organisms can be converted to acetyl-CoA and finally to acetate, thereby generating ATP via substrate-level phosphorylation (38, 41). In fact, the scenario described above is proved to be true only for the natural fresh-water habitats of methanogens. In sulfate-rich anoxic environments such as marine sediments, the breakdown of organic matter is primarily dependent on sulfate-reducing bacteria (44). They oxidize all classical fermentation products to CO_2 with the concomitant of reducing SO_4^{2-} H_2S (45).

Taking into account all information, it is evident that methanogenesis represents the terminal step in the anaerobic breakdown of organic matter under sulfate-limiting conditions indicating the important function of these organisms in recycling of carbon components. Moreover, it is important to note that methanogens created most of the natural gas (fossil fuel) reserves which are trapped as energy sources for domestic or industrial use. On the other hand, methanogenesis also has a severe effect on the global ecology. Methane is the most important greenhouse gas after carbon dioxide and contributes to 16% of the greenhouse effect (46). Only 30% of total methane emission (about 1 billion tons per year) originates from natural sources, whereas 70% is linked to human activities, including expanded cultivation of rice and ruminant life stock (47). Part of the methane released from anaerobic habitats is used by aerobic methylotrophic bacteria as carbon and energy sources. However, the major portion (60–90%) of the methane is emitted from flooded rice paddies to the atmosphere because it is transported through the aerenchyma of the rice plants and cannot be attacked by bacteria (48, 49). Methane production in the digestive tract of production animals is estimated to be responsible for 22% of the anthropogenic sources (46). In this respect, it is important to note that previous studies on ice cores and recent *in situ* measurements have shown that CH_4 has increased from about 0.75 to 1.73 μmol/mol air during the past 150 years (50). This rise is clearly due to anthropogenic sources of methane emissions. Thus, besides the CO_2 content, we must also consider future atmospheric CH_4 budgets in the context of the stabilization of greenhouse gas concentrations.

However, there is one more important aspect of methanogenesis. It is thought that this process is involved or even responsible for the formation of the so-called methane hydrates (51). Gas hydrates are solids, composed of rigid cages of water molecules that trap gas molecules (52). These ice-like, methane-bearing structures are formed and are stable at moderately high pressures and low temperatures, conditions found on land in permafrost regions and within ocean floor sediments at water depths greater than about 500 m. The hydrate deposits themselves may be several hundred meters thick. Accordingly, global estimates of the amount of the methane hydrates exceed 10^{16} kg, which represents one of the largest sources of hydrocarbon on Earth (53). Methane incorporated into the hydrates is believed to have been generated by microorganisms in organic-rich sediments or may migrate from deeper gas deposits (54). Both sources of methane are most likely microbially generated originating from methanogenic archaea. Speculations about large releases of methane to the ocean-atmosphere system have raised questions about the involvement of methane hydrates in climate change (55). Furthermore, the gas hydrate story has provided an almost inexhaustible supply for press reports concerning greenhouse effects, landslides, global warming, and mysterious events such as the loss of aircraft in the "Bermuda Triangle." On the other hand methane hydrates represent a potentially enormous natural gas resource. If it is determined that the production of methane from hydrates is possible and economically viable, the long-term energy security would be ensured, and environmental quality would be improved (56).

B. Taxonomy and Cellular Characteristics of Methanogens

Methanogens are the most diverse group within the kingdom Euryarchaeota indicated by their genomic DNA G+C content which ranges from 26 to 68 mol%. The diversity is also dramatically illustrated by morphological features, because a wide variety of shapes and sizes were found, including regular and irregular cocci, rods of varying length, spirilla, and irregular unusual flattened plates. Some species can aggregate in clusters forming sarcina packages. Moreover, there is a considerable diversity among the methanogens with respect to their cell wall composition. They are surrounded by pseudomurein, heteropolysaccharides, or protein subunits (57). A typical murein layer, the peptidoglycan of Bacteria, is not produced. Therefore, methanogens are insensitive to antibiotics that inhibit the synthesis of cell walls in Bacteria. More than 80 methanogenic species have been described thus far (20) including psychrophilic (*Methanogenium frigidum*), mesophilic (most species), thermophilic (e.g., *Methanothermobacter marburgensis* and *Methanothermobacter thermautotrophicus*), and extremely thermophilic (e.g., *Methanococcus jannashii*, *Methanopyrus kandleri*, and *Methanothermus fervidus*) organisms. They are classified into five different orders, each of which are as distantly phylogenetically related to the other as humans are to slime molds. Members of the orders *Methanobacteriales, Methanococcales, Methanomicrobiales, and Methanopyrales* are referred to as obligate hydrogenotroph organisms able to use $H_2 + CO_4$ as substrate. Most of them can also oxidize formate to form methane. However, they are not able to utilize methylated C_1 compounds or acetate (42). The only exception is *Methanosphaera stadtmaniae* which deviates from the typical mode of hydrogenotrophic metabolism in that the organism catalyzes a H_2-dependent methanol reduction to CH_4 (58). A number of hydrogenotrophic methanogens can form methane in the presence of CO_2 and certain alcohols as hydrogen donors (for review, see Ref. 22). The *Methanosarcinales* are most versatile with respect to their substrate spectrum (22) because many species use $H_2 + CO_2$, acetate, methanol, and other methylated C_1 compounds such as methylamines (mono-, di-, or trimethylamine) and methylated thiols (dimethylsulfide, methanethiol, or methylmercaptopropionate). Others are restricted to acetate (genus *Methanosaeta*) or are unable to use molecular hydrogen as an electron source and can, therefore, not grow on $H_2 + CO_2$ (e.g., *Methanosarcina thermophila*). Another important aspect of the methanogenic metabolism is the detoxification of certain man-made chemicals. *Methanosarcina* stains have been used for dechlorination of tetrachlorethylene (59), chloroform (60), or dehalogenation of trichlorofluoromethane (61). Other methanogens have the capacity to oxidize furfural to furfuryl alcohol in the presence of $H_2 + CO_2$ (62) and nitroaromatic compounds in the presence of formate (63, 64).

III. Biochemistry of Methanogenesis

During the last 3 decades, the metabolic pathway leading to methane formation has been elucidated in detail mainly by the work from the laboratories of Ralph Wolfe, Gottfried Vogels, Rolf Thauer, Gerhard Gottschalk, and Greg Ferry. Of course, many other scientists have contributed to the understanding of certain aspects of the unusual metabolic pathways. It is not possible to mention them all by name. Hence, for a comprehensive overview of the papers published before 1993 the reader is referred to the book "Methanogenesis: Ecology, Physiology, Biochemistry and Genetics" edited by Greg Ferry (65).

A. Unusual Cofactors

Several unusual coenzymes and prosthetic groups have been discovered that are involved in methane formation (66). The conversion of methanogenic substrates to methane proceeds via C_1 intermediates bound to methanofuran (MFR) (67, 68), tetrahydromethanopterin (H_4MPT) (69), and coenzyme M (HS-CoM). Methanofuran is a C_4-substituted furfurylamine able to bind carboxyl and formyl groups (Fig. 1). H_4MPT is a C_1-carrier coenzyme with a central pterin ring

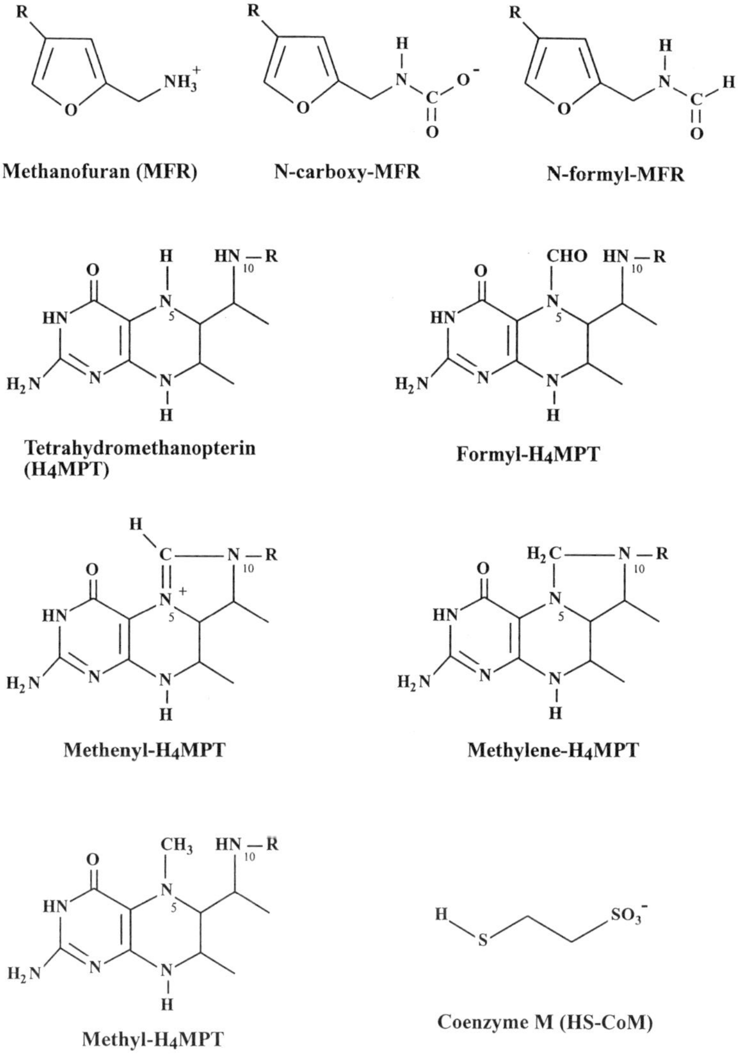

FIG. 1. C_1 carrier in methanogenic archaea.

system able to carry C_1 fragments between formyl and methyl oxidation levels (Fig. 1). Structurally, it resembles tetrahydrofolate, but the two carriers are not functionally equivalent (70).

Methanogens can be recognized by their strong autofluorescence under a fluorescence microscope. This phenomenon is mainly caused by coenzyme F_{420}, which has an absorption maximum of 420 nm and emits light at a wavelength

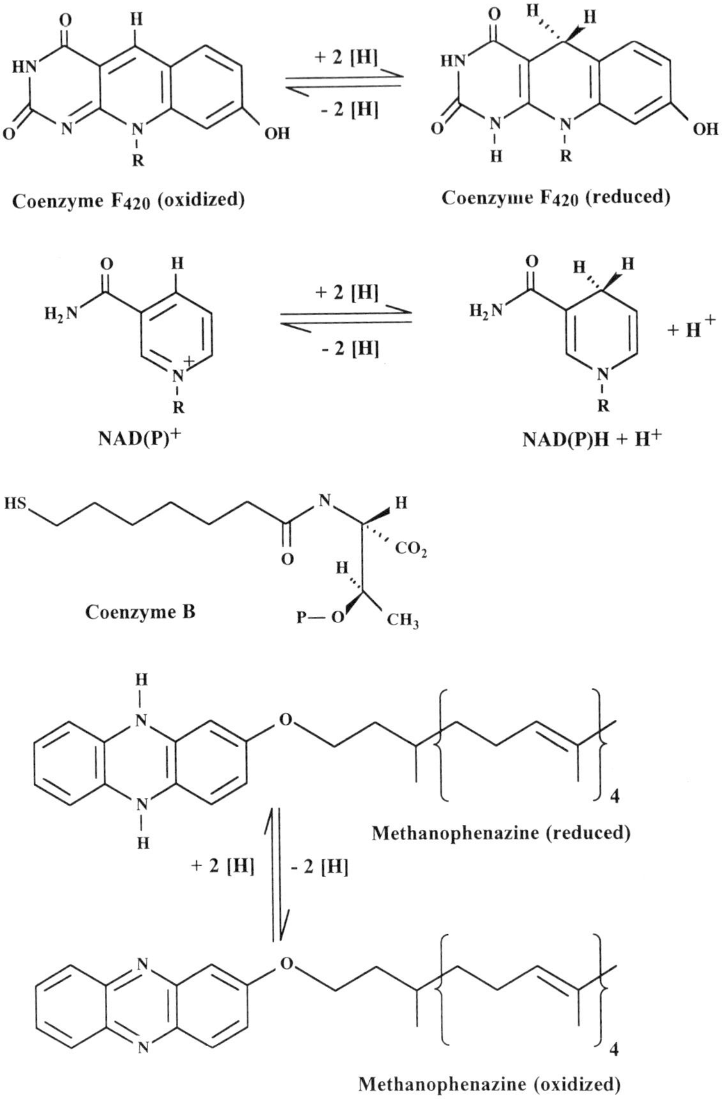

FIG. 2. Electron carrier in methanogenic archaea.

of about 520 nm (71). F_{420} is a deazaflavine derivative with a midpoint potential of −360 mV that functions as a central electron carrier in the cytoplasm of methanogens (72). At first sight, it looks like a flavin derivative, but chemically it is more related to nicotinamides (Fig. 2). A closer look at the active site at carbon atom 4 reveals that the chemical structure in this part of the molecule is identical to the one in pyridine nucleotides. Thus, like NAD^+ and $NADP^+$, coenzyme F_{420} accepts or donates hydride ions in the course of enzymatic reactions (Fig. 2).

The side chains of MFR, H_4MPT, and coenzyme F_{420} are very complex and contain phosphorylated sugars (e.g., ribityl-phosphate, ribose phosphate), organic acids (e.g., lactyl residues, hydroxyglutarate), and amino acids (e.g., glutamate). Several variations of the cofactors have been described (73). It was found that hydrogenotrophic methanogens characteristically contain the typical cofactors H_4MPT and MFR as well as a F_{420} derivative with two glutamate residues. In contrast, the cofactors of methylotrophic methanogens generally possess additional glutamates in their side chains. Examples are methanosarcinapterin and MFR-b which differ from their counterparts in containing one additional glutamate at the very end of the molecule (73).

Besides F_{420}, coenzyme B (HS-CoB) also acts as an electron carrier in the metabolism of methanogens (Fig. 2). The structure of HS-CoB was determined as N-7-mercaptoheptanoyl-L-threonine phosphate (74) and confirmed by chemical synthesis (75). Kobelt et al. (76) have demonstrated that only the L form of HS-CoB is active. A detailed functional description of the cofactor is given in Section III,E.

The most simple methanogenic cofactor is coenzyme M (HS-CoM) which is a ubiquitous methyl-group carrier in the pathway of methanogenesis (Fig. 1). HS-CoM was the first of the unusual coenzymes of methanogens to be discovered and characterized as 2-mercaptoethane-sulfonate (77). The methylated form ((2-methylthio)ethanesulfonate) is the substrate for the methyl-CoM reductase which catalyzes the terminal step of methanogenesis (see Section III,E). This enzyme contains another unusual and unique methanogenic cofactor called coenzyme F_{430} (78). From the chemical point of view, it is a tetrapyrole ring system which, in contrast to all other porphinoids, coordinates a nickel ion (79, 80). Another striking feature is the state of reduction of the pyrole ring system. The cofactor has only five double bonds, of which two pairs are conjugated; thus, it represents the most highly reduced tetrapyrol found in nature (81).

Electron transport in respiratory chains is usually mediated by quinones. In the cytoplasmic membrane of methanogens; however, thus far, only tocopherolquinones in very low concentration, which have obviously no function in the electron transport chain, were identified (82). Thus, the question arose whether other [H] carriers exist, which differ in their chemical structure from quinones and can take over the role of these cofactors (83). To search for any other redox-active, lipid-soluble components, washed membranes of Ms. mazei Göl were lyophilized and extracted with isooctane. Examination of the isooctane extract by analytical HPLC indicated the presence of one major UV-absorbing component. Detailed NMR analysis revealed the aromatic structure as a phenazine derivative (Fig. 2) connected at C-2 to an unsaturated side chain via an ether bridge (83, 84). The lipophilic side chain is responsible for the anchorage in the membrane and consists of five isoprenoid units linked to each other in a head-to-tail manner. Unlike the saturated C_5 unit, which is directly linked to the 2-phenazinyl residue, the remaining four units are unsaturated. The redox-active natural product referred to as methanophenazine is the first phenazine whatsoever isolated form *Archaea*.

Some years ago, it was believed that the coenzymes described above were restricted to methanogenic archaea. However, during the last decade it became evident that some of them are also present in other *Archaea* and even in *Bacteria* or *Eukarya*. The first example is the sulfate-reducing archaeon *Archaeoglobus fulgidus* which in part performs a metabolism similar to the one in methylotrophic methanogens (85). Hence, it was not a surprise that H_4MPT, MFR, and F_{420} were identified in this hyperthermophilic organism, which by the way is phylogenetically closely related to members of the order *Methanosarcinales*. The cofactors in *Archaeoglobus fulgidus* are involved in methyl-group oxidation and are essential during growth on lactate, pyruvate, etc. Interestingly, H_4MPT has also been found in several aerobic methyltrophic bacteria such as *Methylobacterium extorquens*, and evidence for the presence of MFR in these organisms has been presented (86, 87). The cofactors participate in a novel formaldehyde-oxidation pathway during methanol conversion (88). But these are not the only examples for "methanogenic cofactors" in bacteria: Recent studies of propylene metabolism in the *Xanthobacter* strain Py2, a Gram-negative proteobacterium, and *Rhodococcus rhodochrous*, a Gram-positive actinomycete, have demonstrated the presence of coenzyme M which is involved in propylene degradation (89, 90). During studies on the biodegradation of picrate and 2,4-dinitrophenol, coenzyme F_{420} and a NADP:F_{420} oxidoreductase were isolated from *Norcardioides simplex*. In this organism, the components participate in the initial attack for ring cleavage of the xenobiotics (91). In particular, reduced F_{420} is important for hydride transfer from NADPH to the nitroaromatic compounds. In *Streptomyces* species, some steps in tetracycline and lincomycin biosynthesis require F_{420} (92, 93). *Mycobacterium* species are known to contain F_{420} and F_{420}-dependent glucose-6-phosphate dehydrogenases (94, 95). In addition, F_{420} has been found in both the cynaobacterium *Anacystis nidulans* (96) and the eukaryotic green algae *Scenedesmus acutus* (97). The cofactor is described as a target for new antitubercular drugs in *Mycobacterium tuberculosis* (98) and is believed to inhibit growth of some malignant cell lines (99, 100).

B. Methanogenesis from $H_2 + CO_2$ and Formate

The overall process of methanogenesis from $H_2 + CO_2$ can be described according to

$$CO_2 + 4H_2 \longrightarrow CH_4 + 2H_2O \quad (\Delta G'_o = -131 \text{ kJ/mol CH}_4). \quad \text{(i)}$$

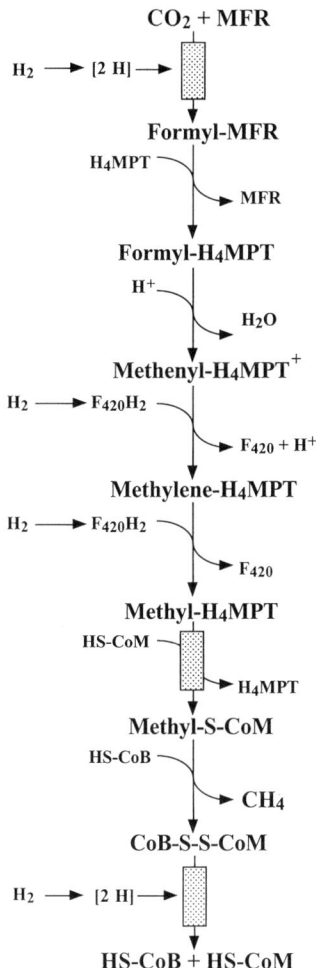

FIG. 3. Pathway of methanogenesis from $H_2 + CO_2$. Reaction catalyzed by membrane-bound enzyme complexes are boxed.

The series of reactions involved in methane formation from the above-mentioned substrate are initiated by the formylmethanofuran dehydrogenase (Fig. 3). The enzyme catalyzes the formation of N-carboxymethanofuran (Fig. 1) from methanofuran and CO_2 (101). The nucleophilic attack of the uncharged amino group of MFR upon CO_2 results in the production of the carbamate in a spontaneous and reversible manner. In a second endergonic reaction N-carboxymethanofuran is reduced to formyl-MFR (Fig. 1) which is the first stable intermediate in the pathway. Reducing equivalents for this reductive process are provided by

H_2 via a hydrogenase (Section V,B,1) or by formate via a formate dehydrogenase. The formyl group is then transferred to H_4MPT and stepwise reduced to N^5-methyl-H_4MPT. Intermediates of this fully reversible process are N-formyl-H_4MPT, N^5,N^{10}-methenyl-H_4MPT, and N^5,N^{10}-methylene-H_4MPT (Fig. 3). The reactions are catalyzed by the formyl-MFR:H_4MPT formyltransferase (102–105), methenyl-H_4MPT cyclohydrolase (106, 107), methylene-H_4MPT dehydrogenase (108, 109), and methylene-H_4MPT reductase (109–112) which are all located in the cytoplasm of methanogens. Reduced coenzyme F_{420} is the source of reducing equivalents for the reduction of N^5,N^{10}-methenyl-H_4MPT and N^5,N^{10}-methylene-H_4MPT (Section III,A). F_{420} is reduced with the help of molecular hydrogen by the F_{420}-reducing hydrogenase (113–117). The cytoplasmic enzyme is composed of three different subunits and contains FAD and several iron–sulfur clusters. Alternatively to the F_{420}-dependent methylene-H_4MPT dehydrogenase, some obligate hydrogenotrophic methanogens contain a second dehydrogenase which is able to use molecular hydrogen as electron donor (118–121). Thus, the protein acts like a hydrogenase despite the fact that it does not contain iron–sulfur clusters or Ni/Fe clusters found in all other hydrogenases (Section V,B,1). Recently, evidence was presented that the metal-free hydrogenase does posses a low-molecular-mass, thermolabile cofactor that is tightly bound to the enzyme. However, there was no indication that the cofactor contains a redox-active transition metal (122).

As mentioned above, the first part of methanogenesis from $H_2 + CO_2$ leads to the formation of methyl-H_4MPT (Fig. 3). In the next reaction, a membrane-bound methyltransferase catalyzes the transfer of the methyl moiety to HS-CoM, thereby producing methyl-S-CoM (123, 124). This exergonic process is coupled to sodium ion extrusion resulting in the generation of an electrochemical sodium ion gradient (Section V,G) (125–130). In the final step methyl-S-CoM is reductively cleaved by the methyl-CoM reductase which uses coenzyme B as electron donor (76, 131–135). In summary, this reaction leads to the formation of methane and a mixed disulfide from coenzyme M and coenzyme B which is called heterodisulfide (CoM–S–S–CoB) (131, 132). The reduction of CoM–S–S–CoB as intermediate of this metabolic pathway is described in detail in Section V,B,3.

About half of all methanogens can grow on formate, including species of the genera *Methanococcus* (Mc.), *Methanothermobacter,* and *Methanospirillum* (Eq. ii). It is not used for growth by any *Methanosarcina* (Ms.) species:

$$4HCOOH \longrightarrow 3CO_2 + CH_4 + 2H_2O \ (\Delta G'_0 = -106 \text{ kJ/mol } CH_4). \quad \text{(ii)}$$

Methanogenesis from formate involves oxidation of the substrate to produce CO_2 and a reduced electron carrier. The reaction is catalyzed by a formate dehydrogenase. The enzyme has been isolated from *Methanobacterium formicicum* and

Mc. vannielii (*136, 137*) and is composed of two subunits containing molybdenum and Fe/S clusters. Both proteins couple the oxidation of formate with the reduction of coenzyme F_{420}. The second product CO_2 enters the carbon dioxide reduction pathway outlined above (Fig. 3) (*138*). Reduced F_{420} serves as electron donor for the reduction of methenyl-H_4MPT, methylene-H_4MPT, and CoM–S–S–CoB (*22, 117*). The electron-transfer route from formate to formyl-MFR is not yet completely clear.

C. Methane Formations from Methylated C_1 Compounds

The most important methanogenic substrates are $H_2 + CO_2$ and acetate, but in marine and brackish environments methylated compounds are thought to predominate. The methylamines and methylthiols arise from the anaerobic breakdown of common cellular osmolytes (choline derivatives, betaine, trimethylamine-*N*-oxide, and dimethylsulfoniopropionate) of many marine phytoplankton and certain plants. Members of the family *Methanosarcinaceae* are able produce methane from methyl compounds such as methanol, methylamines, methylmercaptopropionate, or dimethylsulfide (*22, 139, 140*). Most of our current knowledge derives from studies using *Methanosarcina* strains growing on methanol. In the absence of H_2 the methylotrophic methanogens are able to convert methanol according to Eq. (iii):

$$4CH_3OH \longrightarrow 3CH_4 + 1CO_2 + 2H_2O$$
$$(\Delta G_0' = -106 \text{ kJ/mol } CH_4) \quad \text{(iii)}$$

$$3CH_3OH + 6[H] \longrightarrow 3CH_4 + 3H_2O \quad \text{(iv)}$$

$$1CH_3OH + 1H_2O \longrightarrow 1CO_2 + 6[H]. \quad \text{(v)}$$

The process can be described as a disproportionation event of four methanol molecules (Fig. 4). Three methyl groups are reduced to methane (Eq. iv) and the fourth methyl moiety is oxidized to CO_2 (Eq. v). Also in this methanogenic pathway methyl-S-CoM is an important intermediate. Methyl transfer from methanol to HS-CoM is catalyzed by a soluble methyltransferase system composed of three polypeptides: MtaA, MtaB, and MtaC, the latter of which harbors a corrinoid prosthetic group (*141, 142*). In the reductive branch of the pathway, three out of four methyl-CoM molecules are reductively demethylated to methane. The conversion of methyl-CoM to methane is identical to the one described above (Section III,B). Thus, also in this pathway the activity of the methyl-CoM reductase gives rise to the formation of CoM–S–S–CoB and CH_4 using HS-CoB as the reductant (Fig. 4).

In the oxidative branch of the pathway, one out of four methyl groups is oxidized stepwise to carbon dioxide (Fig. 4) by the reversed CO_2-reduction route

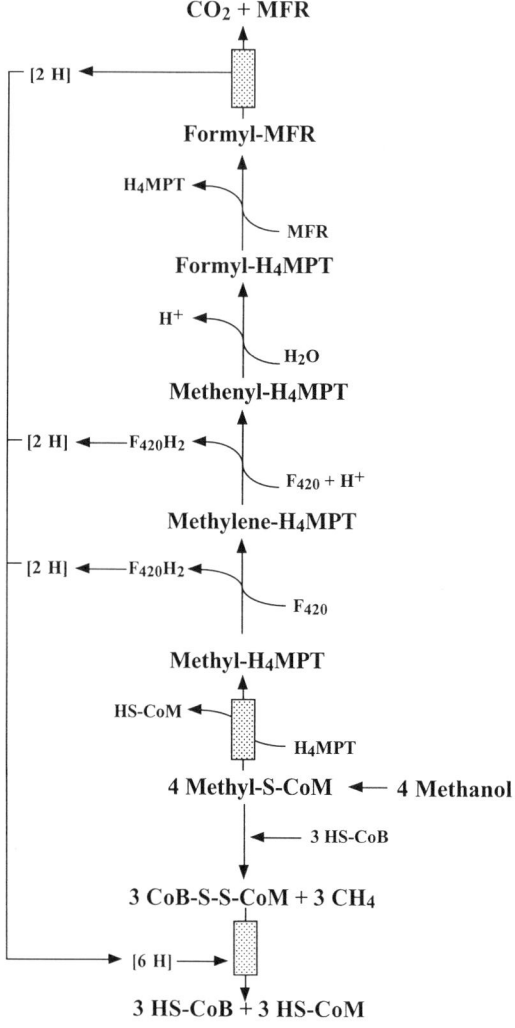

FIG. 4. Pathway of methanogenesis from methanol. Reaction catalyzed by membrane-bound enzyme complexes are boxed.

as described in Section III,B (*143–145*). In this case the methyl transfer from methyl-S-CoM to H$_4$MPT is an endergonic reaction ($\Delta G'_0 = 30$ kJ/mol) and is driven by a electrochemical sodium ion gradient (Section V,G) (*125, 146*). Reducing equivalents derived from the oxidation of N^5-methyl-H$_4$MPT and N^5,N^{10}-methylene-H$_4$MPT (Fig. 4) are used for F$_{420}$ reduction. At the end of the pathway the formyl-MFR dehydrogenase catalyzes the oxidation of CHO–MFR leading to the release of CO_2 and to the regeneration of MFR. The electron acceptor of the protein is still under debate. The electron-transfer reaction is an exergonic process and is probably coupled to the formation of a transmembrane ion gradient (Section V,H).

Many *Methanosarcina* strains can grow on trimethylamine, dimethylamine, and monomethylamine, the substrates being metabolized to CH_4, CO_2, and NH_3 (147–150). Furthermore, evidence has been presented that a *Methanococcoides* species is able to use tetramethylammonium as substrate (151). The pathway for the degradation of methylamines is almost identical to the process of methane formation from methanol as described above (152). The only difference is the transfer of methyl groups to HS-CoM which can be summarized as (Eqs. vi–viii) (27).

$$(CH_3)_3\,NH^+ + H\text{–}S\text{–}CoM \longrightarrow (CH_3)_2\,NH_2^+ + CH_3\text{–}S\text{–}CoM \qquad \text{(vi)}$$

$$(CH_3)_2\,NH_2^+ + H\text{–}S\text{–}CoM \longrightarrow CH_3\text{–}NH_3^+ + CH_3\text{–}S\text{–}CoM \qquad \text{(vii)}$$

$$CH_3\text{–}NH_3^+ + H\text{–}S\text{–}CoM \longrightarrow NH_4^+ + CH_3\text{–}S\text{–}CoM. \qquad \text{(viii)}$$

The methyltransferases catalyzing the reactions have been purified and characterized from *Methanosarcina barkeri*. For the conversion of trimethylamine to dimethylamine and methyl-S-CoM, three proteins are required (153); a 52-kDa protein (MttB), a 26-kDa protein (MttC), and a 36-kDa protein (MtbA or MtaA). The MttC protein contains a corrinoid group, which accepts the methyl group from trimethylamine in a reaction catalyzed by MttB (Eq. ix). From the methylated corrinoid protein, the methyl group is transferred to coenzyme M (154). This reaction is catalyzed by MtbA or MtaA (Eq. x) (142, 155):

$$(CH_3)_3\,NH^+ + MttC \longrightarrow (CH_3)_2\,NH_2^+ + CH_3\text{–}MttC \qquad \text{(ix)}$$

$$CH_3\text{–}MttC + H\text{–}S\text{–}CoM \longrightarrow MttC + CH_3\text{–}S\text{–}CoM. \qquad \text{(x)}$$

For methyl-coenzyme M formation from dimethylamine and coenzyme M, three proteins are also required. The methyltransferase MtbB catalyzes the transfer of the methyl moiety from dimethylamine to the corrinoid protein MtbC (Eq. xi). The second methyl-transfer reaction from the methylated corrinoid to HS-CoM is mediated by MtbA (Eq. xii) (156–158):

$$(CH_3)_2\,NH_2^+ + MtbC \longrightarrow CH_3\text{–}NH_3^+ + CH_3\text{–}MtbC \qquad \text{(xi)}$$

$$CH_3\text{–}MtbC + HS\text{–}CoM \longrightarrow MtbC + CH_3\text{–}S\text{–}CoM \qquad \text{(xii)}$$

A set of methyltransferases, which are referred to as MtmB (catalyzing reaction xiii) and MtmC (the corrinoid protein), are also available for methyl-CoM production from monomethylamine (159). Again MtbA is involved in methyl-group transfer from methyl-MtmC to HS-CoM (Eq. xiv):

$$CH_3-NH_3^+ + MtmC \longrightarrow NH_4^+ + CH_3-MtmC \quad \text{(xiii)}$$

$$CH_3-MtmC + HS\text{-}CoM \longrightarrow MtmC + CH_3\text{-}S\text{-}CoM. \quad \text{(xiv)}$$

As evident all pathways lead to the formation of methyl-S-CoM which is then reduced to methane or oxidized to CO_2 as described above.

D. Conversion of Acetate to Carbon Dioxide and Methane

Despite the fact that the major part of methane production in nature originates from the methyl group of acetate, only two genera (*Methanosarcina* and *Methansaeta*) have been described which are able to use this substrate for methanogenesis and growth. In *Methanosarcina* strains, the so-called aceticlastic pathway of methane formation from acetate starts with the activation of the carboxyl group by phosphorylation as catalyzed by an acetate kinase (*160–162*). The resulting acetyl-phosphate is converted to acetyl-CoA by the catalytic activity of a phosphotransacetylase (Fig. 5) (*163*):

$$CH_3-COO^- + H^+ \longrightarrow CO_2 + CH_4$$
$$(\Delta G_0' = -36 \text{ kJ/mol } CH_4) \quad \text{(xv)}$$

$$CH_3-COO^- + ATP \longrightarrow CH_3CO_2PO_3^{2-} + ADP \quad \text{(xvi)}$$

$$CH_3CO_2PO_3^{2-} + HS\text{-}CoA \longrightarrow CH_3\text{-}COSCoA + P_i. \quad \text{(xvii)}$$

In obligatory acetotrophic methanogens of the genus *Methanosaeta*, the activation of acetate is performed by acetyl-CoA synthetase (*164*):

$$CH_3-COO^- + HS\text{-}CoA + ATP \longrightarrow CH_3\text{-}CO\text{-}SCoA + AMP + PP_i. \quad \text{(xviii)}$$

The CO dehydrogenase/acetyl-CoA synthase complex (from here on referred to as CO dehydrogenase) is the key enzyme of the aceticlastic pathway and functions to cleave the C–C and the C–S bond of acetyl-S-CoA (*165, 166*). Subsequently, the enzyme transfers the methyl group to H_4MPT and oxidizes CO to CO_2. Electrons derived from the oxidation are used for ferredoxin reduction (for details the reader is referred to Refs. *167–170*). The last steps of methanogenesis are identical in the CO_2-reducing and the methylotrophic pathways. The methyl group of CH_3-H_4MPT is transferred to HS-CoM. Finally, CoM–S–S–CoB and CH_4 are formed from methyl-CoM and HS-CoB (Fig. 5) (*171*).

E. Pathways of Methanogenesis: A Summary

As described in the former sections, methyl-coenzyme M is formed from all methanogenic substrates and is the central intermediate in the metabolism of methanogenic archaea. The methyl-S-CoM reductase (MCR) is responsible for the microbial formation of methane and catalyzes the reduction of the

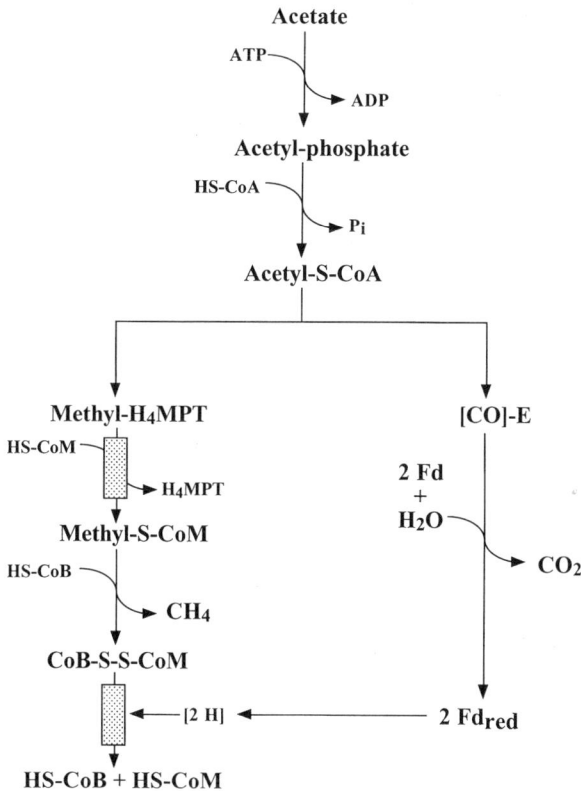

FIG. 5. Methane formation from acetate. Energy-transducing enzymes are boxed.

methylthioether with HS-CoB to CH_4 and CoM–S–S–CoB (132, 172). The 300-kDa protein is organized as a hexamer in an α_2, β_2, γ_2 arrangement (171, 173–176). The crystal structures of the enzyme from *Methanothermobacter marburgensis* (formerly *Methanobacterium thermoautotrophicum*) has been determined by R. K. Thauer and co-workers at 1.45 angstrom resolution (134). The three-dimensional structures revealed that two molecules of the nickel porphinoid coenzyme F_{430} are embedded between the subunits forming two identical active sites. Furthermore, the electron density map indicated five modified amino acids in the α subunit near the active site region. Four of these modifications are C-, N-, and S-methylations; the fifth is an unusual thioglycine (177, 178). A narrow channel formed by the protein allows methyl-CoM to enter one of the active sites which is locked after binding of the second substrate coenzyme B. A reaction mechanism is proposed that involves a nickel organic compound and radical intermediates (25). The crystal structures of the methyl-CoM reductase are also known from *Ms. barkeri* and *Methanopyrus kandleri* (178) and are very similar to the one in *Methanothermobacter marburgensis* (formerly *Methanobacterium thermoautotrophicum* strain Marburg).

The formation of the heterodisulfide marks the end of the processes leading to methane formation. Thus, the question arises as to how this intermediate is reduced in order to regenerate HS-CoM and HS-CoB for the next reaction cycles. When synthetic heterodisulfide became available, it was found that CoM–S–S–CoB is the terminal electron acceptor of a branched respiratory chain in these organisms. The enzyme reducing the heterodisulfide is the membrane-bound heterodisulfide reductase that functions as a terminal respiratory reductase (Section V,B,3). Hence, it is evident that the whole process of methanogenesis is only performed to produce an organic disulfide compound functioning as an electron acceptor for anaerobic respiration. In this respect, methanogenic metabolism resembles the process of sulfur respiration as found in many bacteria and archaea. These organisms use polysulfide as an external electron acceptor for respiration which is formed from elemental sulfur in the presence of H_2S (*31*). With molecular hydrogen as electron donor, the reaction can be summarized according to Eq. (xix). Also in the heterodisulfide reduction an S–S bond is cleaved as shown in Eq. (xx) (see Section V,B,3) indicating the similarity of the reactions:

$$H_2 + S_n^{2-} \longrightarrow HS^- + S_{n-1}^{2-} + H^+ \quad (\Delta G_0' = -31 \text{ kJ/mol } H_2) \quad \text{(xix)}$$

$$H_2 + \text{CoB–S–S–CoM} \longrightarrow \text{HS-CoB} + \text{HS-CoM} \quad (\Delta G_0' = -40 \text{ kJ/mol } H_2). \quad \text{(xx)}$$

In the absence of molecular hydrogen methanogens are forces to use alternative electron donors for the reduction of CoM–S–S–CoB. During growth on methylated C_1 compounds reduced coenzyme F_{420} ($F_{420}H_2$) is formed (Section III,C) and acts as an electron donor. Reduced ferredoxin has a similar function when acetate is utilized (Section III,D). In recent years it turned out that the oxidation of $F_{420}H_2$ and of reduced ferredoxin as well as the reduction of CoM–S–S–CoB is catalyzed by membrane-bound electron transport chains that couple the redox reaction with the translocation of protons (*30, 31*). The energy-conservation systems involved show interesting new features and will be described in detail in the following sections.

IV. Mechanism of ATP Synthesis in Methanogenic Archaea

The principal mechanisms of ATP generation fall into just two classes, namely, substrate-level phosphorylation (SLP) and ion-gradient-driven ATP synthesis. SLP occurs during the degradation of organic substrates where a small number of intermediates is formed containing high-energy phosphoryl bonds (e.g., acetyl phosphate, phosphoenolpyruvate, and 1,3-bisphosphoglycerate). Further metabolism of such compounds is coupled to the transfer of the phosphate group to ADP. The second mechanism of ATP synthesis is effected by the interaction of ion-translocating enzymes and ATP synthases, both tightly associated with the inner mitochondrial membrane in *Eukarya* or the cytoplasmic membrane of *Bacteria* and *Archaea*. According to the chemiosmotic theory (*179*) the proton-motive force (Eqs. xxi and xxii) consisting of a proton gradient (ΔpH) and a membrane potential ($\Delta\Psi$), is the driving force for ATP synthesis from ADP + P_i as catalyzed by ATP synthases (see Section V,I):

$$\Delta\tilde{\mu}_{H^+} = RT \ln([H_i^+]/[H_o^+]) + F\Delta\Psi \qquad \text{(xxi)}$$

$$\Delta p = \Delta\tilde{\mu}_{H^+}/F = \Delta\Psi - Z\Delta pH. \qquad \text{(xxii)}$$

$$Z = 2.3 \, RT/F$$

Later it was shown that some membrane-residing systems can take advantage of sodium ions instead of protons as coupling ions for the formation of an electrochemical potential ($\Delta\mu_{Na^+}$) (for review, see Ref. *180*).

Ion-translocating proteins can be divided into two classes: (i) The transfer of protons or sodium ions across the membrane is coupled to redox reactions as catalyzed by membrane-bound electron transport systems (respiratory chain and photosynthesis). (ii) The generation of an ion motive force is independent from primary redox reaction. Examples for this kind of energy conservation are proton/product symporters (*181*), decarboxylases functioning as primary Na^+ pumps (*180, 182*), and Na^+-translocating methyltransferases (*130*).

Although methanogens are nutritionally rather similar and employ almost identical pathways for methanogenesis, they differ significantly with respect to the components involved in the electron transport chain and, therefore, most likely employ different mechanisms to generate the proton gradient. Since most of our current knowledge derives from studies using methylotrophic methanogens, in particular, *Ms. barkeri* and *Ms. mazei*, the next sections focus on these organisms.

Taking a closer look at the processes of methanogenesis, it is obvious that there is no site for ATP regeneration by substrate-level phosphorylation. Moreover, in the course of the aceticlastic pathway, one or even two ATP molecules are spent for acetyl-CoA formation. Therefore, it has been assumed for quite some time that electron transport processes might be coupled to ion translocation and that the ion-motive force generated might be taken advantage of for ATP synthesis (*183*). However, experimental proof was missing for a long time. Finally, through analysis of the hydrogen-dependent methanol conversion in *Ms. barkeri*, it became evident that ATP is formed by a chemiosmotic mechanism (*184, 185*). Further details of the energy metabolism were obtained when methanogenesis from methyl-CoM and H_2 was studied using subcellular preparations of *Ms. mazei* (*186*). In contrast to other *Methanosarcina* species, this strain lacks a heteropolysaccharide layer and does only possess a proteinous cell wall. By treatment with proteases the protein layer can be digested resulting in the formation of protoplasts (*187*). After gentle French pressure treatment a cell-free lysate was obtained containing closed membrane structures with 90% inside-out orientation (*188*). Later on, it was shown that inverted vesicles, free of cytoplasmic components, are capable of energy transduction (*189*). The great advantage of such a system is that the active centers of the enzymes involved in membrane-bound electron transport face the outside and are accessible for highly charged substrates such as coenzyme F_{420} and the heterodisulfide.

V. Energy-Conserving Systems in *Methanosarcina* Strains

The deep phylogenetic difference between the *Methanosarcinales* and obligate hydrogenotrophic methanogens is reflected by their different energy metabolism which will be described in the following sections. From the thermodynamic point of view, it is obvious that only the reduction of the heterodisulfide (−40 kJ/mol), the oxidation of formyl-MFR (−16 kJ/mol), and the methyl transfer from H_4MPT to HS-CoM (−29 kJ/mol) are sufficiently exergonic to be coupled to energy conservation (*29*). All these reactions are catalyzed by membrane-bound enzyme complexes (Sections V,B–V,H). This fact is in agreement with a chemiosmotic mode of energy transduction because membrane-integral proteins are a stringent necessity for the generation of electrochemical ion gradients.

A. Redox-Driven Proton Translocation in *Methanosarcina mazei*

During the last years, the energy-conserving systems of the methylotrophic methanogen *Ms. mazei* have been analyzed comprehensively (*29, 30*). It was shown that the organism possesses two membrane-bound electron transport systems, both of which are able to use the heterodisulfide as electron acceptor and either H_2 or $F_{420}H_2$ as electron acceptor (Eqs. xxiii and xxiv) (*189–191*):

$$H_2 + CoM\text{–}S\text{–}S\text{–}CoB \longrightarrow HS\text{-}CoM + HS\text{-}CoB$$
$$(\Delta G_0' = -40 \text{ kJ/mol}) \quad\quad \text{(xxiii)}$$

$$F_{420}H_2 + CoM\text{–}S\text{–}S\text{–}CoB \longrightarrow HS\text{-}CoM + HS\text{-}CoB + F_{420}$$
$$(\Delta G_0' = -30,9 \text{ kJ/mol}). \quad\quad \text{(xxiv)}$$

Taking advantage of the above-described inverted vesicles, it has been shown that these preparations catalyze a H_2-dependent CoM–S–S–CoB reduction, whereas methyl-CoM or other disulfide components did not serve as electron acceptors (190). The electron transport system is referred to as H_2:heterodisulfide oxidoreductase (Eq. xxiii) and is involved in methanogenesis from $H_2 + CO_2$ (Fig. 6A). The redox reactions are catalyzed by a F_{420}-nonreducing hydrogenase (Section V,B,1) and a heterodisulfide reductase (Section V,B,3) which are connected by the membrane-integral cofactor methanophenazine.

During growth on methanol or methylamines, part of the methyl groups of the substrates are oxidized, and reducing equivalents are transferred to F_{420} (Section III,C). Using washed inverted vesicles from *Ms. mazei*, it was shown that $F_{420}H_2$ is reoxidized in the presence of CoM–S–S–CoB by a membrane-bound electron transport named $F_{420}H_2$:heterodisulfide oxidoreductase (192). The system consists of an $F_{420}H_2$ dehydrogenase (Section V,B,2) and a heterodisulfide reductase (Fig. 6B). Electron transport between the enzymes is mediated by methanophenazine. An $F_{420}H_2$:heterodisulfide oxidoreductase system has also been found in *Methanolobus tindarius* (192). Furthermore, the genome sequence of *Ms. barkeri* revealed that this organism contains all genes necessary for the formation of the above-mentioned electron transport system (see http://spider.jgi-psf.org for further detail).

It has been shown that electron transfer as catalyzed by the $F_{420}H_2$:heterodisulfide oxidoreductase and the H_2:heterodisulfide oxidoreductase is coupled to proton translocation into the lumen of the inverted vesicles of *Ms. mazei* (189, 190). Each anaerobic respiratory chain yields stoichiometries of 4 mol of translocated protons/mol of reduced heterodisulfide, respectively. The addition of the protonophore SF 6847 or of the ATP synthase inhibitor DCCD resulted in effects that resemble the phenomenon of respiratory control as described for mitochondria (193). The electrochemical transmembrane ion gradient generated in the course of redox-driven H^+ translocation is the driving force for ATP synthesis from $ADP + P_i$. The reaction is catalyzed by an A_1A_o-type ATP synthase (Section V,I). H_2:heterodisulfide oxidoreductases have also been identified in both *Ms. barkeri* and *Methanothermobacter marburgensis;* however, a redox-driven ion translocation by these complexes has not been observed thus far (194–196).

Methanogenesis from acetate is coupled to the smallest change of free energy ($\Delta G^{o'} = -36$ kJ/mol CH_4) of all methanogenic substrates. Since in *Methanosarcina* strains 1 ATP is invested in acetate activation, it is necessary that methanogenesis from acetyl-CoA gives rise to the synthesis of more than 1 ATP per CH_4 formed. Therefore, the organisms must possess efficient energy-conserving systems. As mentioned above, the reaction of the CO-dehydrogenase/acetyl-CoA synthase results in the formation of methyl-H_4MPT, reduced ferredoxin (Fd_{red}), and CO_2. The former intermediate is converted to CoM–S–S–CoB by

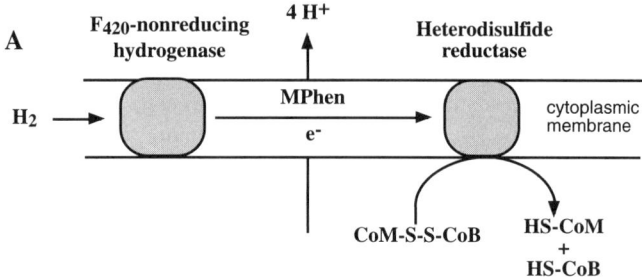
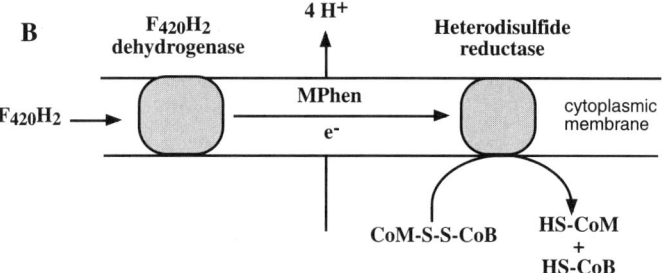
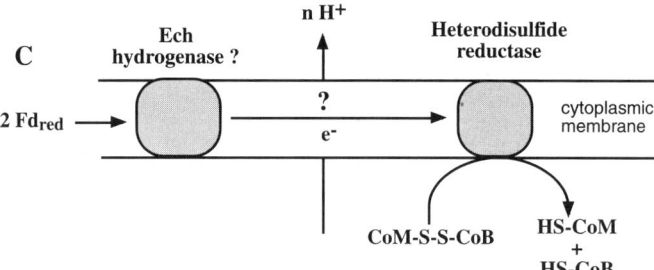

FIG. 6. Overview of anaerobic respiratory chains in *Methanosarcina* strains. (A) H2:heterodisulfide oxidoreductase system, (B) $F_{420}H_2$:heterodisulfide oxidoreductase system, (C) reduced ferredoxin:heterodisulfide oxidoreductase. The question mark indicates that the mechanisms of electron transport and proton translocation are unclear.

the methyl-H_4MPT:HS-CoM methyltransferase and methyl-S-CoM reductase. The methyl-group transfer from CH_3–H_4MPT to HS-CoM leads to Na^+ transfer across the cytoplasmic membrane, resulting in the formation of an electrochemical sodium ion gradient (Fig. 1, Section V,G). The remaining intermediates CoM–S–S–CoB and Fd_{red} are further metabolized by a third membrane-bound electron transport system which can be described as a Fd_{red}:heterodisulfide oxidoreductase (Fig. 6C):

$$2Fd_{red} + CoM\text{–}S\text{–}S\text{–}CoB \longrightarrow 2Fd_{ox}\ HS\text{-}CoM + HS\text{-}CoB$$
$$(\Delta G'_0 \sim -40\ kJ/mol). \qquad (xxv)$$

The composition of this system is still a matter of debate. However, strong evidence has been provided by the group of R. Hedderich (University of Marburg, Germany) that a so-called Ech hydrogenase and the heterodisulfide reductase are involved in electron-transfer and proton translocation (Fig. 6C) (31).

B. Components of the H^+-Translocating Systems

1. HYDROGENASES

Several types of hydrogenases have been described in methanogenic archaea. With the exception of the H_2-forming methylene H_4MPT dehydrogenase, all enzymes belong to the superfamily of nickel–iron hydrogenases (Fig. 7) (197). Generally, the core of these enzymes is composed of a small electron-transfer subunit and a large catalytic subunit harboring a nickel–iron center which is ligated to the protein by one motif in the amino-terminal portion of the polypeptide (RxCxxCxxxH) and a second one at the carboxyl-terminal region (DPCxxCxxH/R) (198–201) In some hydrogenases the last cysteine residue in the latter sequences is replaced by a selenocysteine (116). It has been shown that the metals in the bimetallic reaction center are directly involved in H_2 cleavage or H_2 formation in the reverse reaction:

$$H_2 \rightleftharpoons 2H^+ + 2e^-. \qquad (xxvi)$$

The small subunit contains two to three Fe/S clusters and is responsible for electron transport from or to the catalytic center (202). Besides the core subunits, methanogenic [NiFe] hydrogenases are equipped with several other polypeptides depending on their physiological function.

a. F_{420}-Reducing Hydrogenase. As mentioned in Section III,B, the F_{420}-dependent hydrogenase reduces coenzyme F_{420}, the central electron carrier in methanogenic archaea (66). In addition to the small and large subunits (FrhG and FrhA; Fig. 7A and 7D) it contains a third, FAD-carrying polypeptide (FrhB) that is able to bind and to reduce F_{420} (115, 203–205). In the methylotrophic *Ms. mazei*, the enzyme was found to be located in the cytoplasm and is, therefore, most likely not involved in ion translocation (206). In contrast, the F_{420}-reducing hydrogenase from *Mc. voltae* and other obligate hydrogenotrophic methanogens is bound to the cytoplasmic membrane and might interact with membrane-integral electron carriers (117, 207). The operon encoding the F_{420}-reducing hydrogenase contains a fourth gene called *frhD* (Fig. 7D). The corresponding

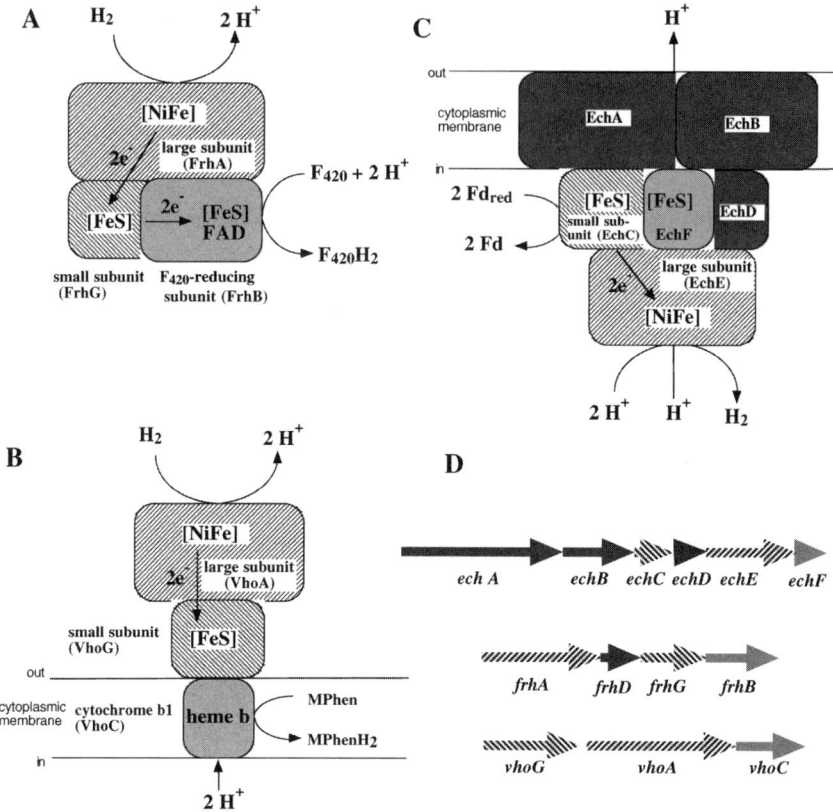

FIG. 7. NiFe hydrogenases in methanogenic archaea. (A) F_{420}-reducing hydrogenase, (B) F_{420}-nonreducing hydrogenase, (C) Ech hydrogenase, (D) structure of operons encoding NiFe hydrogenases. Genes encoding the large subunit (containing the catalytic center) and the small subunit (containing iron–sulfur clusters) are hatched. Genes encoding the electron accepting subunit are shown in grey.

polypeptide FrhD is not copurified with the catalytic enzyme. It is probably involved in the processing of the large subunit (116). F_{420}-reducing hydrogenases were also identified in several other methanogens. In *Mc. voltae* and *Mc. jannaschii*, for example, two operons were characterized, respectively, that code for F_{420}-reducing hydrogenases of the NiFe type (*frcADGB*) and the NiFeSe type (*fruADGB*; 114, 208).

b. *F_{420}-Nonreducing Hydrogenase.* The second type of hydrogenase from methanogens is not able to reduce F_{420}. Hence, it is referred to as F_{420}-nonreducing hydrogenase (Fig. 7B). In *Methanosarcina* strains, the enzyme is membrane associated and has been purified from the particulate fraction using detergents for solubilization (206, 209). The structural genes of the F_{420}-nonreducing hydrogenases from Ms. mazei Göl were cloned and sequenced (210, 211). The genes, arranged in the order vhoG and vhoA, were identified as those encoding the small and the large subunits of the NiFe hydrogenases (Fig. 7D). Northern blot analysis revealed that the structural genes form an operon, containing one additional open-reading frame (vhoC) which codes for a membrane-spanning cytochrome b. It is reasonable to assume that the b-type cytochrome acts as the primary electron acceptor of the core hydrogenase in Ms. mazei and possibly in other species of the family Methanosarcinaceae (191, 212, 213).

The comparison of the sequencing data of vhoG to the experimentally determined N terminus of the small subunit indicates the presence of a 48-amino acid leader peptide in front of the polypeptide (210). It contains a typical twin-arginine motif (214), whereas the large subunit lacks any known targeting signal for export. Several Bacteria contain highly homologous hydrogenases with respect to amino acid sequences and subunit composition (215). Genetic and biochemical data support the assumption that the large subunit is cotranslocated with the small subunit across the cytoplasmic membrane by the Sec-independent, twin-arginine pathway (214). In accordance, electron microscopic immunogold-labeling experiments revealed that at least part of the large subunit of the enzyme from Ralstonia eutropha is exposed toward the periplasm (216). Therefore, it is most possible that the active center of the enzyme is located at the periplasmic face of the membrane. Since the membrane-bound hydrogenase from R. eutropha is homologous to the corresponding enzyme from Ms. mazei Göl, the methanogenic enzyme might have the same orientation (Fig. 7B).

Direct evidence for the involvement of F_{420}-nonreducing hydrogenase in energy conservation was obtained during electron transport studies in Ms. mazei using the hydrophilic methanophenazine homolog 2-hydroxyphenazine (2-OH-phenazine) (213, 217). It was found that the enzyme catalyzes the oxidation of molecular hydrogen and transfers electrons via a b-type cytochrome to 2-OH-phenazine. Furthermore, it has been shown that the reaction is coupled to proton translocation exhibiting stoichiometries of about $2H^+/2e^-$. Later studies revealed that methanophenazine also serves as electron acceptor to the F_{420}-nonreducing hydrogenase (84, 218).

The elucidation of the three-dimensional structure of the periplasmic NiFe hydrogenase from Desulfovibrio gigas (202) together with a wealth of biophysical, biochemical, and genetic studies (201) allowed speculation on a possible reaction mechanism of H_2 oxidation. Accordingly, the active center in the large subunit contains the bimetallic Ni/Fe center that catalyzes the heterolytic split of the hydrogen molecule. The iron–sulfur clusters in the small subunit accept the electrons and transfer them to the respective electron acceptor which is located at the surface of the protein (219, 220). Furthermore, it has been suggested that several histidine and glutamate residues participate in a specific channel within the large subunit that transfer protons from the active site to the surface where they are released into the periplasmic space (*219, 221*). Since the F_{420}-nonreducing hydrogenase from *Ms. mazei* and the enzyme from *Desulfovibrio gigas* are highly homologous and since the critical histidine and glutamate residues are conserved (*210*), it is tempting to speculate that the mechanisms of electron transfer and proton transfer should be similar in the methanogenic enzyme. In the case of the F_{420}-nonreducing hydrogenase, the electrons would be transferred from the Ni/Fe active site via the FeS cluster to the cytochrome *b* subunit. The remaining protons are released outside of the membrane. To complete the reaction cycle, the membrane-integral cytochrome *b* subunit should accept two protons from the cytoplasm for the reduction of methanophenazine Thus, the overall reaction would lead to the production of two scalar protons (Fig. 7B) (*217*).

c. Ech Hydrogenase. Recently, a novel hydrogenase (Ech) was discovered in acetate-grown cells of *Ms. barkeri* (*222, 223*) which shows homologies to hydrogenases 3 and 4 from *E. coli* (*224–226*) and to the CO-induced hydrogenase from *Rhodospirillum rubrum* (*227*). The purified enzyme catalyzed the H_2-dependent reduction of a 2[4Fe–4S]ferredoxin from *Ms. barkeri* and is also able to perform the reverse reaction, namely, hydrogen formation from reduced ferredoxin (Fig. 7C) (*223*). Biochemical and genetic analyses revealed that the Ech hydrogenase is composed of six subunits (EchABCDEF) which are encoded by the *ech* operon (*222*). The polypeptide EchE represents a homolog of the large subunit of the classical [NiFe] hydrogenases. However, the overall sequence similarities between the respective subunits are very low and are mainly restricted to motifs involved in the formation of the bimetallic catalytic center. The sequence alignment of EchC with amino acid sequences from small subunits of [NiFe] hydrogenases indicated that in EchC only one of two to three iron–sulfur-binding motifs is conserved. Despite this fact, it is believed that the polypeptide functions as an electron-transfer subunit in the Ech hydrogenase and is evolutionary related to VhoG and FrhG (*31*) from the F_{420}-reducing and F_{420}-nonreducing hydrogenases, respectively. Hydropathy plots indicate that EchA and EchB are membrane-integral subunits, whereas EchC, EchD, EchE, and EchF reveal a hydrophilic nature and are probably membrane associated (Fig. 7C). All polypeptides of the *ech* hydrogenase show sequence similarities to subunits of the energy-conserving NADH dehydrogenase from eukaryotes and prokaryotes (Section V,B,2). Because of these homologies it was proposed that the enzyme functions as a proton pump (*31*).

The Ech hydrogenase is most likely involved in membrane-bound electron transfer of acetate-grown *Methanosarcina* strains, since Hedderich *et al.* (*31*) showed that Fd_{red} produced by the CO dehydrogenase is reoxidized by the Ech hydrogenase (Fig. 7C). It is thought that electrons derived from this reaction are used for proton reduction leading to the formation of molecular hydrogen.

Furthermore, it was suggested that the overall reaction is coupled to the formation of $\Delta\tilde{\mu}_{H^+}$ (228). The evidence in favor of this hypothesis is the fact that in Ms. barkeri the oxidation of CO to $CO_2 + H_2$ ($\Delta G^{o\prime} = -20$ kJ/mol) is coupled to proton transfer across the cytoplasmic membrane (229, 230). In Rhodospirillum rubrum this reaction is also coupled to the generation of an electrochemical ion gradient (227). Since the CO dehydrogenase of this organism is located in the cytoplasm (231) the membrane-bound CO-induced hydrogenase is most likely the site of energy conservation.

2. $F_{420}H_2$ DEHYDROGENASE

$F_{420}H_2$ dehydrogenases have been purified from the cytoplasmic membrane of Ms. mazei, Methanolobus tindarius, and the sulfate-reducing archaeon Archaeoglobus fulgidus (232–235). The protein from Ms. mazei is composed of five different polypeptides with molecular masses of 40, 37, 22, 20, and 16 kDa and contains non-heme iron, acid-labile sulfur, and FAD (232). The specific activity was 17 U/mg protein using $F_{420}H_2$ as electron donor and methylviologen and metronidazole as electron acceptors.

The subunits are encoded by the fpo cluster that comprises 12 genes which were designated fpo A, B, C, D, H, I, J, K, L, M, N, O (Fig. 8) (236). Northern blot analysis indicated that the genes are organized in one operon. The deduced N-terminal amino acid sequences from fpo D, B, C, and I were identical to the N termini of four subunits (40, 22, 20, and 16 kDa) of the purified $F_{420}H_2$ dehydrogenase. The gene encoding the 37-kDa subunit (Fpo F) of the purified enzyme is not part of the operon and is located at a different site on the chromosome (234, 236). Taking into account that the fpo operon contains 12 genes and that the purified $F_{420}H_2$ dehydrogenase was composed of only 5 subunits, it is obvious that only a subcomplex of protein has been purified. The gene products of fpo A, H, J, K, L, M, and N were not found in the homogeneous protein preparation, and it is most likely that they were lost during purification (236). However, the core enzyme composed of FpoBCDFI showed catalytic activity with $F_{420}H_2$ as electron donor and several artificial dyes as electron acceptors but could not reduce methanophenazine. Hence, it is reasonable to assume that the missing subunits are necessary for the reduction of the physiological electron acceptor.

Recently, the genes encoding the $F_{420}H_2$ dehydrogenase from Archaeoglobus fulgidus have been sequenced in the course of the genome project of this hyperthermophilic archaeon (237). The operon encoding the enzyme is very similar to the one found in Ms. mazei with the exception that fqoF—the counterpart of fpoF—is located at the end of the operon from A. fulgidus (238). The deduced amino acid sequences of fqoF and fpoF reveal similarities to distinct subunits of F_{420}-reducing hydrogenases (114) and F_{420}-dependent formate dehydrogenases (208, 277). Therefore, it has been supposed that FqoF

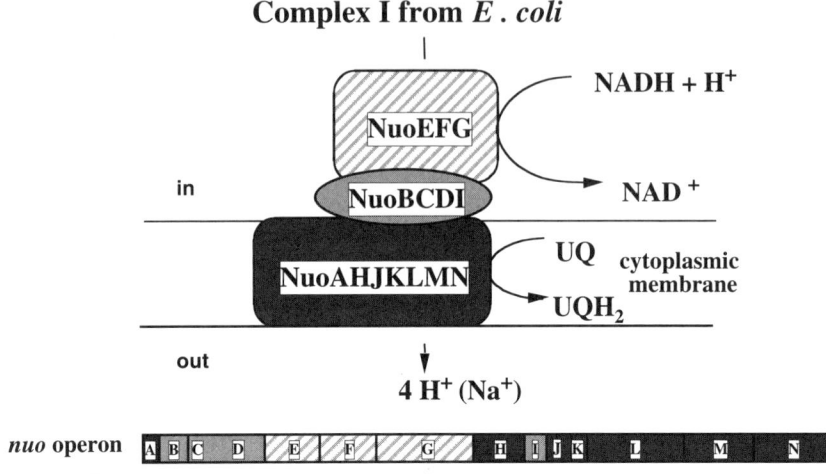

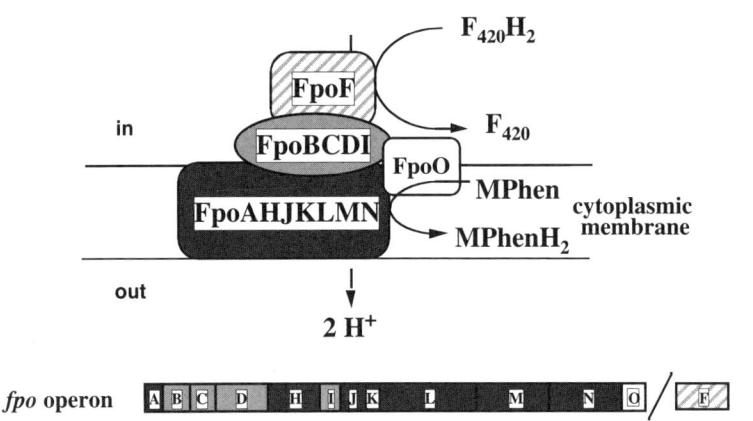

FIG. 8. Tentative models of the $F_{420}H_2$ dehydrogenase from *Ms. mazei* and the NADH dehydrogenase 1 from *E. coli*.

and FpoF represent the subunits within the $F_{420}H_2$ dehydrogenases which are able to oxidize reduced F_{420}. Recently, FpoF from *A. fulgidus* was overproduced in *E. coli* in a functional state (238). As expected from the amino acid sequence, the purified protein contained two [4Fe–4S] clusters and one FAD molecule. The subunit was highly active exhibiting a specific activity of 64 μmol $F_{420}H_2$ oxidized/mg protein at 78°C in the presence of viologen dyes. Taking together these results, it is evident that FqoF and FpoF form the $F_{420}H_2$-oxidizing device of the $F_{420}H_2$ dehydrogenases from *A. fulgidus* and *Ms. mazei*, respectively.

The deduced primary sequences of the *Ms. mazei* $F_{420}H_2$ dehydrogenase subunits were compared to those of other organisms and to those of some phylogenetically related enzymes. For 11 proteins encoded by *fpoA* to *fpoN*, related counterparts exist in bacterial and eukaryotic NADH dehydrogenases (*236*). Interestingly, the highest homology scores were obtained for gene products of higher plants and algae. On the other hand, the *fpo* genes are arranged in the same order as the bacterial genes encoding NADH dehydrogenases (e.g., *ndh/nqo* from *E. coli*, *Thermus thermophilus*, and *Rhodobacter capsulatus* (*239–241*). The proton-translocating NADH dehydrogenase from bacteria is composed of 14 different subunits. In *E. coli*, the enzyme (NDH-1) is encoded by the *nuo* operon (*239*) and the corresponding polypeptides form the three different subcomplexes (*242, 243*): (i) The NADH-oxidizing module is composed of subunits NuoEFG. It contains most of the redox-active prosthetic groups. One noncovalently bound FMN and at least five EPR detectable iron–sulfur clusters form a long electron-transfer chain guiding electron transfer from NADH to FeS clusters present in the membrane-associated module (see below). (ii) NuoAHJKLMN form the membrane-integral module and are involved in quinone reduction and proton translocation. (iii) The membrane-associated module is made from NuoBCDI that connect the above-mentioned subunits and catalyze electron transfer from module 1 to 2 (*244*). In contrast to other bacterial NADH dehydrogenases, subunits NuoC and NuoD are fused in the *E. coli* enzyme (*242*). Very recently, evidence was presented that NuoI contains two tetranuclear FeS clusters similar to 8Fe-ferredoxins (*245*). In addition a second EPR-detectable FeS cluster is found in the connecting fragment (cluster N2) that deserves special attention because it is located close to the membrane and has a pH-dependent midpoint potential (*246, 247*). Several hypotheses concerning the catalytic mechanism of proton-translocating NADH dehydrogenases have been published (*248–250*). Unfortunately, the three-dimensional structure of the complex is not known, and therefore it is difficult to predict electron/proton-transfer reactions. However, it is well established that reduced N2 successively injects single electrons into the membranous subcomplex, thereby activating a serial array of quinones which are directly involved in H^+ translocation (*251*). In agreement with this hypothesis is the discovery of a novel redox group in the membrane-integral module. It is speculated that this electron carrier has a quinoid structure and may be involved in electron transfer from cluster N2 to ubiquinone (*245*).

The native structure of the $F_{420}H_2$ dehydrogenase is still unknown. However, the homology to bacterial NADH dehydrogenases allows us to draw a tentative model (Fig. 8). As mentioned above, the gene product FpoF forms the input module, which oxidizes $F_{420}H_2$ by hydride transfer. FAD present in this subunit catalyzes a two-electron/one-electron switch to reduce the [Fe4–S4] clusters. From the $F_{420}H_2$-oxidizing device, the electrons are then channeled to the amphipatic-connecting fragment composed of FpoBCDI which is highly homologous to the corresponding module of the proton-translocating NADH dehydrogenase. All iron–sulfur signatures of the bacterial enzyme are conserved in FpoB and FpoI from the $F_{420}H_2$ dehydrogenase of Ms. mazei. It is, therefore, reasonable to assume that a FeS cluster comparable to N2 is present in one of the archaeal subunits. In analogy to complex I, cluster N2 should transfer electrons to the membrane-integral module. Interestingly, the composition of the membranous part of the $F_{420}H_2$ dehydrogenase (FpoA, H, J, K, L, M, and N) and NADH dehydrogenase (NuoA, H, J, K, L, M, and N) is identical. However, electron transport through the membrane-integral module of the $F_{420}H_2$-dependent enzyme is difficult to predict because methanogenic archaea do not contain quinones. Therefore, the reaction mechanism of the $F_{420}H_2$ dehydrogenase must be different at this point and must involve the electron carrier methanophenazine (see below).

Using water-soluble analogs of methanophenazine, the catalytic activity of the $F_{420}H_2$ dehydrogenase could be analyzed in more detail. It has been shown that the membrane-bound enzyme is able to reduce 2-OH-phenazine at the expense of $F_{420}H_2$:

$$F_{420}H_2 + \text{2-OH-phenazine} \longrightarrow F_{420} + \text{dihydro-2-OH-phenazine}. \qquad \text{(xxvii)}$$

Moreover, it became evident that the $F_{420}H_2$ dehydrogenase from Ms. mazei is directly involved in the generation of an electrochemical proton gradient. In the course of electron transfer from $F_{420}H_2$ to 2-OH-phenazine as catalyzed by washed inverted vesicles from Ms. mazei (252), about two protons were translocated per $2e^-$ transported (Fig. 8) (236). The midpoint potential of the electron donor and the electron acceptor was determined to be -360 (72) and -255 mV (253), respectively. Thus, the change of free energy ($\Delta G^{o\prime}$) coupled to the $F_{420}H_2$-dependent 2-OH-phenazine reduction is only -20.2 kJ/mol compared to a $\Delta G^{o\prime}$ of -80.9 kJ/mol for the NADH-dependent reduction of ubiquinone. These thermodynamic facts are reflected by the coupling efficiencies of the enzymes since the maximal $H^+/2e^-$ ratio of the $F_{420}H_2$ dehydrogenase is 2.0, in contrast to complex I which translocates four or even more protons across the membrane per reaction cycle.

Despite the aforementioned differences, the $F_{420}H_2$ dehydrogenase from Ms. mazei resembles eukaryotic and bacterial proton-translocating NADH dehydrogenases in many ways. The enzyme from the methanogenic archaeon functions as a complex I homolog and is equipped with an alternative electron-input unit for the oxidation of reduced cofactor F_{420} and a modified-output module adopted to the reduction of phenazine derivatives. Furthermore, the

$F_{420}H_2$ dehydrogenase is a novel proton pump contributing to the generation of the electrochemical proton gradient in the methanogenic organism (Fig. 8).

3. HETERODISULFIDE REDUCTASE

The heterodisulfide reductase catalyzes the final step in the anaerobic respiratory chain of methanogens. This reaction involves the two-electron reduction of CoM–S–S–CoB to the free thiols HS-CoB and HS-CoM (254). The heterodisulfide reductase (Fig. 9) has been purified from the methylotrophic methanogens *Ms. barkeri* (194, 195, 255) and *Ms. thermophila* (256) and consists of two subunits (HdrD and HdrE). The large subunit HdrD from the latter organism contains two distinct [4Fe–4S] clusters with midpoint potentials of −100 and −400 mV and harbors the active site for disulfide reduction. The membrane-integral subunit HdrE represents a *b*-type cytochrome and contains a low-spin and a high-spin heme with midpoint potentials of −180 and −23 mV, respectively. Evidence has been presented that the physiological

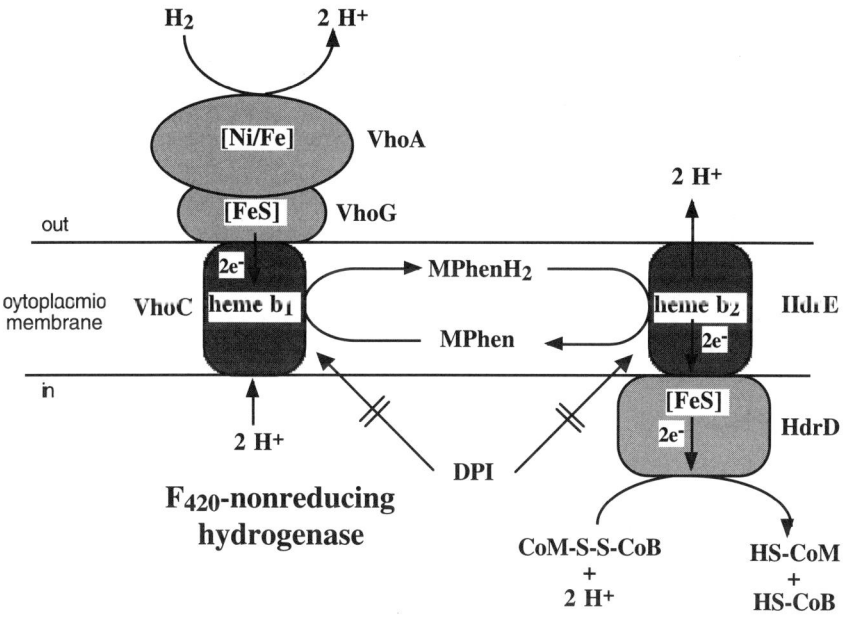

Heterodisulfide reductase

FIG. 9. Tentative scheme of the proton-translocating H_2:heterodisulfide oxidoreductase of *Methanosarcina* strains. Interrupted arrows indicate reactions which are inhibited by DPI (diphenyleniodonium chloride).

electron donor of the heterodisulfide reductase is the membrane-bound cofactor methanophenazine (*84*). The water-soluble analog 2-OH-phenazine is an effective substitute for methanophenazine (*252, 257*). Washed inverted vesicles of *Ms. mazei* were able to couple the dihydro-2-OH-phenazine-dependent reduction of the heterodisulfide with the transfer of two protons across the cytoplasmic membrane (*236*).

The mechanism of energy conservation might be based on scalar proton transfer as in case of the F_{420}-nonreducing hydrogenase (Section V,B,1; *217*). It is reasonable to assume that *in vivo* electrons from dihydromethanophenazine are transferred to the one-electron-accepting prosthetic groups of the enzyme and scalar protons are released at the outer phase of the cytoplasmic membrane (Fig. 9). In a second step, the electrons are channeled to the reactive center, and CoM–S–S–CoB is reduced to HS-CoM and HS-CoB. It has been proposed that protons necessary for this reaction are derived from the cytoplasm and are transferred to the active site by a proton-conducting channel built around a selected number of polar amino acid side chains as well as bound water molecules (*217*).

Recently, the pathway of electron transfer from reduced 2-OH-phenazine to the heterodisulfide was studied by steady-state and transient kinetics using the purified heterodisulfide reductase from *Ms. thermophila* (*257*). Stopped flow experiments and inhibitor studies indicated that only the low-potential heme participates in reduction of CoM–S–S–CoB because the reduction of the second heme was significantly slower than the turnover number for the enzyme. Disruption of the iron–sulfur clusters by mersaryl acid inhibited heme reduction by 2-OH-phenazine, indicating that one of the [4Fe–4S] centers is the initial electron acceptor for the reduced cofactor. Furthermore, spectroscopic and kinetic studies with the inhibitor diphenylene iodonium indicated that only the high-potential cluster is involved in electron transfer to CoB–S–S–CoM. Thus the proposed intramolecular electron transfer pathway is

$$\text{2-OH-phenazine}_{red} \longrightarrow [Fe_4S_4]_{high} \longrightarrow \text{heme}b_{low} \longrightarrow \text{CoB-S-S-CoM}.$$

The central problem that needs to be solved is how a one-electron donor, the low-potential heme, can carry out a concerted reaction involving reductive cleavage of the disulfide substrate. In this connection, it is of note that the reductive cleavage of S–S bonds in organic molecules is widespread in the metabolism of living cells and is usually catalyzed by pyridine nucleotide:disulfide oxidoreductases. Usually, these enzymes contain FAD in the active center which mediates a two-electron transfer from NAD(P)H to the disulfide bound. However, the reaction mechanism of the heterodisulfide reductases from methylotrophic methanogens has to be different because the enzymes do not contain flavin (*255*). There is only one other example of a disulfide reductase lacking FAD which is the ferredoxin–thioredoxin reductase from plants and cyanobacteria.

This enzyme catalyzes the reduction of thioredoxin with reduced ferredoxin by two one-electron transfer reactions (258). As evident from EPR spectra and redox titrations, a one-electron-reduced intermediate might also be produced by the heterodisulfide reductase from Ms. barkeri in the course of CoM–S–S–CoB reduction (31, 259), indicating a similar mechanism as the ferredoxin–thioredoxin reductase.

C. Further Analysis of the Membrane-Bound Electron Transport Systems of Ms. mazei

Detailed spectroscopic analysis revealed that the proton-translocating electron transport systems ($F_{420}H_2$:heterodisulfide oxidoreductase and H_2:heterodisulfide oxidoreductase) of Ms. mazei are inhibited by diphenyleneiodonium chloride (DPI) indicated by IC_{50} values of 20 and 45 nmol DPI/mg protein, respectively (213). These effects are due to a complex interaction of DPI with key enzymes of the electron transport chains. It was found that 2-OH-phenazine-dependent reactions as catalyzed by F_{420}-nonreducing hydrogenase, $F_{420}H_2$ dehydrogenase, and heterodisulfide reductase were inhibited. Analysis of the redox behavior of membrane-bound cytochromes indicated that DPI affected CoB–S–S–CoM-dependent oxidation of reduced cytochromes and H_2-dependent cytochrome reduction. Since DPI and phenazines are structurally similar with respect to their planar configuration, it is assumed that the inhibitor is able to bind to positions where interactions between phenazines and components of the electron transport systems occur. Thus, electron transfer from H_2 to cytochrome b_1 as subunit of the membrane-bound hydrogenase and from reduced 2-OH-phenazine to cytochrome b_2 as part of the heterodisulfide reductase is affected by DPI (Fig. 9). In the case of the $F_{420}H_2$ dehydrogenase electron transport from either FAD or FeS centers to 2-OH-phenazine is inhibited (213).

Very recently the structure and chemical synthesis of methanophenazine, the first phenazine from Archaea, have been published (Fig. 2) (84). Experiments on the function of this new cofactor in Ms. mazei Göl demonstrate that it is also the first phenazine whatsoever involved in the membrane-bound electron transport in biological systems. After completion of the total synthesis, the first experiments on the biological function could be performed with methanophenazine. Washed cytoplasmic membranes of Methanosarcina mazei Göl were combined with the chemically synthesized cofactor, and the activities of the respective enzymes were determined. The results clearly indicated that methanophenazine serves as an electron acceptor to both the membrane-bound hydrogenase and the $F_{420}H_2$ dehydrogenase if H_2 and $F_{420}H_2$ were added, respectively. In addition, the heterodisulfide reductase uses the reduced form of methanophenazine as an electron donor for the heterodisulfide reduction. Therefore, methanophenazine is able to mediate the electron transport between the membrane-bound enzymes (84).

D. Proton Translocating during Growth on $H_2 + CO_2$ or Methylated C_1 Compounds: A Summary

As mentioned above, the composition of the respiratory chain of methylotrophic methanogens is variable and depends on the pathway used for methanogenesis. During growth on $H_2 + CO_2$ the H_2:heterodisulfide oxidoreductase is of great importance (Fig. 9). The membrane-bound F_{420}-nonreducing hydrogenase channels electrons via its b-type cytochrome to the heterodisulfide reductase which reduces the terminal electron donor. As mentioned above, electron transport between the complexes is mediated by methanophenazine. Both partial reactions are coupled to the transfer of protons across the cytoplasmic membrane ($2\ H^+/2e^-$). The results indicate that these organisms possess two different proton-translocating segments in the H_2:heterodisulfide oxidoreductase system (Fig. 9).

$$H_2 + \text{methanophenazine} \longrightarrow \text{dihydro-methanophenazine}$$
$$(\Delta G^{o\prime} = -31.8\ \text{kJ/mol}) \quad \text{(xxviii)}$$

$$\text{dihydro-methanophenazine} + \text{CoM-S-S-CoB} \longrightarrow \text{methanophenazine}$$
$$+ \text{HS-CoM} + \text{HS-CoB}\ (\Delta G^{o\prime} = -10.6\ \text{kJ/mol}) \quad \text{(xxix)}$$

When cells are grown on methanol part of the methyl groups are oxidized to CO_2 and reducing equivalents are transferred to coenzyme F_{420}. In *Methanosarcina*-strains $F_{420}H_2$ is reoxidized by the membrane-bound $F_{420}H_2$ dehydrogenase (Fig. 8) which is part of the $F_{420}H_2$:heterodisulfide oxidoreductase.

As in the H_2-dependent system, electrons are channeled to the heterodisulfide reductase resulting in the reduction of CoM–S–S–CoB and in the formation of an electrochemical proton gradient. The partial reactions are catalyzed by the $F_{420}H_2$ dehydrogenase and the heterodisulfide reductase which are both coupled to the translocation of $2H^+$:

$$F_{420}H_2 + \text{methanophenazine} \longrightarrow \text{dihydro-methanophenazine} + F_{420}$$
$$(\Delta G^{o\prime} = -31.8\ \text{kJ/mol}) \quad \text{(xxx)}$$

$$\text{dihydro-methanophenazine} + \text{CoM-S-S-CoB} \longrightarrow \text{methanophenazine}$$
$$+ \text{HS-CoM} + \text{HS-CoB}\quad (\Delta G^{o\prime} = -10.6\ \text{kJ/mol}). \quad \text{(xxxi)}$$

E. Proton Translocating during Growth on Acetate

In the course of the aceticlastic pathway of methanogenesis, acetate is converted to methane and carbon dioxide, thereby producing CoM–S–S–CoB and reduced ferredoxin. These intermediates are recycled by a membrane-bound electron transport system which can be defined as reduced ferredoxin:heterodisulfide oxidoreductase. However, the exact composition and the reaction mechanism of the system are still a matter of debate (Fig. 10):

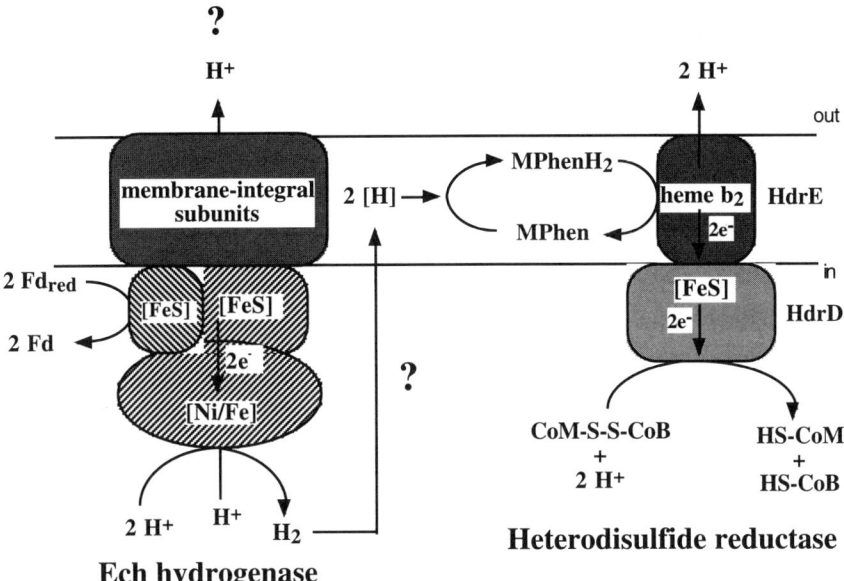

FIG. 10. Schematic representation of the Fd_{red}:heterodisulfide oxidoreductase system. The question marks indicate (i) the mechanism of electron transfer is still unresolved and (ii) the proton-translocating activity of the Ech hydrogenase has not yet been shown.

(1) Hedderich *et al.* *(31)* proposed that the Ech hydrogenase and the heterodisulfide oxidoreductase are involved in this process. In this case the oxidation of reduced ferredoxin would be catalyzed by the Ech hydrogenase resulting in the release of molecular hydrogen which is then reoxidized by the F_{420}-nonreducing hydrogenase and channeled via methanophenazine to the heterodisulfide reductase. An argument against this model is that the electron transport system would comprise three proton-translocating segments: (i) the Ech hydrogenase (Section V,B,1), (ii) the F_{420}-nonreducing hydrogenase (Section V,B,1), and (iii) the heterodisulfide reductase (Section V,B,3). Hence, the described mechanism of an "intraspecies" hydrogen cycle is very unlikely because of thermodynamic reasons.

(2) Another hypothesis is that H_2 is not an intrinsic intermediate. Instead, the electrons derived from reduced ferredoxin would rather be transferred from the Ech hydrogenase directly to methanophenazine. The reduced cofactor is then reoxidized by the heterodisulfide reductase. In this respect, it is important to note that the purified Ech hydrogenase catalyzes an Fd_{red}-dependent 2-OH-phenazine reduction only at very low rates *(223)*, indicating that phenazines may not be suitable electron acceptors of the enzyme. However, whether methanophenazine may be the physiological carrier mediating electron transfer between the Ech hydrogenase and the heterodisulfide reductase has not yet been tested.

(3) According to Ferry (27) a CO-dependent heterodisulfide reductase system from acetate-grown Ms. thermophila can be reconstituted with purified CO dehydrogenase, ferredoxin, membranes, and purified heterodisulfide reductase (260). It was postulated that ferredoxin transfers electrons from the CO dehydrogenase complex to membrane-bound electron carriers, including cytochrome b, that are required for electron transfer to the heterodisulfide reductase which reduces CoM–S–S–CoB. Recently, evidence was presented that an iron–sulfur flavoprotein (Isf) was linked to this electron transport chain (261, 262). Purified CO dehydrogenase complex slowly reduced Isf with CO; however, the rate was stimulated severalfold by the addition of ferredoxin, the electron acceptor of the CO dehydrogenase (263). The results suggest that ferredoxin is a physiological electron donor to Isf. Becker et al. (261) stated that Isf functions to couple electron transfer from ferredoxin to membrane-bound electron carriers, such as methanophenazine and/or b-type cytochromes. In summary, it is obvious that further studies are necessary to elucidate the composition of the Fd_{red}:heterodisulfide oxidoreductase.

F. Proton-Translocating Pyrophosphatases

Genome sequence data from the Ms. mazei strain Göl revealed the existence of two different open-reading frames encoding proton-translocating pyrophosphatases (PPases; Deppenmeier and Gottschalk, unpublished results). They are linked by a 750-bp intergenic region containing TC-rich stretches and are transcribed in opposite directions. The corresponding polypeptides are referred to as Mvp1 and Mvp2 consisting of 671 and 676 amino acids, respectively. Both enzymes represent extremely hydrophobic, membrane-integral proteins with 15 predicted transmembrane segments. Multiple alignments revealed that the PPase isoenzymes from Ms. mazei might have different phylogenetic backgrounds. Mvp1 is closely related to eukaryotic PPases, whereas Mvp2 belongs to the bacterial PPase lineage. Northern blot experiments using RNA from methanol-grown cells harvested in the mid-log growth phase indicated that only Mvp2 was produced under these conditions. The enzyme had a specific activity of 0.34 U/mg protein when washed membranes were analyzed. Proton-translocation experiments with inverted membrane vesicles prepared from methanol-grown cells showed that pyrophosphate hydrolysis was coupled to the translocation of about one proton across the cytoplasmic membrane. Several core biosynthetic pathways generate PP_i (e.g., amino acid activation, polysaccharide synthesis, DNA and RNA assembly, and formation of fatty acyl-CoA). Thus, pyrophosphate may function as a high-energy substrate of the membrane-bound pyrophosphatases that saves a portion of the free energy of PP_i hydrolysis for the generation of an electrochemical proton gradient (264). This assumption is also feasible from the thermodynamic point of view: The free energy of PP_i hydrolysis in the cytoplasm was calculated to be -30 kJ/mol (265). Taking into account that the membrane potential Δ_p is about -150 mV (266), the energy of -14.6 kJ/mol would be necessary to translocate one proton across the cytoplasmic membrane. Thus, PP_i hydrolysis coupled to the translocation of 1 mol of protons/mol of PP_i would still be an exergonic process (-15.4 kJ/mol) capable of driving biosynthetic reactions toward completion. The advantage of a membrane-integral PPase is that some of the energy is conserved as a transmembrane electrochemical proton gradient in comparison to soluble PPases which merely thermally dissipate all free energy. The pyrophosphatases of *Ms. mazei* Göl represent the first examples of this class of enzymes in methanogenic archaea and might be part of their energy-conserving system.

G. The Membrane-Bound Methyltransferase in Methanogenic Archaea: A Primary Sodium Ion Pump

Early investigations already revealed that growth of methanogenic archaea depends on the presence of sodium ions (267). Comprehensive analyses of resting cell suspensions using different substrate combinations led to the identification of the methyl-H_4MPT:HS-CoM methyltransferase (Fig. 11) as the sodium ion-requiring protein (125, 268, 269). As mentioned in Section III, the membrane-bound enzyme catalyzes the methyl-group transfer from H_4MPT to HS-CoM, indicating its essential function in the central methanogenic pathway. Therefore, Na^+ transport is obligatory for methane formation. Using

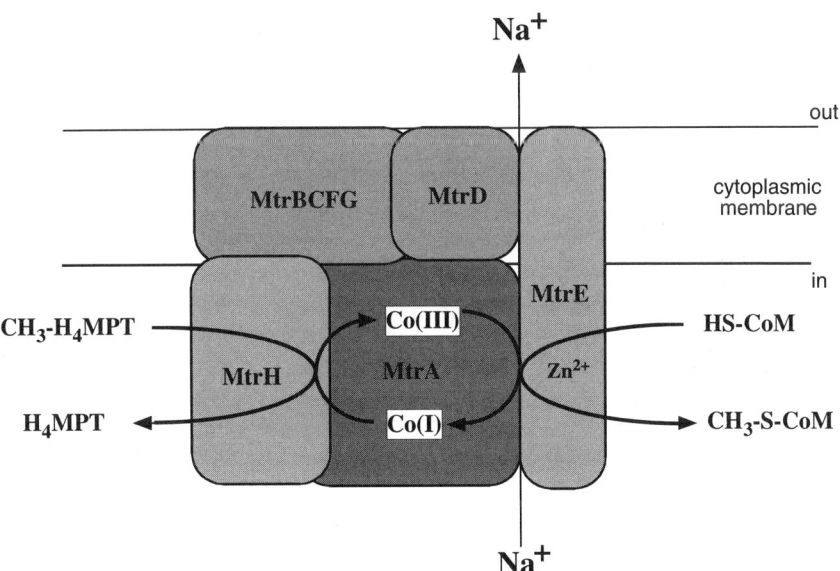

FIG. 11. Hypothetical scheme of the organization of the membrane-bound methyl-H_4MPT:HS-CoM methyltransferase.

inverted vesicle preparations of *Ms. mazei*, it was finally shown that the protein functions as a primary sodium ion pump (*126, 127*). In fact, the purified enzyme reconstituted in liposomes catalyzed vectorial Na$^+$ translocation (*129*). A ratio of 1.7 mol sodium ions translocated/mol of methyl group transferred was calculated. Thus, the enzyme represents a new class of methyltansferases able to couple the methyl-group transfer reaction with ion transport across a membrane. The protein generates a sodium ion potential during methanogenesis from CO_2 or acetate in the course of methyl-S-CoM formation from methyl-H_4MPT. In contrast, during methanogenesis from methylated C_1 compounds the methyltransferase takes advantage of the sodium ion potential in order to drive the endergonic methyl transfer from CH_3–S–CoM to H_4MPT (*130*).

Biochemical and genetical analyses indicated that the membrane-bound methyltransferase (Fig. 11) of *Methanothermobacter marburgensis* and *Ms. mazei* is composed of eight different subunits (MtrA, B, C, D, E, F, G, and H) (*123, 124, 129, 270, 271*). Further characterization revealed the presence of one [4Fe–4S] cluster with an $E^{0\prime}$ of −215 mV and a 5-hydroxybenzimidazolyl cobamide (*272–274*) with a standard reduction potential of −426 mV for the $Co^{2+/1+}$ couple (*275*). Moreover, it has been suggested that a zinc ion is bound to the enzyme and is involved in the reaction cycle (*276*). The corresponding genes of the methyltransferase are organized in the transcription unit *mtrED CBAFGH* (*124, 271*). The same number of genes and the same organization were later found in *Methanococcus jannaschii* (*208*), *Methanobacter thermoautotrophicus* (*277*), and *Ms. barkeri* (*130*).

Recent studies on the structural and catalytic properties of the membrane-bound methyltransferase allowed us to draw a tentative scheme of the reaction mechanism (Fig. 11). According to Gottschalk and Thauer (*130*), the subunits MtrA and MtrH are located at the inner site of the cytoplasmic membrane bound to the membrane-integral subunits MtrC, MtrD, and MtrE, as well as to the membrane-associated subunits MtrB, MtrF, and MtrG. MtrA is the corrinoid-harboring subunit of the protein complex (*278–280*) which is methylated and demethylated in the cycle by the catalytic activity of MtrH and MtrE, respectively. The methylation of the cob(I)amide is believed to be associated with the ligation of an additional histidine residue to the methylcob(III)amide. In the second partial reaction, methyl-Co(III) is subjected to a nucleophilic attack, probably by a thiolate anion of HS-CoM, to give rise to methyl-CoM and regenerated Co(I) (*278, 130*). Evidence as been presented that zinc ions bound to the enzyme are involved in stabilizing the thiolate at neutral pH, indicating the essential role of the ions in the reaction mechanism (Fig. 11) (*281*):

$$CH_3–H_4MPT + \text{cobal(I)amine} \longrightarrow H_4MPT + CH_3\text{–cobal(III)amine}$$
$$\Delta G^{0\prime} = -15\,\text{kJ/mol} \quad \text{(xxxii)}$$

$$HS\text{-CoM} + CH_3\text{–cobal(III)amine} \longrightarrow CH_3\text{–S–CoM} + \text{cobal(I)amine}$$
$$\Delta G^{0\prime} = -15\,\text{kJ/mol.} \quad \text{(xxxiii)}$$

In summary, it is evident that the demethylation results in the dissociation of the histidine residue from cob(I)amide and leads to a dramatic conformational change in the hydrophilic part of the enzyme. Since demethylation is dependent on sodium ions (128, 274), the conformational change associated with this step probably couples with the vectorial translocation of this cation via membrane-integral subunits (282). Recent studies indicated that MtrD and MtrE directly participate in Na^+ transport across the membrane (130, 283). The function of the other subunits remains to be analyzed.

H. The Formyl-Methanofuran Dehydrogenase System

As mentioned in Section III, the formylmethanofuran dehydrogenase (formyl-MFR dehydrogenase) from methanogenic archaea catalyzes the reversible conversion of CO_2 and methanofuran to formylmethanofuran, which is an intermediate in methanogenesis from CO_2 and methylated C_1 compounds (Figs. 3 and 4). The enzyme from *Ms. barkeri* was purified and shown to be composed of five different subunits. It contained iron–sulfur clusters and a molybdopterin dinucleotide as prosthetic groups (284). The corresponding genes (in the order *fmdEFACDB*) have been cloned and sequenced (Fig. 12) (285). Strong evidence was presented that FmdB harbors the active site. Thus, it is the catalytic subunit containing the molybdopterin dinucleotide and probably a [4Fe–4S] cluster. FmdF was shown to be a polyferredoxin indicated by the presence of eight motifs for the formation of [4Fe–4S] clusters. It might function as the direct electron carrier for the catalytic subunit FmdB. The other subunits of the purified enzyme are encoded by FmdA, FmdC, and FmdD. The function of these polypeptides is unknown. The corresponding polypeptides of the sixth gene *fmdE* was not copurified and is obviously not important for enzymatic activity.

In *Methanothermobacter marburgensis* and *Methanothermobacter wolfei* two formyl-MFR dehydrogenases were identified (Fig. 12), one being a molybdenum iron–sulfur protein (Fmd) (286, 287) and the other a tungsten iron–sulfur protein (Fwd). Moreover, in *Methanopyrus kandleri* two tungsten formyl-MFR dehydrogenases were found, one of them containing a selenocysteine in its active site (288). The enzymes from *Methanothermobacter marburgensis* were purified and shown to contain two subunits (FmdAB) and four subunits (FwdABCD), respectively. The genes encoding the tungsten- and the molybdenum-containing enzymes have been sequenced and are organized in the order *fwdHFGDACB* and *fmdECB*, respectively (Fig. 12) (289, 290). By sequence comparison, FwdB and FmdB were identified to be the subunits containing the active sites which harbor the pterin dinucleotide cofactor. The *fmdC* gene encodes a protein with sequence similarity to FwdC in its N-terminal domain and with sequence similarity to FwdD in the C-terminal part. In fact, the polypeptide represents a fusion of FwdC and FwdD. Later on it was found that subunits FmdA and FwdA are identical and encoded by the same gene, namely, *fwdA* in the *fwd* operon.

molybdenum formylmethanofuran dehydrogenase

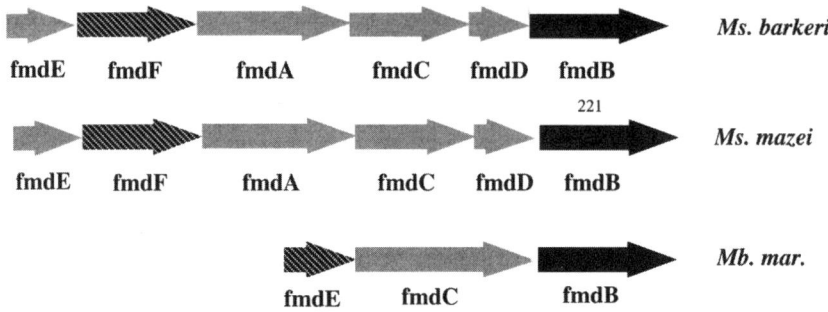

tungsten formylmethanofuran dehydrogenase

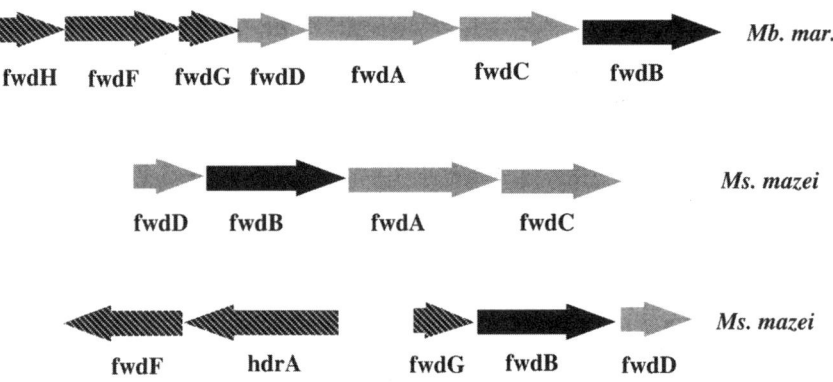

FIG. 12. Operons encoding formylmethanofuran dehydrogenases in methanogens. The genes encoding the catalytic subunit are shown in black. Genes encoding subunits that are involved, in electron transfer are hatched. All other genes are shown in grey. *Mb. mar.*, *Methanothermobacter marburgensis*, *Ms.*, *Methanosarcina*.

Thus, the assembly of an intact molybdenum enzyme requires the expression of both the *fmdECB* and the *fwdHFGDACB* operon (*290*). In accordance, northern blot analysis of RNA revealed that the *fwdHFGDACB* gene cluster is transcribed constitutively, whereas the *fmdECB* gene cluster was only transcribed when molybdate was added. FmdE and FwdF found in the molybdenum and tungsten enzymes of *Methanothermobacter marburgensis*, respectively, are homologous to FmdE from *Ms. barkeri* and also represent polyferredoxins containing eight motifs for the formation of [4Fe–4S] clusters. In addition, FwdH and FwdG are predicted to be ferredoxin-like proteins both containing two [4Fe–4S] clusters.

New insights on the number and structure of formyl-MFR dehydrogenases were obtained from the genome project of *Ms. mazei* (Deppenmeier and Gottschalk, unpublished results). In this organism, three gene clusters were identified that potentially encode formyl-MFR dehydrogenases (Fig. 12). All clusters contain a gene encoding the catalytic active subunit B. The first cluster *fmdEFACDB* is highly homologous to the corresponding operon found in *Ms. barkeri* and is most likely responsible for the formation of the molybdenum formyl-MFR dehydrogenase. The genes encompassed in cluster *fwd*1 (*fwdDBAC*) show the highest homologies to the tungsten enzyme from *Methanothermobacter marburgensis*. Interestingly, counterparts of the iron–sulfur proteins FwdG, F, and H are missing in the *Ms. mazei* enzyme. It is most likely that other proteins are responsible for electron transfer within the predicted enzyme complex. The third cluster *fwd*2 has a very unusual arrangement. The genes *fwdG, B*, and *D* are transcribed in one direction, whereas *fwdF* is oriented in the other direction (Fig. 12). The presence of *hdrA*, which is located between these genes, is very peculiar because the corresponding polypeptide is part of the heterodisulfide reductase from obligate hydrogenotrophic methanogens and is not found in the corresponding *Methanosarcina* enzyme (see Section VI). HdrA probably contains flavin and FeS centers and could function as an electron carrier in the enzyme complex. It is evident that there are still many questions concerning the expression and the function of the gene clusters encoding formyl-MFR dehydrogenases in *Ms. mazei*. Further research work will be necessary to elucidate these questions.

The formyl-MFR dehydrogenase cannot use H_2 for CO_2 reduction directly, indicating that a hydrogenase has to be involved in the initial reaction of the CO_2-reducing pathway of methanogenesis. Based on the free-energy changes associated with the formation of an amide bond the midpoint potential of the CO_2 + MFR/formyl-MFR couple has been calculated to be between -500 and -530 mV (287). The midpoint potential of the H^+/H_2 couple under biochemical standard conditions is equal to -414 mV. Therefore, H_2-dependent CO_2 reduction is an endergonic process ($\Delta G^{o'} = 16.5$ kJ/mol). This value is even higher ($\Delta G' = 43$ to 47 kJ/mol) when the low H_2 partial pressures in the natural habitats of methanogens are taken into account. The mechanism of formyl-MF formation is still a matter of debate. Kaesler and Schönheit proposed that the electrochemical sodium ion gradient generated by the methyl-H_4MPT:HS-CoM methyltransferase is the driving force for H_2-dependent CO_2 reduction (291, 292). On the other hand, it should be noted that formaldehyde oxidation as catalyzed by *Ms. mazei* Göl is not sodium ion dependent (293). These findings favor a proton-motive formyl-MFR dehydrogenase reaction. Very recently the groups of R. Hedderich (University of Marburg, Germany) and W. Metcalf (University of Illinois, USA) achieved a major breakthrough to solve this question. They created a mutant of *Ms. barkeri* by the recently developed genetic

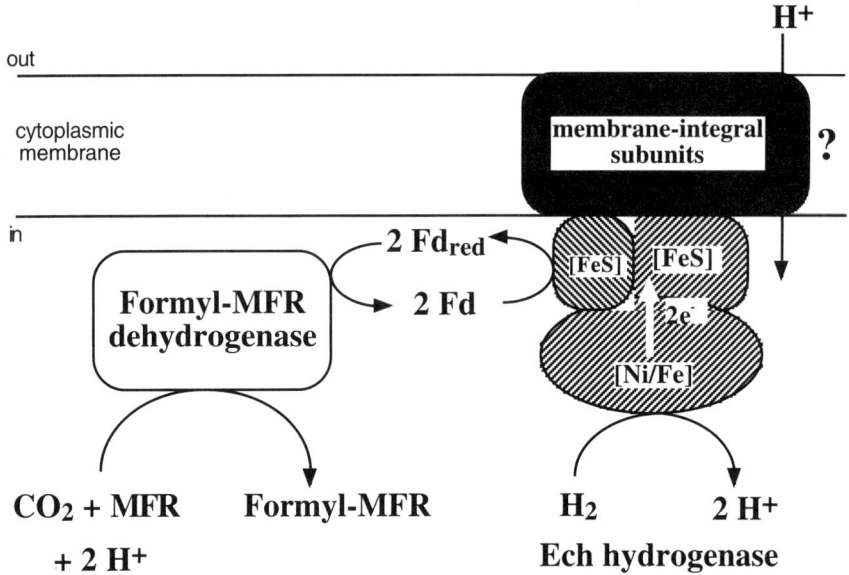

FIG. 13. Tentative scheme of CO_2 reduction as catalyzed by the formyl-MFR dehydrogenase system.

system for *Methanosarcina* strains (294). The mutant was defective in the Ech hydrogenase (Section V,B,1) and was not able to grow on $H_2 + CO_2$ (295). However, cell extracts catalyzed the conversion of formyl-MRF to methane indicating an essential role of the Ech hydrogenase in the CO_2-reducing reaction. Moreover, it has been shown that both the formyl-MFR dehydrogenase and the Ech hydrogenase are able to interact with ferredoxins (295). From these findings, it is reasonable to assume that formyl-MFR is produced in the following manner (Fig. 13): In the first reaction, H_2 is oxidized by the membrane-bound Ech hydrogenase, and electrons are transferred to a ferredoxin. Taking into account the redox potential of the CO_2 + MFR/formyl-MFR couple, the midpoint potential of the electron acceptor should be in the range of -500 mV to allow MFR–COO$^-$ reduction as catalyzed by formyl-MFR dehydrogenase. The redox potential difference between H_2 (-414 mV) and the low-potential ferredoxin could be overcome by the influx of protons through the Ech hydrogenase. Thus, the endergonic reduction of CO_2 would finally be driven by the electrochemical proton gradient.

During growth on methylated compounds formyl-MFR is formed in the course of methyl-group oxidation (Section III,C). In this case, the hydrogenase/formyl-MFR dehydrogenase system should catalyze the reverse reaction:

Formyl-MFR would be oxidized to CO_2 and part of the free energy released would be stored by reducing the low potential ferredoxin. The reduced form of ferredoxin then functions as electron donor of the Ech hydrogenase which in turn produces molecular hydrogen and probably transfers proton across the membrane. The remaining H_2 molecule would be used as electron donor for the H_2:heterodisulfide oxidoreductase system which reduces the heterodisulfide formed in the reductive branch of the methyltrophic pathway of methanogenesis.

The final support for this hypothesis is still missing, because ferredoxins with a potential of -500 mV have not yet been characterized in methanogens. However, it is noteworthy that more than 20 open-reading frames were identified which potentially code for ferredoxins in the genome of *Ms. mazei* (Deppenmeier and Gottschalk, unpublished results). Thus, the presence of such a low-potential ferredoxin cannot be excluded.

I. ATP Synthases

Membrane-bound ATPases are essential for every organism, because the enzymes couple the synthesis or hydrolysis of ATP to the translocation of protons or sodium ions across the cell membrane. Three different classes of ATPases exist in nature which are designated F_1F_o-, V_1V_o-, and A_1A_o-ATPases (*296, 298, 299*). F-type ATP synthases occur in mitochondria, chloroplasts, and most bacteria. With the exception of obligate fermentative prokaryotes, the enzymes usually function in the synthesis mode; i.e., they use a transmembrane ion gradient to drive ATP synthesis. This is also true for the A-type ATP synthase found in archaeal organisms (*298*). In contrast, V-type ATPases which are located in endomembranes and some plasma membranes of eukaryotes exhibit only ATP hydrolase activity and translocate ions across the lipid bilayers (*299*). Based on subunit composition and primary structures of the subunits, the archaeal A-type ATPases and the eukaryal V-type ATPases are closely related (*298, 300*). Despite the structural similarities, A_1A_o-ATPases differ significantly from the V_1V_o-ATPase by their ability to synthesize ATP, a feature shared with F_1F_o-ATPases. Structurally, all ATPases are composed of two domains: a membrane-embedded, ion-translocating module (F_o, A_o, and V_o and a membrane-peripheral, ATP hydrolyzing/synthesizing headpiece (F_1, A_1, and V_1). The best studied proteins of this enzyme class are the F_1F_o-ATPases from mitochondria and *E. coli*. An overwhelming number of original articles and reviews are available on the structure and the reaction mechanism of this enzyme. For a detailed description of the ATP synthase, the reader is referred to excellent collections of reviews published in *Biochim. Biophys. Acta* 1458 issue 2–3, *J. Bioenerg. Biomembr.* **32**, issue 4, and *J. Exp. Biol.* **203**, issue 1.

X-ray crystallography and NMR spectroscopy revealed details about the active site and subunit relationships of F-type ATPases (Fig. 14A). The F_1 part of

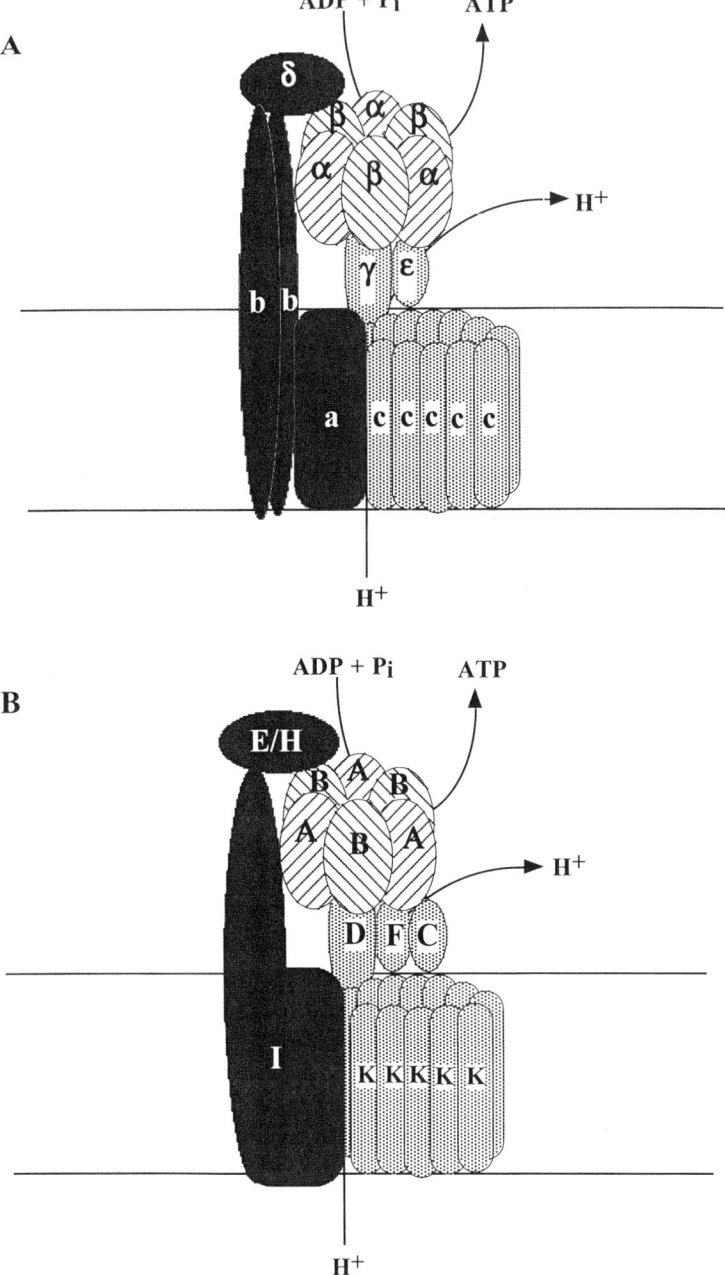

FIG. 14. Structure of the F_1F_o-type ATPase in comparison to the A_1A_o-type ATPase.

the bacterial protein is made of five different subunits in a $\alpha_3\beta_3\gamma\delta\varepsilon$ stoichiometry (301–303). The ion-conducting F_o portion is composed of three different subunits that form the $a_1b_2c_{10-14}$ subunit complex (304–306). The number of c subunits might not be the same for all species and/or growth conditions. In fact, rings of 10 to 14 c subunits have been observed (307–309). Strong evidence has been presented that the F_1F_o-ATPase is functionally bipartite and can be divided into a stator and a rotor part. According to the current model, the enzyme is divided into a headpiece formed by three α and three β subunits of the F_1 part of the enzyme. Together with subunits a, b_2, and δ (310), the hexamer forms the stator of the enzyme (Fig. 14A). The rotor portion comprises subunits c, ε, and γ. It is believed that the centrally located γ subunit extends from the top of the F_1 part to below the bottom of its headpiece and is able to rotate in the center of the $\alpha_3\beta_3$ ring. Together with the ε subunit it builds the "traditional" stalk that connects the F_1 and the F_o part (Fig. 14A) and can be identified in electromicroscopic images. The "second stalk" comprised of subunit δ and two copies of subunit b holds the $\alpha\beta$-hexamer in place (310, 311). The F_o part has an asymmetric appearance, consistent with the idea that the a subunit is attached outside of a symmetric ring of c subunits. The analysis of the three-dimensional structure of the mitochondrial F_1 domain (302, 312, 313) and the finding that it functions as a molecular motor (314, 315) gave rise to speculations on the reaction mechanism of this fascinating enzyme. According to a widely accepted model, the electrochemical ion gradient across the membrane is the driving force for the rotation of a cylindrical rotor assembled from membrane-integral c subunits. It has been proposed that subunit a is involved, forming one channel or two half-channels that allow protons (or sodium ions) to gain access to buried carboxyl groups on subunit c and to exit this site to the opposite aqueous compartment (316, 337). Furthermore, it has been proposed that the protonation and deprotonation of carboxyl groups in subunit c trigger the rotation of the entire c ring (316, 317). The rotational energy generated by ion flux through the motor is mechanically transferred via the γ and ε subunits to the catalytic sites in F_1. The rotation of subunit γ within the headpiece causes conformational changes of the catalytic sites present in the β subunits resulting in the release of preformed ATP in one of the three active sites. At the same time, ADP and P_i are bound and occluded at the remaining β subunits (310, 318). For every four protons translocated, the rotor moves by 120° inside the $\alpha\beta$ hexamer. Hence, the γ subunit acts like a crankshaft that is coupled to the opening and closing of the catalytic nucleotide-binding sites according to the binding-change mechanism (319) In summary, all studies mentioned above revealed a unique mechanism of the F-type ATPase which transforms an electrochemical ion gradient ($\Delta\tilde{\mu}_{H^+}$ or $\Delta\tilde{\mu}_{Na^+}$) into chemical energy (ATP) through the rotation of a subunit assembly.

Evidence for the presence of F-type ATPases was also found in methanogenic organisms. Inhibitor studies indicated the simultaneous presence of both H^+-dependent A_1A_o-ATPases and Na^+-translocating F_1F_o ATP synthases in *Ms. mazei* (320). Very recently, it became evident from the genome-sequencing project of this organism that it contains an A_1A_o-ATPase, but genes encoding a F-type ATPase are not present on the chromosome. In accordance, the published genome sequences of two methanogens, *M. jannaschii* (208) and *Methanothermobacter thermautotrophicus* (formerly *Methanobacterium thermoautotrophicum* ΔH) (277), revealed that they are able to form the A_1A_o-ATPase but lack genes coding for an F-type ATPase. In another organism, *M. barkeri* MS, a gene cluster encoding an F_1F_o-ATPase has been identified (321); however, a gene-encoding subunit δ could not be identified, the deduced amino acid sequence of the γ subunit is very unusual, and the corresponding polypeptide is presumably nonfunctional. In addition, mRNA transcripts could not be detected in cells grown on methanol, indicating that the F_1F_o-like genes are not expressed in *M. barkeri* (V. Müller, personal communication). Most likely, the F_1F_o-ATPase genes present at least in *M. barkeri* MS have arisen from horizontal gene transfer and are not functional.

It is important to note that methanogens are the only microorganisms (with the exception of few bacteria) known to generate two primary ion gradients, $\Delta\tilde{\mu}_{Na^+}$ and $\Delta\tilde{\mu}_{H^+}$, at the same time (Sections V,A and V,G). Hence, they must be able to couple both ion gradients to the synthesis of ATP (29, 32, 322). How this is achieved is still a matter of discussion. The most reasonable explanation is that the electrochemical sodium ion gradient established by the methyltransferase reaction is converted to a secondary proton gradient that drives synthesis of ATP via a H^+-translocating A_1A_o-ATPase (Fig. 14B) (291, 292). The universal device that couples these H^+ and Na^+ movements across biological membranes of *Bacteria*, *Archaea*, and *Eukarya* are Na^+/H^+ antiporters. Much biochemical and genetic evidence has been presented that such enzymes are present in methanogens. It is believed that these enzymes are able to transform the $\Delta\tilde{\mu}_{Na^+}$ generated by the primary sodium ion pump into a proton gradient which is then used for ATP synthesis.

As outlined above, ATP, the universal biological energy currency, is formed in methanoarchaea by a chemiosmotic mechanism. The transmembrane ion gradient generated by the respiratory chain, the membrane-bound methyltransferase, or the formyl-MF dehydrogenase is the driving force for ATP synthesis from ADP and inorganic phosphate. Since a F_1F_o-type ATP synthase is obviously not involved in this process in methanogens, the A_1A_o-type ATP synthase is the only remaining candidate which could form ATP (Fig. 14B). A-type ATPases have been enriched from various methanogens, but unfortunately they are very unstable (298). In every case reported so far, only incomplete enzymes with two to six subunits were purified which were not in a functional, ATP-synthesizing state. The H^+-translocating A_1A_o-ATPase from the methanogenic archaeon *Ms. mazei* has been purified, and the corresponding genes have been sequenced by V. Müller and co-workers (University of Munich, Germany). It is the best investigated specimen of the unique class of A-type ATPases. The molecular data revealed that the enzyme is composed of 10 different subunits (Fig. 14B). The corresponding genes, arranged in the order *H, I, K, E, C, F, A, B, D*, and *G*, are organized in the *aha* operon (*300*). Sequence homologies, size comparison, and secondary structure predictions revealed that most of the Aha subunits have a counterpart in the F_1F_o-ATPase. According to Müller *et al.* (*298*) the hydrophilic subunit AhaA is homologous to the β subunit of F-type ATPases and is believed to be the catalytic subunit for ATP synthesis/hydrolysis in the methanogenic A_1A_o-ATPase. The other polypeptide in the hexagonal headpiece of the enzyme is the α subunit homolog AhaB. Very recently, the hydrophilic subunits A, B, C, D, and F of the enzyme from *Ms. mazei* were overproduced in *E. coli* in a functional ATP-hydrolyzing state (Müller, personal communication). A low-resolution structure obtained from small-angle neutron scattering of the purified subcomplex indicated that it is composed of five different subunits with a A_3B_3CDF stoichiometry (*323*). The overall quaternary structure of the A_1-ATPase showed two well-defined domains, with a knob-like region containing the alternating three copies each of subunits A and B and a stalk presumably made from AhaD (in the center) and AhaC and AhaF (in the periphery of the stalk), as revealed by limited tryptic digests. These findings are in accordance with sequence comparisons predicting that AhaD and AhaF are counterparts of subunits γ and ε, respectively, which are located in the rotary stalk of F-type ATPase. The stator in the archaeal enzyme might be built by AhaE and AhaI. The former polypeptide could be similar to the δ subunit. Aha I is a two-domain subunit, and it is very tempting to speculate that the hydrophobic C-terminal part and the hydrophilic N-terminal part have distinct functions which are separated in F_1F_o-ATPases on subunits *a* and *b*, respectively (Fig. 14B). The highly hydrophobic subunit AhaK is referred to as the proteolipid and is a homolog of subunit *c* (*300*).

From the comparison of the subunits of the ATP-forming enzymes it can be concluded that the A-type ATPases have essentially the same structural composition compared to V_1V_o- or F_1F_o-type ATPases. Although many details of the molecular structure of archaeal ATP synthases remain unknown, a tentative model of the A_1A_o-ATPase from *Ms. mazei* was recently published (*298*). As mentioned above, the headpiece of the enzyme is formed by an alternating hexagonal arrangement of three A and three B subunits. In analogy to the γ subunit of F_1F_o-ATPases the centrally located subunit D interacts asymmetrically with the catalytic A subunits, thereby promoting tight ADP + P_i binding and ATP release. The energy for the movement of AhaD derives from ion translocation through A_o which led to a rotation of the symmetric ring of AhaK subunits. The central stalk composed of Aha D, AhaF, and AhaC allows the contact to the AhaK subunits and transfers the rotational energy to the A subunits. In this respect, the proposed reaction mechanism of the A_1A_o-ATP synthase (298) is very similar to the one predicted for the F-type ATPase.

There is still one important question to discuss that concerns the ability of F- and A-type ATPases to function as reversible machines able to hydrolyze and to produce ATP, whereas V_1V_o-ATPases are unable to synthesize ATP. The latter enzyme exhibits only ATP hydrolyase activity and serves to energize certain membranes by generating a ΔpH. Until recently, it was thought that this difference is caused by the structure of the proteolipid that forms the ion-driven rotor ring system in the membrane. Proteolipids (subunit c/subunit K) from several A_1A_o- and F_1F_o-ATPases have been purified. They have a molecular mass of about 8 kDa and are arranged in the membrane in a hairpin-like fashion with two transmembrane helices (306). In contrast, the corresponding protein from V-type ATPases possesses M_r of about 16,000 that is caused by a duplication and subsequent fusion of the original genes encoding the proteolipid (325). Thus, the subunits are predicted to form four transmembrane helices. This obvious difference was believed to be the reason for the apparent inability of V_1V_o-ATPases to synthesize ATP. However, the recently published genome sequences of methanogens indicate that the proteolipid of *Mb. thermoautotrophicus* has a mass of 15.6 kDa with four transmembrane helices, and that of *Mc. jannaschii* is predicted to be 21.3 kDa with six transmembrane helices. The molecular masses of the proteolipids have been verified by biochemical methods and by mass spectroscopy (283, 326). Hence, the duplication of the proteolipid genes is not accompanied by a failure of the ATPase to synthesize ATP. The important conclusion is that the size of the proteolipid cannot be the reason for the inability of V_1V_o-ATPases to synthesize ATP (326). Further investigations are inevitable for determining the structural necessities for ATP synthesis in addition to hydrolysis.

VI. Energy Conservation in Obligate Hydrogenotrophic Methanogens

All methanogenic archaea are characterized by their ability to produce methane. Accordingly, the central methanogenic pathways are almost identical in all methanogens, and the enzymes involved in this process are very similar (Section III,B). However, in recent years it turned out that this is not the case for the energy-conserving systems. Detailed biochemical analysis of the enzymes involved in energy transduction and the information derived from the genome sequences of *Mb. thermotrophicus* (277), *Mc. jannaschii* (208), *Ms. barkeri* (http://spider.jgi-psf.org), and *Ms. mazei* (http://www.g2l.bio.uni-goettingen.de) indicate that methanogens can be divided into two classes, the obligate

BIOCHEMISTRY OF METHANOGENS 271

hydrogenotrophic methanogens (*Methanococcales, Methanobacteriales, Methanomicrobiales*) and the methylotrophic methanogens (*Methanosarcinales*). First, evidence that these groups are different with respect to their electron carriers was already given in the late 1970s when it became clear that b-type and c-type cytochromes are present in members of the genera *Methanosarcina, Methanolobus, Methanosaeta,* and *Methanococcoides* (327, 328). These carriers could not be found in the chemolithotrophic *Methanothermobacter marburgensis, Methanothermobacter bryantii, Methanobrevibacter arboriphilus, Methanococcus vannelli, Methanosphaera stadtmanie, Methanoplanus limicola,* and *Methanospirillum hungatei* (58, 328, 329). Because of these findings, it was suggested that cytochromes are directly involved in the first step of methyl-group oxidation (330). Later on it was found that the heme-containing electron carriers participate directly in the anaerobic respiratory chain of *Methanosarcina* strains. As mentioned in Section V,B,1, one b-type cytochrome ($cytb_1$) acts as a primary-electron acceptor of the F_{420}-nonreducing hydrogenases, a second cytochrome b ($cytb_2$) has been identified as part of the heterodisulfide reductase (Fig. 9; Section V,B,3) (191).

Since these cytochromes are absent in obligate hydrogenotrophic methanogens, it is evident that redox-driven energy conservation has to be different from that in methylotrophic methanogens. Furthermore, recent studies revealed that methanophenazine, the central electron carrier in the cytoplasmic membrane of *Ms. mazei,* is obviously not present in *Methanothermobacter marburgensis* and *Mc. marispaludis,* indicating that obligate hydrogenotrophic methanogens lack this cofactor. In contrast, strong evidence was obtained that methanophenzine is produced by the methylotrophic methanogens *Ms. thermophila* and *Methanolobus tindarius* (Deppenmeier, unpublished results). In summary, it is evident that cytochromes and methanophenazine, which represent intrinsic elements of the respiratory chain of methylotrophic methanogens, are not formed by obligate hydrogenotrophic methanoarchaea. On the other hand, all key enzymes of the electron transport systems from *Methanosarcina* strains described in Section V,B are also present in members of the genera *Methanococcus* and *Methanothermobacter*. However, all of them differ in structure and composition from their counterparts found in methylotrophic methanogens.

The heterodisulfide reductase and the F_{420}-nonreducing hydrogenase from *Methanothermobacter* strains have been analyzed in detail and can serve as models of these enzymes in hydrogenotrophic methanogens (196, 331, 332). The heterodisulfide reductase was purified, and it was shown that it is an iron–sulfur flavoprotein composed of the subunits HdrA (80 kDa), HdrB (36 kDa), and HdrC (21 kDa) (333). Interestingly, HdrC and HdrB are homologous to the N-terminal third and the C-terminal two-thirds of HdrD from *Ms. barkeri,* respectively. These findings and sequence comparisons with heterodisulfide reductases from *Methanopyrus kandleri* and *Mc. jannaschii* indicate that subunits HdrB and HdrC form the active site (255). According to Madadi-Kahkesh *et al.* (259), the mechanism of S–S-bond cleavage of CoM–S–S–CoB is similar in the heterodisulfide reductases from *Ms. barkeri* and *Methanothermobacter marburgensis*. However, the pathway of electron transfer to the active site has to be different because heme groups found in the heterodisulfide reductase from *Methanosarcina* species are absent from the enzyme of hydrogenotrophic methanogens. The primary electron donor of the latter enzymes could be subunit HdrA that contains FeS clusters and FAD. Thus, it might be responsible to transfer electrons to the active site in HdrCD.

F_{420}-nonreducing hydrogenases from *Methanothermobacter* strains are composed of three different subunits (331) and are encoded by the *mvhDGAB* operon (332). The small subunit (MvhG) and the large subunit (MvhA) are homologous to the corresponding polypeptides found in all [NiFe] hydrogenases (Section V,B,1). MvhD, which copurifies with the F_{420}-nonreducing hydrogenase, is not essential for hydrogen oxidation and its function is unknown. Sequence analysis and biochemical characterization revealed that MvhG is a polyferredoxin containing eight Fe_4S_4 clusters (334). It might act as electron acceptor of the core enzyme and could be a substitute for the *b*-type cytochromes found in F_{420}-nonreducing hydrogenases from methylotrophic methanogens. Another important difference is that a signal sequence identified in the gene encoding the small subunit of the F_{420}-nonreducing hydrogenase from *Ms. mazei* (210) is not present in *mvhG*. Hence, it is assumed that the latter enzyme is not transported across the cytoplasmic membrane and that its active site is not oriented toward the periplasm. This assumption is in accordance with the fact that the enzyme from hydrogenotrophic methanogens was purified from the soluble cell fraction without using detergents, indicating that the major part of the enzyme is located in the cytoplasm (205, 331). In contrast, the F_{420}-nonreducing hydrogenase from *Ms. mazei* and *Ms. barkeri* was found to be bound to the cytoplasmic membrane (206, 209). The localization of the hydrogenase has a major impact on the mechanism of energy conservation in hydrogenotrophic methanogens. If it turns out to be true that the active site of the enzyme is orientated toward the cytoplasm, the production of scalar protons, as observed in methylotrophic methanogens (Section V,B,1), is not possible, and it is hard to imagine how the hydrogenase reaction could be coupled to the generation of an electrochemical ion gradient in hydrogenotrophic methanogens.

Interestingly, the F_{420}-nonreducing hydrogenase and the heterodisulfide reductase from *Methanothermobacter marburgensis* can be purified in one protein complex from the soluble fraction after cell lysis (196, 331). The complex contained several other proteins with unknown functions and catalyzed the H_2-dependent reduction of CoM–S–S–CoB at high rates (196). Further analysis indicated that the activity of the H_2:heterodisulfide oxidoreductase complex could be stimulated by the addition of a membrane protein with a size of 5–7 kDa that could function as a membrane anchor (196). However, evidence for an ion translocating activity has not yet been found.

Another important difference between hydrogenotrophic and methylotrophic methanogens concerns the role and the utilization of coenzyme F_{420} in energy metabolism. In *Methanosarcina* strains, the $F_{420}H_2$ dehydrogenase is one of the essential components of the respiratory chain (Section V,B,2). In contrast, there is no indication for the presence of such an enzyme from the analysis of the genomes from *Mc. jannaschii* (208) and *Methanothermobacter thermautotrophicus* (227). Further evidence that hydrogenotrophic methanogens lack an $F_{420}H_2$ dehydrogenase came from the analysis of *Mc. voltae*. The organism contains a membrane-bound $F_{420}H_2$-oxidizing enzyme and a heterodisulfide reductase which remains partly membrane bound after cell lysis (117). However, a distinct $F_{420}H_2$ dehydrogenase was not found in this organism. Instead, several lines of evidence indicate that $F_{420}H_2$ oxidation is catalyzed by a membrane-associated F_{420}-reducing hydrogenase (114). In contrast to methylotrophic methanogens, the enzyme seems to be tightly membrane associated in *Mc. voltae* (205, 206). Also the F_{420}-reducing hydrogenase from *Methanothermobacter* strains might be connected to the cytoplasmic membrane via linker particles of 10 to 20 kDa (207). From the results presented (117), it seems reasonable that the membrane-bound F_{420}-reducing hydrogenase catalyzes the oxidation of $F_{420}H_2$ in *Mc. voltae* and channels the electrons via unknown membrane-integral electron carriers to the heterodisulfide reductase resulting in the reduction of CoB–S–S–CoM. In parallel to the anaerobic respiratory chain found in methylotrophic methanogens, membrane-bound electron transport should also be coupled to ion translocation.

Further evidence for electron transfer-driven energy transduction in *Mc. voltae* has been presented by Dybas and Konisky (335, 336). The authors stated that methanogenesis from methyl-S-CoM and H_2 is coupled to the translocation of sodium ions across the cytoplasmic membrane. Furthermore, it has been proposed that the resulting $\Delta\tilde{\mu}_{Na^+}$ can be used for ATP synthesis by a Na^+-dependent ATP synthase. Hence, *Mc. voltae* would possess at least one primary sodium ion pump besides the membrane-bound methyl-H_4MPT:HS-CoM methyltransferase. According to Dybas and Konisky (336), this could be a H_2-dependent heterodisulfide reductase. If this hypothesis is correct, there would be another major difference between the two groups of methanogenic archaea: The bioenergetics of *Mc. voltae* and probably other hydrogenotrophic methanogens would be based mainly on a sodium ion cycle across the membrane. In contrast, methylotrophic methanogens take advantage of both $\Delta\tilde{\mu}_{H^+}$ and $\Delta\tilde{\mu}_{Na^+}$ for energy transduction. However, the physiological differences of these groups have to be considered while attempting to depict a universal model of the use of coupling ions. One has to clearly differentiate between marine (such as *Mc. voltae*) and fresh-water (such as *Ms. mazei*) organisms when talking about Na^+ as a coupling ion used in energy-transducing systems of these organisms.

In summary, the mechanisms of membrane-bound electron transfer and ion translocation of hydrogenotrophic methanogens are still unclear. Deeper insights into the energy-conserving systems of hydrogenotrophic methanogens have been prevented by the fact that inverted plasma membrane vesicles are not accessible. A final conclusion about the mechanisms of ATP synthesis has to await experiments with defined systems such as intact complexes of energy-transducing membrane proteins that are incorporated into artificial liposomes.

Taking into account all aspects mentioned above, I hope the reader agrees that the phylogenetic differences of methylotrophic methanogens and hydrogenotrophic methanogens are reflected in the composition of their respiratory systems. In contrast, the central methanogenic pathways are very similar, indicating that these organisms are closely related with respect to the origin of enzyme and coenzymes involved in CH_4 formation.

ACKNOWLEDGMENTS

Research in the author's laboratory has been supported by grants of the Deutsche Forschungsgemeinschaft and by Fonds der Chemischen Industrie. The author is indebted to many students who have worked in the laboratory over the years. The following deserve special mention for their contribution to our research on the metabolism of methanogens: H.-J. Abken, S. Bäumer, J. Brodersen, H. Brüggemann, F. Falinski, K. Hoffmann, and T. Ide. The author expresses his thanks to Professor G. Gottschalk, Göttingen, for support and stimulating discussion, and to Professor V. Müller (University of Munich) for his critical reading of the manuscript. Finally, I thank A. Johann and C. Jacobi from the Göttingen Genomics Laboratory for providing me with unpublished sequence information from *Methanosarcina mazei*.

REFERENCES

1. C. R. Woese, *Microbiol. Rev.* **51,** 221 (1987).
2. C. R. Woese, O. Kandler, and M. L. Wheelis, *Proc. Natl. Acad. Sci. U.S.A.* **87,** 4576 (1990).
3. G. J. Olsen and C. R. Woese, *FASEB J.* **7,** 113 (1993).
4. J. R. Brown and W. F. Doolittle, *Microbiol. Mol. Biol. Rev.* **61,** 457 (1997).
5. S. M. Barns, C. F. Delwiche, J. D. Palmer, and N. R. Pace, *Proc. Natl. Acad. Sci. U.S.A.* **93,** 9188 (1996).
6. H. P. Klenk and W. F. Doolittle, *Curr. Biol.* **4,** 920 (1994).
7. D. Langer, J. Hain, P. Thuriaux, and W. Zillig, *Proc. Natl. Acad. Sci. U.S.A.* **92,** 5768 (1995).
8. I. K. O. Cann and Y. Ishino, *Genetics* **152,** 1249 (1999).
9. M. A. Andrade, C. Ouzounis, C. Sander, J. Tamames, and A. Valencia, *J. Mol. Evol.* **49,** 551 (1999).

BIOCHEMISTRY OF METHANOGENS 275

10. N. Iwabe, K. Kuma, M. Hasegawa, S. Osawa, and T. Miyata, *Proc. Natl. Acad. Sci. U.S.A.* **86**, 9355 (1989).
11. J. R. Brown and W. F. Doolittle, *Proc. Natl. Acad. Sci. U.S.A.* **92**, 2441 (1995).
12. C. Ouzounis and N. Kyrpides, *FEBS Lett.* **390**, 119 (1996).
13. N. Glansdorff, *Mol. Evol.* **49**, 432 (1999).
14. J. Castresana and D. Moreira, *J. Mol. Evol.* **49**, 453 (1999).
15. A. Lazcano and S. L. Miller, *J. Mol. Evol.* **49**, 424 (1999).
16. N. Kyrpides, R. Overbeek, and C. Ouzounis, *J. Mol. Evol.* **49**, 413 (1999).
17. A. Gambacorta, A. Trincone, B. Nicolaus, L. Lama, and M. DeRosa, *System. Appl. Microbiol.* **16**, 518 (1994).
18. M. Nishihara and Y. Koga, *J. Biochem.* **122**, 572 (1997).
19. J. L. Garcia, *FEMS Microbiol. Rev.* **87**, 297 (1990).
20. J. L. Garcia, B. K. C. Patel, and B. Ollivier, *Anaerobe* **6**, 205 (2000).
21. W. B. Whitman, T. L. Bowen, and D. R. Boone, *in* "The prokaryotes" (A. Balows, H. G. Trüper, M. Dworkin, W. Harder, and K. J. Schleifer, eds.), p. 719. Springer Verlag, New York, 1991.
22. D. R. Boone, W. B. Whitman, and P. E. Rouviere, *in* "Methanogensis" (J. G. Ferry, ed.), p. 35. Chapman & Hall, New York/London, 1993.
23. M. Blaut, *Antonie van Leewenhoek* **66**, 187 (1994).
24. R. K. Thauer, *Antonie van Leewenhoek* **71**, 21 (1997).
25. R. K. Thauer, *Microbiology* **144**, 2377 (1998).
26. J. G. Ferry, *Critic. Rev. Biochem. Mol. Biol.* **27**, 473 (1992).
27. J. G. Ferry, *FEMS Microbiol. Rev.* **23**, 13 (1999).
28. V. Müller, M. Blaut, and G. Gottschalk, *in* "Methanogenesis" (J. G. Ferry, ed.), p. 360. Chapman & Hall, New York/London, 1993.
29. U. Deppenmeier, V. Müller, and G. Gottschalk, *Arch. Microbiol.* **165**, 149 (1996).
30. U. Deppenmeier, T. Lienard, and G. Gottschalk, *FEBS Lett.* **457**, 291 (1999).
31. R. Hedderich, O. Klimmek, A. Kröger, R. Dirmeier, M. Keller, and K. O. Stetter, *FEMS Microbiol. Rev.* **22**, 353 (1999).
32. G. Schäfer, M. Engelhard, and V. Müller, *Microbiol. Mol. Biol. Rev.* **63**, 570 (1999).
33. J. N. Reeve, *Annu. Rev. Microbiol.* **46**, 165 (1992).
34. J. N. Reeve, J. Nolling, R. M. Morgan, and D. R. Smith, *J. Bacteriol.* **179**, 5975 (1997).
35. R. I. Eggen, *FEMS Microbiol. Rev.* **15**, 251 (1994).
36. J. Leigh, *Curr. Opin. Microbiol.* **2**, 131 (1999).
37. D. L. Tumbula and W. B. Whitman, *Mol. Microbiol.* **35**, 697 (2000).
38. R. Conrad, *FEMS Microbiol. Ecol.* **8**, 193 (1999).
39. G. Diekert, and G. Wohlfarth, *Anton. Leeuw. Int. J.* **66**, 209 (1994).
40. H. L. Drake, S. L. Daniel, K. Kusel, C. Matthies, C. Kuhner, and S. Braus-Stromeyer, *Biofactors* **6**, 13 (1997).
41. B. Schink, *Microbiol. Mol. Biol. Rev.* **61**, 262 (1997).
42. S. H. Zinder, *in* "Methanogenesis" (J. G. Ferry, ed.), p. 128. Chapman & Hall, New York/London, 1993.
43. M. J. Wolin and T. L. Miller, *ASM News* **48**, 561 (1982).
44. L. Raskin, B. E. Rittmann, and D. A. Stahl, *Appl. Environ. Microbiol.* **62**, 3847 (1996).
45. F. Widdel, *in* "Biology of Anaerobic Microorgansims" (A. J. B. Zehnder, ed.). Wiley, New York, 1988.
46. D. Demeyer and V. Fievez, *Annales de Zootechnie* **49**, 95 (2000).
47. R. Conrad, *Microbiol. Rev.* **60**, 609 (1996).
48. M. S. Aulakh, J. Bodenbender, R. Wassmann, and H. Rennenberg, *Nut. Cycl. Agroecosyst.* **58**, 357 (2000).

49. W. Liesack, S. Schnell, and N. P. Revsbech, *FEMS Microbiol. Rev.* **24**, 625 (2000).
50. M. A. K. Khalil and R. A. Rasmussen, *Chemosphere* **29**, 833 (1994).
51. J. A. Marchesi, A. J. Weightman, B. A. Cragg, R. J. Parkes, and J. C. Fry, *FEMS Microbiol. Ecol.* **34**, 221 (2001).
52. T. S. Collett, *Bull. Can. Petr. Geol.* **45**, 317 (1997).
53. B. A. Buffett and O. Y. Zatsepina, *Marine Geol.* **164**, 69 (2000).
54. M. Kastner, K. A. Kvenvolden, and T. D. Lorenson, *Earth Planet Sci. Lett.* **156**, 173 (1998).
55. J. Laherrere, *Energ. Explor. Exploit.* **18**, 349 (2000).
56. K. A. Kvenvolden, *Ann. NY Acad. Sci.* **912**, 17 (2000).
57. W. J. Jones, D. P. Nagle, and W. B. Whitman, *Microbiol. Rev.* **51**, 135 (1987).
58. T. L. Miller and M. J. Wolin, *Arch. Microbiol.* **141**, 116 (1985).
59. B. Z. Fathepure and S. A. Boyd, *Appl. Environ. Microbiol.* **54**, 2976 (1988).
60. M. D. Mikesel and S. A. Boyd, *Appl. Environ. Microbiol.* **56**, 1198 (1990).
61. U. E. Krone and R. K. Thauer, FEMS *Microbiol. Lett.* **90**, 201 (1992).
62. N. Belay, R. Boopathy, and G. Voskuilen, *Appl. Environ. Microbiol.* **63**, 2092 (1997).
63. T. Daussmann, A. Aivasidis, and C. Wandrey, *Eur. J. Biochem.* **248**, 889 (1997).
64. R. Boopathy, *Arch. Microbiol.* **162**, 167 (1994).
65. J. G. Ferry (ed), "Methanogenesis: Ecology, Physiology, Biochemistry and Genetics." Chapman & Hall, New York/London, 1993.
66. R. S. Wolfe, *TIBS* **10**, 396 (1985).
67. J. C. Escalante-Semerena, J. C. Leigh, and K. L. Rinehart, Jr., and R. S. Wolfe, *Proc. Natl. Acad. Sci. U.S.A.* **81**, 1976 (1984).
68. J. A. Leigh, K. L. Rinehart, and R. S. Wolfe, *Biochemistry* **24**, 995 (1985).
69. M. I. Donnely and R. S. Wolfe, *J. Biol. Chem.* **261**, 16,653 (1986).
70. B. E. Maden, *Biochem. J.* **350**, 609 (2000).
71. L. D. Eirich, G. D. Vogels, and R. S. Wolfe, *Biochemistry* **17**, 4583 (1978).
72. C. Walsh, *Acc. Chem. Res.* **19**, 216 (1986).
73. L. G. M. Gorris and C. Van der Drift, *Biofactors* **4**, 139 (1994).
74. K. M. Noll, R. K. Rinehart, R. S. Tanner, and R. S. Wolfe, *Proc. Natl. Acad. Sci. U.S.A.* **83**, 4238 (1986).
75. K. M. Noll and R. S. Wolfe, *Biochem. Biophys. Res. Commun.* **145**, 204 (1987).
76. A. Kobelt, A. Pfaltz, D. Ankel-Fuchs, and R. K. Thauer, *FEBS Lett.* **214**, 265 (1987).
77. C. D. Taylor and R. S. Wolfe, *J. Biol. Chem.* **249**, 4886 (1974).
78. W. L. Ellefson, W. B. Whitman, and R. S. Wolfe, *Proc. Natl. Acad. Sci. U.S.A.* **79**, 3707 (1982).
79. A. Pfaltz, B. Jaun, A. Fassler, A. Eschenmoser, R. Jaenchen, R. Gilles, G. Diekert, and R. K. Thauer, *Helv. Chim. Acta* **65**, 828 (1982).
80. D. A. Livingston., A. Pfaltz, J. Schreiber, A. Eschenmoser, D. Ankel-Fuchs, J. Moll, R. Jaenchen, and R. K. Thauer, *Helv. Chim. Acta* **67**, 334 (1984).
81. D. Ankel-Fuchs, R. Hüster, E. Mörschel, S. P. J. Albracht, and R. K. Thauer, *System. Appl. Microbiol.* **7**, 383 (1986).
82. P. E. Hughes and S. B. Tove, *J. Bacteriol.* **151**, 1397 (1982).
83. H. J. Abken, M. Tietze, J. Brodersen, S. Bäumer, U. Beifuss, and U. Deppenmeier, *J. Bacteriol.* **180**, 2027 (1998).
84. U. Beifuss, M. Tietze, S. Bäumer, and U. Deppenmeier, *Angew. Chem.* **112**, 2583 (2000).
85. J. Vorholt, J. Kunow, K. O. Stetter, and R. K. Thauer, *Arch. Microbiol.* **163**, 112 (1995).
86. B. K. Pomper, J. A. Vorholt, L. Chistoserdova, M. E. Lidstrom, and R. K. Thauer, *Eur. J. Biochem.* **261**, 475 (1999).
87. J. A. Vorholt, L. Chistoserdova, S. M. Stolyar, R. K. Thauer, and M. E. Lidstrom, *J. Bacteriol.* **181**, 5750 (1999).

88. L. Chistoserdova, J. A. Vorholt, R. K. Thauer, and M. E. Lidstrom, *Science* **281,** 99 (1998).
89. J. R. Allen, D. D. Clark, J. G. Krum, and S. A. Ensign, *Proc. Natl. Acad. Sci. U.S.A.* **96,** 8432 (1999).
90. J. G. Krum and S. A. Ensign, *J. Bacteriol.* **182,** 2629 (2000).
91. S. Ebert, P. G. Rieger, and H. J. Knackmuss, *J. Bacteriol.* **181,** 2669 (1999).
92. J. R. D. McCornick and G. O. Morton, *J. Am. Chem. Soc.* **104,** 4014 (1982).
93. J. H. Coats, G. P. Li, M. S. Kuo, and D. A. Yurek, *J. Antibiot.* **42,** 472 (1989).
94. E. Purwantini, T. P. Gillis, and L. Daniels, *FEMS Microbiol. Lett.* **146,** 129 (1997).
95. E. Purwantini and L. Daniels, *J. Bacteriol.* **180,** 2212 (1998).
96. A. P. M. Eker, P. Kooiman, J. K. C. Hessels, and A. Yasui, *J. Biol. Chem.* **265,** 8009 (1990).
97. A. P. M. Eker, J. K. C. Hessels, and J. Van der Velde, *Biochemistry* **17,** 1758 (1988).
98. C. K. Stover, P. Warrener, D. R. VanDevanter, D. R. Sherman, T. M. Arain, M. H. Langhorne, S. W. Anderson, J. A. Towell, Y. Yuan, D. N. McMurray, B. N. Kreiswirth, C. E. Barry, and W. R. Baker, *Nature* **405,** 962 (2000).
99. P. C. Raemakers-Franken, R. A. De-Abreu, J. G. Willems, C. Van der Drift, and G. D. Vogels, *Biochem. Pharmacol.* **41,** 561 (1991).
100. P. C. Raemakers-Franken, M. E. Bracke, N. Van-Larebeke, B. Vyncke, and M. M. Mareel, *Anticancer Res.* **12,** 547 (1992).
101. S. Bartoschek, J. A. Vorholt, R. K. Thauer, B. H. Geierstanger, and C. Griesinger, *Eur. J. Biochem.* **267,** 3130 (2000).
102. J. Breitung and R. K. Thauer, *FEBS Lett.* **275,** 226 (1990).
103. J. Kunow, S. Shima, J. A. Vorholt, and R. K. Thauer, *Arch. Microbiol.* **165,** 97 (1996).
104. S. Shima, R. K. Thauer, H. Michel, and U. Ermler, *Proteins* **26,** 118 (1996).
105. U. Ermler, M. C. Merckel, R. K. Thauer, and S. Shima, *Structure* **5,** 635 (1997).
106. M. Vaupel, H. Dietz, D. Linder, and R. K. Thauer, *Eur. J. Biochem.* **236,** 294 (1996).
107. W. Grabarse, V. Vaupel, J. A. Vorholt, S. Shima, R. K. Thauer, A. Wittershagen, G. Bourenkov, H. D. Bartunik, and U. Ermler, *Struct. Fold. Des.* **7,** 1257 (1999).
108. M. Enssle, C. Zirngibl, D. Linder, and R. K. Thauer, *Arch. Microbiol.* **155,** 483 (1991).
109. B. W. J. Te Brömmelstroet, W. J. Geerts, J. T. Keltjens, C. Van der Drift, and G. D. Vogels, *Biochim. Biophys. Acta* **1079,** 293 (1991).
110. K. Ma and R. K. Thauer, *FEMS Microbiol. Lett.* **70,** 119 (1990).
111. B. W. J. Te Brömmelstroet, C. M. H. Hensgens, J. T. Keltjens, C. Van der Drift, and G. D. Vogels, *J. Biol. Chem.* **265,** 1852 (1990).
112. S. Shima, E. Warkentin, W. Grabarse, M. Sordel, M. Wicke, R. K. Thauer, and U. Ermler, *J. Mol. Biol.* **300,** 935 (2000).
113. M. Vaupel and R. K. Thauer, *Arch. Microbiol.* **169,** 201 (1998).
114. S. Halboth and A. Klein, *Mol. Gen. Genet.* **233,** 217 (1992).
115. R. Michel, C. Massanz, S. Kostka, M. Richter, and K. Fiebig, *Eur. J. Biochem.* **233,** 727 (1995).
116. O. Sorgenfrei, S. Müller, M. Pfeiffer, I. Sniezko, and A. Klein, *Arch. Microbiol.* **167,** 189 (1997).
117. J. Brodersen, G. Gottschalk, and U. Deppenmeier, *Arch. Microbiol.* **171,** 115 (1999).
118. C. Afting, A. Hochheimer, and R. K. Thauer, *Arch. Microbiol.* **169,** 206 (1998).
119. C. Afting, E. Kremmer, C. Brucker, A. Hochheimer, and R. K. Thauer, *Arch. Microbiol.* **174,** 225 (2000).
120. B. H. Geierstanger, T. Prasch, C. Griesinger, G. Hartmann, G. Buurman, and R. K. Thauer, *Angew. Chem. Int. Edit.* **37,** 3300 (1998).
121. G. C. Hartmann, E. Santamaria, V. M. Fernandez, and R. K. Thauer, *J. Biol. Inorg. Chem.* **1,** 446 (1996).
122. G. Buurman, S. Shima, and R. K. Thauer, *FEBS Lett.* **485,** 200 (2000).
123. E. Stupperich, A. Juza, M. Hoppert, and F. Mayer, *Eur. J. Biochem.* **217,** 115 (1993).

124. U. Harms, D. S. Weiss, P. Gärtner, D. Linder, and R. K. Thauer, *Eur. J. Biochem.* **228**, 640 (1995).
125. V. Müller, M. Blaut, and G. Gottschalk, *Eur. J. Biochem.* **172**, 601 (1988).
126. B. Becher, V. Müller, and G. Gottschalk, *J. Bacteriol.* **174**, 7656 (1992).
127. B. Becher, V. Müller, and G. Gottschalk, *FEMS Microbiol. Lett.* **91**, 239 (1992).
128. D. Weiss, P. Gärtner, and R. K. Thauer, *Eur. J. Biochem.* **226**, 799 (1994).
129. T. Lienard, B. Becher, M. Marschall, S. Bowien, and G. Gottschalk, *Eur. J. Biochem.* **239**, 857.
130. G. Gottschalk and R. K. Thauer, *Biochim. Biophys. Acta,* **1505**, 28 (2001).
131. T. A. Bobik, K. D. Olson, K. M. Noll, and R. S. Wolfe, *Biochem. Biophys. Res. Commun.* **149**, 455 (1987).
132. J. Ellermann, R. Hedderich, R. Böcher, and R. K. Thauer, *Eur. J. Biochem.* **172**, 669 (1988).
133. P. L. Hartzell, J. C. Escalante-Semerena, T. A. Bobik, and R. S. Wolfe, *J. Bacteriol.* **170**, 2711 (1988).
134. U. Ermler, W. Grabarse, S. Shima, M. Goubeaud, and R. K. Thauer, *Science* **278**, 1457 (1997).
135. D. F. Becker and S. W. Ragsdale, *Biochemistry* **37**, 2639 (1998).
136. J. Barber, L. M. Siegel, N. L. Schauer, H. D. May, and J. G. Ferry, *J. Biol. Chem.* **258**, 10,839 (1983).
137. J. B. Jones and T. C. Stadtman, *J. Biol. Chem.* **256**, 656 (1981).
138. R. Sparling, and L. Daniels, *J. Bacteriol.* **172**, 1464 (1990).
139. K. Finster, Y. Tanimoto, and F. Bak, *FEMS Microbiol. Ecol.* **74**, 295 (1990).
140. M. J. E. C. Van der Maarel, M. Jansen, and T. A. Hansen, *Appl. Environ. Microbiol.* **61**, 48 (1995).
141. K. Saue, and R. K. Thauer, *Eur. J. Biochem.* **261**, 674 (1999).
142. K. Sauer, U. Harms, and R. K. Thauer, *Eur. J. Biochem.* **243**, 670 (1997).
143. J. T. Keltjens and G. D. Vogels, in "Methanogensis" (J. G. Ferry, ed.), p. 209. Chapman & Hall, New York/London, 1993.
144. R. Fischer and R. K. Thauer, *FEBS Lett.* **269**, 368 (1990).
145. A. Mahlmann, U. Deppenmeier, and G. Gottschalk, *FEMS Microbiol. Lett.* **61**, 115 (1989).
146. V. Müller and G. Gottschalk, in "Acetogenesis" (H. L. Drake, ed.), p. 127. Chapman & Hall, New York, 1992.
147. P. J. Weimer and J. G. Zeikus, *Arch Microbiol.* **119**, 49 (1978).
148. H. Hippe, C. Caspari, K. Fiebig, and G. Gottschalk, *Proc. Natl. Acad. Sci. U.S.A.* **76**, 494 (1979).
149. R. Walther, K. Fahlbusch, R. Sievert, and G. Gottschalk, *J. Bacteriol.* **148**, 371 (1981).
150. K. Fahlbusch, H. Hippe, and G. Gottschalk, *FEMS Microbiol. Lett.* **19**, 103 (1983).
151. S. Asakawa, K. Sauer, W. Liesack, and R. K. Thauer, *Arch. Microbiol.* **170**, 220 (1998).
152. E. Naumann, K. Fahlbusch, and G. Gottschalk, *Arch. Microbiol.* **138**, 79 (1984).
153. L. Paul, D. J. Ferguson, Jr., and J. A. Krzycki, *J. Bacteriol.* **182**, 2520 (2000).
154. D. J. Ferguson, Jr. and J. A. Krzycki, *J. Bacteriol.* **179**, 846 (1997).
155. G. M. LeClerc and D. A. Grahame, *J. Biol. Chem.* **271**, 18725 (1996).
156. S. A. Burke and J. A. Krzycki, *J. Bacteriol.* **177**, 4410 (1995).
157. S. A. Burke and J. A. Krzycki, *J. Biol. Chem.* **272**, 16,570 (1997).
158. D. J. Ferguson, Jr., N. Gorlatova, D. A. Grahame, and J. A. Krzycki, *J. Biol. Chem.* **275**, 29,053 (2000).
159. S. A. Burke, S. L. Lo, and J. A. Krzycki, *J. Bacteriol.* **180**, 3432 (1998).
160. M. T. Latimer and J. G. Ferry, *J. Bacteriol.* **175**, 6822 (1993).
161. K. A. Buss, C. Ingram-Smith, and J. G. Ferry, *Protein Sci.* **6**, 2659 (1997).
162. K. A. Buss, D. R. Cooper, C. Ingram-Smith, J. G. Ferry, D. A. Sanders, and M. S. Hasson, *J. Bacteriol.* **183**, 680 (2001).

163. M. E. Rasche, K. S. Smith, and J. G. Ferry, *J. Bacteriol.* **179**, 7712 (1997).
164. M. S. M. Jetten, A. J. M. Stams, and A. J. B. Zehnder, *J. Bacteriol.* **171**, 5430 (1989).
165. S. A. Raybuck, S. E. Ramer, D. R. Abbanat, J. W. Peters, W. H. Orme-Johnson, J. G. Ferry, and C. T. Walsh, *J. Bacteriol.* **173**, 929 (1991).
166. D. R. Abbanat and J. G. Ferry, *Proc. Natl. Acad. Sci. U.S.A.* **88**, 3272 (1991).
167. J. G. Ferry, *Annu. Rev. Microbiol.* **49**, 305 (1995).
168. J. G. Ferry, *Biofactors* **6**, 25 (1997).
169. J. C. Fontecilla-Camps and S. W. Ragsdale, *Adv. Inorg. Chem.* **47**, 283 (1999).
170. E. Kocsis, M. Kessel, E. DeMoll, and D. A. Grahame, *J. Struct. Biol.* **128**, 165 (1999).
171. P. E. Jablonski and J. G. Ferry, *J. Bacteriol.* **173**, 2481 (1991).
172. T. A. Bobik and R. S. Wolfe, *J. Bacteriol.* **171**, 1423 (1989).
173. J. Ellermann, S. Rospert, R. K. Thauer, M. Bokranz, A. Klein, M. Voges, and A. Berkessel, *Eur. J. Biochem.* **184**, 63 (1989).
174. W. L. Ellefson and R. S. Wolfe, *J. Biol. Chem.* **256**, 4259 (1981).
175. S. Rospert, J. Breitung, K. Ma, C. Zirngibl, R. K. Thauer, D. Linder, R. Huber, and K. O. Stetter, *Arch. Microbiol.* **156**, 49 (1991).
176. M. Hoppert and F. Mayer, *FEBS Lett.* **267**, 33 (1990).
177. T. Selmer, J. Kahnt, M. Goubeaud, S. Shima, W. Grabarse, U. Ermler, and R. K. Thauer, *J. Biol. Chem.* **275**, 3755 (2000).
178. W. Grabarse, F. Mahlert, S. Shima, R. K. Thauer, and U. Ermler, *J. Mol. Biol.* **303**, 329 (2000).
179. P. Mitchell, *Nature* **191**, 144 (1961).
180. P. Dimroth, *Biochim. Biophys. Acta* **1318**, 11 (1997).
181. W. N. Konings, S. J. Lokema, H. W. van Veen, B. Poolman, and A. J. M. Driessen, *Antonie van Leeuwenhook* **71**, 117 (1997).
182. P. Dimroth and B. Schink, *Arch. Microbiol.* **170**, 69 (1998).
183. R. K. Thauer, K. Jungermann, and K. Decker, *Bact. Rev.* **41**, 100 (1977).
184. M. Blaut and G. Gottschalk, *Eur. J. Biochem.* **141**, 217 (1984).
185. M. Blaut, V. Müller, and G. Gottschalk, *FEBS Lett.* **215**, 53 (1987).
186. S. Peinemann, M. Blaut, and G. Gottschalk, *Eur. J. Biochem.* **186**, 175 (1989).
187. A. Jussofie, F. Mayer, and G. Gottschalk, *Arch. Microbiol.* **146**, 245 (1986).
188. F. Mayer, A. Jussofie, M. Salzmann, M. Lübben, M. Rohde, and G. Gottschalk, *J. Bacteriol.* **169**, 2307 (1987).
189. U. Deppenmeier, M. Blaut, A. Mahlmann, and G. Gottschalk, *Proc. Natl. Acad. Sci. U.S.A.* **87**, 9449 (1990).
190. U. Deppenmeier, M. Blaut, and G. Gottschalk, *Arch. Microbiol.* **155**, 272–277 ((1991).
191. B. Kamlage, and M. Blaut, *J. Bacteriol.* **174**, 3921 (1992).
192. U. Deppenmeier, M. Blaut, A. Mahlmann, and G. Gottschalk, *FEBS Lett.* **261**, 199 (1990).
193. F. M. Harold, "The Vital Force: A Study of Bioenergetics." Freeman, New York, 1986.
194. S. Heiden, R. Hedderich, E. Setzke, and R. K. Thauer, *Eur. J. Biochem.* **213**, 529 (1993).
195. S. Heiden, R. Hedderich, E. Setzke, and R. K. Thauer, *Eur. J. Biochem.* **221**, 855 (1994).
196. E. Setzke, R. Hedderich, S. Heiden, and R. K. Thauer, *Eur. J. Biochem.* **220**, 139 (1994).
197. J. N. Reeve and G. S. Beckler, *FEMS Microbiol. Rev.* **87**, 419 (1990).
198. A. E. Przybyla, J. Robbins, N. Menon, and H. D. Peck, *FEMS Microbiol. Rev.* **88**, 109 (1992).
199. G. Voordouw, *Adv. Inorg. Chem.* **38**, 397 (1992).
200. P. M. Vignais and B. Toussaint, *Arch. Microbiol.* **161**, 1 (1994).
201. S. P. J. Albracht, *Biochim. Biophys. Acta* **1188**, 167 (1994).
202. A. Volbeda, M. H. Charon, C. Piras, E. C. Hatchikian, M. Frey, and J. C. Fontecilla-Camps, *Nature* **373**, 580 (1995).
203. J. A. Fox, D. J. Livingston, W. H. Orme-Johnson, and C. T. Walsh, *Biochemistry* **26**, 4219 (1987).

204. S. F. Baron and J. G. Ferry, *J. Bacteriol.* **171**, 3854 (1989).
205. E. Muth, E. Mörschel, and A. Klein, *Eur. J. Biochem.* **169**, 571 (1987).
206. U. Deppenmeier, M. Blaut, B. Schmidt, and G. Gottschalk, *Arch. Microbiol.* **157**, 505 (1992).
207. I. J. Braks, M. Hoppert, S. Roge, and F. Mayer, *J. Bacteriol.* **176**, 7677 (1994).
208. C. J. Bult, O. White, G. J. Olsen, L. Zhou, R. D. Fleischmann, G. G. Sutton, J. A. Blake, L. M. FitzGerald, R. A. Clayton, J. D. Gocayne, A. R. Kerlavage, B. A. Dougerty, J.-F. Tomb, M. L. Adams, C. I. Reich, R. Overbeek, E. F. Kirkness, K. G. Weinstock, J. M. Merrick, A. Glodek, J. L. Scott, N. S. M. Geoghagen, J. F. Weidman, J. L. Fuhrmann, D. Nguyen, T. R. Utterback, J. M. Kelley, J. D. Peterson, P. W. Sadow, M. C. Hanna, M. D. Cotton, K. M. Roberts, M. A. Hurst, B. P. Kaine, M. Borodovsky, H.-P. Klenk, C. M. Fraser, H. O. Smith, C. R. Woese, and J. C. Venter, *Science* **273**, 1058 (1996).
209. J. M. Kemner and J. G. Zeikus, *Arch. Microbiol.* **161**, 47 (1994).
210. U. Deppenmeier, M. Blaut, S. Lentes, C. Herzberg, and G. Gottschalk, *Eur. J. Biochem.* **227**, 261 (1995).
211. U. Deppenmeier, *Arch. Microbiol.* **164**, 370 (1995).
212. J. M. Kemner, J. A. Krzycki, R. C. Prince, and J. G. Zeikus, *FEMS Microbiol. Lett.* **48**, 267 (1987).
213. J. Brodersen, S. Bäumer, H. J. Abken, G. Gottschalk, and U. Deppenmeier, *Eur. J. Biochem.* **259**, 218 (1999).
214. L. F. Wu, A. Chanal, and A. Rodrigue, *Arch Microbiol.* **173**, 319 (2000).
215. L. F. Wu and M. A. Mandrand, *FEMS Microbiol. Rev.* **10**, 243 (1993).
216. K. Eismann, K. Mlejnek, D. Zipprich, M. Hoppert, H. Gerberding, and F. Mayer, *J. Bacteriol.* **177**, 6309 (1995).
217. T. Ide, S. Bäumer, and U. Deppenmeier, *J. Bacteriol.* **181**, 4076 (1999).
218. U. Beifuss and M. Tietze, *Tetrahedron Lett.* **41**, 9759 (2000).
219. J. C. Fontecilla-Camps and S. W. Ragsdale, *Adv. Inorg. Chem.* **47**, 283 (1999).
220. F. Dole, A. Fournel, V. Magro, E. C. Hatchikian, P. Bertrand, and B. Guigliarelli, *Biochemistry* **36**, 7847 (1997).
221. E. Garcin, Y. Montet, A. Volbeda, C. Hatchikian, M. Frey, and J. C. Fontecilla-Camps, *Biochem. Soc. Trans.* **26**, 396 (1998).
222. A. Künkel, J. A. Vorholt, R. K. Thauer, and R. Hedderich, *Eur. J. Biochem.* **252**, 467 (1998).
223. J. Meuer, S. Bartoschek, J. Koch, A. Künkel, and R. Hedderich, *Eur. J. Biochem.* **265**, 325 (1999).
224. R. Böhm, M. Sauter, and A. Böck, *Mol. Microbiol.* **4**, 231 (1990).
225. M. Sauter, R. Böhm, and A. Böck, *Mol. Microbiol.* **6**, 1523–1532 ((1992).
226. S. C. Andrews, B. C. Berks, J. McClay, A. Ambler, M. A. Quail, P. Golby, and J. R. Guest, *Microbiology* **143**, 3633 (1997).
227. J. D. Fox, R. L. Kerby, G. Roberts, and P. W. Ludden, *J. Bacteriol.* **178**, 1515 (1996).
228. S. P. J. Albracht and R. Hedderich, *FEBS Lett.* **485**, 1 (2000).
229. M. Bott, B. Eikmanns, and R. K. Thauer, *Eur. J. Biochem.* **159**, 393 (1986).
230. M. Bott and R. K. Thauer, *Eur. J. Biochem.* **179**, 469 (1989).
231. D. Bonam and P. W. Ludden, *J. Biol. Chem.* **262**, 2980 (1987).
232. H. J. Abken and U. Deppenmeier, *FEMS Lett.* **154**, 231 (1997).
233. P. Haase, U. Deppenmeier, M. Blaut, and G. Gottschalk, *Eur. J. Biochem.* **203**, 527 (1992).
234. D. J. Westenberg, A. Braune, C. Ruppert, V. Müller, C. Herzberg, G. Gottschalk, and M. Blaut, *FEMS Microbiol. Lett.* **170**, 389 (1999).
235. K. Kunow, D. Linder, K. O. Stetter, and R. K. Thauer, *Eur. J. Biochem.* **223**, 503 (1994).
236. S. Bäumer, T. Ide, C. Jacobi, A. Johann, G. Gottschalk, and U. Deppenmeier, *J. Biol. Chem.* **275**, 17,968 (2000).

237. H.-P. Klenk, R. A. Clayton, J. F. Tomb, O. White, K. A. Nelson, R. J. Dodson, M. Gwinn, E. K. Hickey, J. D. Peterson, D. L. Richardson, A. R. Kerlavage, D. E. Graham, N. C. Kyrpides, R. D. Fleischmann, J. Quackenbusch, N. H. Lee, G. G. Sutton, S. Gill, E. F. Kirkness, B. A. Dougherty, K. McKenny, M. D. Adams, B. Loftus, and J. C. Venter, *Nature* **390**, 364 (1997).
238. H. Brüggemann and U. Deppenmeier, *Eur. J. Biochem.* **267**, 5810 (2000).
239. U. Weidener, S. Geier, A. Ptock, T. Friedrich, H. Leif, and H. Weiss, *J. Mol. Biol.* **233**, 109 (1993).
240. T. Yano, T., S. S. Chu, V. D. Sled, T. Ohnishi, and T. Yagi, *J. Biol. Chem.* **272**, 4201 (1997).
241. A. Dupuis, M. Chevallet, E. Darrouzet, H. Duborjal, J. Lunardi, and J. P. Issartel, *Biochim. Biophys. Acta* **1364**, 147 (1998).
242. T. Friedrich, *Biochim. Biophys. Acta* **1364**, 134 (1998).
243. T. Friedrich and D. Scheide, *FEBS Lett.* **479**, 1 (2000).
244. T. Friedrich and H. J. Weiss, *Theor. Biol.* **187**, 529 (1997).
245. T. Friedrich, B. Brors, P. Hellwig, L. Kintscher, T. Rasmusen, D. Scheide, U. Schulte, W. Mäntele, and H. Weiss, *Biochim. Biophys. Acta* **1459**, 305 (2000).
246. T. Yagi, Y. Takahiro, S. DiBernardo, and A. Matsuno-Yagi, *Biochim. Biophys. Acta* **1364**, 125 (1998).
247. V. D. Sled, T. Friedrich, H. Leif, H. Weiss, S. W. Meinhardt, Y. Fukumori, M. W. Calhoun, R. B. Gennis, and T. Ohnishi, *J. Bioenerg. Biomembr.* **25**, 347 (1993).
248. A. Vinogradov, *J. Bioenerg. Biomembr.* **25**, 367 (1993).
249. M. Degli Esposti and A. Ghelli, *Biochim Biophys. Acta* **1187**, 116 (1994).
250. U. Brandt, *Biochim. Biophys. Acta* **1318**, 79 (1997).
251. P. Hellwig, D. Scheide, S. Bungert, W. Mäntele, and T. Friedrich, *Biochemistry* **39**, 10884 (2000).
252. S. Bäumer, E. Murakami, J. Brodersen, G. Gottschalk, S. W. Ragsdale, and U. Deppenmeier, *FEBS Lett.* **428**, 295 (1998).
253. S. Mann, *Arch. Mikrobiol.* **71**, 304 (1970).
254. R. Hedderich, A. Berkessel, and R. K. Thauer, *FEBS Lett.* **255**, 67 (1989).
255. A. Künkel, M. Vaupel, S. Heim, R. K. Thauer, and R. Hedderich, *Eur. J. Biochem.* **244**, 226 (1997).
256. M. Simianu, E. Murakami, J. M. Brewer, and S. W. Ragsdale, *Biochemistry* **37**, 10027 (1998).
257. E. Murakami, U. Deppenmeier, and S. W. Ragsdale, *J. Biol. Chem.* **276**, 2432 (2001).
258. C. R. Staples, E. Gaymard, A. L. Stritt-Etter, J. Telser, B. M. Hoffman, P. Schurmann, D. B. Knaff, and M. K. Johnson, *Biochemistry* **37**, 4612 (1998).
259. S. Madadi-Kahkesh, E. C. Duin, S. Heim, S. P. J. Albracht, M. K. Johnson, and R. Hedderich, *Eur. J. Biochem.* **268**, 2566 (2001).
260. C. W. Peer, M. H. Painter, M. E. Rasche, and J. G. Ferry, *J. Bacteriol.* **176**, 6974 (1994).
261. D. F. Becker, U. Leartsakulpanich, K. K. Surerus, J. G. Ferry, and S. W. Ragsdale, *J. Biol. Chem.* **273**, 26,462 (1998).
262. U. Leartsakulpanich, L. Mikhail, L. Antonkine, and James G. Ferry, *J. Bacteriol.* **182**, 5309 (2000).
263. M. T. Latimer, M. H. Painter, and J. G. Ferry, *J. Biol. Chem.* **271**, 24023 (1996).
264. P. A. Rea and D. Sanders, *Physiol. Plantarum* **71**, 131 (1987).
265. J. M. Davis, R. J. Poole, and D. Sanders, *Biochim. Biophys. Acta* **1141**, 29 (1993).
266. S. Peinemann, PhD Thesis, University of Göttingen (1989).
267. H. J. Perski, P. Schönheit, and R. K. Thauer, *FEBS Lett.* **143**, 323 (1982).
268. M. Blaut, V. Müller, K. Fiebig, and G. Gottschalk, *J. Bacteriol.* **164**, 95 (1985).
269. V. Müller, C. Winner, and G. Gottschalk, *Eur. J. Biochem.* **178**, 519 (1988).

270. P. Gärtner, A. Ecker, R. Fischer, D. Linder, G. Fuchs, and R. K. Thauer, *Eur. J. Biochem.* **213**, 537 (1993).
271. T. Lienard and G. Gottschalk, *FEBS Lett.* **425**, 204 (1998).
272. C. M. Poirot, S. W. M. Kengen, E. Valk, J. T. Keltjens, C. van der Drift, and G. D. Vogels, *FEMS Microbiol. Lett.* **40**, 7 (1987).
273. H. Schulz, S. P. J. Albracht, J. M. C. C. Coremans, and G. Fuchs, *Eur. J. Biochem.* **171**, 589 (1988).
274. P. Gärtner, D. S. Weiss, U. Harms, and R. K. Thauer, *Eur. J. Biochem.* **226**, 465 (1994).
275. W. P. Lu, B. Becher, G. Gottschalk, and S. W. Ragsdale, *J. Bacteriol.* **177**, 2245 (1995).
276. K. Sauer and R. K. Thauer, *Eur. J. Biochem.* **267**, 2498 (2000).
277. D. R. Smith, L. A. Doucette-Stamm, C. Deloughery, H. Lee, J. Dubios, T. Aldrege, R. Bashirzahdeh, D. Blakely, R. Cook, K. Gilbert, D. Harrison, L. Hoang, P. Keagle, W. Lumm, B. Pothier, D. Qiu, R. Spadafora, R. Vicaire, Y. Wang, J. Wierzbowski, R. Gibson, N. Jiwani, A. Caruso, D. Bush, H. Safer, D. Patwell, S. Prabhakar, S. McDougall, G. Shimer, A. Goyal, S. Pietrokovski, G. M. Church, C. J. Daniels, J. Mao, P. Rice, J. Nölling, and J. N. Reeve, *J. Bacteriol.* **179**, 7135 (1997).
278. U. Harms and R. K. Thauer, *Eur. J. Biochem.* **235**, 653 (1996).
279. U. Harms and R. K. Thauer, *Eur. J. Biochem.* **250**, 783 (1997).
280. K. Sauer and R. K. Thauer, *Eur. J. Biochem.* **253**, 698 (1998).
281. R. G. Matthews and C. W. Goulding, *Curr. Opin. Chem. Biol.* **1**, 332 (1997).
282. K. Sauer and R. K. Thauer, *Eur. J. Biochem.* **261**, 674 (1999).
283. C. Ruppert, R. Schmid, R. Hedderich, and V. Müller, *FEMS Microbiol. Lett.* **195**, 47 (2001).
284. M. Karrasch, G. Börner, M. Enssle, and R. K. Thauer, *FEBS Lett.* **253**, 226 (1989).
285. J. A. Vorholt, M. Vaupel, and R. K. Thauer, *Eur. J. Biochem.* **236**, 309 (1996).
286. R. A. Schmitz, M. Richter, D. Linder, and R. K. Thauer, *Eur. J. Biochem.* **207**, 559 (1992).
287. P. A. Bertram and R. K. Thauer, *Eur. J. Biochem.* **226**, 811 (1994).
288. J. A. Vorholt and R. K. Thauer, *Eur. J. Biochem.* **248**, 919 (1997).
289. A. Hochheimer, R. A. Schmitz, R. K. Thauer, and R. Hedderich, *Eur. J. Biochem.* **234**, 910 (1995).
290. A. Hochheimer, D. Linder, R. K. Thauer, and R. Hedderich, *Eur. J. Biochem.* **242**, 156 (1996).
291. B. Kaesler and P. Schönheit, *Eur. J. Biochem.* **186**, 309 (1989).
292. B. Kaesler and P. Schönheit, *Eur. J. Biochem.* **184**, 223 (1989).
293. C. Winner and G. Gottschalk, *FEMS Microbiol. Lett.* **65**, 259 (1989).
294. W. W. Metcalf, J. K. Zhang, E. Apolinario, K. R. Sowers, and R. S. Wolfe, *Proc. Natl. Acad. Sci. U.S.A.* **94**, 2626 (1997).
295. J. Meuer, H. C. Kuettner, W. W. Metcalf, and R. Hedderich, "Biospektrum," special issue, p. 88. Spektrum Verlag, Berlin/New York, 2001.
296. P. L. Pedersen and E. Carafoli, *Trends Biochem.* **12**, 146 (1987).
297. G. Schäfer and M. Meyering-Vos, *Biochim. Biophys. Acta* **1101**, 232 (1992).
298. V. Müller, C. Ruppert, and T. Lemker, *J. Bioenerg. Biomembr.* **31**, 15 (1999).
299. M. Forgac, *J. Biol. Chem.* **274**, 12,951 (1999).
300. C. Ruppert, S. Wimmers, T. Lemker, and V. Müller, *J. Bacteriol.* **180**, 3448 (1998).
301. W. A. Catterall and P. L. Pedersen, *Biochm. Soc. Spec. Publ.* **4**, 63 (1974).
302. J. P. Abrahams, A. G. W. Leslie, R. Lutter, and J. E. Walker, *Nature* **370**, 621 (1994).
303. M. A. Bianchet, J. Hullihen, P. L. Pedersen, and L. M. Amzel, *Proc. Natl. Acad. Sci. U.S.A.* **95**, 11,065 (1998).
304. J. C. Greie, G. Deckers-Hebestreit, and K. Altendorf, *J. Bioenerg. Biomembr.* **32**, 357 (2000).
305. K. Altendorf, W. D. Stalz, J. C. Greie, and G. Deckers-Hebestreit, *J. Exp. Biol.* **203**, 19 (2000).
306. R. H. Fillingame, W. Jiang, and O. Y. Dmitriev, *J. Exp. Biol.* **203**, 9 (2000).
307. D. Stock, A. G. W. Leslie, and J. E. Walker, *Science* **286**, 1700 (1999).

308. H. Seelert, A. Poetsch, N. A. Dencher, A. Engel, H. Stahlberg, and D. J. Müller, *Nature* **405,** 418 (2000).
309. R. A. Schemidt, J. Qu, J. R. Williams, and W. S. A. Brusilow, *J. Bacteriol.* **180,** 3205 (1998).
310. W. Junge, H. Lill, and S. Engelbrecht, *Trends Biochem. Sci.* **22,** 420 (1997).
311. S. Wilkens and R. A. Capaldi, *Biochim. Biophys. Acta* **1365,** 93 (1998).
312. J. P. Abrahams, S. K. Buchanan, M. J. Van Raaij, I. M. Fearnley, A. G. W. Leslie, and J. E. Walker, *Proc. Natl. Acad. Sci. U.S.A.* **93,** 9420 (1996).
313. M. J. Van Raaij, J. P. Abrahams, A. G. W. Leslie, and J. E. Walker, *Proc. Natl. Acad. Sci. U.S.A.* **93,** 6913 (1996).
314. H. Noji, R. Yasuda, M. Yoshida, and K. Kinosita, Jr., *Nature* **386,** 299 (1997).
315. D. Sabbert, S. Engelbrecht, and W. Junge, *Nature* **381,** 623 (1996).
316. P. Dimroth, G. Kaim, and U. Matthey, *Biochim. Biophys. Acta* **1365,** 87 (1998).
317. R. H. Fillingame, *Nat. Struct. Biol.* **7,** 1002 (2000).
318. K. Kinosita, Jr., R. Yasuda, H. Noji, S. Ishiwata, and M. Yoshida, *Cell* **93,** 21 (1998).
319. P. D. Boyer, *Biochim. Biophys. Acta* **1140,** 215 (1993).
320. B. Becher and V. Müller, *J. Bacteriol.* **176,** 2543 (1994).
321. M. Sumi, M. Yohda, Y. Koga, and M. Yoshida, *Biochem. Biophys. Res. Commun.* **241,** 427 (1997).
322. M. Blaut, V. Müller, and G. Gottschalk, *J. Bioenerg. Biomemb.* **24,** 529 (1992).
323. G. Grüber, D. I. Svergun, U. Coskun, T. Lemker, M. H. J. Koch, H. Schägger, and V. Müller, *Biochemistry* **40,** 1890 (2001).
324. J. P. Gogarten and L. Taiz, *Photosynth. Res.* **33,** 137 (1992).
325. H. Arai, G. Terres, S. Pink, and M. Forgac, *J. Biol. Chem.* **263,** 8796 (1988).
326. C. Ruppert, H. Kavermann, S. Wimmers, R. Schmid, J. Kellermann, F. Lottspeich, H. Huber, K. O. Stetter, and V. Müller, *J. Biol. Chem.* **274,** 25,281 (1999).
327. W. Kühn, F. Fiebig, R. Walther, and G. Gottschalk, *FEBS Lett.* **105,** 271 (1979).
328. W. Kühn, K. Fiebig, H. Hippe, R. A. Mah, B. A. Huser, and G. Gottschalk, *FEMS Microbiol. Lett.* **20,** 407 (1983).
329. A. Jussofie and G. Gottschalk, *FEMS Microbiol. Lett.* **37,** 15 (1986).
330. M. Blaut and G. Gottschalk, *Trends. Biol. Sci.* **10,** 486 (1986).
331. R. Hedderich, A. Berkessel, and R. K. Thauer, *Eur. J. Biochem.* **193,** 255 (1990).
332. J. N. Reeve, G. S. Beckler, D. S. Cram, P. T. Hamilton, J. W. Brown, J. A. Krzycki, A. F. Kolodziej, L. Alex, W. H. Orme-Johnson, and C. T. Walsh, *Proc. Natl. Acad. Sci. U.S.A.* **86,** 3031 (1989).
333. R. Hedderich, J. Koch, D. Lindner, and R. K. Thauer, *Eur. J. Biochem.* **225,** 253 (1994).
334. R. Hedderich, S. P. J. Albracht, D. Linder, J. Koch, and R. K. Thauer, *FEBS Lett.* **298,** 65 (1992).
335. M. Dybas and J. Konisky, *J. Bacteriol.* **171,** 5866 (1989).
336. M. Dybas and J. Konisky, *J. Bacteriol.* **174,** 5575 (1992).
337. P. Dimroth, U. Matthey, and G. Kaim, *Biochim. Biophys. Acta* **1459,** 506 (2000).

A History of Poly A Sequences: From Formation to Factors to Function

MARY EDMONDS

Department of Biological Sciences
University of Pittsburgh
Pittsburgh, Pennsylvania 15260

I. Introduction	287
II. From Polymerases to Poly A(+) mRNA	290
III. Sequences Required for Polyadenylation	291
A. Processing of Adenovirus Primary Transcripts	291
B. The AAUAAA Signal	292
C. Sequences Downstream of the Poly A Site	294
IV. The Biochemistry of Polyadenylation	296
A. Transcription and Processing of Pre-mRNAs in Isolated Nuclei	297
B. Initiation of Transcription in Cell Extracts	298
C. Transcription Coupled to 3′ End Processing	299
D. mRNA 3′ End Formation in Nuclear Extracts	299
E. The Mechanism of Cleavage/Polyadenylation	300
F. Polyadenylation Complexes	303
G. Roles of snRNPs in 3′ End-Processing Complexes	305
V. Cleavage/Polyadenylation Proteins	306
A. Separating and Enumerating the Components—A Time of Confusion	306
B. Fractionating Nuclear Extracts	306
C. UV Crosslinking of Proteins to Poly A Sites	309
D. Sequences at Poly A Sites Crosslinked to Specific Proteins	310
E. Assembly of Complexes from Nuclear Fractions	311
VI. The Core Components of Cleavage/Polyadenylation	313
A. The Cleavage Stimulation Factor (CstF)	313
B. The Cleavage and Polyadenylation Specificity Factor (CPSF)	314
C. Poly A Polymerase (PAP)	317
D. Poly A-Binding Protein (PABII)	318
VII. Cloning, Sequencing, and Expressing the Core Proteins	320
A. Poly A Polymerases	322
B. The CstF Complex	326
C. The CPSF Complex	329
D. The Poly A-Binding Protein, PABII (PAB2)	332
E. $CF1_m$ and $CF11_m$ Cleavage Factors	333
VIII. Regulation of Polyadenylation	335
A. Phosphorylation of Poly A Polymerase	335
B. Regulation of Poly A Site Choice	337
C. Processing Multiple Poly A Sites in 3′ Noncoding Exons	337
D. Processing Multiple Poly A Sites in Different Coding Exons	339
E. Down-Regulation of mRNA 3′ End Processing by a Viral Protein	348

IX. Polyadenylation in Yeast 351
    A. Yeast Poly A Signals 352
    B. Yeast mRNA 3' End-Processing Factors 355
    C. Genes for Yeast mRNA 3' End-Processing Factors 360
X. Polyadenylation in *E. coli* 364
    A. Poly A Sequences 364
    B. A Poly A Polymerase Gene................................... 367
    C. Poly A Sequences Target RNA Decay.......................... 370
    D. The Search for a Second Poly A Polymerase Gene................ 373
XI. Polyadenylation in *Vaccinia Virus* 375
    A. A *Vaccinia* Poly A Polymerase 376
    B. Poly A Signals in *Vaccinia* mRNA 378
    C. VP39 is a 5' Cap Methylase 379
    References... 381

Biological polyadenylation, first recognized as an enzymatic activity, remained an orphan enzyme until poly A sequences were found on the 3' ends of eukaryotic mRNAs. Their presence in bacteria viruses and later in archeae (ref. 338) established their universality. The lack of compelling evidence for a specific function limited attention to their cellular formation. Eventually the newer techniques of molecular biology and development of accurate nuclear processing extracts showed 3' end formation to be a two-step process. Pre-mRNA was first cleaved endonucleolytically at a specific site that was followed by sequential addition of AMPs from ATP to the 3' hydroxyl group at the end of mRNA. The site of cleavage was specified by a conserved hexanucleotide, AAUAAA, from 10 to 30 nt upstream of this 3' end. Extensive purification of these two activities showed that more than 10 polypeptides were needed for mRNA 3' end formation. Most of these were in complexes involved in the cleavage step. Two of the best characterized are CstF and CPSF, while two other remain partially purified but essential. Oddly, the specific proteins involved in phosphodiester bond hydrolysis have yet to be identified. The polyadenylation step occurs within the complex of poly A polymerase and poly A-binding protein, PABII, that controls poly A length. That the cleavage complex, CPSF, is also required for this step attests to a tight coupling of the two steps of 3' and formation. The reaction reconstituted from these RNA-free purified factors correctly processes pre-mRNAs. Meaningful analysis of the role of poly A in mRNA metabolism or function was possible once quantities of these proteins most often over-expressed from cDNA clones became available. The large number needed for two simple reactions of an endonuclease, a polymerase and a sequence recognition factor, pointed to 3' end formation as a regulated process. Polyadenylation itself had appeared to require regulation in cases where two poly A sites were alternatively processed to produce mRNA coding for two different proteins. The 64-KDa subunit of CstF is now known to be a regulator of poly A site choice between two sites in the immunoglobulin heavy chain of B cells. In resting cells the site used favors the mRNA for a membrane-bound protein. Upon differentiation to plasma cells, an upstream site is used the produce a secreted form of the heavy chain. Poly A site choice in the

calcitonin pre-mRNA involves splicing factors at a pseudo splice site in an intron downstream of the active poly site that interacts with cleavage factors for most tissues. The molecular basis for choice of the alternate site in neuronal tissue is unknown. Proteins needed for mRNA 3′ end formation also participate in other RNA-processing reactions: cleavage factors bind to the C-terminal domain of RNA polymerase during transcription; splicing of 3′ terminal exons is stimulated by cleavage factors that bind to splicing factors at 3′ splice sites; nuclear export of mRNAs is linked to cleavage factors and requires the poly A II-binding protein. Most striking is the long-sought evidence for a role for poly A in translation in yeast where it provides the surface on which the poly A-binding protein assembles the factors needed for the initiation of translation. This adaptability of eukaryotic cells to use a sequence of low information content extends to bacteria where poly A serves as a site for assembly of an mRNA degradation complex in *E. coli*. Vaccinia virus creates mRNA poly A tails by a streamlined mechanism independent of cleavage that requires only two proteins that recognize unique poly A signals. Thus, in spite of 40 years of study of poly A sequences, this growing multiplicity of uses and even mechanisms of formation seem destined to continue.
© 2002, Elsevier Science (USA).

I. Introduction

The poly A story began 40 years ago with the discovery of an enzyme in calf thymus nuclear extracts that polymerized ATP, but not ADP, into poly A (*1*). This substrate requirement distinguished this activity from the only enzyme known at the time to synthesize polyribonucleotides, i.e., polynucleotide phosphorylase (*2*). Neither enzyme required the DNA template assumed to be necessary for the transfer of genetic information. A DNA template-directed RNA polymerase was found shortly thereafter in extracts of *Escherichia coli* (*3, 4*) and rat liver (*5*). Poly A polymerases were purified and characterized over the next decade from many tissues, cells, and viruses. However, they remained interesting curiosities, until poly A sequences were found in both messenger RNA (mRNA) and in their nuclear precursors (hnRNAs) (*6–8*). This discovery diverted attention from the poly A polymerases and refocused it on the properties and function of the poly A sequences and especially on the RNA molecules containing them, as it was quickly realized that these sequences would be very useful for studying mechanisms of eukaryotic gene expression. To begin with, poly A sequences provided the basis for the long-sought route for mRNA purification, and even more importantly their 3′ end location supplied an ideal site for the priming of reverse transcription of mRNAs by the enzyme that at the time had just been discovered in retroviruses (*9, 10*). The experimental protocols devised at that time remained standard methods for both purifying mRNAs and generating the cDNAs and the probes derived from them on which so many studies of gene expression continue to depend.

However, although it is clear that poly A has done a lot for molecular biologists, it is only recently that clues to what poly A does for the cell have emerged. The earliest evidence obtained in the mid-1970s came from experiments showing that poly A tails stabilized mRNAs injected into frog eggs (11, 12). This reasonable, but rather obvious, function of protecting mRNAs from nucleolytic degradation had always seemed more of an explanation by default for lack of credible evidence for other functions. Although a protective role for poly A has now been amply supported [see review by E. J. Butler (13)], interesting possibilities for other roles can now be considered. Foremost among these is a role in translation, but evidence is also accumulating for a need for polyadenylation for nuclear export of mRNA and surprisingly for the removal of introns from the terminal exon of pre-mRNAs.

A role for poly A in translation was first indicated by the coincidence in timing of poly A tail elongation of a specific mRNA and its recruitment into polysomes for immediate translation during maturation of both frog (14) and mouse (15) oocytes. The participation of poly A in the stimulation of translation of the poly A(+) mRNAs of yeast came first from genetic studies showing translation was sharply reduced by mutations in the gene for poly A-binding protein (PABP) (16). The suppression of such mutations by several *E. coli* strains with mutations in genes for proteins in the 60S ribosomal subunit implicated the poly A-binding protein and by extension poly A in translation, presumably as a PABP–poly A(+) mRNA complex in cytoplasm. The molecular mechanisms underlying these observations in *Saccharomyces cereviscae* are now being explored with an improved yeast *in vitro* translation system (17) that has shown the PABP-poly A complex to accelerate formation of the translational initiation complex by increasing the binding of the 40S ribosomal subunit to poly A(+) mRNA (18). The role of poly A in translation and its regulation are described from several different perspectives in three chapters of a monograph on translational control (19–21). These include a wide-ranging consideration of developmental decisions involving translational control that emphasizes the role of 3' untranslated regions of mRNA including their poly A tails (19); another summarizes the evidence for the participation of poly A in translation, in general, that leads to a prediction of an interaction between the 5' capped and 3' poly A ends of mRNA during translation initiation that is summarized in a closed loop model (20); and finally there is a chapter characterizing cytoplasmic mRNA polyadenylation and deadenylation during oocyte maturation and embryogenesis and its relationship to translational activation during development (21).

The long-sought role for poly A in export of mRNA from the nucleus has proved to be difficult to study because of the lack of appropriately processed deadenylated substrates. A novel solution to this problem of providing deadenylated mRNAs (22) has resulted in experimental support for a need for pre-mRNA polyadenylation (although not necessarily poly A itself) for the nuclear export of mRNA (23).

A role for polyadenylation in the removal of introns from terminal exons first predicted by the exon definition model for splicing is supported by current evidence for the interaction of splicing and polyadenylation factors across 3′ terminal exons (24).

In spite of this progress in defining roles for poly A in processes likely to occur at sites for the regulation of gene expression, molecular mechanisms defining poly A function in all of these systems are poorly understood. What is clear is that poly A and polyadenylation signal sequences serve as staging sites for the assembly of various proteins and protein complexes that not only function catalytically but also may regulate mRNA processing, transport, or translation. One of the long-standing impediments to understanding poly A function had been the ignorance of the pathways and components involved in polyadenylation itself. This ignorance has been dispelled over the past decade by biochemical approaches that have shown an unexpectedly large number of proteins to assemble in a complex that carries out not only site-specific polyadenylation but also site-specific cleavage. The tight linkage of mRNA 3′ polyadenylation with site-specific cleavage was a major conceptual advance that could be successfully exploited once the two reactions were separated *in vitro*. A review of this progress made in the early years (25) was followed several years later by a comprehensive summary of the biochemistry of mRNA 3′ end formation (26). Two recent reviews cover the rapid expansion of this information (27, 27a) that has revealed several mechanisms that regulate mRNA 3′ end formation and decay.

This review will present a historical approach that stresses the impact of new methodology on the experimental evidence that has led to an understanding of polyadenylation and its function. Needless to say, this path of discovery has been decidedly uneven, especially in the early years where an initial unexpected finding would be followed by a gradually dwindling interest, once the available technologies had been exploited. Such was the case for the discovery of the poly A polymerases in the 1960s and for polyadenylated RNAs in the 1970s. These early periods have been fully reviewed for both polymerases (28, 29) and polyadenylated mRNAs (30) and will be noted only where relevant here. Since the mid-1980s, polyadenylation studies have increased steadily so that most of the core components and their genes are fully characterized. This has opened up a new era of discoveries of the role of polyadenylation in gene expression that are now being exploited with new techniques.

The scope of this review will include not only those biochemical studies largely from mammalian cells and their purified components but also studies, more recently, from *S. cereviscae* that have led to current views of polyadenylation. Studies in bacteria and viruses that have long been an important, but often ignored part of the polyadenylation story will also be reviewed by focusing on *E. coli* long-overlooked, but now revealing unanticipated roles for poly A in prokaryotes and on vaccinia virus with its more completely characterized polyadenylation system.

II. From Polymerases to Poly A(+) mRNA

The revived interest in poly A polymerases that followed the discovery of poly A tails on mRNAs led to some highly purified preparations from several sources. Study of these enzymes waned in the late 1970s, when none of them showed an appropriate substrate specificity. Each could add AMP residues to the 3′ ends of RNAs that were not polyadenylated *in vivo*, such as transfer RNAs and ribosomal RNAs. This suggested either that these enzymes were not those that added poly A *de novo* to primary transcripts or that a factor that conferred specificity on the selection of precursor mRNAs was missing from these preparations. The first possibility, implying the existence of multiple poly A polymerases in a cell, was suggested by reports of the separation of distinct polymerases from some cells (28, 29). A role for a separate cytoplasmic polymerase was suggested by the reports of turnover of the 3′ ends of the poly A sequences of cytoplasmic mRNAs (31, 32). A similar turnover of short segments at the 3′ ends of poly A on nuclear RNAs was also reported (33). The existence of such "repair" poly A polymerases has not been established and now appears unlikely. However, the second possibility, i.e., the removal of a specificity factor, seemed the more attractive explanation at that time in view of the ability of the then recently discovered sigma factor to confer a promoter site-recognition function on the *E. coli* RNA polymerase holoenzyme (34). As will be described later, this view has been amply confirmed in recent years by the separation of a protein complex from the polymerase that imparted specificity not only to the separated polymerase but also to nonspecific purified polymerases stored for many years (35, 36). Although at the time it was logical to push the analogy with sigma factor even further and to assume there was a sequence common to mRNAs that determined polyadenylation, it would be many years before such a sequence requirement was clearly established. At that time, there was little sequence information for mRNAs, and prospects for increasing it were limited by the lack of sufficient quantities of specific mRNAs as well as by the laborious techniques used for classical RNA sequencing. In spite of these problems, a few specific mRNAs abundant in differentiated cells or tissues or from simple viruses were partially sequenced from cDNA copies. Reverse transcription in the presence of oligo dT primers assured that at least the 3′ ends of mRNAs would be represented in the cDNA population, since reverse transcriptases of that period rarely produced full-length copies of larger mRNAs. From the 3′ end fragments of six different mRNAs sequenced by a classical single-strand DNA-sequencing strategy, it was possible to spot the crucial sequence element now known to be required for mRNA polyadenylation, the hexanucleotide, AAUAAA (37). In addition, it appeared from these few cases that the distance of the hexanucleotide from the 3′ poly A addition site was also relatively conserved at 10 to 30 nucleotides.

Messenger RNAs with mutations in this sequence would be needed to establish its role in polyadenylation. Some of the obstacles to creating such mutant mRNAs at the time have already been noted, but they were soon to be greatly diminished by the nearly simultaneous advent of molecular cloning and rapid gel read-out techniques for sequencing DNA. Within the next few years, the presence of AAUAAA was found in more than 30 mRNA species at similar distances from the poly A site, although a minor variant, AUUAAA, was present in a few cases. As this hexanucleotide was the only sequence element common to these mRNAs, it seemed certain that it was involved in some general feature of mRNA metabolism and that polyadenylation was a likely candidate. The conserved location of AAUAAA had practical uses as well, as it provided a marker for locating the 3' ends of mRNAs within their genomic sequences.

III. Sequences Required for Polyadenylation

A. Processing of Adenovirus Primary Transcripts

The first clues to the basic mechanism of mRNA 3' end formation emerged in the late 1970s from a detailed set of experiments made possible by the cloning of restriction fragments of the adenovirus genome that could be used to map transcripts expressed at different periods of the infectious cycle of the virus. Particularly revealing were studies of the major late ∼25-knt transcript of Adenovirus type 2 which is transcribed from a single promoter from which five major polyadenylated RNA families are derived (38). Members of each family were related by similar sequences at their 3' ends that were later found to be splicing intermediates from each of the five major transcript classes. Hybridization analyses of pulse-labeled RNA with specific DNA fragments showed that use of these poly A sites occurred at ratios of about 1:2:3:2:2 in late stages of infection (38). This was not the result expected for 3' end formation by a transcriptional termination event. Assuming a roughly constant rate of elongation, it would be expected that transcripts closer to the promoter would be more abundant than for those more distal. Rather, these data suggested that 3' ends of each family were formed by a specific cleavage at a site that was subsequently polyadenylated. This was surprising, as 3' ends of bacterial transcripts were known to be formed by transcription termination. That transcription extended even well beyond the fourth and last poly A site, where a cleavage event would not actually be needed, suggested further that cleavage might be part of a more general mechanism of 3' end formation for RNA polymerase II transcripts (39). Evidence was already at hand showing transcription of the SV40 late region also extended beyond the poly A addition site (40), and it was soon shown to be the case for two of the early adenovirus transcription units (*41*). That cleavage was not merely a feature of the processing of viral transcripts with multiple processing sites was made clear when B-globin mRNA was also shown to be derived by cleavage of a 3' extended transcript (*42*). The advent of rapid sequencing techniques would soon make it possible to identify sequences in primary transcripts that specify cleavage site selection and/or polyadenylation.

B. The AAUAAA Signal

The first compelling evidence for a role for AAUAAA in polyadenylation came from an elegant study of *SV40* mutants in which small deletions were made in the vicinity of the poly A site of *SV40* late mRNA (*43*). Analysis of several viable mutants showed each had retained the AAUAAA sequence, suggesting it was likely to be required for virus viability. Interestingly, in mutants in which sequences downstream of AAUAAA had been deleted, poly A addition occurred at a new downstream site approximately equivalent to the length of the deletion. Obtaining direct evidence for the role of AAUAAA with this approach required some ingenuity since its deletion was not likely to result in viable virus. To ensure viability, a plasmid was constructed with a tandem duplication of a 240-nt segment of this region from which RNAs terminated at both poly A sites were produced in infected cells. However, when a 16-nt segment that included AAUAAA was deleted from either one of the two sites, only the RNA that terminated at the other site was produced. From evidence obtained with other mutants from which portions of the 16-nt sequence other than AAUAAA deleted were without effect on polyadenylation, it was concluded that the critical nucleotides for polyadenylation were AAUAAACA, but that most likely only AAUAAA was needed as the 3' CA is not conserved.

The telling evidence for this function came from point mutations created by a clever selection technique that exploited the presence of the UAA stop codon embedded within the sequence (*44*). Insertion of a small DNA fragment containing AAUAAA into the reading frame for the B-galactosidase *lacZ* gene should cause premature translational termination resulting in lack of expression of the enzyme in cells infected with such mutant phages. Restoration of expression should be achieved by base changes within the stop codon that would restore ribosome readthrough of the B-galactosidase mRNA. To create such mutations, the fragment containing the AAUAAA sequence was inserted within the *lacZ* gene carried by the M13 phage. After UV irradiation, colonies from bacteria infected with these mutagenized phages were then examined for the blue color characteristic of B-galactosidase activity. The DNA insert recovered from four such colonies was sequenced and each showed the predicted mutations within the UAA stop codons that had restored ribosome readthrough as detected by expression of B-galactosidase.

The effect of each of these mutations (AUUAAA, AACAAA, AAUGAA, and AAUACA) on 3' end formation of mRNA was tested with an oocyte injection technique (44). It had been shown previously that plasmid DNA containing mammalian genes when injected into the nucleus of frog oocytes could be correctly processed into functional polyadenylated mRNAs (45). The plasmid PBR322 containing the 3' untranslated region of *SV40* virus late mRNA and about 90 additional nucleotides downstream of the poly A site ($-141/+90$) was transcribed and correctly processed in these oocytes. The replacement of the AAUAAA of this construct with one of the mutant sequences shown above resulted in each case in drastic reductions of polyadenylated RNA. The S1 nuclease mapping technique that distinguished processed from unprocessed RNA in these tests showed that amounts of total RNA from the mutants were similar to that from the native construct, but most of it remained unprocessed. However, where small amounts of cleaved products did accumulate with some of the mutants, they were also polyadenylated. That AAUAAA might be related to cleavage rather than to polyadenylation was not an entirely unexpected outcome. In the previous year, it had been shown that replacement of AAUAAA by AAGAAA in the adenovirus early mRNA *E1A* gene resulted in an elongated transcript in HeLa cells infected with this viral mutant (46). Again the small amount of correctly cleaved *E1A* mRNA recovered was polyadenylated.

Evidence of a quite different kind came from a study of a mutation in the α-*globin* gene that also led to the conclusion that AAUAAA was involved in the cleavage step of mRNA 3' end formation. Analysis of the α-*globin* gene of an individual with α-thalessemia revealed a point mutation, AAGAAA, within the hexanucleotide that apparently accounted for the greatly reduced amounts of α-*globin* mRNA in reticulocytes (47). When this mutant DNA was transfected into HeLa cells on a recombinant plasmid, most of the mRNA products extended far beyond the poly A site.

It was clear from the earliest sequence data that AAUAAA could not be the sole determinant for mRNA 3' end formation, as several mRNAs had more than one AAUAAA within their 3' untranslated region (3' UT), yet polyadenylation was detected at only one site. It was also recognized very early on that some mRNAs were processed at each of several poly A sites within their 3'UT sequence. A striking example was the pre-mRNA for dihydrofolate reductase (DHFR), where several mRNAs were produced that differed only in the length of their 3' UT (48). As yet no differences in the coding function of these DHFR mRNAs processed at different poly A sites have been reported (49). More interesting cases of poly A site selection arose when two poly A sites in different exons of a mRNA could each express a different protein. It was unlikely that these early observations on the role of choice among poly A sites (for reviews, see Refs. 50 and 51) could be interpreted until the sequence determinants for selection of a single poly A site were known.

C. Sequences Downstream of the Poly A Site

The requirement for a specific cleavage of an elongated transcript extended the range of possible sites within regions of the 3′ ends of pre-mRNA where other critical sequences might be found. Those surrounding the poly A site were obvious candidates, although evidence for a highly conserved element near the poly A site was lacking. The results of Fitzgerald and Shenk (43) had shown that although the distance separating AAUAAA from the poly A site was important, the specific sequence of this spacer was not. That sequences downstream of the poly A addition site were required for correct 3′ end formation of a functional mRNA was first clearly shown by transfecting human cells with several plasmids bearing the adenovirus E2A gene with deletions distal to the poly A addition site (52). As long as 35 nt remained downstream of the poly A addition site, expression of the E2A mRNA and the 72-kDa protein coded by it were unaffected, but when this sequence was reduced to 20 nt, the protein was not produced and polyadenylated E2A mRNA could not be detected on Northern blots. The E2A RNAs recovered from such mutants were shown by S1 nuclease analysis to extend well beyond the poly A addition site, thereby implicating sequences between 20 and 35 nt downstream of the poly A site in the cleavage reaction. Failure to detect these elongated species among the poly A(+) RNAs probed by northern blot analysis supported the view that cleavage was closely linked to polyadenylation, as had been forecast by the similar results seen with a point mutation within the AAUAAA sequence noted earlier (44, 46).

Downstream sequences were soon reported to be needed for the efficient production of SV40 late mRNA, as a 60-nt deletion of the sequence flanking the 3′ side of AAUAAA markedly reduced the amounts of mRNA and increased the accumulation of readthrough transcripts (53). Similar effects on the levels of B-globin mRNA expression were seen in cells transfected with plasmids bearing the B-globin gene from which sequences from +3 to +31 nt downstream of the poly A site were deleted (54).

More refined mapping of the end points of these deletions were soon reported for the E2A gene and for the SV40 early gene where an essential downstream region had also been detected (55). The E2A deletion (+20/+38) included a tandem repeat of T (A/G) TTTTT of which only one was needed for normal 3′ end formation. The boundaries of the SV40 early mRNA deletions that extended from +5 to +18 nt beyond the poly A site had a GT-rich sequence (GT_2GTG_2T). In spite of this lack of homology, the T-rich E2A sequence restored normal cleavage and polyadenylation when inserted into this inactive SV40 deletion mutant. Thus, it seemed clear that sequences within this rather short deletion need not be highly specified. To avoid the possibility that these terminal deletions disrupted critical distance constraints for cleavage, a less intrusive oligonucleotide replacement approach was developed in which chemically synthesized oligonucleotides were inserted within the deletion (56). Again the critical downstream oligonucleotides of both *E2A* and *SV40* pre-mRNAs were able to restore efficient cleavage to the *SV40* mutant lacking the downstream element.

The question of base specificity was then examined with a set of single base substitutions at each site within each of the synthetic oligonucleotides. Most such replacements were well tolerated within the T-rich element of the *E2A* gene, while several within the *SV40* GT-rich element caused major, but not complete loss of cleavage of the *SV40* pre-mRNA. Replacement of a dinucleotide was reported in only one case for each sequence. Here, replacing the central GT in the GT-rich oligonucleotide with a TA-abolished cleavage of the *SV40* pre-mRNA, but replacing the sole GT of the *E2A* T-rich element with an AG reduced cleavage only 50%. Replacing a single G with a T residue actually enhanced cleavage two- or threefold in the case of three of the four G's of the GT-rich element, showing that the native sequence was not necessarily optimal for cleavage. These results showed that two very different downstream elements are functionally interchangeable at least for the expression of the *SV40* early mRNA. While specific nucleotides were shown to be important, especially in the case of one dinucleotide in the GT-rich element, a stringent base requirement was lacking for the T-rich element where each single T and the one G could be replaced without effects on cleavage.

Similar oligonucleotide replacements within the 35 nucleotides downstream of the poly A site of the rabbit *B-globin* gene previously shown to be needed for efficient processing of the 3' ends of *B-globin* mRNA showed that the active element included both a GT- and a T-rich sequence separated by only two nucleotides (57). Each alone gave low levels of cleaved RNA but gave normal levels when combined. Removal of the dinucleotide separating them had no effect, but a separation of more than 7 nt abolished activity, as did moving the entire bipartite element 16 nt downstream. Similar distance constraints had also been noted for the *E2A* and early *SV40* downstream elements, which were inactive when moved 40 nucleotides downstream (55). A later confirmation of the bipartite nature of the *B-globin* downstream element was reported for the mouse gene where two similar sequences at sites corresponding to the rabbit gene were needed for maximal activation of cleavage (58). Cluster mutations of random CA containing oligonucleotides that replaced short regions across the entire bipartite element had little effect on cleavage, strongly emphasizing the functional redundancy of significant portions of these downstream sequences. Significant cleavage activity (about 30%) survived even when the entire downstream element was eliminated (58).

A similar redundancy was noted for the bovine growth hormone gene that was first shown by deletions of its downstream elements to result in displacements of the cleavage site (59). Replacement mutations in this downstream region that lacks any GU- or U-rich elements only marginally reduce polyadenylation efficiency, although a sequence from 18 to 27 nt downstream from the cleavage site is required for correct cleavage (60).

The view of downstream elements derived from these early transfection experiments would shortly be supported with data from the nuclear extracts that carry out site-specific polyadenylation (61, 62). Although U- or GU-rich elements in regions 10 to 50 nt downstream of the poly A site were important for accurate cleavage of these few well-studied transcripts, gene bank searches showed that many transcribed genes lacked such sequences (63). More recent searches of 180 mammalian genes showed that 27% had oligo U sequences of four or more U residues within 25 nt of the cleavage site, but none showed the consensus GT-rich element, YGTGTTYY, in this region (64).

This redundancy that accounted for much of the confusion in defining downstream elements in the past did, however, present the possibility that such sequences could determine the relative strengths of different poly A sites and account for differences in levels of expression of specific mRNAs and for differential use of multiple poly A sites within a single gene. These elaborate mutational studies of the 3' regions of pre-mRNA genes not only clearly showed that 3' ends of mRNAs are formed by a specific cleavage that depends on a downstream as well as an upstream AAUAAA but also provided crucial sequence information for designing those substrates that would be used to usher in the biochemical era of polyadenylation.

IV. The Biochemistry of Polyadenylation

By the mid 1980s, the addition of poly A tails to mRNAs could be seen as a simple two-step reaction with predictable protein components. The proteins known at the time that were likely to be key components of the pathway were the poly A polymerases, although none of these preparations had ever been shown to be involved in polyadenylation of mRNA *in vivo*. Moreover the lack of substrate specificity of these polymerases had led to early predictions that a specificity factor would be needed to discriminate among the different cellular RNAs that could receive a poly A tail (65). The eventual demonstration of the requirement for the AAUAAA sequence for site-specific polyadenylation provided a suitable recognition element for this predicted specificity factor. Finally, the discovery that the poly A addition site was formed by cleavage of an elongated pre-mRNA introduced the need for a nuclease, most likely an endonuclease, into the scheme. This endonuclease would produce the terminal 3' hydroxyl group of the mRNA onto which single AMPs would be added by a poly A polymerase. Thus polyadenylation would require (i) the recognition of a poly A site within pre-mRNA, (ii) cleavage at or near that site, and (iii) polymerization of AMPs onto the site. Identification of the proteins that carry out this overall reaction would ultimately require their separation from cell extracts that polyadenylate mRNAs at specific sites.

A. Transcription and Processing of Pre-mRNAs in Isolated Nuclei

The development of *in vitro* systems for cleavage/polyadenylation lagged considerably behind the more intensive efforts devoted to achieving correctly initiated transcription in cell extracts. As might be expected, the eventual achievement of the latter provided important insights as well as experimental guidelines for similar attempts to develop systems that correctly spliced and polyadenylated these transcripts. The successful transcription of exogenous DNA in a soluble eukaryotic system required nearly 10 years of effort that began with the discovery of the eukaryotic RNA polymerases in 1969 (66). The major impediment, as was also the case for polyadenylation, was the lack of appropriate substrates. At the time, much had yet to be learned about eukaryotic genes, most importantly their intervening sequences, but even the location and sequence of promoters were essentially unknown. The latter information would obviously be critical for designing exogenous DNA substrates for transcription systems. While cell-free transcription initiated on exogenous DNA remained an elusive goal, transcription of endogenous DNA in isolated nuclei did not. Large amounts of labeled RNA could be synthesized in nuclei, although only very small amounts of any specific cellular pre-mRNAs were made. However, many more viral-specific transcripts were produced in nuclei isolated from cells infected with DNA viruses (67, 68). Once again the collection of cloned DNA fragments from defined regions of the major late transcription unit of adenovirus provided hybridization probes to show that the endogenous adenovirus DNA in nuclei from infected cells was accurately transcribed and at least some of the transcripts were correctly initiated and capped at their 5' ends. However, for the most part, such isolated nuclei initiate poorly *in vitro* and did not reinitiate. This property provides the basis for the nuclear run-off technique that measures the level of specific transcripts that have been initiated *in vivo* but remain incomplete at the time of isolation of the nuclei. The RNA polymerase, which remains bound to its DNA template during this isolation, can resume transcription during a subsequent *in vitro* incubation. The number of labeled nucleotides needed to complete the transcripts of any specific gene will be related to its length and to the number of molecules already initiated *in vivo*. This last number is often used to compare levels of transcription of a specific gene at different times or environments for the cell.

The retention of such processing functions as capping and 3' end formation by isolated nuclei was foreshadowed by earlier evidence for the synthesis of a collection of polyadenylated RNAs that resembled those of intact cells (69, 70). It was never clear how many of these poly A sequences were added to

RNA transcribed *in vitro* or to endogenous, possibly nonspecific, RNA. Unfortunately, biotinylated ribonucleoside triphosphates were not then available to make RNA *in vitro* that could be separated from endogenous RNAs by avidin chromatography. At least one case of a successful separation was reported in which mercurated nucleotides were incorporated into the transcripts made in isolated nuclei (68). The mercurated RNA recovered on sulfhydryl sepharose was faithfully transcribed from the correct strand of the endogenous adenovirus DNA. Most attempts to isolate mercurated polyadenylated RNAs that were transcribed and processed in isolated nuclei were less successful (70), no doubt because of technical problems associated with the anomalous behavior of mercurated RNAs on solid chromatographic supports (71).

The ability to initiate accurate transcription and to process 3' ends of adenoviral mRNAs accurately although limited showed that nuclei isolated by aqueous fractionation techniques retained those components needed for cleavage/polyadenylation, suggesting such preparations could be a source for a soluble system. Furthermore, it appeared that polyadenylation could be uncoupled from transcription, as isolated nuclei in the presence of actinomycin D or α-amanitin still correctly processed the 3' ends of those pre-mRNAs transcribed prior to the addition of inhibitors (67). This result held out the attractive possibility that exogenous pre-mRNA substrates would be processed in nuclear extracts.

B. Initiation of Transcription in Cell Extracts

Again, it was the major late transcription unit of adenovirus with its strong promoter that allowed the first correct initiation and faithful transcription of DNA added to cell extracts. The first of these cell-free systems was created by supplementing purified RNA polymerase II with the high-speed supernatant (the S100 fraction) of a homogenate of HeLa cells (73), and another was a high-salt extract of HeLa cells (74). Accurate transcription with these preparations, even though very inefficient, was somewhat surprising, as both were deficient in nuclear components, as each was designed to reduce levels of endogenous DNA that could compete with or inhibit the initiation of transcription on the exogenous specific DNA. However, the recovery of a 5' capped oligonucleotide from the transcripts made in the whole cell extracts that was identical in sequence to the 5' end of the major late adenovirus pre-mRNA established the accuracy of initiation of *in vitro* transcription (73, 74). These whole cell extracts were not only a source of transcription initiation factors but also provided endogenous transcripts for the first *in vitro* RNA-processing systems. Furthermore, success with these cell extracts encouraged a widespread search for conditions that would support splicing of those pre-mRNAs transcribed from exogenous DNA. Some of these experiments have been described in an earlier volume of this series (75).

C. Transcription Coupled to 3' End Processing

It was soon shown that minor modifications of the conditions that allowed a small amount of transcription initiation in these whole cell extracts enabled them to polyadenylate the transcribed RNA as well (76). Before describing these results, it is worth noting that new techniques were available that would soon vastly improve the efficiency of all RNA-processing reactions in cell-free systems. Among the most important was a simple soluble transcription system that used a purified bacteriophage RNA polymerase to copy a piece of DNA inserted downstream of the highly specific bacteriophage promoter contained in a plasmid with multiple cloning sites (77). In contrast to the inefficient eukaryotic and prokaryotic transcription systems then available, this system could produce essentially unlimited quantities of RNA from any clonable DNA sequence. Thus the substrate availability problem that had so severely restricted studies of RNA synthesis and processing quickly disappeared.

The second important development was an extract that overcame one of the major deficiencies of the whole cell extract, e.g., its low concentration of nuclear components. This new preparation from the HeLa cells, which was a concentrated high-salt extract of isolated nuclei from which chromatin and nuclear membranes had been removed, soon supplanted other *in vitro* systems and remained the standard extract for *in vitro* RNA processing (78).

Sequence-specific cell-free polyadenylation was first achieved, however, without the benefit of these two breakthroughs. Two systems were used, both based on the transcription extracts prepared from whole cells and with RNA substrates generated from the major late adenovirus promoter, but differed in the mode of delivery of that RNA to the extract. In the first case, RNA transcribed from exogenous DNA in these same extracts was isolated and added to another extract to be polyadenylated (76). In the other, pre-mRNA was generated *in situ* from the *L3* gene of the adenovirus major late transcription unit (79).

It was clear that correct polyadenylation of exogenous RNAs would greatly simplify the subsequent biochemical analyses. Prospects for success were, in fact, encouraged by the earlier evidence that cleavage and polyadenylation occurred in the absence of transcription in isolated nuclei (67). However, the early *in vitro* results favored the coupled transcription-processing system, as exogenous RNA was not cleaved in this whole-cell extract, although it was polyadenylated as long as the 3' end of the pre-mRNA was not too far downstream of a poly A signal. Sequence specificity was retained, however, as RNAs lacking sequences around the poly A site that included the AAUAAA sequence were not polyadenylated (76).

D. mRNA 3' End Formation in Nuclear Extracts

The coupled transcription/polyadenylation system, on the other hand, specifically cleaved and polyadenylated the endogenous *L3* pre-mRNA, but this same

RNA was not cleaved when added directly to these extracts, although it was polyadenylated (79) in agreement with the first study (76). That polyadenylation was independent of ongoing transcription was seen when the cleaved RNAs continued to be polyadenylated even in the presence of inhibitors of transcription (79). Although expectations were raised by these results, the apparent need to generate pre-mRNA substrates by transcription *in situ* in order to sustain the cleavage reaction was bound to limit further progress. However, this limitation was overcome soon after it was shown that processing, in this case RNA splicing, was greatly enhanced when the nuclear extracts discussed above replaced these whole-cell extracts and when exogenous pre-mRNA made with a bacteriophage transcription system was the substrate (80). These innovations removed the major obstacles to the biochemical analysis of both 3′ end formation and RNA splicing. This nuclear extract from HeLa cells remains the standard assay system for RNA processing and the primary source of the processing complexes and their protein and ribonuclear protein components (78).

It should not be assumed that the experimental uncoupling of polyadenylation from transcription *in vitro* mimics the situation *in vivo*. A later look at this coupling has shown that pre-mRNAs generated by transcription in whole-cell extracts were more efficiently polyadenylated than when added exogenously to these same extracts (81). Nor was the cleavage reaction in these extracts restricted to pre-mRNAs with 3′ ends close to the cleavage site as is seen with model exogenous RNA substrates. Aside from considerations of efficiency, the coupled system was recommended for examining the well-known requirement of poly A signals for efficient transcription termination (82) as well as for examining a possible communication between promoter sites and polyadenylation signals discussed in this paper (81). The long-sought molecular basis for such coupling is now evident (83, 84). The deletion of many of the 55 heptad repeats of the C-terminal domain (CTD) of the large subunit of the mouse RNA polymerase II had little effect on transcription initiation but marked effects on RNA processing. Splicing, 3′ end processing, and transcription termination were each inhibited independently in transfected cells expressing the truncated CTD. Most satisfying was evidence that the cleavage/polyadenylation factors, CstF and CPSF, are involved in linking transcription with RNA processing. Each of these complexes, but not PAP, binds to an immobilized CTD fragment and also copurifies with RNA Pol II in a high-molecular-weight complex that was dubbed "the RNA factory" (83, 84).

E. The Mechanism of Cleavage/Polyadenylation

The biochemistry of polyadenylation was clearly revealed with a set of experiments in which nuclear extracts were shown by a simple sensitive RNase protection assay to cleave and correctly polyadenylate a small RNA fragment consisting of the 3′ untranslated region and sequences downstream of the poly A site of the *L3* adenovirus pre-mRNA (79, 85, 86). Although cleavage had been clearly seen in transfection experiments, any downstream products disappeared too rapidly *in vivo* to be identified. Apparently the relative inefficiency of cleavage/polyadenylation in these extracts allowed the accumulation of a fragment of appropriate length that gave a RNase T1 fingerprint that indicated it was derived from sequences 3′ of the poly A site (85). Since the polyadenylated fragment was also shown by RNA fingerprints to be the upstream cleavage fragment, these results confirmed that the site of polyadenylation was produced by cleavage. An important finding of this set of experiments was the uncoupling of cleavage from polyadenylation (85). This was done by substituting the methylene analogs of ATP for ATP in these extracts. The AMPP(CH$_2$)P analog with its nonhydrolyzable B–γ bond was utilized as well as ATP, but the α–B analog AMP(CH$_2$)PP, as expected, was not. Although poly A was not made, the cleavage products described above were produced. The analog 3′dATP (cordycepin triphosphate) also allowed cleavage, but in this case although poly A polymerase was not inhibited, addition of a single 3′dAMP to the 3′ hydroxyl group at the cleavage site terminated elongation. This eliminated the possibility that the polyadenylation reaction *in vivo* involved the ligation of a preformed poly A to mRNA.

Using these inhibitors, it was possible to recover the cleavage products from nuclear extracts for a detailed analysis of their termini (86). The upstream sequence ended with a 3′ hydroxyl group on a terminal adenosine residue, while the downstream fragment bore a 5′ phosphate group on the nucleotide known to be adjacent to the poly A site of this mRNA. These experiments defining a mechanism for endonucleolytic cleavage also revealed the terminal 3′ hydroxyl group on mRNA required by the poly A polymerases characterized in the past. With the biochemical mechanism of polyadenylation seen in Fig. 1 now clearly established, it seemed that applying classical protein fractionation techniques

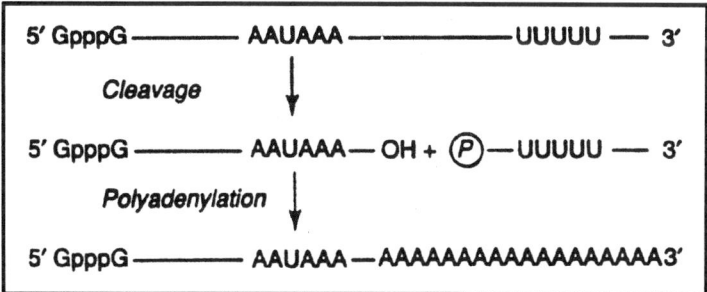

FIG. 1. The two steps in mRNA 3′ end processing. Endonucleolytic cleavage between the AAUAAA sequence and the downstream element shown as an oligo U sequence generates an upstream fragment with a 3′ OH group and a downstream fragment with a 5′ phosphate group. The upstream fragment is polyadenylated and the downstream fragment is degraded. Reprinted from *Trends in Biochemical Sciences*, **21**, p. 247. E. Wahle and W. Keller, Biochemistry of polyadenylation, Copyright (1996), with permission from Elsevier Science.

to these nuclear extracts would lead quickly to the identification of the components involved in pre-mRNA recognition, cleavage, and polyadenylation. Quite unsuspected at this time were the large number of components now known to be involved in such well-established reactions.

While the identification of those proteins that participate in mRNA 3' end formation was at that time expected to be difficult, it was relatively easy to establish the sequence specificity for polyadenylation in nuclear extracts. This involved merely recloning into the newly available plasmid transcription vectors those same poly A sites already cloned in plasmids used earlier to establish the *in vivo* sequence requirements. Using these now readily synthesized RNA substrates, it was reassuring to find that both the AAUAAA sequence (85, 87) and the downstream sequences (61) were needed for efficient 3' end formation *in vitro*. As is usually the case, the *in vitro* system revealed more subtle features of the reactions that were only suggested by the *in vivo* data, such as the tight coupling of cleavage to polyadenylation. The early *in vivo* experiments had shown that the AAUAAA sequence was required for cleavage but not necessarily for polyadenylation, as the few substrates with mutations within the AAUAAA that were cleaved were also polyadenylated.

A direct *in vitro* test of the role of AAUAAA in polyadenylation was done with RNA substrates with 3' ends at the poly A site, the so-called precleaved RNAs (88). These RNAs were readily polyadenylated in nuclear extracts but not if other bases replaced those within the AAUAAA sequence. This surprising result might have been anticipated by results with the whole-cell extracts that polyadenylate but do not cleave pre-mRNAs. Here, however, in such cases poly A was added only to those uncleaved RNAs with an intact AAUAAA sequence (76). The insights derived from this apparent need for AAUAAA for polyadenylation in the absence of cleavage, aside from the demonstration that cleavage and polyadenylation could be uncoupled *in vitro*, suggested that some component that recognizes AAUAAA contacts not only a component involved in cleavage but also one involved in polyadenylation. A prediction made at the time was that the other components could be the endonuclease and the poly A polymerase joined in a complex with a recognition factor (88).

A nice confirmation of the sequences critical for mRNA 3' end formation as deduced from these mutational studies was obtained with a modification interference technique similar to that originally developed to identify those sites in DNA that interacted with specific proteins (89). Here RNAs were chemically treated to give approximately one base modification per molecule, before adding to nuclear extracts that cleave and polyadenylate. After incubation, both processed and unprocessed RNAs are then chemically cleaved at the modified sites to produce the nested set of fragments that are then resolved by gel electrophoresis to produce an RNA-sequencing ladder. If modification at a particular site has prevented cleavage, then the RNA fragments normally derived from the site would be missing from the sequencing ladder of the processed but not that of the unprocessed RNA. These would be seen as gaps in the sequence ladder that correspond to those sites in the RNA that are critical for processing. Modification of any of the bases of AAUAAA appeared to prevent cleavage as well as polyadenylation in the case of precleaved RNA. Other base modifications in regions surrounding the poly A site gave cleavage products identical to those of the unmodified RNA. This included bases downstream of the poly A site shown to be needed for cleavage by mutational analysis, thus reaffirming the earlier *in vivo* evidence for sequence redundancy in the downstream element needed for cleavage.

F. Polyadenylation Complexes

It was logical to assume that the recognition, cleavage, and polyadenylation reactions required for 3' end formation of mRNA would occur within a complex as is the case for most multistep cellular processes. Especially influential at the time was the recognition of the spliceosome complex as the site of RNA splicing in these same nuclear extracts (90). Early evidence for such complexes for cleavage/polyadenylation came from an indirect experimental approach that predicted several of the properties of such complexes quite accurately. An oligonucleotide/RNaseH protection technique that assayed the availability of the AAUAAA sequence for hybridization to a complementary deoxyoligonucleotide during the course of pre-mRNA processing in nuclear extracts provided evidence for complex formation (91). If AAUAAA was available for hybridization, the sequence would become a substrate for the enzyme RNaseH that cleaves the RNA strand of an RNA:DNA hybrid. It was shown that the time during the incubation at which the anti-deoxyoligonucleotide was no longer accessible to the AAUAAA sequence of the pre-mRNA coincided with the onset of the cleavage reaction (92). This protection displayed the same sequence requirements already established for the cleavage reaction, e.g., an intact AAUAAA sequence and a downstream element. Polyadenylated mRNA added to the extracts was not protected, suggesting that, once processed, mRNA was released from such complexes (91).

A particularly useful finding of these experiments was the ability of 3' dATP to stabilize the putative complex that was detected as a prolonged inaccessibility of the AAUAAA sequence to the anti-oligonucleotide after cleavage had occurred which was not seen in the presence of ATP (91). As noted earlier, 3' dATP does not inhibit cleavage but limits AMP addition by poly A polymerase to a single terminal 3' dAMP at the cleavage site. The availability of stabilized complexes did allow substrate competition experiments that showed that cleavage was saturable since the reaction was competed by exogenous pre-mRNA, but only if the poly A signals were intact (91). This showed that cleavage factors were not only titratable in nuclear extracts but also were able to recycle during the reaction, since a reaction saturated for cleavage could still cleave labeled pre-mRNA added during a subsequent incubation. This recycling was not observed in reactions where 3′dATP was present, further attesting to the stability of such complexes. Finally, poly A(+) RNA did not compete for cleavage as would be expected from its failure to form a complex in which AAUAAA was protected. These protection experiments suggested a simple working model for complex formation that postulated a factor in nuclear extracts that recognizes the AAUAAA sequence in pre-mRNA and becomes physically associated with it, resulting in a complex that also requires sequences downstream of the poly A site. Once formed, cleavage occurs rapidly with polyadenylation following immediately. After poly A addition, the factors are released and reassemble on other pre-mRNAs.

A satisfactory understanding of complex formation and function would of course require complex isolation and reconstitution from separated components with designated functions. Progress toward this goal was soon reported by several laboratories (92–95) that adopted a technique that had been used for the separation of spliceosome complexes from the same nuclear extracts (96). The approach is based on a readily detected decrease in electrophoretic mobility that RNA (or DNA) molecules undergo, when bound to protein, the so-called mobility shift assay. In the case of RNA present in crude nuclear extracts, the isolation is done in the presence of an excess of a large polyanion, in this case heparin, to reduce the binding of nonspecific proteins. Such extracts applied directly to nondenaturing polyacrylamide gels and electrophoresed at low ionic strength produce a limited number of labeled RNA bands. The naked pre-mRNA migrates most rapidly, followed by a heterodisperse smear of pre-mRNAs complexed with a limited subset of nuclear proteins that include those characteristics of the well-known hnRNP structures. The latter form immediately upon addition of RNA to a nuclear extract and are independent of RNA sequence. When ATP is included in extracts that carry out 3′ end processing, one or two more slowly moving compact bands containing labeled RNA appear (92–95). Kinetic studies with the different pre-mRNA substrates used by each group showed that complex formation preceded the appearance of cleavage products, while two studies showed that the onset of cleavage coincided with the appearance of a slightly faster moving complex when polyadenylation was suppressed by 3′dATP (92, 93).

This latter complex was enriched in the 5′ cleaved half of the pre-mRNA that was absent from the complex appearing earlier that contained only pre-mRNA (92). In addition to the separation of pre- and postcleavage complexes, a third faster moving complex was observed only under conditions that allowed polyadenylation (93). This poly A(+) RNA-rich complex was assumed to be a postprocessing complex that formed on poly A(+) RNA released from the cleavage/polyadenylation complex. The separation of precleavage and postcleavage complexes with an appropriate RNA content established the existence of a

3′ end-processing complex. The poly A site was protected during complex formation and processing was demonstrated by the isolation of a 67-nt fragment of the adenovirus *L3* pre-mRNA from an RNase T1 digest of a nuclear extract early in the processing reaction. The "protected" fragment included the poly A site and both upstream and downstream poly A signals (*93*).

The use of heparin to identify complexes from nuclear extracts raised questions about their biological authenticity. It was reassuring to observe that complexes formed in density gradients without heparin treatment (*97, 98*). Accumulation of a 50S complex during pre-mRNA processing was shown to require ATP, AAUAAA, and the downstream sequences needed for 3′ end processing. Heparin-treated fractions from the isolated 50S complex had electrophoretic mobilities similar to those recovered directly from heparin-treated extracts. Both gradient and gel-isolated complexes contained poly A(+) RNAs and smaller amounts of precursor RNA. Heparin treatment of extracts prior to sedimentation resulted in more slowly sedimenting complexes of 20–25S, but with RNA contents similar to the 50S complexes (*97*). Isolated 50S complexes did not process pre-mRNAs, suggesting these might be the postprocessing complexes noted earlier, rather than active intermediates, although this inactivity may also have resulted from experimental manipulations required for the assay.

G. Roles of snRNPs in 3′ End-Processing Complexes

The marked reduction in sedimentation velocity of the complex seen after heparin treatment was attributed at the time to the removal of snRNPs from the 50S complex. Specific snRNPs were already known to be involved in both splicing and processing of preribosomal RNA and of the 3′ end of histone mRNAs that are not polyadenylated. Thus, at the time it was widely anticipated that snRNPs would participate in many RNA-processing reactions including polyadenylation. A direct role for SnRNPs in cleavage/polyadenylation had been supported by immunological evidence obtained from early *in vitro* experiments (*79*). A direct role for those UsnRNAs involved in splicing, however, appeared unlikely to be involved in polyadenylation when it was shown that the oligonucleotide-directed RNaseH inactivation of the U1-, U2-, and U4- plus U6SnRNAs that abolished splicing had no effect on polyadenylation in these same extracts (*99*). Attention was ultimately focused on a specific snRNP, U11SnRNP, that copurified with a nuclear fraction that conferred specificity on cleavage/polyadenylation (*100, 101*). The upcoming discussion of polyadenylation factors will note that U11SnRNP was eventually separated from these factors without loss of its function, thereby eliminating a direct role for U11SnRNP in 3′ end formation (*103*). It was then doubtful that any other RNAs were directly involved in the basic mechanism of cleavage/polyadenylation, as they could be reconstituted from RNA-free components (*104*). However, a less direct role for an snRNP protein in polyadenylation has emerged from evidence that terminal intron removal is closely linked to polyadenylation efficiency (24), apparently through an interaction between the U1A protein of the U1SnRNP and the 160-kDa subunit of the cleavage/polyadenylation specificity factor (CPSF) that binds to the poly A site (105).

V. Cleavage/Polyadenylation Proteins

A. Separating and Enumerating the Components—A Time of Confusion

Although the discovery of polyadenylation complexes was a significant advance, it was difficult to exploit until the individual components were identified. One of these was certain to be a poly A polymerase, although even here it could not be assumed that it would resemble those nonspecific poly A polymerases purified in the past. There were no obvious candidates for the putative specificity factors or for a nuclease specific for cleavage. Despite these uncertainties, prospects for separating such activities were favorable by the mid-1980s. Several fractions active in RNA splicing had recently been separated from these same nuclear extracts. Most encouraging was the uncoupling of cleavage from polyadenylation that could be achieved in these extracts. This work had provided simple assays for measuring cleavage in the absence of polyadenylation, either by omitting Mg^{2+} or replacing ATP with the chain terminator, 3'dATP. Specific polyadenylation could also be assayed in the absence of cleavage, as precleaved mRNAs were effective substrates if they retained the canonical hexanucleotide. An abundance of pre-mRNA substrates (and their mutant forms) along with these simple assays for individual steps of what appeared to be a straightforward set of reactions encouraged several groups to begin the fractionation of nuclear extracts using classical protein separation techniques. Although the tedium of this approach made it an uncongenial one, particularly for molecular biologists, the potential rewards for the purification of even a small amount of a protein had increased greatly by this time because of the new technologies of gene cloning and monoclonal antibody production. However, it would require a large effort by several laboratories before most of these rewards could be realized, as an unexpectedly large number of proteins were found to be involved in mRNA 3' end formation.

B. Fractionating Nuclear Extracts

The unexpected complexity of cleavage/polyadenylation that was compounded by the different fractionation protocols used by different groups as well as by the lack of a standard nomenclature for individual components produced a literature that did not make easy reading. Nonetheless, in 1988 several groups reported the separation of a poly A polymerase activity from the cleavage activity by anion-exchange chromatography (*100, 102, 106, 107*). This activity, which was only weakly bound by DEAE sepharose, added poly A to RNAs with the lack of specificity that had characterized the poly A polymerases studied in the past. However, when this fraction was combined with a more tightly bound fraction, specific polyadenylation was seen if the precleaved RNA substrate retained an intact AAUAAA sequence. This specificity was lost if Mn^{2+} replaced Mg^{2+} in the reaction, an effect often seen for both RNA and DNA polymerases. The fraction conferring specificity also contained a cleavage activity that was specific for poly A sites with an intact AAUAAA sequence. For most pre-mRNA substrates, specific cleavage required the polymerase fraction as well.

A more extensive fractionation of this cleavage/specificity activity separated another fraction needed only for cleavage (CF) from a second fraction (CPF) that retained the activity needed for both sequence-specific cleavage and polyadenylation (*100*). Only when CF and CPF were combined with the polymerase activity (PAP) did site-specific cleavage occur. The CF fraction had no effect on the specific polyadenylation of precleaved RNAs. This identification of three fractions, each of which retained one of the three activities needed for the generation of correctly processed 3′ ends, i.e., sequence recognition or specificity (CPF), cleavage (CF), and polyadenylation (PAP), raised expectations that a straightforward purification of each fraction would lead to purified components from which 3′ end formation could be reconstituted (*100*).

It was somewhat surprising then when another more extensive fractionation by a different route led to the separation from a crude cleavagespecificity fraction four activities, all needed for cleavage in addition to the PAP activity (*108*). Among them were two fractions separated from the CF activity noted above that were designated CFI and CFII, as both were needed for cleavage, but neither was needed for specific polyadenylation. This suggested that CFI and CFII participated in the cleavage reaction after the cleavage site had been selected by a specificity fraction designated by this group as SF, which resembled the CPF activity noted above (*100*).

The existence of another fraction that strongly stimulated cleavage (CstF) when combined with CFI, CFII, SF, and PAP had gone unrecognized earlier, as it apparently had copurified with PAP that is also needed for cleavage. In this protocol PAP was separated early in the scheme as a low-molecular-weight component by gel filtration, leaving the four activities needed for cleavage among the larger components to be separated by anion-exchange chromatography (*108*).

A fraction that resembled this CstF fraction in chromatographic behavior designated as CF1 was separated from similar extracts with yet another protocol devised by a third group (*106*) which identified a second fraction also needed for cleavage, CF2, that was apparently similar to the cleavage activity described as CF (*100*). Neither CF1 nor CF2 had any effect on polyadenylation, but both were required for cleavage in the presence of a specificity fraction and

TABLE I
MAMMALIAN CLEAVAGE AND POLYADENYLATION FACTORS

Factor	Subunits (kDa)	Function
CPSF (CPF, SF, PF2)[a]	160, 100, 73, 30	Cleavage and polyadenylation; binds AAUAAA
CstF (CF1)[a]	77, 64, 50	Cleavage; binds downstream element
CFl$_m$ (CF2)[a]	(72, 68, 59)[b], 25	Cleavage; binds RNA
CFll$_m$ (CF2)[a]	Unknown	Cleavage
PAP (PF1)	82	Cleavage and polyadenylation Catalyzes AMP polymerization
PABP2 (PAB11)[a]	33	Poly A extension stimulates PAP; controls poly A length

[a] Alternative designations in use circa 1988–1992.
[b] One of these three interacts with the 25-kDa subunit to give a functional heterodimer.

a polymerase. Two other fractions separated in this study were required for specific polyadenylation of precleaved RNA (107). One was a polymerase activity, PF1, while the other, PF2, conferred specificity on both the polymerase in the presence of Mg^{2+} and the cleavage reaction. Most interesting at this stage was the finding of each study that the cleavage step alone required the specificity fraction and the polymerase in addition to the three separate cleavage activities. This suggested that a complex must assemble before any processing occurs, a conclusion consistent with the earlier identification of precleavage complexes formed very soon after the addition of pre-mRNAs to nuclear extracts (see above).

Table I includes synonyms no longer in use derived from the early experiments on complexes and individual proteins required for efficient cleavage and polyadenylation of mammalian pre-mRNAs *in vitro*. The current nomenclature derived from well-defined evidence is used in subsequent discussions of these factors and their subunits.

A striking observation revealed by density gradient sedimentation analyses of each cleavage fraction was the high molecular weight estimated for each active fraction. Assuming a globular structure, values of greater than 300 kDa were estimated for CPSF, nearly 200 kDa for CstF, and greater than 100 kDa for both CFI and for CFII. Poly A polymerase was estimated at only 50 to

60 kDa, values often reported for PAPs (29). It was duly noted that the sum of these molecular weights was close to that estimated earlier for those 25S polyadenylation complexes recovered from these extracts after heparin treatment (97, 98).

The restoration of function on recombining these partially purified fractions brightened prospects for reconstituting 3' end formation from purified components. Further purification would eventually identify the active proteins within each fraction, which dispelled the confusion that had arisen from these different fractionation schemes. Defining the function of each component, however, proved to be more difficult. The polymerase was most easily purified because of a simple enzymatic assay for both nonspecific and specific polyadenylation. Interestingly, none of the other separated fractions involved in cleavage were able to cleave pre-mRNA alone but did so only when combined, emphasizing the need for a multicomponent cleavage complex. Even now the role of individual cleavage proteins in phosphodiester bond hydrolysis is not clear. However, one protein considered likely to be involved in pre-mRNA sequence recognition had already been identified in unfractionated nuclear extracts by UV crosslinking to the pre-mRNA substrate (112).

C. UV Crosslinking of Proteins to Poly A Sites

The early detection of a potential specificity factor for cleavage/polyadenylation illustrates the particular virtue of the UV crosslinking technique (112). It can be used in crude extracts or even in whole cells if an acceptable signal-to-noise ratio can be attained. Its success depends on the assumption that proteins recognizing a specific RNA sequence will interact with it or a surrounding sequence at some time during the reaction and that this interaction can be stabilized by crosslinking the two components by UV irradiation. This assumption had by this time been amply supported by the crosslinking of relevant proteins to transfer (109) and to ribosomal RNAs (110). Later those core proteins known to be associated with hnRNA in hnRNP particles had also been UV crosslinked to polyadenylated hnRNA in these nuclear extracts (111).

The UV irradiation of nuclear extracts carrying out the polyadenylation of each of three well-studied pre-mRNA substrates, i.e., adenovirus L3, SV40 early, and SV40 late pre-mRNA constructs, resulted in the crosslinking of a predominant 64-kDa protein to each (112). Crosslinking was abolished when the canonical AAUAAA sequence was converted to AAGAAA. Each of these RNA substrates competed for the crosslinking of the 64-kDa protein to any other single substrate as long as it retained the AAUAAA signal. Partial proteolysis of the 64-kDa protein crosslinked to each substrate gave similar digestion products, indicating that the same protein was bound by each RNA. Another study found this same protein (designated in this case as p68) could be crosslinked to the adenovirus L3 poly A site if the AAUAAA sequence was intact (113). Exogenous substrate did not compete for crosslinking if its hexanucleotide was mutated to AAGAAA. Nor was this mutant substrate able to compete for the formation of the specific polyadenylation complexes reported earlier by this group (95). Thus, crosslinking of the 64-kDa protein and complex formation both required the AAUAAA signal. Others reported the UV crosslinking of larger proteins of 155 kDa (113) and 130 and 170 kDa to the L3 pre-mRNA substrate (114) with an intact AAUAAA signal. A follow-up study of UV crosslinking with more purified fractions of CPSF and CstF (115) showed these larger proteins were from the CPSF fraction, while the crosslinked 64-kDa protein was in the CstF fraction. Both factors were required for efficient crosslinking of p64, emphasizing again that the assembly of a large complex was needed to bring those proteins into close contact with RNA.

D. Sequences at Poly A Sites Crosslinked to Specific Proteins

Early studies had indicated that the crosslinking of the 64-kDa protein from extracts had required the intact AAUAAA poly A signal in the pre-mRNA (112, 113) but not sequences downstream of the poly A site (112). However, another study (115) found crosslinking of the 64-kDa protein was achieved with an AdL3 substrate with a AACAAA poly A signal, but not if the downstream sequence was replaced with vector sequences. These discrepancies arising from attempts to define the sites of crosslinking to RNA remained controversial until more sophisticated techniques for analysis of UV crosslinking clarified, at least in the case of the 64-kDa protein, the sites to which it binds (64). However, all of the early studies reported the strong binding of the 64-kDa protein and its dependence on poly A signals as well as a need for multiple factors for achieving efficient UV crosslinking.

It was at this point that crosslinking studies converged with protein fractionation efforts, leading two research groups to a fruitful collaboration that began with attempts to crosslink RNAs to each of the four subfractions that were needed for cleavage (116). Surprisingly, pre-mRNAs could not be UV crosslinked to any of the four fractions separately but required the presence of CPSF and CstF (116), a finding shortly confirmed with equivalent fractions (PF2 and CF1) from another purification scheme (115). Again it appeared that multiple protein:protein interactions create the specific site for UV crosslinking of p64 as well as for cleavage of pre-mRNA. This structure apparently allows close contact of p64 with pre-mRNA when CstF and CPSF are present. At the time, it seemed likely that p64 would be a component of CPSF since this complex, which is required for both cleavage and polyadenylation, is dependent on the AAUAAA sequence for both reactions. It had appeared from early evidence that crosslinking of p64 to pre-mRNA also required an intact AAUAAA sequence. However, as already noted, others failed to confirm this requirement and found downstream sequences to be needed instead (115, 117). Such anomalies would eventually be resolved by the large-scale systematic fractionation of nuclear extracts by three separate groups that produced purified components to be examined with improved UV crosslinking techniques (64).

E. Assembly of Complexes from Nuclear Fractions

The 50S putative polyadenylation complexes sedimented from nuclear extracts did not display 3′ end-processing activity (97). However, an early achievement of the biochemical fractionation approach was the reconstitution of a cleavage/polyadenylation activity from fractions separated from nuclear extracts (100, 102, 106). The names used by different groups to describe the complexes and their subunits during this period along with the current designations are listed in Table I. An early kinetic study focused on complex formation with each of four partially purified fractions that had been found necessary for the reconstitution of cleavage/polyadenylation activity (114). The two fractions sufficient for sequence-specific polyadenylation of precleaved mRNA were designated as PF1, a poly A polymerase, and PF2, a specificity factor for recognition of the AAUAAA sequence. PF2, but not PF1, formed an unstable complex with pre-mRNA that was unaffected by the presence of PF1. Neither of the two fractions required for cleavage, i.e., CF1 (CstF) or CF2, formed a stable complex with pre-mRNA either separately or in combination. However, CF1 (CstF) when combined with PF2 (CPSF) formed a far more stable complex with pre-mRNA that was designated as the commitment complex (B) that migrated more slowly than complex A in nondenaturing gels and in contrast persisted in the presence of competitor pre-mRNA.

The sequence requirements for complex formation and complex stability paralleled those for cleavage and polyadenylation. The formation of the A complex of PF2 (CPSF) with the AdL3 pre-mRNA required an intact AAUAAA signal but not a downstream sequence element, while complex B formation of PF2 + CF1 requires both (114). A mutation of two bases of the downstream element of SV40 pre-mRNA resulted in an 85% reduction in stability of this B complex, suggesting that a component of the CF1 (CstF) complex, possibly the 64-kDa protein that so readily crosslinks to pre-mRNA, recognizes the G+U-rich downstream element.

From these data a multistep pathway depicting the key reactions of mRNA 3′ end formation was proposed (Fig. 2) (114, 117). The first step is the recognition of the poly A signal by the PF2 (CPSF) specificity factor to form a relatively unstable complex that is then stabilized by the binding of CF1 (CstF) to the G+U-rich downstream element of the pre-mRNA. This commits this

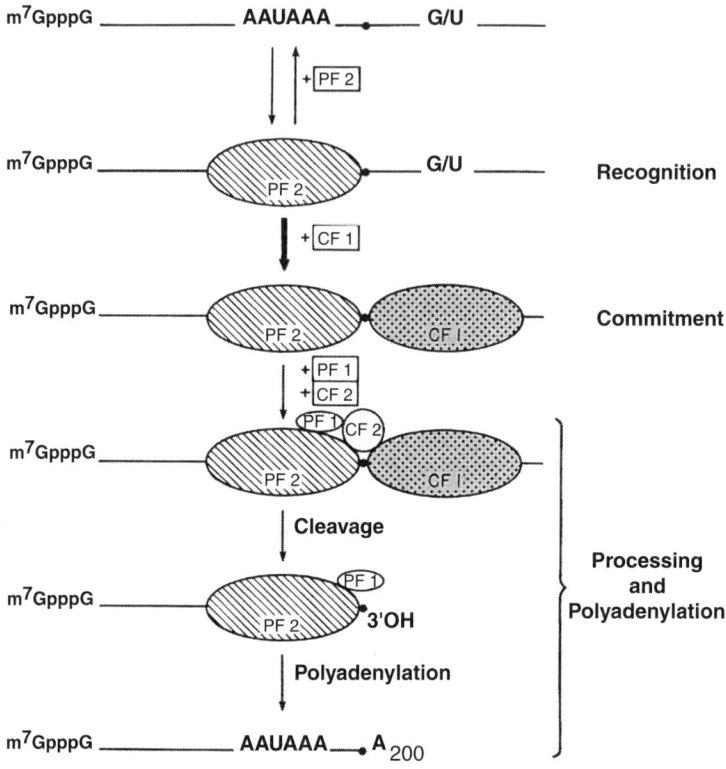

FIG. 2. A schematic depiction of the RNA:protein interactions at the poly A site that result in mature mRNAs. The current designations for these complexes and factors is shown in Table I. Reprinted from E. A. Weiss, G. M. Gilmartin, and J. R. Nevins, Poly A site efficiency reflects the stability of complex formation involving downstream element, *EMBO J.* (1991) **10**, 215, by permission of Oxford University Press.

larger complex to the cleavage event that occurs only if cleavage factor(s) CF2 and poly A polymerase are present. Once cleaved, the 5′ mRNA portion bound to the PF2 specificity factor is polyadenylated at the 3′ hydroxyl group by the polymerase.

This scheme, generally supported by the results of others (*118, 119*), provided a useful framework for sorting out the potential roles of those fractions separated from nuclear extracts. However, it was already clear from on-going purification efforts that the key complexes were a mix of several proteins. The structural analysis and particularly the cloning of cDNA for each of them would become primary focuses of research on cleavage/polyadenylation in the early 1990s.

VI. The Core Components of Cleavage/Polyadenylation

The fractionation of nuclear extracts separated an unexpectedly large number of fractions needed for cleavage alone (Table I). Two of these fractions after purification turned out to be complexes of equimolar amounts of three or four single polypeptides. Two other cleavage factors were large proteins, but neither was needed for polyadenylation. A poly A polymerase was separated from cleavage factors that was a single polypeptide of about 60 kDa. A new activity, not previously recognized as participating in poly A synthesis, was later identified as a poly A-binding protein (PABPII) that converts polymerization from a distributive to a processive reaction after several AMPs have been added to the 3' end of mRNA (*120*) (see Table I).

A. The Cleavage Stimulation Factor (CstF)

The first isolation of a cleavage factor, CstF (*116*), was soon followed by the purification of a cleavage factor, CF1, by another group (*114*). That these were the same complex became clear when it was noted that the most purified fraction was separated into three similar sized polypeptides in equimolar amounts, when electrophoresed in a denaturing gel. Interestingly, one of these was 64 kDa while the others were 77 and 50 kDa.

One of these groups made effective use of monoclonal antibodies to characterize the CstF complex (*121*). Antibodies were created by injecting purified CstF into mice. Hybridomas created by fusing spleen cells with myeloma cells gave several clones that produced monoclonal antibodies for the 50- or the 64-kDa subunits of CstF. Initially, each monoclonal was used to immunodeplete a crude CstF fraction of cleavage activity which could in each case be restored by adding back highly purified CstF. The important question of whether the 64-, 50-, or 77-kDa polypeptides were merely cofractionating or were part of a stable complex was addressed by immunoprecipitation of ^{125}I-labeled CstF with these mAb. All three subunits were precipitated with each antibody, although not in equimolar amounts in all cases. These same antibodies were also used to answer the question of whether the 64-kDa protein that so readily crosslinks in UV light to pre-mRNA in nuclear extracts is the 64-kDa subunit of CstF. Extracts exposed to UV during polyadenylation of ^{32}P-pre-mRNA were digested with RNase before treating either with the 50- or the 64-kDa mAb. A 64-kDa protein crosslinked to a ^{32}P-labeled RNA fragment was recovered after immunoprecipitation of CstF with either mAb. This localization of the 64-kDa crosslinking protein within the CstF complex was somewhat unexpected as it had been assumed, incorrectly as noted above, that the specificity fraction, SF (CPSF), which also required an intact AAUAAA for polyadenylation of precleaved mRNA and was subsequently shown to interact with it (103), would include the p64 subunit. However, the localization of p64 in CstF was consistent with the downstream sequence requirements for complex formation and the RNA crosslinking observed for the analogous CF1 complex described above (115, 117).

In summarizing these early successful purification efforts, it should be stressed that further purification of the remaining components needed for cleavage and polyadenylation posed some major difficulties, given the very small amounts of purified CstF that had been recovered from large volumes of cultured HeLa cells. The characterization of CstF had depended heavily on the availability of monoclonal antibodies for two of the subunits. An obvious way around this protein scarcity was to purify the components from mammalian tissues in the expectation that increased amounts would allow development of either a heterologous system or an entirely mammalian one. An encouraging sign that this would be possible had been given by an experiment in which a poly A polymerase purified from calf thymus was able to replace the human polymerase of HeLa nuclear extracts for sequence-specific polyadenylation (122). This predicted that the putative cleavage/specificity proteins of calf thymus would also confer substrate specificity on a human poly A polymerase.

B. The Cleavage and Polyadenylation Specificity Factor (CPSF)

1. THE STRUCTURE OF CPSF

That the CPSF factor was a critical component for organizing the reactions of 3′ end processing was evident, once it was established that CPSF was needed for specifying both cleavage and polyadenylation (100, 102, 106). It was the only factor, aside from the polymerase, that was needed for specific polyadenylation of precleaved mRNAs (103). A new large-scale purification of poly A polymerase (PAP) from calf thymus (124) had shown it to lack the specificity for mRNA typical of all PAPs purified much earlier as well as for a PAP recently purified from HeLa nuclear extracts (101). That a CPSF fraction from HeLa cells conferred specificity on this bovine PAP confirmed the prediction that this function would be conserved between humans and cows.

CPSF was purified from calf thymus using this ability to confer specificity on the calf thymus PAPase as an assay (103). From its sedimentation velocity in density gradients, CPSF activity appeared to be in a large component of ~500 kDa, most likely a multisubunit complex. This indeed proved to be the case, as purified preparations from both HeLa nuclear extracts and calf thymus could be separated into three large and one small polypeptide during electrophoresis in SDS gels (103). The fact that preparations from both sources had similar specific activities and contained similar subunits in roughly equimolar amounts lent strong support for the characterization of CPSF as a complex of these four polypeptides.

The electrophoretic mobilities of 160-, 100-, and 73-kDa subunits were unaffected by denaturation in disulfide-reducing agents, but the smallest 30-kDa subunit disappeared, a finding that may have been related to the failure of others to detect this protein in their purified preparations of CPSF (*118*). Moreover, the three largest polypeptides conformed closely to the size of three polypeptides among several that were immunoprecipitated with a human SLE serum of the Sm serotype from a partially purified ^{125}I-labeled specificity factor (PF2) that was equivalent to CPSF (*115*).

An active CPSF complex also purified from calf thymus with subunits of 165, 105, and 70 kDa, but lacking the smallest 30-kDa subunit, also lacked RNA, which again cast doubt on a direct role for an SnRNP in 3' end formation (*118*).

2. CPSF SUBUNIT INTERACTIONS

Each of the purified CPSF complexes showed the highly cooperative interaction with CstF that produced the stable commitment complex described earlier for the PF2 (CPSF) and CF1 (CstF) fractions shown in Fig. 2 (*115, 117*). The failure to detect a complex between a second CPSF preparation and pre-mRNA (*118*) as reported for the other (*103*) may have reflected the general sensitivity of such recognition complexes to variations under assay conditions or in contaminating proteins.

The reaction sequence proposed in Fig. 2 for the assembly of the recognition complex that begins with the interaction of CPSF with pre-mRNA remains somewhat arbitrary, as an initial interaction of CstF with the downstream element is also possible, since a weak complex was reported to form with CstF on pre-mRNA in the absence of CPSF (*118*). The strong cooperative interaction between these two factors in stabilizing the commitment complex for cleavage casts doubt on the significance of these unstable complexes detected *in vitro* for the cleavage reaction in the cell.

In addition to its role in organizing the commitment complex for cleavage, CPSF was also needed for the subsequent polyadenylation of AAUAAA-dependent mRNA by poly A polymerase. CPSF fractions conferred both specificity and increased activity on purified PAPs for precleaved substrates, i.e., those with 3' ends at the poly A addition site (*126*).

3. AN SnRNP IN CPSF

That CPSF had originally been reported to contain a SnRNP epitope was consistent with (*79, 85, 100, 114, 127*), although not always observed in (*102*), inhibition of polyadenylation in nuclear extracts by both anti-Sm- and/or U1SnRNP antibodies. An obvious candidate that emerged early on was the U11SnRNP (*100*) which, however, disappeared upon further purification (*103*). Nor was U11SnRNA or any RNA detected in another purified CPSF complex (*118*). Since correct mRNA 3' end processing (*108, 118*) can be reconstituted in the absence of RNA, it was concluded that SnRNPs per se are not needed for basic cleavage/polyadenylation. The earlier evidence for an SnRNP involvement may have been related to the recently observed stimulation of polyadenylation that occurs during the splicing of 3′ terminal exons from pre-mRNA (*24*). The molecular details that gradually emerged for this coupling have implicated an interaction between U1SnRNP and CPSF. The stimulation of polyadenylation by U1SnRNP was found to correlate with the binding of U1RNA to several sites upstream of the AAUAAA signal on *SV40* and *Ad L3* polyadenylation substrates (*128*). The A protein of U1SnRNP was then found to interact with sequence elements upstream of the *SV40* late poly A signal that enhance polyadenylation, while suppression of this interaction inhibited *in vitro* polyadenylation. Later, this A protein of the U1SnRNP was co-immunoprecipitated with the 160-kDa subunit of CPSF with polyclonal antibodies for a 100 kDa fragment of the 160-kDa protein (*105*). Thus, although the core reactions of cleavage/polyadenylation occur in the absence of SnRNPs, U1SnRNP appeared to coordinate terminal exon splicing with 3′ polyadenylation through an interaction of its A protein with the 160-kDa subunit of CPSF.

4. CPSF BINDING TO RNA

The early observation that CPSF depended on an intact AAUAAA sequence for the polyadenylation of precleaved mRNA clearly predicted an interaction of CPSF with the hexamer (*100, 101, 106*). It was then shown with modification interference experiments in which purines were modified with diethylpyrocarbonate and pyrimidines with hydrazine that the nucleotides of the AAUAAA were essential for polyadenylation and complex formation with CPSF (*123*). Modification of the other sites of the RNA were without effect. Similar effects were seen with controlled 2′-*O*-methylation of ribose where only modification of nucleosides of the hexamer interfered with polyadenylation or complex formation (*130*). However, in this case more subtle effects of methylation on the individual nucleosides of AAUAAA were seen. For example, 2′-*O*-methylation of the ribose of the second adenosine led to marked reductions in polyadenylation that were not seen for a modified ribose of the uridine residue even though replacement of the U with a G abolished both polyadenylation and complex formation. Such differential effects stressed the likelihood that CPSF makes highly specific contacts with individual structural features of both the bases and the ribose of the AAUAAA.

The 160-kDa subunit of CPSF emerged as the likely candidate to make such contacts with AAUAAA as some (*113, 114*), but not all (*112, 121*), of the early studies had reported a protein of this size to be among the proteins of nuclear extracts to be UV crosslinked to RNA polyadenylation substrates. The two different preparations of CPSF described above gave contradictory results for the UV crosslinking of the 160-kDa subunit to RNA. One group observed proteins of 160 and 35 kDa crosslinked to RNA (*123*). This binding could be readily competed away with excess unlabeled RNA substrate, but not if this RNA had a point mutation in the AAUAAA poly A signal. Another group with a more purified CPSF complex that lacked a 30-kDa protein did not observe crosslinking of the 160-kDa subunit (*118*). Compelling evidence for the binding of the 160-kDa subunit to AAUAAA containing RNAs was later obtained with separate antibodies raised against the 100- and 160-kDa purified subunits of CPSF (*131*). Either of these antibodies co-immunoprecipitated from nuclear extracts a complex of four proteins with electrophoretic mobilities indistinguishable from those UV crosslinked in the presence of the RNA substrate; each antibody co-immunoprecipitated the 160- and 30-kDa proteins crosslinked to the RNA substrate.

In spite of these unexplained differences in RNA crosslinking, it is clear that it is the 160-kDa subunit of CPSF that recognizes the AAUAAA poly A signal and binds to pre-mRNA. Studies with a recombinant form of the 160-kDa protein confirmed this binding (*132*) but also suggest that the 160-kDa protein alone binds with less specificity for the AAUAAA poly A signal than it does within the CPSF complex. This study also reported the specific binding of the 160-kDa subunit to the 77-kDa protein of the CstF complex, providing a molecular basis for the cooperativity of the interaction of CPSF with CstF to form the stable complex committed to specific pre-mRNA cleavage.

C. Poly A Polymerase (PAP)

Poly A polymerase is now among the best characterized proteins comprising the core components of cleavage/polyadenylation. It originally was the unexpected specificity of the polymerase for ATP that had first suggested to some that poly A sequences might be present in cells, although early on others thought the activity could be an *in vitro* artifact, at least until poly A sequences were found a decade later to be covalently attached to mRNAs and their nuclear precursors (*6–8*). This finding led to renewed efforts to purify the polymerase that resulted in several highly purified preparations. Such efforts along with descriptions of the properties of poly A polymerases have been described in reviews that antedate the modern era of sequence-specific mRNA polyadenylation covered in this history (*28, 29, 133*). The most disappointing feature of these preparations had been their lack of RNA substrate specificity, a property also shared by the polymerase purified later from HeLa nuclear extracts (*101*). Specificity factors such as CPSF recovered from HeLa extracts were, however, able to confer substrate specificity on both the HeLa and the calf thymus polymerases, suggesting that the two enzymes shared common sequences (*124*). This encouraged the development of a new protocol for large-scale purification of PAP from calf thymus (*124*). Larger quantities of protein were needed not only for biochemical studies but also even more importantly for providing oligopeptides from which sequence information could be obtained for the synthesis of deoxyoligonucleotide probes for screening cDNA libraries for clones that express PAP. The yields and purity of the PAP purified from HeLa cells up to this point were hardly adequate for these undertakings.

The highly purified nonspecific polymerase from calf thymus acquired specificity for pre-mRNA substrates in the presence of the HeLa specificity factor CPSF and Mg^{2+} (124). Thus, it was finally clear that those calf thymus polymerases purified long ago were in fact responsible for the polyadenylation of pre-mRNAs *in vivo*. A study of this enzyme confirmed that it retained most of the properties described for these earlier preparations. Some important characteristics of the enzyme protein were, however, clarified, such as its monomeric structure, shown by the similarity of molecular weights for the native and denatured protein. The molecular weights calculated from sedimentation and gel filtration data were in agreement with values of about 60 kDa reported for several earlier preparations (29) but now known from cloned cDNA sequences to be a large underestimate (125, 126).

A detailed kinetic analysis using oligoA$_{12-18}$ or poly A$_{65}$ primers in the presence of Mg^{2+} or Mn^{2+} provided an explanation for the often reported stimulatory effects of Mn^{2+} on the catalytic activity (124). The K_M for 3' ends of both primers was reduced about 30- to 100-fold when Mn^{2+} replaced Mg^{2+}, but the V_{max} was relatively insensitive to this substitution. In an excess of primer, the reaction rates were similar for the two cations, suggesting that Mn^{2+} increased the affinity of the enzyme for the 3' end of the primer. Thus it would appear that *in vivo*, where Mg^{2+} serves as divalent cation, the polymerase has an intrinsically low affinity for the 3' ends of RNA that can be greatly increased in the presence of the specificity factor CPSF that recognizes the AAUAAA signal.

These studies also established the mechanism of nonspecific poly A chain elongation as a distributive rather than a processive reaction. This was deduced from analysis of the size distribution of the polyadenylated products accumulating during the course of the reaction. With a distributive mechanism, the polymerase dissociates from the substrate after each catalytic event. In a processive one, the enzyme remains bound to substrate as it extends the length of the sequence to some specified end point. These mechanisms are readily distinguished in reactions with a large molar excess of primer over enzyme. Here a uniform increase in length of all RNAs specifies a distributive reaction as in the case of nonspecific polyadenylation noted above, while a processive mechanism is characterized by both fully elongated and unreacted RNAs with few observable intermediates. Polyadenylation in the presence of purified CPSF, however, also appeared to be distributive in these reactions (124).

D. Poly A-Binding Protein (PABII)

The lack of processivity noted above for the sequence-specific reaction catalyzed by PAP even in the presence of the purified CPSF complex was soon shown to have been the result of the removal of a factor from CPSF during its purification (134). A protein of 49 kDa (actually 33 kDa from the cDNA sequence) was subsequently purified from these extracts that in the presence of a purified CPSF and PAP gave a strikingly different distribution of polyadenylated products from those seen with the purified CPSF alone (134). Instead of the gradual elongation of all the RNAs, a population of RNAs with about 10 AMP residues accumulated early in the reaction as did mRNAs with tails of about 200 AMP residues. This latter population continued to accumulate, while RNAs with the short poly A tails disappeared. RNAs with intermediate lengths of poly A did not accumulate. This biphasic reaction induced apparently by this 49-kDa protein had already been observed in a kinetic analysis of polyadenylation in nuclear extracts where a similar distribution of products accumulated over time (135). It was also reported that the early stages of the reaction required substrates with an intact AAUAAA signal as expected, but once reaction products with close to 10 AMPs had accumulated this signal was no longer needed for further elongation. It was suggested that a short oligo A tail provides the recognition element for a factor in nuclear extracts that accelerates polyadenylation processively (135). This was tested later with a set of mRNAs bearing different lengths of oligo A tails and either an AAUAAA or an AAGAAA signal. Substrates with at least 9 AMP residues were efficiently elongated to the poly A length found *in vivo* in spite of the mutated poly A signal, while those with less than 9 residues required AAUAAA for further elongation (136).

This protein was a likely candidate to mediate suppression of a mutated poly A signal as it binds specifically and tightly to poly A and also stimulated poly A tail extension in the absence of AAUAAA as long as the 3′ end of the mRNA had a short oligo A tail. It was designated poly A-binding protein II (PABII) to distinguish it from PABI from which it was chromatographically separated during its purification to near homogeneity from calf thymus (136). Although the two proteins differ in size and sequence, they have similar interactions with poly A in terms of the minimal binding site of 12 nt and the binding of 23 AMP residues per protein monomer. The latter value has been corrected to 13 AMPs for PABII based on a revised size of 33 kDa (138). That they have different cellular functions is indicated by immunofluorescence analysis showing PABII to be concentrated in the cell nucleus (137) while PABI is in cytoplasm.

That CPSF would have no role in the second processive phase of the reaction initiated by the presence of PABII seemed likely. Surprisingly, elongation was far more efficient, and longer poly A tails, resembling those attained *in vivo*, were made when it was present (119), suggesting that the polymerase remains in contact with CPSF as well as with PABII during tail extension. None of these three components alone forms stable complexes with RNA, although CPSF and PABII each forms weak complexes with RNA that are stabilized by PAP leading to somewhat processive reactions that become fully processive when all three components are present producing at least 100 AMP residues for each polymerase-binding event (*119, 120*).

A comparative study of the elongation of mRNA poly A tails of different lengths showed that tails shorter than 200 nt were each elongated rapidly and processively, but mRNA substrates with tails of 250 nt were elongated slowly and distributively even in the presence of PABII and CPSF (*138*). Presumably this switch from a rapid processive to a slow distributive reaction, once lengths similar to those of native poly A tails are reached, results from a disruption of the polyadenylation complex. How this length is sensed by the complex is unclear, but it would seem to be related to the number of PABII molecules that can be bound within a complex that is elongating the poly A tail of an RNA.

VII. Cloning, Sequencing, and Expressing the Core Proteins

Much of the stimulus for the large-scale purification of the polyadenylation core components came from the need for sequence information that could be used to clone their specific cDNAs. By this time, cloning only required a few short sequenced oligopeptides derived from the purified protein to design the DNA probes for screening cDNA libraries. In the late 1980s, the advent of automated chemical synthesis of DNA could readily supply the pairs of short DNA primers needed for amplification of cDNA by the polymerase chain reaction (PCR) that had also recently been automated. Selected sequences could then be chemically synthesized to provide single-stranded probes for screening colonies of bacteria expressing cloned cDNA libraries. From one or more of these cDNA sequences, a complete amino acid sequence of the protein could usually be deduced. Verification of course required evidence that the cDNA expressed a protein with the properties of the native protein. Such evidence was obtained either from the *in vitro* translation product of the mRNA transcribed from the cDNA or from the recovery of a recombinant protein from the extracts of the cells expressing the cloned cDNA. Such recombinant proteins are often identified with antibodies against the native protein, but only a direct comparison of the properties of the native and the recombinant protein can establish this identity with certainty. Retention of biological activity of a recombinant protein or its truncated form greatly expands the possibilities for either biochemical or structural studies, although even an inactive protein may retain antigenic activity.

By 1991, protein fractionation techniques had identified most of the proteins involved in 3′ end formation, albeit in most cases only as a band on a gel. These allowed the assignment of apparent molecular weights for each component, and in some cases enough protein was available to raise specific antibodies, as noted above in the case of the heterotrimeric protein complex CstF. Future progress was likely to be limited to the generation of antibodies for other purified components that could be useful in the further characterization of the subunit composition of the more stable complexes such as CPSF. At this point, the situation was not unlike what it had been in the mid-1970s when several nearly homogeneous preparations of PAP were available and at least one PAP antibody had been reported (*139*). Nevertheless, the availability of these purified polymerases proved not to be particularly useful until sequence-specific polyadenylation was achieved with nuclear extracts nearly 10 years later.

By 1990, it was clear, however, that much could be learned from the cloning of cDNAs for the core components of cleavage/polyadenylation. The primary sequence of an open reading frame not only can establish the correct size of the protein, as proved to be important in the case of PAP and PABII, but also can reveal homologies to known functional motifs. Such motifs within proteins involved in RNA processing had gradually emerged from the ever-expanding sequence databases. Foremost among these were the RNA-binding domains (RBD) of several proteins present in both hnRNP and snRNP ribonuclear protein complexes. The most common motif is characterized by two short conserved sequences of six (RNP2) and eight (RNP1) amino acids separated by sequences about 40 to 60 amino acids that include several highly conserved hydrophobic amino acids (*140*). The presence of this motif suggests a direct interaction of the protein with RNA that is testable by measuring the effects of mutagenesis of these sites on such properties as binding or crosslinking to RNA substrates. Even more intriguing are sequence homologies to supposedly unrelated proteins that often give unexpected insights into possible functions shared by both proteins. A particularly startling example was the realization that the gene for the *E. coli* PAP was identical to a *pcn* gene already known to be associated with the control of a plasmid copy number in *E. coli* (*141*). (See Section X.)

Another obvious benefit of cloning is the possibility of overexpression of a cDNA in bacteria or other hosts that can enhance the low yields of proteins often recovered by the classic purification protocols for eukaryotic cells or tissues. Major improvements of the purification of recombinant proteins from bacterial extracts had been achieved by the expression of the protein from a genetically engineered cDNA containing a short selectable oligonucleotide, such as a run of histidines that allows the recombinant protein to be specifically selected from lysates by immobilized nickel ion chromatography (*142*), which has the added benefit of avoiding the selection of endogenous forms of the protein. These innovations in the recovery of pure recombinant proteins combined with the more efficient eukaryotic expression systems typified by those Baculovirus-recombinant DNA vectors that infect cultured insect cells (*143*) accelerated progress to the point where cDNAs for nearly all of the core proteins have been cloned and expressed. Quantities sufficient for crystallization have recently allowed the X-ray diffraction analysis of the bovine (*154*) and yeast PAPs (*253*).

A. Poly A Polymerases

1. cDNA CLONES AND MULTIPLE PAPases

PAP was an early candidate for cloning, as purified preparations stored for many years were available to obtain the sequenced tryptic oligopeptides needed to design oligonucleotide probes for probing cDNA libraries. Enzymatic activity also simplified its detection in expression systems. In 1991, two groups (*125, 126*) reported the isolation of cDNA clones for poly A polymerase from different bovine tissue libraries. Each group sequenced a cDNA that contained essentially identical open reading frames (ORF) for a protein of 740 amino acids with a predicted size of 83 kDa. One group purified a PAP activity associated with an 82-kDa protein in the lysate of the bacterial clone expressing this cDNA (*126*). The other group translated in a reticulocyte lysate an mRNA, transcribed from a slightly smaller cDNA, into a protein of 77 kDa (*125*). Both of these recombinant proteins carried out nonspecific as well as AAUAAA-dependent polyadenylation. The failure of recombinant PAP to distinguish mutated from wild-type AAUAAA sequences in the absence of CPSF finally ruled out a direct role for PAP in the recognition of poly A signals.

The surprising differences in the size of these recombinant bovine polymerases from the most purified bovine PAP (*124*), as well as others reported in the past to be about 60 kDa, most likely resulted from their degradation during purification probably at one or a few susceptible sites. That such sites are likely to be in the C-terminal region is suggested by the finding that deletions of the C-terminal 150 amino acids of the ORF still allowed its expression *in vitro* as a truncated protein that catalyzed both nonspecific and specific polyadenylation (*125*).

One of the bovine cDNA libraries yielded two classes of PAP cDNA clones. One designated as type II is the cDNA with the 740-amino acid ORF noted above. The other cDNA designated as type I codes for a 77-kDa protein of 689 amino acids that except for the terminal 50 amino acids is identical to the type II cDNA. The enzymatic activity of PAPI and PAPII expressed from these cDNAs could not be distinguished as would be expected given the dispensability of the C-terminal region for such activity. The cDNA libraries from human cells (*144*) and frog oocytes (*145*) contained PAP cDNAs that were 97 and 83% identical to the bovine PAPII. The ORF of a yeast PAP cDNA shares a 47% identity to the ORF of the N-terminal half of PAPII but bears no resemblance to its C-terminal region, which may account for its inability to replace the mammalian PAP in HeLa nuclear extracts (*146*).

Each of these cDNA libraries contained truncated forms similar to one from bovine libraries designated as PAPIII which lacks the 3′ half of PAP but is otherwise identical. More recently, other truncated mRNAs in addition to PAPIII have been detected in mouse tissue designated as PAPV and PAPVI, each of which arises from alternative processing of the transcript of a poly A polymerase gene to be described below (*147*). In contrast to the long forms of PAP, protein products have not been detected for these truncated mRNAs (*147*). One possible exception is an mRNA of frog oocytes that resembles the mammalian PAPIII. When expressed in *E. coli* from a cDNA as a glutathione fusion protein it showed a weak poly A polymerase activity (*148*).

The elongated functional mRNAs for PAPI, II, and IV arise by alternative splicing of the PAP gene transcript and show tissue-specific expression that could account for early evidence for multiple cellular PAPs (*29*). The significance of multiple forms, however, remains unknown. The early speculations on the need for more than one PAP were based in part on the apparent distinction between *de novo* poly A synthesis in the nucleus and poly A tail elongation in the cytoplasm (*30*). A cytoplasmic function is unlikely for any of the three forms of PAP described above, as each contains the identical bipartite NLS. At present, there is little evidence for a distinct cytoplasmic PAP even in *Xenopus* oocytes where a few specific mRNAs with a specific U-rich recognition sequence are polyadenylated in cytoplasm during oocyte maturation (*148*). Purified PAP and CPSF from calf thymus nuclei could replace the *Xenopus* polyadenylation factors in this specific cytoplasmic reaction (*149*).

2. Genomic Clones

The sequence of a PAP gene from a mouse genomic library clarified the processing pathways that could generate all of the long and truncated forms of PAP detected in cDNA libraries (*147*). Splicing of the 22 exons of the gene would generate PAPII while skipping exon 20, and substituting a downstream exon for exon 22 would produce PAPI. Alternative splicing again by skipping exon 20 would give PAPIV an elongated mRNA recently detected along with PAPII by reverse transcription–PCR of total mouse brain RNA. PAPIV predominated over PAPII in mouse brain but not in other mouse tissues or other species, while PAPI has only been detected in bovine libraries (*147*).

Antibodies raised against PAPI detected the three elongated forms of PAP expressed in baculovirus extracts on Western blots and more importantly in lysates of HeLa cells. Recombinant forms of the truncated PAPIII, V, and VI were expressed in baculovirus-infected cells, but not in HeLa cell lysates. They had no PAP activity nor did they competitively inhibit PAPI activity or display nuclease activity on the RNA substrate (*147*).

3. Sequence Motifs

The cDNAs of the bovine PAP showed several recognizable sequence motifs (*125*, *126*) such as a putative RNA-binding domain (RBD) near the N terminus with similarity to the RNP1 and RNP2 sequences and their spacing, a catalytic domain with several well-spaced modules that are characteristic of

RNA-dependent polymerases and designated as the polymerase or PM module, and a nuclear localization signal (NLS) in the C-terminal region that was followed by a long serine/threonine-rich or S/T domain. The expression of PAP activity with a truncated cDNA lacking the S/T region that is often a site of phosphorylation showed it was not needed for PAP activity. Verification of the function of these and other sequences of PAP was somewhat problematic. Not only were ambiguities encountered in interpreting the effects of certain mutations on these recognizable motifs but also results from two major mutational studies of PAP (*150, 151*) have led to some different conclusions about the function of these motifs. This is hardly surprising, as one group focused on conservative replacements of one or a few amino acids within the conserved regions of each of the motifs (*150*), while the other relied largely on C-terminal or C-internal deletions and, in the case of the catalytic site, on "carboxylate scans" in which individual aspartate or glutamates of PAP were replaced with alanines or histidines (*151*). The earlier study (*150*) measured effects on both nonspecific and specific PAP activity, while the more recent study (*151*) examined the kinetic parameters of the enzyme reaction and the binding of mutant PAPs to RNA by gel retardation and UV crosslinking assays.

a. The NLS Signal. The putative bipartite NLS motif is needed to localize PAP in the nucleus, as several nonconservative amino acid substitutions within each of two conserved sites showed the NLS to be necessary but surprisingly not sufficient for efficient localization of PAP to the nucleus. A second bipartite NLS about 140 amino acids downstream of NLS1 also participated in nuclear localization. This need appeared from observations that an expressed epitope-tagged truncated PAP that retained NLS1, but lacked a C-terminal segment, was distributed over cytoplasmic and nuclear regions of the cell, while the wild-type PAP was essentially nuclear. Mutations of both NLS1 and NLS2 regions resulted in a largely cytoplasmic fluorescence of cells treated with the anti-epitope-tagged PAP antibody which established the need for both signals for efficient nuclear localization (*150*).

b. The Catalytic Site. The function of the putative PM module in PAP that is characteristic of several RNA-dependent RNA polymerases could not be demonstrated with single amino acid replacements within three of the four short conserved, but widely separated, motifs of the PM domain. A deletion of three conserved amino acids within the A module closest to the N terminus did abolish PAP activity, suggesting the catalytic site included this region but not as part of a classic PM motif (*150*). A "carboxylate scan" of the first 520 of the 740 amino acids of PAP in which individual aspartates and a few glutamates were replaced by alanine or histidine detected three aspartate residues critical for PAP activity, Asp_{113}, Asp_{115}, and Asp_{167} (*151*). Mutations of these residues were later shown to greatly reduce crosslinking efficiency of ATP analogs to this catalytically important site (152) that shares homologies with many nucleotidyl transferases (153).

The catalytic site has now been viewed in PAP crystals from both bovine and yeast cells in a complex with 3' dATP, an ATP analog that PAP adds to the 3' OH end of the primer but cannot extend (154, 253). This historic breakthrough has revealed an overall PAP structure that deviates from the "palm of the hand" models that characterize template-dependent polymerases (153) but retains a U shape in which the catalytic domain spreads across one side of the cavity rather than at the bottom of the U as with template-dependent polymerases. The structure is most conveniently designated as composed of three domains beginning from the N terminus as the catalytic, central, and RNA-binding domain. The latter is highly compact and is presumed to interact with the catalytic domain, but the placement of the RNA will only be known when a structure of a PAP:RNA primer cocrystal can be determined. The solution of this structure marks the culmination of 40 years of study, albeit intermittent, of PAPs that began with a description of a novel ATP-polymerizing activity for which no cellular product was then known (1). A detailed comparison of the bovine and yeast PAP crystal structures has been published (337).

c. An RNA-Binding Domain. The tentative assignment of a function to the RBD motif in the N-terminal region of PAP was based on the effects of nonconservative replacements of conserved amino acids of the RBD on specific and nonspecific polyadenylation (150). Single amino acid replacements in RNP1 had rather small effects, but a triple mutation in RNP1 abolished enzymatic activity, and substitution of a pair of conserved amino acids in the RNP2 motif also reduced activity. Direct RNA-binding assays were not attempted because of the known instability of PAP:RNA complexes.

A different region of PAP was designated as an RNA-binding domain in a later study that analyzed the effects of C-terminal or short C-internal amino acid deletions on PAP functions (151). The deletion of a 25-amino acid sequence in the C-terminal region that partly overlaps NLS1 resulted in sharp decreases of specific and nonspecific PAP activities that were characterized by large increases in K_M but small decreases in K_{cat}. These losses in enzymatic activity were well correlated with losses in mutant PAP binding to RNA as measured in gel-retardation assays where high levels of PAP had been used to overcome the inherent instability of PAP:RNA complexes originally reported by this group (119). Fragments of PAP of about 100 amino acids that included the amino acids critical for PAP activity also formed gel-retarded complexes with RNA. Complex formation was also well correlated with UV crosslinking of these fragments to RNA at amino acids 488–508. Interestingly, these residues that are at the C-terminal end of the truncated PAP crystal noted above folded into a compact domain resembling the fold in an RNA recognition motif (154). As noted, crystals of PAP complexed to polyadenylated RNA will be needed to define the relationship between the catalytic site and this RNA-binding site and ultimately sites of interactions of PAP with CPSF 160 and PABII.

d. The Serine/Threonine-Rich Domain. That the S/T-rich region at the C terminus of PAP was not required for polyadenylation was clear when a truncated form of PAP lacking about 150 C-terminal amino acids retained specific and non specific PAP activities (125). It was noted that several conserved motifs that were potential sites for phosphorylation by the cyclic AMP-dependent protein kinase (Cdk2) were present in the S/T domain. This domain was subsequently assigned a regulatory function when full-length PAPs expressed *in vivo* were phosphorylated while a truncated form lacking the S/T domain was not (150).

Nuclear forms of human (144) and *Xenopus* oocyte PAP (145) were also shown to be phosphorylated *in vivo*. During oocyte maturation, phosphorylated forms of nuclear PAP increased but decreased rapidly upon fertilization, even though total PAP protein and PAP activity remained relatively unchanged over this time period (145). These observations of a cycle of phosphorylation–dephosphorylation for PAP raised questions of whether polyadenylation is directly regulated or whether such changes in phosphorylation are under the control of the cell cycle. (See Section VIII.) In any case it appeared that changes in the level of phosphorylation of PAP could be an important aspect of the regulation of mRNA formation.

B. The CstF Complex

1. cDNA Clones

The first of the accessory polyadenylation factors to be cloned was the 64-kDa subunit of CstF (155) which was also the first to be identified, primarily for its propensity to crosslink to substrate RNA (112). A cDNA clone was soon reported for the 50-kDa subunit (156), while a cDNA clone for a 77-kDa protein was sequenced and expressed later (157). The molecular weights calculated for all three cDNAs ORFs were in close agreement with sizes assigned earlier to each subunit from gel mobilities.

All three cDNAs for CstF subunits have been expressed in cultured insect cells infected with recombinant baculoviruses (157). The coexpressed 50- and 64-kDa subunits were each immunoprecipitated from extracts of these cells only by their specific antibodies. However, coexpression in these cells of either one with the 77-kDa protein allowed it to be immunoprecipitated by its specific antibody in a complex with the 77-kDa protein. Coexpression of all three subunits gave cell extracts active in cleavage/polyadenylation from which the CstF heterotrimer could be precipitated by either the 50- or 64-kDa-specific antibody.

This central role of the 77-kDa subunit in creating the CstF heterotrimer was reinforced with *in vitro* experiments using antibodies for the 77-kDa protein that were not available in the earlier studies. This was achieved by insertion of a flu virus epitope into the 77-kDa cDNA that could be expressed as a flu-epitope-tagged 77-kDa protein precipitable by anti-flu antibodies. Combinations of mRNAs transcribed from this cDNA and the cDNAs for 50- and 64-kDa proteins were translated in reticulocyte lysates, and their protein products were examined for complex formation with the anti-flu antibody specific for the 77-kDa subunit (*157*). Cotranslation of either mRNA along with the epitope-tagged mRNA for the 77-kDa protein gave a complex of the two proteins that was precipitable by the anti-flu antibody. Cotranslation of the mRNAs for the 50- and 64-kDa proteins gave no complex, but only the single protein precipitated by its specific antibody. This *in vitro* evidence for separate interactions of the 77-kDa protein with both the 50- and 64-kDa proteins that do not interact with each other consolidated the evidence from earlier experiments with the 50- and 64-kDa antibodies that the 77-kDa subunit organizes the heterotrimeric CstF complex (*121*). The cDNA sequences for CstF subunits have several recognizable motifs and some surprising homologies to other proteins. However, mutational analyses that could establish a function for these domains have yet to be reported.

a. The 64-kDa Subunit. An RNP-type RNA-binding domain highly homologous to an RBD consensus sequence correlates well with the UV crosslinking of this protein to polyadenylation substrates, as does the crosslinking of an N-terminal fragment that contains this putative RBD domain (*157*). The CstF complex in the presence of CPSF and an AAUAAA containing pre-mRNA had been presumed to interact through the 64-kDa subunit with the downstream sequence element essential for cleavage of pre-mRNA. This was deduced from mutations within this element that reduced UV crosslinking of the 64-kDa protein to RNA (*115*) and destabilized the CstF:CPSF pre-mRNA complex (*117*). The actual site of 64-kDa binding had, however, remained unclear until a more sensitive RNA-mapping technique was used to localize its crosslinking to a short U-rich downstream site on the SV40 L poly A substrate (*64*). The 64-kDa protein protected this region from hybridizing to a complementary deoxyoligonucleotide that confers RNaseH susceptibility on the RNA portion of the DNA:RNA hybrid. The protected RNA sequence was UV crosslinked to the bound protein and immunoprecipitated with the 64-kDa antibody.

A recombinant form of the 64-kDa subunit in the absence of other factors was also crosslinked to this RNA, but at a number of sites both upstream and downstream of the cleavage site (*64*). These data reemphasize the role of the cleavage/polyadenylation complex in restricting the binding of the 64-kDa protein to a site that favors the interaction of CstF with the CPSF complex on the AAUAAA sequence to form a complex committed to cleavage.

This 64-kDa protein has no recognizable nuclear localization signal, although immunofluorescent microscopy has shown it to be localized in the nucleus (121). The suggestion that it enters the nucleus in a CstF complex where one of the other subunits supplies the signal seems likely in view of the detection of an NLS signal in the 77-kDa subunit (157).

A unique 5-amino acid unit (MEARA/G) repeated 12 times in the C terminus can be folded into an unusually long α-helix of 60 amino acids in which the charged amino acids are all on the external surface creating a potential for exposing it to solvent. The proline and glycine residues surrounding this rigid rod-like helix would allow flexibility for the interaction with the 77-kDa subunit of CstF noted above.

b. The 50-kDa Subunit. The most striking feature of this subunit that lacks RBD and NLS motifs is a set of seven homologous C-terminal repeats, each of about 40 amino acids (156). Similar repeats are present in the B subunit of human and bovine G proteins (transducins) (158) which share a 20% identity and a 36% similarity with the 50-kDa subunit. The homology is concentrated in the C-terminal segments of the repeat. The heterotrimeric structure of G proteins suggested that CstF might be a structural analog but was ruled out by a lack of sequence similarity of the α and γ subunits to either the 64- or the 77-kDa subunits of CstF. A group of proteins with functions not obviously related to G proteins also have transducin repeats such as the prp4 subunit of the U4SnRNP complex of yeast. A deletion of the central repeat of PRP4 greatly reduced its interaction with prp3 protein of the U4SnRNP, that in turn destabilized the U4/U6 complex (159), which in the presence of U5 SnRNP constitutes the catalytic core of the spliceosome.

More recently, a yeast two-hybrid screen using CstF50 as bait selected a nuclear protein, BARD1 (160), known to associate *in vivo* with the breast cancer tumor suppressor, BRCA1. This BRCA1-associated ring domain protein formed a complex with CstF50 that inhibited polyadenylation *in vitro* (160). Since BARD1 as well as CstF50 interacts with the CTD domain of RNA Pol II, it was suggested that the inhibitory effect of BARD1 on polyadenylation may prevent inappropriate 3' end formation during transcription pausing or at sites of DNA repair.

c. The 77-kDa Subunit. Immunoprecipitation experiments with the recombinant forms of each of the CstF subunits as discussed earlier showed the 77-kDa protein forming separate complexes with each of the other two subunits. When all three were present, an active complex formed. The lack of nuclear localization signals on the 50- and 64-kDa subunits that had led to the prediction that a NLS would be found on the remaining subunit proved to be correct. A sequence homologous to the bipartite NLS is present in the 77-kDa protein (157) that may allow the CstF complex to be transported into the nucleus, although only the 64-kDa protein has been localized there thus far.

A striking feature of the 77-kDa sequence is its similarity to the suppressor of forked gene of *Drosophila melanogaster* to which it is 56% identical and 69% similar. One of the well-studied phenotypes of *Drosophila* suppressed by a mutated *Su(f)* gene is the result of an insertion of the gypsy transposon into an intron of the heat-shock 82 gene. This results in the accumulation of a truncated *hsp82* mRNA (*161*) produced by termination of transcription within the LTR of the transposon (*162*). A *Su(f)* mutation restores the production of the normal *hsp82* mRNA. The *Su(f)* gene had already been implicated in RNA processing from its shared homology with the *RNA14* gene of yeast (*163*). This gene along with a related gene, *RNA15*, was originally believed to be involved in yeast mRNA stability (*163*), but both have now been shown to function in mRNA 3' end formation (*164*). (See Section IX.)

The reversal by the *Su(f)* mutation of the effects of the gypsy transposon insertion into an intron of the *hsp82* gene can now be understood as a modulation of the cleavage/polyadenylation reaction that differentially affects the use of multiple poly A sites in this pre-mRNA. The low levels of the 77-kDa subunit of CstF, the apparent *Su(f)* gene product, would favor the use of stronger poly A signals of the *hsp82* pre-mRNA rather than those of the weaker site in the upstream LTR of the inserted transposon, thus restoring wild-type *hsp82* mRNA production. However, wild-type levels of the 77-kDa subunit that result in high levels of the CstF complex allow use of the upstream poly A site that produces the truncated mRNA which thereby precludes use of the normal downstream site in the *hsp82* pre-mRNA.

The cloning, sequencing, and expression of each of the CstF subunits has clarified the organization of the heterotrimer and its specific binding site on pre-mRNA. The similarity of the 77-kDa subunit to the *Drosophila Su(f)* gene product suggested CstF could be a regulator of poly A site choice. Recent evidence has confirmed such a role for CstF 64 in the mechanism of poly A site choice that characterizes alternative processing of the immunoglobin heavy chain mRNA (*165*). (See Section VIII.)

C. The CPSF Complex

1. CLONES

The four predominant proteins separated from CPSF during electrophoresis in denaturing gels (*103, 119*) have now been confirmed as authentic subunits by a series of co-immunoprecipitation experiments similar to those used to characterize the CstF complex. Monoclonal antibody specific for each of the large subunits immunoprecipitated the entire CPSF complex, including the controversial 30-kDa subunit that was precipitated with an antibody for the 160-kDa protein (166). These antibodies were critical for monitoring the recombinant proteins expressed in bacteria transfected with cDNA clones of each of the three large subunits and for establishing their specific antigenic properties. For example, monoclonal antibodies for the native 73-kDa subunit immunoprecipitated the recombinant protein expressed in bacteria from a cDNA for the 73-kDa protein, while polyclonal antibodies raised with the recombinant 73-kDa protein recognized the native 73-kDa subunit in a purified CPSF complex (167).

cDNA clones described for the 160-kDa (132, 166), the 100-kDa (168), the 73-kDa (167), and the 30-kDa (169) subunits of CPSF have open reading frames with calculated molecular weights of 161.8, 88.5, 77.5, and 31-kDa, respectively. The originally designated size of the 100-kDa subunit now appears to be an overestimate.

2. SEQUENCE MOTIFS AND HOMOLOGS

a. The 160-kDa Subunit. Sequences derived from clones of bovine (166) and human cDNAs (132) containing an intact open reading frame are 96% identical. An identifiable bipartite nuclear localization signal would account for the localization of the 160-kDa protein in the nucleoplasm of HeLa cells analyzed by indirect immunofluorescence (132). Surprisingly, a recognizable RNA-binding motif was lacking, since much evidence already cited here showed that the 160-kDa subunit recognizes only AAUAAA-containing pre-mRNAs to which it can be crosslinked with UV irradiation (166). This suggested that an unrecognized variant of the RNP motif exists. Functional analysis of domains of CPSF, however, have yet to be reported.

The 160-kDa protein had no strong similarities to known proteins including a variety of peptides derived from the 30-kDa subunit that coprecipitates in the CPSF complex with antibodies for either the 160- or the 100-kDa subunit which eliminated the possibility that the 30-kDa subunit was a degradation product of the 160-kDa protein (166).

b. The 100-kDa (88.5-kDa) Subunit. A single cDNA cloned from a bovine heart library apparently has a complete open reading frame of 88.5 kDa as its *in vitro* translation product comigrated with the 100-kDa subunit of CPSF during electrophoresis (168). This coincidence of mobilities suggests an anomalous electrophoretic behavior of the protein originally designated as 100 kDa.

Polyclonal antibodies raised against a large C-terminal region of the recombinant 100-kDa protein were able to deplete CPSF activity from HeLa extracts and to detect the 100-kDa subunit of CPSF on immunoblots. These antibodies also inhibited polyadenylation, but not cleavage, while a monoclonal antibody for the 100-kDa protein inhibited cleavage, but not polyadenylation, suggesting separate functional domains of the 100-kDa protein for each reaction for which these antibodies would be useful probes (168). The 100-kDa subunit had no recognized motifs or similarities to proteins in the databases except for a significant identity (23%) and similarity (49%) to its companion subunit, the 73-kDa protein of CPSF.

c. The 73-kDa Subunit. A cDNA clone expressed in *E. coli* was identified with monoclonal antibodies for the 73-kDa subunit as a polypeptide with an ORF of 684 amino acids and a predicted mass of 77.5 kDa. A C-terminal recombinant fragment was used to raise polyclonal antibodies that recognized the 73-kDa subunit of a CPSF functional complex on immunoblots and was able to immunoprecipitate a CPSF complex that had been purified by immunoprecipitation with a monoclonal antibody for the 100-kDa subunit (167).

The sequence similarities of this recombinant protein to the 100-kDa subunit have already been noted, but it was a similarity to a gene of *S. cereviscae* that was particularly striking. This gene designated earlier as *BRR5* (170) and more recently as *YSH1* (171) codes for a 779-amino acid polypeptide that is essential for yeast viability and shares a 53% identity and a 73% similarity to the 73-kDa subunit of CPSF. Disruption of this yeast gene results in 3′-end-processing defects seen as an accumulation of elongated pre-mRNAs both *in vivo* and in yeast extracts active for cleavage/polyadenylation (171). Antibodies raised with a C-terminal recombinant fragment of the yeast Ysh1 protein were used to localize the Ysh1 protein in the PF1 fraction of yeast extracts (167). PF1, a polyadenylation factor, was one of three yeast fractions that include cleavage factors CF1 and CF11 that in combination with poly A polymerase could reconstitute cleavage/polyadenylation *in vitro* (172). A 150-kDa protein in the cleavage factor CF1 fraction of *S. cerevisciae,* designated as Cft1, also shares sequences with the 160-kDa subunit of the mammalian CPSF complex (173). The finding of yeast homologs for two of the subunits of CPSF was somewhat unexpected, as CPSF requires the AAUAAA signal that is not essential for 3′ end formation in yeast. These findings suggested a closer evolutionary relationship of mammalian cleavage/polyadenylation systems to *S. cerevisciae* than had been forecast earlier primarily from differences in poly A signal sequences.

d. 30-kDa Subunit. The earlier uncertainty of a role for a 30-kDa protein as a functional component of the CPSF cleavage and polyadenylation complex was eliminated by the isolation of a cDNA clone that expressed a 30-kDa protein in *E. coli* (169). Its essential role in mRNA 3′ end formation was established when it restored cleavage and polyadenylating activity to cell extracts and to CPSF preparations immunodepleted of the 30-kDa protein. The cDNA codes for a 31-kDa protein with five consecutive zinc-binding repeats and a carboxy-terminal "zinc knuckle." These structures have strong homology to similar proteins in *Drosophila, C. elegans,* and yeast. An essential yeast gene, *YTH1,* is 40%

identical to the 30-kDa subunit of CPSF, and its expressed protein is present in the PF1 polyadenylation factor of yeast (169).

The CPSF30 protein has recently been implicated in the export of mRNAs from the nucleus (208). Inactivation of CPSF30 by a protein of an infecting virus markedly reduces cellular mRNA nuclear export but not export of viral mRNAs whose 3′ ends are processed by a different mechanism. (See Section VIII.)

D. The Poly A-Binding Protein, PABII (PAB2)

1. CLONES

The sequences of seven different cDNAs cloned from several bovine tissue cDNA libraries were used to reconstruct an open-reading frame identical to a PABII cDNA transcribed from poly A(+) mRNA. This cDNA amplified by PCR with PABII primers was expressed in *E. coli* as a recombinant protein with the properties of PABII, most notably by conferring processivity on poly A tail elongation by poly A polymerase (174).

The ORF of this cDNA specified a 32-kDa protein rather than the 50 kDa assigned to PABII on the basis of electrophoretic mobility and sedimentation velocity (136). A reconsideration of the properties of the native PABII protein that included mass spectrometry sequencing and analysis of cyanogen bromide-cleaved fragments confirmed the lower molecular weight of 32 kDa (174). The higher values originally reported were ascribed to a dimerization of PABII during velocity sedimentation at concentrations used in earlier studies. In this reappraisal, a value of close to 33 kDa was obtained when sedimentation was measured in a denaturing solvent. A largely sequenced genomic clone of 17 kB contained each of the seven exons comprising the ORF and the 3′ and 5′ UTRs of the cDNA and the six introns (174).

2. SEQUENCE MOTIFS AND HOMOLOGS

The PABII sequence contains a close match to an RNP-type RNA-binding domain with RNP1 and RNP2 consensus sequences. Although both immunofluorescence and electron microscopy have shown PABII to localize almost exclusively in the nucleus (137), no consensus NLS signals were seen in the cDNA sequence. Nor did the cDNA resemble other known proteins, although expressed sequence tags homologous to PABII have been mapped to two different human chromosomes, which may represent additional genes or pseudo genes. The highly basic C-terminal domain of the ORF contains many arginines, but many appear to be modified as half of them remained undetected in a sequenced C-terminal peptide. Although dimethyl arginine is a likely candidate, it was not detected (174).

PABII, as well as CPSF 30-kDa protein, bind the NS1 influenza viral protein that blocks nuclear export of polyadenylated mRNA. The nuclear accumulation of mRNAs with short poly A tails that characterizes this inhibition results from the inactivation of PABII when bound to the effector domain of NS1. A mechanism for this regulation of nuclear export of mRNA is described under Section VIII.

E. $CF1_m$ and $CF11_m$ Cleavage Factors

Although $CF1_m$ and $CF11_m$ were recognized early on as essential mammalian cleavage factors in fractions separated from nuclear extracts, their purification had been delayed undoubtably because of complex multiple interactions among subunits of $CF1_m$ about to be described. Little has been reported on the structure of the $CF11_m$ factor at this time. Purified $CF1_m$, surprisingly, contained major proteins of 68, 59, and a minor one of 72 kDa (*175, 176*). (Table I). The three major proteins could be UV crosslinked to a pre-mRNA cleavage substrate at an unknown site that did not include the CPSF-binding site that surrounds the AAUAAA signal. The p68 and p25 proteins were cloned and expressed as functional proteins that together were sufficient to restore normal cleavage in the presence of purified CPSF, CstF, and PAP and $CF11_m$ (*176*). This activity of the p65:p25 heterodimer suggested that p72 and p59 subunits of $CF1_m$ were alternative forms of p65, since p59 shares extensive homology with the N- and C-terminal sequences of p65. Sequence information for the less abundant p72 subunit has been limited by lack of material.

The sequence of 68p has several unusual motifs including an RNA-binding domain of the RNP type in the N-terminal domain that would account for UV crosslinking of 68p to pre-mRNA. The central domain of 200 amino acids is proline rich (47%) with two to five prolines interspersed between noncharged residues, usually glycines. Although not previously seen in mRNA-processing factors, such motifs are known to promote protein:protein interactions. The C-terminal domain includes a 60-amino acid sequence of arginines alternating with either glutamate or aspartate along with several SR dipeptides. Again such domains are predicted to be involved in protein:protein interactions.

The $CF1_m$ heterodimer was shown to accelerate both the rate of assembly of an active cleavage complex on pre-mRNA and the subsequent cleavage reaction. The authors suggested that the $CF1_m$ heterodimer may resemble those SR proteins such as the U1 snRNP 70-kDa protein that activate a commitment complex to initiate splicing (*176*).

With the exception of $CF11_m$, proteins needed for cleavage/polyadenylation of mammalian mRNAs have been cloned, sequenced, and expressed as recombinant proteins or large recombinant fragments in a few cases. The antibodies raised with these proteins have already provided the crucial reagents for probing the structural organization of CPSF, CstF, and most recently the $CF1_m$ complex. This array of reagents from genes to antibodies will be indispensable for the further exploration of the role of mRNA-processing reactions in the regulation of gene expression.

A current overview of the essential proteins and their many interactions involved in mRNA 3' end formation is modeled in Fig. 3. To compare this to

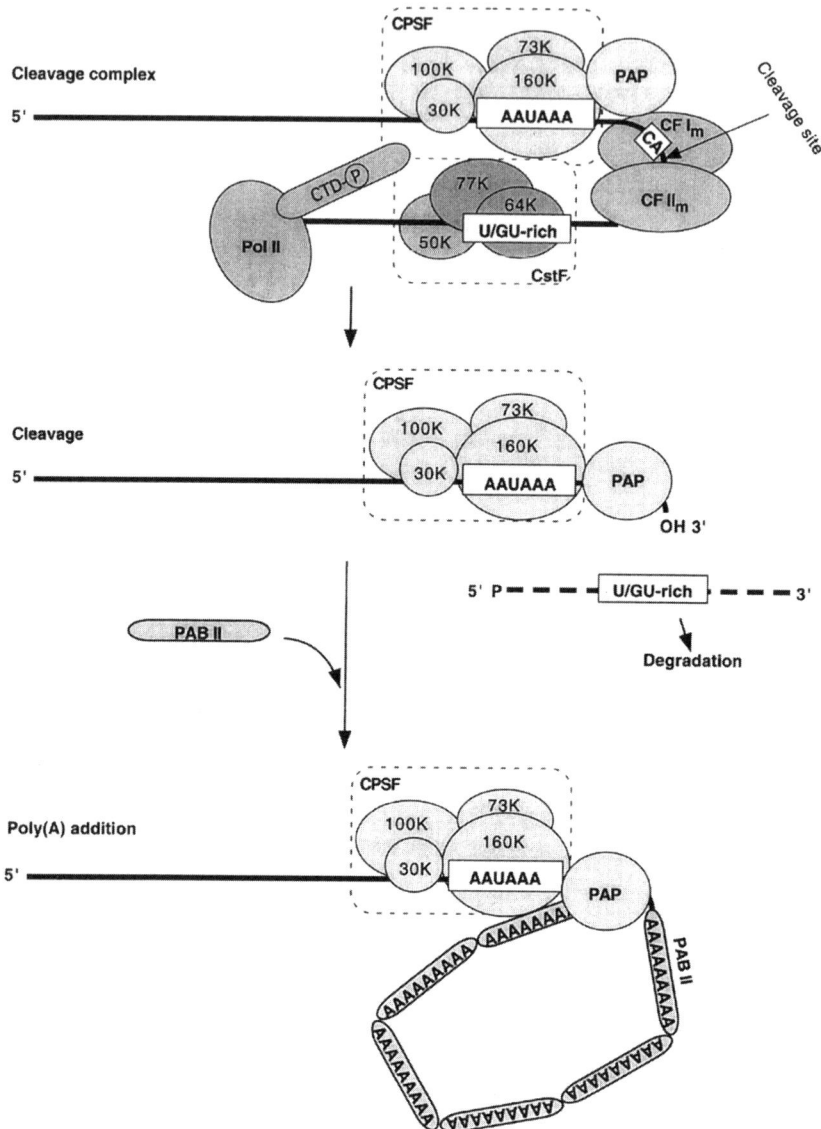

FIG. 3. Schematic representation of the mammalian 3′ end-processing complex. The cleavage complex assembles through a cooperative binding of CPSF at the AAUAAA signal and CstF at the U/GU-rich sequence. CPSF160 interacts directly with CstF77 and PAP. The arrangement of $CF1_m$ and $CF11_m$ is not known. After cleavage, CPSF and PAP remain bound to the cleaved RNA, and the poly A tail is elongated in the presence of PABII. Reprinted from *Microbiol. Mol. Biol. Rev.*, **63**, 405 (1999), J. Zhao, L. Hyman, and C. Moore, with permission from American Society for Microbiology/DC.

the earlier model diagramed in Fig. 2 is not only to be impressed with the experimental achievements of the intervening decade but also to appreciate the value of the simple, but essentially accurate early model that served as the paradigm guiding subsequent experiments. This eventually reduced the initial confusion arising from the multiplicity of proteins required for a simple two-step reaction and led to the recognition that several of these proteins have roles in regulating mRNA levels by altering mRNA 3' end formation and function. Some of these mechanisms are described in the subsequent sections.

VIII. Regulation of Polyadenylation

In view of the surprising complexity of the basic cleavage/polyadenylation reaction, it is now obvious that early efforts to explore the regulation of polyadenylation were hopelessly premature. All that could be studied at the time were alterations in the activity of the polymerase and in the levels of poly A(+) mRNA. Very few cDNA probes were available to quantitate specific mRNAs. Despite such limitations, the effects of hormones, of stages of embryonic development, and of the cell cycle on polyadenylation were all reported in the 1970s (29). It was logical to expect that the presence of poly A or variations in its length would affect mRNA translatability or stability. However, little interpretable data came from such experiments that could be ascribed to a regulatory role (or any role for that matter) for polyadenylation. There was one possible exception that, although it remains unconfirmed, may now be understood in light of the recently described properties of the mammalian PAP (see below).

A. Phosphorylation of Poly A Polymerase

An early *in vivo* study of the effects of hormonal stimulation on PAP activity of the perfused rabbit heart (177) reported that perfusion with either noradrenaline or dibutyryl cyclic AMP, which mediates the effects of noradrenaline, resulted in rapid (less than 5 min) and large (12-fold) increases in PAP activity. When a pulse of labeled ribose was included in the perfusate, the microsomal RNA of the noradrenaline-treated heart contained more label than the RNA from the untreated heart. Essentially all of the increase of label was in the AMP of the ribonucleotides of the hydrolyzed RNA which suggested a correlation between enhanced PAP catalytic activity and synthesis of cytoplasmic poly A. From these *in vivo* results, the authors postulated that the increased PAP activity resulted from its phosphorylation by a cyclic AMP-dependent kinase activated by noradrenaline (177).

Somewhat later, a low level of phosphorylation at serine and threonine residues of PAP recovered from isolated nuclei incubated with ATP was reported (178). The purified PAP incubated *in vitro* with ATP and with a nuclear protein kinase also was phosphorylated (*179*). Both preparations were reported to have increased initial rates of poly A synthesis relative to controls. The relevance of those phosphorylations to the regulation of PAP is now problematic in light of recent structural studies of PAP showing that most, if not all, of the purified mammalian PAPs of that era lacked the large serine/threonine-rich C-terminal domain. Although dispensable for both nonspecific and specific polyadenylation, this region has recently been established as a major site of phosphorylation (*149*). Both cyclic AMP-dependent and cyclin B-cell-cycle-dependent kinase consensus sites, first seen in cloned cDNAs of PAPI and PAPII (*125*), forecast a regulatory role for phosphorylation–dephosphorylation of the C-terminal domain. Evidence for phosphorylation was first detected in forms of PAP that migrated more slowly during electrophoresis than PAPI and PAPII, but which comigrated with them after phosphatase treatment (*144, 145*).

A regulatory role for phosphorylation–dephosphorylation was first suggested by a progressive increase in phosphorylation of PAP during progesterone-induced maturation of *Xenopus* oocytes that was rapidly reversed on fertilization (*145*). The phosphorylation of PAP that occurred during the transition of maturing oocytes from interphase to metaphase ensured that PAP entered mitosis in a hyperphosphorylated state. This suggested that it was likely that phosphorylation–dephosphorylation of PAP was controlled by the cell cycle rather than by some mechanism specific for the regulation of polyadenylation. It was also noted that major changes in PAP catalytic activity or PAP protein were not detected during this oocyte maturation.

Strong evidence linking phosphorylation of PAP to the cell cycle has recently been seen as a marked accumulation of hyperphosphorylated PAP in human cells arrested in mitosis, in contrast to asynchronous cells where several underphosphorylated forms of PAP are present (*180*). Direct linkage to the cell cycle was established by a dependence of this hyperphosphorylation on the maturation/mitosis-promoting factor (MPF), which includes a cyclin B subunit and the cyclin-dependent kinase, p34 *cdc2*. This dependence was established by coinfecting sf21 insect cells with the baculoviruses that included a his-tagged recombinant PAP and two baculoviruses, each expressing one of the MPF subunits. A single hyperphosphorylated recombinant PAP was expressed in such cells, while several underphosphorylated forms of PAP were expressed in cells infected with the recombinant PAP virus alone.

A physiological role for this MPF-dependent phosphorylation was indicated when a similar hyperphosphorylated PAP accumulated in HeLa cells arrested in mitosis (*180*). That this accumulation depended on one or more of the three cyclin-dependent kinase consensus sites (CDK) in the S/T-rich C-terminal domain was clear when cdk^- PAP, in which a serine of each of the three CDK sites were converted to alanine, was expressed in HeLa cells, as several hypophosphorylated forms. The PAP expressed from this cdk^- PAP mRNA was clearly less phosphorylated than PAP expressed from wild-type mRNA injected into Xenopus oocytes, although a significant phosphorylation of cdk^- PAP still occurred showing other sites of PAP are phosphorylated.

That the hyperphosphorylated PAP accumulating in the M phase of the cell cycle is biologically relevant was also indicated by the greatly reduced *in vitro* enzymatic activity of hyperphosphorylated forms that could be restored by phosphatase treatment. Furthermore, both specific and nonspecific polyadenylating activities were reduced, indicating that it is PAP that is altered by phosphorylation rather than its interaction with a component(s) of CPSF. Although the details of the mechanism are unknown, it is clear that PAP activity can be downregulated by an MPF-dependent phosphorylation that coincides with a general shutdown of RNA and protein synthesis during mitosis.

B. Regulation of Poly A Site Choice

The concept of poly A site choice as a regulatory mechanism arose from studies of the processing of the single major late Adenovirus transcription unit briefly described here earlier. A key observation was the identification of five distinct classes of polyadenylated RNA designated from the 5' end of the major late transcript as L1 through L5 that code for specific proteins. The accumulation of poly A(+) RNAs in late infection in each class in fixed molar ratios of 1:2:3:2:2 had pointed to a regulation of RNA processing (*177*) and was supported by the observation that only L1 and L2 transcripts were produced in early infection (*178*), apparently the result of an early termination of transcription (*179*). The general lack of understanding of the biochemistry of RNA processing coupled with the tendency of multiple processing reactions to produce overlapping size products discouraged continued analysis of this large adenoviral transcription unit. However, the advent of rapid cloning and sequencing of cDNAs soon revealed the existence of multiple poly A sites in several cellular mRNAs that underwent alternative processing. Most of these alternative forms of specific mRNAs differed in the length of their 3' untranslated regions but retained identical coding regions.

C. Processing Multiple Poly A Sites in 3' Noncoding Exons

1. Dihydrofolate Reductase

Among the earliest of such mRNAs to be characterized was dihydrofolate reductase, (DHFR) (*48*), for which 11 poly A sites in the 3' untranslated region could be used, although the 4 sites closest to the coding region were used most often (*49*). The sequence of the DHFR gene showed all 11 sites were in a single terminal exon containing the 3' UT. Early studies did not find major differences in the distribution of the most abundant forms within different tissues (*184*), nor was there evidence for their differential stability, at least in cytoplasm where their distribution was similar to that in the nucleus (*185*). It had been observed earlier that the DHFR protein increased greatly after a shift from resting to growing cells (*186*). This was correlated with increases of DHFR mRNA that did not involve a major increase in its rate of transcription (*187, 188*). Strong evidence that the increase was due to post-transcriptional events involving sequences 3′ of the coding region, presumably at poly A sites, was obtained with clones of Chinese hamster cells stably transformed with DHFR cDNAs differing in their poly A site sequences (*189*). The poly A sites of the DHFR mRNAs were clearly needed to observe the growth-dependent increase of DHFR mRNA, as the increase disappeared when functional poly A sites from other cellular mRNAs replaced DHFR sites. The distribution of mRNAs processed from the four major poly A sites of the cloned DHFR cDNAs was, however, similar in resting and growing cells (*189*).

2. Translation Initiation Factor eIF2α

Although these early studies of multiple DHFR mRNAs with different 3′ ends failed to find significant changes in their distribution in tissues or in different cellular environments, many such mRNAs have since been found for which the distribution of the alternatively processed mRNAs does differ, either among tissues or in cells at different stages of growth or differentiation. A recent compilation of nearly 100 mRNAs expressed from multiple poly A sites in noncoding 3′ exons notes that about one-third of these show such differences (*190*). Seven of those clearly showed tissue differences in translatability or stability of the alternative forms that could be expected to alter levels of the protein expressed. A recent well-documented example is the 4.2- and 1.6-kb mRNAs for the protein synthesis initiation factor, eIF2α, that are processed from two widely separated poly A sites within a single noncoding terminal exon of the mRNA (*191*). These two mRNAs show marked differences in distribution in a variety of human tissues that is similar to their distribution in the same tissues of the mouse. A third eIF2α mRNA of 1.7 kb, processed from a poly A site close to the 1.6-kb site, was expressed only in human and mouse testis, further emphasizing the evolutionary conservation of the distribution patterns. Both mRNAs were efficiently translated *in vivo* and apparently *in vitro* as both were efficiently bound to functioning polyribosomes. However, striking differences in their stability were noted during mitogen-induced T-cell differentiation where the level of the 1.6-kb mRNA fell steadily over several hours while the 4.2-kb mRNA level remained unchanged. This clearly suggests that the marked difference in distribution is not due to differences in poly A site strength but to metabolic changes in the T cells during differentiation that presumably alter factors involved in 3′ end processing. Thus, while it is clear from these data that different forms of mRNA that code for the same protein can be metabolized differently, the significance of such differences for cell function is unknown.

D. Processing Multiple Poly A Sites in Different Coding Exons

A less common, but more interesting, possibility for the regulation of poly A site choice exists if alternative poly A sites are in different coding exons so that choice can result in the expression of different proteins with distinct biological functions. In 1980, two such mRNAs were described that have remained among the most studied models for the regulation of poly A site choice and that also involve alternative splicing. These are the mRNAs for the hormone, calcitonin (192), and for the immunoglobulin (Ig) heavy chain proteins (193) that are of particular interest, as the use of a poly A site determines the structure of the protein expressed which in each case generates proteins with very different cellular functions.

1. CALCITONIN/CGRP PRE-mRNA

The mRNA resulting from the use of an upstream poly A site of the calcitonin gene transcript codes for the circulating Ca^{2+}-activated calcitonin polypeptide that is produced in thyroid and other tissues, but not in neuronal tissues, while a larger mRNA with two additional downstream exons is processed at a poly A site further downstream in neuronal tissue (Fig. 4). The larger protein expressed from this mRNA is the calcitonin gene-related peptide (CGRP) which has several neurotrophic activities. Its restricted tissue expression suggested that transacting factors specific to neuronal tissues regulate this alternative processing, while the calcitonin mRNA of most other tissues is processed by a default pathway (194).

a. Exon Skipping Model. The model favored at that time predicted that a neuronal specific factor(s) commits calcitonin/CGRP pre-mRNA to a splicing pathway that skips exon 4 bearing the upstream poly A site, and splices exon 3 to exon 5 followed by exon 6 which contains the downstream poly A site from which CGRP mRNA is processed. This model emphasized alternative splicing via exon skipping rather than competition between poly A sites as the basis for the differential processing of the calcitonin/CGRP transcript. Although the molecular basis for the skipping of exon 4 in neuronal cells is unclear, surprising new findings have revealed a novel regulatory mechanism for the CT/CGRP-processing pathway in nonneuronal cells that uses an intron enhancer downstream of exon 4 to activate cleavage at the exon 4 poly A site to enhance the formation of the CT mRNA.

b. An Intron Enhancer of Poly A Site Choice. The intronic enhancer 168 nt downstream of the poly A site of exon 4 was shown to be required for the inclusion of exon 4 in calcitonin mRNA *in vivo* (195). A novel sequence element in the middle of the 127-nt enhancer that is critical for its function includes a consensus 3' splice site followed immediately by a conserved 5' splice site.

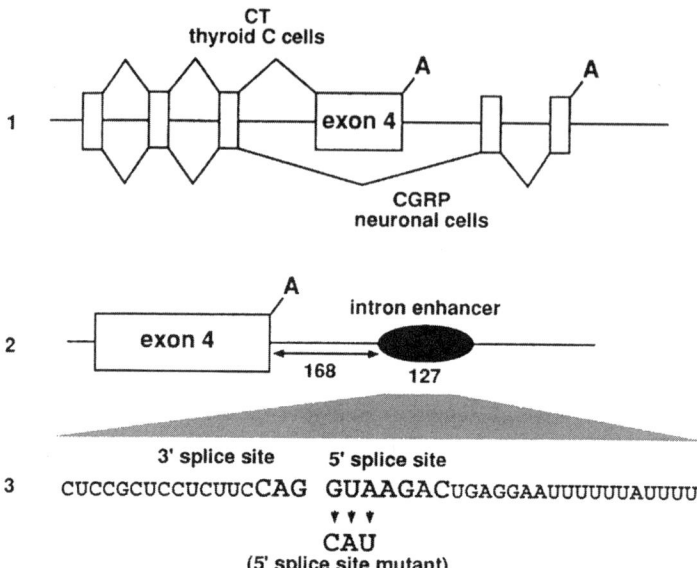

FIG. 4. *In vivo* recognition of exon 4 requires a pseudo 5′ splice site within a downstream intron. Line 1, A diagram of the *CT/CGRP* gene showing alternative processing of the pre-mRNA in thyroid and neuronal tissue. Line 2, Location of the 127-nt intron enhancer 168 nt downstream of the exon 4 poly A site and 346 nt upstream of exon 5. Line 3, Core sequence of the enhancer and a 5′ splice site mutant. Reprinted from *Genes and Development* **10**, 208 (1996), H. Lou, R. F. Gagel, and S. M. Berget, with permission from Cold Spring Harbor Laboratory Press.

This orientation is characteristic of splice sites surrounding internal exons, but oddly, in this case, the internal exon sequences are missing, making it unlikely that these sites are used for splicing. In HeLa cells, a point mutation in the pseudo 5′ splice site of this enhancer resulted in the skipping of exon 4 and the splicing of exon 3 to exon 5 the pathway normally restricted to neuronal cells (*195*). This suggested that the enhancer activates either the 3′ splice site or the 3′ poly A site of exon 4, both known to be weak sites, to favor its inclusion into calcitonin mRNA. The need for the pseudo 5′ splice site for enhancer activity suggested that components that recognize 5′ splice sites, such as U1SnRNPs might participate in exon 4 inclusion. An *in vitro* poly A site cleavage assay using a modified substrate containing the 3′ half of exon 4 and the downstream intron enhancer showed that mutations of either the enhancer pseudo 5′ splice site or the adjacent polypyrimidine tract reduced cleavage significantly at the exon 4 poly A site (*196*).

The participation of components that recognize standard splice sites was established using the usual techniques for observing effects of their depletion or inactivation on enhancer-activated cleavage and their UV crosslinking to the enhancer. The U1RNA of the U1SnRNP was essential as enhancer-activity was abolished by an oligo nucleotide directed RNase cleavage of the 5' end of U1RNA. U1RNA also protected the 5' splice site of the enhancer from RNaseH inactivation. The splicing factor ASF/SF2 that binds 5' splices could also be UV crosslinked to the wild-type enhancer, but not to one with mutations in the 5' splice site. However, U2AF and U2SnRNA that bind sequences at 3' splice sites including the branch point could not be crosslinked to the enhancer. However, an hnRNP polypyrimidine tract protein (PTB) that binds pyrimidine tracts, most notably those upstream of 3' splice sites, was UV crosslinked to the wild type but not to the enhancer with a mutated 5' splice site. Competition for PTB binding with oligonucleotides optimal for enhancer PTB-binding sites decreased PTB binding and reduced polyadenylation of exon 4 *in vitro*. Competition with oligoribonucleotides optimal for U2AF binding to polypyrimidine tracts at 3' splice sites had no effect on either PTB binding to the enhancer or exon 4 inclusion, emphasizing the specificity of PTB enhancer binding.

That PTB positively regulated inclusion of exon 4 was unexpected, as PTB had been reported to promote the exclusion of several alternatively spliced exons presumably by binding to the polypyrimidine tract at 3' splice sites, thereby displacing core-splicing factors such as U2AF (see Ref. *197*, for example). In the case of exon 4, this positive role for PTB-stimulated exon 4 inclusion in CT mRNA was established by cotransfecting both nonneuronal and neuronal cells with a CT/CGRP minigene and a plasmid expressing the PTB gene (*197*). In each cell type, increased levels of PTB correlated with enhanced inclusion of exon 4 into CT mRNA. Furthermore, the increase required the wild-type pyrimidine tract of the enhancer since replacement with a polypyrimidine tract optimal for U2AF^{65} protein abolished enhancement and polyadenylation *in vitro*.

This dual function of PTB as both a negative and a positive regulator of exon 4 inclusion in CT mRNA was depicted in a model (Fig. 5). Here PTB is a dimer in which one subunit binds to the polypyrimidine tract of the enhancer to suppress splicing, while the other binds to a pyrimidine tract 150 nt upstream of the poly A site of exon 4 (*197*). This could bring enhancer complexes with their U1SnRNP and SR protein(s) into contact with the cleavage/polyadenylation complexes to enhance processing at the poly A site of exon 4. Although the nature of these interactions and the function of PTB at the poly A site are unknown, PTB has emerged as a key regulator of alternative exon splicing that in this case is positively mediated by binding an enhancer within an intron downstream of the exon. This discovery of intron-enhancer-mediated regulation of poly A site choice enlarges the number of mechanisms available to cells for alternative processing of pre-mRNAs. The frequency of its deployment among primary transcripts is not yet known.

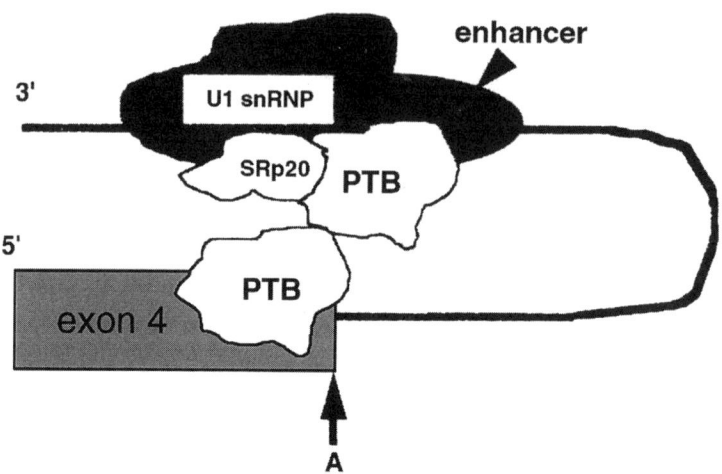

FIG. 5. A model depicting the role of PTB in the intron enhancer-mediated facilitated polyadenylation of CT exon 4. Reprinted from *Mol. Cell. Biol.* **19,** 78 (1999), H. Lou, D. M. Helfman, R. F. Gagel, and S. M. Berget, with permission from American Society for Microbiology/DC.

2. THE IMMUNOGLOBIN HEAVY-CHAIN PRE-mRNA

Alternative processing generates two different immunoglobulin heavy-chain mRNAs in B cells. During B-cell differentiation into antibody-secreting plasma cells, the level of the mRNA processed from the upstream poly A site of the Ig pre-mRNA heavy chain that expresses the secreted protein increases greatly (Fig. 6). The mRNA processed from the downstream poly A site includes two additional exons as well as the 3' end of the last constant region exon. The other upstream exons of the two mRNAs are identical. The additional exons of this mRNA result in the expression of an Ig heavy chain with C-terminal sequences that allow binding to the plasma membrane, while the shorter Ig heavy chains expressed from the Ig pre-mRNA processed at the upstream poly A site are secreted in plasma cells.

a. Competition for Poly A Sites. A consensus has now been reached that the processing of the Ig heavy-chain transcription unit of both the μ- and the γ-Ig heavy-chain genes of the mouse is regulated primarily at the level of poly A site choice. Over a decade of study was needed to restrict, but not necessarily eliminate, other possible mechanisms that might control this switch from about equal amounts of membrane and secreted forms of Ig mRNAs in resting cells from 50- to 100-fold increases in the secreted form of Ig mRNA of differentiated B cells or plasma cells. The experimental evidence that minimizes but does not necessarily eliminate other mechanisms for this shift, such as

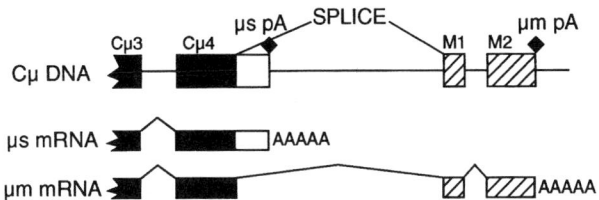

FIG. 6. Alternative processing pathways of the immunoglobulin heavy-chain pre-mRNA that produce either the secreted or the membrane forms of the *Ig* heavy-chain mRNA. 3′ End processing at the promoter proximal μs poly A site eliminates the splicing reaction, while splicing of the Cμ4 exon to the downstream M1 exon precludes use of the Cμ4 exon μs poly A site. Both the mouse *Ig-μ* and the *Ig-γ* heavy-chain gene transcripts undergo similar alternative processing. Reprinted from "Handbook of B and T Lymphocytes" (E. C. Snow, ed.), p. 321. M. L. Peterson, RNA Processing and Expression of Immunoglobulin Genes (1994) with permission from Academic Press, San Diego, CA.

altered rates of transcription initiation, changes in transcription termination, or in either the efficiency of splicing or the mRNA turnover, is well-documented in a comprehensive review (*198*).

A central feature of this regulation is the potential competition that exists between splicing of the last constant region exon Cμ4 to exon M1 and a cleavage at the proximal poly A site or secretory site (designated a μs or γs for the μ- and γ-*Ig* genes) in this same constant region exon (Fig. 6). These are mutually exclusive reactions, since splicing of this exon to exon M1 eliminates the secreted poly A site, while cleavage at this upstream site eliminates the downstream acceptor splice site along with the membrane poly A site. Since splicing and 3′ polyadenylation can occur simultaneously on the same transcript, the increase of the secreted form of Ig mRNA during B-cell differentiations could result either from a reduced efficiency of splicing that allows the secretory site to persist or from a more efficient cleavage at the secretory site that would eliminate the splicing reaction and the competing downstream membrane poly A site.

Studies over a number of years failed to detect any major differences in the efficiency of the 3′ end-splicing reaction between B cells and plasma cells or in their tumor counterparts, lymphomas and myeloma cells (see Ref. *198*). However, the alternative possibility of an increased efficiency of use of the poly A_{sec} site in plasma cells has recently received support from several sources. Experiments leading up to these findings required development of the concept of poly A site strength that underlies the mechanism of poly A site choice. One site is often used more than the other when two different sites are present in a single RNA (*199*), while two identical sites in a single transcript usually result in a greater use of the upstream promoter proximal site (*200*). However, this inherent advantage of promoter proximity can be overcome by a stronger downstream poly A site. Experiments that switched the location of the two competing poly A sites of the

Ig heavy-chain transcript have designated the downstream membrane site as stronger than the secretory poly A site for both μ (201) and γ genes (202).

b. Stability of Complexes at Different Poly A Sites. It has already been noted here that poly A site strength correlates with the stability of polyadenylation complexes found on the RNA. A point mutation within a downstream element that reduced *in vitro* specific cleavage also greatly destabilized the polyadenylation complex formed with this RNA, leading the authors to conclude that stability of the polyadenylation complex determines the efficiency of a poly A site (117). They further predicted that differential complex stability provides the basis for choice between two competing poly A sites, such as those in the pre-mRNA for Ig heavy-chain mRNA. The site that forms the most stable complex will be favored under conditions where components of the CstF or CPSF complexes are limited. When in excess, a weak site will compete more effectively with a strong site. Evidence for this prediction was obtained by assaying for the CstF complex in cells infected with adenovirus (203). The switch in mRNA processing from the adenovirus L1 mRNA poly A site in early infection to the L3 mRNA poly A site later that had been reported many years earlier (181) was accompanied by a decrease in the level of CstF complex in late infection. According to the prediction this would define the L3 poly A site as the stronger site that could be used at reduced levels of CstF while the weaker L1 site could not. As predicted the L3 mRNA formed a more stable complex *in vitro* with CstF and CPSF than did the L1 mRNA. In a follow-up study (204), the complexes on heavy-chain IgM pre-mRNAs with either μs or μm were equally stable in HeLa nuclear extracts, but in lymphoma nuclear extracts, RNA with the μs site was far less stable than the complex associated with the RNA with the μm site. The disappearance of this instability when purified CstF and CPSF replaced the B-cell extract suggested an inhibitor of complex formation had been removed. Evidence for an inhibitor was obtained when a partially purified fraction from the B-cell extract destabilized the μs RNA complex but had no effect on the μm RNA complex. It was proposed that resting B cells have an inhibitor that blocks access of polyadenylation factors to the μs but not to the μm site. Upon B-cell differentiation, this inhibitor either disappears or is displaced by components of polyadenylation that stabilize an active processing complex at the μs site.

c. Role of CstF64 Levels in Poly A Site Choice. Evidence that changes in levels of cleavage/polyadenylation factors may account for the increased use of the weaker secretory poly A site of the IgM heavy-chain pre-mRNA during B-cell differentiation was soon reported by others. In one case, the efficiency of UV crosslinking of the 64-kDa subunit of CstF to IgM pre-mRNA heavy chain was 5- to 10-fold greater in nuclear extracts from several plasmacytoma cell lines than from extracts of early-stage B-cell extracts (205). However, somewhat unexpected was the finding that amounts of CstF-64 and its phosphorylated forms were similar in early- and late-stage B cells. The authors proposed an activator protein in late-stage cells that promotes an increase in RNA-binding affinity of the CstF complex to allow weaker poly A sites to compete more effectively with stronger sites. Others suggested that the failure to detect differences in CstF-64 levels of early and late B cells reflected the transformed state of both types of cultured B-cell lines where CstF-64 levels were already derepressed (*165*).

That levels of CstF-64 are a key factor in the increased expression of the secreted form of the IgM heavy-chain characteristic of differentiated B cells was shown more directly when mitogen-induced differentiation of primary mouse spleen B cells showed large increases in CstF-64, while the CstF-77 subunit of CstF and several SR proteins involved in splicing remained unchanged (*165*). The secreted form of the IgM heavy-chain mRNA also exceeded that of the membrane form by 10-fold as is characteristic of differentiated B cells *in vivo*.

These effects of the induction of cell proliferation on CstF-64 levels are not limited to B cells but have been seen in 3T3 fibroblasts stimulated by serum to enter the S phase from a G_o resting state (*206*). CstF-64 levels increased 5-fold from low levels in resting cells, while levels of other protein subunits of the CstF and CPSF complexes were unchanged from their considerably higher levels in resting cells. It appeared from these studies that CstF-64, an essential core protein for cleavage/polyadenylation, could be playing a special role in gene expression that includes regulation of the IgM heavy-chain switch. Such effects are seen in several chicken B-cell clones that overexpress chicken CstF-64 and have a 10-fold increased output of the secreted form of IgM heavy-chain mRNA compared to cells transformed with CstF-64 cDNAs lacking the RNA-binding domain. Increased levels of CstF-64 were associated with increased levels of the CstF complex (Fig. 7A) as measured by increased levels of CstF-77 within the complex, suggesting that CstF-64 is limiting in resting B cells (*165*). The greater strength of the μm poly A site was also indicated in these experiments when the complex formed on an Ig pre-mRNA with only the μm site was more stable *in vitro* than that formed with a Ig pre-mRNA with only the μs site (*165*).

A reduced expression of CstF-64 was also achieved with these same chicken spleen B cells by replacing the *CstF64* gene with a *CstF-64* transgene with a tetracycline-regulated promoter (*207*). Levels of tetracycline that reduced CstF-64 expression 10-fold not only reduced the IgM heavy chain 20-fold but also shifted the ratio of $mRNA_{sec}$:$mRNA_{mem}$ from about 10 to 1. Reductions in the expression of actin and several other less abundant mRNAs were not seen; neither was transcription of the IgM H-chain gene reduced nor cell viability affected. However, further reduction of these low levels of CstF-64 that arrest these B cells at the G_o/G_1 boundary of the cell cycle caused them to enter apoptosis, evidence that *CstF-64* is needed for cell survival. That B cells can

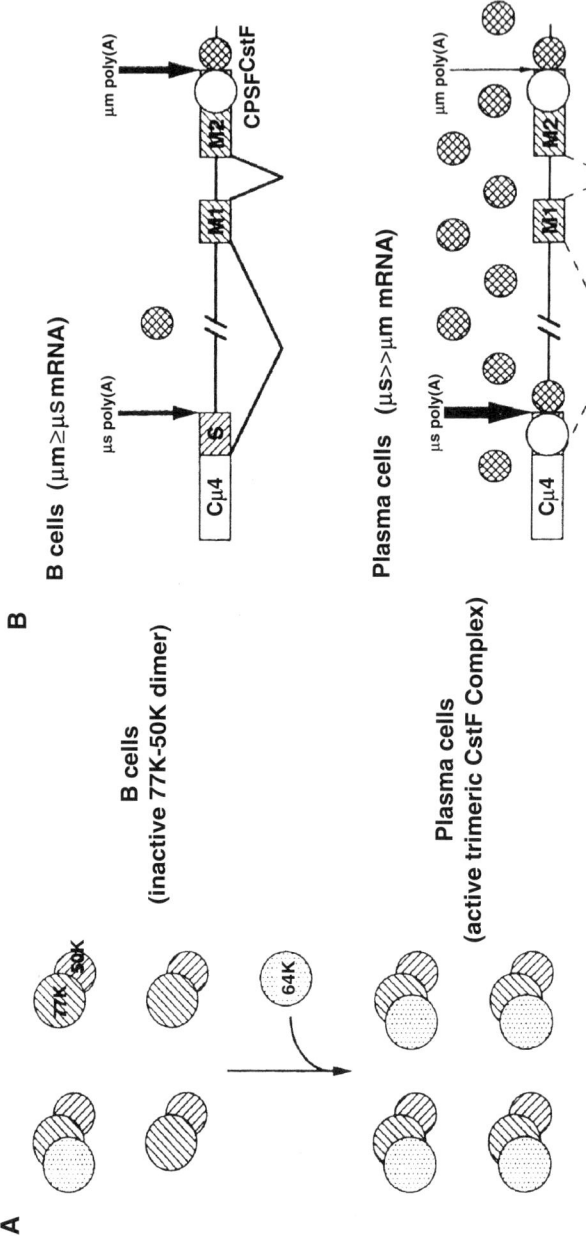

FIG. 7. A model for the regulation of the *IgM-μ* heavy-chain gene by CstF64 protein. (A) CstF 64 is limiting for CstF complex formation. At the final stage of B-cell differentiation (plasma cells) CstF64 expression is increased to increase formation of active CstF complexes (hatched circles). (B) Enhanced CstF levels switches μ gene mRNA expression from the μm to the μs poly A site. In undifferentiated B cells, the stronger μm poly A site competes effectively with the weaker μs site. In plasma cells, higher levels of CstF allow enhanced use of the μs poly A site to give higher levels of μs mRNA and ultimately higher levels of secreted form of the Ig heavy-chain protein. Thickness of arrows reflects the relative use of the two poly A sites. Reprinted from *Cell* **87**, Y. Takagaki, R. L. Seipelt, M. L. Peterson, and J. L. Manley, The polyadenylation factor CstF64 regulates alternative processing of Ig-M heavy chain pre-mRNA during B cell differentiation, p. 941, Copyright (1996), with permission from Elsevier Science.

tolerate levels of CstF-64 either 10-fold higher or lower than normal proliferating B cells with only minor affects on viability or rates of transcription makes CstF-64 an attractive candidate for regulating gene expression at the level of 3′ end formation where differences in poly A site strengths among pre-mRNAs would allow their differential expression.

A simple model (Fig. 7) based on levels of CstF-64 as the primary factor in control of the switch from the membrane-bound to the secreted form of IgM protein has been proposed (165). Here low levels of CstF-64 in resting B cells reduce formation of active CstF complexes, although adequate levels of CstF-77 and CstF-50 are apparently present as inactive dimers. At these low levels, the splicing of exon Cμ4 to the downstream exon M1 competes effectively with the weaker poly A site of exon 4 and allows use of the stronger downstream μm poly A site to produce μm Ig mRNA. Upon differentiation to plasma cells, the increased levels of CstF-64 allow increased cleavage at the weaker μs site which eliminates the competing splicing reaction by removing the downstream acceptor splice site along with the downstream poly A site. The increased cleavage at the promoter proximal site eliminates formation of mRNA$_{mem}$ at the stronger downstream poly A site, thereby increasing the RNA$_{sec}$-to-RNA$_{mem}$ ratio to favor expression of the secreted form of IgM protein.

That increased levels of CstF-64 are sufficient to activate the switch from membrane-bound to secreted IgM during B-cell differentiation has been questioned by experiments in which CstF-64 levels were measured under conditions which allowed the proliferation and differentiation stages of B cells to be examined separately (206). Previous experiments depended on mitogen stimulation of resting B cells that induces both their proliferation and differentiation to assess the role of cleavage/polyadenylation factors such as CstF-64 in the immunoglobulin switch in differentiated B cells. It is clear that the shift from the G$_o$ to S phase will induce multiple changes in cell components with unknown effects on the differentiation process. To clarify this situation, human resting B cells were stimulated to replicate but not to differentiate with a factor, CD40 ligand, secreted by fibroblasts layered beneath the B-cell suspension. B-cell proliferation was accompanied by fivefold increases in CstF-64 protein with only small increases in several other protein subunits of the CstF and CPSF cleavage complexes. However, there was no increase in either the IgM-secreted protein or the ratio of secreted mRNA to membrane mRNA. Thus, the large increase in CstF-64 in these proliferating B cells did not activate a switch to increase IgM protein secretion. However, inclusion of a lymphokine, IL-10, along with the CD40 ligand did induce synthesis of the IgM secreted protein and a threefold shift in the μ_{sec} versus μ_{mem} mRNA ratio. IL-10 in the absence of the CD40 ligand had no effect on CstF-64 levels or on either the secretion of IgM proteins or the ratio of secretory to membrane-bound mRNA.

Other evidence that CstF-64 levels alone are insufficient to activate the switch to IgM protein secretion was obtained with a continuously growing B-cell line (SKW6.4) that produces only low levels of IgM-secreted protein. Treatment with lymphokine IL-6 resulted in an eightfold increase in secreted IgM and a sixfold increase in the ratio of secreted to membrane-bound mRNA. However, CstF64 levels remained unchanged (206). These experiments suggest that the mechanism that activates this switch to IgM secretion is more complex than a simple response to increased levels of the protein CstF64. This conclusion is now supported by the recent discovery that the hnRNP F protein competes with CstF64 for binding of sequences downstream of the Ig-secretory poly A site and inhibits the cleavage reaction in nuclear extracts of plasma cells (206a). Protein levels of hnRNP F relative to other related hnRNP proteins were higher in memory B cells than in plasma cells. Overexpression of hnRNP F in plasma cells significantly decreased the ratio of the secreted to membrane heavy-chain mRNA as well as the total mRNA. This resulted in relative levels of the two forms of heavy-chain mRNA characteristic of undifferentiated B cells. Thus hnRNP F would fulfill the role of a negative regulatory factor once predicted to restrict access or activity of a processing complex at the Ig μs poly A site (204).

In summary, it is clear that 2 decades of study of alternative processing of the IgM heavy-chain and the *calcitonin/CTRG* gene transcripts have not only validated the role of poly A site choice as a vehicle for regulating specific gene expression but also revealed distinctive regulatory mechanisms that include competition for specific cleavage–polyadenylation factors and a novel regulatory mechanism that uses a downstream intron enhancer sequence that binds specific proteins to stimulate polyadenylation of an upstream alternative exon to produce CT mRNA in nonneuronal tissues. A clarification of the role of the specific components and their interactions, which are needed to understand each of these elegant regulatory systems for alternative mRNA processing, is now well underway.

E. Down-Regulation of mRNA 3' End Processing by a Viral Protein

Recently both the 30-kDa subunit of CPSF (208) and the poly A-binding protein (PABll) (209) have been linked to the inhibitory effects of influenza viral infection on the nuclear export of polyadenylated cellular mRNA that is mediated by NSl, an abundant viral protein. Separate yeast two-hybrid screens use NSl as bait-selected cDNAs for each protein.

Antibodies for CPSF-30 immunoprecipitated the 30-kDa subunit from extracts of viral-infected cells in a complex with NS1 (208). NS1 antibodies co-immunoprecipitated the 30-kDa subunit along with CPSF-160 and CPSF-73. As none of these CPSF subunits alone were immunoprecipitated by NSl antibodies, it was apparent that the 30-kDa subunit interaction with NSl occurred within the CPSF complex. NS1 was then shown to inhibit not only cleavage and polyadenylation both *in vitro* and *in vivo* but also the binding of CPSF to pre-mRNA *in vitro*. Each of these effects involved the activation domain of NS1 known to be required for its inhibitory effects in virus-infected cells (*210*).

Blocking the access of CPSF to pre-mRNA was a critical feature of a model for the inhibition of mRNA 3' end processing that is induced by NS1 binding to CPSF (*208*). This block would account for the accumulation of unprocessed cellular mRNAs in the nucleus, while allowing viral mRNA nuclear export since viral mRNA 3' ends are processed by a different mechanism involving the viral transcriptase. Thus NS1 enhances access of viral mRNAs to the translation system by reducing nuclear export of cellular mRNA.

This model was soon found to be inadequate to account for some correctly cleaved cellular mRNAs with oligo A tails of ~12 nt that did accumulate in the nucleus of infected cells. It suggested that factors in addition to CPSF that affect poly A tail length are involved in mRNA nuclear export. A likely candidate, PAB11, needed for the processive elongation by PAP of poly A tails, was then selected in a different yeast two-hybrid screen baited with NS1. NS1 was then shown to inhibit poly A tail elongation *in vitro*, while virus-infected cells accumulated B-actin mRNAs with short poly A tails in the nucleus, but only if NS1 retained a functional effector domain. Influenza virus strains with mutations in this NS1 domain allowed the usual accumulation of *B-actin* mRNA with normal-length poly A tails in the cell cytoplasm. Both PAB11 and CPSF bind to the effector domain of NS1, but apparently at nonoverlapping sites as neither competed with the other for NS1 binding. All three proteins form a ternary complex *in vitro*. The simultaneous binding of CPSF and PAB11 to NS1, as well as the binding of CPSF via the 30-kDa subunit to PAB11 in the absence of NS1, led to a more elaborate two-stage model for the inhibition of mRNA nuclear export by NS1 that includes a major role for PAB11-dependent poly A tail elongation (Fig. 8). A role for additional polyadenylating factors had been forecast earlier by evidence that polyadenylating activity and not merely elongated poly A tails was needed for polyadenylated mRNA export from the nucleus (*23*).

The current model for NS1 inhibition of nuclear mRNA export invokes a dual function for NS1 (Fig. 8) in which cleavage of some pre-mRNAs is inhibited by binding of NS1 to the 30-kDa subunit of CPSF which blocks access of CPSF to the pre-mRNA. Such a block was supported by the *in vitro* inhibition of CPSF binding to pre-mRNA in the presence of NS1. Those pre-mRNAs that continue to be correctly cleaved are retained in the nucleus of infected cells, but with short poly A tails, since PAB11 that allows processive polyadenylation by poly A polymerase has also been inactivated by binding to NS1. Such mRNAs apparently cannot be exported from the nucleus. Thus, NS1 provides a fail-safe mechanism for ensuring viral replication by an inactivation of both steps of cellular mRNA

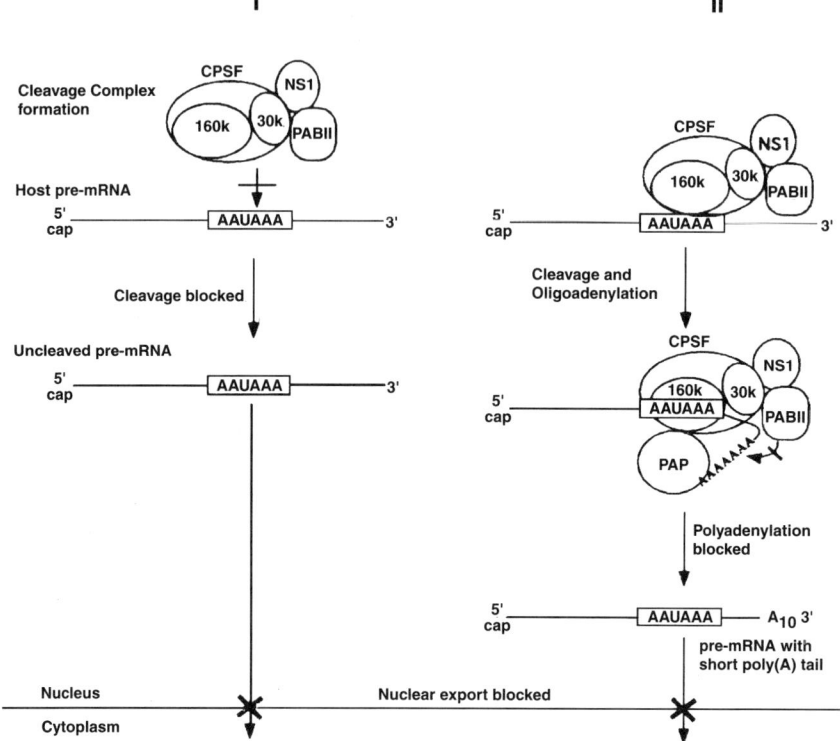

FIG. 8. Proposed two-pronged mechanism by which the NS1 protein of *Influenza* A virus inhibits cellular mRNA 3' end processing in infected cells. Pathway I: Binding NS1 protein (and PAB11) to 30-kDa subunit of CPSF blocks the cleavage of some pre-mRNAs. Uncleaved mRNAs remain in the nucleus. Pathway II: CPSF binding to the AAUAAA sequences of other mRNAs is not blocked, although NS1 and PAB11 are bound to the CPSF30 subunit. A short poly A tail is added to these cleaved pre-mRNAs by PAP in a CPSF-dependent reaction, but elongation is blocked by the binding of NS1 and PAB11. These mRNAs with short poly A tails accumulate in the nucleus of *Influenza* virus-infected cells. Reprinted from *EMBO J.*, **18**, 2273 (1999), Z. Chen, Y. Li, and R. M. Krug, with permission from Oxford University Press.

3' end processing that does not affect viral mRNA polyadenylation and its nuclear export, thereby favoring access of viral mRNAs to the translational system.

The study of the down-regulation of cellular mRNA 3' end formation by an infecting influenza virus has not only uncovered interactions among the core components of mRNA 3' end-processing complexes but also suggested potential reactions and/or sites for their function. Examples are the stimulation of poly A tail elongation by PAB11 that is apparently closely linked to the nuclear export of the mRNA, while the elusive CPSF30 has now been implicated as a potential regulator of the cleavage reaction itself. Thus, viruses which have historically provided fruitful systems for the study of RNA processing since the early recognition of the role of polyadenylation in the processing of the major late transcription unit of adenovirus continue to provide important evidence to the current effort of understanding the role of multiple 3′ end mRNA-processing factors in the regulation of gene expression.

IX. Polyadenylation in Yeast

Early studies of polyadenylation in yeast closely paralleled those of mammalian systems. Soon after poly A tails had been detected in the latter, poly A tails of about 50 nt were found on yeast mRNAs (211, 212). A poly A polymerase was partially purified from S. cerevisae shortly afterwards (213, 214). For the next decade, the lull that followed the early discoveries of polyadenylation in animal cells extended to studies in yeast. However, genetic approaches launched in the late 1970s would eventually prove to be important. These characterized two conditional mutants displaying aberrant poly A(+) mRNA metabolism (215). Since both mutants retained wild-type poly A polymerase activity, attention was deflected from possible defects in mRNA 3′ end formation until many years later when sequences of these genes showed homologies to the subunits of the mammalian CstF cleavage complex (164).

Early evidence that the mechanism of mRNA 3′ end formation might resemble that of mammalian cells came from studies of a yeast cytochrome c mutant (CYC1–512) lacking a 38-bp sequence at the 3′ end of the gene that produced low levels of CYC1 mRNA (216). When mutational analysis failed to detect any single, short regions within this sequence that were essential for correct 3′ end formation such as an AAUAAA hexanucleotide, it was concluded that the deleted sequence functioned in transcription termination. A subsequent analysis of revertants of the CYC1–512 mutant revealed that multiple signals in the 3′UTR could restore normal CYC1 mRNA 3′ end formation (217), a redundancy that continued to impede the analysis of the mechanism of yeast mRNA 3′ end formation.

However, the development of extracts that correctly cleaved and polyadenylated yeast mRNAs clarified the situation in yeast by revealing its similarity to the mechanism already established for mammalian cells (218). These extracts were eventually separated into four distinct fractions, two of which were required for cleavage, another for polyadenylation, and a fourth fraction with poly A polymerase activity (219). The latter which had recently been purified, cloned, and sequenced was 47% identical in sequence to the mammalian PAP (146). However, it could not replace the latter in mammalian nuclear extracts, suggesting that the functional domains of yeast PAP that interact with accessory factors were likely to differ from mammalian factors.

Recently newer genetic approaches such as the yeast two-hybrid system and searches of the sequenced yeast genome for homologs of mammalian polyadenylation factors have accelerated the identification of yeast factors to the point where their number at least equals those needed for mammalian mRNA 3' end formation (220). The continued comparisons of the structure and function of proteins from the two sources will undoubtedly add to the current evidence supporting multiple roles for polyadenylation in the regulation of gene expression.

A. Yeast Poly A Signals

For many years the marked degeneracy and redundance of yeast signals for mRNA 3' end formation confounded attempts to define or localize them. It was even predicted that such signals did not exist except for a general AU richness within the 3' untranslated regions of yeast mRNAs. However, once a larger sample of mRNAs with a variety of 3' UT mutations was analyzed for effects, not only on amounts of correctly processed mRNAs but also on the precise location of poly A sites, did a common motif emerge that in spite of a marked degeneracy does accommodate the sequence requirements (221). This motif includes three separate short-sequence elements upstream of the poly A site: (i) an A-rich positioning element 10 to 30 nt upstream of the poly A site, (ii) an AU-rich efficiency element at a variable distance upstream of the positioning element, and (iii) the poly A site. The order of three elements is important, but the spacing between them can vary. The key experiments that have led to this current view of yeast poly A signals are presented in a well-illustrated recent review (222) and will be described only briefly here. A diagram from this review that summarizes the sequences and properties of those three elements of yeast mRNA is shown in Fig. 9.

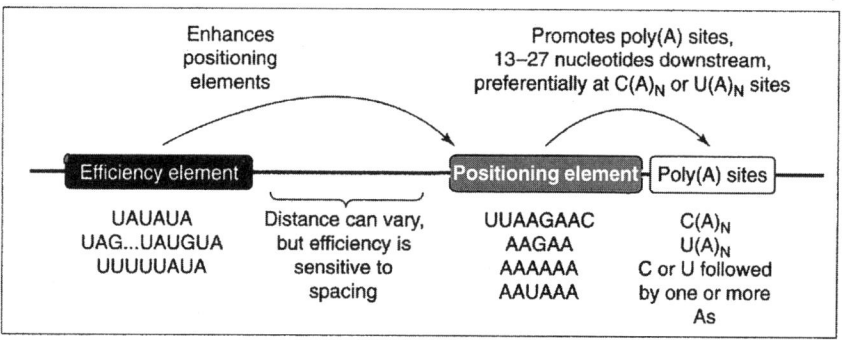

FIG. 9. The sequences and properties of three major elements specifying 3' end formation of *S. cereviscae* mRNAs. Reprinted from *Trends in Biochemical Sciences*, **21**, Z. Guo, and F. Sherman, 3' end-forming signals of yeast mRNA, p. 477, Copyright (1996), with permission from Elsevier Science.

The first sequence to be recognized as essential for 3' end processing of a yeast mRNA was a 38-nt segment 8 nt upstream of the poly A site of the CYC1 mRNA coding for iso1-cytochrome c. A mutant lacking this sequence produced only low levels of elongated polyadenylated mRNAs with multiple 3' ends (216). This deletion lacks a mammalian AAUAAA poly A signal as do the 3' UTRs of more than 50% of yeast mRNAs. Deletions or mutations of the AAUAAA sequence in the 3' UTR of other mRNAs such as the ADH2 mRNA had little effect on the efficiency of 3' end processing (223, 224), although the actual site of the 3' end was not reported in these studies. Deletion of AAUAAA from the 3' end of ADH2 mRNA, however, gave a low level of elongated transcripts that foreshadowed a potential role for AAUAAA as a positioning element (223). It was clear, however, that the mammalian poly A signal was not essential for 3' end formation in yeast. Analysis of several revertants showed two separated sequences within the 38-nt element that functioned together to restore 3' mRNA end processing (225).

A major experimental breakthrough that clarified the basic mechanism of mRNA 3' end formation in yeast was achieved when yeast extracts correctly processed several yeast pre-mRNAs including that for the CYC1 gene. The 3' ends of each mRNA were created by a site-specific endonucleolytic cleavage in the 3' UT of a larger transcript (218, 227). These ends were then polyadenylated with the 60 to 80 AMPs typical of cellular yeast mRNAs. The recognition that the mechanism of 3' end formation of yeast mRNAs resembled that of higher eukaryotes rather than that of bacterial transcription termination provided the rationale for searching these extracts for cleavage and polyadenylation factors (219).

Meanwhile, the advent of site-directed mutagenesis and PCR-amplification techniques prompted not only more detailed sequence analysis of the original CYC1 revertants but also newly created CYC1 deletions and point mutants. These studies (221, 226, 228) established a functional distinction between the two sequence elements originally proposed for CYC1 mRNA (225). Mutations of A-rich sequences 10–30 nt upstream of the poly A site resulted in the disappearance of the normal poly A site with its reappearance at a preferred downstream site (228). These mutations, however, had little effect on the total level of CYC1 mRNA, suggesting that this region near the poly A site was a positioning element, while the site further upstream was an efficiency element required for 3' end formation both *in vitro* and *in vivo* (226). That the two elements function together was clear from the observation that not only lack of a strong efficiency element reduced processing to low levels but also the mRNAs had multiple 3' ends *in vivo* (228). Although both elements are AU rich, they are highly degenerate as evident from comparison of each putative element in the 3' UTR of seven yeast pre-mRNAs (226).

One attempt to define a consensus sequence for the efficiency element was based on an approach that avoids the marked redundancy that has made such definitions so difficult. The method capitalized on the discovery that the 3' UTR of a plant cauliflower mosaic virus mRNA could replace the 3' UTR of the ADH1 yeast gene without altering the levels of ADH1 mRNA (229). Deletion of a 9-nt sequence from this element about 40 nt upstream of the poly A site abolished 3' end processing of the ADH1 pre-mRNA. A saturation mutagenesis of this sequence, TAGTATGTA, showed only the last 6 nt were needed for activity. A consensus sequence, TAYRTA, was derived from the remaining 6 nt of which only the first and fifth T were invariant, while TATATA was the most effective hexamer. The latter sequence had already been recognized among the CYC1–512 revertants (225) and $(AT)_6$ was known to be the efficiency element for the Gal 7 pre-mRNA processing (230).

A similar search for a consensus sequence for the positioning element was carried out on a modified construct of the CYC1–512 mRNA deletion (226). Several constructs, each with a TATATA sequence replacing the normal efficiency element, received one of several A-rich putative positioning elements. Several of these were effective in restoring cleavage at the preferred poly A site. The mammalian AAUAAA sequence was among the most effective, although saturation mutagenesis showed it to be highly degenerate in that all single-nucleotide replacements were tolerated except for substitution of a G for the first A.

The third element showing some degree of sequence conservation is the poly A site. Although its location is specified by the positioning element which in turn requires enhancement by the efficiency element, the 3' ends of yeast mRNAs are either at a C or at a U residue followed by one or more A's. Cleavage occurs 3' to the adenosine residue of this $Py(A)_n$ consensus that receives the poly A tail (231).

The critical question of the *in vivo* relevance of these three elements was answered when they were combined in a construct that directed correct 3' end processing of an mRNA in yeast regardless of surrounding or intervening sequences (232). This was shown by creating a 3' UTR with the optimum sequences for efficiency (UAUAUA), for positioning (AAUAAA), and as a poly A site (CAAA). The elements were separated by suitably placed computer-generated random sequences presumably irrelevant for mRNA 3' end formation. This construct inserted either into the coding region or in the 3' UTR of the *LacZ* gene efficiently directed correct 3' end formation at the poly A sites as specified by these synthetic signals.

In summary, all three yeast sequence elements show a marked degeneracy (Fig. 9) that differentiates them at least in part from those of mammals where the major poly A signal, AAUAAA, is highly conserved. Although the location of the yeast efficiency element differs from that of the "downstream sequence element" of mammalian pre-mRNAs making it uncertain whether they are functionally equivalent, both efficiency elements show marked degeneracy, while the positioning elements of yeast does not display the sequence specificity of the mammalian AAUAAA signal. Nonetheless, it now appears that mRNA 3' end signals of yeast and mammals do not differ as widely as was once assumed.

B. Yeast mRNA 3' End-Processing Factors

The search for processing factors could only begin when active yeast extracts became available (218, 227). The separation of four fractions from such an extract, that when recombined reconstituted correct 3' end formation, provided the basis for the eventual isolation of the individual active components (219). A role for many of the proteins purified from the CF1 and CF11 fractions has now been established by genetic approaches. The proteins expressed from the cloned *RNA14* and *RNA15* genes, for example, were used to raise antibodies that identified analogous proteins in the CF1 complex (233). For simplicity, the individual proteins of the yeast complexes and their cognate genes will be considered separately here as biochemical and genetic approaches were pursued independently, until evidence from new genetic approaches confirmed the identity of the proteins that function in mRNA 3' end formation. Both genetic and biochemical data for mammals and yeast proteins are compared in a recent review (27a).

Extensive purification of the yeast fractions needed for cleavage, CF1 and CF11, showed each to include multiple polypeptides that copurify over many purification steps in roughly equimolar amounts, supporting the view that they exist within complexes (233–235). The fractions assigned to a specific role in polyadenylation include PF2 which was quickly (219) identified as the poly A polymerase. This is apparently the sole activity as recombinant yeast PAP replaced PF2 in a reconstituted reaction (219). Defining the components of the PF1 fraction generated some confusion as three proteins that are the yeast homologs of the mammalian CPSF cleavage complex were purified within a tetrapeptide complex from the CF11 fraction (234), while these same three proteins were recovered in a much larger complex that included PF1 components from the PF1 fraction (235) that will be described below.

1. THE CF1 COMPLEX

The CF1 fraction was separated into two fractions, CF1A and CF1B (Fig. 10). The purified CF1A fraction has four proteins of 76, 70, 50, and 38 kDa. The products of the *RNA14* and *RNA15* genes detected earlier in the CF1 fraction (164) were the 76- and 38-kDa proteins, respectively, of the purified CF1A complex (233). (See section on the two-hybrid system below.) The RNA 15p was the only CF1 subunit to be UV crosslinked to mRNA as might have been expected from its classic RNA-binding domain (233). The 70-kDa protein of CF1A was shown to be the product of the *PCF11* gene (polyadenylation cleavage factor 1) that was selected with a two-hybrid system from a yeast library in which either *RNA14* or *RNA15* proteins served as the bait (237). The properties of the 50-kDa protein were not reported, but a yeast gene, *CLP1*, that codes for the

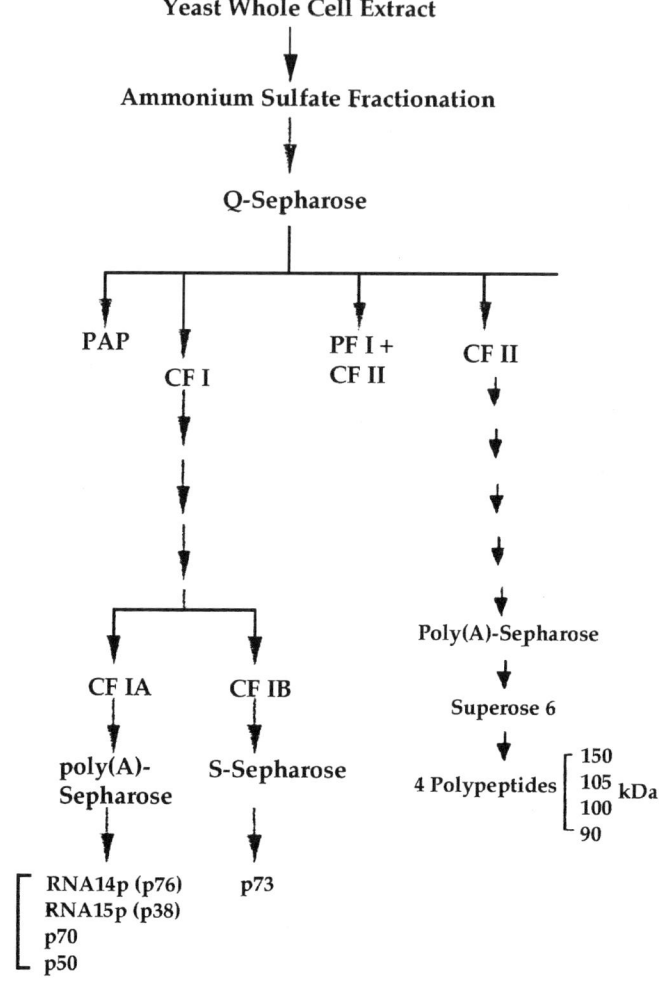

FIG. 10. Purification schemes for yeast cleavage factors CF1 and CF11. A composite diagram reproduced with modification from Refs. 233 and 234. Unlabeled arrows depict individual steps in protocol omitted from original references. Reproduced and modified from *J. Biol. Chem.*, **271**, 27,167 (1996); and *J. Biol. Chem.* **272**, 10,831, with permission from American Society for Biochemistry and Molecular Biology.

50-kDa protein in CF1 has been identified (238). Somewhat later, a fifth protein of 70 kDa was reported to cofractionate with the CF1A complex (252) that was the yeast cytoplasmic poly A-binding protein, Pab1, that functions in translation initiation (18) and in mRNA turnover (251). Pabp1 mutants were first recognized by the elongated tails on mRNA (250) that are also produced in extracts of *Pabp1* mutants and are restored to normal lengths by addition of recombinant Pabp1 (*252*). The weak association of Pabp1 with CF1A and its likely comigration with the 70-kDa subunit of CF1A would account for its apparent absence from some CF1A preparations. A weak association could favor the ready dissociation of Pabp1 from the nuclear CF1A complex for export to the cytoplasm where it functions in translation and mRNA turnover.

The CF1B, which in combination with CF1A reconstituted CF1 activity *in vitro*, showed a single gel band migrating as a 73-kDa protein. Sequenced peptides from it (*239*) were a perfect match to a yeast gene for a heterogeneous ribonuclear protein (HRP1) that was first isolated as a suppressor of a mutation in a gene required for the export of messenger RNA from the cell nucleus (NPL3) (*240*). Other evidence showed that CF1B/hrp1 protein is not essential for pre-mRNA cleavage either *in vivo* or *in vitro* but is needed to ensure accurate cleavage rather than the multiple cleavage sites generated in its absence (*241*). It does so in a concentration-dependent manner that is reminiscent of the function of the hnRnpA1 protein in alternative 5′ splice site selection in mammals (*242*). Both proteins share the property of cycling between nuclear and cytoplasmic compartments.

2. THE CF11 COMPLEX

Four polypeptides of 150, 105, 100, and 90 kDa were separated from a highly purified CF11 complex (*234*) (see Fig. 10). Microsequencing of the 105- and the 100-kDa gel bands showed the 100-kDa protein to be identical to a 87-kDa protein expressed from a previously cloned yeast gene, *BRR5/YSH1*, that is a homolog of the 73-kDa subunit of the mammalian CPSF complex (*167, 171*). The 105-kDa protein band had a calculated molecular mass of 94 kDa and shared a 24% identity and a 43% similarity with the 100-kDa subunit of CPSF. This protein designated as Cft2 was the first yeast factor to depend on intact yeast poly A signals for its UV crosslinking to the pre-mRNA substrate (*234*). The 150-kDa subunit of CF11 was identified as the product of the *Cft* gene that had been identified earlier as a homolog of the mammalian 160-kDa subunit of CPSF (*173*). Immunoblots with antibodies for the Cft1 protein expressed from the cloned *Cft1* gene showed it was the 150-kDa gel band of the CF11 complex (*173*). The clones of the gene for the remaining 90 kDa band showed it was the *PTA1* gene product needed for tRNA processing (*243*).

The CF11 complex appeared to be the yeast equivalent of the mammalian CPSF complex as it shared homologies with its three largest subunits. Others had assigned a yeast protein (Ysh1) a homolog of the 73-kDa subunit of bovine CPSF to the PF1 fraction based on its cofractionation with the fip1 protein, a marker for PF1 (*167*). The immunodepletion of PF1 activity with antibodies for Ysh1 could be restored with a purified PF1 fraction (*167*). The later purification of PF1 found it contained other yeast homologs of bovine CPSF subunits (*235*) that are described below.

3. THE PF1 FRACTION

The first accessory protein for yeast 3′ end mRNA formation was recovered in a two-hybrid screen as a factor interacting with poly A polymerase (fip1). fip1 mutants showed reduced levels of poly A(+) mRNAs with shortened poly A tails (*236*). Immunodepletion of fip1 from cell extracts abolished polyadenylation but reduced cleavage only moderately. The localization of fip1 in PF1 was compatible with the polyadenylating activity ascribed to PF1 for which fip1 has served as a marker. The properties of fip1 resembled that of the mammalian poly A-binding protein (PAB11) but unlike PAB11, fip1 did not bind poly A. However, fip1 did interact with the RNA14 protein of CF1 in a two-hybrid screen, suggesting it could also function in the cleavage reaction complex (*235*).

Such a multiprotein complex with PF1 activity was extensively purified from yeast extracts using as an assay restoration of activity to PF1 extracts that had been inactivated by a mutation in *fip1*. This complex included fip1 and PAP and all four subunits of the CF11 complex as well as the yeast homolog of the CPSF smallest 30-kDa subunit, Yth1, and two unidentified proteins of 58 and 53 kDa that are coded by the yeast PFS1 and PFS2 genes. The composition of this complex was confirmed by its immunoprecipitation from whole yeast cells with antibodies for fip1 or PAP1 (*27a*). The functional significance of the separate CF11 complex that is also part of this larger complex from PF1 is unclear. It could result from *in vitro* manipulations that detach it from the larger complex, or CF11 may exist in both free and bound forms. The cleavage and polyadenylation activity of this PF1 complex that include homologs of the mammalian CPSF complex establishes a functional resemblance of PF1 to CPSF that is also used for cleavage and polyadenylation. Figure 11 summarizes a recent view of the apparent location of the interacting proteins within the two cleavage complexes and the PF1 complex during the polyadenylation step of the yeast mRNA 3′ end.

4. THE PF11 FRACTION IS POLY A POLYMERASE

The sole activity of PF11 was that of PAP that had already been characterized enzymatically (*244*) and cloned and sequenced (*146*). Structure–function analysis by truncation of nonconserved sequences at both ends of PAP identified putative RNA-binding domains near the C terminus and an 18-nt sequence at the N terminus required for specific, but not nonspecific, polyadenylation that designated likely regions for interactions with other proteins (*245*). The mechanism of processivity of polyadenylation was examined later by mapping fip1-binding sites on PAP1 using the yeast two-hybrid system with N- or C-terminal truncations of PAP1 as bait (*246*). fip1-binding sites were found at both ends of PAP1, but the role of fip1 in processivity was restricted to the site at the C terminus. This site overlapped an RNA-binding site (RBS) known to be needed for processivity. Binding of fip1 blocked access of RNA primer to this RBS resulting in an

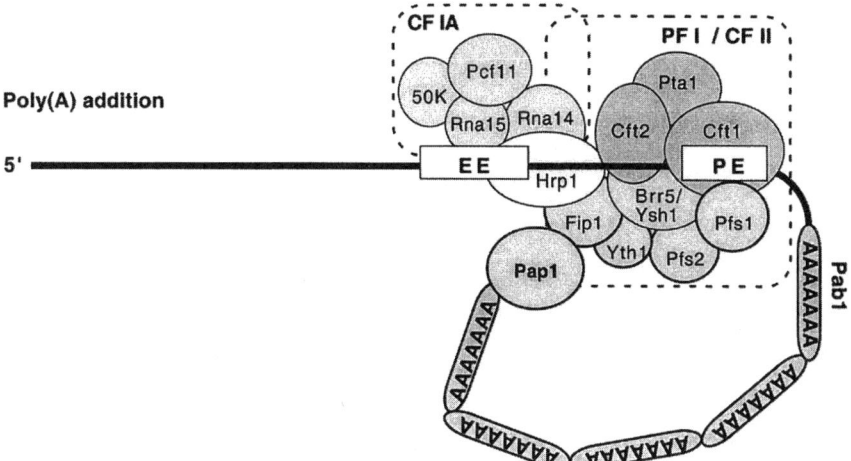

FIG. 11. Schematic representation of yeast 3′ end processing. Depicted is the post-cleavage-processing complex undergoing poly A tail addition. Cleavage requires the CF1A, Hrp1, and CF11, while polyadenylation requires the PF1 complex (see text) and Pap1 and Pab1. EE is the mRNA efficiency element; and PE, the positioning element. Reproduced with modification from *Microbiol. Mol. Biol. Rev.*, **63**, 405 (1999), J. Zhao, L. Hyman, and C. Moore, with permission from American Society for Microbiology/DC.

activity shift from processive to a distributive mode *in vitro*. Thus, fip1 appears to activate this switch by limiting the interaction of PAP1 with RNA primer.

This same study examined effects of various structural features of NTPs and of nucleotides at the 3′ end of the primer on PAP1 activity. Interestingly the unique specificity of PAP1 for adenosine (ATP) was not confined to catalytic site contacts but included interactions with the 3′ ends of RNA primers. From these biochemical data, a simple model of PAP1 was developed that included two functional domains. One in the N-terminal region included a catalytic pocket common to other polymerases, while that of the C-terminal region included an RNA-binding site near the C terminus. Although the RNA primer bound this site, its 3′ end made contacts with the catalytic site in the N-terminal domain.

It was soon possible to compare this model with three-dimensional structures of a yeast PAP1 obtained by X-ray diffraction analysis of a crystal both alone and as a cocrystal complexed with 3′dATP (*253, 337*). PAP1 has three domains that encircle a large cleft. The N-terminal and its adjoining domain form one wall of the cleft along which the catalytic site extends, while the C-terminal domain forms the opposite wall to which the RNA primer binds. Modeling of the dATP contacts at the catalytic site of the cocrystal showed that two molecules of 3′dATP bind per molecule, one for the incoming base and the other presumably for the base at the 3′ end of the primer. Both the bovine and the yeast PAPs will require the cocrystallization of an RNA primer complexed to PAP to provide a full description of the core components of the simple polymerization reaction. Beyond this looms the formidable task of unraveling such a complex with the accessory factors needed for specific RNA recognition and ultimately for its regulation of mRNA levels.

C. Genes for Yeast mRNA 3′ End-Processing Factors

1. CONDITIONAL MUTANTS

Genetic approaches to mRNA 3′ end processing began more than 20 years ago when yeast cells were screened for sensitivity to 3′ deoxyadenosine (cordycepin), an inhibitor of poly A(+) mRNA production in cultured animal cells (215). Two temperature-sensitive independent mutants were selected (cor1 and cor2) that had greatly reduced levels of poly A(+) mRNA at the nonpermissive temperature, but activities of PAP were unaffected at this higher temperature. A rapid loss of poly A(+) RNA during a pulse–chase experiment at the restrictive temperature suggested that an increased rate of mRNA decay was responsible for the mutant phenotype.

Many years later, but well before active yeast mRNA 3′ end-processing extracts were reported, the wild-type genes for the two mutants, now renamed *RNA14* and *RNA15*, were cloned and sequenced (163). Each, present as a single genomic copy, was required for cell viability. *RNA14* and *RNA15* showed no homologies to other sequenced proteins although *RNA15* had a classic RNA-binding domain at the N terminus. Although each was present on a different chromosome, the fact that a plasmid bearing a wild-type *RNA14* gene suppressed the ts phenotype of an *RNA15*-mutated strain and vice versa suggested that their gene products interacted. As rates of transcription and PAP activities of each mutant were relatively unimpaired at the restrictive temperature (163), the reduced levels of poly A(+) and shortened poly A tails again pointed to accelerated mRNA decay as the primary defect, although it was noted that a defect in mRNA processing had not been excluded.

2. NEW GENETIC APPROACHES

The classic genetic selection of conditional mutants with poly A(+) mRNA-processing defects which ultimately led to the isolation of the *RNA14* and *RNA15* genes was eventually augmented by newer approaches such as synergistic lethality assays and the yeast two-hybrid system and by searches of the yeast genome for mammalian homologs that participate in mRNA 3′ end formation.

3. SYNERGISTIC (SYNTHETIC) LETHALITY IDENTIFIES RELATED GENE PRODUCTS

In vivo evidence that RNA 14p and RNA 15p might be involved in pA(+) mRNA formation rather than in its decay came from a genetic test that suggested both might participate with poly A polymerase in 3' end formation. Here, a ts-mutated *PAP* gene (*164*) was introduced into either an RNA14 or an RNA15 ts mutant strain to create a double mutant in which the endogenous *PAP* had been inactivated. These double mutants were no longer viable at the permissive temperature in contrast to cells bearing only one of these mutations. This so-called synergistic lethality is explained as the enhancement of one conditional defect by a second conditional mutation which often means that the two genes participate in a common reaction (*247*). This was confirmed when extracts of RNA14 and RNA15 mutant strains each showed marked defects in cleavage and polyadenylation. Wild-type extracts immunodepleted of RNA14 or RNA15 proteins were also defective in both cleavage and polyadenylation but could be reactivated with a partially purified CF1 fraction that was shown by immunoblotting to contain both *RNA14* and *RNA15* proteins (*164*). Somewhat later, RNA 14p and RNA 15p were identified by immunoprecipitation as the 76- and 38-kDa polypeptides, respectively, of the purified CF1 complex (*233*). An earlier view that CF1 was the yeast counterpart of the mammalian CPSF complex since both are required for cleavage and polyadenylation had to be modified when *RNA14* and *RNA15* genes showed sequence homologies with the 77- and 64-kDa proteins of the mammalian and *Drosophila* CstF complexes (*157*; see Section VII,A,4).

4. GENE ISOLATION WITH YEAST TWO-HYBRID SYSTEM

The two-hybrid system of yeast is based on the separation of a transcription activator protein (the Gal4 protein) into two distinct functional domains (*248*). One domain binds to a specific site in the promoter region, the DNA-binding domain (DB), while the other domain binds to a site that activates transcription at the promoter (the AD domain) to initiate transcription of the downstream gene. A key feature of this system is that transcriptional activation can occur even if the two domains are on separate molecules as long as the two protein domains can be brought into a proper alignment for transcriptional activation of the DNA. The interaction is not limited to the Gal4 protein sequences but may be achieved by any two interacting polypeptides. In practice, the method requires the *in vivo* expression of translational fusion proteins expressed from plasmids bearing a gene for a known protein fused to a segment of the DB domain and a second plasmid with a gene fused to the AD domain. An interaction between the two expressed fusion proteins can initiate transcription of a downstream reporter gene, usually the Z gene for B-galactosidase, whose activity can be measured colorimetrically in cell extracts. This activity provides a measure of the strength of the interaction between the two fusion proteins, although interaction through a third protein is also possible. The method is well suited for cloning genes from a library of DNA fragments fused either to a Gal4 activation domain or to the DNA-binding domain. Success in either case depends on the selection of the appropriate bait expressed as the fusion protein that can interact with a fusion protein expressed from a clone fished out of the plasmid genomic library. Several recent successful isolations of yeast genes involved in mRNA 3′ end formation are discussed here.

a. The FIP1 Gene. The first use of the two-hybrid system for the isolation of mRNA 3′ end-processing genes was limited to gene products that could interact with PAP, the only cloned yeast polyadenylation factor available then as the bait. Yeast cells transformed with a plasmid bearing a *PAP* gene fused to the Gal4 DB domain and a yeast genomic plasmid library fused to the Gal4 AD domain were screened for clones expressing B-galactosidase (236). One such clone expressing a 1.3-kb ORF was used to screen the yeast genome where a 5.4-kb fragment with an ORF of 327 amino acids was found. This gene was designated as a factor interacting with poly A polymerase (*FIP1*). Extracts of conditional mutants of *FIP1* obtained by chemical mutagenesis were deficient in polyadenylation but not in cleavage. Wild-type polyadenylation was restored with a PFl fraction of the yeast extract. The fip lp expressed from the cloned *FIP1* gene formed a 1:1 complex with PAP *in vitro* as predicted from the results of the two-hybrid analysis. Antibodies for fip1p also bound fip 1p in a complex with RNA 14p but not with RNA 15p. In contrast, antibodies for PAP did not immunoprecipitate RNA 14p nor co-immunoprecipitate it with PAP unless fip lp was also present suggesting that fip lp links PAP and RNA 14p (a CFI subunit) within a larger complex that could account for the role of the CF1 cleavage complex in the polyadenylation step as well (233).

b. The PCF11 Gene (First Protein of CF1). Using either *RNA14* or *RNA15* as bait, a two-hybrid system selected a clone from a yeast genomic library with an ORF of 626 amino acids with an estimated size of 70 kD (237). This gene that is required for cell viability was implicated in mRNA 3′ end processing when cells with ts mutations of *PCF11* were crossed with ts mutants of either *RNA14* or *RNA15* to give double mutants that displayed synergistic lethality. The mutant strains of *PCF11* had reduced mRNA levels *in vivo* and the mRNA of some mutants had shortened poly A tails. Extracts of these mutants did not add poly A tails to precleaved mRNA and produced only low levels of cleaved products. Antibodies generated from recombinant pcf 11p detected RNA 14p and RNA 15p in the same chromatographic fraction, making it likely that the 70-kDa protein band found in the purified CF1 complex (233) was the *PCF11* gene product.

The two-hybrid system can identify the sites of interactions by introducing specific mutations into proteins that can be fused either to Gal4 BD or AD domains. As noted above, this approach identified specific fip1-binding sites on yeast Pap1 (246). These few early examples that are now rapidly multiplying illustrate how the use of two-hybrid analysis has transformed what had become a bewildering collection of components involved in 3′ end formation in yeast into a more orderly arrangement of interacting components modeled in Fig. 11. More precise localizations will require other approaches.

5. PROBING THE YEAST GENOME FOR MAMMALIAN HOMOLOGS

The growing number of sequenced cDNAs for mammalian cleavage/polyadenylation factors coupled with the sequenced yeast genome with relatively few introns has allowed shortcuts to the identification of yeast homologs of the mammalian factors. This is nicely illustrated by the discovery of the *CFT1* yeast gene product (*173*). It began with a search of a yeast protein sequence database for homologs of all the sequenced mammalian polyadenylation factors. An ORF of 1357 amino acids was detected on yeast chromosome IV with a 24% identity and a 51% similarity to the 160-kDa subunit of CPSF. A large COOH terminal gene fragment (889 amino acids) was amplified by PCR, cloned, and expressed in *E. coli*. Antibodies to the expressed fragment localized the protein in a crude CF11 fraction. As it was needed for cleavage/polyadenylation, it was designated as the first protein of the cleavage factor two complex (Cft1). These same antibodies later showed that the 150-kDa protein of the purified CF11 complex was Cft1 that is the yeast homolog of the 160-kDa subunit of CPSF as are the 105- and 100-kDa subunits of CF11 (*234*).

a. Summary. The understanding of yeast mRNA 3′ end formation began with biochemical approaches used earlier for mammalian cells. Poly A polymerase was characterized initially, while poly A tails were found on mRNA 3′ ends. The genetic approach that followed identified two temperature-sensitive mutants with reduced levels of total mRNA with shortened poly A tails. At that time, the genes were predicted to be involved in mRNA turnover as poly A polymerase activity was unaffected, a reasonable assumption at a time when it would have been hard to imagine the multiplicity of factors (including the products of these two genes) that were needed for mRNA 3′ end formation.

Establishing a mechanism for mRNA 3′ end formation in yeast was hindered by both the confusion engendered by the redundancy of yeast poly A signals and the delayed availability of extracts capable of sequence-specific polyadenylation. Once resolved, techniques of protein fractionation and of cloning and expression of relevant genes were pursued quite independently. However, once antibodies were raised from the recombinant proteins expressed from these genes, the two approaches were merged to identify possible functions for individual proteins needed for cleavage and/or polyadenylation. That many of these proteins were homologs of the mammalian CPSF and CstF complexes dispelled the view that the mechanism of mRNA 3′ end formation would differ greatly from that of animal cells. The yeast two-hybrid system has continued to provide insights into the organization of factors for 3′ end formation in yeast as well as in mammalian factors. The recent solutions of crystal structures for both yeast and bovine poly A polymerases complexed with 3′ dATP have launched the dissection of the molecular details of the polyadenylation reaction and its regulation.

X. Polyadenylation in *E. coli*

Although, from the beginning, studies of polyadenylation were almost entirely focused on eukaryotes and their viruses, by the mid-1970s, a few parallel studies in bacteria provided an essentially similar, albeit more rudimentary, view of polyadenylation. This included a poly A polymerase with properties resembling the mammalian enzymes (256–261) and solid evidence for poly A sequences covalently attached to the 3′ ends of a heterogeneous collection of RNA molecules of *E. coli* with properties expected for mRNA (263, 264).

In spite of this resemblance, further studies of poly A in prokaryotes lagged, primarily because of the marked instability of the mRNAs of *E. coli*. Even more restrictive was the lack of any specific mRNAs analogous to the globin- and ovalbumin-polyadenylated mRNAs widely studied in the 1970s as models for RNA processing in mammalian cells. Thus, although poly A sequences continued to be detected throughout the 1980s in several species of bacteria, this was largely a period of consolidation and confirmation of earlier results. Nimila Sarkar and her colleagues who had begun to apply the emerging techniques of molecular biology to *Bacilli* and *E. coli* RNA were largely responsible for sustaining interest in prokaryotic polyadenylation from the mid-1970s up to their breakthrough observations on a cloned DNA for the poly A polymerase of *E. coli*. A surprising near identity of this sequence with a previously characterized gene of *E. coli* immediately renewed interest in prokaryotic polyadenylation (141). These novel findings and the observations that led up to them are described here and in two recent reviews (254, 255).

A. Poly A Sequences

Poly A had been reported in yeast and bacteria (211) shortly after it was found in the RNA of nuclear and cytoplasmic fractions of Ehrlich ascites tumor cells (262). Since both studies preceded the discovery of the covalent linkage of poly A to RNA, the RNAs were not necessarily subjected to the RNase treatments developed later to release the RNase-resistant poly A sequence from the RNA. Even without such treatment, sequences highly enriched in AMP (75%) were recovered from *E. coli* RNA on oligo dT cellulose that undoubtedly were the result of endogenous RNase activity during RNA isolation (211). The size of these AMP-rich species estimated from their electrophoretic comigration with the poly A from animal cell RNAs as 150 to 200 nt was certainly an overestimate. The size of those poly A sequences recovered from RNAs treated with exogenous RNase T1 that were 98% AMP were not reported in this early study.

Polyadenylation was reexamined in *E. coli* after the poly A(+) RNAs and their poly A sequences had been characterized in animal cells (*263, 264*). Low levels of poly A(+) RNA of a size capable of coding for proteins and with the rapid turnover characteristic of *E. coli* mRNA were found (*263*). These poly A sequences ranging from 30 to 50 nt appeared to be covalently linked to the 3' ends of such RNAs. Aside from differences in length, these poly A(+) mRNAs resembled those of eukaryotes. This, along with the functional similarity of their poly A polymerases, made it likely that polyadenylation bestowed some as yet unknown, but probably similar, function on the mRNAs of both prokaryotes and eukaryotes.

In spite of the genetic approaches afforded by *E. coli* and its transducing phages and plasmids, these findings on polyadenylated RNAs in *E. coli* evoked little interest. Nor did they stimulate a follow-up of the much earlier observations on the specific and rapid decline in PAP activity following infection of *E. coli* with the T-even bacteriophages that was predicted to result from the synthesis of a phage-specified protein inhibitor of PAP (*265*). One follow-up study attempted to demonstrate the source of the gene for the inhibitor by infecting *E. coli* with 16 different T4 phage amber mutants, but none were able to reverse the inhibition (*266*). Although these early inhibition data remain unexplained, they sustained over time my own view that polyadenylation could be a significant feature of RNA metabolism well before poly A had been found covalently linked to RNA.

Although neglected in *E. coli,* poly A sequences were studied in *Bacilli* by Sarkar and co-workers over a period of years, presumably because more poly A(+) RNA could be recovered from *Bacilli* than from *E. coli* where less than 1% of pulse-labeled RNA bound to oligo dT cellulose as opposed to 15 to 25% in *B. subtilis* (*267*). The increased yields were ascribed to an improved one-step isolation of RNA from cell lysates that depended on a controlled proteinase K digestion (*268*). These studies gave a generally similar view of poly A sequences in *Bacilli* as reported earlier for *E. coli*. They also included evidence that poly A(+) RNA might enhance mRNA translation from its strong stimulation of amino acid incorporation in an *E. coli* translation system (*269*) as well as evidence that poly A(+) RNA of *B. subtilis* could serve as a template for reverse transcription, but only in the presence of an oligo dT primer (*270*). Not only was the latter an independent confirmation of the presence of poly A sequences in *Bacilli* but also the resulting cDNAs were of sufficient length to suggest that the poly A would be close to the 3' end of the RNA. This work also prepared the way for the creation of cDNA libraries that could be cloned into *E. coli*. Several years later 30 unique clones of recombinant cDNAs of *B. brevis* were subcloned into an M13 phage to obtain single-stranded cDNAs for sequencing (*271*). Six of eight clones in the correct orientation for the sense strand had oligo dTs ranging from 4 to 19 residues abutting the vector sequences along with upstream sequences of 70 to 90 nt of RNA. The fact that a termination codon was present in at least one of the three possible reading frames of each of these RNA sequences suggested that they were derived from 3' ends of mRNAs with varied, and in some cases extensive, untranslated regions. Although the poly A sequences were short, at least two of them (A19 and A13) were longer than the longest present in the Genbank DNA sequences of bacteria available at that time, suggesting they were added post-transcriptionally.

The usual benefits derived from cloned cDNA libraries could not be fully realized for bacteria until specific mRNAs became available. This occurred in 1980 when a stable mRNA for a small abundant outer membrane lipoprotein (*lpp*) of *E. coli* was sequenced largely by the classical RNA methods, although by this time the sequenced RNA fragments could be ordered by examining overlap with sequenced restriction fragments of the *lpp* gene (272). The *lpp* RNA did not end with poly A but with a putative stem–loop structure followed by a stretch of UMP residues characteristic of the rho-independent termination sites of bacteria. A plasmid expression vector carrying the *lpp* gene was then constructed that could be transcribed at high levels from an inserted *lac*-inducible promoter in *E. coli* with a mutant *lpp* gene (273).

The absence of a poly A tail on this sequence of the *lpp* mRNA did not deter Sarkar and colleagues from using *lpp* probes derived from this vector to search *E. coli* RNA for a polyadenylated *lpp* mRNA (274). The earlier success of this group in obtaining cDNAs from polyadenylated RNAs of *B. subtilis* led them to similar attempts to isolate 3′ regions of the abundant *lpp* mRNA of *E. coli*. The approach, however, was thwarted by the low yields of cDNA obtained with oligo dT primed reverse transcription of *E. coli* RNA, presumably the result of excessive 3′ exoribonuclease activity. Instead the *lpp* phage expression vector was used in an affinity chromatographic approach (274). A single-stranded DNA obtained by subcloning a restriction fragment from this vector into an M13 bacteriophage was covalently linked to a solid support for selection of *lpp* mRNA. Of the 1% of total *E. coli* RNA selected, about 40% bound to oligo dT cellulose. RNase-resistant fragments of 10 to 15 nt were recovered by gel electrophoresis of the digested oligo dT-bound RNA. These oligomers acquired label when the putative poly A(+) RNA was treated with RNA ligase in the presence of 32pCp. Upon hydrolysis all of the label appeared in AMP, suggesting these oligomers were localized at 3′ ends of an *lpp* mRNA terminated by an adenosine residue with a free 3′ OH group. Other less direct evidence was obtained with two enzymatic reactions, each dependent on the presence of an oligo dT:oligo A duplex. These included the dependence of reverse transcription of the pA(+) *lpp* mRNA on an oligo dT primer and of RNaseH (an endonuclease that cleaves the RNA of RNA:DNA hybrids) removal of the poly A sequence of *lpp* RNA in the presence of added oligo dT.

A second less stable mRNA of *E. coli*, coding for the α-subunit of the biosynthetic enzyme tryptophan synthetase, that had been cloned into a plasmid expression vector gave similar evidence for 3′ polyadenylation when examined by these same procedures (275). This included the binding of a sizable fraction of the TrpA mRNA to oligo dT cellulose and similarities in length and location of short oligo A's at 3′ ends. Although the critical sequence data for both mRNAs were lacking, it was likely that these two E. coli mRNAs had poly A tails. Improved conditions for preparing cDNA libraries from E. coli when combined with PCR amplification of specific cDNAs would eventually provide the confirming sequence evidence as well as some clues on mechanisms of 3′ end processing of E. coli mRNAs (276).

Although these experiments showed that obstacles to the study of polyadenylation in E. coli were not insurmountable, they failed to attract the attention of investigators who were then beginning to focus on mechanisms of mRNA turnover in E. coli. This probably stemmed from the generally accepted view that 3′ ends of mRNA in E. coli were formed by transcription termination either at stem–loop structures followed by a short stretch of UMPs (rho-independent termination) or at sites shortly downstream of the stem–loop where the rho protein participates. Neither mechanism included any role for post-transcriptional poly A addition. Furthermore, the failure to find poly A in the original isolates of lpp mRNA and TrpA mRNA (a concern that Sarkar and colleagues were unable to account for), undoubtedly discouraged a search for a link between polyadenylation and 3′ end formation, despite the evidence then emerging from mammalian systems that these reactions were tightly coupled. Thus few references to prokaryotes appeared in the literature of polyadenylation that was rapidly accumulating for eukaryotes in the late 1980s.

B. A Poly A Polymerase Gene

This neglect of prokaryotes may well have persisted had it not been for Sarkar's decision to clone the gene for the E. coli poly A polymerase. Given the existing stalemate in the study of prokaryotic polyadenylation, it made sense to shift from studies of mRNA structure to an analysis of PAP structure, the only protein then known to be involved in prokaryotic polyadenylation. At the very least it would be of interest to compare bacterial PAP sequences with those of calf thymus and yeast that had just been reported (125, 126). Furthermore, the original protocols for purifying the E. coli PAP could, with some updating, be used to obtain an enzyme sufficiently pure to provide sequenced tryptic peptides for designing DNA probes needed for screening and amplifying PAP sequences in E. coli genomic libraries.

As it turned out, a N-terminal tryptic peptide from PAP containing 25 amino acids provided sufficient sequence information to make cloning of the gene unnecessary. Degenerate deoxyoligonucleotide probes derived from this N-terminal peptide were used to amplify a 74-bp fragment in an E. coli genomic library that encoded the 25 amino acids near the N terminus of PAP (141). A gene bank search uncovered an identical sequence within a gene already known to control the copy number of certain colicin-like plasmids such as pBR322, then widely used as a cloning vector (277). The specific function of this *pcnB* gene was unknown, but its gene sequence coded for a protein of about 47 kDa (278, 279) and was expressed in minicells harboring a *pcnB*$^+$ plasmid (278). To prove that the *pcn* gene encoded a poly A polymerase, *E. coli* mutant at the *pen* locus was transformed with an inducible *pcn*$^+$ plasmid (141). Extracts from these cells had 150-fold greater PAP activity than did control cells. Several truncated proteins formed by deletion of either 5' or 3' regions of the open-reading frame of the *pen* gene were also expressed in extracts but were enzymatically inactive.

1. PAP1 REGULATES PLASMID DNA REPLICATION

How host-cell poly A polymerase could regulate copy number of a restricted group of plasmids, presumably at the level of DNA replication, was not immediately obvious, as PAPs were only known to add AMPs to 3' ends of mRNA. An important clue came from studies that had established a correlation between the control of the copy number of certain colicin-related plasmids by the *pcn* gene and the mechanism of their replication (280). Replication is under the control of a 108-nt anti-sense RNA designated as RNAI. This was the first naturally occurring anti-sense RNA known to exert biological control, in this case by binding to and inactivating the plasmid-specified RNA primer of its DNA replication, designated as RNAII (281, 282). RNAI is transcribed in the opposite direction from RNAII and includes a complementary region of the DNA near the origin of replication. This creates a region at the 5' end of RNAI that is complementary to a segment of the 555-nt transcript specifying the RNAII primer. The interaction of RNAI with this 5' region of the RNAII primer prevents it from binding to the DNA template, thereby inhibiting the initiation of replication.

Early evidence that this *pen* gene product interacted with RNA had come from its sequence that had several regions of strong homology to the *E. coli* tRNA nucleotidyl transferase gene (279). This enzyme that adds the CCA 3' terminal sequence to all tRNAs presumably requires recognition of the three-dimensional folded structure shared by all tRNAs. Since RNAI folds into a similar structure (see Fig. 12), it was predicted that the *pcn* gene product would bind to RNA1 and prevent it from inactivating RNA11 (279). Although this prediction of a direct interaction between the *pcn* gene product and RNA1 proved to be correct, there was no evidence that would have linked anti-sense RNAl to polyadenylation at the time. Few investigators may have been aware of polyadenylation in bacteria in spite of the long-standing evidence for it. Even the evidence for poly A sequences at 3' ends of at least two specific *E. coli* would not have been seen as relevant to the function of RNAl that suppressed plasmid DNA replication without apparent involvement of an mRNA. Nor were there well-documented examples of polyadenylated RNA other than mRNAs. Thus, it came as a surprise that the interaction of the *pcn* gene product with RNA1 was that of a poly A polymerase with its substrate which meant that poly A tails could be added to RNA1, an RNA not known to be polyadenylated.

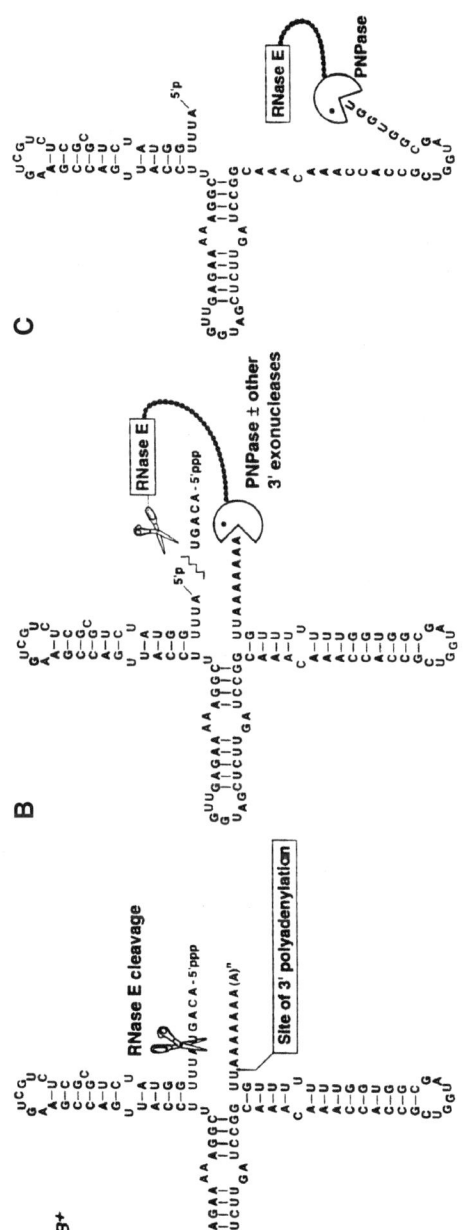

FIG. 12. A model depicting the pathway of decay for the anti-sense RNA1 repressor of ColE1-type plasmids. A poly A tail serves as a site for binding a ribonuclease complex, the degradosome, that degrades RNA1 and other mRNAs (see text). Reprinted from *Cell*, **80**, S. N. Cohen, Surprises at the 3′ end RNA, p. 829, Copyright (1995), with permission from Elsevier Science.

C. Poly A Sequences Target RNA Decay

1. THE ANTI-SENSE RNA1

A search for polyadenylated RNA1 quickly became a top priority for those probing the mechanism of RNA1 control of plasmid copy number (283, 284). The search was facilitated by ongoing studies that were already focused on other factors, ostensibly unrelated to the *pcn* gene, that participated in the regulation of plasmid DNA replication by RNA1. Foremost among these were two RNase-dependent processes that controlled the rate of degradation of RNA1, an obvious mechanism for relieving RNA1 suppression of plasmid DNA synthesis. One such process removes five nucleotides from the 5' end of RNA1 by an endonucleolytic cleavage by RNaseE (280), an endonuclease already known to process *E. coli* ribosomal RNA transcripts. The product of this cleavage, RNA1-5, is then rapidly degraded by 3' exonucleases (Fig. 12). It is the rapid decay of this cleavage product that is the critical step in the relief of suppression of plasmid DNA replication. To circumvent the experimental problem posed by this rapid decay of the cleavage product, a more stable form replaced the RNA1 of another colicin-related plasmid, pSL-C1O1 (280). This RNA is identical in sequence to the RNaseE cleavage product of RNA1 except for a triphosphate group at the 5' end that marks the *in vivo* transcription start site. This pppRNA1-5 is 10 times more stable than the native RNaseE cleavage product, pRNA1-5, suggesting incidentally that 5' ends may also be involved in RNA1 degradation. More important, however, was the retention of the anti-sense function of this RNA1 analog through its interaction with the RNA11 primer of plasmid DNA replication.

The effect of the *pcn* gene on the structure and stability of RNA1 and its decay intermediates was examined during a rifampicin chase in *E. coli* transfected with plasmids expressing either the native or this modified *RNA1* gene (283). Elongated forms of both *RNA1* and pppRNA1-5 were seen in *pcn*[+], but not *pcn*[−] cells. The 5' end of each RNA was ruled out as the site of this elongation by evidence that both bore triphosphate ends as expected for transcription start sites. The suspected 3' elongation of RNA1 by a poly A sequence was confirmed by several indirect approaches that compared properties of RNA1 from *pcn*[+] and *pcn*[−] cells for binding to oligo dT cellulose and for oligo dT priming of reverse transcription.

To confirm this indirect evidence for polyadenylation, the 3' regions of *RNA1* genes were amplified, cloned, and sequenced (283). Of the 11 clones sequenced from 30 RNA1 clones isolated, none contained an oligo A comparable in length to those previously characterized either in specific mRNAs or in total *E. coli* RNA. Interestingly, each of the 11 clones had a different 3' end, all generated from multiple sites within the stem–loop structure at the 3' end of RNA1. Two of the 11 sequences, however, showed sites where adenosine had replaced other nucleosides to create a short stretch of AMPs (A6 was the longest) at the

3′ end of these clones that is not present in RNA1, making it likely that a post-transcriptional addition of AMP had occurred at these processed 3′ ends. It was hardly surprising that longer oligo A's were not seen, given the small amounts of elongated polyadenylated RNAl relative to unprocessed RNA1 found in these pcn^- cells. pcn^- cells accumulate large amounts of decay intermediates shorter than RNA1 or RNA1–5 not seen in pcn^+ cells which attests to the increased stability of RNA1 lacking poly A tails. This accumulation of RNA1 enhances the inactivation of RNAlI, the primer of plasmid DNA replication that is recognized as a reduction of plasmid copy number.

The effects of the *pcn* gene now known as the *PAPI* gene on RNA1 structure and turnover revealed some unsuspected features of polyadenylation. First among them was that RNAs other than mRNAs are polyadenylated. More importantly, this poly A tail regulates the lifetime of this anti-sense RNA, not by protecting it from degradation, as has often been proposed as a major function for poly A, but, in this case, by targeting RNA1 for degradation.

2. *E.COLI* mRNA Decay

The surprising effects of polyadenylation on the stability of RNAl, a plasmid-encoded transcript, immediately raised questions of similar effects on *E. coli* mRNA. A role for poly A in prokaryotic mRNA turnover had not been considered during the previous decade, although considerable progress in understanding mRNA decay was achieved largely with the development of RNase-deficient mutants. In fairness, it should be noted that this failure could have been justified at the time on the basis of scarcity. Only about 1% of total pulse-labeled *E. coli* RNAs have detectable poly A which contrasts with eukaryotes where nearly all mRNAs are highly polyadenylated. In the case of RNA1, it is the very presence of poly A that enhances this scarcity by accelerating the rate of degradation of poly A(+) RNA1.

3. RNases That Degrade mRNAs

Among the achievements of the previous decade was the identification and characterization of RNases involved in mRNA degradation (reviewed in Ref. 285). The isolation of *E. coli* mutants for each of several such RNases and the combination of two or more of these RNase mutations in a single viable strain ultimately provided the resource needed to implicate polyadenylation in prokaryotic mRNA decay. Among four such RNases was RNaseE, already noted for cleaving a pentanucleotide from the 5′ end of RNA1 (280). RNaseE also makes specific endonucleolytic cleavages within the 3′ noncoding regions of several mRNAs (287, 289). RNaseIII, an already well-known endonuclease that cleaves within double-stranded, often intercistronic regions of polycistronic mRNA, was shown to have negligible effects on the rates of decay of monocistronic mRNAs (285). It is the 3′ exoribonucleases, polynucleotide phosphorylase, PNPase, and

RNaseII that are largely responsible for the phosphorolytic and hydrolytic cleavages, respectively, that degrade E. coli mRNAs. A double mutant lacking the PNP gene and bearing a temperature-sensitive RNaseII gene was found to raise the poly A(+) RNA level from the 1% typical of wild-type strains to more than 5% in the mutant confirming the role of these 3′ exonucleases in mRNA decay (276).

One of the groups that had developed several of the E. coli RNase mutants for study of rates and patterns of mRNA decay reported that a quadruple mutant of all four of the RNases described above was still able to degrade substantial amounts of mRNA, leading them to conclude that "important aspects of mRNA decay are still not understood" (285). One such aspect turned out to be polyadenylation.

4. PAP1 Stimulated mRNA Decay

After a PAP1 mutation had been implicated in the destabilization of RNA1, total pulse-labeled E. coli RNA was shown to have an increased stability when a PAP1 mutation was included in cells mutant for RNaseE and the 3′ exonucleases RNaseII and PNP (288). In cells mutant only for PAP1, the lpp mRNA showed a longer half-life compared to the wild-type cells, but half-lives of the stable OmpA mRNA coding for an outer membrane protein and the unstable thioredoxin mRNA were unchanged, although patterns of decay intermediates were altered. However, including a PAP1 mutation in the triple-RNase mutant increased the stability of all three mRNAs (288).

Further convincing support for the stabilization of mRNA in E. coli strains lacking PAP was reported by a group studying the processing of rpsO mRNA that codes for the S15 protein of the small ribosomal subunit (289). This gene is transcribed as a polycistronic mRNA along with the PNPase gene. The rpsO mRNA is processed from the transcript into an abundant monocistronic mRNA whose decay begins with an endonucleolytic cleavage by RNaseE just before a terminal stem–loop structure. In strains lacking RNaseE and the 3′ exonucleases PNPase and RNaseII, elongated forms of rpsO mRNA that were not seen in an isogenic strain bearing a PAP1 mutation accumulated during a rifampicin chase. Evidence that polyadenylation was responsible for this elongation was supported by an oligo dT-dependent priming of reverse transcription of the rpsO mRNA from the PAP+, but not from the PAP− strain. Regions surrounding the putative mRNA–poly A junction of these elongated cDNAs were amplified by PCR and cloned into E. coli. The sequences of 32 of 33 selected clones showed poly A sequences of 18–20 nt linked to the same site at the 3′ end of the terminator hairpin of each mRNA clone. Recognizing that pitfalls in the determination of mRNA decay rates could arise from the use of E. coli strains with multiple RNase deficiencies in addition to PAP1, the effects on mRNA stability of a defective PAP gene were examined in a strain that retained both of the 3′ exonucleases, but one that lacked RNaseE (289). The total amount of poly A(+) RNA was reduced as would be expected in such strains, but the half-life of the *rpsO* mRNA was still extended, in this case from 4 to >15 min showing an effect of PAP1 on the decay of this mRNA even as poly A tails were being removed or shortened by 3′ exonucleases.

New approaches using plasmids bearing an inducible *PAP1* gene have confirmed and greatly extended the evidence for a role for PAP1 in mRNA decay (296). The ability to control levels of PAP1 without reducing growth or cell viability is an experimental advance over previous dependence on the null mutants that implicated PAP1 in mRNA decay. Here increased levels of PAP1 were correlated with reduced half-lives of the same mRNAs used in the earlier PAP1 mutant studies. Of special note is the contrast between these mRNAs and those coding for RNaseE and PNPase whose half-lives were actually extended, resulting in increased levels of these nucleases that paralleled the overexpression of PAP1 activity, showing PAP1 stimulated mRNA decay to be a well-regulated process. However, other pathways for mRNA decay in *E. coli* must exist, as cell viability is retained in PAP1 mutants and a significant fraction of polyadenylated mRNAs is insensitive to PAP1 mutations (297). Evidence for a second PAP activity in *E. coli* extracts (297) raised the possibility that a second *PAP* gene could account for the retention of significant viability of PAP1 mutants (see below).

5. A POLY A TAIL DEGRADATION COMPLEX—THE DEGRADOSOME

The stabilization of the several selected mRNAs observed in PAB1 mutants by others (288) had also lacked RNaseE, while stabilization of the *rpsO* transcripts was only seen in PAP1 mutants that also lacked RNaseE that cleaves near mRNA 3′ ends usually leaving a stable stem–loop structure. This is the case for RNA1 as well, where it was suggested (290) that a single-stranded poly A extension would be a favorable structure for the assembly of an RNA-degradation complex such as that formed by RNaseE and polynucleotide phosphorylase (291) (see Fig. 12). Attention is currently focused on a multienzyme degradation complex of *E. coli*, the degradosome, that includes those two nucleases and an RNA helicase and other proteins, but not PAP1 (292, 293) (for a recent review, see Ref. 294). Recently those mRNAs with 3′ ends embedded in a stable stem loop structure were found to be resistant to degradosome attack *in vitro* until exposed 3′ ends were lengthened with five or more AMP residues (295). While supporting the model of poly A as a site for anchoring the degradosome, it may be single strandness rather than base specificity that is the significant functional element as other generic oligomers could replace oligo A, but only if lacking G residues (295).

D. The Search for a Second Poly A Polymerase Gene

The continued viability of PAP1 mutants had raised the possibility of a second *PAP* gene that could account for the persistence of PAP enzymatic activity and the continued synthesis of poly A tails on mRNA (141). A second PAP activity was then purified from a PAP1 deletion strain. The designated protein differed in size, i.e., 35 versus 55 kDa for PAP1 and in primer specificity (297).

A strategy for isolating the gene for a PAPII was based on the assumption that a PAP1 deletion strain would require a second functional PAP to retain viability. Such a potential double-PAP mutant was created in a PAP1 deletion strain by disrupting the putative *PAPII* gene with the *tn10*-transposon-insertion technique (298). Viability of such a double mutant would require the acquisition of a *PAP1* gene from a transforming plasmid which could be activated by an arabinose inducible promoter for the *PAP1* gene. Sequences of 80- to 90-bp fragments replicated from DNA of the *tn10*-transposon-disrupted gene of the viable strain exactly matched an unassigned 36-kDa ORF designated f310 in the *E. coli* gene bank that had no sequence similarities to *E. coli PAP1* or any other eukaryotic or viral *PAP* genes.

The critical evidence for assigning a PAP function to the *f310* gene came from comparisons of levels of PAP activity in PAP1 deletions transformed with plasmids either overexpressing or lacking the *f310* gene (299). One PAP assay measured radioactivity from ATP that accumulated over a 36-kDa gel band separated from proteins of cells transformed with *f310* gene. Another assay measured PAP activity in an eluate from a nickel sepharose column in which the major component was a 36-kDa protein, presumably the hexahistidine f310 fusion protein overexpressed in the transformed cells. The apparent enhanced PAP activity and ATP binding associated with overexpression of the *f310* gene in those two assays led to the conclusion that *f310* was a second poly A polymerase gene that could sustain the viability of PAP1 mutants.

This conclusion now seems to have been premature as a recent more detailed and quantitative analysis of the effects of overexpression of the *f310* gene in a variety of wild-type and PAP1 mutants failed to show any effects on poly A metabolism (300). Varied levels of overexpressed *f310* mRNA *in vivo* had minimal effects on poly A content or length. Nor was the expected decrease in poly A content of the double mutant, PAP1$^-$, f310$^-$ observed. Levels of PAP enzymatic activity were also not increased in strains overexpressing the *f310* protein. The diversity of this evidence failing to support a PAP function for *f310* overshadows the earlier supporting evidence derived from PAP enzymatic assays that did not include a characterization of the reaction product (299). Two consensus ATP-binding sites in the *f310* ORF were proposed to account for the radioactivity associated with the overexpressed 36-kDa protein on the membrane blot assay for PAP of the earlier study (300).

The elimination of a PAP function for f310 has led to renewed efforts to locate the source of the residual PAP activity of PAP1 mutants (301). Although an obvious potential candidate, polynucleotide phosphorylase (PNP) had seemed an unlikely one in view of the levels of inorganic phosphate in *E. coli* that would favor the reverse reaction, i.e., the phosphorolysis of phosphodiester bonds from

RNA 3' ends. This view may now have to be modified to accommodate newer evidence that PNP may account for the residual PAP activity of PAP1 mutants (*301*). Coupling these data with the elimination of the leading candidate, i.e., the *f310* gene product, currently leaves PAP1 as the sole poly A polymerase in E. coli. However, poly A addition to RNA 3' ends may under appropriate conditions be catalyzed by polynucleotide phosphorylase, an enzyme lacking base specificity that polymerizes NDPs rather than NTPs into polyribonucleotides.

The study of polyadenylation in bacteria was ignored for many years until a small plasmid-coded anti-sense RNA that regulates ColE1-type plasmid DNA replication was found to have a 3' poly A tail that accelerated its degradation (*283, 284*). The connection between the rate of decay of RNA1 and polyadenylation was established when the *pcn* gene that regulates plasmid copy number in E. coli was found to be the *PAP1* gene. RNA1 was then shown to have a poly A tail that accelerated its rate of decay as did the poly A tail of several E. coli mRNAs.

These surprising aspects of polyadenylation in E. coli established the universality of polyadenylation as a mechanism for modifying mRNA function in prokaryotes as well as eukaryotes and their viruses. Poly A is present in archeae as well (*338*). They clearly show a role for poly A directed destablization of certain E. coli mRNAs. Decay is mediated by the binding to poly A tails of a multienzyme complex, the degradosome, that includes RNaseE and PNPase, a helicase and other proteins. In mammals, poly A had often been viewed as stabilizing mRNAs presumably through a poly A-binding protein (*302*). The mRNA-destabilizing effects of poly A tails have now been seen in eukaryotes as well. The decay of a major fraction of yeast mRNAs is initiated by a shortening of the poly A tail that activates mRNA 5' end decapping that in turn allows 5' exonucleolytic degradation (*303*). A similar poly A shortening initiates mRNA decay in mammalian cells (*304, 305*) that can now be reproduced in cytoplasmic extracts (*306*). Deadenylation and subsequent mRNA degradation in these extracts were both stimulated by the presence of AU-rich sequences in the 3' UTRs of labile mRNA substrates as has long been known for such mRNAs *in vivo* (*307*). Current studies are now focused on characterizing specific deadenylases (*308*) and the RNases for mRNA turnover in mammals that are likely to be regulated by complex mechanisms if those of yeast and E. coli are relevant examples.

XI. Polyadenylation in *Vaccinia* Virus

Some of the earliest evidence for the existence of polyadenylated RNAs came from studies of macromolecular synthesis of vaccinia virus. *Vaccinia*, a large DNA virus that replicates in cell cytoplasm, encodes the gene products needed for its replication, transcription, and RNA processing that are packaged within virions that retain these activities *in vitro*. Our early studies of poly A sequences in animal cells (262) were encouraged by a study of RNA synthesis in vaccinia virions noting that a detergent treatment of virions abolished the incorporation of UTP, but not of ATP, into an acid-insoluble product (RNA) even in the presence of the four nucleotides (309). It was tentatively concluded that this ATP incorporation was likely to result from poly A synthesis, although the labeled product was not analyzed. In retrospect, this striking difference in the effects of detergent treatment of virions on UTP and ATP incorporation predicted that RNA and poly A synthesis were catalyzed by different enzymes. At that time, however, it was necessary to show that poly A sequences were being made by *Vaccinia* virions. This was soon verified when poly A sequences of 150 to 200 nt presumably covalently attached to large *Vaccinia* RNAs were synthesized in permeable *Vaccinia* core particles (310). When the synthesis of this poly A was inhibited by actinomycin D and other DNA-intercalating molecules, it was concluded that they arose by transcription even though there was no evidence that Vaccinia DNA contained such long poly dT sequences. The fact that *E. coli* RNA polymerase synthesized large amounts of poly A *in vitro* from Vaccinia DNA also favored a transcriptional mechanism, although the lengths of poly A were not reported. A slippage mechanism for poly A synthesis seemed to be ruled out as poly A was made in the presence of all four nucleotides, while poly A synthesis by slippage occurred primarily with only ATP with *E. coli* RNA polymerase (311).

A. A *Vaccinia* Poly A Polymerase

Somewhat later, after transcription had essentially been ruled out for cellular poly A synthesis and poly A had been localized to 3′ ends of cellular mRNAs, a poly A polymerase activity was detected in extracts of purified vaccinia virions (312). This preparation was stimulated by short riboadenylate oligomers, while another more purified polymerase was activated with either ribo- or deoxyribopolynucleotides (313).

Priming by poly dA:dT renewed the possibility that *Vaccinia* poly A was synthesized by transcription, but this was soon eliminated by the detection of poly A tails covalently attached to 3′ hydroxyl ends of both ribo- and deoxyribopolynucleotides (314). A highly purified 80-kDa virion PAP on denaturation gave a 35- and a 51-kDa protein on gels (315), but in amounts insufficient to examine their function which would only be clarified many years later.

A PAP activity from a cytoplasmic fraction of HeLa cells infected with vaccinia closely resembled this virion PAP and was activated by ribo- but not by deoxypolyribonucleotides (316, 317). Later a vaccinia-induced PAP from HeLa cell cytoplasm that added up to 100 AMPs to *Vaccinia* mRNAs *in vivo* was purified. It appeared to be identical to the virion *Vaccinia* PAP and distinct from HeLa PAP in some catalytic properties (318).

These studies of *Vaccinia* poly A polymerases coincided with the close of this enzymological phase of polyadenylation research. Little was to be learned until molecular cloning technology made available larger amounts of RNA substrates a decade later. The discovery of specific poly A signals in mRNAs and the preparation of cell-free mammalian extracts active in cleavage/polyadenylation revived interest in mRNA 3' end formation in *Vaccinia*. Soluble extracts of virions that correctly initiated and terminated transcription of early mRNA genes were developed that also capped 5' and polyadenylated 3' ends of these transcripts (*319*). In contrast to cellular systems, the poly A tails were added directly to sites of transcription termination. Termination of early *Vaccinia* mRNAs required a TMP-rich sequence TTTTTNT, usually about 30 to 50 nt upstream of the termination site (*320*). No consensus sequences specify the site of termination for mammalian cells and although cleavage/polyadenylation signals are required, they are insufficient for pre-mRNA termination (*321*).

Among the first questions to be answered upon resumption of studies of polyadenylation in *Vaccinia* was the nature of the genes for the two proteins that copurified as the poly A polymerase heterodimer. Several sequenced oligopeptides recovered from tryptic digests of each purified subunit allowed specific oligonucleotide probes to be synthesized for screening a *Vaccinia* DNA library where separate genes for each protein were located within open reading frames on separate DNA fragments (*322*). Molecular weights of 55 and 39 kDa calculated from their amino acid sequences agreed well with those that had been assigned earlier on the basis of gel mobilities (*315*). Although active forms of each subunit could not be expressed from their cloned genes transfected into *E. coli*, denatured forms were recovered that elicited specific antibodies in rabbits for VP55 and for VP39 that did not cross-react. Immunoprecipitation of Vaccinia-infected cell extracts showed all of VP55 was in the heterodimer along with equimolar amounts of VP39, but similar amounts of VP39 were also present as the monomer (*322*). Dissociation of the dimer was achieved by treating it with VP55 antibody immobilized on a solid matrix that bound VP55, but not VP39. Fortunately polymerase activity was retained even while complexed with antibody attached to the gel, as it could not be released without inactivation. The soluble unbound VP39 had no polymerase activity, but it stimulated the bound VP55 to synthesize much longer poly A tails. VP39 also bound specifically to poly A, but VP55 did not. These properties of VP39 bore a strong resemblance to those of the mammalian protein described at about this same time that stimulated poly A tail elongation by the bovine polymerase that was designated as PABII, a 33-kDa nuclear protein (*120*).

Both VP39 and VP55 were eventually obtained as functional monomers by cloning each gene in a vector that could be expressed in mammalian cells. Such a system had already been successfully deployed in the cloning of *Vaccinia* transcription factors and depended on the coinfection of HeLa cells with two recombinant *Vaccinia* viral vectors (*323*). One vector was designed to express an inserted gene from a T7 RNA promoter, while the other expressed T7 RNA polymerase continuously from a constitutive promoter of Vaccinia. The properties of the recombinant forms of VP55 and VP39 were studied separately and in the reconstituted dimer (*324, 325*). The recombinant *Vaccinia* VP55 monomer differed from cellular PAPs in its catalytic function, as it catalyzed a rapid burst of processive polyadenylation that abruptly slowed to a nonprocessive reaction after about 35 AMPs had been added to mRNA (*324*). This slower phase could be reversed to the more rapid processive reaction upon the addition of VP39 (*325*) much as PABII stimulated processive elongation of poly A tails by the calf thymus polymerase (*120*). The cause of this abrupt shift with the VP55 catalyzed reaction was examined by comparing the priming activity of several *Vaccinia* mRNA substrates bearing oligo A tails of different lengths (*326*). For all RNA substrates, with tails that ranged from 0 to 25 AMPs, the switch was observed when a length of about 35 nt had been reached, suggesting that oligo A tail length specified the transition rather than the number of reactions catalyzed by the polymerase. VP55 did polyadenylate oligo A primers, but slowly and nonprocessively (*324*), suggesting that the early processive phase of the reaction depended on other sequences in the 3′ ends of *Vaccinia* mRNAs.

B. Poly A Signals in *Vaccinia* mRNA

In contrast to animal cells where even precleaved mRNAs require the AAUAAA signal for polyadenylation, *Vaccinia* mRNAs have no such specific sequence requirements although the length of this 3′ terminal segment is critical as sequences less than 30 nt were inactive (*326*). Analysis of the effects of extensive sequence scrambling within a 40-nt segment at the 3′ end of the VGF early mRNA failed to eliminate polyadenylation or binding of VP35 to the mRNA (*326*). Surprisingly VP55 activity and binding were not indifferent to base composition as reduction of the UMP content of this segment below 30% abolished priming, while elimination of any one of the other three nucleotides or replacement of all of three with deoxyribose analogs had no effect on VP55 binding or polyadenylation. A more precise localization of the functional oligo U sequences was achieved with a set of DNA:RNA chimeras in which a single tetra–oligo U sequence was systemically inserted across a 34-nt sequence in which oligo dC's served as an inert background (*327*). Two separate tetra UMP inserts at 10 and 25 nt upstream of the RNA terminal 3′OH group most effectively bound VP55. Combining these two groups at similar sites in a single sequence restored optimal VP55 binding. From these and other nucleotide replacement experiments a consensus motif emerged, $(rU_2)–(N_{15})–rU$ in which N_{15} could be any deoxy- or ribonucleotide (*328*). The optimal placement of the motif was about 9 nt from the 3′ end of the mRNA. If the motif was the sole source of UMPs, processivity was sustained for the addition of only three to five AMPs. The presence of additional UMPs both upstream and downstream of the motif restored normal processivity to about 50 AMPs indicating that other UMPs are recognized by VP55 during this initial burst of processive activity.

Thus length, UMP content, and UMP location are the key structural elements in VP55 RNA binding and activity. Effects of uracil and ribose analog substitutions on VP55 binding further defined the minimum structural requirements as an unsubstituted pyrimidine ring and a ribose 2′ hydroxyl group (328). Other well-recognized oligo U-binding proteins had no sequence similarity to VP55.

1. THE MECHANISM OF PROCESSIVE POLYADENYLATION

The involvement of two well-separated UMP-binding sites suggested a translocation of VP55 along the mRNA by a typical inchworm mechanism (326, 327). In this model the RNA-binding domain of VP55 "locks on" to the more distal UMP site, while the catalytic domain binds to the UMP site proximal to the 3′ end and initiates addition of AMP residues. Processive elongation could be achieved either by a conformational extension of VP55 or by a folding of the RNA sequence spanning the conserved UMP sites. Either mechanism could define the limits for either VP55 elongation or RNA shortening and thereby the length of the oligo A added to mRNA during the initial processive phase. In the presence of VP39 processivity is retained with a heteroduplex of VP39 and VP55 that stabilizes VP55 binding to the elongating poly A tail. The mechanism for this stabilization is unclear, but includes binding sites for both subunits. VP55 binds both the upstream rU_2 site and the downstream rU site of the optimal consensus sequence for *in vivo* processive polyadenylation, while VP39 binds only the rU site of an $(rU_2)-N_{25}-rU$ sequence (329). Note the 10-nt extension of this spacer region for the heterodimer oligo U-binding sites (N_{25}) versus the N_{15} length for the VP55 monomer-binding sites noted above. This extension is seen as an RNA loop in the heterodimer:RNA complex of Fig. 13.

Stabilization of the vaccinia PAP by VP39 is similar to that attained with PABII for the bovine poly A polymerase (120), although here processivity is attained only after a short oligo A tail has been added nonprocessively to mRNA by the bovine PAP in the presence of accessory factors of the CPSF complex. Thus *Vaccinia* VP55 alone carries out substrate specific processive polyadenylation, while a mammalian PAP is neither specific nor processive. Although VP39 mimics the poly A-binding and processivity functions of PABII, it has no homologous sequences, nor can it replace PABII in mammalian extracts or vice versa (P. D. Gershon, Personal communication).

C. VP39 is a 5′ Cap Methylase

A quite unexpected property of VP39 emerged when its sequence was found to be identical to a well-studied *Vaccinia* transmethylase that methylates the

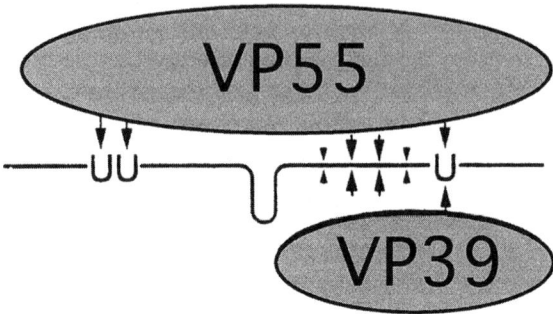

FIG. 13. Summary of proposed interactions within the VP55–VP39-primer ternary complex (arrows denote protein–RNA contacts). The internal RNA loop accounts for the ~10 nt increased spacing between the two uridylate-binding sites for the heterodimer compared to the monomeric VP55 spacer. Reproduced from L. Johnson and P. D. Gershon, RNA binding characteristics and overall topology of the Vaccinia poly A polymerase processivity primer complex, *Nucleic Acids Res.* (1999), **72**, 2708, by permission of Oxford University Press.

2′ hydroxyl group of the penultimate nucleotide of the 5′ cap of *Vaccinia* mRNAs (*330*). Each activity had been purified by a different route, but a battery of biochemical assays confirmed their identity as predicted from sequence analysis. This finding raised the possibility that VP39 linked activities at the 5′ and 3′ ends of *Vaccinia* mRNAs. Some evidence existed that ends of *Xenopus* mRNAs might interact during the polyadenylation of those mRNAs recruited to polysomes during oocyte maturation (*331*). Here cytoplasmic poly A tail elongation was temporally linked with the specific 2′-O-methylation of the 5′ cap of the mRNA. Later, however, 3′ poly A tail elongation was uncoupled from the 5′ cap methylation of the maturation specific c-mos mRNA, that was injected into *Xenopus* oocytes treated with a 2′-O-methylase inhibitor (*332*). Although cap methylation of c-mos mRNA and its subsequent translation were blocked, polyadenylation of c-mos mRNA and protein synthesis in general were unaffected. A detailed *in vitro* study with new assays for activities at the 5′ and 3′ ends of *Vaccinia* mRNAs also failed to find any effects on reactions at one end of the mRNA by those at the other (*333*). Neither the presence or absence of 2′-O-methylation of the 5′ cap, nor the presence or absence of a 5′ capping site had any effect on the elongation of the mRNA oligo A tail promoted by VP39, while caps of mRNAs were still methylated in the absence of VP55, the poly A polymerase.

A rationale for the apparent independence of the two functions of VP39 may be found in a 1.85 Å resolution crystal of VP39 that is a single oblate spherical structure with two major faces on opposite sides of the molecule (*334, 335*). One site contains sites for each of the several functions expected for the 2′-O-methylase reaction, while the opposite site has a heterodimeric interaction domain for the interaction with VP55. It was not clear that *VP55* binds solely to this site, but the separation of the two sites could allow VP39 to carry out both functions simultaneously or independently. The apparent molar excess of the monomeric form of VP39 *in vivo* (322) would allow it to function independently of VP55. The dual function of VP39 *in vivo*, however, may allow a joining of 5′ and 3′ ends of *Vaccinia* mRNAs to stimulate translation initiation much as the poly A-binding protein does in yeast (18). In yeast, Pabp1 functions within a complex system in which a translational initiation factor eIF4G associates with both a cap-binding protein and Pabpl to stimulate initiation of translation of a circularized poly A(+) mRNA (336).

a. Summary. When compared to mammalian or yeast polyadenylation systems, that of *Vaccinia* virions is quite rudimentary. Much of this simplification is the result of the generation of 3′ ends of *Vaccinia* mRNAs by transcription termination rather than by a cleavage reaction that in eukaryotic cells involves many proteins, while only two proteins carry out sequence and length-specific polyadenylation here. VP55 retains the sequence recognition function along with the catalytic function, while a poly A-binding protein VP39 allows the polymerase to remain bound to the growing poly A tail as it is processively elongated. The fact that both viral and cellular systems need a poly A-binding protein for processive polyadenylation emphasizes the importance of this long overlooked feature of basic polyadenylation.

The *Vaccinia* poly A polymerase heterodimer has provided a simple compact system for uncovering the molecular details of mRNA sequence recognition and poly A tail elongation in *Vaccinia*. The surprising additional role for the VP39 processivity factor as a 2′-*O*-methylase for the 5′ caps of RNA suggested that *in vivo* VP39, most likely within the heterodimer with VP55, may coordinate processing of the 5′ and 3′ ends of *Vaccinia* mRNAs and potentially activate translation. Although attempts to demonstrate such a functional coordination between 5′ cap methylation and 3′ poly A tail elongation have not been successful, further exploration of the molecular details of RNA processing and its relationship to gene expression in this relatively simple system may produce other novel insights into the processing of 3′ ends of mRNAs.

REFERENCES

1. M. Edmonds and R. Abrams, *J. Biol. Chem.* **235**, 1142 (1960).
2. M. Grunberg-Manago and S. Ochoa, *J. Am. Chem. Soc.* **77**, 3165 (1955).
3. A. Stevens, *Biochem. Biophys. Res. Commun.* **3**, 92 (1960).
4. J. Hurwitz, A. Bresler, and R. Diringer, *Biochem. Biophys. Res. Commun.* **3**, 15 (1960).
5. S. B. Weiss, *PNAS* **46**, 1020 (1960).

6. M. Edmonds, M. Vaughan, and H. Nakazato, *PNAS* **68,** 1336 (1971).
7. J. E. Darnell, R. Wall, and R. J. Tushinski, *PNAS* **68,** 1331 (1971).
8. S. Lee, J. Mendecki, and G. Brawerman, *PNAS* **68,** 1331 (1971).
9. D. Baltimore, *Nature* **226,** 1209 (1970).
10. H. Temin and S. Mitsutani, *Nature* **226,** 1211 (1970).
11. G. Huez, G. Marbaix, F. Hubert, Y. Cleuter, M. Leclerq, H. Chantrenne, R. Devos, H. Soreq, Un. Nudel, and U. Littauer, *Eur. J. Biochem.* **59,** 589 (1975).
12. G. Huez, G. Marbaix, D. Gallwitz, E. Weinberg, R. Devos, E. Hubert, and Y. Cleuter, *Nature* **271,** 572 (1978).
13. E. J. Butler, *in* "Control of Messenger RNA Stability" (J. Belasco and G. Brawerman, eds.), p. 367. Academic Press, 1993.
14. L. L. McGrew, E. Dworkin-Rastle, M. B. Dworkin, and J. D. Richter, *Genes Dev.* **3,** 803 (1989).
15. J. Huarta, D. Belin, A. Vassalli, S. Strickland, and J. D. Vassalli, *Genes Dev.* **1,** 1201 (1987).
16. A. Sachs and R. Davis, *Cell* **58,** 857 (1989).
17. N. Iizuka, L. Najita, A. Fraususoff, and P. Sarnow, *Mol. Cell. Biol.* **14,** 7322 (1994).
18. S. Z. Tarun and A. B. Sachs, *Genes Dev.* **9,** 2997 (1995).
19. M. Wickens, J. Kimble, and S. Strickland, *in* "Translational Control" (J. W. Hershey, M. B. Matthews, and N. Sonenberg, eds.), Monograph 30, p. 411. Cold Spring Harbor Press, Cold Spring Harbor, New York, 1996.
20. A. Jacobson, *Ibid.* p. 451.
21. J. D. Richter, *Ibid.* p. 481.
22. R. W. Eckner, W. Ellmeier, and M. L. Birnstiel, *EMBO J.* **10,** 3513 (1991).
23. Y. Huang and G. C. Carmichael, *Mol. Cell Biol.* **16,** 1534 (1996).
24. S. Berget, *J. Biol. Chem.* **270,** 2411 (1995).
25. J. L. Manley, *Biochim. Biophys. Acta* **950,** 1 (1988).
26. E. Wahle and W. Keller, *Annu. Rev. Biochem.* **61,** 419 (1992).
27. E. Wahle and U. Kuhn, *Prog. Nucleic Acid Res. Mol. Biol.* **57,** 41 (1997).
27a. J. Zhao, L. Hyman, and C. Moore, *Microbiol. Mol. Biol. Rev.* **63,** 405 (1999).
28. M. Edmonds and M. A. Winters, *Prog. Nucleic Acid Res. Mol. Biol.* **17,** 149 (1976).
29. M. Edmonds, *The Enzymes,* **XV,** p. 217 (1982).
30. G. Brawerman, *Crit. Rev. Biochem.* **10,** 1 (1981).
31. J. Diez and G. Brawerman, *PNAS* **71,** 4041 (1974).
32. G. Brawerman and J. Diez, *Cell* **5,** 271 (1975).
33. S. Sawicki, W. Jelinek, and J. E. Darnell, **113,** 219 (1977).
34. A. A. Travers and R. Burgess, *Nature* **222,** 537 (1969).
35. V. J. Bardwell, D. Zarkower, M. Edmonds, and M. Wickens, *Mol. Cell. Biol.* **10,** 846 (1990).
36. T. Raabe, F. J. Bollum, and J. L. Manley, *Nature* **353,** 229 (1991).
37. N. J. Proudfoot and G. Brownlee, *Nature* **263,** 211 (1976).
38. J. R. Nevins and J. E. Darnell, *Cell* **15,** 1477 (1978).
39. N. W. Fraser, J. R. Nevins, E. Ziff, and J. E. Darnell, *J. Mol. Biol.* **129,** 643 (1979).
40. J. Ford and M. T. Hsu, *J. Virol.* **28,** 795 (1978).
41. J. R. Nevins, J. M. Blanchard, and J. E. Darnell, *J. Mol. Biol.* **144,** 377 (1980).
42. E. Hofer and J. E. Darnell, *Cell* **23,** 585 (1981).
43. M. Fitzgerald and T. Shenk, *Cell* **24,** 251 (1981).
44. M. Wickens and P. Stephenson, *Science* **226,** 1045 (1983).
45. M. P. Wickens and J. B. Gurdon, *J. Mol. Biol.* **163,** 1 (1983).
46. C. Montell, E. F. Fisher, M. H. Caruthers, and A. Berk, *Nature* **305,** 600 (1983).
47. D. R. Higgs, S. E. Goodbourn, J. Lamb, J. B. Clegg, D. J. Weatherall, and N. J. Proudfoot, *Nature* **306,** 398 (1983).

48. D. R. Setzer, M. McGrogan, and R. T. Schimke, *J. Biol. Chem.* **257,** 5143 (1982).
49. A. Hook and R. E. Kellems, *J. Biol. Chem.* **263,** 2337 (1988).
50. S. E. Leff, M. G. Rosenfeld, and R. M. Evans, *Annu. Rev. Biochem.* **55,** 1091 (1986).
51. F. R. Blattner and P. W. Tucker, *Nature* **307,** 417 (1984).
52. M. A. McDevitt and M. J. Imperiale, H. Ali, and J. R. Nevins, *Cell* **37,** 993 (1984).
53. M. Sadowski and J. C. Alwine, *Mol. Cell Biol.* **4,** 1460 (1984).
54. A. Gil and N. J. Proudfoot, *Nature* **312,** 473 (1984).
55. R. P. Hart, M. A. McDevitt, H. Ali, and J. R. Nevins, *Mol. Cell Biol.* **5,** 2975 (1986).
56. M. A. McDevitt, R. P. Hart, W. Wong, and J. R. Nevins, *EMBO J.* **5,** 2907 (1986).
57. A. Gil and N. J. Proudfoot, *Cell* **49,** 399 (1987).
58. J. S. Chen and J. Nordstrom, *NAR* **20,** 2565 (1992).
59. R. P. Woychik, R. H. Lyons, L. Post, and F. M. Rottman, *PNAS* **81,** 3944 (1984).
60. E. C. Goodwin and F. M. Rottman, *J. Biol. Chem.* **267,** 330 (1992).
61. R. P. Hart, M. A. McDevitt, and J. R. Nevins, *Cell* **43,** 677 (1985).
62. T. L. Green and R. P. Hart, *Mol. Cell. Biol.* **8,** 1839 (1988).
63. J. McLaughlan, D. Gaffney, J. L. Whitton, and J. B. Clements, *Nucleic Acids Res.* **13,** 1347 (1985).
64. C. C. McDonald, J. Wilusz, and T. Shenk, *Mol. Cell. Biol.* **14,** 6647 (1994).
65. M. A. Winters and M. Edmonds, *J. Biol. Chem.* **248,** 4763 (1973).
66. R. G. Roeder and W. J. Rutter, *Nature* **224,** 234 (1969).
67. J. L. Manley, P. A. Sharp, and M. L. Gefter, *PNAS* **76,** 160 (1979).
68. V. W. Yang and S. J. Flint, *J. Virol.* **32,** 394 (1979).
69. W. R. Jelinek, *Cell* **2,** 197 (1974).
70. R. M. Kieras, R. J. Almendinger, and M. Edmonds, *Biochemistry* **17,** 3221 (1978).
71. K. Shafer, *NAR* **4,** 4465 (1977).
72. J. L. Manley, P. A. Sharp, and M. L. Gefter, *J. Mol. Biol.* **159,** 581 (1982).
73. P. A. Weil, D. S. Luse, J. Segal, and R. G. Roeder, *Cell* **18,** 469 (1979).
74. J. L. Manley, A. Fire, A. Cano, P. A. Sharp, and M. L. Gefter, *PNAS* **77,** 3855 (1980).
75. J. L. Manley, *Prog. Nucleic Acid Res. Mol. Biol.* **30,** 195 (1983).
76. J. L. Manley, *Cell* **33,** 595 (1983).
77. D. A. Melton, P. A. Krieg, M. R. Rebagliati, T. Maniatis, K. Zinn, and M. R. Green, *Nucleic Acids Res.* **12,** 7035 (1984).
78. J. D. Dignam, R. M. Lebowitz, and R. G. Roeder, *Nucleic Acids Res.* **11,** 1475 (1983).
79. C. L. Moore and P. A. Sharp, *Cell* **36,** 581 (1984).
80. A. R. Krainer, T. Maniatis, B. Ruskin, and M. R. Green, *Cell* **36,** 993 (1984).
81. R. C. Mifflin and R. E. Kellems, *J. Biol. Chem.* **266,** 19,593 (1991).
82. N. J. Proudfoot, *Trends Biochem. Sci.* **14,** 105 (1989).
83. S. McCracken, N. Fong, K. Yankulov, S. Ballentyne, G. Pan, J. Greenblatt, S. D. Patterson, M. Wickens, and D. L. Bentley, *Nature* **385,** 357 (1997).
84. J. C. Dantonel, K. G. Murthy, J. L. Manley, and L. Tora, *Nature* **389,** 399 (1997).
85. C. L. Moore and P. A. Sharp, *Cell* **41,** 845 (1985).
86. C. L. Moore, H. Skolnik-David, and P. A. Sharp, *EMBO J.* **5,** 1929 (1986).
87. J. L. Manley, H. Yu, and L. Ryner, *Mol. Cell. Biol.* **5,** 373 (1985).
88. D. Zarkower, P. Stephenson, M. Sheets, and M. Wickens, *Mol. Cell. Biol.* **6,** 2317 (1986).
89. L. Conway and M. Wickens, *EMBO J.* **6,** 4177 (1987).
90. P. J. Grabowski, S. R. Seiler, and P. A. Sharp, *Cell* **42,** 345 (1985).
91. D. Zarkower and M. Wickens, *EMBO J.* **6,** 177 (1987).
92. D. Zarkower and M. Wickens, *EMBO J.* **6,** 4185 (1987).
93. T. Humphrey, G. Christofori, V. Lucijanic, and W. Keller, *EMBO J.* **6,** 4159 (1987).
94. F. Zhang and C. N. Coles, *Mol. Cell. Biol.* **7,** 3277 (1987).

95. H. Skolnik-David, C. L. Moore, and P. A. Sharp, *Genes Dev.* **1,** 672 (1987).
96. M. M. Konarska and P. A. Sharp, *Cell* **46,** 845 (1986).
97. C. L. Moore, H. Skolnick-David, and P. A. Sharp, *Mol. Cell. Biol.* **226,** (1988).
98. J. E. Stefano and D. E. Adams, *Mol. Cell. Biol.* **8,** 205 (1988).
99. S. M. Berget and B. L. Robberson, *Cell* **46,** 691 (1988).
100. G. Cristofori and W. Keller, *Cell* **54,** 875 (1988).
101. G. Cristofori and W. Keller, *Mol. Cell. Biol.* **9,** 193 (1991).
102. Y. Takagaki, L. C. Ryner, and J. L. Manley, *Cell* **52,** 731 (1988).
103. S. Bienroth, E. Wahle, C. Suter-Crazzolara, and W. Keller, *J. Biol. Chem.* **266,** 19, 768 (1991).
104. W. Keller, *Cell* **81,** 829 (1995).
105. C. S. Lutz, K. G. K. Murthy, N. Shek, J. P. O'Connor, J. L. Manley, and J. C. Alwine, *Genes Dev.* **10,** 325 (1995).
106. G. M. Gilmartin, M. A. McDevitt, and J. R. Nevins, *Genes Dev.* **2,** 578 (1988).
107. M. A. McDevitt, G. M. Gilmartin, W. H. Reeves, and J. R. Nevins, *Genes Dev.* **2,** 588 (1988).
108. Y. Takagaki, L. C. Ryner, and J. L. Manley, *Genes Dev.* **3,** 1711 (1989).
109. H. J. Schoenmaker and P. R. Schimmel, *J. Mol. Biol.* **84,** 503 (1974).
110. B. Ehresmann, E. Backendorf, and J. P. Ebel, *FEBS Lett.* **78,** 261 (1977).
111. G. Dreyfuss, Y. D. Choi, and S. A. Adam, *Mol. Cell. Biol.* **4,** 1104 (1984).
112. J. Wilusz and T. Shenk, *Cell* **52,** 221 (1988).
113. C. L. Moore, J. Chen, and J. Whoriskey, *EMBO J.* **7,** 3159 (1988).
114. G. M. Gilmartin and J. R. Nevins, *Genes Dev.* **3,** 2180 (1989).
115. G. M. Gilmartin and J. R. Nevins, *Mol. Cell. Biol.* **11,** 2432 (1991).
116. J. Wilusz, T. Shenk, Y. Takagaki, and J. L. Manley, *Mol. Cell. Biol.* **10,** 1244 (1990).
117. E. A. Weiss, G. M. Gilmartin, and J. R. Nevins, *EMBO J.* **10,** 215 (1991).
118. K. G. K. Murthy and J. L. Manley, *J. Biol. Chem.* **267,** 14, 804 (1992).
119. S. Bienroth, W. Keller, and E. Wahle, *EMBO J.* **12,** 585 (1993).
120. E. Wahle, *Cell* **66,** 759 (1991).
121. Y. Takagaki, J. L. Manley, C. C. McDonald, J. Wilusz, and T. Shenk, *Genes Dev.* **4,** 2112 (1990).
122. V. J. Bardwell, D. Zarkower, M. Edmonds, and M. Wickens, *Mol. Cell. Biol.* **10,** 846 (1990).
123. W. Keller, S. Bienroth, K. M. Lang, and G. Christofori, *EMBO J.* **10,** 4241 (1991).
124. E. Wahle, *J. Biol. Chem.* **266,** 3131 (1991).
125. T. Raabe, F. J. Bollum, and J. L. Manley, *Nature* **353,** 229 (1991).
126. E. Wahle, G. Martin, E. Schlitz, and W. Keller, *EMBO J.* **10,** 4251 (1991).
127. C. Hashimoto and J. A. Steitz, *Cell* **45,** 581 (1986).
128. K. Wassarman and J. A. Steitz, *Genes Dev.* **7,** 647 (1993).
129. C. S. Lutz and J. C. Alwine, *Genes Dev.* **8,** 576 (1994).
130. V. J. Bardwell, M. Wickens, S. Bienroth, W. Keller, R. S. Sproat, and A. I. Lamond, *Cell* **65,** 125 (1991).
131. A. Jenny, H. P. Hauri, and W. Keller, *Mol. Cell. Biol.* **14,** 8183 (1994).
132. K. G. K. Murthy and J. L. Manley, *Genes Dev.* **9,** 2672 (1995).
133. S. T. Jacob and K. Rose, "Enzymes of Nucleic Acid Synthesis and Modification," Vol. 2, p. 135. CRC Press, Boca Raton, FL, 1983.
134. E. Wahle, *Cell* **66,** 759 (1991).
135. M. D. Sheets and M. Wickens, *Genes Dev.* **3,** 1401 (1989).
136. E. Wahle, A. Lustig, P. Jeno, and P. Mauer, *J. Biol. Chem.* **268,** 2937 (1993).
137. S. Kraus, S. Fakan, K. Weis, and E. Wahle, *Exp. Cell Res.* **214,** 75 (1994).
138. E. Wahle, *J. Biol. Chem.* **270,** 2800 (1995).
139. D. A. Stetler and S. T. Jacobs, *J. Biol. Chem.* **259,** 7239 (1984).

140. C. G. Burd and G. Dreyfuss, *Science* **265**, 615 (1994).
141. G. Cao and N. Sarkar, *PNAS* **89**, 10, 380 (1992).
142. R. Gentz, C. Chen, and C. A. Rosen, *PNAS* **86**, 821 (1989).
143. D. W. Miller, P. Safer, and L. K. Miller, "Genetic Engineering," Vol. 8, p. 277. Plenum Press, New York, 1986.
144. A. Thuresson, J. Astrom, A. Astrom, K. Gronvik, and A. Virtanen, *PNAS* **91**, 979 (1994).
145. S. Ballantyne, A. Bilger, J. Astrom, A. Virtanen, and M. Wickens, *RNA* **1**, 64 (1995).
146. J. Lingner, J. Kellermann, and W. Keller, *Nature* **354**, 496 (1991).
147. W. Zhao and J. L. Manley, *Mol. Cell. Biol.* **16**, 2378 (1996).
148. F. Gebauer and J. D. Richter, *Mol. Cell. Biol.* **15**, 1422 (1995).
149. A. Bilger, C. A. Fox, E. Wahle, and M. Wickens, *Genes Dev.* **8**, 1106 (1994).
150. T. Raabe, K. G. K. Murthy, and J. L. Manley, *Mol. Cell Biol.* **14**, 2946 (1994).
151. G. Martin and W. Keller, *EMBO J.* **15**, 2593 (1996).
152. G. Martin, P. Jeno, and W. Keller, *Protein Sci.* **8**, 2380 (1999).
153. C. M. Joyce and T. A. Steitz, *Annu. Rev. Biochem.* **63**, 777 (1994).
154. G. Martin, W. Keller, and S. Doublie, *EMBO J.* **19**, 4193 (2000).
155. Y. Takagaki, C. C. McDonald, T. Shenk, and J. L. Manley, *PNAS* **89**, 1403 (1992).
156. Y. Takagaki and J. L. Manley, *J. Biol. Chem.* **267**, 471 (1992).
157. Y. Takagaki and J. L. Manley, *Nature* **372**, 471 (1994).
158. M. I. Simon, M. P. Strathmann, and N. Grutam, *Science* **252**, 802 (1991).
159. J. Hu, Y. Xu, K. Schappert, T. Harrington, A. Wang, R. Braga, J. Mogridge, and J. Friesen, *Nucleic Acids Res.* **22**, 1724 (1994).
160. F. E. Kleinman and J. L. Manley, *Nature* **285**, 1576 (1999).
161. D. Dorsett, A. Viglianti, B. J. Rutledge, and M. Meselson, *Genes Dev.* **3**, 454 (1989).
162. A. Mitchelson, M. Simonvelig, C. Williams, and K. O'Hare, *Genes Dev.* **7**, 241 (1993).
163. L. Minvielle-Sebastia, B. Winsor, N. Bonneaud, and F. LaCroute, *Mol. Cell. Biol.* **11**, 3075 (1991).
164. L. Minvielle-Sebastia, P. J. Preker, and W. Keller, *Science* **266**, 1702 (1994).
165. Y. Takagaki, R. L. Seipelt, M. L. Peterson, and J. L. Manley, *Cell* **87**, 941 (1996).
166. A. Jenny and W. Keller, *Nucleic Acids Res.* **23**, 2629 (1995).
167. A. Jenny, L. Minvielle-Sebastia, P. J. Preker, and W. Keller, *Science* **274**, 1514 (1996).
168. A. Jenney, H. P. Hauri, and W. Keller, *Mol. Cell. Biol.* **14**, 8183 (1994).
169. S. M. Barbino, W. Hubner, A. Jenny, L. Minvielle-Sebastia, and W. Keller, *Genes Dev.* **11**, 1703 (1997).
170. S. M. Noble and C. Guthrie, *Genetics* **143**, 67 (1996).
171. G. Chanfreau, S. Noble, and C. Guthrie, *Science* **274**, 1511 (1996).
172. J. Chen and C. Moore, *Mol. Cell. Biol.* **12**, 3470 (1992).
173. G. Stumpf and H. Domdey, *Science* **274**, 1517 (1996).
174. A. Nemeth, S. Krause, D. Blank, A. Jenny, P. Jemo, A. Lustig, and E. Wahle, *Nucleic Acids Res.* **23**, 4034 (1996).
175. U. Ruegsegger, K. Beyer, and W. Keller, *J. Biol. Chem.* **271**, 6107 (1996).
176. U. Ruegsegger, D. Blank, and W. Keller, *Mol. Cell* **1**, 243 (1998).
177. A. Casti, A. Corti, N. Reali, G. Mezzetti, G. Orlandini, and C. M. Caldarera, *Biochem. J.* **168**, 333, 341 (1977).
178. K. M. Rose and S. T. Jacob, *J. Biol. Chem.* **254** (10), 256 (1979).
179. K. M. Rose and S. T. Jacob, *Biochemistry* **19**, 1472 (1980).
180. D. F. Colgan, K. Murthy, C. Prives, and J. L. Manley, *Nature* **384**, 282 (1996).
181. J. R. Nevins and M. C. Wilson, *Nature* **290**, 113 (1981).
182. A. R. Shaw and E. B. Ziff, *Cell* **22**, 905 (1980).
183. G. Akusjarvi and H. Person, *Nature* **292**, 420 (1981).

184. D. R. Setzer, M. McGrogan, J. H. Nunberg, and R. T. Schimke, *Cell* **22,** 361 (1980).
185. J. J. Yen and R. E. Kellems, *Mol. Cell Biol.* **7,** 3732 (1987).
186. F. W. Alt, R. E. Kellems, and R. T. Schimke, *J. Biol. Chem.* **251,** 3063 (1976).
187. S. L. Hendrickson, J. R. Wu, and L. F. Johnson, *PNAS* **77,** 5140 (1980).
188. E. J. Leys and R. E. Kellems, *Mol. Cell Biol.* **1,** 96, 1 (1981).
189. R. J. Kaufman and P. A. Sharp, *Mol. Cell Biol.* **3,** 1602 (1983).
190. G. Edwalds-Gilbert, K. L. Veraldi, and C. Milcarek, *Nucleic Acids Res.* **25,** 2547 (1997).
191. S. Miyamoto, J. Chiorini, E. Ureelay, and B. Safer, *Biochemistry* **315,** 791 (1996).
192. M. G. Rosenfeld, S. G. Amara, B. A. Roos, E. S. Ong, and R. M. Evans, *Nature* **290,** 63, (1981).
193. J. Rogers, P. Early, C. Carter, K. Calame, M. Bond, L. Hood, and R. Wall, *Cell* **20,** 303 (1980).
194. S. E. Leff, M. G. Rosenfeld, and R. M. Evans, *Cell* **48,** 517 (1987).
195. H. Lou, Y. Yang, G. J. Cote, S. M. Berget, and R. F. Gagel, *Mol. Cell Biol.* **15,** 7135 (1995).
196. H. Lou, R. F. Gagel, and S. M. Berget, *Genes Dev.* **10,** 208 (1996).
197. H. Lou, D. M. Helfman, R. F. Gagel, and S. M. Berget, *Mol. Cell. Biol.* **19,** 78 (1999).
198. M. L. Peterson, "Handbook of B and T Lymphocytes," (E. C. Snow, ed.), p. 321. Academic Press, San Diego, 1994.
199. N. Levitt, D. Briggs, A. Gil, and N. J. Proudfoot, *Genes Dev.* **3,** 1019 (1989).
200. R. M. Denome and C. N. Cole, *Mol. Cell. Biol.* **8,** 4829 (1988).
201. G. Galli, J. Guise, M. A. McDevitt, P. W. Tucker, and J. R. Nevins, *Genes Dev.* **1,** 471 (1987).
202. C. R. Lasman and C. Milcarek, *J. Immunol.* **148,** 2578 (1992).
203. K. R. Mann, E. A. Weiss, and J. R. Nevins, *Mol. Cell Biol.* **13,** 2411 (1993).
204. D. H. Yan, E. A. Weiss, and J. R. Nevins, *Mol. Cell Biol.* **15,** 1901 (1995).
205. G. Edwalds-Gilbert and C. Milcarek, *Mol. Cell Biol.* **15,** 6420 (1995).
206. K. Martincic, R. Campbell, G. Edwalds-Gilbert, L. Souan, M. T. Lotze, and C. Milcarek, *PNAS* **95,** 11,095 (1998).
206a. K. L. Veraldi, G. K. Arhin, K. Martincic, L. H. Chung-Ganster, J. Wilusz, and C. Milcarek, *Mol. Cell. Biol.* **21,** 1228 (2001).
207. Y. Takagaki and J. L. Manley, *Mol. Cell* **2,** 761 (1998).
208. M. E. Nemeroff, S. M. Barabino, Y. Li, W. Keller, and R. M. Krug, *Mol. Cell* **1,** 991 (1998).
209. Z. Chen, Y. Li, and R. M. Krug, *EMBO J.* **18,** 2273 (1999).
210. X. Qian, F. Alonso-Caplan, and R. M. Krug, *J. Virol.* **68,** 2433 (1994).
211. M. Edmonds and D. W. Kopp, *Biochem. Biophys. Res. Commun.* **41,** 1531 (1970).
212. C. S. McLaughlin, J. R. Warner, M. Edmonds, H. Nakazato, and M. H. Vaughan, *J. Biol. Chem.* **248,** 1466 (1973).
213. L. A. Haff and E. B. Keller, *Biochem. Biophys. Res. Commun.* **51,** 704 (1973).
214. L. A. Haff and E. B. Keller, *J. Biol. Chem.* **250,** 1838 (1975).
215. J. C. Bloch, F. Perrin, and F. Lacroute, *Mol. Genet.* **165,** 123 (1978).
216. K. S. Zaret and F. Sherman, *Cell* **28,** 563 (1982).
217. K. S. Zaret and F. Sherman, *J. Mol. Biol.* **177,** 107 (1984).
218. J. S. Butler and T. Platt, *Science* **242,** 1270 (1988).
219. J. Chen and C. L. Moore, *Mol. Cell. Biol.* **12,** 3470 (1992).
220. W. Keller and L. Minvielle-Sebastia, *Curr. Opin. Cell Biol.* **9,** 329 (1997).
221. P. Russo, W.-Z. Li, Z. Guo, and F. Sherman, *Mol. Cell. Biol.* **13,** 7836 (1993).
222. Z. Guo and F. Sherman, *Trends Biochem. Sci.* **21,** 477 (1996).
223. L. E. Hyman, S. H. Seiler, J. Whoriskey, and C. L. Moore, *Mol. Cell. Biol.* **11,** 2004 (1991).
224. S. Henikoff, J. D. Kelly, and E. H. Cohen, *Cell* **33,** 607 (1983).
225. P. Russo, W.-Z. Li, D. M. Hampsey, K. S. Zaret, and F. Sherman, *EMBO J.* **10,** 563 (1991).
226. Z. Guo and F. Sherman, *Mol. Cell. Biol.* **15,** 5983 (1995).

227. J. S. Butler, P. P. Sadhale, and T. Platt, *Mol. Cell. Biol.* **10,** 2599 (1990).
228. Z. Guo, P. Russo, D.-F. Yun, S. Butler, and Fisherman, *PNAS* **92,** 4211 (1995).
229. S. Irniger and G. H. Braus, *PNAS* **91,** 257 (1994).
230. A. Abe, Y. Hiraoka, and T. Fukasawa, *EMBO J.* **9,** 3691 (1990).
231. S. Heidmann, C. Schinewolf, and H. Domdey, *Mol. Cell. Biol.* **14,** 4633 (1994).
232. Z. Guo and F. Sherman, *Mol. Cell. Biol.* **16,** 2772 (1996).
233. M. M. Kessler, J. Zhao, and C. L. Moore, *J. Biol. Chem.* **271,** 27,167 (1996).
234. J. Zhao, M. M. Kessler, and C. L. Moore, *J. Biol. Chem.* **272** (10), 831 (1997).
235. P. J. Preker, M. Ohnacker, L. Minvielle-Sebastia, and W. Keller, *EMBO J.* **16,** 4727 (1997).
236. P. J. Preker, J. Lingner, and L. Minvielle-Sebastia, *Cell* **81,** 379 (1995).
237. N. Amrani, M. Minet, F. Wyers, M. E. Dufour, L. P. Aggerbeck, and F. Lacroute, *Mol. Cell. Biol.* **17,** 1102 (1997).
238. L. Minvielle-Sebastia and W. Keller, *Curr. Opin. Cell. Biol.* **11,** 352 (1999).
239. M. M. Kessler, M. F. Henry, E. Shen, J. Zhao, S. Gross, P. A. Silver, and C. L. Moore, *Genes Dev.* **11,** 2545 (1997).
240. M. Henry, C. Z. Borland, M. Bossie, and P. A. Silver, *Genetics* **142,** 103 (1996).
241. L. Minvielle-Sebastia, K. Beyer, A. M. Krecic, R. E. Hector, M. S. Swanson, and W. Keller, *EMBO J.* **17,** 7454 (1998).
242. A. Mayeda and A. R. Krainer, *Cell* **68,** 365 (1992).
243. P. O'Connor and C. Peebles, *Mol. Cell. Biol.* **12,** 3843 (1992).
244. J. Lingner, I. Radtke, E. Wahle, and W. Keller, *J. Biol. Chem.* **266,** 8741 (1991).
245. A. M. Zhelkovsky, M. M. Kessler, and C. L. Moore, *J. Biol. Chem.* **270,** 26,715 (1995).
246. A. Zhelkovsky, M. M. Kessler, and C. L. Moore, *Mol. Cell. Biol.* **18,** 5942 (1998).
247. L. Guarente, *Trends Genet.* **9,** 369 (1993).
248. C.-T. Chien, P. L. Bartel, R. Sternglanz, and S. Fields, *PNAS* **88,** 9578 (1991).
249. N. Amrani, M. Minet, M. LeGovar, F. Lacroute, and F. Wyers, *Mol. Cell. Biol.* **17,** 3694 (1997).
250. A. Sachs and R. W. Davis, *Cell* **58,** 857 (1989).
251. G. Caponigro and R. R. Parker, *Genes Dev.* **9,** 2421 (1995).
252. L. Minvielle-Sebastia, P. J. Preker, T. Wiederkehr, Y. Strahm, and W. Keller, *PNAS* **94,** 7897 (1997).
253. J. Bard, A. M. Zhelkovsky, S. Helmling, T. N. Earnest, C. L. Moore, and A. Bohm, *Science* **289,** 1346 (2000).
254. N. Sarkar, *Microbiology* **142,** 3125 (1996).
255. N. Sarkar, *Annu. Rev. Biochem.* **66,** 173 (1997).
256. J. T. August, P. Ortiz, and J. Hurwitz, *J. Biol. Chem.* **237,** 3786 (1962).
257. M. E. Gottesman, Z. N. Canellakis, and E. S. Canellakis, *Biochem. Biophys. Acta* **61,** 34 (1962).
258. S. J. S. Hardy and C. G. Kurland, *Biochemistry* **5,** 3668 (1966).
259. A. E. Sippel, *Eur. J. Biochem.* **37,** 31 (1973).
260. M. Terzi, A. Cascino, and C. Urbani, *Nature* **226,** 1052 (1970).
261. H. Sano and G. Faix, *Eur. J. Biochem.* **71,** 577 (1976).
262. M. Edmonds and M. G. Caramela, *J. Biol. Chem.* **244,** 1314 (1969).
263. H. Nakazato, S. Venkatesan, and M. Edmonds, *Nature* **256,** 144 (1975).
264. P. R. Srinivasan, M. Ramanarayanan, and E. Rabani, *PNAS* **72,** 2910 (1975).
265. P. Oritz, J. T. August, M. Watanabe, A. M. Kaye, and J. Hurwitz, *J. Biol. Chem.* **240,** 423 (1965).
266. B. Goz, *Biochim. Biophys. Acta* **204,** 267 (1970).
267. N. Sarkar, D. Langley, and H. Paulus, *Biochemistry* **17,** 3468 (1978).
268. Y. Gopalkrishnan, D. Langley, and N. Sarkar, *Nucleic Acids Res.* **9,** 3545 (1981).

269. Y. Gopalkrishnan and N. Sarkar, *Biochemistry* **21**, 2724 (1982).
270. Y. Gopalkrishnan and N. Sarkar, *J. Biol. Chem.* **257**, 2747 (1982).
271. P. Karnik, Y. Gopalkrishnan, and N. Sarkar, *Gene* **49**, 161 (1986).
272. K. Nakamura, R. M. Pirtle, I. L. Pirtle, K. Takeishi, and M. Inouye, *J. Biol. Chem.* **255**, 210 (1980).
273. K. Nakamura and M. Inouye, *EMBO J.* **1**, 771 (1982).
274. J. Taljanidisz, P. Karnik, and N. Sarkar, *J. Mol. Biol.* **193**, 507 (1987).
275. P. Karnik, J. Taljanidisz, M. Sasvari-Szekely, and N. Sarkar, *JMB* **196**, 347 (1987).
276. G. J. Cao and N. Sarkar, *PNAS* **89**, 7546 (1992).
277. J. Lopilato, S. Bortner, and J. Beckwith, *Mol. Gen. Genet.* **205**, 285 (1986).
278. J. Liu and J. S. Parkinson, *J. Bact.* **171**, 1254 (1989).
279. M. Masters, J. B. March, I. R. Oliver, and J. F. Collins, *Mol. Gen. Genet.* **220**, 341 (1990).
280. S. Lin-Chao and S. N. Cohen, *Cell* **65**, 1233 (1991).
281. R. M. Lacatena and G. Cesarini, *Nature* **294**, 623 (1981).
282. J. Tomizawa and T. Itoh, *PNAS* **78**, 6096 (1981).
283. F. Xu, S. Lin-Chao, and S. N. Cohen, *PNAS* **90**, 6756 (1993).
284. L. He, F. Soderbom, E. Gerhart, N. Binus, and M. Masters, *Mol. Microbiol.* **9**, 1131 (1993).
285. P. Babitzke, L. Granger, J. Olszewski, and S. R. Kushner, *J. Bact.* **175**, 229 (1993).
286. E. A. Mudd, H. M. Krisch, and C. F. Higgins, *Mol. Microbiol.* **4**, 2127 (1990).
287. P. Regnier and E. Hajnsdorf, *J. Mol. Biol.* **217**, 283 (1991).
288. E. B. O'Hara, J. A. Chekanova, C. A. Ingle, Z. R. Kushner, E. Peters, and S. R. Kushner, *PNAS* **92**, 1807 (1995).
289. E. Hajnsdorf, F. Braun, J. Haugel-Nielsen, and P. Regnier, *PNAS* **92**, 3973 (1995).
290. S. N. Cohen, *Cell* **80**, 829 (1995).
291. A. J. Carpousis, G. Van Houwe, G. Ehretsmann, and H. M. Krisch, *Cell* **76**, 889 (1994).
292. B. Py, H. Causton, E. A. Mudd, and C. F. Higgins, *Mol. Microbiol.* **14**, 717 (1994).
293. B. Py, C. F. Higgins, H. M. Krisch, and A. J. Carpousis, *Nature* **381**, 169 (1996).
294. G. A. Coburn and G. A. Mackie, *Prog. Nucleic Acid Res. Mol. Biol.* **62**, 55 (1999).
295. E. Blum, A. J. Carpousie, and C. F. Higgins, *JBC* **274**, 4009 (1999).
296. B. K. Mohanty and S. R. Kushner, *Mol. Microbiol.* **34**, 1094 (1999).
297. M. P. Kalapos, G. J. Cao, S. R. Kushner, and N. Sarkar, *Biochem. Biophys. Res. Commun.* **198**, 459 (1994).
298. N. Kleckner, J. Bender, and S. Gottesman, *Methods in Enzymol.* **204**, 10,473 (1991).
299. G. J. Cao, J. Pogliano, and N. Sarkar, *PNAS* **93**, 1158 (1996).
300. B. K. Mohanty and S. R. Kushner, *Mol. Microbiol.* **34**, 1109 (1999).
301. B. K. Mohanty and S. R. Kushner, *PNAS* **97**, 11,966 (2000).
302. P. Bernstein, S. W. Peltz, and J. Ross, *Mol. Cell. Biol.* **9**, 659 (1989).
303. C. J. Decker and R. Parker, *Genes Dev.* **7**, 1632 (1993).
304. C.-Y. A. Chen, N. Xu, and A.-B. Shyu, *Mol. Cell. Biol.* **15**, 5777 (1995).
305. N. Xu, C.-Y. A. Chen, and A.-B. Shyu, *Mol. Cell. Biol.* **17**, (1997).
306. L. F. Ford, J. Watson, J. D. Keene, and J. Wilusz, *Genes Dev.* **13**, 188 (1999).
307. G. Shaw and R. Kamen, *Cell* **46**, 659 (1986).
308. C. G. Korner and E. Wahle, *J. Biol. Chem.* **272**, 10,448 (1997).
309. J. H. Subak-Sharpe, T. H. Pennington, J. F. Szilagyi, M. C. Timbury, and J. F. Williams, "The Effect of Rifampicin on Mammalian Viruses and Cells," Vol. 1, p. 260. Lepetit Coll. Biol. Med., North-Holland, Amsterdam, 1970.
310. J. Kates and J. Beeson, *J. Mol. Biol.* **50**, 19 (1970).
311. M. Chamberlin and P. Berg, *PNAS* **48**, 81 (1962).
312. M. K. Brown, J. W. Dorson, and F. J. Bollum, *J. Virol.* **12**, 203 (1973).
313. B. Moss, E. N. Rosenblum, and E. Paolettti, *Nature NB* **245**, 59 (1973).

314. B. Moss and E. N. Rosenblum, *J. Virol.* **14,** 86 (1974).
315. B. Moss, E. N. Rosenblum, and A. Gershowitz, *J. Biol. Chem.* **250,** 4722 (1975).
316. C. Brakel and J. Kates, *J. Virol.* **14,** 715 (1974).
317. C. Brakel and J. Kates, *J. Virol.* **14,** 724 (1974).
318. J. Nevins and W. K. Joklik, *JBC* **252,** 6939 (1977).
319. G. Rohrman, L. Yuen, and B. Moss, *Cell* **46,** 1029 (1986).
320. L. Yuen and B. Moss, *PNAS* **84,** 6417 (1987).
321. E. Whitelaw and N. J. Proudfoot, *EMBO J.* **5,** 2915 (1986).
322. P. Gershon, B.-Y. Ahu, M. Garfield, and B. Moss, *Cell* **66,** 1269 (1991).
323. B. Moss, P. Earl, and N. Cooper, "Current Protocols in Molecular Biology," Vol. 2, p. 16.15–16.98. Greene and Wiley, New York, 1991.
324. P. Gershon and B. Moss, *Genes Dev.* **6,** 1575 (1992).
325. P. Gershon and B. Moss, *J. Biol. Chem.* **268,** 2203 (1993).
326. P. Gershon and B. Moss, *EMBO J.* **12,** 4705 (1993).
327. L. Deng and P. Gershon, *EMBO J.* **16,** 1103 (1997).
328. L. Deng, L. Beigelman, J. Matulic-Adamie, A. Karpeisky, and P. D. Gershon, *J. Biol. Chem.* **272** (31), 542 (1997).
329. L. Johnson and P. D. Gershon, *Nucleic Acids Res.* **72,** 2708 (1999).
330. B. S. Schnierle, P. D. Gershon, and B. Moss, *PNAS* **89,** 2897 (1992).
331. H. Kuge and J. D. Richter, *EMBO J.* **14,** 6301 (1995).
332. H. Kuge, G. G. Brownlee, P. D. Gershon, and J. D. Richter, *Nucleic Acids Res.* **26,** 3208 (1998).
333. P. D. Gershon, X. Shi, and A. E. Hodel, *Virology* **246,** 253 (1998).
334. A. E. Hodel, P. D. Gershon, X. Shi, and F. A. Quiocho, *Cell* **85,** 247 (1996).
335. P. D. Gershon, *Seminars Virol.* **8,** 343 (1998).
336. S. E. Wells, P. E. Hilmer, R. D. Vale, and A. B. Sachs, *Mol. Cell.* **2,** 135 (1998).
337. P. D. Gershon, *Nature Struct. Biol.* **7,** 819 (2000).
338. J. W. Brown and J. N. Reeve, *J. Bact.* **166,** 686 (1986).

A Growing Family of Guanine Nucleotide Exchange Factors Is Responsible for Activation of Ras-Family GTPases

LAWRENCE A. QUILLIAM,
JOHN F. REBHUN, AND
ARIEL F. CASTRO

*Department of Biochemistry and Molecular
Biology and Walther Oncology Center
Indiana University School of Medicine
Indianapolis, Indiana 46202*

I. Introduction	392
II. Regulation of *in Vivo* Ras-GTP Levels by Inhibition of GTPase-Activating Proteins	394
III. Early Identification of Ras-Family GEFs	395
IV. GEF Structure and the Nucleotide Exchange Reaction	398
V. Dominant Inhibitory Ras Proteins Target GEFs	404
VI. Biological Assays for GEF Activity	406
VII. Ras-Family GEFs	407
A. Sos1 and 2	407
B. GRF1 and 2	409
C. The GRP/CalDAG-GEF Family	411
D. GRASP-1	414
E. C3G	414
F. The Epac/cAMP-GEFs	416
G. MR-GEF (KIAA0277, GFR, or Repac)	418
H. PDZ-GEFs	418
I. Phospholipase Cε	420
J. Other Putative Rap1 GEFs	422
K. The RalGDS Family	422
L. RalGPS	423
M. BCAR3 and Related Putative GEFs	424
N. Smg GDS	425
VIII. GEFs and Disease	427
IX. Are There More GEFs in Our Future?	428
References	428

Abbreviations: DAG, diacylglycerol; ERK, extracellular receptor-activated protein kinase; GAP, GTPase-activating protein; GEF, guanine nucleotide exchange factor; PI3-kinase, phosphatidylinositol 3-kinase; PKA, protein kinase A or cyclic AMP-dependent protein kinase; RA, Ras association or RalGDS/AF6 homology domain; RBD, Ras-binding domain (a subgroup of RA domains).

GTPases of the Ras subfamily regulate a diverse array of cellular-signaling pathways, coupling extracellular signals to the intracellular response machinery. Guanine nucleotide exchange factors (GEFs) are primarily responsible for linking cell-surface receptors to Ras protein activation. They do this by catalyzing the dissociation of GDP from the inactive Ras proteins. GTP can then bind and induce a conformational change that permits interaction with downstream effectors. Over the past 5 years, approximately 20 novel Ras-family GEFs have been identified and characterized. These data indicate that a variety of different signaling mechanisms can be induced to activate Ras, enabling tyrosine kinases, G-protein-coupled receptors, adhesion molecules, second messengers, and various protein-interaction modules to relocate and/or activate GEFs and elevate intracellular Ras-GTP levels. This review discusses the structure and function of the catalytic or CDC25 homology domain common to almost all Ras-family GEFs. It also details our current knowledge about the regulation and function of this rapidly growing family of enzymes that include Sos1 and 2, GRF1 and 2, CalDAG-GEF/GRP1–4, C3G, cAMP-GEF/Epac 1 and 2, PDZ-GEFs, MR-GEF, RalGDS family members, RalGPS, BCAR3, Smg GDS, and phospholipase Cε.
© 2002, Elsevier Science (USA).

I. Introduction

Ras proteins are molecular switches that couple extracellular stimuli to cellular response machinery. The archetypal *Ras* genes were identified in rodent retroviruses (*1*), but by the mid to late 1980s, a large family of mammalian cDNAs that encoded a superfamily of related GTPases had been cloned. Based on either structure or function, Ras proteins can be divided into a limited number of subfamilies (*2*). The classic Ras-subfamily proteins primarily regulate mitogenic-signaling pathways (*3, 4*); the Rho subfamily is predominantly involved in cytoskeletal regulation (*5–7*); ARFs and the largest Rab-family control membrane trafficking (*8, 9*), while Ran controls nuclear trafficking and mitotic spindle formation (*10, 11*). All have in common a core ∼160-residue GTP-binding domain, a hypervariable domain, and, with the exception of ARF, Rit and Rin, Rerg, and the Rad/GEM family, a lipid-modified C terminus that is required for membrane targeting (ARFs are N terminally myristoylated) (*12*). Each of these proteins acts as a regulated switch that can exist in an either active GTP- or inactive GDP-bound state. Conversion between these states can occur through intrinsic nucleotide exchange or GTP hydrolysis (*13–15*). However, with half-times typically in the order of minutes to hours, this is not of physiological consequence. *In vivo* Ras is rapidly turned on (Ras-GTP) or off (Ras-GDP) by regulatory molecules (*13, 16*). Guanine nucleotide exchange factors (GEFs) promote the release of GDP to enable loading with GTP while GTPase-activating proteins (GAPs) coax the Ras-protein-intrinsic GTPase activities to rapidly hydrolyze their bound GTP, thus returning them to the inactive GDP-bound state (Fig. 1). Either inhibition of GAPs or activation of GEFs will

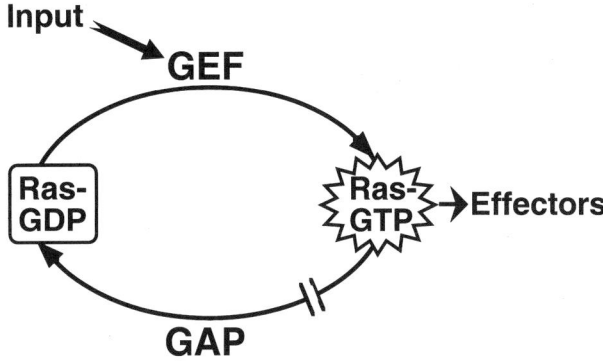

FIG. 1. The Ras GDP/GTP cycle. Ras normally exists in a GDP-bound state in resting cells. Upon external stimulation, guanine nucleotide exchange factors are activated or relocated resulting in their association with Ras and the release of bound nucleotide. GTP is typically in a 10- to 20-fold excess over GDP in the cytoplasm of cells and binds to the Ras-GEF complex, displacing the exchange factor. The binding of GTP induces a conformational in Ras leading to its ability to bind and activate downstream effector proteins. GTPase-activating proteins (GAPs) enhance the intrinsic GTPase activity of Ras causing Ras to adopt its inactive GDP-bound conformation. However, oncogenic mutations at residues 12, 13, and 61 of Ras prevent the intrinsic and GAP-stimulated GTPase activity, thus locking Ras in its active GTP-bound state and constitutively activating effector proteins/pathways.

lead to a shift in equilibrium in favor of the active GTP-bound state. Several examples of ligand-induced GAP regulation have been recorded in the literature (see Section II); however, the most common links between cell-surface ligand/receptor engagement and Ras protein activation are GEFs.

Here we will focus on activators of the Ras subfamily that is made up of ~20 members (see Fig. 2). The founding members, Harvey, Kirsten, and neuroblastoma (H-, K-, and N-) Ras are major conduits of mitogenic signaling between growth factor receptors and the nuclear transcription machinery. Point mutations at codons 12, 13, and 61 block basal and GAP-stimulated GTPase activity resulting in accumulation of increased Ras-GTP and constitutive downstream signaling to effectors such as Raf, phosphatidylinositol (PI) 3-kinase, and RalGDS (4). These oncogenic Ras mutations have been found in approximately 30% of human tumors (17). The closely related TC21/R-Ras2 and M-Ras/R-Ras3 play similar roles, but unlike the classic Ras proteins, activating mutants have not yet been found in primary human tumor samples (18). R-Ras and Rap1 have been implicated in integrin action/cell attachment/spreading (19, 20). Rap1 and related proteins such as NOEY2 may also function as tumor suppressors (21). Ral is a downstream effector of Ras and contributes to gene expression and transformation (22). Details on these and other Ras-family GTPases may be found in the following references (9, 23–27).

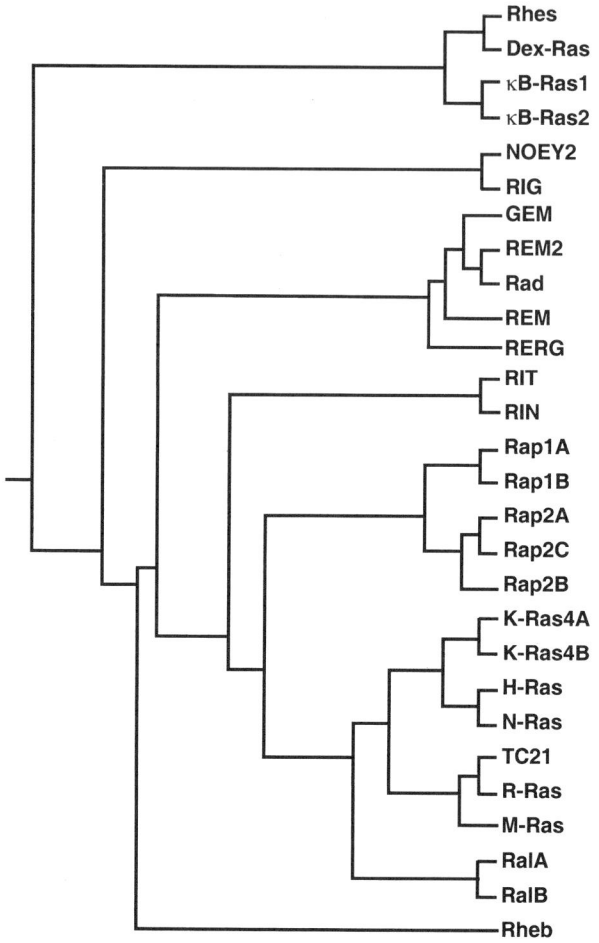

FIG. 2. Homology of Ras-family GTPases. Mammalian Ras-family GTPases were aligned using the ClustalW program and a dendogram generated to indicate their relative homology. Accession numbers for Rap2C (AL049685) and Rig (AY056037) are provided, as these have not been described elsewhere in the literature.

II. Regulation of *in Vivo* Ras-GTP Levels by Inhibition of GTPase-Activating Proteins

Despite extensive literature indicating that GEFs are predominantly responsible for coupling activated receptors to Ras protein activation, a number of examples of Ras activation via inhibition of their cognate GAP(s) have been reported. Since this fact is often overlooked, some examples are described here before focusing on GEFs.

An early report in platelets suggested that erythropoeitin-induced activation of Ras correlated with inhibition of p120 Ras GAP (28). Subsequently, chemotractive peptides have been found to inhibit p120 GAP activity in neutrophils (29). Since PI3-kinase inhibitors abrogate Ras-GTP accumulation via loss of GAP activity (30), D-3-phosphoinositide-mediated inhibition of RasGAP may contribute to Ras activation. The activity of the GTPase Rap1 also appears to be modulated via GAP regulation, and its interaction with heterotrimeric G proteins has been associated with both increased and decreased *in vivo* Rap1-GTP levels. M2-muscarinic-receptor-induced activation of G_i induced the translocation of the Rap1GAPII isoform to the membrane, resulting in reduced Rap1-GTP levels. This decrease in Rap-GTP results in enhanced activation of ERK by Ras (31). Alternatively, the inactive forms of G_i or G_o may sequester Rap1GAP (32). In this mode, it would be the release of Rap1GAP upon G_o activation that would result in suppression of Rap1-GTP levels leading to increased Ras-Raf-MEK ERK signaling. Finally, Carey *et al.* have shown that the ability of the T-cell coreceptor, CD28, to inhibit Rap1-GTP levels to enhance T-cell-receptor signaling via Ras (33) can occur in Lck-mediated Rap1 GAP activation (34). Thus, although GAP regulation may not be the major mechanism of Ras activation, regulation of both Ras and Rap GAPs can promote increased ERK activation.

III. Early Identification of Ras-Family GEFs

The first GEF to be identified was CDC25p, which was found to be genetically upstream of RAS in bakers yeast, *Saccharomyces cerevisiae* (35). Several other proteins sharing homology with CDC25 were also identified in yeast, including STE6, SDC25, LTE1, and BUD5 (16). Although biochemical confirmation that CDC25 was indeed a nucleotide exchange factor did not come until 1991 (36), the advent of PCR led to a flurry of attempts to clone its human homolog(s). Despite these efforts, the first mammalian GEF to be identified was Smg GDS, purified first as an activity that promoted GTP loading of Rap1 (37) and subsequently several other GTPases including members of the Rho subfamily of GTPases (38, 39). Upon cloning, Smg GDS was found to have little or no homology with CDC25 (40) (Fig. 3). In the early 1990s a CDC25-related protein, Sos, was identified in *Drosophila* (41, 42) followed by its mammalian counterparts, Sos1 and 2 (43, 44). These discoveries coincided with the identification of the Ras GEF, $CDC25^{Mm}$/GRF1 (45, 46–48), the Ral GEF, RalGDS (49), and later the Rap1 GEF, C3G (50, 51), all of which contained a catalytic domain sharing homology with *S. cerevisiae*. CDC25p (Figs. 3 and 5). The discovery of

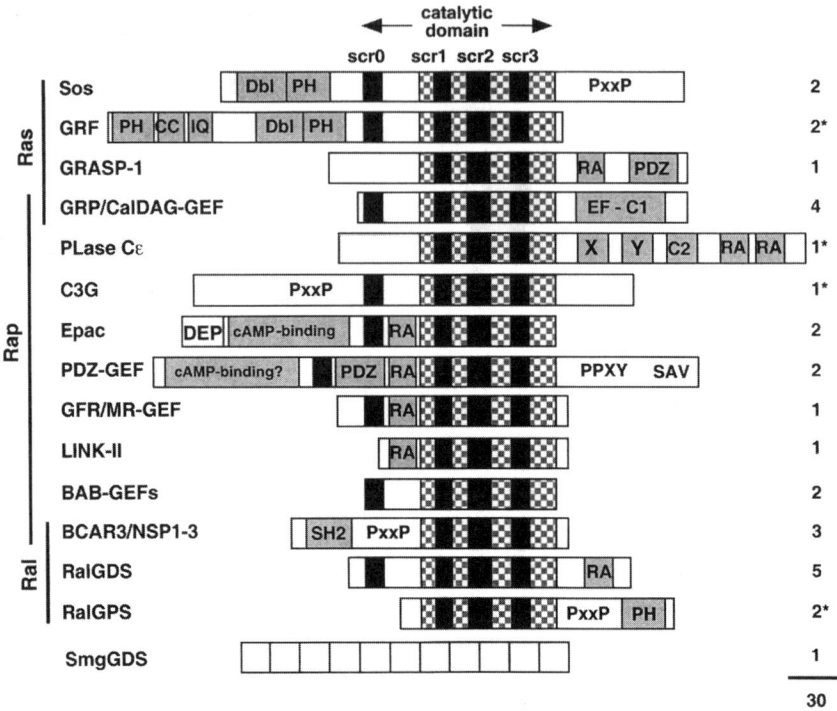

FIG. 3. Diagrams of mammalian Ras-family GEFs indicating the presence of functional domains. The structurally conserved regions (scr) 0, 1, 2, and 3 found in the catalytic domain are shown in black; and the core CDC25 homology region, in checkered gray. Other regulatory domains include Dbl homology (DH) domains that function as Rac GEFs, pleckstrin homology (PH) domains, proline-rich regions that bind SH3 (PXXP) or WW (PPXY) domains, Ca^{2+}-binding EF hands, diacylglycerol- and phorbol ester-binding C1 domains, PDZ peptide-binding domains, and hydrophobic (Ser-Ala-Val) PDZ-binding pepides, Ras-associating (RA) domains, cAMP-binding domains, membrane-binding DEP, coiled–coil regions (CC), IQ calmodulin-binding domains, phosphotyrosine-binding SH (Src homology) two domains, phospholipase C catalytic domain (X and Y domains) and lipid-binding C2 domain. See text for more information on the roles of these regions. The numbers to the right indicate the total number of distinct family members, and an asterisk indicates the presence of splice variants. BAB-GEFs are putative GEFs (GenBank accession numbers BAB31080 and BAB31297). Link II is a putative Rap1 GEF (GenBank accession number AF117946).

these molecules together with the unexpected broad specificity and weak activity of Smg GDS resulted in its fading from interest.

The large number of Ras-subfamily members suggested to many scientists in the mid-1990s that many more GEFs must exist to regulate these GTPases. This notion was fueled by the knowledge that the similarly sized Rho subfamily of GTPases was regulated by nearly 30 Dbl homology (DH) domain-containing

Rho GEFs (reviewed in Refs. 6, 52, 53). Probably in excess of 50 Dbl family proteins have now been identified. Work by several groups over the last 3–4 years has upheld the notion that there is also a large family of upstream Ras regulators; nearly 20 new Ras-family GEFs/CDC25 homology proteins have recently been described (Fig. 4). The majority of these exchange factors were identified by the homology-based searching of the National Center for Biotechnology Information (NCBI) expressed sequence tag (EST) database, while others have been cloned following functional assays, yeast two-hybrid screens, or coprecipitation with their regulators. This study will focus on summarizing the structure, specificity, function, and regulation of each of the Ras-family exchange factors.

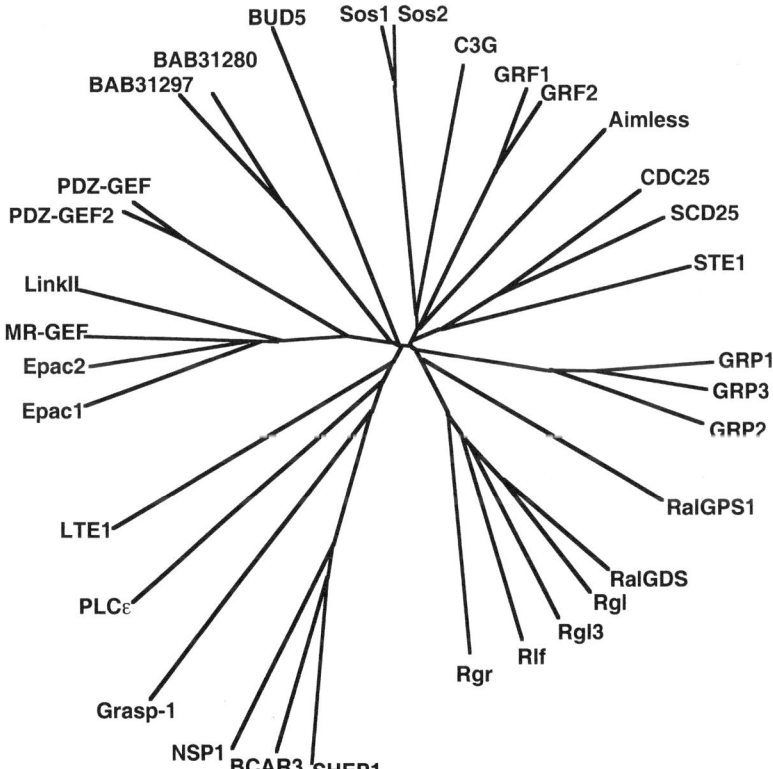

FIG. 4. Homology of Ras-family GEFs. The core CDC25 homology domains of mammalian as well as a selection of yeast (CDC25, SDC25, LTE1, STE6, and BUD5) and *Dicteostelium* (Aimless) GEFs were aligned using the ClustalW program, and a dendogram was generated to indicate their relative homologies.

IV. GEF Structure and the Nucleotide Exchange Reaction

The common feature of Ras GEFs is their catalytic or CDC25 homology domain. The minimal ~250-residue functional region of the yeast CDC25 was determined by Powers and colleagues (54). Three structurally conserved regions (scr1–3) were noted in this domain by sequence alignment of Sos, GRF, RalGDS, and multiple-yeast GEFs (16). Subsequently, two other regions of sequence conservation have been noted by alignment of additional mammalian GEFs and termed scr4 and 5 (55) (see Fig. 5). Solution of the Sos catalytic domain cocrystallized with apoRas (56) demonstrated that scr1–3 form the Ras-binding pocket of the GEF.

Upon switching from its inactive GDP-bound to active GTP-bound state, just two regions of Ras undergo significant conformational change. These are generally referred to as switch 1 and switch 2 (57). Switch 1 (Ras residues 30–38) approximately encompasses the effector-binding loop that is responsible for Ras-GTP interaction with downstream target molecules such as Raf1 and PI3-kinase (13). Switch 2 encompasses loop 4 (including the catalytic Q61 residue) and the following α-helix 2 and is adjacent to the gamma phosphate. This region undergoes the biggest conformational change upon GTP binding (13, 57). A number of previous Ras mutagenesis studies had shown a dependence on the switch 1, switch 2, and α-helix 3 (residues 102–107) of Ras for activation by GEFs. Mutations in each of these regions render the wild-type protein inactivatable or unable to interact with Sos or GRF1 (reviewed in Ref. 14). Each of these regions is on the same surface of Ras and not surprisingly turn out to represent the main GEF contact points on Ras (56).

The switch 2 region of Ras is held in a very tight embrace by Sos with almost all its external side chains interacting with the GEF (56). Meanwhile, an extended antiparallel pair of helices encompassing scr3 and 4 displaces switch 1, and thus pries open the nucleotide-binding pocket. Mutagenesis of switch 2 residues showed that only a few side chains affect Sos binding, with the most important contact being mediated by tyrosine 64 that buries into a hydrophobic pocket of Sos in the Ras–Sos complex (58). Substitutions of Sos side chains that insert into the Mg^{2+} and phosphate-binding site of switch 2 (Sos Leu938, and Glu942) had no effect on the catalytic function of Sos. Additionally an A59G mutation that prevented the Ala methyl group from occupying the Mg^{2+}/phosphate pocket did not affect Sos-mediated GDP release (58). These results indicate that the interaction of Sos with switch 2 is necessary for tight binding but is not the critical driving force for GDP displacement. Disruption of switch 1's interaction with the scr3 portion of the helical hairpin structure of Sos did inhibit GEF action, suggesting that interaction with switch 2 mediates the anchoring of Ras to Sos, whereas the interaction with switch 1 leads to disruption

	scr1		
Sos1	LLTLHPIEIARQITLLESDIYRAVQPSEIVGSVWTKE	....	DKEINSPNLKMIRHTTNLILMFEKCIVETE.
Sos2	IMTLHPIEIARQITLLESDIYRKVQPSEIVGSVWTKE	....	DKEINSPNLKMIRHTTNLIMFEKCIVEAE.
C3G	LHDFHSHEIAEQLTLLDAELFYKIEIPEVL..LVAKE	....	QNEEKSPNITQFTEHFNNMSYWVRSIMLQE.
GRF	FENHSAMEIAEQILLDHIVFKSIPYEEFFGQGMMKA.	....	DKNERTPYIMKTTRHFHIHSNLIASEITRNE.
GRF2	FETISAMELAEQILLDHIVFRSIPFEIYRNGQGMMKL	....	DKNERTPYIMKTSOHFNEMSNLVASQIMNYA.
Aimless	IYDIDEEIARQITLFEFIYRNIKPPELINQSWNKT	....	KLKSRAPNLVKMIDRFNSVSMWVATMIQTT.
CDC25	LLDIDPYTYAIIQLTVLEHDLYRIITMFECLDRAVGTK	....	YCN.MGGSPNITKFIANANTLINFVSHTIVKQA.
SCD25	ILAVDPVLFATQOILTEHEIYCEITIFDCLQKIWKNK.	....	YTKSYGASPGNEFISFANKLINFISYSVVKEA.
STE6	VLLIPPREIAKQICILFLFQSFSHISRIQFITKIWDN.	....	LNRFSPKEKTSTFYLSNHLVNFVTETIVQEE.
GRP1	FDHLEPELSEHLTYLEFKSFRRISFSDYQNYIVNSCV	KEN.....	PTMERSIALCNGISQWVQLMVISRP.
GRP2	FDHLEPMELAEHLTYIEYRSFCKILFQDYHSFVTHGCT	VDN.....	PVIERFISLFNVSRQWVQLMLISKP.
GRP3	FDHLEPIEIAEHLTFLEHKSFRRISFTDYQSYIVHGCL	ENN.....	PTIERSIALFNGISKWVQLMVISKP.
RalGDS	LLLFPPDLVADQFILMDAELFKKVPYHCLGSIWSQRA.	....	KKGKEHLAPTIRATVAQFNVANCVITTCLGDQ.
Rgl	FTNFSEDIVAEQITYMDAQLFKHVHHCLGCIVSQRD.	....	KKENKHLAPTIRATISQFNTLTKCVVSTVLGDR.
Rlf	VIVFLADHIAEQILILDAELFLNIIPSQCIGGLWGHRD	....	RPGHSHLCPSVRATVTQFNKVAGAVSSVIGATSIGEGPREVT
Rg13	.DFSVDDVAPQLTLMVELFLRVRSCECLGSMWSQRD	....	RPGAAGISPTVRATVAQFNTVGCVLGSVLAAP.
Rgr	ITAFSPRTMAEQILLDHDLFQKVPFHCLGSWTKSN.	....	RKGKEHLASTVHATVQQFNCVTNCVVTTCLGDP.
Epac1	LDLVSAKDLAGQIIDHDWSLFNSIHQVELHYALVGP.	....	QHLRDVTTANLERFMRRFNELQYWVTTCLCLCPV.
Epac2	FELMSSKDIAYQITIYDWELFNCIHELELIYHTG.	....	RHNFKKTTANIDLFLRRFNEIQFWVTTEICLCSQ.
MR-GEF	ILGMNTWDIALEIMNFWSLFNSIHEQELIYFFS.	....	RQGSGEHTANLSLLQRCNEQLWVAATEILCSQ.
Link II	IHRVEPEDVANHLTAFHWELFRCMELEFVDYVFHG.	....	ERGR.RETANIELLQRCSEVHWVATEWILCEA.
PDZ-GEF	ILQLSTEVATQISMRNFEFLFRNIEPTEYIDDLFK..	....	LRSKTSCANIKRFEEVINQETFWVASEITEAN.
PDZ-GEF2	MLQLSTEVATQISMRDFDLFRNIEPTEYIDDLFK..	....	LNSKTGNTHIKRFEDIVNQETFWVASEITETAN.
BAB31280	GVCSDPLVIAQQLTHIELERVNSIRPEDLMQIISHMDSL.	....	DNHRCRGDMTKTFSLEAYDNWFNCSMIVATEVCRVVK.
BAB31297	GVCSDPYTIAQQLTHVELERLRHIGPEEFVQAFVNKDPL.	....	AGTKPR.FSDKTNNVEAYVKWFNRLCYILVATEICMPAK.
SHEP-1	VTKVDCIVARILGVTKEMQTLMGVWGMELITIPHG.	....	RQLRLDLIERFHTMSIMAVDIGCTG.
BCAR3	VLSMDCVARILGVEMRRNMGVSSGLELITIPHG.	....	HQLRLDIIERHNTMAIGIAVDIGCTG.
NSP1	LLLVDCOATGLLGVTRDRGNMGVSSGLELITIPHG.	....	HHLRLELLERHQTIALAGALAVIGCSG.
RalGPS1	VLKVTPEEFASQITLMDIPVFRAIQPELASCGNSKKE	....	KHSLAPNVAFTRFNQVSFWVRELITAQT.
BUD5	ALNVSPWSTAKILTLSSSLYLDIETIEFTRHFKHN..	....	DTTIDSVFTLSNQLSSYWLETTIQQT.
LTE1	ILMYDSLSVAQQTLHKEILGEIDWKDLLDLKMKHEGP.	....	QVISWLQLLVRNETLSGIDLAISRFNILTVDWIISEILLTKS.
PLCe	....IQFPPEVASITMEQEQTIYRRVLPVDYICFITRDLGTPECQSSLPCLKASISASLITTQNGEHNAIEDLVMRFNEVSSWVTWLITTAGS.
consensus	---*--**-*-**--	---	-- --*-*----*-

FIG. 5. Alignment of CDC25 homology domains. Core CDC25 homology domains of the proteins shown in Fig. 4 were aligned using the ClustalW program. Following minor manual realignment, residues showing greater than 60% conservation or similarity were highlighted in black or gray, respectively, using the program MacBoxShade. Two completely conserved residues (in scr3 and 5) are indicated in bold type and by a (!) on the consensus line. Structurally conserved regions (scr) found in most GEFs and involved in Ras binding or structural stability (see text) are boxed. GRASP-1 was omitted due to its poor alignment with other GEFs.

(*continues*)

```
                                                                        scr2
Sos1        ....NLEERVAVVSRIEFLQVFQ..EINNENGVLEVVSAMNSSPVYRIDHTFEEQIPSRQKKILEEAHELSEDH..YKKYL............
Sos2        ....NFERVAISRIIELQVFR..DNNFNGVLEIVSVSVYRLDHTFEAIQERKRKILDEAVELSQDH..FKKYL............
C3G         ....KAQDRERLLLKFIKMKHLR..KLNNFNSYLAILSIDSAPIRRL..EWQKQTSEG...LAEYCTLIDSSSSFRAYR............
GRF         ....EVSARASTEKWVADICR...CLHNYNAVLEITSSINRSAIFRIKKTHLKTSKQTKSLFDKIQKLVSSDGRFKNLR..........
GRF2        ....DISSRPNAIEKWVADICR..CLHNYNGVLEITSALNRSPIYRLYKTWAKVSKQTKALMDKIQKTVSSEGRFKNLR..........
Aimless     ....KVKARARMTRFIKIADHLK..NNNYNSMAIIAGNFSSVYRLYKTREELSAQTMRTYSDFEKIMNSEGSFKTY3..........
CDC25       ....DVKIRSKITQYFVTVAQHCK..ELNNESIYVAILSSPIYRLTKDLLKNINLMDSKRNFVKYRE............
SCD25       ....DKSKRAKLISHFIFIAEYCR..KFNNFSSMTAIISALYSSPIYRIEKTKTQAVIPQTRDLLQSINKIMDPKKNFINYRN.........
STE6        ....EPRRTNVLAYFIQVCDYLR..EINNFASIFSIIAINSSPIHRLRKTWANINSKTLASFELINNLTEARKFNSNYRD.........
GRP1        ....TPQIRAEVFIKFIHVAQKLH..QLQNFNTIMAVIGLCHSSISRIKETSSHVPHEINKVLGEVTELLSSCRNYDNYRRAYGECTHFK..........
GRP2        ....TAPQRALVITHKFIHAEKLL..QLQNFNTMAVVGISHSSISRLIKETHSHVSPETIKLWEGTELIVTATGNVRRLAACVGFR..........
GRP3        ....TPQCRAEVITHKFINVAKKLL..QLQNFNTMAVVGISHSSISRLIKETHSHVSVSEVTRNWNEMELIVSSNGNYCNYRKAFADCDGFK..........
RalGDS      ....SMKAPDRARRVEHWIEVARECI..ALKNFSSIYAILSAIQSNAIHRLRKTIWEEVSRDSFRVFQKISEIFSDENNYSLSRELLIKEGTSKFATLEMNPR
Rgl         ....ELKTQCRARIEKWINIAHECR..IKKNFESIRAIISALQSNSIYRLRKKAAVPKDRMLMFEEISDIFSDHNNLLTSRELLMKEGTSKFANLDSSVK
Rlf         VRPLREPQRARLLEKWVADICR..EYRLNESIYAIISALQSSPIHRLRAAMGETTRDSLRVFSSICQIFSEEDNYSQSRELLTQEVRKFQPPVEPHSKK
Rgl3        ....GLAASCRAQREKWIRIAQRCR..ELRNFESIRAIISALQSNPIYRLRSMGAVSHEPLSVFRKISQIFSDEDNHLSSRAILSQEETEDDCPSG..
Rgr         ....SMKATRARVEHWIKVAKECC..TLRNVSSAHAIITLIQTSPICRLQETWGEVSSKSCKKFFERLCEKDNG....LSRDRLTKKGAFKLAARENPQ
Epac1       ....PGFRAQLLRKFIIKAAHLK..EQKNLNESFFAVMPGLSNSAIRAHTWERLPHKVRKLSALERLLDPSWNHRVYR..LALAKLS...........
Epac2       ....LSKRVQLLRKFIKIAAHCK..AEQRNLNESFFAIVMGLNTASVSRIAIQTWEKLPSKFKKFYAEIFESLMDPSRNHRAYR..LTVAKLE...........
MR-GEF      ....AQRNLNESFFAIVMGLNTASVSRIAIQTWEKIPSKFKKFLFESILTDPSINHRAYR..DAFKKMK...........
Link II     ....PGKRAQLLKKFIKIAALCK..QNQLLSFAVVMEDNAAVSRIRLTWEKIPGKFKNLFRKFENLTDPCRNHKSYR..EVISKMK............
PDZ-GEF     ....QLKRMKLIKHEFIKIALHCR..ECKENFSMFAIISGINLNAPVARIRTTWEKLPNKYEKLFQDLFDPSRNMAKYRNVLNSQNLQ.
PDZ-GEF2    ....QLKRMKLIKHFIKIALHCR..ECKENFNSIFAIISGLNLASVARIRGTWEKIPSKYEKHLQDIQDIFDPSRNMAKYRNILSSQSMQ.
BAB31280    ....KKHRTRMLEFFIVARECF..NIGENFNSMAIISGMNLSPVARIRKTWEKTSKYK...TAKFFIEHHMDPSSNFCNYRTALQGATQR.
BAB31297    ....KKQRAQVIEFFIDVARECF..NIGENFNSMAIISGMNMSPVSRIRTKKTWAKYK...TAKFFIEHQMDPTGNFCNYRTALRGAAHRS.
SHEP-1      ....SAEERAALLHKTIQLAAELR..GTMGNMFSFAAVMGAIEMAQISREIEQTMTIRQRHTEGAILYEKKLKFLKSLNEGKE..GPPLSN.
BCAR3       ....TLEDRAATLSKIIQVAVELK..DSMGDLYSFSALMKALEMPOITRLEKTWALRHQYTQTALLYEKQLKPFSKLLHEGRESTCVPNN.
NSP1        ....PLEERAAAIRGLVELALIRPGAAGDLPGAAVMGALIMPQVSRIEHTWQLRRSHTEAALAFEQELKPLMRALDEGAG..PCDPGE.
RalGPS1     ....LKIRAELISHFYKIAKKLL..ELNNLHSMSVVSALQSAPIFRLTKTWALINRKDKTTFEKIDYLMSKEDLKRTREYIRSLKMVP.
BUD5        ....HTISYWLQVALACL..YLENLNSIASITTSIQNHSIERIS...LPIFDVKSDHLFQRLIEPGDLLTWEEIKKIFSLDRNLSTIRNLLNSVNPLVG.
LTE1        ....SKMKRNVLQRFIHVADHCR..TFQNFNTIMEILALSSSVQKFTDAVLIEPGDLLTWEEIKKIFSLDRNLSTIRNLLNSVNPLVG.
PLCe        ....MEEKREWFSYLMVHVAKCCWN..MGNYNAVMEFTACERS...RKVLLKMVQFMDQSDIETMRSLKDAMAQHESSCEYRKVVTRALHPG.
consensus   -----*--------**----*----*--**-*-*-*----*-*--*-*-------*------*----------------*-------------------
```

FIG. 5. (continued)

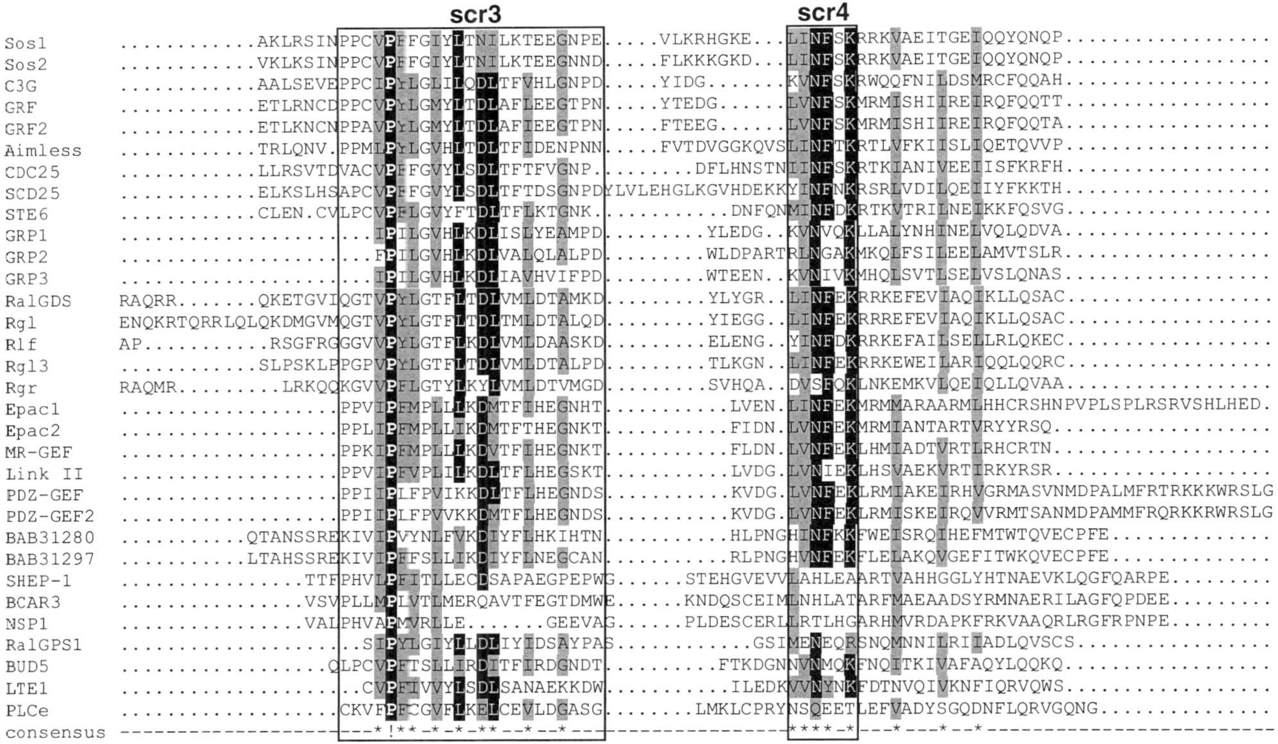

FIG. 5. *(continued)*

```
                                                                                          scr5
Sos1       ...........................YCLRVESDIKRFFENLNPMGNSM.EKEFTDYLFNKSLEIEPR
Sos2       ...........................YCLRIEPDMRFFENLNPMGSAC.EKEFTDYLFNKSLEIEPR
C3G        ...........................YDMRRNDDINFFNDFS.......DHLAEEAWELSLKIKPR
GRF        ...........................YKIEQPKVTQYLVDET.......FVLDDESIYEASLRIEPK
GRF2       ...........................YRIDQQPKVIQYLLDKA.......LVIDEDSLYELSLKIEPR
Aimless    ...........................YNLQPVHQIQEFLLNIRSDLKAHTLDQYQQELYRESLKIEPK
CDC25      ...........................YKLKRLDDIQTVIEASLE......NVPHIEKQYQLSLQWEPR
SCD25      ...........................YDFTKDRTVIECISNSLE......NIPHIEKQYQLSLIEPK
STE6       ...........................YMFNPINEVQELLNEVIS......RERNTNNYQRSLTVEPR
RasGRP     ...........................PPLDANKDLVHLLTLSLD......LYYTEDEIYELSYAREPR
RapGRP     ...........................PPVQANPDLLSLLTVSLD......QYQTEDELYQLSLQREPR
GRP3       ...........................HHLEPNMDLINLLTLSLD......LYHTEDDIYKLSLVEPR
RalGDS     ...........................NNYSIAPEEHFGTWFRAM......ERLSEAESYTISCELEPP
Rg1        ...........................NSYCMGPDQKFIQWFQRQ......QLSEEESYALSCEIEAA
Rlf        ...........................RGYDLRPNSDIQQWLQGL......QPLTEAQSHRVSCEVEPP
Rg13       ...........................QRYSLSPRPPLAALRAQ......RQLSEEQSYRVSRVIEPP
Rgr        ...........................SKYYFKRDEEFGVWFRSV......EFLTEKESYALSRQIEPP
Epac1      .............SQVARISTCSEQSLSTRSPASTWAYVQQLKVIDNQREISRISREIEEP
Epac2      .............QFG...DLSPKEHQELKSYVNHLYVIDSQQAEFEISHRIEPR
MR-GEF     .............QFNPDAAQANKNHQDVRSYVRQLNVIDNQRTISQMSHRIEPR
Link II    SLSQGSTNAITVLDVAQTGGHKKRVRRSSFLNAKKLYEDAQMARKVKQYLSNLELEMDEESLQTISLQCEFA
PDZ-GEF    SLSQGSTNSNMLDVQGG.AHKKRARRSSLLNAKKLYEDAQMARKVKQYLSSLDVETDEEKFQMMSLQWEPA
PDZ-GEF2   .............PLCDMEASNHLQTKAYVRQFQVIDNQNILFEISYKLEAN
BAB31280   .......................KDKKIQSYVL.........TAPIYSEEALFIASFESEGP
BAB31297   .......................QDPSITHYLY.........TAPIFSEDGLYLASYESESP
SHEP-1     .......................LLEVFSTEFQMRLLWGSQGANSSQAWRYEKFDKVITALSHKIEPA
BCAR3      .......................MNEICKTEFQMRLLWGSKGAQVNQTERYEKFNQITALSRKIEPP
NSP1       .......................LREALTTGFVRRLLWGSRGAGAPRAERFEKFQRVLGVLSQRIEPD
RalGPS1    .......................YDHLTTLPHVQKYLKSVRYIEELQKFVEDDNYKLSLRIEPG
BUD5       .......................YEDIHCSNTTARSLLGAMIKVHTLYNDNKDRAYQVSIAKVPR
LTE1       .......................KFYTFKVNHELLSKCVY....ISTLTQEEINELST
PlCe       .......................LKNSEKESTVNSIFQVIRSCNRSLETEDEDSPSEGNSSRKSSLKDKSR
consensus  .........................................................*:..*.**
```

FIG. 5. (continued)

of the nucleotide-binding site and GDP dissociation (58, 59). These recent data dispel the speculation (discussed in Ref. 60) that Ras GEFs act primarily by disrupting the interaction with the nucleotide phosphates and Mg^{2+} and are supported by the finding that GRF1 activity is detectable even in the presence of EDTA (61).

Although the structures of Rho- and Rab-family GEFs are quite distinct from the CDC25 domains, mutation of residues within switch 1, 2, and α3 of these GTPases suggested that they likely contact their GEFs in a similar fashion (62). This has been borne out by recently solved crystal structures of Tiam1–Rac1 (63) and ARF1–Sec7 (64) complexes. The overall structural features of the H-Ras–Sos, ARF1–Sec7, and Rac1–Tiam1 complexes share similarities in that in each the GEF interacts extensively with switch 2 and induces the displacement of switch 1. However, although Rac1 residue Tyr64 is again a key site of interaction with its GEF, the postulated exchange mechanisms for the Sec7 and Tiam1 appear to differ from that observed for Sos.

Although the core catalytic domain of CDC25p functioned as a GEF *in vitro*, a larger region of ~450 residues was required *in vivo*. This larger sequence encompassed a ~50-residue N-terminal homology region (54) (see Fig. 6). This

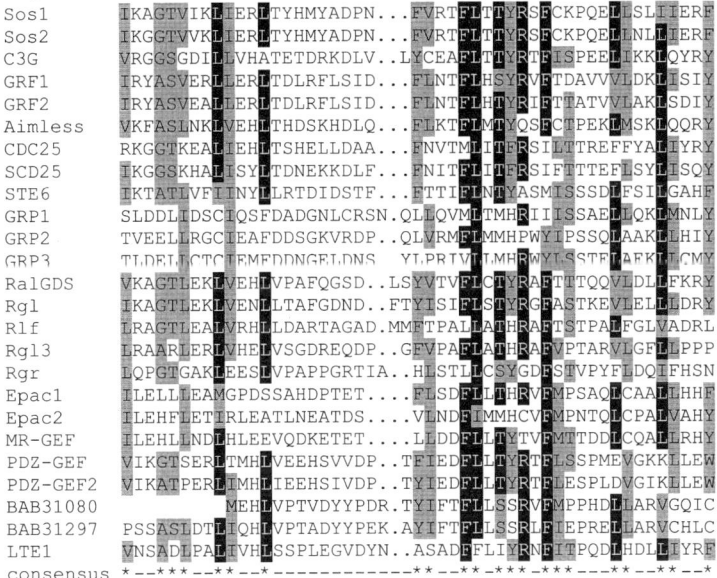

FIG. 6. Alignment of structurally conserved region 0. A region outside of the core catalytic domain, alternately termed REM (Ras exchange motif), CNC (conserved noncatalytic), LTE1 homology region or scr0, is required for efficient GEF activity. This region of Sos1 interacts with and stabilizes the core catalytic domain (see Section IV). scr0 is aligned using ClustalW, and residues sharing >60 conservation or similarity are shaded in black or gray, respectively, using the program MacBoxShade.

has been variably described as the REM (Ras exchange motif), CNC (conserved noncatalytic), LTE1 homology region (after the yeast GEF, LTE1) or scr0 (being N terminal to scr1) in the literature (54, 56, 65). scr0 is present in most but not all GEFs (see Fig. 3) and, based on the crystal structure of Sos1, plays a key role in stabilizing the large helical hairpin structure that pries open the GTP-binding pocket (56). In Sos1 the N domain containing scr0 forms a hydrophobic pocket that accommodates residues I956 and F958 present on the outer surface of the hairpin structure. These residues form the highly conserved hydrophobic I/VNF motif, described as scr4 in Refs. 55 and 66. In GEFs that lack an scr0, the scr4 is considerably more charged (see BUD5, RalGPS, phospholipase Cε, and BCAR3 in Fig. 5). This is consistent with the anticipated surface exposure of these residues. It will be interesting to determine if an alternative, charged, domain stabilizes the hairpin structure of these atypical GEFs.

V. Dominant Inhibitory Ras Proteins Target GEFs

Dominant inhibitory mutants of Ras have been very helpful in dissecting signaling pathways and implicating the involvement of Ras. The most commonly used mutant is H-RasS17N (67). This and other inhibitory H–Ras mutants that include 15A (68) and 57Y (69) act by binding to GEFs with a higher affinity than normal, endogenous Ras, and so prevent its activation (reviewed in Refs. 14, 67). This is due to GEF binding with much higher affinity to nucleotide-free Ras, the intermediate state during the nucleotide exchange reaction. Interestingly, a 57N mutant was found in one allele of the Rho-family protein, Rac2, in the blood cells of an immunodeficient patient. This mutant acts in a dominant inhibitory fashion resulting in inhibition of Rac activation (70).

These dominant inhibitory mutants block the GEFs rather than the Ras proteins; thus, if a GEF regulates multiple targets, e.g., Sos1 activates Ras, M-Ras, and TC21 (Table I), activation of each GTPase will be blocked by H-Ras17N. Thus, these mutants are not GTPase specific. While equivalent mutations have been utilized in many GTPases, not all successfully block Ras or GEF activity (67). Confusingly, Rap1-17N has been shown to be effective in some instances (33, 71, 72), but not in others (73), at blocking Rap-signaling events/RapGEF action, suggesting that other cellular events might contribute to its GEF-binding ability. Alternative means of blocking Ras signaling include the use of a GTP-bound mutant that cannot target to the plasma membrane: H-Ras61L/186S (74), overexpression of specific GAPs (33), or the isolated Ras-binding domains of effector proteins such as those of Raf or Rgl (75, 76). However, here too there is an issue of GTPase specificity.

TABLE I
Ras GEF Substrates and Interacting Molecules

GEF and aliases	Known substrates	Activators/interacting molecules
Sos1	Ras, TC21, M-Ras (15, 44, 93, 126)	Grb2 (44, 92, 93, 100, 285)
Sos2	Ras?	Grb2 (286)
GRF1	Ras, R-Ras, M-Ras (45, 46–48,125, 126)	Ca^{2+} (129, 131)
GRF2	Ras (65, 127)	Ca^{2+} (65)
GRP1 RasGRP CalDAG-GEF II	Ras, R-Ras, TC21 (15, 147, 148, 150)	Ca^{2+}, phorbol esters (147, 148, 150)
GRP2 CalDAG-GEF I	Rap1, N-Ras, K-Ras, R-Ras, TC21 (15, 148, 151)	Ca^{2+}, phorbol esters (148, 151)
GRP3 CalDAG-GEF III KIAA0846	Ras, Rap1, M-Ras (15, 66, 146)	Phorbol esters (287)
GRP4	Ras (Reuther and Der, unpublished)	
Epac1 cAMP-GEF I	Rap1, Rap2 (192)	Cyclic AMP (190–192)
Epac2 cAMP-GEF II	Rap1, Rap2 (192)	Cyclic AMP (190–192) Ras (JFR and LAQ, unpublished)
C3G	Rap1, R-Ras, TC21 (15, 125, 162)	Crk, p130Cas (50, 51, 227)
GRASP-1	Ras (161)	GRIP (161)
PDZ-GEF1 KIAA0313 RA-GEF NRapGEP CNrasGEF	Rap1, Rap2 (66, 141, 209–211)	Rap1, 2 (66, 208, 210), S-SCAM (211, 212), cAMP and NEDD4 (141)
RA-GEF2 PDZ-GEF2	Rap1, Rap2 (216)	M-Ras (216)
MR-GEF GFR KIAA0277 Repac	Rap1 and Rap2 (15, 66, 192, 207)	M-Ras (66)
Link II	Rap1?	Ras?
BAB31080	?	
BAB31297	?	
Phospholipase Cε	Rap1, Ras (225)	Ras (223, 224)
RalGDS	Ral (49)	Ras, Rap, M-Ras, R-Ras (83, 227, 236, 288, 289)
Rgl	Ral (227)	Ras (83, 227)
Rlf Rgl2	Ral (228, 229)	Ras, Rap1 (228, 229)

(continued)

TABLE I (Continued)

GEF and aliases	Known substrates	Activators/interacting molecules
Rgl3	Ral (230, 231)	M-Ras, Rit, Ras (230, 231)
Rgr	Ral (232, 233)	
RalGPS1 A/B RalGEF2 KIAA0351	Ral (55, 250)	Nck, GRB2 (55)
RalGPS2	?	
BCAR3 NSP2 AND-34	Ral, Rap1, R-Ras (260)	p130Cas (256, 260)
Chat NSP3 SHEP1	Can bind Rap1 and R-Ras (257)	p130Cas/HEF1 (257, 259)
NSP1	?	p130Cas (258)
Smg GDS	Rap1, K-Ras, RhoA, Rac1, Rac2, Ral (37–39)	SMAP (274)

VI. Biological Assays for GEF Activity

A number of *in vivo* assays for GEF activity have been devised based on the measurement of Ras-GTP. Initially, metabolic labeling of the GTP/GDP pools by incubation of cells with $^{32}PO_4$ was performed. Ras was then immunoprecipitated, bound nucleotides were separated by thin-layer chromatography, and the percentage of bound GTP was detected (77). The β- and γ-phosphates rapidly turn over and incorporate label, so the percentage of GTP can be calculated as $GTP/(GTP + 0.5 \times GDP)$. This technique has been refined by Boss and colleagues, who measured GTP and GDP mass from immunoprecipitated GTPases making an involved but much more sensitive assay (78). These assays are the most reliable and quantitative, but they require the use of significant amounts of radioactivity or specialized reagents/equipment.

Ras-GTP binding was brought to the masses by taking advantage of the fact that the Ras-binding domains (RBDs) of Ras effectors bind to Ras-GTP with at least a 100-fold higher affinity than that to Ras-GDP. Thus a GST–RBD fusion protein can extract Ras-GTP from a cell lysate (79–82) and be detected by immunoblotting. Since different Ras binding or RA (Ras association or Ral-GDS/AF6 homology) domains have differing affinities or specificities for Ras proteins (83), careful selection can improve the quality of results. Typically the RafRBD is used for Ras, while the RA domain of RalGDS is used for Rap1.

Mutation of the RafRBD may also improve its affinity for Ras-GTP resulting in an improved probe (84, 85). A limitation with Ras-GTP pull-down assays has been encountered in cotransfection assays (66). Since activated Ras promotes increased expression of genes driven by CMV (and other) promoters, increased Ras activity can drive up GEF expression leading to artifactual results. Nonlinearity of western blots can make normalization for total Ras protein expression less than satisfactory. This is not such a problem in the $^{32}PO_4$-metabolic-labeling assays, where the percentage rather than the mass of Ras-GTP is being measured.

Most recently, fluorescence resonance energy transfer (FRET) between variants of the green fluorescent protein has been used to measure Rac1 and Rap1 activation in real time in living cells (86, 87). This technique also enables subcellular location of GTPase activation to be monitored.

VII. Ras-Family GEFs

A. Sos1 and 2

As indicated in Section III, Sos was the first CDC25 homology protein identified in higher eukaryotes (41, 42). There are two closely related isoforms in mammals: Sos1 and Sos2 (43). Sos1 is ubiquitously expressed and likely to be the most commonly used Ras GEF. Consistent with its major role in cellular signaling, a Sos1 knockout resulted in embryonic lethality. In contrast, deletion of Sos2 resulted in no significant phenotype (88, 89). Sos2 was found to be more sensitive to ubiquitination than Sos1 (90), suggesting that its expression may be more transient. Both Sos isoforms consist of a central CDC25 homology domain flanked by C-terminal proline-rich regions and by N-terminal Dbl homology (DH) and pleckstrin homology (PH) domains (Fig. 3).

This GEF was originally identified as son of sevenless, a downstream mediator of the receptor protein tyrosine kinase, sevenless (41, 42), involved in the formation of the omatidia that makes up the fly compound eye. Subsequently, the fly and mammalian enzymes were found to respond to many protein tyrosine kinases that coupled to Grb2 (91), an SH3–SH2–SH3 domain-containing adapter protein (92–96). Receptor activation and *trans*-phosphorylation create docking sites for the SH2 domain of Grb2. Meanwhile the SH3 domains of Grb2 interact constitutively with the proline-rich, PXXP motifs in the Sos C terminus. Consequently, association of Grb2 with growth factor receptors results in increased proximity of Sos to the plasma membrane and, hence, Ras. Present within the proline-rich SH3-binding region of the Sos C terminus are several consensus sites for proline-directed kinases such as the ERK family of mitogen-activated protein kinases. Phosphorylation of these sites can influence the interaction of Grb2 with Sos and may represent a negative-feedback loop by ERKs to shut down receptor-induced Ras activation (97).

Not all growth factor receptors contain the phospho-tyrosine-X-asparagine-X consensus sequence necessary for Grb2 SH2 domain binding, but this is frequently overcome by placing an additional coupling protein into the complex. This may be another adapter protein, such as Shc or Gab1, that contains SH2 and PTB domains that can bind alternate phosphotyrosine sequences on upstream receptors (98, 99). Tyr-X-Asn-X motifs present on these proteins then become phosphorylated and recruit the SH2 domain of Grb2. Other SH2-containing proteins such as the tyrosine phosphatase SHP2 (100) can also double as adapters to couple receptors with Grb2. In this way, SHP2 can function as both an effector and an inhibitor of phosphotyrosine signaling. Many reviews have been dedicated to describing the role of Sos in PTK signaling (96, 101).

Many studies have demonstrated that Sos is also a downstream effector of integrins and G-protein-coupled receptors. However, this typically requires that these other receptor systems can couple to protein tyrosine kinases such as FAK (102), PYK2 (103), Src, or growth factor receptors, and signal to Sos via SH2-domain-containing adapter proteins. These pathways are reviewed in Ref. 104.

Sos was reported to specifically activate H-, K-, and N-Ras (105) but not R-Ras, Rap, or Ral (93). More recently, it was shown to also act on TC21/R-Ras2 and M-Ras/R-Ras3 (15, 66). This is significant, since it indicates that Sos activates GTPases other than the classic Ras proteins and so may activate additional downstream effector pathways to Ras. This may result in differential pathway activation depending on what Ras proteins are expressed in a given cell type. Also upon introduction of the dominant inhibitory H-Ras17N mutant, which binds to GEFs, the activation of multiple GTPases in addition to H-Ras will be blocked.

It was initially proposed by Buday and Downward, based on their finding that membrane and cytoplasmic Sos had similar Ras GEF activity (93), that the role of Grb2 was to target Sos to the plasma membrane where it could activate Ras. This notion was supported by data from three groups where membrane targeting of the isolated CDC25 homology domain (106) or full-length GEF to the plasma membrane (107, 108) enhanced Ras activation. Deletion of C-terminal Sos1 sequences increased GEF activity, suggesting that this region played an inhibitory role in addition to Grb2 binding (107, 109, 110). Studies by Lowy and co-workers (111) and Bar-Sagi and co-workers (110, 112) indicated that sequences N terminal to the CDC25 homology domain also provided negative regulation and that the PH domain was critical for membrane and growth factor receptor binding. Indeed, the isolated Sos1 PH domain could target to the plasma membrane and antagonize serum-induced activation of the Ras-signaling pathway in a dominant inhibitory fashion (112). The Sos PH domain can bind to phosphatidylinositol 4,5-bisphosphate (PIP_2), enabling it to bind to lipid vesicles (112). However, the binding of Sos to membranes *in vivo* was not dependent on

PIP$_2$. Interestingly, PIP$_2$ was reported to inhibit the loading of Ras with GTP *in vitro* (*113*).

In addition to influencing the catalytic activity and membrane localization of Sos, the DH/PH region functions as a GEF for the Rho family GTPase Rac (*114*). Since PI3 kinase is a downstream effector of Ras (*115*) and its product phosphatidylinositol 3, 4, 5-trisphosphate (PIP$_3$) binds to the PH domain of Sos leading to Dbl activation, the Ras GEF activity of Sos may ultimately regulate its Rac GEF capabilities. Thus, similar to the DH/PH Rac GEF, and Vav1, (*116*), the substrate (PIP$_2$) and product (PIP$_3$) of PI3-kinase may differentially regulate exchange activities of Sos1. Based on the crystal structure of the Sos1 DH/PH domain (*117*), the PH domain may occlude the Rac-binding pocket until PIP$_3$ binding induces a conformational change to permit Rac access.

An additional layer of regulation is provided by the adapter proteins Eps8 and Abi1/E3b1 that complex with Sos upon receptor protein tyrosine kinase activation and are also responsible for promoting Rac versus Ras activation by Sos. While immunoprecipitated Sos or its isolated DH/PH region failed to activate Rac *in vitro*, a Sos1–Eps8–E3b1 ternary complex was able to load Rac with GTP (*118*). A C-terminal region of Eps8 is responsible for binding to Sos and promoting Rac GEF activity. Eps8 also binds to F-actin and so targets Sos to the site of Rac activation (*119*).

B. GRF1 and 2

Guanine nucleotide releasing factor 1 (GRF1, also referred to as CDC25^{Mm} for mammalian homolog of the yeast CDC25) was identified independently by multiple groups (*45–48*). A closely related family member, GRF2, was subsequently isolated (*65, 120*). The CDC25 homology domain is located at the extreme C terminus in both proteins, similar to their *Saccharomyces* homolog. Reports from Lowy and Santos and co-workers indicate that there are multiple splice variants of GRF1 (*48, 121, 122*) having variable-length N termini. While full-length, 140-kDa GRF1 is predominantly expressed in brain, GRF2 and the N terminally truncated forms of GRF1 are more widely distributed.

Like Sos, both GRF1 and 2 are GEFs for Ras. A clear preference for H- versus K- or N-Ras was noted in one *in vivo* study (*123*); however, another group has reported equal activation of H-, K4A, K4B, and N-Ras (*124*). GRF1 can also activate R-Ras, TC21/R-Ras2, and M-Ras/R-Ras3 (*15, 125, 126*). Interestingly, GRF2 failed to activate R-Ras due to it being post-translationally gernaylgeranylated rather than farnesylated (*127*). Deleting or swapping the post-translational modification with H-Ras permitted GRF2 to regulate R-Ras *in vitro*. A divergent 30-amino acid sequence in GRF2 was shown to be responsible for this substrate preference (*127*).

As shown in Fig. 3, GRFs have multiple functional domains. In addition to containing a Ras GEF domain, like Sos, they also possess tandem DH/PH domains in their N terminus. Similarly to Sos, this GEF domain specifically activates the Rho-family GTPase, Rac (*128*), again providing coordinated regulation of downstream-signaling pathways. Additionally, GRFs have a second, more N-terminal, PH domain preceded by an IQ (ilimaquinone) domain and a coil–coil region (*65, 129*). A PEST sequence containing a destruction box has also been noted between scr0 and scr1 motifs of GRF2 (*130*).

The activity of GRF1 is markedly enhanced by elevation of intracellular Ca^{2+} levels *in vivo* (*131*). This effect is dependent on calmodulin binding to the IQ motif, as mutations in either protein that disrupt their interaction blocked GEF activation. Mutation of either the Dbl or the adjacent PH domain also prevented GRF1 activation by Ca^{2+} *in vivo* (*132*). Since GRF1 can dimerize with GRF2 (*133*), the above mutations may have disrupted this interaction. Indeed, a point mutant that disrupts dimerization completely destroys Ras GEF activity of GRF1 (*133*). Further, since Ca^{2+} could activate Raf in a Ras-independent manner, this cation may also promote Rac GEF activity, since Rac can promote Raf phosphorylation, increasing its sensitivity to Ras (*134*). Intriguingly, oligomerization of the prototypic Dbl protein has also been implicated in its activation (*135*), suggesting that this may be a common regulatory event. Since yeast CDC25 can also dimerize (*136*), clearly a Dbl domain is not necessary for this function but may be the tool of choice in higher eukaryotes. The N-terminal PH domain, IQ, and coil–coil region function cooperatively to activate GRF; deletion of the PH domain resulted in translocation to the cytosol and loss of Ca^{2+} sensitivity (*129*). Further, the coil–coil and IQ domains as well as the PH domain were necessary to efficiently target the GEF to the membrane. These data suggest that multiple components need to interact with the individual domains of GRF to promote its activation.

In addition to its regulation by various modular domains, GRF activity is further controlled by phosphorylation. This may be the means by which this GEF is regulated by G-protein-coupled receptors, since the G_i and G_o inhibitor, pertussis toxin, effectively blocked the hyperphosphorylation of GRF (*137, 138*). The muscarinic receptor agonist, carbachol, promotes the phosphorylation of the GRF1 residue Ser916 (*139*). Although this residue may also be phosphorylated by PKA, carbachol-induced phopshorylation occurred through a protein kinase A-independent pathway. Full-length Ras-GRF1 that contains an alanine 916 mutation was only partially activated by carbachol, suggesting that phosphorylation at residue 916 is necessary for full activation. Phosphorylation of serine 916 in response to forskolin treatment did not, however, increase the activity of Ras-GRF1, indicating that it is not sufficient for activation (*139*). PKA was found to be much more effective than protein kinase C, calmodulin kinase II, or casein kinase II at phosphorylating purified GRF1 (*140*). Ser745 and Ser822 were found to be the most heavily phosphorylated sites, and an S822D mutation strongly inhibited the exchange activity of $CDC25^{Mm}$ on Ha-Ras. This suggests that PKA may negatively regulate GRF.

Sos2 is considerably less effective at inducing transformation than Sos1 due to becoming ubiquitinated and rapidly degraded by the proteosome (90). The Rap-specific PDZ-GEF (described belowin Section VII,H) also associates with the E3 ubiquitin ligase NEDD4 (141). As noted above, a destruction box is located within the PEST motif of GRF2 that is found between scr0 and scr1. GRF2 protein levels are dramatically decreased upon GRF activation or the expression of dominant inhibitory Ras (130). Although deletion of the destruction box reduces this downregulation, mutation of the CDC25 domain to prevent Ras binding completely eliminates degradation and results in the accumulation of a high-molecular-weight ubiquitinated protein. Thus, binding of Ras likely induces a conformational change in GRF2 that exposes the destruction box to ubiquitination with the resultant GEF downregulation. Whether Ras or Rap regulates the stability of Sos2 or PDZ-GEF in a similar manner remains to be determined. Interestingly, a deubiquitinating enzyme was recently isolated in association with GRF1, adding an additional layer of complexity to the regulation of Ras GEF levels (142).

Although GRF1 mRNA is detectable in embryonic stem cells (120), its expression increases sharply in the brain postnatally (143) and is present at synaptic junctions (144), suggesting that it plays a role in mature brain function rather than in development. Consistent with this notion, knockout of the GRF1 gene resulted in no overt defect in development (145). However, these mice do not learn fear responses that require the function of the amygdala and show defects in the delayed phase of memory formation. These results implicate GRF-Ras-ERK signaling in synaptic events leading to formation of long-term memories.

C. The GRP/CalDAG-GEF Family

A family of four GEFs containing calcium and diacylglycerol (DAG)-binding motifs has recently been described by numerous investigators (66, 146 151). These molecules each contain an amino-terminal scr0 motif followed by a potential nuclear localization signal (146) and central CDC25 homology domain (Fig. 3). This domain is highly conserved among GRPs and is quite distinct from that of other Ras family exchange factors (Fig. 5). C terminal to the catalytic domain are two tandem Ca^{2+}-binding "EF hands" that are homologous to those found in calmodulin. The carboxyl-terminal region contains a DAG/phorbol ester-binding C1 domain similar to that present in the classical and atypical protein kinases C (152). As noted above, coiled–coil domains in Ras-GRF1 cooperate with its IQ and pleckstrin homology domains to promote maximal exchange activity. Although coiled–coil domains are present in a couple of GRPs (151), they are found in different regions and are probably not a part of a common ancestral molecule.

The founding member of the GRP family was originally described by Stone and colleagues and designated Ras guanine–nucleotide-releasing peptide (GRP1150). It was isolated from a rat cDNA library as a weakly transforming complement to v-H-Ras in rat2 fibroblasts. It activates H-Ras *in vitro* and cooperates with H-, K-, and N-Ras to activate MAP kinase and transform NIH 3T3 fibroblasts (*147, 150*). The C-terminal DAG domain of GRP1 appears to regulate GRP1 activity by affecting intracellular location. It binds to phorbol esters with high affinity and recruits GRP1 from the cytosol to the plasma membrane (*150, 153*). Consequently, the phorbol ester, PMA, enhances GRP1's ability to activate H-Ras *in vivo* (*150*). GRP1 mutants lacking the C1 domain do not activate MAP kinase as efficiently as wild type, and they lose the ability to transform NIH 3T3 fibroblasts (*147, 150*). By adding a membrane-localizing prenylation signal to these mutants, NIH 3T3-cell transformation and MAP-kinase-activation capacity are restored, presumably through the efficient activation of Ras at the plasma membrane (*147*). Calcium's role in regulating GRP1 is not well understood. The "EF hands" of GRP1 coordinate the binding of calcium to GRP1, the second EF hand binding more effectively (*150*); however, GRP1 mutants lacking the EF hands remain capable of transforming NIH 3T3 cells (*147*).

Several lines of evidence link GRP1 to Ras activation in the growth and development of the immune system. GRP1 is expressed in hematopoietic organs as well as in a number of lymphoid-derived cell lines (*148, 150, 154, 155*). Antibodies against GRP1 or phospholipase C-γ-1 inhibitors which decrease DAG levels diminish Ras activation following T-cell-receptor (TCR) stimulation in Jurkat T cells (*155*). Overexpression of kinase-dead DAG kinase zeta, which normally catabolizes DAG to phosphatidic acid, prolongs Ras activation in TCR-stimulated Jurkat T cells (*156*). Additionally, GRP1-knockout mice fail to develop mature, single-positive (CD4 + CD8− and CD4−CD8 +) thymocytes (*157*). Thus, GRP1 plays an important role in Ras protein activation and T-cell development.

Shortly after the discovery of GRP1, a second family member, CalDAGI/GRP2, was isolated as a striatum-enriched transcript (*148*). In this review, RasGRP/CalDAGII and RapGRP2/CalDAGI are referred to as GRP1 and GRP2, respectively. GRP2 activated Rap1 but did not activate H-Ras, R-Ras, or RalA *in vivo*. It cooperated with the calcium ionophore, A23187, and the phorbol ester PMA for maximal exchange activity, and coadministration of A23187 and PMA had additive effects on Rap1 stimulation (*148*). Hancock's group (*151*) further characterized GRP2 by presenting additional N-terminal sequence not previously described. This sequence contains N-terminal myristoylation and palmitoylation signals that target GRP2 to the plasma membrane, changing its cellular location as well as its substrate specificity. At the plasma membrane, GRP2 was found to interact with and activate N- and K-Ras but not H-Ras (*151*). While it was suggested that microdomains within the plasma membrane may isolate GRP2 from H-Ras (*151*), subsequent *in vitro* data suggest that GRP2 has a structural specificity for N- and K-Ras and not H-Ras (*146*). Interestingly, Clyde-Smith *et al.* (*151*) observed that Ca^{2+} inhibited GRP2's activation of N-Ras *in vitro* suggesting that Ca^{2+} binding causes steric inhibition of GRP2. In contrast, Rap1 activation by GRP2 is enhanced by Ca^{2+}. Thus, the activation of Rap- versus Ras-specific pathways via GRP2 may be regulated by Ca^{2+} concentration. The observation of GRP2 in the target nuclei of striatal termini and the presence of Rap1 in synaptosomes and synaptic vesicles (*158*) suggest the coordinate activation of Ras/Rap via GRP2 for synaptic function. Since Ca^{2+} is released from internal store and can diffuse through the cytoplasm, it provides a means for Rap1, which is predominantly expressed in a perinuclear/Golgi localization, to be rapidly activated by cell-surface stimuli.

Other exchange factors in this family also demonstrate dual specificity for Ras and Rap proteins. GRP3 (KIAA0846/CalDAGIII) originally isolated as part of the Kazusa human cDNA sequencing project (*159, 160*) (http://www.Kazusa.or.jp/huge/) was recognized by several groups as a GEF similar to GRP1 and GRP2 (*15, 66, 146*). GRP4, another family member, acts as an exchange factor for Ras (G. Reuther and C. Der, personal communication). *In vivo*, GRP3 efficiently activates Rap1A, Rap2A, H-Ras, R-Ras, and TC21, and to a lesser extent M-Ras (*15, 66, 146*). *In vitro* assays using purified proteins devoid of additional cellular components supported these findings. Using this assay, Yamashita *et al.* (*146*) made a direct comparison of GRP3's ability to activate different Ras-family GTPases. They observed that GRP3 activated H-Ras and R-Ras more efficiently than GRP1 and M-Ras and TC21 as efficiently as GRP1 (*146*). While GRP3 was extremely effective in activating Rap1A *in vivo*, it was not as effective as N terminally truncated GRP2 in activating Rap1A *in vitro*. Thus, GRP3 may be a more effective Ras exchange factor than a Rap GEF. Comparing known physiological outcomes of Ras or Rap stimulation may be a means of ascertaining the major substrate of GRP3.

GRPs are predominantly expressed in the brain, kidney, and hematopoetic organs. Previously, little overlap was observed between the different GRPs and specific cell types. In the brain, GRP1 was expressed in the pyramidal cells of the hippocampus and the purkinje cells of the cerebellum, while GRP2 was most highly expressed in the basal ganglia (*148*) and in the cell bodies, axons, and axon terminal of striatal projection neurons (*158*). GRP3 was primarily found in the glial cells of the cerebral and cerebellar white matter, primarily in the oligodendroglia (*146*). Further diverse expression patterns are observed in the kidney where GRP1 is expressed in the epithelium of the distal convoluted tubules and collecting tubules, while GRP2 is found in the interstitial cells, and GRP3 is found in the mesangial cells of the glomeruli (*146*). More recently, within the striatum, GRP1 and GRP2 were observed to have a nearly identical distribution pattern: they are coexpressed in striatal projection neurons that give rise to the direct and indirect pathways of the basal ganglia (*158*). Thus, although GRPs are typically expressed in different cell types, there are instances where they are coexpressed.

D. GRASP-1

GRIP (glutamate receptor-interacting protein) is a 7-PDZ-domain-containing adapter protein associated with AMPA-type glutamate receptors. To identify downstream targets of this adapter protein, various fragments were used in a yeast two-hybrid screen (*161*). A central portion interacted with a Rap2-interacting protein while the C-terminal-most PDZ domain bound to a putative Ras GEF termed GRASP-1 (GRIP-associated protein). This GEF proved difficult to align with other family members (and was omitted from Fig. 5) due to weak conservation outside of scr1 and perhaps to different spacing between these regions (*161*). In fact, the original report inadvertently aligned GRASP-1 with a region outside the scr2 of Rlf. An *in vitro* exchange assay using a GST-fused catalytic domain demonstrated the ability of GRASP-1 to act as an effective GEF for H-Ras but did not test its ability to regulate other GTPases. GRASP-1 contains a C-terminal PDZ domain that is responsible for its interaction with GRIP and interestingly also possesses a nearby putative RA domain (Fig. 3; no Ras-binding or specificity studies have been published so far), indicating that Ras proteins may coregulate the GEF activity of GRASP-1.

E. C3G

C3G was originally isolated as a protein that interacts with the SH3 domain of the SH2/SH3-containing adaptor protein, Crk (C3G stand for Crk SH3 domain-binding guanine nucleotide exchange factor) (*50, 51*). It was first identified as a Rap1 exchange factor (*162*) but was later shown to also promote the exchange reaction of R-Ras, TC21, and Rap2 (*15, 125*). It is ubiquitously expressed in adult and fetal tissues (*50*), although a splice variant with more tissue-specific distribution was identified in rat cells (*163*). The catalytic CDC25 homology domain that is responsible for the exchange activity is located in the C terminus. The central region contains four proline-rich sequences, which are capable of individually interacting with the N-terminal SH3 domain of Crk (*51*). Further, the N terminus contains an additional proline-rich sequence that interacts with the SH3 domain of the docking protein p130Cas (*164*). In the absence of stimulation, the N-terminal region of C3G negatively regulates its GEF activity (*165*). Upon receptor tyrosine kinase activation, the SH2 domain of Crk binds to phosphorylated tyrosine and recruits C3G to the receptor. This results in the phosphorylation of C3G on tyrosine 504 and represses the negative-N-terminal regulation of its GEF activity (*165*). Subsequent tyrosine phosphorylation of Crk appears to decrease its binding affinity to C3G and induces dissociation of the Crk–C3G complex (*166*). A similar mechanism regulates the interaction of the Ras exchange factor SOS with GRB2 where serine/threonine phosphorylation of SOS results in the dissociation of the complex (*97*) (discussed in Section VII,A).

The generation of C3G-knockout mice established that C3G is required in early embryogenesis; embryos carrying a C3G deficiency died after implantation, containing only degenerating embryonic tissue (*167*). The lethality of the C3G deficiency indicated that other GEFs could not complement C3G during development. This correlates with the ubiquitous expression of C3G compared to other RapGEFs and with its ability to be activated by many mitogens and signaling pathways (*71, 168–175*). Perhaps more relevant to the defective development of C3G$^{-/-}$ embryos, C3G has been also implicated in integrin-mediated cell adhesion and migration. Adhesion of NIH 3T3 fibroblasts to extracellular matrix proteins induces tyrosine phosphorylation of C3G (*176*). In addition, C3G was found to form complexes with p130Cas and Crk upon integrin ligand binding (*177*) and increased adhesion or migration in hematopoietic cells (*178, 179*). Accordingly, C3G-deficient fibroblast cells showed decreased adhesion and spreading that correlate to the lack of Rap1 activation in these cells. In contrast, they showed increased cell migration that is suppressed by Rap1 activation (*167*). This discrepancy may reflect cell-type differences. As an additional controversy, dissociation of the Crk–C3G complex and Rap1 inactivation was correlated to adhesion in Chinese hamster ovary cells (*180*).

C3G has also been implicated in T-cell-receptor signaling. It was shown that activation of the T-cell receptor induces the recruitment of Crk complexes with tyrosine-phosphorylated Cbl and C3G (*181*). In the absence of costimulation, the activation of the T-cell receptor induces anergy, a phenomenon characterized by an inability to induce the expression of interleukin 2 when it is restimulated with antigen (*182*). Rap1-GTP elevation was detected in anergic T cells, and Rap1 activation negatively regulated their ability to transcribe interleukin 2, indicating that this GTPase might, in part, mediate the anergic process (*183*). Since recruitment of C3G with signaling complexes also occurs in anergic T cells, it was suggested that C3G is responsible for the activation of Rap1 (*183*). Interaction of C3G with Crk has been also reported in B cells, although a correlation to B-cell-receptor engagement or activation of any Ras GTPase have not been established (*184, 185*).

While the regulation of the MAPK activity by C3G seems to depend on Rap1 activation, C3G mediates Jun kinase (JNK) activation through R-Ras, a pathway that involves the mixed-lineage kinase (MLK) family of proteins, MLK3, and dual-leucine zipper kinase (DLK) (*186*). The activation of JNK by C3G appears to be one of the signaling events downstream of the *v-Crk* oncogene. In fact, C3G enhances JNK activity as well as anchorage-dependent and -independent growth of *v-Crk* NIH 3T3-transformed cells. In addition, a dominant-negative form of C3G, lacking the CDC25 homology domain, prevented *v-Crk* JNK activation (*187*). The participation of R-Ras in JNK activation downstream of *v-Crk* and C3G was suggested using constitutively activated R-Ras (38V) and the dominant-negative mutant R-Ras(43N) (*188*). However, it remains to be investigated under what physiological conditions R-Ras is a preferable substrate for C3G to induce the activation of JNK. For instance, it was shown that Rap1 rather than R-Ras was the main C3G substrate upon cell attachment, as only Rap1 activation was abolished by C3G deficiency (*167*).

F. The Epac/cAMP-GEFs

The laboratories of Bos and Graybiel independently identified this new class of cAMP-regulated exchange factors. These represent the third class of cAMP-binding proteins to be identified, the others being cAMP-dependent protein kinases (PKA) and cyclic-nucleotide-gated ion channels (*189*). De Rooij *et al.* found Epac1/cAMP-GEF I searching the genomic database for genes with sequence homology to both CDC25- and cAMP-binding domains (*190*). Meanwhile Kawasaki *et al.* isolated Epac1/cAMP-GEF I and Epac2/cAMP-GEF II by differential display RT-PCR while they were searching for novel brain-enriched signaling genes with second messenger motifs (*191*). They were initially described as Rap1GEFs, but later it was shown that they also activate the small GTPase Rap2 (*192*).

Like the GRP/CalDAG GEF family, the Epacs fit into the category of exchange factors whose activity is regulated by second messengers rather than by adapter proteins. They have a cAMP-binding domain, which is located N terminally to their CDC25 homology domain (Fig. 3). These domains share closest homology with the cAMP-binding sites of the PKA regulatory subunits, RI and RII (*192*). Epac2, like the *C. elegans* Epac, contains a second cAMP-binding site closer to its NH2 terminus. The isolated N-terminal domain possesses only low affinity for cAMP (*192*), making it unlikely that physiological cAMP concentrations could regulate Epac2 via this domain. The other cAMP-binding site appears to function as an inhibitory regulatory domain. In fact, Epac1 only promoted nucleotide exchange on Rap1 *in vitro* if the cAMP-binding domain was removed or cAMP was added to the reaction (*192*). Moreover, the isolated cAMP-binding site inhibited the activity of the isolated GEF catalytic domain (*192*), indicating that interaction between these two domains results in the inhibition of the GEF activity and that this inhibition is relieved by adding cAMP. Accordingly, *in vivo* studies showed that cAMP induces Rap1 activation by Epac1 (*190, 191, 192*).

Epacs also contain a dishevelled, Egl–10, Pleckstrin (DEP) domain that is involved in membrane localization (*192*). This is not regulated by cAMP, indicating that translocation to the membrane is not, like in adapter protein-regulated GEFs, a mechanism of activation in Epacs. Whether they associate with the membrane by direct interaction with lipids or with a membrane-associated protein is unknown. Epac2 possesses an RA domain immediately N terminal to the core CDC25 homology domain that can bind to H-Ras-GTP (unpublished);

thus, similarly to RalGDS, PDZ-, and MR-GEFs as well as GRASP-1 and phospholipase Cε, Epac2 may be regulated by Ras proteins, providing a means of pathway crosstalk. Although there is some homology in this region with Epac1, it is likely that the latter GEF is not able to interact with activated Ras family members.

Epac1 is widely expressed, being particularly abundant in kidney, ovary, and thyroid; while Epac2 is highly expressed in brain and adrenal glands (*190, 191*). Thus, the identification of these cAMP-regulated GEFs may help explain a number of cAMP-induced events that are independent of PKA (not inhibitable by PKA inhibitors). The control of endocrine cell functions by pituitary hormones provides several good examples (*193*). These hormones bind to cell-surface G-protein-coupled receptors (GPCRs) inducing cAMP production via $G_{s\alpha}$ and adenylyl cyclase. Thyrotropin (TSH) stimulates proliferation and differentiated gene expression through cAMP in thyroid cells (*194, 195*). It seems that Rap1 contributes to TSH-regulated differentiation and stimulated Akt/PKB phosphorylation (*196*). Since TSH activates Rap1 through a cAMP-mediated but PKA-independent mechanism, it is suggested that Epacs play a key role in these events (*196, 197*). A similar link may apply to FSH signaling, which appears to induce phosphorylation of Akt and Sgk (serum and glucocorticoid-induced kinase) by a PKA-independent pathway in ovarian granulosa cells (*198*). In pancreatic beta cells, glucagon-like peptide-1(GLP-1) potentiates insulin gene transcription, glucose-induced insulin secretion, and cell growth through the elevation of intracellular cAMP levels (*199, 200*). Some of these events are also PKA independent (*201*). Since beta cells express Epac1 and Epac2 (*202*), it is likely that they play a role in GLP-1-mediated beta-cell growth and differentiation. Epac2 may also be involved in exocytosis; its interaction with Rim2 (a Rab3 small GTPase-interacting molecule), a putative regulator of vesicle fusion to the plasma membrane, mediates cAMP-induced, Ca^{2+}-dependent secretion, suggesting that Epac2 is a direct target of cAMP-regulated exocytosis (*203*). However, whether activation of Rap GTPases is involved in this process was not investigated.

Epacs might not always be the link between cAMP- and PKA-independent events. Indeed, cAMP promotes the activation of ERK by a Rap1- and PKA-independent mechanism that involves Ras activation in melanocytes cells (*204*). It was suggested that an unidentified cAMP-regulated Ras GEF is responsible for the activation of ERK. It would be interesting to test whether the novel PDZ-GEF (Section VII,H), described as a Ras exchange factor under cAMP regulation (*141*), is implicated in this system. More intricate, a recent report showed that cAMP inhibits ERK and Akt activities in a PKA-independent manner by reducing Rap1 activation in rat glioma cells (*205*). As expected, Epac was not involved. To the best of our knowledge, this is the first report that links cAMP-induced inhibition of ERK with Rap1 inactivation. In fact, it was generally accepted that Rap1 activation mediates the inhibition of ERK by cAMP (206). Thus, this novel observation needs further characterization before attempting any speculation.

G. MR-GEF (KIAA0277, GFR, or Repac)

This exchange factor was initially described by Ichiba and colleagues (207), who demonstrated that the KIAA0277 cDNA, cloned by the Kazusa DNA Research Institute, shared homology with the catalytic domain of C3G. They determined that this GEF activated Rap1 but not H-Ras and named it GFR for guanine nucleotide exchange factor for Rap (207). It has since been shown to activate Rap1 and Rap2 but not other GTPases (66, 192). A FLAG-tagged GFR was reported to localize in the nucleus of transfected Hela cells (207); however, we observed a green fluorescent protein fusion to be almost exclusively cytoplasmic (unpublished observation). The reason for this discrepancy is not known. The message for this GEF is highly expressed in the brain and is also detectable in various other tissues including heart, spleen, kidney, liver, placenta, and lung (66).

Although identified in a screen using C3G, GFR shares closest identity to Epac2 (Figs. 4–6) and was named Repac (related to Epac) by de Rooij and colleagues, although it lacks a cAMP-binding domain (192). We independently characterized the KIAA0277 cDNA and found the encoded protein to be regulated by M-Ras-GTP due to the presence of an RA domain N terminal to scr1 (66). We consequently named it MR-GEF (M-Ras regulated). M-Ras GTP reduced the ability of MR-GEF to activate Rap1 by approximately 50% when GEF, Rap, and M-Ras cDNAs were cotransfected. Whether this was due to direct inhibition of the catalytic domain by M-Ras binding or sequestration of MR-GEF away from the Golgi, where Rap1 predominantly resides, to the plasma membrane, where M-Ras is located, was not determined. However, as discussed in Section VII,H, another M-Ras-regulated Rap exchange factor, PDZ-GEF2 or RA-GEF2, is regulated by subcellular redistribution (208).

H. PDZ-GEFs

This exchange factor was independently identified by numerous groups, thus, has accrued multiple names: PDZ-GEF1, RA-GEF, nRap-GEP, or CNrasGEF (66, 141, 209–211). It was identified by the Bos and Quilliam labs by searching the EST database for genes sharing sequence homology with CDC25 (66, 209). A full-length cDNA (KIAA0313) was isolated and sequenced by the Kazusa DNA Research Facility (http://www.Kazusa.or.jp/huge/) and has been the source of full-length cDNA for all studies. de Rooij et al., who identified two PDZ-GEF isoforms, were searching for GEFs that might be regulated by second messengers. They noted the presence of a putative N-terminal cAMP-binding domain similar to that present in the regulatory subunits of cAMP-dependent protein kinase and later identified in the Epac/cAMP-GEFs (described above). One report demonstrated association of PDZ-GEF1 with cyclic nucleotide-immobilized agarose (*141*); however, several other groups have reported that PDZ-GEF1 does not bind to either cAMP or cGMP using more traditional assay methods (*209–211*). This is consistent with the absence of key residues in the cyclic nucleotide-binding domain of PDZ-GEF1 that are required for cAMP binding to the regulatory subunit of PKA (*192, 209, 211*). Similar to Epac1, deletion of the N-terminal cAMP-binding motif activates PDZ-GEF (*192*), suggesting that, even though PDZ-GEF lacks regulation by cyclic nucleotides, the autoinhibitory role of this domain is conserved.

Some confusion also exists regarding the tissue distribution of PDZ-GEF1. Two reports (*141, 211*) indicated that it was predominantly expressed in brain; however, one of these groups has since noted a broader tissue distribution (*212*). Another found the protein to be ubiquitously expressed in both tissues and cultured cell lines (*209*). We generated an antibody to an N-terminal peptide that was very sensitive in detecting N terminally epitope-tagged recombinant PDZ-GEF1 but not the untagged protein, leading us to believe that the predicted initiating ATG may be further 5′ to that actually utilized (unpublished observation).

PDZ-GEF1 is an exchange factor for Rap1 and Rap2 (*66, 209–211*), however Pham *et al.* described a switch in specificity from Rap1 to Rap1 plus Ras upon cAMP stimulation (*141*). As might be expected from its name, PDZ-GEF also possesses a PDZ (PSD95/Dgl/ZO-1) domain, a protein-interaction module that typically binds to hydrophobic C-terminal peptide sequences or other PDZ domains (*213*). This domain is located between scr0 and the RA domain (Fig. 3). Although binding partners that interact with this domain have not yet been described, the C-terminal SAV sequence has been found to interact with the PDZ domain of the synaptic scaffolding protein S-SCAM and related epithelial tight junction protein MAGI-1/BAP1 (*211, 212*). Thus Rap1, via PDZ-GEF, may play a role in cell–cell interaction. This would be consistent with the observation that deletion of Rap1 in *Drosophila* disrupts cell shape and migration (*214*). As noted in Section VII,B, PDZ-GEF1 contains a proline-rich sequence that enables it to bind the ubiquitin ligase, NEDD4, creating a possible means of regulating its degradation by the proteosome (*141*).

PDZ-GEF was named RA-GEF by Kataoka's group, due to the presence of an RA (Ras association/RalGDS-AF6 homology) domain similar to that found in Ras effectors (*210, 215*). This domain has been found to bind to GTP-bound Rap1 and 2 (*66, 210*), and an intact RA domain is required for the subcellular localization and efficient Rap1 activation by PDZ-GEF1 *in vivo* (*208*). Very recently, Kataoka and colleagues further determined that PDZ- or RA-GEF2 is also regulated by M-Ras. As we had observed with MR-GEF (*66*), they noted a decrease in total Rap-GTP upon introducing activated M-Ras into cells that expressed PDZ-GEF2 and confirmed that this was due to translocation of the GEF to the cell periphery (216). Interestingly, they found that although total Rap1-GTP levels were decreased, there was an increase in the activation of Rap1 at the plasma membrane. Therefore, the RA domains present in the Epac/MR-GEF/PDZ-GEF subfamily of Ras exchange factors each appear to play a role in directing them to different subcellular locations where they can activate a specific pool of Rap1 and/or 2. Despite high-sequence homology between PDZ-GEFs 1 and 2, they may play quite different roles due to being activated by different GTPases in different parts of the cell; PDZ-GEF1 is involved in a feed-forward amplification mechanism to activate Rap1 and 2 in the Golgi, while PDZ-GEF2 is involved in M-Ras/Rap pathway crosstalk at the plasma membrane. Based on the Sos crystal structure (56), the close proximity of the RA domain to scr1 likely places it on the opposite face of the GEF to the catalytic pocket. Therefore Ras/Rap-GTP may orient the GEFs at the membrane to efficiently catalyze nucleotide exchange on adjacent Rap molecules. A number of other signaling molecules possess putative RA domains (215), suggesting that Ras family GTPases might regulate the subcellular distribution of a myriad of target proteins.

I. Phospholipase Cε

Since Ras was coupled to cyclic AMP generation in yeast, early studies on Ras entertained the possibility that this GTPase may be the pertussis toxin-insensitive G protein responsible for coupling cell-surface receptors to phospholipase C activation in higher eukaryotes (217, 218). Although later studies suggested this was not the case (219, 220), there has been continued interest in the connection between Ras activation and lipid metabolism. Recent studies by three groups identified a novel phospholipase that, in addition to sharing homology with the catalytic subunit (X and Y domains, Fig. 3) and C2 lipid-binding motif characteristic of phospholipases C, also contains a CDC25 homology domain and two RA domains (Figs. 3 and 7), potentially placing it both upstream and downstream of Ras (221–226). One of the initial studies reported that PLCε activated Ras and downstream ERKs (225), but a subsequent report, supported by our unpublished *in vivo* data, found that PLCε activated Rap1- but not other Ras-family GTPases including H-Ras and Rap2 (222). Activation of ERK via B-Raf but not Raf1 was consistent with this observation.

Phospholipase Cε was first identified as a Ras-binding protein, PLC210, in *Caenorhabditis elegans* due to its containing two C terminally located RA domains (221). Human and rat phospholipase Cε bound to Ras- and Rap1-GTP and oncogenic Ras induced membrane translocation and stimulated its PIP2-hydrolyzing activity (223, 224). This GTP-dependent interaction was primarily mediated by the C-terminal RA domain. Interestingly, the E37G effector binding domain mutant of Ras stimulated phospholipase Cε activity (223, 224). This mutant has previously been used to implicate the Ras effector RalGDS

```
AF6-C     SGGTLRIYASSLKPNIPYKAIILSLTDPADFAVAEAIEKYGLEKENPKDYCLARVLPPGAQHSDE.KGAIEIIDDDECPLQLFREWPSDKGILVFQLKRPP.
Rg1       DCRIIRVQMFLGEDGSIYKSIIVTSQDKAESVIFSRVLKKNNRDSAVASEFLVQILP........VFYAMDG..ASHDFLIQRRR.
R1f       DCRIIRVQMFLGEDGSIYKSIIVTSQDKAESVIFSRVLKKNNRDSAVASEFLVQILP........VFYAMDG..ASHDFLIQRRR.
RalGDS    DCCIIRVSLEVD.NGNIYKSIIVTSQHKAPAVIRKAMDKHNLEEEPEDYELQILS........DDRKLKIPENAN..VFYAMNST.ANYDFVLKKRTF.
Rg13      EARVIRVSINNN.HGNLYRSIIJTCQHKAFSVQHAIEHNVPQPWARDVFQLIP........VFYAMSPA.APGDFLRKMEG.
PLCe-1    CSQTHKVTHGVPGIEPFAVFTINEGTKAKQLLQQIILAVDQDTKLTAADYFLMEKHFISKEKNECRKQPFQRAGPEEIIVQLLNSWFPEEIVGRAVLKPQQE


AF6-N     .HGVMRFYFQDKAAGNFATKCIRVSSTATQDVIETLAEKFRP..DMRMLSSPKYSLYEIHVSGERRLDIEKPLVVQLNWNKDDREGRFVLKNENDAIPPKAQSN
PDZGEF1   ....DRVFKADQQS....RYIMISKDMTAKEVIQAIREFA....VTATPDQYSLCEISVTPEGVIKQRRLPDQLSKIADRIQLSGRYVLKNNMETETLCSDED
PDZGEF2   PDQVIEHVKVDQQS....CYIIISKDMTAKEVFHAVHEFG....LTGASDTYSLCEISVTPEGVIKQRRLPDQFSKIADRIQLNGRYVLKNNMETETLCSDED
EPAC2     DEVLFKVYCMDHTY....TTIVFVAHSVKEVISAJAAPKLG...SGE.........GLIVKMSSGGEKVLIKPNDVSUFTTHTINGRLFACPREQFDSLTLPE
LinkII    DEIFCKVYMPDHSY....VTIRSRLSASVQDIIGSVTEKLQ....YSEEPAGREDSEILVAVSSSGEKVLIQPTEDCVFTAIGINSLFACTRDSYEALVPLPE
MRGEF     EEIFCKVITEHSY.....VSVKAKVSSIAQEIKVAEKIQ....YAEED.....IALVIATFSGEKHELQPNDLVISKSLEASGRIYVYRKDLADTLNPFAE
PLCe-1    ESFFQVHDVSPEQPR...TVIKAPRVSJAQVIQQTLCKAKYSYSILNNPNPCDYVLIEEVMKDAPNKKSSTPKSSQRLLLDQECVFQAQSKWKGAGKFILKLKE
GRASP1    GQRIEKKGSAVLKDLKRQLHIERKRADKLQERQEILTNSK.....SRTGIEEIVLSEMNSPSRTQTGDSSVSFSYREILKEKESSAIPARSLSSS
```

FIG. 7. Alignment of Ras-association domains. RA (Ras-association or RalGDS/AF6 homology) domains (215, 284) that interact with GTP-bound Ras proteins are found in multiple GEFs (see Fig. 3). These regions have been aligned using the ClustalW program. Conserved residues indicated by gray or black shading as in Fig. 5. Due to the poor conservation of sequence, the RA domains of RalGDS family members and most other GEFs were aligned separately. The two RA domains of AF6 are included for comparison. The RA domain of PDZ GEF1 binds Rap1 and 2 (66, 210); that of phospholipase Cε binds to Ras and Rap1 (223, 224); while those of PDZ-GEF2 and MR-GEF bind M-Ras (66, 208) and Epac2 associates with H-Ras (unpublished observation).

in cellular-signaling processes. Thus phospholipase Cε may be responsible for some of the biological effects previously attributed to RalGDS.

Three classes of phospholipase C, β, γ, and δ, that are regulated by heterotrimeric G proteins, tyrosine kinases, and calcium, respectively, have previously been characterized. An exciting discovery is that PLCε is activated by G$_{\alpha 12}$ (225), suggesting that Rap1 too is a mediator of G-protein-coupled receptor signaling. It remains to be established whether G$_{\alpha 12}$ can directly activate Ras or Rap1 via PLCε. Upon stimulation of cells with EGF, translocation of phospholipase Cε to both the plasma membrane and the perinuclear region was observed (222, 223), due to the association of its RA domains with Ras- and Rap-GTP, respectively. Intact RA domains were required for both efficient subcellular localization and GEF activity, suggesting that the Ras-GTP binding plays an important role in substrate targeting.

J. Other Putative Rap1 GEFs

A partial cDNA clone, "Link II," entered into the GenBank database by Graybiel and colleagues (accession number AF117946), shares the closest identity to MR-GEF and is described in the database entry as a Rap1 GEF. Like several of the above-mentioned Rap GEFs, Link II contains an RA domain immediately N terminal to its scr1. cDNA clones, with accession numbers BAB31080 and BAB31297, also encode for proteins that share a high degree of homology with the MR-GEF/PDZ-GEF/Epac family and share a particularly high identity with each other (Figs. 4 and 5). Both were isolated in the same screen for full-length cDNAs but lack RA domains and share no discernible regulatory domains with other proteins. Based on their sequence, it is most likely that they will be (at least) Rap1 GEFs. However, thus far, no functional characterization has been reported.

K. The RalGDS Family

RalGDS was originally isolated in a cDNA library screened using a CDC25 probe to identify novel Ras GEFs (49). It contains a central CDC25 homology domain and a C-terminal RA domain that binds to Ras-GTP. Perhaps because of the C-terminal location of its RA domain resulting in high abundance in oligo dT-primed cDNA libraries, RalGDS family members have frequently been isolated in yeast two-hybrid screens that used Ras-GTP as bait. Currently four additional family members have been cloned: Rgl (RalGDS-like) (227), Rgl2/Rlf (RalGDS-like factor, also Arg-Leu-Phe/RLF are its C-terminal-most three residues) (228, 229), Rgl3/RPM (Ras pathway modulator)(230, 231), and Rgr (RalGDS-related)(232–235). All these proteins appear to be specific GEFs for the Ral A and B proteins. Although it is difficult to pick out residues that are unique to Ral GEFs, there is significant homology between RalGDS family members, including an extension prior to scr3 (Fig. 5).

The RA domains of the various RalGDS family members have different binding affinities for various GTPases. For example, RalGDS preferentially binds Rap1 over H-Ras while Rlf associates only poorly with Ras. Similarly, M-Ras associates with Rlf > Rgl2 > RalGDS (236, and unpublished). Several binding constants can be found in Ref. 83. Although various GTPases including Ras, Rap1, Rap2, M-Ras, R-Ras, and TC21 can interact with one or more RalGDS isoforms, it is currently not known whether many of these GTPases can activate Ral. Rap1 can induce GTP loading of Ral when components are overexpressed in Cos cells, but activation of endogenous Rap1 in fibroblasts does not result in Ral activation (237). The role of the RA domain of RalGDS is one believed to predominantly target GEFs to the membrane, since replacement of this domain with a K-Ras C terminus to constitutively direct to the plasma membrane results in greatly enhanced activity of several family members (238, 239).

RalGDS cooperates with other Ras effectors such as Raf to induce cellular transformation, and constitutive membrane targeting of Rlf allows growth of cells in low serum (239–241). However, RalGDS has frequently been found to be more effective than an activated Ral23V mutant at inducing biological events (240, 242). This has led to speculation that RalGDS might have another substrate or function, such as localizing Ral to a specific subcellular site. Another possibility is that Ral needs to rapidly turn over its bound GTP. A rapid turnover of nucleotide has been shown to be important for strong Rho protein activity (243). Introduction of an N115I mutation into Rab proteins reduces their affinity for nucleotide and so increases their GDP/GTP turnover (244). However, an equivalent 124I RalA mutant did not exhibit increased biological activity in our hands (unpublished observation).

Ras-independent activation of Ral, which is apparently dependent on Ca^{2+} elevation, has been reported (245, 246). This suggested the presence of additional Ral GEFs. However, it is possible that the Ca^{2+} dependence is due to the fact that the C terminus of Ral can associate with calmodulin (247). This is also true of another Ras family member, Rin (248, 249); however, how this association might lead to nucleotide exchange or by what GEF(s) remains to be determined.

L. RalGPS

A family of Ras-independent Ral GEFs was recently independently identified by two groups (55, 250). These GEFs, referred to as RalGPS (GEFs with PH domain and SH3-binding motif), or RalGEF2 (a second family of Ral GEFs, distinct from RalGDS), lack an scr0 region but contain a C-terminal pleckstrin homology domain sharing closest identity to that of the *Drosophila* protein *Still life* and the N-terminal PH domain of TIAM-1, both Rho-family GEFs (55). Two splice variants of RalGPS1 exist: RalGDP 1A and 1B (RalGDP1A and 1B) (55). The 1A isoform was cloned using EST fragments and rapid amplification of cDNA ends (RACE) (55), while the 1B isoform was another clone sequenced by the Kasuza DNA research facility, KIAA0351 (55, 250). RalGPS1B differed by having two internal insertions including one in loop 3 of the PH domain and an alternative, shorter, C-terminal tail (55). The alternate sequence of the PH domain might alter its ligand-binding specificity. A potential cAMP-dependent protein kinase phosphorylation site is also present in this region of RalGPS1B that might influence its GEF localization/activity. Despite interaction of the N-terminal PH domain of TIAM-1 with inositol phospholipids (251), these lipids had no effect on RalGPS1 activity or membrane localization (250), and basal activity was not suppressed by the PI3K inhibitor LY294002 (unpublished observation). This domain does, however, appear to be important for membrane association, since the loss of activity that occurred following removal of the PH domain could be rescued by addition of a Ras-membrane-targeting motif (55, 250). RalGPS1 also has a short proline-rich sequence PPxPRxRxxS that matches the consensus-binding sequence of the Grb2 and Nck adapter proteins (252, 253), both of which could be co-immunoprecipitated from cell coexpressing Flag-tagged RalGPS 1B (55). The proline-rich sequence of RalGPS2 contains a minimal PxxP motif that is generally required for SH3 domain binding but does meet other criteria required for adapter protein interaction (252, 253). A full-length sequence of RalGPS2 has not been described. Future studies are required to confirm that it is a Ral exchange factor, although based on sequence homology with RalGPS1, which is most likely.

M. BCAR3 and Related Putative GEFs

Using retroviral insertion to identify genes that contribute to estrogen-independent growth of breast cancer cells, van Agthoven *et al.*, identified several genes whose overexpression produced such a phenotype. These included breast cancer anti-estrogen-resistance gene *1, BCAR1* (p130Cas) (254), and *BCAR3* (255). Although not recognized in the original study, the C terminus of *BCAR3* shares moderate homology with the CDC25 homology region of Ras GEFs (Fig. 5). Additionally, *BCAR3* has an N-terminal SH2 domain and a central proline-rich region that may interact with SH3 domains. A murine homolog of *BCAR3*, AND-34, was identified in thymic stromal cells following the induction of CD4+CD8+ thymocyte apoptosis (256). An additional family member, SHEP1, was found to be associated with the EphB2 protein tyrosine kinase (257, 258) and is identical to Chat, a Cas/HEF1-associated signal transducer (259). A third family member, NSP1 (novel SH2-containing protein), was found to be associated with p130Cas in response to EGF or insulin stimulation and its overexpression induced JNK activation (258). Due to sequence homology with NSP1, BCAR3/AND-34 and SHEP1/Chat were also cloned as NSP2 and 3, respectively (258). All three family members associate with p130Cas implicating them in cell adhesion. NSP2-BCAR3/AND-34 was most abundantly expressed in heart and skeletal muscle (255, 258); NSP3/SHEP1/Chat was broadly distributed (257–259); while NSP1 message levels were only weakly expressed compared to its family members (258).

Although several nucleotide-free Ras proteins can associate with AND-34 or SHEP1 (257; and our unpublished observations), we were unable to demonstrate activation of these GTPases *in vivo* using GST-RBD pull downs or $^{32}PO_4$ metabolic labeling of GDP/GTP pools under identical conditions to those described in Ref. 260. Similarly, Bos and colleagues failed to detect *in vitro* GEF activity associated with BCAR3 (20). In contrast, in the hands of Gotoh *et al.*, AND-34 promoted moderate nucleotide exchange on Ral, Rap1, and R-Ras but not H-Ras (260). We have not been able to resolve this discrepancy but suspect that it may be related to the phosphorylation state of AND-34 and/or p130Cas in our strain of Cos cells. Interestingly the C-terminal CDC25 homology region of Chat or AND-34 is responsible for its associates with p130Cas (259; and our unpublished observation), suggesting that Cas may regulate GEF activity. Since this family of GEFs lacks the scr0 that is important for stabilizing the catalytic domains of many GEFs, it is tempting to speculate that p130Cas might serve such a function for the NSP family.

Elevated p130Cas (BCAR1) levels have been implicated in the progression of clinical breast cancer (261). Since Ral has been shown to activate Src (262) which in turn phosphorylates p130Cas and the transcription factor STAT3 (262), a signaling complex involving BCAR3 and p130Cas might contribute to the elevated Src and STAT3 activities associated with breast and colon cancer.

N. Smg GDS

The small-molecular-weight G protein guanine nucleotide-dissociation stimulator (Smg GDS) was originally purified as an activity that promoted nucleotide exchange on Rap1/Smg p21 (37). It was subsequently found to regulate several GTPases *in vitro* that include K-Ras, Rac, and Rho (38, 39). Smg GDS did not activate H- or N-Ras. A *Xenopus* homolog was more recently reported to interact with X-Ral (263), and the bovine protein has been found to weakly activate Ral when overexpressed *in vivo* (66). However, the *in vivo* GEF activity of Smg GDS is poor on all reported Ras- and Rho-family substrates (unpublished observation). Smg GDS is the only known GEF to act on Ras- and Rho-family GTPases, although it has not been definitively shown that a common domain of Smg GDS is responsible for activation of both Ras subfamilies. Demonstration that a dominant inhibitory Rho19N mutant, which should bind tightly to Smg GDS (264), could inhibit, e.g., Rap1 activation might help address this issue.

Smg GDS does not share homology with *S. cerevisiae* CDC25 or other Ras-family GEFs, but instead was found to consist almost exclusively of eleven ~41-residue armadillo (Arm) repeats similar to those found in the catenins and various other cellular proteins (265). These domains contain three helices, and the Arm repeats stack to form a super helix of helices with a binding groove (266). Transport of material across the nuclear membrane requires the interaction of the Ras-superfamily GTPase, Ran, with importin β proteins. The N-terminal Ran-binding portion of importin β is made up of 10 Arm repeats and has been crystallized with Ran-GTP (267). The Arm repeats form a groove that is occupied by Ran and may represent a model for Ras/Rho interaction with Smg GDS. Similar to the Ras–Sos and Rac–TIAM-1 structures (56, 63), contact with importin β involves the switch 2 and α-helix 3 regions of Ran (267). Although there was no direct contact with switch 1, it was found to be in close proximity to the first Arm repeat that may play a more significant role in nucleotide exchange in Smg GDS.

A common feature of all Smg GDS substrates is the presence of a polybasic region in the hypervariable region adjacent to the lipid-modified C terminus. Both prenylation of the C terminus (38, 268) and phosphorylation of a PKA consensus site adjacent to the prenylated cysteine (269) have been shown to increase the *in vitro* exchange activity of Smg GDS for Rap1A/B, suggesting an important role for this region in GEF regulation. Rap1 phopshorylation affects both its *in vivo* activation and ability to regulate Akt (196, 270). For heterotrimeric G proteins, interaction of the C-terminal helix with their GEFs, the heptahelical receptors, is believed to be required for exchange (271). However, it is not clear how an interaction with the C terminus of Ras would promote GDP release. Although many crystal structures of Ras proteins have been solved, most have been of residues 1–166 or 172 that lack the C terminus. Alternatively, due to lack of a fixed structure, the C terminus has failed to be detected; therefore, we cannot really address its function as other than a flexible spacer to distance the GTPase from the membrane.

Similarly to classic Ras (CDC25 homology) GEFs, nucleotide exchange occurs via a two-step reaction where the GEF stabilizes a nucleotide-free intermediate state (39). This is supported by the finding that dominant inhibitory Ras proteins that become locked in the apo state have the highest affinity for Smg GDS (264). However, several differences in Smg versus Dbl or CDC25 have been noted (272, 273). Given the low *in vivo* exchange activity and broad substrate specificity of Smg GDS, it is possible that its biological role is more complex than mere nucleotide exchange. Takai's group has performed screens for Smg GDS-interacting proteins and identified SMAP (Smg-associated protein) (274). As is common with Arm-repeat-containing proteins, SMAP also contains such repeats. It is a Src substrate and is identical to KIF3, a microtubule-based ATPase motor for organelle transport. Further, a SMAP-interacting protein may regulate mitotic spindle formation (274) suggesting that Smg GDS links GTPases to microtubules.

Mice lacking Smg GDS often die of heart failure shortly after birth due to apoptosis of cardiomyocytes. Cells from other tissues are also prone to cell death, suggesting that Smg GDS plays a role in cell survival. Interestingly, the phenotypes observed in Smg GDS$^{-/-}$ mice are similar, albeit weaker, to those observed in K-Ras-deficient mice (275, 276), suggesting that Smg GDS may contribute to basal K-Ras-GTP levels. A *Dicteostelium* homolog of Smg GDS, Darlin (*Dictyostelium* armadillo-like protein), has been described (277). Loss of Darlin results in defective starvation-induced aggregation, suggesting that it may regulate a signaling pathway that modulates the chemotactic response during early development. Loss of Darlin did not affect other developmental processes. Study of Darlin may shed further light on the physiological function of this fascinating black sheep of the Ras and Rho GEF superfamilies.

VIII. GEFs and Disease

Several Ras exchange factors were identified based on their ability to induce cellular transformation upon overexpression. For example, GRP1 was isolated following the expression of a T-cell lymphoma cDNA library in fibroblasts (147), and the overexpression of many other GEFs can lead to transformation, indicating the potential for GEFs to contribute to cancer development. While Ras mutations are found in ~30% of human tumors, WT Ras can also contribute to transformation as a result of its overexpression or by mediating the growth-promoting effects of other upstream oncoproteins such as ErbB2 or BCR-Abl (95, 278, 279). Sos is also part of this pathway, and overexpression of this GEF in renal carcinoma cell lines may lead to increased Ras activation (280). Sos amplification may also play an important role in the carcinogenesis of human bladder cancer (281).

Retroviral insertional mutagenesis in BXH-2 recombinant inbred mice induces a high incidence of acute myeloid leukemia (AML), and one of the genes amplified by this technique was the Rap exchange factor, *GRP2*/CalDAG-GEF I (282). Subsequent overexpression of *GRP2* resulted in transformation of cultured fibroblasts and potentially implicates Rap1 signaling in myeloid leukemia. *GRP2* maps to human chromosome 11q13 (149), a region in which both translocations and gene amplifications have been associated with a variety of malignancies. These include breast adenocarcinoma, bladder carcinoma, gastric cancer, B-cell follicular lymphoma, and oral squamous cell carcinoma (282). Interestingly, this study also identified a truncated variant of *GRP2*, formed by alternate polyadenylation, that encoded only the scr0 and adjacent sequence. Whether this protein is normally expressed and what its biological function might be remain to be determined.

Two fusion proteins resulting from chromosomal translocations have been identified containing Ras-family GEFs, but it has not been established which half, if either, contributes to cancer development. The N terminus of the Ral GEF, Rgr, which encoded the CDC25 but not an RA domain, was isolated as a fusion protein with rHR 23A (a rabbit homolog of yeast Rad 23) from a squamous cell carcinoma. This fusion protein was called Rsc (*232*). Although no full-length sequence of Rgr has been isolated, the catalytic domain has been reported to stimulate Ras- and Rho-activated pathways (*234*). A fusion of Smg GDS with the *nucleoporin 98* gene was also found in several cases of adult T-cell acute lymphocytic leukemia (*283*). Although the *NUP98*-Smg GDS fusion protein would be in frame and encode almost the entire Smg GDS protein, and thus potentially retain GEF activity, *NUP98* has been found in several chromosomal translocations and is likely to represent the transforming component of the protein.

Finally as discussed in Section VII,M, BCAR3 is elevated in breast cancers that display resistance to antiestrogens such as tamoxifen (*255, 261*). Its association with BCAR1/p130Cas, Ras-family GTPases and potential regulation of Src and STAT3 make it a likely contributor to breast and other malignancies (*255, 260–262*).

IX. Are There More GEFs in Our Future?

The lack of GEFs for some Ras-family members such as Rit and Rin suggests that there may be more exchange factors to be found. However, given the near completion of the human genome project, it is likely that there are fewer Ras- than Rho-family-exchange factors, which now exceed 50 in number. Weak homology of, e.g., BCAR3 and GRASP-1 with other GEFs suggests that we may just need to devise better searches for the scr (structurally conserved regions) within the CDC25 domain in order to expand this family. Determining how the recently identified GEFs are coupled to extracellular stimuli and identifying the targets of these proteins may provide useful biochemical tools to establish the function of novel Ras-protein-signaling pathways and identify targets for therapeutic intervention of the diseases that relay information through them.

Acknowledgments

The authors' research was supported by Research Project Grant 00-125-01-TBE from the American Cancer Society, the R.W & G.M. Showalter Trust Fund, and the Walther Cancer Institute. JFR was supported by Training Grant T32 07774 from the National Heart Lung and Blood Institute.

References

1. J. C. Lacal and S. R. Tronick, The ras oncogene. *in* "The Oncogene Handbook" (E. P. Reddy, A. M. Skalka, and T. Curran, eds.), pp. 257–304. Elsevier Science Publishers, Amsterdam, 1988.

2. A. Valencia, P. Chardin, A. Wittinghofer, and C. Sander, The ras protein family: Evolutionary tree and role of conserved amino acids. *Biochemistry* **30**, 4637–4648 (1991).
3. J. M. Shields, K. Pruitt, A. McFall, A. Shaub, and C. J. Der, Understanding ras: 'it ain't over 'til it's over.' *Trends Cell. Biol.* **10**, 147–154 (2000).
4. S. L. Campbell, R. Khosravi-Far, K. L. Rossman, G. J. Clark, and C. J. Der, Increasing complexity of Ras signaling. *Oncogene* **17**, 1395–1413 (1998).
5. A. Hall, Rho GTPases and the actin cytoskeleton. *Science* **279**, 509–514 (1998).
6. I. M. Zohn, S. L. Campbell, R. Khosravi-Far, K. L. Rossman, and C. J. Der, Rho family proteins and Ras transformation: The RHOad less traveled gets congested. *Oncogene* **17**, 1415–1438 (1998).
7. K. Kaibuchi, S. Kuroda, and M. Amano, Regulation of the cytoskeleton and cell adhesion by the Rho family GTPases in mammalian cells. *Annu. Rev. Biochem.* **68**, 459–486 (1999).
8. P. Chavrier and B. Goud, The role of ARF and Rab GTPases in membrane transport. *Curr. Opin. Cell Biol.* **11**, 466–475 (1999).
9. Y. Takai, T. Sasaki, and T. Matozaki, Small GTP-binding proteins. *Physiol. Rev.* **81**, 153–208 (2001).
10. Y. Azuma and M. Dasso, The role of Ran in nuclear function. *Curr. Opin. Cell Biol.* **12**, 302–307 (2000).
11. P. R. Clarke and C. Zhang, Ran GTPase: A master regulator of nuclear structure and function during the eukaryotic cell division cycle? *Trends Cell. Biol.* **11**, 366–371 (2001).
12. H. W. Fu and P. J. Casey, Enzymology and biology of CaaX protein prenylation. *Rec. Prog. Horm. Res.* **54**, 315–342 (1999).
13. A. Wittinghofer, Signal transduction via Ras. *Biol. Chem.* **379**, 933–937 (1998).
14. L. A. Quilliam, R. Khosravi-Far, S. Y. Huff, and C. J. Der, Guanine nucleotide exchange factors: activators of the Ras superfamily of proteins. *Bioessays* **17**, 395–404 (1995).
15. Y. Ohba, N. Mochizuki, S. Yamashita, A. M. Chan, J. W. Schrader, S. Hattori, K. Nagashima, and M. Matsuda, Regulatory proteins of R-Ras, TC21/R-Ras2, and M-Ras/R-Ras3. *J. Biol. Chem.* **275**, 20,020–20,026 (2000).
16. M. S. Boguski and F. McCormick, Proteins regulating Ras and its relatives. *Nature* **366**, 643–654 (1993).
17. G. J. Clark and C. J. Der, Ras proto-oncogene activation in human malignancy. *in* "Cellular Cancer Markers" (C. Garrett and S. Sell, eds.), pp. 17–52. Humana Press, Totowa, NJ, 1995.
18. K. T. Barker and M. R. Crompton, Ras-related TC21 is activated by mutation in a breast cancer cell line, but infrequently in breast carcinomas in vivo. *Br. J. Cancer* **78**, 296–300 (1998).
19. A. J. Self, E. Caron, H. F. Paterson, and A. Hall, Analysis of R-Ras signalling pathways. *J. Cell. Sci.* **114**, 1357–1366 (2001).
20. J. L. Bos, J. de Rooij, and K. A. Reedquist, Rap1 signalling: Adhering to new models. *Nat. Rev. Mol. Cell. Biol.* **2**, 369–377 (2001).
21. Y. Yu, F. Xu, H. Peng, X. Fang, S. Zhao, Y. Li, B. Cuevas, W. L. Kuo, J. W. Gray, M. Siciliano, G. B. Mills, and R. C. Bast, Jr., NOEY2 (ARHI), an imprinted putative tumor suppressor gene in ovarian and breast carcinomas. *Proc. Natl. Acad. Sci. U.S.A.* **96**, 214–219 (1999).
22. D. O. Henry, S. A. Moskalenko, K. J. Kaur, M. Fu, R. G. Pestell, J. H. Camonis, and M. A. White, Ral GTPases contribute to regulation of cyclin D1 through activation of NF-kappaB. *Mol. Cell. Biol.* **20**, 8084–8092 (2000).
23. G. W. Reuther and C. J. Der, The Ras branch of small GTPases: Ras family members don't fall far from the tree. *Curr. Opin. Cell Biol.* **12**, 157–165 (2000).
24. B. S. Finlin, C. L. Gau, G. A. Murphy, H. Shao, T. Kimel, R. S. Seitz, Y. F. Chiu, D. Botstein, P. O. Brown, C. J. Der, F. Tamanoi, D. A. Andres, and C. M. Perou, Rerg is a novel ras-related, estrogen-regulated and growth-inhibitory gene in breast cancer. *J. Biol. Chem.* **276**, 42259–42267 (2001).

25. B. S. Finlin, H. Shao, K. Kadono-Okuda, N. Guo, and D. A. Andres, Rem2, a new member of the Rem/Rad/Gem/Kir family of Ras-related GTPases. *Biochem. J.* **347**, Pt 1, 223–231 (2000).
26. J. D. Falk, P. Vargiu, P. E. Foye, H. Usui, J. Perez, P. E. Danielson, D. L. Lerner, J. Bernal, and J. G. Sutcliffe, Rhes: A striatal-specific Ras homolog related to Dexras1. *J. Neurosci. Res.* **57**, 782–788 (1999).
27. C. Fenwick, S. Y. Na, R. E. Voll, H. Zhong, S. Y. Im, J. W. Lee, and S. Ghosh, A subclass of Ras proteins that regulate the degradation of IkappaB. *Science* **287**, 869–873 (2000).
28. M. Torti, K. B. Marti, D. Altschuler, K. Yamamoto, and E. G. Lapetina, Erythropoietin induces p21ras activation and p120GAP tyrosine phosphorylation in human erythroleukemia cells. *J. Biol. Chem.* **267**, 8293–8298 (1992).
29. L. Zheng, J. Eckerdal, I. Dimitrijevic, and T. Andersson, Chemotactic peptide-induced activation of Ras in human neutrophils is associated with inhibition of p120-GAP activity. *J. Biol. Chem.* **272**, 23,448–23,454 (1997).
30. I. Rubio and R. Wetzker, A permissive function of phosphoinositide 3-kinase in Ras activation mediated by inhibition of GTPase-activating proteins. *Curr. Biol.* **10**, 1225–1228 (2000).
31. N. Mochizuki, Y. Ohba, E. Kiyokawa, T. Kurata, T. Murakami, T. Ozaki, A. Kitabatake, K. Nagashima, and M. Matsuda, Activation of the ERK/MAPK pathway by an isoform of rap1GAP associated with G alpha(i). *Nature* **400**, 891–894 (1999).
32. J. D. Jordan, K. D. Carey, P. J. Stork, and R. Iyengar, Modulation of rap activity by direct interaction of Galpha(o) with Rap1 GTPase-activating protein. *J. Biol. Chem.* **274**, 21,507–21,510 (1999).
33. K. A. Reedquist, E. Ross, E. A. Koop, R. M. Wolthuis, F. J. Zwartkruis, Y. van Kooyk, M. Salmon, C. D. Buckley, and J. L. Bos, The small GTPase, Rap1, mediates CD31-induced integrin adhesion. *J. Cell. Biol.* **148**, 1151–1158 (2000).
34. K. D. Carey, T. J. Dillon, J. M. Schmitt, A. M. Baird, A. D. Holdorf, D. B. Straus, A. S. Shaw, and P. J. Stork, CD28 and the tyrosine kinase lck stimulate mitogen-activated protein kinase activity in T cells via inhibition of the small G protein Rap1. *Mol. Cell. Biol.* **20**, 8409–8419 (2000).
35. D. Broek, T. Toda, T. Michaeli, L. Levin, C. Birchmeier, M. Zoller, S. Powers, and M. Wigler, The S. cerevisiae CDC25 gene product regulates the RAS/adenylate cyclase pathway. *Cell* **48**, 789–799 (1987).
36. S. Jones, M. L. Vignais, and J. R. Broach, The CDC25 protein of Saccharomyces cerevisiae promotes exchange of guanine nucleotides bound to ras. *Mol. Cell. Biol.* **11**, 2641–2646 (1991).
37. T. Yamamoto, K. Kaibuchi, T. Mizuno, M. Hiroyoshi, H. Shirataki, and Y. Takai, Purification and characterization from bovine brain cytosol of proteins that regulate the GDP/GTP exchange reaction of smg p21s, ras p21-like GTP-binding proteins. *J. Biol. Chem.* **265**, 16,626–16,634 (1990).
38. T. Mizuno, K. Kaibuchi, T. Yamamoto, M. Kawamura, T. Sakoda, H. Fujioka, Y. Matsuura, and Y. Takai, A stimulatory GDP/GTP exchange protein for smg p21 is active on the post-translationally processed form of c-Ki-ras p21 and rhoA p21. *Proc. Natl. Acad. Sci. U.S.A.* **88**, 6442–6446 (1991).
39. T. H. Chuang, X. Xu, L. A. Quilliam, and G. M. Bokoch, SmgGDS stabilizes nucleotide-bound and -free forms of the Rac1 GTP-binding protein and stimulates GTP/GDP exchange through a substituted enzyme mechanism. *Biochem. J.* **303**, 761–767 (1994).
40. K. Kaibuchi, T. Mizuno, H. Fujioka, T. Yamamoto, K. Kishi, Y. Fukumoto, Y. Hori, and Y. Takai, Molecular cloning of the cDNA for stimulatory GDP/GTP exchange protein for smg p21s (ras p21-like small GTP-binding proteins) and characterization of stimulatory GDP/GTP exchange protein. *Mol. Cell. Biol.* **11**, 2873–2880 (1991).
41. L. Bonfini, C. A. Karlovich, C. Dasgupta, and U. Banerjee, The Son of sevenless gene product: A putative activator of Ras. *Science* **255**, 603–606 (1992).

42. M. A. Simon, D. D. Bowtell, G. S. Dodson, T. R. Laverty, and G. M. Rubin, Ras1 and a putative guanine nucleotide exchange factor perform crucial steps in signaling by the sevenless protein tyrosine kinase. *Cell* **67**, 701–716 (1991).
43. D. Bowtell, P. Fu, M. Simon, and P. Senior, Identification of murine homologues of the Drosophila son of sevenless gene: Potential activators of ras. *Proc. Natl. Acad. Sci. U.S.A.* **89**, 6511–6515 (1992).
44. P. Chardin, J. H. Camonis, N. W. Gale, L. van Aelst, J. Schlessinger, M. H. Wigler, and D. Bar-Sagi, Human Sos1: A guanine nucleotide exchange factor for Ras that binds to GRB2. *Science* **260**, 1338–1343 (1993).
45. E. Martegani, M. Vanoni, R. Zippel, P. Coccetti, R. Brambilla, C. Ferrari, E. Sturani, and L. Alberghina, Cloning by functional complementation of a mouse cDNA encoding a homologue of CDC25, a Saccharomyces cerevisiae RAS activator. *EMBO J.* **11**, 2151–2157 (1992).
46. C. Shou, C. L. Farnsworth, B. G. Neel, and L. A. Feig, Molecular cloning of cDNAs encoding a guanine-nucleotide-releasing factor for Ras p21 [see comments]. *Nature* **358**, 351–354 (1992).
47. W. Wei, R. D. Mosteller, P. Sanyal, E. Gonzales, D. McKinney, C. Dasgupta, P. Li, B. X. Liu, and D. Broek, Identification of a mammalian gene structurally and functionally related to the CDC25 gene of Saccharomyces cerevisiae. *Proc. Natl. Acad. Sci. U.S.A.* **89**, 7100–7104 (1992).
48. H. Cen, A. G. Papageorge, R. Zippel, D. R. Lowy, and K. Zhang, Isolation of multiple mouse cDNAs with coding homology to Saccharomyces cerevisiae CDC25: identification of a region related to Bcr, Vav, Dbl and CDC24. *EMBO J.* **11**, 4007–4015 (1992).
49. C. F. Albright, B. W. Giddings, J. Liu, M. Vito, and R. A. Weinberg, Characterization of a guanine nucleotide dissociation stimulator for a ras-related GTPase. *EMBO J.* **12**, 339–347 (1993).
50. S. Tanaka, T. Morishita, Y. Hashimoto, S. Hattori, S. Nakamura, M. Shibuya, K. Matuoka, T. Takenawa, T. Kurata, K. Nagashima, and a. et, C3G, a guanine nucleotide-releasing protein expressed ubiquitously, binds to the Src homology 3 domains of CRK and GRB2/ASH proteins. *Proc. Natl. Acad. Sci. U.S.A.* **91**, 3443–3447 (1994).
51. B. S. Knudsen, S. M. Feller, and H. Hanafusa, Four proline-rich sequences of the guanine-nucleotide exchange factor C3G bind with unique specificity to the first Src homology 3 domain of Crk. *J. Biol. Chem.* **269**, 32,781–32,787 (1994).
52. R. A. Cerione and Y. Zheng, The Dbl family of oncogenes. *Curr. Opin. Cell Biol.* **8**, 216–222 (1996).
53. I. P. Whitehead, S. Campbell, K. L. Rossman, and C. J. Der, Dbl family proteins. *Biochim. Biophys. Acta* **1332**, F1–F23 (1997).
54. C. C. Lai, M. Boguski, D. Broek, and S. Powers, Influence of guanine nucleotides on complex formation between Ras and CDC25 proteins. *Mol. Cell. Biol.* **13**, 1345–1352 (1993).
55. J. F. Rebhun, H. Chen, and L. A. Quilliam, Identification and characterization of a new family of guanine nucleotide exchange factors for the Ras-related GTPase Ral. *J. Biol. Chem.* **275**, 13,406–13,410 (2000).
56. P. A. Boriack-Sjodin, S. M. Margarit, D. Bar-Sagi, and J. Kuriyan, The structural basis of the activation of Ras by Sos. *Nature* **394**, 337–343 (1998).
57. M. V. Milburn, L. Tong, A. M. deVos, A. Brunger, Z. Yamaizumi, S. Nishimura, and S. H. Kim, Molecular switch for signal transduction: Structural differences between active and inactive forms of protooncogenic ras proteins. *Science* **247**, 939–945 (1990).
58. B. E. Hall, S. S. Yang, P. A. Boriack-Sjodin, J. Kuriyan, and D. Bar-Sagi, Structure-based mutagenesis reveals distinct functions for ras switch 1 and switch 2 in Sos-catalyzed guanine nucleotide exchange. *J. Biol. Chem.* **276**, 27,629–27,637 (2001).
59. F. Wittinghofer, Ras signalling. Caught in the act of the switch-on. *Nature* **394**, 317, 319–320 (1998).

60. J. Y. Pan and M. Wessling-Resnick, GEF-mediated GDP/GTP exchange by monomeric GTPases: A regulatory role for Mg^{2+}?. *Bioessays* **20**, 516–521 (1998).
61. C. Lenzen, R. H. Cool, H. Prinz, J. Kuhlmann, and A. Wittinghofer, Kinetic analysis by fluorescence of the interaction between Ras and the catalytic domain of the guanine nucleotide exchange factor Cdc25Mm. *Biochemistry* **37**, 7420–7430 (1998).
62. G. J. Day, R. D. Mosteller, and D. Broek, Distinct subclasses of small GTPases interact with guanine nucleotide exchange factors in a similar manner. *Mol. Cell. Biol.* **18**, 7444–7454 (1998).
63. D. K. Worthylake, K. L. Rossman, and J. Sondek, Crystal structure of Rac1 in complex with the guanine nucleotide exchange region of Tiam1. *Nature* **408**, 682–688 (2000).
64. J. Goldberg, Structural basis for activation of ARF GTPase: Mechanisms of guanine nucleotide exchange and GTP-myristoyl switching. *Cell* **95**, 237–248 (1998).
65. N. P. Fam, W. T. Fan, Z. Wang, L. J. Zhang, H. Chen, and M. F. Moran, Cloning and characterization of Ras-GRF2, a novel guanine nucleotide exchange factor for Ras. *Mol. Cell. Biol.* **17**, 1396–1406 (1997).
66. J. F. Rebhun, A. F. Castro, and L. A. Quilliam, Identification of guanine nucleotide exchange factors (GEFs) for the rap1 GTPase. Regulation of MR-GEF by M-Ras-GTP interaction. *J. Biol. Chem.* **275**, 34,901–34,908 (2000).
67. L. A. Feig, Tools of the trade: Use of dominant-inhibitory mutants of ras-family GTPases. *Nat. Cell Biol.* **1**, E25–E27 (1999).
68. S. Y. Chen, S. Y. Huff, C. C. Lai, C. J. Der, and S. Powers, Ras-15A protein shares highly similar dominant-negative biological properties with Ras-17N and forms a stable, guanine-nucleotide resistant complex with CDC25 exchange factor. *Oncogene* **9**, 2691–2698 (1994).
69. V. Jung, W. Wei, R. Ballester, J. Camonis, S. Mi, L. Van Aelst, M. Wigler, and D. Broek, Two types of RAS mutants that dominantly interfere with activators of RAS. *Mol. Cell. Biol.* **14**, 3707–3718 (1994).
70. D. A. Williams, W. Tao, F. Yang, C. Kim, Y. Gu, P. Mansfield, J. E. Levine, B. Petryniak, C. W. Derrow, C. Harris, B. Jia, Y. Zheng, D. R. Ambruso, J. B. Lowe, S. J. Atkinson, M. C. Dinauer, and L. Boxer, Dominant negative mutation of the hematopoietic-specific Rho GTPase, Rac2, is associated with a human phagocyte immunodeficiency. *Blood* **96**, 1646–1654 (2000).
71. R. D. York, H. Yao, T. Dillon, C. L. Ellig, S. P. Eckert, E. W. McCleskey, and P. J. Stork, Rap1 mediates sustained MAP kinase activation induced by nerve growth factor. *Nature* **392**, 622–626 (1998).
72. E. Caron, A. J. Self, and A. Hall, The GTPase Rap1 controls functional activation of macrophage integrin alphaMbeta2 by LPS and other inflammatory mediators. *Curr. Biol.* **10**, 974–978 (2000).
73. N. van den Berghe, R. H. Cool, G. Horn, and A. Wittinghofer, Biochemical characterization of C3G: An exchange factor that discriminates between Rap1 and Rap2 and is not inhibited by Rap1A(S17N). *Oncogene* **15**, 845–850 (1997).
74. D. W. Stacey, L. A. Feig, and J. B. Gibbs, Dominant inhibitory Ras mutants selectively inhibit the activity of either cellular or oncogenic Ras. *Mol. Cell. Biol.* **11**, 4053–4064 (1991).
75. T. R. Brtva, J. K. Drugan, S. Ghosh, R. S. Terrell, S. Campbell-Burk, R. M. Bell, and C. J. Der, Two distinct Raf domains mediate interaction with Ras. *J. Biol. Chem.* **270**, 9809–9812 (1995).
76. M. Okazaki, S. Kishida, H. Murai, T. Hinoi, and A. Kikuchi, Ras-interacting domain of Ral GDP dissociation stimulator like (RGL) reverses v-Ras-induced transformation and Raf-1 activation in NIH3T3 cells. *Cancer Res.* **56**, 2387–2392 (1996).
77. J. B. Gibbs, Determination of guanine nucleotides bound to Ras in mammalian cells. *Methods Enzymol.* **255**, 118–125 (1995).
78. J. S. Scheele, J. M. Rhee, and G. R. Boss, Determination of absolute amounts of GDP and GTP bound to Ras in mammalian cells: Comparison of parental and Ras-overproducing NIH 3T3 fibroblasts. *Proc. Natl. Acad. Sci. U.S.A.* **92**, 1097–1100 (1995).

79. S. J. Taylor and D. Shalloway, Cell cycle-dependent activation of Ras. *Curr. Biol.* **6**, 1621–1627 (1996).
80. J. de Rooij and J. L. Bos, Minimal Ras-binding domain of Raf1 can be used as an activation-specific probe for Ras. *Oncogene* **14**, 623–625 (1997).
81. M. van Triest, J. de Rooij, and J. L. Bos, Measurement of GTP-bound Ras-like GTPases by activation-specific probes. *Methods Enzymol.* **333**, 343–348 (2001).
82. S. J. Taylor, R. J. Resnick, and D. Shalloway, Nonradioactive determination of Ras-GTP levels using activated ras interaction assay. *Methods Enzymol.* **333**, 333–342 (2001).
83. D. Esser, B. Bauer, R. M. Wolthuis, A. Wittinghofer, R. H. Cool, and P. Bayer, Structure determination of the Ras-binding domain of the Ral-specific guanine nucleotide exchange factor Rlf. *Biochemistry* **37**, 13,453–13,462 (1998).
84. M. Fridman, H. Maruta, J. Gonez, F. Walker, H. Treutlein, J. Zeng, and A. Burgess, Point mutants of c-raf-1 RBD with elevated binding to v-Ha-Ras. *J. Biol. Chem.* **275**, 30,363–30,371 (2000).
85. M. Fridman, F. Walker, B. Catimel, T. Domagala, E. Nice, and A. Burgess, c-Raf-1 RBD associates with a subset of active v-H-Ras. *Biochemistry* **39**, 15,603–15,611 (2000).
86. V. S. Kraynov, C. Chamberlain, G. M. Bokoch, M. A. Schwartz, S. Slabaugh, and K. M. Hahn, Localized Rac activation dynamics visualized in living cells. *Science* **290**, 333–337 (2000).
87. N. Mochizuki, S. Yamashita, K. Kurokawa, Y. Ohba, T. Nagai, A. Miyawaki, and M. Matsuda, Spatio-temporal images of growth-factor-induced activation of Ras and Rap1. *Nature* **411**, 1065–1068 (2001).
88. X. Qian, L. Esteban, W. C. Vass, C. Upadhyaya, A. G. Papageorge, K. Yienger, J. M. Ward, D. R. Lowy, and E. Santos, The Sos1 and Sos2 Ras-specific exchange factors: Differences in placental expression and signaling properties. *EMBO J.* **19**, 642–654 (2000).
89. L. M. Esteban, A. Fernandez-Medarde, E. Lopez, K. Yienger, C. Guerrero, J. M. Ward, L. Tessarollo, and E. Santos, Ras-guanine nucleotide exchange factor sos2 is dispensable for mouse growth and development. *Mol. Cell. Biol.* **20**, 6410–6413 (2000).
90. K. H. Nielsen, A. G. Papageorge, W. C. Vass, B. M. Willumsen, and D. R. Lowy, The Ras-specific exchange factors mouse Sos1 (mSos1) and mSos2 are regulated differently: mSos2 contains ubiquitination signals absent in mSos1. *Mol. Cell. Biol.* **17**, 7132–7138 (1997).
91. E. J. Lowenstein, R. J. Daly, A. G. Batzer, W. Li, B. Margolis, R. Lammers, A. Ullrich, E. Y. Skolnik, D. Bar-Sagi, and J. Schlessinger, The SH2 and SH3 domain-containing protein GRB2 links receptor tyrosine kinases to ras signaling. *Cell* **70**, 431–442 (1992).
92. S. E. Egan, B. W. Giddings, M. W. Brooks, L. Buday, A. M. Sizeland, and R. A. Weinberg, Association of Sos Ras exchange protein with Grb2 is implicated in tyrosine kinase signal transduction and transformation. *Nature* **363**, 45–51 (1993).
93. L. Buday and J. Downward, Epidermal growth factor regulates p21ras through the formation of a complex of receptor, Grb2 adapter protein, and Sos nucleotide exchange factor. *Cell* **73**, 611–620 (1993).
94. N. W. Gale, S. Kaplan, E. J. Lowenstein, J. Schlessinger, and D. Bar-Sagi, Grb2 mediates the EGF-dependent activation of guanine nucleotide exchange on Ras. *Nature* **363**, 88–92 (1993).
95. A. M. Pendergast, L. A. Quilliam, L. D. Cripe, C. H. Bassing, Z. Dai, N. Li, A. Batzer, K. M. Rabun, C. J. Der, J. Schlessinger, *et al.*, BCR-ABL-induced oncogenesis is mediated by direct interaction with the SH2 domain of the GRB-2 adaptor protein. *Cell* **75**, 175–185 (1993).
96. S. E. Egan and R. A. Weinberg, The pathway to signal achievement [news]. *Nature* **365**, 781–783 (1993).
97. S. Corbalan-Garcia, S. S. Yang, K. R. Degenhardt, and D. Bar-Sagi, Identification of the mitogen-activated protein kinase phosphorylation sites on human Sos1 that regulate interaction with Grb2. *Mol. Cell. Biol.* **16**, 5674–5682 (1996).

98. M. Rozakis-Adcock, J. McGlade, G. Mbamalu, G. Pelicci, R. Daly, W. Li, A. Batzer, S. Thomas, J. Brugge, P. G. Pelicci, J. Schlessinger, and T. Pawson, Association of the Shc and Grb2/Sem5 SH2-containing proteins is implicated in activation of the Ras pathway by tyrosine kinases. *Nature* **360,** 689–692 (1992).
99. M. Holgado-Madruga, D. R. Emlet, D. K. Moscatello, A. K. Godwin, and A. J. Wong, A Grb2-associated docking protein in EGF- and insulin-receptor signalling. *Nature* **379,** 560–564 (1996).
100. W. Li, R. Nishimura, A. Kashishian, A. G. Batzer, W. J. Kim, J. A. Cooper, and J. Schlessinger, A new function for a phosphotyrosine phosphatase: Linking GRB2-Sos to a receptor tyrosine kinase. *Mol. Cell. Biol.* **14,** 509–517 (1994).
101. J. Schlessinger, How receptor tyrosine kinases activate Ras. *Trends Biochem. Sci.* **18,** 273–275 (1993).
102. D. D. Schlaepfer and T. Hunter, Signal transduction from the extracellular matrix—a role for the focal adhesion protein-tyrosine kinase FAK. *Cell Structure Function* **21,** 445–450 (1996).
103. I. Dikic, G. Tokiwa, S. Lev, S. A. Courtneidge, and J. Schlessinger, A role for Pyk2 and Src in linking G-protein-coupled receptors with MAP kinase activation. *Nature* **383,** 547–550 (1996).
104. K. L. Pierce, L. M. Luttrell, and R. J. Lefkowitz, New mechanisms in heptahelical receptor signaling to mitogen activated protein kinase cascades. *Oncogene* **20,** 1532–1539 (2001).
105. E. Porfiri, T. Evans, P. Chardin, and J. F. Hancock, Prenylation of Ras proteins is required for efficient hSOS1-promoted guanine nucleotide exchange. *J. Biol. Chem.* **269,** 22,672–22,677 (1994).
106. L. A. Quilliam, S. Y. Huff, K. M. Rabun, W. Wei, W. Park, D. Broek, and C. J. Der, Membrane-targeting potentiates guanine nucleotide exchange factor CDC25 and SOS1 activation of Ras transforming activity. *Proc. Natl. Acad. Sci. U.S.A.* **91,** 8512–8516 (1994).
107. A. Aronheim, D. Engelberg, N. Li, N. al-Alawi, J. Schlessinger, and M. Karin, Membrane targeting of the nucleotide exchange factor Sos is sufficient for activating the Ras signaling pathway. *Cell* **78,** 949–961 (1994).
108. L. J. Holsinger, D. M. Spencer, D. J. Austin, S. L. Schreiber, and G. R. Crabtree, Signal transduction in T lymphocytes using a conditional allele of Sos. *Proc. Natl. Acad. Sci. U.S.A.* **92,** 9810–9814 (1995).
109. W. Wang, E. M. Fisher, Q. Jia, J. M. Dunn, E. Porfiri, J. Downward, and S. E. Egan, The Grb2 binding domain of mSos1 is not required for downstream signal transduction. *Nat. Genet.* **10,** 294–300 (1995).
110. S. Corbalan-Garcia, S. M. Margarit, D. Galron, S. S. Yang, and D. Bar-Sagi, Regulation of Sos activity by intramolecular interactions. *Mol. Cell. Biol.* **18,** 880–886 (1998).
111. X. Qian, W. C. Vass, A. G. Papageorge, P. H. Anborgh, and D. R. Lowy, N terminus of Sos1 Ras exchange factor: critical roles for the Dbl and pleckstrin homology domains. *Mol. Cell. Biol.* **18,** 771–778 (1998).
112. R. H. Chen, S. Corbalan-Garcia, and D. Bar-Sagi, The role of the PH domain in the signal-dependent membrane targeting of Sos. *EMBO J.* **16,** 1351–1359 (1997).
113. A. B. Jefferson, A. Klippel, and L. T. Williams, Inhibition of mSOS-activity by binding of phosphatidylinositol 4,5-P2 to the mSOS pleckstrin homology domain. *Oncogene* **16,** 2303–2310 (1998).
114. A. S. Nimnual, B. A. Yatsula, and D. Bar-Sagi, Coupling of Ras and Rac guanosine triphosphatases through the Ras exchanger Sos. *Science* **279,** 560–563 (1998).
115. P. Rodriguez-Viciana, P. H. Warne, A. Khwaja, B. M. Marte, D. Pappin, P. Das, M. D. Waterfield, A. Ridley, and J. Downward, Role of phosphoinositide 3-OH kinase in cell transformation and control of the actin cytoskeleton by Ras. *Cell* **89,** 457–467 (1997).
116. J. Han, K. Luby-Phelps, B. Das, X. Shu, Y. Xia, R. D. Mosteller, U. M. Krishna, J. R. Falck, M. A. White, and D. Broek, Role of substrates and products of PI 3-kinase in regulating activation of Rac-related guanosine triphosphatases by Vav. *Science* **279,** 558–560 (1998).

117. S. M. Soisson, A. S. Nimnual, M. Uy, D. Bar-Sagi, and J. Kuriyan, Crystal structure of the Dbl and pleckstrin homology domains from the human Son of sevenless protein. *Cell* **95,** 259–268 (1998).
118. G. Scita, J. Nordstrom, R. Carbone, P. Tenca, G. Giardina, S. Gutkind, M. Bjarnegard, C. Betsholtz, and P. P. Di Fiore, EPS8 and E3B1 transduce signals from Ras to Rac. *Nature* **401,** 290–293 (1999).
119. G. Scita, P. Tenca, L. B. Areces, A. Tocchetti, E. Frittoli, G. Giardina, I. Ponzanelli, P. Sini, M. Innocenti, and P. P. Di Fiore, An effector region in Eps8 is responsible for the activation of the Rac-specific GEF activity of Sos-1 and for the proper localization of the Rac-based actin-polymerizing machine. *J. Cell. Biol.* **154,** 1031–1044 (2001).
120. L. Chen, L. J. Zhang, P. Greer, P. S. Tung, and M. F. Moran, A murine CDC25/ras-GRF-related protein implicated in Ras regulation. *Dev. Genet.* **14,** 339–346 (1993).
121. H. Cen, A. G. Papageorge, W. C. Vass, K. E. Zhang, and D. R. Lowy, Regulated and constitutive activity by CDC25Mm (GRF), a Ras-specific exchange factor. *Mol. Cell. Biol.* **13,** 7718–7724 (1993).
122. C. Guerrero, J. M. Rojas, M. Chedid, L. M. Esteban, D. B. Zimonjic, N. C. Popescu, J. Font de Mora, and E. Santos, Expression of alternative forms of Ras exchange factors GRF and SOS1 in different human tissues and cell lines. *Oncogene* **12,** 1097–1107 (1996).
123. M. K. Jones and J. H. Jackson, Ras-GRF activates Ha-Ras, but not N-Ras or K-Ras 4B, protein in vivo. *J. Biol. Chem.* **273,** 1782–1787 (1998).
124. K. H. Nielsen, L. Gredsted, J. R. Broach, and B. M. Willumsen, Sensitivity of wild type and mutant ras alleles to Ras specific exchange factors: Identification of factor specific requirements. *Oncogene* **20,** 2091–2100 (2001).
125. T. Gotoh, Y. Niino, M. Tokuda, O. Hatase, S. Nakamura, M. Matsuda, and S. Hattori, Activation of R-Ras by Ras-guanine nucleotide-releasing factor. *J. Biol. Chem.* **272,** 18,602–18,607 (1997).
126. L. A. Quilliam, K. R. Graham, A. F. Castro, C. B. Martin, C. J. Der, and C. Bi, M-Ras/R-Ras3, a novel transforming Ras protein regulated by Sos1 and p120 GAP, interacts with the putative Ras effector AF6. *J. Biol. Chem.* **274,** 23,850–23,857 (1999).
127. T. Gotoh, X. Tian, and L. A. Feig, Prenylation of target GTPases contributes to signaling specificity of Ras-guanine nucleotide exchange factors. *J. Biol. Chem.* **276,** in press (2001).
128. W. T. Fan, C. A. Koch, C. L. de Hoog, N. P. Fam, and M. F. Moran, The exchange factor Ras-GRF2 activates Ras-dependent and Rac-dependent mitogen-activated protein kinase pathways. *Curr. Biol.* **8,** 935–938 (1998).
129. R. Buchsbaum, J. B. Telliez, S. Goonesekera, and L. A. Feig, The N-terminal pleckstrin, coiled-coil, and IQ domains of the exchange factor Ras-GRF act cooperatively to facilitate activation by calcium. *Mol. Cell. Biol.* **16,** 4888–4896 (1996).
130. C. L. de Hoog, J. A. Koehler, M. D. Goldstein, P. Taylor, D. Figeys, and M. F. Moran, Ras binding triggers ubiquitination of the Ras exchange factor Ras-GRF2. *Mol. Cell. Biol.* **21,** 2107–2117 (2001).
131. C. L. Farnsworth, N. W. Freshney, L. B. Rosen, A. Ghosh, M. E. Greenberg, and L. A. Feig, Calcium activation of Ras mediated by neuronal exchange factor Ras-GRF. *Nature* **376,** 524–527 (1995).
132. N. W. Freshney, S. D. Goonesekera, and L. A. Feig, Activation of the exchange factor Ras-GRF by calcium requires an intact Dbl homology domain. *FEBS Lett* **407,** 111–115 (1997).
133. P. H. Anborgh, X. Qian, A. G. Papageorge, W. C. Vass, J. E. DeClue, and D. R. Lowy, Ras-specific exchange factor GRF: Oligomerization through its Dbl homology domain and calcium-dependent activation of Raf. *Mol. Cell. Biol.* **19,** 4611–4622 (1999).
134. H. Sun, A. J. King, H. B. Diaz, and M. S. Marshall, Regulation of the protein kinase Raf-1 by oncogenic Ras through phosphatidylinositol 3-kinase, Cdc42/Rac and Pak. *Curr. Biol.* **10,** 281–284 (2000).

135. K. Zhu, B. Debreceni, F. Bi, and Y. Zheng, Oligomerization of DH domain is essential for Dbl-induced transformation. *Mol. Cell. Biol.* **21,** 425–437 (2001).
136. C. Camus, M. Geymonat, H. Garreau, S. Baudet-Nessler, and M. Jacquet, Dimerization of Cdc25p, the guanine-nucleotide exchange factor for Ras from Saccharomyces cerevisiae, and its interaction with Sdc25p. *Eur. J. Biochem.* **247,** 703–708 (1997).
137. C. Shou, A. Wurmser, K. L. Suen, M. Barbacid, L. A. Feig, and K. Ling, Differential response of the Ras exchange factor, Ras-GRF to tyrosine kinase and G protein mediated signals. *Oncogene* **10,** 1887–1893 (1995).
138. R. R. Mattingly, V. Saini, and I. G. Macara, Activation of the Ras-GRF/CDC25Mm exchange factor by lysophosphatidic acid. *Cell Signal* **11,** 603–610 (1999).
139. R. R. Mattingly, Phosphorylation of serine 916 of Ras-GRF1 contributes to the activation of exchange factor activity by muscarinic receptors. *J. Biol. Chem.* **274,** 37,379–37,384 (1999).
140. S. Baouz, E. Jacquet, K. Accorsi, C. Hountondji, M. Balestrini, R. Zippel, E. Sturani, and A. Parmeggiani, Sites of phosphorylation by protein kinase A in CDC25Mm/GRF1, a guanine nucleotide exchange factor for Ras. *J. Biol. Chem.* **276,** 1742–1749 (2001).
141. N. Pham, I. Cheglakov, C. A. Koch, C. L. de Hoog, M. F. Moran, and D. Rotin, The guanine nucleotide exchange factor CNrasGEF activates ras in response to cAMP and cGMP [In Process Citation]. *Curr. Biol.* **10,** 555–558 (2000).
142. N. Gnesutta, M. Ceriani, M. Innocenti, I. Mauri, R. Zippel, E. Sturani, B. Borgonovo, G. Berruti, and E. Martegani, Cloning and characterization of mouse-UBPy, a deubiquitinating enzyme that interacts with the Ras guanine nucleotide exchange factor CDC25 Mm/Ras-GRF1. *J. Biol. Chem.* **276,** 39,448–39,454 (2001).
143. R. Zippel, N. Gnesutta, N. Matus-Leibovitch, E. Mancinelli, D. Saya, Z. Vogel, E. Sturani, Z. Renata, G. Nerina, M. L. Noa, M. Enzo, S. Daniella, V. Zvi, and S. Emmapaola, Ras-GRF, the activator of Ras, is expressed preferentially in mature neurons of the central nervous system. *Brain. Res. Mol. Brain Res.* **48,** 140–144 (1997).
144. E. Sturani, A. Abbondio, P. Branduardi, C. Ferrari, R. Zippel, E. Martegani, M. Vanoni, and S. Denis-Donini, The Ras Guanine nucleotide Exchange Factor CDC25Mm is present at the synaptic junction. *Exp. Cell Res.* **235,** 117–123 (1997).
145. R. Brambilla, N. Gnesutta, L. Minichiello, G. White, A. J. Roylance, C. E. Herron, M. Ramsey, D. P. Wolfer, V. Cestari, C. Rossi-Arnaud, S. G. Grant, P. F. Chapman, H. P. Lipp, E. Sturani, and R. Klein, A role for the Ras signalling pathway in synaptic transmission and long-term memory. *Nature* **390,** 281–286 (1997).
146. S. Yamashita, N. Mochizuki, Y. Ohba, M. Tobiume, Y. Okada, H. Sawa, K. Nagashima, and M. Matsuda, CalDAG-GEFIII activation of Ras, R-ras, and Rap1. *J. Biol. Chem.* **275,** 25,488–25,493 (2000).
147. C. E. Tognon, H. E. Kirk, L. A. Passmore, I. P. Whitehead, C. J. Der, and R. J. Kay, Regulation of RasGRP via a phorbol ester-responsive C1 domain. *Mol. Cell. Biol.* **18,** 6995–7008 (1998).
148. H. Kawasaki, G. M. Springett, S. Toki, J. J. Canales, P. Harlan, J. P. Blumenstiel, E. J. Chen, I. A. Bany, N. Mochizuki, A. Ashbacher, M. Matsuda, D. E. Housman, and A. M. Graybiel, A Rap guanine nucleotide exchange factor enriched highly in the basal ganglia. *Proc. Natl. Acad. Sci. U.S.A.* **95,** 13,278–13,283 (1998).
149. D. Kedra, E. Seroussi, I. Fransson, J. Trifunovic, M. Clark, J. Lagercrantz, E. Blennow, H. Mehlin, and J. Dumanski, The germinal center kinase gene and a novel CDC25-like gene are located in the vicinity of the PYGM gene on 11q13. *Hum Genet* **100,** 611–619 (1997).
150. J. O. Ebinu, D. A. Bottorff, E. Y. Chan, S. L. Stang, R. J. Dunn, and J. C. Stone, RasGRP, a Ras guanyl nucleotide-releasing protein with calcium- and diacylglycerol-binding motifs. *Science* **280,** 1082–1086 (1998).
151. J. Clyde-Smith, G. Silins, M. Gartside, S. Grimmond, M. Etheridge, A. Apolloni, N. Hayward, and J. F. Hancock, Characterization of RasGRP2, a plasma membrane-targeted, dual specificity Ras/Rap exchange factor. *J. Biol. Chem.* **275,** 32,260–32,267 (2000).

152. M. G. Kazanietz, Eyes wide shut: Protein kinase C isozymes are not the only receptors for the phorbol ester tumor promoters. *Mol. Carcinog.* **28,** 5–11 (2000).
153. P. S. Lorenzo, M. Beheshti, G. R. Pettit, J. C. Stone, and P. M. Blumberg, The guanine nucleotide exchange factor RasGRP is a high-affinity target for diacylglycerol and phorbol esters. *Mol. Pharmacol.* **57,** 840–846 (2000).
154. P. Pierret, R. J. Dunn, B. Djordjevic, J. C. Stone, and P. M. Richardson, Distribution of ras guanyl releasing protein (RasGRP) mRNA in the adult rat central nervous system. *J Neurocytol* **29,** 485–497 (2000).
155. J. O. Ebinu, S. L. Stang, C. Teixeira, D. A. Bottorff, J. Hooton, P. M. Blumberg, M. Barry, R. C. Bleakley, H. L. Ostergaard, and J. C. Stone, RasGRP links T-cell receptor signaling to Ras. *Blood* **95,** 3199–3203 (2000).
156. M. K. Topham and S. M. Prescott, Diacylglycerol kinase zeta regulates Ras activation by a novel mechanism. *J. Cell. Biol.* **152,** 1135–1143 (2001).
157. N. A. Dower, S. L. Stang, D. A. Bottorff, J. O. Ebinu, P. Dickie, H. L. Ostergaard, and J. C. Stone, RasGRP is essential for mouse thymocyte differentiation and TCR signaling. *Nat. Immunol.* **1,** 317–321 (2000).
158. S. Toki, H. Kawasaki, N. Tashiro, D. E. Housman, and A. M. Graybiel, Guanine nucleotide exchange factors CalDAG-GEFI and CalDAG-GEFII are colocalized in striatal projection neurons. *J. Comp. Neurol.* **437,** 398–407 (2001).
159. M. Suyama, T. Nagase, and O. Ohara, HUGE: A database for human large proteins identified by Kazusa cDNA sequencing project. *Nucleic Acids Res.* **27,** 338–339 (1999).
160. O. Ohara, T. Nagase, K. Ishikawa, D. Nakajima, M. Ohira, N. Seki, and N. Nomura, Construction and characterization of human brain cDNA libraries suitable for analysis of cDNA clones encoding relatively large proteins. *DNA Res.* **4,** 53–59 (1997).
161. B. Ye, D. Liao, X. Zhang, P. Zhang, H. Dong, and R. L. Huganir, GRASP-1: A neuronal RasGEF associated with the AMPA receptor/GRIP complex. *Neuron* **26,** 603–617 (2000).
162. T. Gotoh, S. Hattori, S. Nakamura, H. Kitayama, M. Noda, Y. Takai, K. Kaibuchi, H. Matsui, O. Hatase, and H. Takahashi, T. Kurata, and M. Matsuda, Identification of Rap1 as a target for the Crk SH3 domain-binding guanine nucleotide-releasing factor C3G. *Mol. Cell. Biol.* **15,** 6746–6753 (1995).
163. Shivakrupa, R. Singh, and G. Swarup, Identification of a novel splice variant of C3G which shows tissue-specific expression. *DNA Cell. Biol.* **18,** 701–708 (1999).
164. K. H. Kirsch, M. M. Georgescu, and H. Hanafusa, Direct binding of p130(Cas) to the guanine nucleotide exchange factor C3G. *J. Biol. Chem.* **273,** 25,673–25,679 (1998).
165. T. Ichiba, Y. Hashimoto, M. Nakaya, Y. Kuraishi, S. Tanaka, T. Kurata, N. Mochizuki, and M. Matsuda, Activation of C3G guanine nucleotide exchange factor for Rap1 by phosphorylation of tyrosine 504. *J. Biol. Chem.* **274,** 14,376–14,381 (1999).
166. S. Okada, M. Matsuda, M. Anafi, T. Pawson, and J. E. Pessin, Insulin regulates the dynamic balance between Ras and Rap1 signaling by coordinating the assembly states of the Grb2-SOS and CrkII-C3G complexes. *EMBO J.* **17,** 2554–2565 (1998).
167. Y. Ohba, K. Ikuta, A. Ogura, J. Matsuda, N. Mochizuki, K. Nagashima, K. Kurokawa, B. J. Mayer, K. Maki, J. Miyazaki Ji, and M. Matsuda, Requirement for C3G-dependent Rap1 activation for cell adhesion and embryogenesis. *EMBO J.* **20,** 3333–3341 (2001).
168. K. Yokote, U. Hellman, S. Ekman, Y. Saito, L. Ronnstrand, C. H. Heldin, and S. Mori, Identification of Tyr-762 in the platelet-derived growth factor alpha-receptor as the binding site for Crk proteins. *Oncogene* **16,** 1229–1239 (1998).
169. H. Larsson, P. Klint, E. Landgren, and L. Claesson-Welsh, Fibroblast growth factor receptor-1-mediated endothelial cell proliferation is dependent on the Src homology (SH) 2/SH3 domain-containing adaptor protein Crk. *J. Biol. Chem.* **274,** 25,726–25,734 (1999).
170. Y. Alsayed, S. Uddin, S. Ahmad, B. Majchrzak, B. J. Druker, E. N. Fish, and L. C. Platanias, IFN-gamma activates the C3G/Rap1 signaling pathway. *J. Immunol.* **164,** 1800–1806 (2000).

171. D. Sakkab, M. Lewitzky, G. Posern, U. Schaeper, M. Sachs, W. Birchmeier, and S. M. Feller, Signaling of hepatocyte growth factor/scatter factor (HGF) to the small GTPase Rap1 via the large docking protein Gab1 and the adapter protein CRKL. *J. Biol. Chem.* **275,** 10,772–10,778 (2000).
172. J. Du, Y. M. Alsayed, F. Xin, S. J. Ackerman, and L. C. Platanias, Engagement of the CrkL adapter in interleukin-5 signaling in eosinophils. *J. Biol. Chem.* **275,** 33,167–33,175 (2000).
173. G. Posern, U. R. Rapp, and S. M. Feller, The Crk signaling pathway contributes to the bombesin-induced activation of the small GTPase Rap1 in Swiss 3T3 cells. *Oncogene* **19,** 6361–6368 (2000).
174. S. Kao, R. K. Jaiswal, W. Kolch, and G. E. Landreth, Identification of the mechanisms regulating the differential activation of the mapk cascade by epidermal growth factor and nerve growth factor in PC12 cells. *J. Biol. Chem.* **276,** 18,169–18,177 (2001).
175. Y. Nosaka, A. Arai, N. Miyasaka, and O. Miura, CrkL mediates Ras-dependent activation of the Raf/ERK pathway through the guanine nucleotide exchange factor C3G in hematopoietic cells stimulated with erythropoietin or interleukin-3. *J. Biol. Chem.* **274,** 30,154–30,162 (1999).
176. R. de Jong, A. van Wijk, N. Heisterkamp, and J. Groffen, C3G is tyrosine-phosphorylated after integrin-mediated cell adhesion in normal but not in Bcr/Abl expressing cells. *Oncogene* **17,** 2805–2810 (1998).
177. K. Vuori, H. Hirai, S. Aizawa, and E. Ruoslahti, Introduction of p130cas signaling complex formation upon integrin-mediated cell adhesion: A role for Src family kinases. *Mol. Cell. Biol.* **16,** 2606–2613 (1996).
178. A. Arai, Y. Nosaka, H. Kohsaka, N. Miyasaka, and O. Miura, CrkL activates integrin-mediated hematopoietic cell adhesion through the guanine nucleotide exchange factor C3G. *Blood* **93,** 3713–3722 (1999).
179. N. Uemura and J. D. Griffin, The adapter protein Crkl links Cbl to C3G after integrin ligation and enhances cell migration. *J. Biol. Chem.* **274,** 37,525–37,532 (1999).
180. C. S. Buensuceso and T. E. O'Toole, The association of CRKII with c3g can be regulated by integrins and defines a novel means to regulate the mitogen-activated protein kinases. *J. Biol. Chem.* **275,** 13,118–13,125 (2000).
181. K. A. Reedquist, T. Fukazawa, G. Panchamoorthy, W. Y. Langdon, S. E. Shoelson, B. J. Druker, and H. Band, Stimulation through the T cell receptor induces Cbl association with Crk proteins and the guanine nucleotide exchange protein C3G. *J. Biol. Chem.* **271,** 8435–8442 (1996).
182. R. H. Schwartz, A cell culture model for T lymphocyte clonal anergy. *Science* **248,** 1349–1356 (1990).
183. V. A. Boussiotis, G. J. Freeman, A. Berezovskaya, D. L. Barber, and L. M. Nadler, Maintenance of human T cell anergy: Blocking of IL-2 gene transcription by activated Rap1. *Science* **278,** 124–128 (1997).
184. L. Smit, G. van der Horst, and J. Borst, Sos, Vav, and C3G participate in B cell receptor-induced signaling pathways and differentially associate with Shc-Grb2, Crk, and Crk-L adaptors. *J. Biol. Chem.* **271,** 8564–8569 (1996).
185. R. J. Ingham, D. L. Krebs, S. M. Barbazuk, C. W. Turck, H. Hirai, M. Matsuda, and M. R. Gold, B cell antigen receptor signaling induces the formation of complexes containing the Crk adapter proteins. *J. Biol. Chem.* **271,** 32,306–32,314 (1996).
186. S. Tanaka and H. Hanafusa, Guanine-nucleotide exchange protein C3G activates JNK1 by a ras-independent mechanism. JNK1 activation inhibited by kinase negative forms of MLK3 and DLK mixed lineage kinases. *J. Biol. Chem.* **273,** 1281–1284 (1998).
187. S. Tanaka, T. Ouchi, and H. Hanafusa, Downstream of Crk adaptor signaling pathway: Activation of Jun kinase by v-Crk through the guanine nucleotide exchange protein C3G. *Proc. Natl. Acad. Sci. U.S.A.* **94,** 2356–2361 (1997).

188. N. Mochizuki, Y. Ohba, S. Kobayashi, N. Otsuka, A. M. Graybiel, S. Tanaka, and M. Matsuda, Crk activation of JNK via C3G and R-Ras. *J. Biol. Chem.* **275**, 12,667–12,671 (2000).
189. J. Downward, Signal transduction. New exchange, new target. *Nature* **396**, 416–417 (1998).
190. J. de Rooij, F. J. Zwartkruis, M. H. Verheijen, R. H. Cool, S. M. Nijman, A. Wittinghofer, and J. L. Bos, Epac is a Rap1 guanine-nucleotide-exchange factor directly activated by cyclic AMP. *Nature* **396**, 474–477 (1998).
191. H. Kawasaki, G. M. Springett, N. Mochizuki, S. Toki, M. Nakaya, M. Matsuda, D. E. Housman, and A. M. Graybiel, A family of cAMP-binding proteins that directly activate Rap1. *Science* **282**, 2275–2279 (1998).
192. J. de Rooij, H. Rehmann, M. van Triest, R. H. Cool, A. Wittinghofer, and J. L. Bos, Mechanism of regulation of the Epac family of cAMP-dependent RapGEFs. *J. Biol. Chem.* **275**, 20,829–29,836 (2000).
193. J. S. Richards, New signaling pathways for hormones and cyclic adenosine $3',5'$-monophosphate action in endocrine cells. *Mol. Endocrinol.* **15**, 209–218 (2001).
194. E. Kupperman, W. Wen, and J. L. Meinkoth, Inhibition of thyrotropin-stimulated DNA synthesis by microinjection of inhibitors of cellular Ras and cyclic AMP-dependent protein kinase. *Mol. Cell. Biol.* **13**, 4477–4484 (1993).
195. J. E. Dumont, F. Lamy, P. Roger, and C. Maenhaut, Physiological and pathological regulation of thyroid cell proliferation and differentiation by thyrotropin and other factors. *Physiol. Rev.* **72**, 667–697 (1992).
196. O. M. Tsygankova, A. Saavedra, J. F. Rebhun, L. A. Quilliam, and J. L. Meinkoth, Coordinated regulation of Rap1 and thyroid differentiation by cyclic AMP and protein kinase A. *Mol. Cell. Biol.* **21**, 1921–1929 (2001).
197. S. Dremier, F. Vandeput, F. J. Zwartkruis, J. L. Bos, J. E. Dumont, and C. Maenhaut, Activation of the small G protein Rap1 in dog thyroid cells by both cAMP-dependent and -independent pathways. *Biochem. Biophys. Res. Commun.* **267**, 7–11 (2000).
198. I. J. Gonzalez-Robayna, A. E. Falender, S. Ochsner, G. L. Firestone, and J. S. Richards, Follicle-Stimulating hormone (FSH) stimulates phosphorylation and activation of protein kinase B (PKB/Akt) and serum and glucocorticoid-lnduced kinase (Sgk): evidence for A kinase-independent signaling by FSH in granulosa cells. *Mol. Endocrinol.* **14**, 1283–1300 (2000).
199. D. J. Drucker, Glucagon-like peptides. *Diabetes* **47**, 159–169 (1998).
200. H. C. Fehmann, R. Goke, and B. Goke, Cell and molecular biology of the incretin hormones glucagon-like peptide-I and glucose-dependent insulin releasing polypeptide. *Endocr. Rev.* **16**, 390–410 (1995).
201. G. Skoglund, M. A. Hussain, and G. G. Holz, Glucagon-like peptide 1 stimulates insulin gene promoter activity by protein kinase A-independent activation of the rat insulin I gene cAMP response element. *Diabetes* **49**, 1156–1164 (2000).
202. C. A. Leech, G. G. Holz, O. Chepurny, and J. F. Habener, Expression of cAMP-regulated guanine nucleotide exchange factors in pancreatic beta-cells. *Biochem. Biophys. Res. Commun.* **278**, 44–47 (2000).
203. N. Ozaki, T. Shibasaki, Y. Kashima, T. Miki, K. Takahashi, H. Ueno, Y. Sunaga, H. Yano, Y. Matsuura, T. Iwanaga, Y. Takai, and S. Seino, cAMP-GEFII is a direct target of cAMP in regulated exocytosis. *Nat. Cell. Biol.* **2**, 805–811 (2000).
204. R. Busca, P. Abbe, F. Mantoux, E. Aberdam, C. Peyssonnaux, A. Eychene, J. P. Ortonne, and R. Ballotti, Ras mediates the cAMP-dependent activation of extracellular signal-regulated kinases (ERKs) in melanocytes. *EMBO J.* **19**, 2900–2910 (2000).
205. L. Wang, F. Liu, and M. L. Adamo, Cyclic AMP inhibits extracellular signal-regulated kinase and phosphatidylinositol 3-kinase/Akt pathways by inhibiting Rap1. *J. Biol. Chem.* **276**, 37,242–37,249 (2001).

206. T. Chen, R. W. Cho, P. J. Stork, and M. J. Weber, Elevation of cyclic adenosine 3′,5′-monophosphate potentiates activation of mitogen-activated protein kinase by growth factors in LNCaP prostate cancer cells. *Cancer Res.* **59,** 213–218 (1999).
207. T. Ichiba, Y. Hoshi, Y. Eto, N. Tajima, and Y. Kuraishi, Characterization of GFR, a novel guanine nucleotide exchange factor for Rap1. *FEBS Lett.* **457,** 85–89 (1999).
208. Y. Liao, T. Satoh, X. Gao, T. G. Jin, C. D. Hu, and T. Kataoka, RA-GEF-1, a guanine nucleotide exchange factor for Rap1, is activated by translocation induced by association with Rap1{middle dot}GTP and enhances Rap1-dependent B-Raf activation. *J. Biol. Chem.* **276,** 28,478–28,483 (2001).
209. J. de Rooij, N. M. Boenink, M. van Triest, R. H. Cool, A. Wittinghofer, and J. L. Bos, PDZ-GEF1, a Guanine Nucleotide Exchange Factor Specific for Rap1 and Rap2. *J. Biol. Chem.* **274,** 38,125–38,130 (1999).
210. Y. Liao, K. Kariya, C. D. Hu, M. Shibatohge, M. Goshima, T. Okada, Y. Watari, X. Gao, T. G. Jin, Y. Yamawaki-Kataoka, and T. Kataoka, RA-GEF, a novel Rap1A guanine nucleotide exchange factor containing a Ras/Rap1A-associating domain, is conserved between nematode and humans. *J. Biol. Chem.* **274,** 37,815–37,820 (1999).
211. T. Ohtsuka, Y. Hata, N. Ide, T. Yasuda, E. Inoue, T. Inoue, A. Mizoguchi, and Y. Takai, nRap GEP: A novel neural GDP/GTP exchange protein for rap1 small G protein that interacts with synaptic scaffolding molecule (S-SCAM). *Biochem. Biophys. Res. Commun.* **265,** 38–44 (1999).
212. A. Mino, T. Ohtsuka, E. Inoue, and Y. Takai, Membrane-associated guanylate kinase with inverted orientation (MAGI)-1/brain angiogenesis inhibitor 1-associated protein (BAP1) as a scaffolding molecule for Rap small G protein GDP/GTP exchange protein at tight junctions. *Genes Cells* **5,** 1009–1016 (2000).
213. A. S. Fanning and J. M. Anderson, PDZ domains: Fundamental building blocks in the organization of protein complexes at the plasma membrane. *J. Clin. Invest.* **103,** 767–772 (1999).
214. H. Asha, N. D. de Ruiter, M. G. Wang, and I. K. Hariharan, The Rap1 GTPase functions as a regulator of morphogenesis in vivo. *EMBO J.* **18,** 605–615 (1999).
215. J. Wojcik, J. A. Girault, G. Labesse, J. Chomilier, J. P. Mornon, and I. Callebaut, Sequence analysis identifies a ras-associating (RA)-like domain in the N-termini of band 4.1/JEF domains and in the Grb7/10/14 adapter family. *Biochem. Biophys. Res. Commun.* **259,** 113–120 (1999).
216. X. Gao, T. Satoh, Y. Liao, C. Song, C. D. Hu, K. Kariya Ki, and T. Kataoka, Identification and characterization of RA-GEF-2, a Rap guanine nucleotide exchange factor that serves as a downstream target of M-Ras. *J. Biol. Chem.* **276,** 42,219–42,225 (2001).
217. V. Chiarugi, F. Porciatti, F. Pasquali, and P. Bruni, Transformation of BALB/3T3 cells with EJ/T24/H-ras oncogene inhibits adenylate cyclase response to beta-adrenergic agonist while increases muscarinic receptor dependent hydrolysis of inositol lipids. *Biochem. Biophys. Res. Commun.* **132,** 900–907 (1985).
218. M. J. Wakelam, S. A. Davies, M. D. Houslay, I. McKay, C. J. Marshall, and A. Hall, Normal p21N-ras couples bombesin and other growth factor receptors to inositol phosphate production. *Nature* **323,** 173–176 (1986).
219. J. Downward, J. de Gunzburg, R. Riehl, and R. A. Weinberg, p21ras-induced responsiveness of phosphatidylinositol turnover to bradykinin is a receptor number effect. *Proc. Natl. Acad. Sci. U.S.A.* **85,** 5774–5778 (1988).
220. L. A. Quilliam, C. J. Der, and J. H. Brown, GTP-binding protein-stimulated phospholipase C and phospholipase D activities in ras-transformed NIH 3T3 fibroblasts. *Second Messengers & Phosphoproteins* **13,** 59–67 (1990).
221. M. Shibatohge, K. Kariya, Y. Liao, C. D. Hu, Y. Watari, M. Goshima, F. Shima, and T. Kataoka, Identification of PLC210, a Caenorhabditis elegans phospholipase C, as a putative effector of Ras. *J. Biol. Chem.* **273,** 6218–6222 (1998).

222. T. G. Jin, T. Satoh, Y. Liao, C. Song, X. Gao, K. Kariya Ki, C. D. Hu, and T. Kataoka, Role of the CDC25 homology domain of PLC{epsilon} in amplification of Rap1-dependent signaling. *J. Biol. Chem.* **276,** 30,301–30,307 (2001).
223. C. Song, C. D. Hu, M. Masago, K. Kariyai, Y. Yamawaki-Kataoka, M. Shibatohge, D. Wu, T. Satoh, and T. Kataoka, Regulation of a novel human phospholipase C, PLCepsilon, through membrane targeting by Ras. *J. Biol. Chem.* **276,** 2752–2757 (2001).
224. G. G. Kelley, S. E. Reks, J. M. Ondrako, and A. V. Smrcka, Phospholipase C(epsilon): A novel Ras effector. *EMBO J.* **20,** 743–754 (2001).
225. I. Lopez, E. C. Mak, J. Ding, H. E. Hamm, and J. W. Lomasney, A novel bifunctional phospholipase c that is regulated by Galpha 12 and stimulates the Ras/mitogen-activated protein kinase pathway. *J. Biol. Chem.* **276,** 2758–2765 (2001).
226. P. J. Cullen, Ras effectors: Buying shares in Ras plc. *Curr. Biol.* **11,** R342–R344 (2001).
227. A. Kikuchi, S. D. Demo, Z. H. Ye, Y. W. Chen, and L. T. Williams, ralGDS family members interact with the effector loop of ras p21. *Mol. Cell. Biol.* **14,** 7483–7491 (1994).
228. S. N. Peterson, L. Trabalzini, T. R. Brtva, T. Fischer, D. L. Altschuler, P. Martelli, E. G. Lapetina, C. J. Der, and G. C. White, 2nd, Identification of a novel RalGDS-related protein as a candidate effector for Ras and Rap1. *J. Biol. Chem.* **271,** 29,903–29,908 (1996).
229. R. M. Wolthuis, B. Bauer, L. J. van't Veer, A. M. de Vries-Smits, R. H. Cool, M. Spaargaren, A. Wittinghofer, B. M. Burgering, and J. L. Bos, RalGDS-like factor (Rlf) is a novel Ras and Rap 1A-associating protein. *Oncogene* **13,** 353–362 (1996).
230. H. Shao and D. A. Andres, A novel RalGEF-like protein, RGL3, as a candidate effector for rit and Ras. *J. Biol. Chem.* **275,** 26,914–26,924 (2000).
231. G. R. Ehrhardt, C. Korherr, J. S. Wieler, M. Knaus, and J. W. Schrader, A novel potential effector of M-Ras and p21 Ras negatively regulates p21 Ras-mediated gene induction and cell growth. *Oncogene* **20,** 188–197 (2001).
232. D. R. D'Adamo, S. Novick, J. M. Kahn, P. Leonardi, and A. Pellicer, rsc: A novel oncogene with structural and functional homology with the gene family of exchange factors for Ral. *Oncogene* **14,** 1295–1305 (1997).
233. T. Goi, G. Rusanescu, T. Urano, and L. A. Feig, Ral-specific guanine nucleotide exchange factor activity opposes other Ras effectors in PC12 cells by inhibiting neurite outgrowth. *Mol. Cell. Biol.* **19,** 1731–1741 (1999).
234. I. Hernandez-Munoz, M. Malumbres, P. Leonardi, and A. Pellicer, The Rgr oncogene (homologous to RalGDS) induces transformation and gene expression by activating Ras, Ral and Rho mediated pathways. *Oncogene* **19,** 2745–2757 (2000).
235. M. Malumbres, I. Perez De Castro, M. I. Hernandez, M. Jimenez, T. Corral, and A. Pellicer, Cellular response to oncogenic ras involves induction of the Cdk4 and Cdk6 inhibitor p15(INK4b). *Mol. Cell. Biol.* **20,** 2915–2925 (2000).
236. A. Kimmelman, T. Tolkacheva, M. V. Lorenzi, M. Osada, and A. M. Chan, Identification and characterization of R-ras3: a novel member of the RAS gene family with a non-ubiquitous pattern of tissue distribution. *Oncogene* **15,** 2675–2685 (1997).
237. F. J. Zwartkruis, R. M. Wolthuis, N. M. Nabben, B. Franke, and J. L. Bos, Extracellular signal-regulated activation of Rap1 fails to interfere in Ras effector signalling. *EMBO J.* **17,** 5905–5912 (1998).
238. K. Matsubara, S. Kishida, Y. Matsuura, H. Kitayama, M. Noda, and A. Kikuchi, Plasma membrane recruitment of RalGDS is critical for Ras-dependent Ral activation. *Oncogene* **18,** 1303–1312 (1999).
239. R. M. Wolthuis, N. D. de Ruiter, R. H. Cool, and J. L. Bos, Stimulation of gene induction and cell growth by the Ras effector Rlf. *EMBO J.* **16,** 6748–6761 (1997).
240. T. Urano, R. Emkey, and L. A. Feig, Ral-GTPases mediate a distinct downstream signaling pathway from Ras that facilitates cellular transformation. *EMBO J.* **15,** 810–816 (1996).

241. R. Khosravi-Far, M. A. White, J. K. Westwick, P. A. Solski, M. Chrzanowska-Wodnicka, L. Van Aelst, M. H. Wigler, and C. J. Der, Oncogenic Ras activation of Raf/mitogen-activated protein kinase-independent pathways is sufficient to cause tumorigenic transformation. *Mol. Cell. Biol.* **16**, 3923–3933 (1996).
242. M. Okazaki, S. Kishida, T. Hinoi, T. Hasegawa, M. Tamada, T. Kataoka, and A. Kikuchi, Synergistic activation of c-fos promoter activity by Raf and Ral GDP dissociation stimulator. *Oncogene* **14**, 515–521 (1997).
243. R. Lin, R. A. Cerione, and D. Manor, Specific contributions of the small GTPases Rho, Rac, and Cdc42 to Dbl transformation. *J. Biol. Chem.* **274**, 23,633–23,641 (1999).
244. E. J. Tisdale, J. R. Bourne, R. Khosravi-Far, C. J. Der, and W. E. Balch, GTP-binding mutants of rab1 and rab2 are potent inhibitors of vesicular transport from the endoplasmic reticulum to the Golgi complex. *J. Cell. Biol.* **119**, 749–761 (1992).
245. F. Hofer, R. Berdeaux, and G. S. Martin, Ras-independent activation of Ral by a Ca(2+)-dependent pathway. *Curr. Biol.* **8**, 839–842 (1998).
246. R. M. Wolthuis, B. Franke, M. van Triest, B. Bauer, R. H. Cool, J. H. Camonis, J. W. Akkerman, and J. L. Bos, Activation of the small GTPase Ral in platelets. *Mol. Cell. Biol.* **18**, 2486–2491 (1998).
247. K. L. Wang, M. T. Khan, and B. D. Roufogalis, Identification and characterization of a calmodulin-binding domain in Ral-A, a Ras-related GTP-binding protein purified from human erythrocyte membrane. *J. Biol. Chem.* **272**, 16,002–16,009 (1997).
248. C. H. J. Lee, N. G. Della, C. E. Chew, and D. J. Zack, Rin, a neuron-specific and calmodulin-binding small G-protein, and Rit define a novel subfamily of ras proteins. *J. Neurosci.* **16**, 6784–6794 (1996).
249. P. D. Wes, M. Yu, and C. Montell, RIC, a calmodulin-binding Ras-like GTPase. *EMBO J.* **15**, 5839–5848 (1996).
250. K. M. de Bruyn, J. de Rooij, R. M. Wolthuis, H. Rehmann, J. Wesenbeek, R. H. Cool, A. H. Wittinghofer, and J. L. Bos, RalGEF2, a pleckstrin homology domain containing guanine nucleotide exchange factor for Ral. *J. Biol. Chem.* **275**, 29,761–29,766 (2000).
251. L. E. Rameh, A. Arvidsson, K. L. 3rd Carraway, A. D. Couvillon, G. Rathbun, A. Crompton, B. VanRenterghem, M. P. Czech, K. S. Ravichandran, S. J. Burakoff, D. S. Wang, C. S. Chen, and L. C. Cantley, A comparative analysis of the phosphoinositide binding specificity of pleckstrin homology domains. *J. Biol. Chem.* **272**, 22,059–22,066 (1997).
252. A. B. Sparks, J. E. Rider, N. G. Hoffmann, D. M. Fowlkes, L. A. Quilliam, and B. K. Kay, Distinct ligand preferences of SH3 domains from Src, Yes, Abl, cortactin, p53bp2, PLCγ, Crk and Grb2. *Proc. Natl. Acad. Sci. U.S.A.* **93**, (1996).
253. L. A. Quilliam, Q. T. Lambert, L. A. Mickelson-Young, J. K. Westwick, A. B. Sparks, B. K. Kay, N. A. Jenkins, D. J. Gilbert, N. G. Copeland, and C. J. Der, Isolation of a NCK-associated kinase, PRK2, an SH3-binding protein and potential effector of Rho protein signaling. *J. Biol. Chem.* **271**, 28,772–28,776 (1996).
254. A. Brinkman, S. van der Flier, E. M. Kok, and L. C. Dorssers, BCAR1, a human homologue of the adapter protein p130Cas, and antiestrogen resistance in breast cancer cells. *J. Natl. Cancer Inst.* **92**, 112–120 (2000).
255. T. van Agthoven, T. L. van Agthoven, A. Dekker, P. J. van der Spek, L. Vreede, and L. C. Dorssers, Identification of BCAR3 by a random search for genes involved in antiestrogen resistance of human breast cancer cells. *EMBO J.* **17**, 2799–2808 (1998).
256. D. Cai, L. K. Clayton, A. Smolyar, and A. Lerner, AND-34, a novel p130Cas-binding thymic stromal cell protein regulated by adhesion and inflammatory cytokines. *J Immunol* **163**, 2104–2112 (1999).

257. V. C. Dodelet, C. Pazzagli, A. H. Zisch, C. A. Hauser, and E. B. Pasquale, A novel signaling intermediate, SHEP1, directly couples Eph receptors to R-Ras and Rap1A. *J. Biol. Chem.* **274,** 31,941–31,946 (1999).
258. Y. Lu, J. Brush, and T. A. Stewart, NSP1 defines a novel family of adaptor proteins linking integrin and tyrosine kinase receptors to the c-Jun N-terminal kinase/stress-activated protein kinase signaling pathway. *J. Biol. Chem.* **274,** 10,047–10,052 (1999).
259. A. Sakakibara and S. Hattori, Chat, a Cas/HEF1-associated adaptor protein that integrates multiple signaling pathways. *J. Biol. Chem.* **275,** 6404–6410 (2000).
260. T. Gotoh, D. Cai, X. Tian, L. A. Feig, and A. Lerner, p130Cas regulates the activity of AND-34, a novel Ral, Rap1, and R-Ras guanine nucleotide exchange factor. *J. Biol. Chem.* **275,** 30,118–30,123 (2000).
261. S. van der Flier, A. Brinkman, M. P. Look, E. M. Kok, M. E. Meijer-van Gelder, J. G. Klijn, L. C. Dorssers, and J. A. Foekens, Bcar1/p130Cas protein and primary breast cancer: Prognosis and response to tamoxifen treatment. *J. Natl. Cancer Inst.* **92,** 120–127 (2000).
262. T. Goi, M. Shipitsin, Z. Lu, D. A. Foster, S. G. Klinz, and L. A. Feig, An EGF receptor/Ral-GTPase signaling cascade regulates c-Src activity and substrate specificity. *EMBO J.* **19,** 623–630 (2000).
263. N. Iouzalen, J. Camonis, and J. Moreau, Identification and characterization in Xenopus of XsmgGDS, a RalB binding protein. *Biochem. Biophys. Res. Commun.* **250,** 359–363 (1998).
264. D. Strassheim, R. A. Porter, S. H. Phelps, and C. L. Williams, Unique in vivo associations with SmgGDS and RhoGDI and different guanine nucleotide exchange activities exhibited by RhoA, dominant negative RhoA(Asn-19), and activated RhoA(Val-14). *J. Biol. Chem.* **275,** 6699–6702 (2000).
265. M. Hatzfeld, The armadillo family of structural proteins. *Int. Rev. Cytol.* **186,** 179–224 (1999).
266. A. H. Huber, W. J. Nelson, and W. I. Weis, Three-dimensional structure of the armadillo repeat region of beta-catenin. *Cell* **90,** 871–882 (1997).
267. I. R. Vetter, A. Arndt, U. Kutay, D. Gorlich, and A. Wittinghofer, Structural view of the Ran-Importin beta interaction at 2.3 A resolution. *Cell* **97,** 635–646 (1999).
268. K. Kotani, A. Kikuchi, K. Doi, S. Kishida, T. Sakoda, K. Kishi, and Y. Takai, The functional domain of the stimulatory GDP/GTP exchange protein (smg GDS) which interacts with the C-terminal geranylgeranylated region of rap1/Krev-1/smg p21. *Oncogene* **7,** 1699–1704 (1992).
269. I. Itoh, K. Kaibuchi, T. Sasaki, and Y. Takai, The smg GDS-induced activation of smg p21 is initiated by cyclic AMP-dependent protein kinase-catalyzed phosphorylation of smg p21. *Biochem. Biophys. Res. Commun.* **177,** 1319–1324 (1991).
270. D. L. Altschuler, S. N. Peterson, M. C. Ostrowski, and E. G. Lapetina, Cyclic AMP-dependent activation of Rap1b. *J. Biol. Chem.* **270,** 10,373–10,376 (1995).
271. H. R. Bourne, How receptors talk to trimeric G proteins. *Curr. Opin. Cell Biol.* **9,** 134–142 (1997).
272. H. Yaku, T. Sasaki, and Y. Takai, The Dbl oncogene product as a GDP/GTP exchange protein for the Rho family: its properties in comparison with those of Smg GDS. *Biochem. Biophys. Res. Commun.* **198,** 811–817 (1994).
273. H. Nakanishi, K. Kaibuchi, S. Orita, N. Ueno, and Y. Takai, Different functions of Smg GDP dissociation stimulator and mammalian counterpart of yeast Cdc25. *J. Biol. Chem.* **269,** 15,085–15,091 (1994).
274. K. Shimizu, H. Shirataki, T. Honda, S. Minami, and Y. Takai, Complex formation of SMAP/KAP3, a KIF3A/B ATPase motor-associated protein, with a human chromosome-associated polypeptide. *J. Biol. Chem.* **273,** 6591–6594 (1998).

275. L. Johnson, D. Greenbaum, K. Cichowski, K. Mercer, E. Murphy, E. Schmitt, R. T. Bronson, H. Umanoff, W. Edelmann, R. Kucherlapati, and T. Jacks, K-ras is an essential gene in the mouse with partial functional overlap with N-ras. *Genes Dev.* **11,** 2468–2481 (1997).
276. K. Koera, K. Nakamura, K. Nakao, J. Miyoshi, K. Toyoshima, T. Hatta, H. Otani, A. Aiba, and M. Katsuki, K-ras is essential for the development of the mouse embryo. *Oncogene* **15,** 1151–1159 (1997).
277. K. K. Vithalani, C. A. Parent, E. M. Thorn, M. Penn, D. A. Larochelle, P. N. Devreotes, and A. De Lozanne, Identification of darlin, a Dictyostelium protein with Armadillo-like repeats that binds to small GTPases and is important for the proper aggregation of developing cells. *Mol. Biol. Cell.* **9,** 3095–3106 (1998).
278. F. C. von Lintig, A. D. Dreilinger, N. M. Varki, A. M. Wallace, D. E. Casteel, and G. R. Boss, Ras activation in human breast cancer. *Breast Cancer Res. Treat.* **62,** 51–62 (2000).
279. D. M. Reese and D. J. Slamon, HER-2/neu signal transduction in human breast and ovarian cancer. *Stem Cells* **15,** 1–8 (1997).
280. N. Shinohara, Y. Ogiso, M. Tanaka, A. Sazawa, T. Harabayashi, and T. Koyanagi, The significance of Ras guanine nucleotide exchange factor, son of sevenless protein, in renal cell carcinoma cell lines. *J. Urol.* **158,** 908–911 (1997).
281. T. Watanabe, N. Shinohara, K. Moriya, A. Sazawa, Y. Kobayashi, Y. Ogiso, M. Takiguchi, J. Yasuda, T. Koyanagi, N. Kuzumaki, and A. Hashimoto, Significance of the Grb2 and son of sevenless (Sos) proteins in human bladder cancer cell lines. *IUBMB Life* **49,** 317–320 (2000).
282. A. J. Dupuy, K. Morgan, F. C. von Lintig, H. Shen, H. Acar, D. E. Hasz, N. A. Jenkins, N. G. Copeland, G. R. Boss, and D. A. Largaespada, Activation of the Rap1 guanine nucleotide exchange gene, CalDAG-GEF I, in BXH-2 murine myeloid leukemia. *J. Biol. Chem.* **276,** 11,804–11,811 (2001).
283. D. J. Hussey, M. Nicola, S. Moore, G. B. Peters, and A. Dobrovic, The (4;11)(q21;p15) translocation fuses the NUP98 and RAP1GDS1 genes and is recurrent in T-cell acute lymphocytic leukemia. *Blood* **94,** 2072–2079 (1999).
284. C. P. Ponting and D. R. Benjamin, A novel family of Ras-binding domains. *Trends Biochem. Sci.* **21,** 422–425 (1996).
285. M. Rozakis-Adcock, R. Fernley, J. Wade, T. Pawson, and D. Bowtell, The SH2 and SH3 domains of mammalian Grb2 couple the EGF receptor to the Ras activator mSos1. *Nature* **363,** 83–85 (1993).
286. S. S. Yang, L. Van Aelst, and D. Bar-Sagi, Differential interactions of human Sos1 and Sos2 with Grb2. *J. Biol. Chem.* **270,** 18,212–18,215 (1995).
287. P. S. Lorenzo, J. W. Kung, D. A. Bottorff, S. H. Garfield, J. C. Stone, and P. M. Blumberg, Phorbol esters modulate the Ras exchange factor RasGRP3. *Cancer Res.* **61,** 943–949 (2001).
288. C. Herrmann, G. Horn, M. Spaargaren, and A. Wittinghofer, Differential interaction of the ras family GTP-binding proteins H-Ras, Rap1A, and R-Ras with the putative effector molecules Raf kinase and Ral-guanine nucleotide exchange factor. *J. Biol. Chem.* **271,** 6794–6800 (1996).
289. M. Spaargaren and J. R. Bischoff, Identification of the guanine nucleotide dissociation stimulator for Ral as a putative effector molecule of R-ras, H-ras, K-ras, and Rap. *Proc. Natl. Acad. Sci. U.S.A.* **91,** 12,609–12,613 (1994).

Practical Approaches to Long Oligonucleotide-Based DNA Microarray: Lessons from Herpesviruses

EDWARD K. WAGNER,[*]
J. J. GARCIA RAMIREZ,[†]
S. W. STINGLEY,[*]
S. A. AGUILAR,[*] L. BUEHLER,[‡]
G. B. DEVI-RAO,[*] AND
PETER GHAZAL[†]

[*]*Department of Molecular Biology
and Biochemistry and Center
for Virus Research
University of California
Irvine, California 92717*
[†]*Genomic Technology & Informatics
Centre
University of Edinburgh, Summerhall
Edinburg EH9 1QH, Scotland*
[‡]*Division of Biology
University of California
San Diego, California 92093*

I. A Rationale for Developing DNA Microarrays for Herpesviruses	446
II. Herpes Simplex and Cytomegaloviruses—Two Herpesviruses That Share Features of Productive Infection but Differ Markedly in Patterns of Latency and Reactivation	447
A. HSV—A Well-Characterized Human Pathogen	447
B. CMV—A Less Well-Characterized Human Pathogen	450
III. Design Criteria for Herpesvirus DNA Microarrays	451
A. Microarray Analysis of Herpesvirus Transcription	452
B. PCR Fragment-Based Approach	453
C. Oligonucleotide-Based Approach	453
D. Concluding Remarks on Design of Viral Microarrays	454
IV. The Construction and Validation of an Oligonucleotide-Based Hsv-1 DNA Microarray on Glass Slides	455
A. Transcript Labeling, Hybridization, and Scanning Protocols	455
B. Selection of Oligonucleotides Specific for Viral Transcripts	461
C. Chip Fabrication and Scanning	464
D. Receiver Operating Characteristic (ROC) of Viral Microarrays	465
E. Normalization and Statistical Evaluation of Chip Data	465
F. Representation and Standardization of Experimental Data	470
V. Exemplary Applications	472
A. Genotype Profiling of HCMV	472
B. Profiling Cell-Specific Temporal Viral Gene Expression	474

VI. Conclusions 486
References 487

I. A Rationale for Developing DNA Microarrays for Herpesviruses

The application of robotic microarraying techniques and laser-based image analysis has lead to the development of DNA microarrays as a powerful tool for the global analysis of transcriptional responses of cells and microorganisms to perturbations, such as stress, in their environment. While the equipment necessary to synthesize chips is rapidly changing, designs are freely available and a number of groups have made excellent prototypes (cf. (*1, 2*) and http://arrayit.com). Further, a number of commercially available fabricators are available at an affordable price for generating common-use facilities in major universities and research centers. While it might be argued that the application of a high-throughput technology such as DNA microarray-based global transcriptional analysis is unnecessary for the study of viruses with their circumscribed genomes, it is important to remember that much if not most of our understanding of the molecular basis of gene regulation, as applied to the normal and abnormal developments of human cells, comes directly or tangentially from the study of the aspects of virus replication and of the interaction between viruses and their hosts. The response of specific cellular genes to virus infection and the expression of specific viral functions are the defining factors in controlling viral pathogenesis. Further, it is the interaction between the cell and the specific viral genes related to cellular genes regulating cellular replication that leads to virus-mediated-growth transformation and oncogenesis. DNA microarray analysis affords the best opportunity for studying these variables on a global scale.

In addition to the relevance of increasing our understanding of the function of genes important in characterizing aspects of disease and human morbidity, an understanding of herpesvirus infections, in terms of the functional relationship between viral and host genetic complement, has specific relevance to the use of viruses as tools for gene delivery, therapy, and control of disease. Clearly, understanding those features important in differential gene expression in the context of the viral genome, as well as the consequences of perturbation of normal control on the course of virus infection, is critical in the evaluation of potential vectors and therapeutic agents. Thus, global studies are highly relevant to a broad spectrum of questions concerning specific aspects of cellular gene expression and to the results of defined failures in such.

Effective molecular study of viral pathogenesis requires optimized conditions for studying virus replication in differentiated cells under physiologically relevant conditions. While this is a technical goal, it is an essential one. General patterns of viral gene expression and gene function have been derived from studies where relatively large quantities of cultured cells are infected under selected and/or optimized conditions. As useful as this approach has been and will continue to be, it does not reflect the process of virus infection and spreading in the host. The development of microarray analysis of viral replication, in reactivation from latent infection and in infection of a specific organ or cell type, may be mediated by infection of either one or, at most, very few cells under suboptimal conditions of low multiplicity requiring high levels of sensitivity.

The fact that both abundance and time of appearance of mRNA are controlled during productive infection by herpesviruses argues that regulated gene expression has a central place in viral pathogenesis; yet direct investigation of this has not been extensive. There are a large number of defined regulatory and kinetic mutations of the virus available in which either the timing of expression or the function of a single gene has been modified. Microarray analysis allows the investigation of the role of both kinetics of expression and transcript abundance upon the replication of the subject virus in cultured cells and, ultimately, in animal models. The power of the approach will be realized by assaying an increasing collection of cellular genes (both human genes and those specific for the animal model being studied) chosen to represent those whose expression is modulated by the stress of viral infection. The broad patterns of viral and critical cellular gene expression exhibited under both varying conditions of infection and the influence of defined modification of critical regulatory genes will illuminate potential critical points or "bottlenecks" in the course of virus replication in the whole animal. Those differentiated cells and tissues where there is a critical restriction of virus replication are precisely the points where it can be expected that the full panoply of viral regulatory circuits must operate to optimum effect. They will provide important experimental subjects for further regulatory studies.

II. Herpes Simplex and Cytomegaloviruses—Two Herpesviruses That Share Features of Productive Infection but Differ Markedly in Patterns of Latency and Reactivation

A. HSV—A Well-Characterized Human Pathogen

The study of prototypical neurotropic human herpesvirus (HSV) provides a major resource for understanding the replication and patterns of gene expression and the pathogenesis of all herpesviruses. The relative ease of manipulation of the HSV genome, the ability to test defined modifications of the viral genome in the context of replication in cultured cells of many origins, the availability of a number of animal models emphasizing different aspects of viral pathogenesis, and the ability to use such animals for establishing and maintaining latent infections, as well as the ability to recover virus from such latent infections are powerful arguments in favor of the continued value for this model. Perhaps no other virus or system allows such a spectrum for studying the effects of perturbation and elimination of both *cis*- and *trans*-acting control points on parameters of infection.

HSV replication and pathogenesis are well characterized (see, for example, Refs. 3–5). A rapidly resolving initial acute infection is followed by life-long latent infection interspersed with sporadic reactivation episodes. HSV (as well as other herpesviruses) has a promoter-rich genome. During infection, specific promoters mapping at cognate genes mediate transcript expression. All data suggest that the coordinate regulation of expression of viral transcripts must involve the transcriptional machinery of the cell *in toto*. A major factor in the control of viral gene expression is the differential activity of promoters whose functional architecture is, in a large part, responsible for controlling access to the transcriptional machinery of the cell. A second major factor in the regulation of expression of at least some viral genes is the alteration of post-transcriptional processing and transport of viral transcripts mediated by the activity of the immediate-early $\alpha 27$ gene (6, 7). In this light and in the most general sense, regulation of HSV gene expression can only be understood in light of normal cellular transcription processes. While viral regulatory proteins operate to drastically alter the regulatory environment of the host cell during infection, their effects are manifest through existing cellular transcription factors and enzymes. Two well-studied examples of this are the initial activation of immediate-early viral regulatory genes by the potent *trans*-activator, a-TIF, in conjunction with the action of a cellular adapter, Oct-1 (8, 9). And, the global transcriptional activator ICP4, which is related to the cellular transcriptional activator (10), functions through stabilization of the binding of TFIID (cf. 11).

If productive infection by HSV is characterized by the virus fully capturing the cell's transcriptional and translational machinery in a wide range of cell types and state of differentiation, latent infection by HSV in sensory neurons is the opposite—a highly restricted expression of viral genes resulting in no immune signature during the latent phase. Latency involves three distinct stages: establishment, maintenance, and reactivation (12–16). The first two, in which viral gene expression is rigidly suppressed, are clearly dependent on both the state of differentiation and the physiological context of the infected neuron. The restriction of HSV-productive-cycle gene expression during the establishment and maintenance of latency does not require the expression of any latent phase-specific viral gene (12, 14, 17–19), but a single transcription unit—the latency-associated transcript (LAT) family—is expressed (*20*). In addition to the transcriptional environment within the latently infected neuron, the immune competency of the host plays a major role in limiting viral productive infection in peripheral neurons as well as in spreading the infection to the CNS (*21, 22*).

Viral genomes persisting in sensory neurons provide the reservoir of infectious virus for reactivation; therefore, virus must exit these neural cells to allow productive infection of peripheral cells without causing excessive damage either by virus-induced reactivation or by the host's physiological response to recrudescence. This latter requirement is critical for the laboratory analysis of factors involved in reactivation and limits the usefulness of many animal and tissue culture systems.

Maintenance of latent HSV genomes in the protected and nonreplicating environment requires no viral gene expression. The viral genome is a histone-associated, super-coiled episome during latent infection (*23–25*). This state must preclude frequent sporadic episodes of transcription *via* unmodified cellular processes occurring in latently infected neurons, since low-level expression of viral transcripts, except those specific to the latent phase, is either not detectable or at such low frequency that it requires the use of PCR amplification for detection (*14, 26, 27*). This is an important consideration, since many herpesvirus promoters are active at detectable levels, even in the absence of viral transcriptional activation in both biochemically transformed cells and transient assays (*28–34*).

In both humans and several *in vivo* models, reactivation of HSV results in the appearance of infectious virus at the site of initial primary infection (see Refs. 35 and 36 for reviews). Simply put, reactivation involves either the presentation of a small amount of infectious virus or replication-competent viral genomes from the latent reservoir to peripheral tissues leading to productive replication at the periphery until cleared by the host. While it is, as yet, unclear whether the initial transcriptional events leading to a successful reactivation are regulated, as in productive infection, PCR analysis of the earliest steps in reactivation in several animal models confirms the early expression of immediate-early viral genes (*26, 27*). Unlike the situation with establishment and maintenance of the latent infection, it is clear that expression of the LAT unit plays a significant role in reactivation; however, the precise mechanism of this involvement is unknown, and LAT expression is not an absolute requirement for reactivation, although it greatly increases the efficiency of induced reactivation in several *in vivo* models (*27, 36, 37*). To complicate matters further, the role of LAT expression in several popular (and relatively inexpensive) *in vitro* models is problematic (cf. 26).

The investigation of very early events in reactivation is complicated by the fact that it is necessarily a low-frequency event leading to a low-multiplicity infection, and it is difficult to differentiate those viral functions required for *efficient* productive-phase infection from those *required*. Animal models are an important resource for the rational experimental study of latency and reactivation of neurotropic herpesviruses, but no model fully reflects the actual situation in the natural host (35). Clearly, microchip analysis of changes in both viral and cellular gene expression during the earliest stages of reactivation is an important goal.

B. CMV—A Less Well-Characterized Human Pathogen

Cytomegaloviruses are β-herpesviruses that exhibit a highly restricted host range and efficiently establish latent infection in distinct cell types and in sites throughout their respective hosts. Human cytomegalovirus (HCMV), which replicates exclusively in human cells, is a ubiquitous pathogen that causes severe disease in newborns and in immunosuppressed patients. This strict species specificity, for reasons which are not fully known, has hampered the development of suitable animal model systems for HCMV disease. For this reason, the related but distinct murine cytomegalovirus (MCMV) infection of mice has provided an excellent surrogate animal model system.

As with HSV, the expression of HCMV and MCMV genes in an infection is temporally regulated. The first genes expressed (immediate-early, IE) require host cell factors and mainly encode regulatory or immune modulator factors. The IE regulatory proteins, together with cellular transcription factors, coordinate the next level of gene expression (early genes). The early genes provide an essential source of factors, including viral DNA-replication and DNA-repair enzymes. Late genes are essentially expressed after the onset of viral DNA replication and contribute primarily to the assembly and egress of the virus. The program of temporal gene expression during infection thus represents key checkpoints in modulating productive infection including latency, which is defined as the reversible inhibition of productive infection (38).

HCMV establishes latency at several sites throughout the body including peripheral blood monocytes and macrophages (39). While monocytes are the predominant infected cell type in the circulation, viral gene expression is restricted in these cells (40). In tissue macrophages, however, extensive gene expression can be detected, and it is proposed that macrophage differentiation is a prerequisite for productive HCMV infection (41). MCMV exhibits select sites of latent infection, including macrophages (42). The major site of latency appears to be the lungs, although other sites including salivary glands, adrenal glands, heart (43), bone marrow macrophages (42), spleen, and kidneys (44) can also harbor latent infection. In CMV latency, transcription of its genome is highly restricted. On occasion, minimal transcription can be detected from the IE region with highly sensitive polymerase chain reaction techniques, such as latency-associated transcripts encoding 491 and 579 aa proteins from the HCMV *ie1/ie2* region in granulocyte–macrophage precursors (45). In MCMV infection, transcription appears to be limited to the *ie1* gene, while HCMV transcripts originating from the major *ie1* region have been observed (45). These transcripts are of extremely low abundance, and it is questionable what role they play. They may simply represent sporadic activation events. Thus, at the molecular level, latency is the outcome of cellular restriction of key early checkpoints in the program of viral gene expression (reviewed in Ref. 38).

At the host level, in MCMV latency, immune function is important in quelling reactivation of virus. A hierarchical system of immune control has been proposed with CD8 and NK subsets playing the critical initial roles, followed by help from the CD4 T-cell subset (46). In this connection, IFN-γ potently suppresses reemergent virus from latency (47). The source of IFN-γ is likely due to noncytolytic activity of NK and CD8 cells. Antibodies were also important in preventing dissemination of reactivated MCMV, although they prevent neither primary infection nor recurrence of MCMV infection (48). HCMV undergoes frequent episodes of reactivation and "shedding," so that the virus is present in body fluids and as a consequence continually interacts with immune memory.

Stimuli for reactivation of latent CMV are not understood, although they may include cellular differentiation and hormonal and cytokine (TNF-α) stimulation. The immune system undoubtedly limits the dissemination of reactivated HCMV and MCMV; however, it is unlikely that the actual reactivation events are simply licensed by the loss of immune control (spontaneous reactivation) but also require other exogenous or cellular stimuli (induced reactivation). In this case, changes in cellular factor expression can lead to latent genomes being de-repressed, resulting in an increased probability for reactivation. Recent observations using the MCMV model are consistent with reactivation being an inducible event in which signaling would switch on the productive cycle (49). In this study, the MCMV viral genome was evenly distributed in the latently infected lungs, but a complex mosaic of viral transcriptional activity was focally and randomly distributed in the lungs. These data were used to propose that multiple, sequential checkpoints at both transcription and post-transcriptional levels, which have to be passed on the way from latency to recurrence, exist in the MCMV (49). Clearly, the availability of genome-wide transcriptional profiling for HCMV and MCMV (50) may reveal not only a clearer definition of genome activity modulated by immune modulators but also an exact definition of virus latency.

III. Design Criteria for Herpesvirus DNA Microarrays

Ideally, a DNA microchip designed for the analysis of both viral and cellular gene expression during normal and experimentally perturbed replication cycles should combine ease of construction, stability, specificity for all viral genes, and a wide selection of cellular genes. The DNA microchip should also have the ability to reliably detect both abundant and rare transcripts over a wide range of concentrations and to demonstrate a relative ease in sample preparation.

Cloned DNA fragments are currently the major sources of DNA sequences for gene probes and they are often amplified by PCR and synthetic oligonucleotides. In this section, studies exploring the virus transcription cycle using microarray technology are briefly reviewed. The main advantages and limitations of using PCR fragment and oligonucleotide approaches for large viral genomes are then discussed.

A. Microarray Analysis of Herpesvirus Transcription

As described in greater detail in the previous sections, the expression of herpesvirus genes upon infection is temporally regulated. The first genes expressed (IE) are independent of viral *de novo* protein synthesis and encode either regulatory *trans*-acting factors or immune inhibitory proteins. The next set of genes expressed (early, E) requires the presence of viral regulatory proteins and contributes an essential source of factors, including viral DNA replication, repair enzymes, and other nonstructural proteins, such as those that serve in signal transduction and immune evasion. Late (L) genes are essentially expressed after the onset of DNA replication and contribute primarily to assembly, morphogenesis, and egress of virions. Thus the time of viral gene expression during infection provides an important clue to its functional role. For this reason, viral DNA microarray technology offers an ideal systematic approach for high throughput evaluation of specific viral gene expression in elucidating gene function.

The first generation of a viral DNA chip for genome wide-expression measurements was reported for the human cytomegalovirus (HCMV) genome, the largest member of the herpesvirus family (50). In this study, an HCMV DNA chip was used to both catalogue the temporal class of viral gene expression and characterize the profile of virus transcription upon drug inhibition. Prior to this study, less than 30% of the genome had been transcriptionally mapped, and an excellent correlation between temporal expression class and assigned function was observed. Subsequently, the viral DNA microarray approach was successfully applied to the functional analysis of the well-established herpes simplex (HSV) and the newly discovered Kaposi's sarcoma-associated herpesvirus (KSHV; human herpesvirus 8) (51, 52). In the case of KSHV, only a handful of genes had been previously characterized, and the correlation between gene expression and function has provided key clues for the role of the many KSHV genes yet to be characterized.

The genome wide-mapping of temporal gene expression also provides important information toward understanding and dissecting regulatory pathways of particular virulent genes. For instance, for HCMV sequence compositional analysis of the 5′ noncoding DNA sequences of the temporal classes, using algorithms that automatically search for motifs in unaligned sequences, indicated the presence of potential regulatory elements for a subset of key early and early–late genes.

The ability to rapidly perform parallel gene expression analysis at the whole viral genome level is also beginning to help uncover new aspects of viral biology. For example, a previously unidentified class of viral transcripts, termed virion RNAs, has been discovered using HCMV microarrays (53). These specific transcripts are packaged within virions and allow for the viral genes to be expressed within an infected cell immediately after virus entry. The role of this new class of transcripts is unknown. In other studies, fabricated microarrays of viral genomes are helping to reveal new relationships between lytic and latent gene expression classes (52). In the future, viral DNA microarrays will play an important role in defining classes of virus transcription activity in latency. It is noteworthy that the patterns (signature) of viral gene expression are influenced either directly or indirectly by expression of cellular proteins, and as a consequence quantitative and qualitative differences in profile of viral gene expression have been observed to vary depending on host cell background (51). In this connection, a better understanding of a virus transcription–replication cycle can be appreciated by future studies that aim to integrate host cell expression with viral gene expression. The current approaches and their limitations to the design and use of viral DNA microarry systems are briefly reviewed in the next section.

B. PCR Fragment-Based Approach

In this approach, pioneered by Brown and colleagues (1), PCR is used to amplify segments of DNA from 250 to 2000 bp in length from either cDNA or genomic libraries. These amplified products are deposited onto solid surfaces using high-speed robotic machines. Using such a high-density approach, array elements, from a few hundred to many thousand, can be fabricated in a small area. In general, a specific gene is specified by a single-array element. This approach has been successfully applied to viral genomes including HCMV and KHSV (52–54). The main advantage of this approach is the ability to develop a high-sensitivity array. However, it has the major limitation of not being able to distinguish the polarity of transcription. This can be a real problem in the case of viral genomes where economy of space is a critical feature. The rather extensive amount of labor involved in the characterization, quantitation, and maintenance of such clones is another drawback to their use in the medium-to-small-sized laboratory.

C. Oligonucleotide-Based Approach

The relatively affordable syntheses of oligonucleotides, which are long enough to provide appropriate hybrid stability and specificity (especially strand specificity), provide another approach which falls into two distinct paths: the short-oligonucleotide approach with probe lengths of 25 bases, pioneered by Stephen Fodor of Affymetrix and Ed Southern at the University of Oxford, and the long-oligonucleotide approach with probe lengths in excess of 40 bases.

For the short-oligonucleotide approach, consider the Affymetrix GeneChip arrays, which utilize multiple 25-mer probe sets (16 per gene) that are synthesized directly on the surface of the array using photolithography and combinatorial chemistry. Tens to hundreds of thousands of different oligonucleotide probes are synthesized on each array. Probe arrays are manufactured in a series of cycles. A glass substrate is coated with linkers containing a photolabile protecting group. Then, a mask is applied that exposes selected portions of the probe array to ultraviolet light. Illumination removes the photolabile protecting groups enabling selective nucleoside phosphoramidite addition only at the previously exposed sites. Next, a different mask is applied, and the cycle of illumination and chemical coupling is repeated. By repeating this cycle, a specific set of oligonucleotide probes is synthesized, with each probe type in a known location. These arrays are of both high quality and high expense. The major limitation is that production is strictly limited to syntheses of relatively short-oligonucleotide probes.

1. Long-Oligonucleotide Approach

Alternatively, single-stranded DNA oligonucleotides, complementary to a specific sequence, can be synthesized by more conventional procedures. In this case, it is possible to manufacture defined long-oligonucleotide probes in a microarray format that was first used for the fabrication of microarrays specific for human cytomegalovirus (HCMV—human herpesvirus type 5) using 75-mers (50). This was the first description of a genome-wide chip for a human pathogen. Subsequently, the same general approach was employed to synthesize an oligonucleotide-based DNA microarray for the analysis of herpes simplex virus type 1 (human herpesvirus type 1) of equivalent specificity and sensitivity (51). Since then, this approach has now been extended to other herpesvirus members including MCMV, MHV, and gamma 68 (unpublished observations).

D. Concluding Remarks on Design of Viral Microarrays

For HSV, the extensive analysis of the patterns of viral gene expression carried out on a gene-by-gene basis using other methods such as RNse protection assays, Northern blot analysis, and primer extension (see, for example, Refs. 5, 26, and 55–58) (also see http://darwin.bio.uci.edu/~faculty/wagner/) provides not only the ready ability to optimize conditions and techniques over a relatively large number of samples but also the direct comparison of microarray analysis results to these other methods. On the other hand, the less well-characterized CMV transcription program is clearly a useful application to chip technology.

While there are distinct advantages to these various approaches, the long-oligonucleotide approach is by far the most cost-effective regarding specificity and reliability. For this reason, we recommend long-oligonucleotide-based DNA microarrays as an ideal platform for the global study of patterns of gene expression of large viruses and larger pathogens. The cost of their fabrication is low in comparison to those used for PCR probes, especially when taking into account ease of maintaining inventories and modification of probes for specific purposes. Further, these microarrays provide strand specificity and equivalency under hybridization conditions. The highly detailed information available concerning the patterns of herpes replication and pathogenesis makes them important agents for establishing the basics of the interaction between virus and host. This information can be rapidly and directly applied to the construction and use of DNA microarrays for other neurotropic herpesviruses and provides a model for a concerted and carefully controlled investigation of the interactions between other human herpesviruses and their hosts. The value of such studies for understanding human disease and formulation aspects of health policy cannot be overstated. In the next section, by way of an example, the design considerations for a long-oligonucleotide-based herpes simplex microarray are described.

IV. The Construction and Validation of an Oligonucleotide-Based Hsv-1 DNA Microarray on Glass Slides

A. Transcript Labeling, Hybridization, and Scanning Protocols

Glass-slide-based microarrays require fluorescent-dye-tagged dNTP precursors for cDNA synthesis followed by laser scanning to record hybridization. There are two basic methods for generating mRNA-specific cDNA for hybridization: (i) generating oligo-dT-primed cDNA from total cellular or tissue RNA and (ii) isolating poly(A)-containing RNA from the same material and using random oligomers for cDNA priming. Both approaches have been used with success; however, greater transcript specificity can be attained from the latter. This is because the commercially available fluorescent tags (Cy3 and Cy5) are bulky, and their incorporation into cDNA using reverse transcriptase is not highly efficient, with the average size of cDNA product synthesized averaging only 3–400 nt in length. This places severe restrictions on the location of oligonucleotide probes. In practice, and as described below, the two labeling methods can be combined to provide a high degree of specificity for detecting individual transcripts in overlapping clusters.

The availability of two different fluorescent-tagged deoxynucleotides suggests that double-labeling would be a useful approach toward measuring changes in transcript abundance under various conditions of infection. In practice, however, the approach has a number of limitations. The most significant problem is that it can be shown, using identical cDNA samples, that signals exhibit a differential bias for one of the two labels due to differences in solubility, stickiness, chemical stability, chemical reactivity, and photobleaching. Practically, this means that describing changes in expression of a given transcript or transcripts as differences in the ratios of hybridization signals is subject to a considerable degree of uncertainty, even upon several experimental replicates. When double-labeling using Cy5 is vital, it is important to use a Cy5-dCTP that is as freshly prepared as possible and to carry out more replicate experiments in which both samples are not only labeled separately with both dye-conjugated nucleotides but also assayed at the same time. With particularly important or valuable samples, usually, any cost and time advantages obtained from double-labeling are less important factors than the higher reliability of using Cy3-labeled material.

Currently a promising alternative labeling approach exists where a reactive dialyl-nucleotide derivative is incorporated into cDNA and then the fluorescent tag is added after the cDNA synthesis. It is hoped that such methods and the use of other fluorescent tags would obviate this problem.

We have found that hybridization for 14–18 h in moderately high salt (5 × SSC) with 0.5% SDS at 68°C provides conditions under which there is good hybridization of both the relatively high G + C HSV-1 oligonucleotides and the lower percentage G + C content of the cellular probes included on the chip (51). Unhybridized cDNA is readily removed by a simple rinsing protocol at room temperature. The hybridized glass slides are then scanned using a ScanArray 4000, and fluorescent hybridization signals are quantitated using commercial Quantarray software. Other commercial scanners and software packages are available and should provide equivalent data.

The scanning process samples the fluorescence (expressed in arbitrary units) derived from the spotted oligonucleotide, which is adjusted by subtraction of the background fluorescence of an equivalent area within a concentric ring just outside the spotted sample. The data from individual spots are then expressed using Microsoft Excel. While reduced signals can be directly used, we have recently found that better discrimination of samples displaying low levels of hybridization can be attained by subtracting a second background value, the average of values seen from measuring the reduced fluorescence of a large number (ca. 100) of regions spotted with SSC alone. These values are termed as net (−SSC) values.

With laser scanners, there are two adjustable variables that must be set for each individual scan: the laser power and the photomultiplier gain. While each can be adjusted to any value between 0 and 100%, useful values for the laser power lie between 40 and 90 with the photomultiplier set at some constant value or at a set number of units more or less than those for the laser power. Thus, in practice, only one variable needs to be set. The ratio of fluorescent signal-to-actual sample value is linear only to net (−SSC) values of 40,000 or so. While theoretically this allows a greater than 40-fold discrimination in hybridization

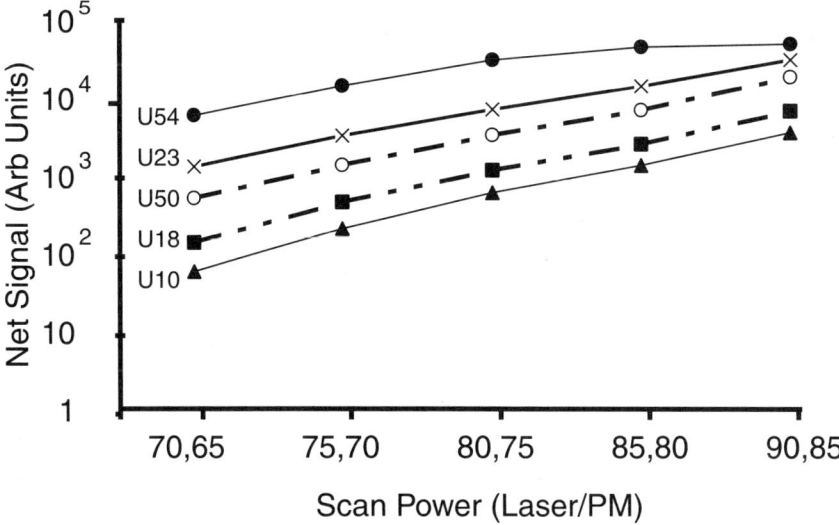

FIG. 1. The effect of laser power on hybridization signal with an HSV-1 oligonucleotide-based DNA microarray. The array was hybridized with Cy-3-labeled cDNA made from poly(A)-containing RNA isolated at 6 h after infection of HFF cells with HSV-1 at a multiplicity of 5 PFU per cell. The chip was scanned as the power and photomultiplier settings indicated, and the total signal for five transcripts known to give different intensity signals was plotted.

values, the fact that weak fluorescent signals are inherently less reliable than strong ones makes the actual range of reciprocity considerably smaller. Doing multiple scans at varying laser power settings can compensate for this if Cy3 tags are used, since these are relatively resistant to photobleaching. Multiple scans of the same microarray at varying laser power are shown in Fig. 1, where it can be seen that fluorescent signals differing by as much as 3 logs can be reliably differentiated.

Multiple scans, which at varying strengths obtain reliable measures of abundance of HSV transcripts under different experimental conditions can be seen in the following experiment, illustrated in Fig. 2. Here, immortalized human foreskin fibroblasts (HFF) (59) were infected at multiplicities of infection of 0.05, 0.5, and 5 PFU of HSV-1 per cell, and poly(A)-containing RNA from infected cells was isolated at 6 h postinfection. Equivalent amounts of this RNA were used to generate oligo-dT-primed cDNA using Cy3-dUTP as a label, and this material was hybridized separately to individual microarrays. Scanning the hybridized microarrays at the same power (laser 75 and photomultiplier 70) clearly shows that the amount of viral RNA detectable at the lowest multiplicity of infection (MOI) is 10-fold or greater reduced as compared to the highest value (Fig. 2A). Despite this, the viral transcripts expressed at the lowest multiplicity are detectable at significant levels when a higher laser power is used—one that would saturate the fluorescence from the more abundant sample (Fig. 2B). Further analysis by measuring the ratios of abundance of diagnostic transcripts demonstrates that the levels of expression of various transcripts are not identical under the two conditions. Indeed, a full statistical analysis of such data demonstrates that

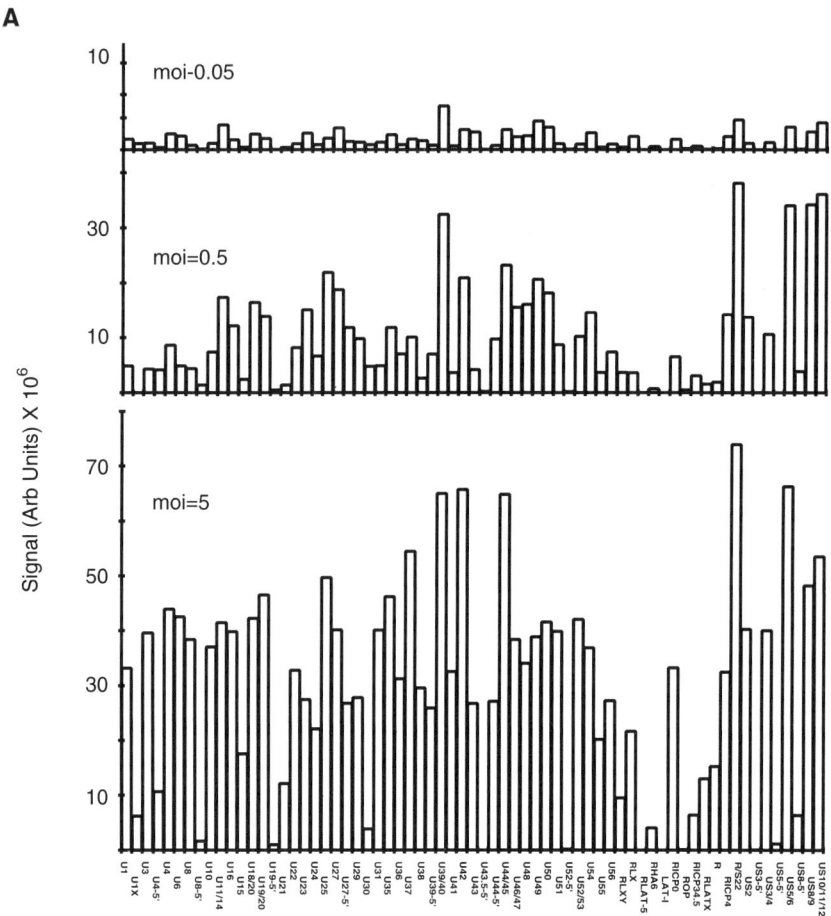

FIG. 2. The effect of laser power on resolution of HSV-1 byrididization signals with an oligonucleotide-based DNA microarray. As described, Cy-3-labeled cDNA, made from poly(A)-containing RNA isolated at 6 h after infection of HFF cells with HSV-1 at the multiplicity of infection shown was hybridized into separate arrays. Data are plotted as total hybridization signal for transcript vs genome position. (A) Each array was scanned at a laser power of 80% and a photomultiplier setting of 75%. (B) The effect of scanning the chip with the lowest overall signal (MOI = 0.05) at a higher laser power than the power used for the more robust signal.

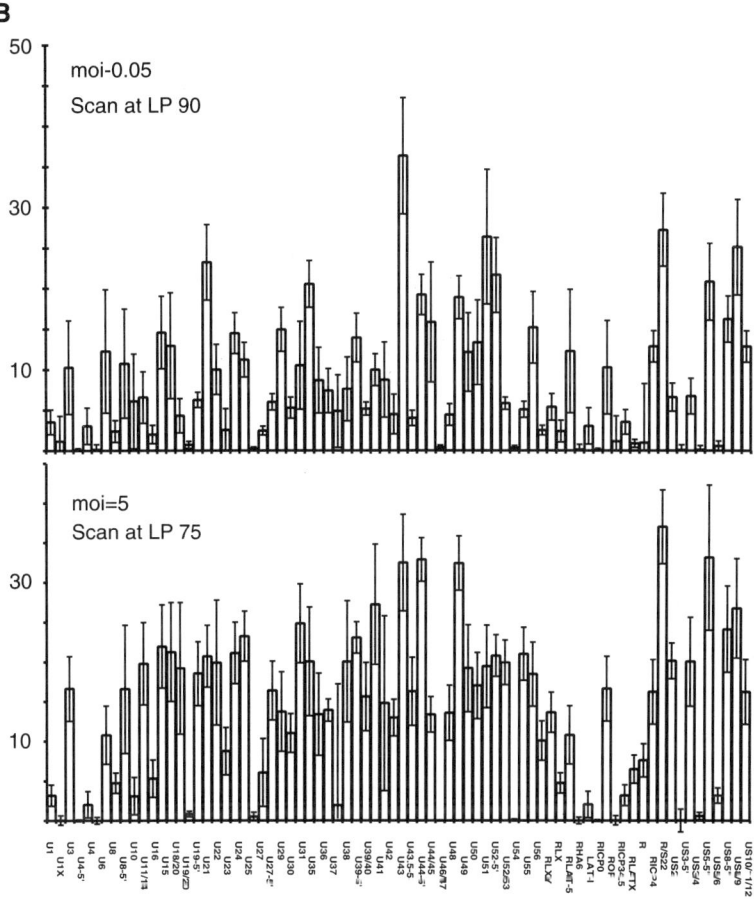

FIG. 2. (*continued*)

the patterns of viral RNA abundance seen in the 0.05 PFU/cell infection resembles those seen at a shorter time postinfection with a high multiplicity of infection (data not shown). In this connection, it is noteworthy that the level of sensitivity should also be taken into consideration. For instance, RNA harvested at late times postinfection with HCMV (Towne) at various MOI demonstrate (after signal strength normalization) an excellent correlation between MOI of 0.5 and 0.05 showing a similar profile of gene expression at high and low MOIs (Fig. 3A). However, this correlation is lost at MOIs below 0.05, most likely due to limitations in the sensitivity of signal detection rather than to different programs of gene expression (Fig. 3B).

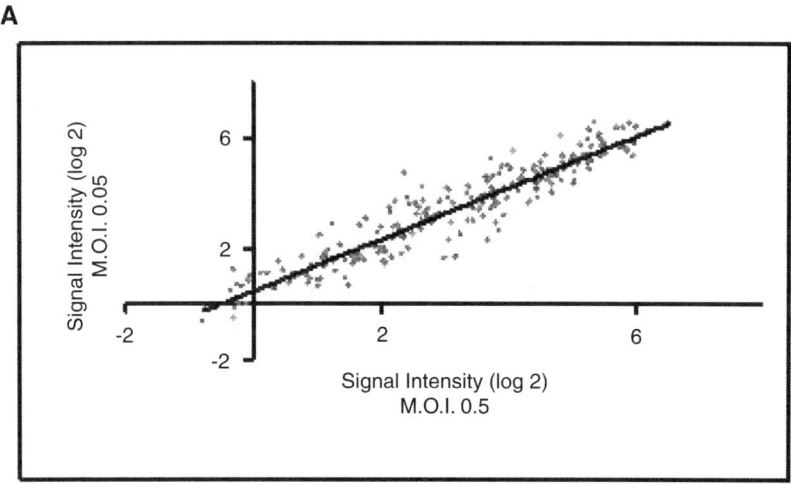

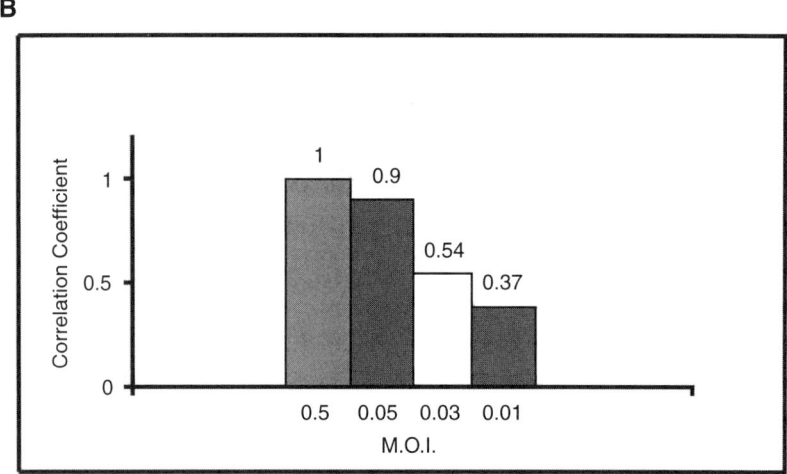

FIG. 3. A. The $\log_2$ of the average of a duplicate experiment of the signal intensity of HCMV-specific oligonucleotide probes hybridized at 72 hr post infection (hpi) with a MOI of 0.5 (x axis) or 0.05 (y axis). Note signal strengths have been normalized. The line of correlation is shown, with a correlation coefficient $R^2 = 0.9$. (B) Correlation coefficient (R^2) analysis at decreasing MOI. The R^2 were calculated using the $\log_2$ of the signal intensities of cells infected at decreasing MOI, shown in the upper-left corner, harvested at 72 hpi. All comparisons were made using MOI of 0.5 as a reference.

B. Selection of Oligonucleotides Specific for Viral Transcripts

For HSV, with its complete transcription map known, it was important to construct a microarray with sufficient probes necessary to uniquely detect all possible HSV transcripts as well as diagnostic host-cell genes, the expression of which is known or expected to be perturbed by viral infection. To maximize efficiency, optimize reliability (especially where low signal is expected), and minimize expense, we used two nonoverlapping probes for each viral transcript wherever possible. Using nonoverlapping probes has the added advantage of increasing the number of independent determinations of hybridization for each gene, which is of great value for statistical analyses (discussed below).

The selection of specific probes was simplified by the fact that each viral transcript is locally controlled by its cognate promoter. Splicing is rare and, where it does occur, not complex. On the other hand, this advantage was offset by the fact that many transcripts occur as nested sets of partially overlapping groups sharing the same 3'-polyadenylation sites. Our criteria for choosing appropriate probes included a position relative to the transcript polyadenylation site, lack of internal repeat or reiterated sequences, a and base composition reasonably close to the average for the region of DNA being transcribed.

It is important in chip design that consideration should also be given to whether sense or antisense probes are to be used. Sense probes should be used studies that involve direct labeling of RNA samples with fluorescent dyes. In contrast, Antisense oligonucleotide probe sets should be prepared for RNA samples that are indirectly labeled, such as with certain amplification protocols for sample RNA. Perhaps ideally, both sense and antisense probes for each region could be spotted, but the extra expense of this approach does not appear to provide commensurate gains in the attainable information.

In our first design, we used 60 sets of sense oligonucleotides for the unique detection of all HSV-1 transcript families. This included a total of 28 HSV-1 transcripts that were uniquely terminated by a cleavage/polyadenylation signal and 24 further transcripts that were expressed as one of two partially overlapping transcripts. To resolve the latter, we subsequently selected probe sets in the 5' (unique) regions of each of these overlapping transcription families to allow a unique measure of expression using random hexamer- (or octamer-) primed cDNA and comparing its hybridization to that observed with oligo-dT-primed material. Thus, our current chip contains 67 unique sets of HSV-1 probes (Table I); with these, 38 individual transcripts are uniquely resolvable (Table II). As shown in Fig. 4, differential patterns of hybridization with the two probes can provide high specificity.

A total of 21 viral transcripts are expressed in one of seven groups of multiple partially overlapping nested transcripts. Only the most distal of each group can

TABLE I
OLIGONUCLEOTIDE PROBES SPECIFIC FOR HSV-1 TRANSCRIPTS

Transcript(s)	poly(A) (R/L)	Probe A	G + C	Probe B	G + C	Transcript(s)	poly(A) (R/L)	Probe A	G + C	Probe B	G + C
UL1 (3')	10945R	10844R	56	10743R	64	UL43.5-5'		95307L	77	95716L	72
UL1X (3')	9635L	9757L	59			UL43 (3')	96063R	95948R	71	95837R	68
UL3 (3')	11717R	11599R	65	11519R	67	UL44 (5')		96711R	68	96451R	63
UL4 (5')		12424L	63	12290L	64	UL44/45 (3')	98663R	98363R	71	98533R	69
UL4/5 (3')	11760L	11961L	73	11896L	56	UL46/47 (3')	98731L	98829L	67	98932L	76
UL6/7 (3')	18037R	17915R	65	17840R	76	UL48 (3')	103542L	103619L	68	103695L	57
UL8/9 (3')	18217L	18296L	52	18406L	67	UL49/49.5 (3')	105467L	105545L	68	105620L	71
UL8 (5')		20469L	65	20144L	69	UL50 (3')	108152R	107969R	64	108038R	68
UL10 (3')	24645R	24544R	68	24407R	64	UL51 (3')	108281L	108357L	64	108432L	71
UL11/14 (3')	24807L	24919L	72	25005L	65	UL52 (5')		109078R	65	109657R	64
UL16L*	30178L	30415L	67	30375L	72	UL52/53 (3')	113443R	113300R	67	113225R	52
UL15R*	34820R	34624R	53	34743R	64	UL54 (3')	115277R	115193R	55	115091R	60
UL18/20 (3')	35028L	35187L	71	35234L	69	UL55 (3')	116098R	115911R	61	115896R	59
UL19/20 (3')	36405L	36690L	69	36483L	65	UL56 (3')	116201L	116341L	61	116282L	59
UL19 (5')		40488L	63	40248L	64	RLXY (3')	118003R	117814R	66		
UL21 (3')	43690R	43485R	72	43611R	59	RLX (3')	118710R	118605R	76		
UL22 (3')	43870L	43971L	63	44061L	56	LAT-5		118881R	68	119251R	71
UL23 (3')	46626L	46706L	59	46816L	69	LAT-I		119540R	64	119777R	68
UL24 (3')	48739R	48622R	67	48548R	65	LAT (RHA6)	120421L	120291L	69		
UL25/26/26.5 (3')	52766R	52681R	53	52571R	75	ICP0 (3')	120693L	120636L	64	120956L	69
UL27 (5')		55525L	65	55304L	55	ORFOP		124939R	59	125680R	63
UL27/28 (3')	53063L	53153L	59	53253L	63	LAT-X		126369R	61	126425R	55
UL29 (3')	58414L	58513L	65	58529L	63	icp34.5	123781R	125616R	80	125073R	54
UL30 (3')	66548R	66357R	64	66218R	63	LAT-3'	127141R	126776R	61	126708R	61
UL31/34 (3')	66382L	66460L	79	66594L	56	ICP4 (3')	127189L	127305L	68	127390L	83
UL35 (3')	70938R	70774R	71	70834R	67	US1 (3')	133941R	133341R	65	133788R	64
UL36/36 (-) (3')	70938L	71119L	64	71182L	68	US2 (3')	134041L	134206L	68	134257L	67
UL37 (3')	80717L	80902L	69	80823L	71	US3 (5')		135226R	63	136157R	61
UL38 (3')	86016R	85936R	68	85820R	67	US3/4 (3')	137508R	137383R	69	137433R	67
UL39 (5')		86553R	64	87813R	60	US5-5'		137731R	64	137871R	69
UL39/40 (3')	90983R	90843R	56	90876R	65	US5/6/7 (3')	141013R	140897R	64	140937R	64
UL41/43.5 (3')*	91121L	91214L	60	91201L	59	US8 (5')		141252R	59	141552R	65
UL42 (3')	94633R	94478R	71	94435R	71	US8/9 (3')	143667R	143514R	61	143592R	57
						US10/11/12 (3')	144139L	144255L	61	144218L	60

TABLE II
Unique Oligonucleotide Probes for HSV-1 Transcripts

Transcript	Function	Transcript	Function
U_L4	Unknown	U_L48	α-TIF
U_L8	part of helicase/primase complex	U_L50	dUTPase
U_L10	gM	U_L51	Unknown
U_L15	Spliced—DNA packaging	U_L52-5′	Helicase/primase complex
U_L21	Auxiliary virion maturation function(?)	U_L54	RNA transport/inhibit splicing
U_L22	gH	U_L55	Unknown—pathogenesis (74, 75)
U_L23	Thymidine kinase	U_L56	Unknown—pathogenesis (74, 75)
U_L24	Unknown—regulated polyA site (76)	ICP0	trans-Activator
U_L29	Single-stranded DNA-binding protein	RHA6	1400 nt 3′ of LAT cap (77)
U_L30	DNA polymerase	ICP34.5	Neurovirulence (72, 73)
U_L35	Capsomer tips	LAT	Latency function
U_L37	Tegument phosphoprotein	LAT-intron	Unknown (4)
U_L38	Efficiency of poly(A) site usage varies with cell type (78, 79)	ICP4	Broad-range trans-activator
U_L39-5′	Large subunit of ribonucleotide reductase	U_S1	Host range
U_L41	vhs, Virion-associated host shutoff protein	U_S2	Unknown
U_L42	Part of helicase/primase complex	U_S3-5′	Protein kinase
U_L43	Unknown (80)	U_S5-5′	gJ
$U_L43.5$	Antisense to U_L42 (81)	U_S8-5′	gE (Fc binding)
U_L44-5′	gC		

be uniquely detected with any cDNA synthesis regimen, but a comparison of hybridization of the 5′-unique transcript to probes specific for interior regions still provides a good measure of sensitivity and specificity. Combined with the fact that, in many cases, the various overlapping transcripts can be distinguished by the time or conditions of the maximum expression, essentially all viral genes are readily assayable in a microarray. This is illustrated in somewhat more detail in a following section.

We used a commercial computer program (Oligo6) and our own databases to select nonoverlapping sets of two or three appropriate 75-nt oligonucleotide sequences within the HSV genome specific for the expression of individual viral transcript sets. About 75% of the predicted sequences hybridized with good efficiency, as listed in Table I. Major criteria for selection included a G + C content ranging between 50 and 70% elimination of homopolymer and inverted repeat sequences. The majority were situated no more than 500 nt 5′ of individual polyadenylation sites to allow the use of oligo-dT-priming of total RNA.

Our most recent chip designs include a total of 144 cellular probes chosen to represent both house-keeping genes (60) and some whose expression can be expected to change under at least some conditions of infection. A preliminary analysis of cellular transcript abundance with these probes has been described

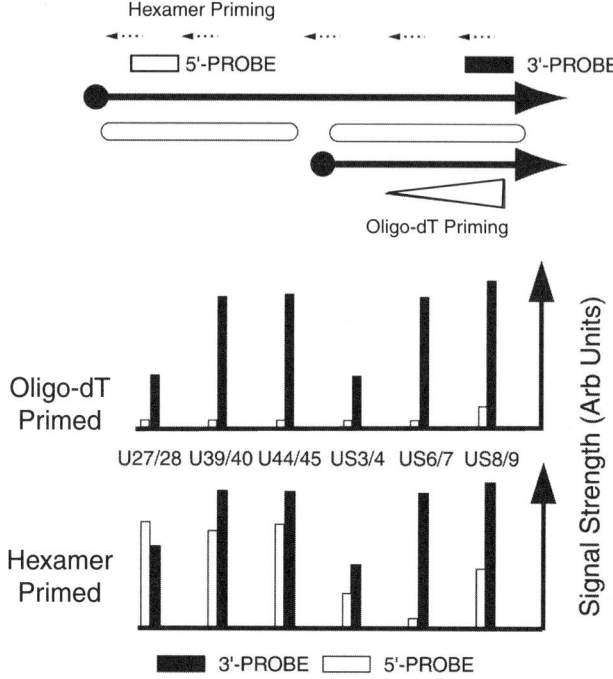

FIG. 4. The effect of labeling protocol on the hybridization signal throughout partially overlapping HSV-1 transcripts. The top panel schematically shows the two cDNA labeling protocols: oligo-dT-priming, which strongly emphasizes sequences near the poly(A) track of the transcript; and random hexamer-priming, which provides a more uniform representation of sequences throughout the cDNA. The bottom panel shows the difference in the relative signal intensities seen for probes at the 3′ and 5′ regions of selected overlapping viral transcript families when the two protocols are used. The location of specific probes are shown in Table 1.

in (51, 61). In the case of HSV, infection generally results in a decline in cellular transcript levels, but a complete statistical analysis of changes in cellular transcript profiles will be carried out on more extensive cellular arrays. The probes present on the viral chips serve mainly to help in chip-to-chip and interchip comparisons.

C. Chip Fabrication and Scanning

Chips were fabricated following the detailed approaches described by Brown and co-workers as applied to the HCMV chip (1, 50) (also described on the microarray web site http://arrayit.com). Oligonucleotides are synthesized commercially and supplied in a 96-multiwell format. The probe sets are then robotically transferred to a 384-well format at a printing (source plate) concentration of

20 μM in a suitable printing buffer. A number of positive and negative control oligonucleotides are also included. We robotically spot three 2- to 4-nl aliquots of each oligonucleotide (1 $\mu g/\mu l$ in 3× SSC) within 12 × 12 arrays on either polylysine or silylated aldehyde-coated glass slides. Polylysine slides are cheaper but suffer from higher background signals in comparison to silylated slides. Each probe element is printed in triplicate on the slide using either the Affymetrix 417 or the Biorobotics MicroGrid II arrayer. The arrays are then stored dry and dust free at constant temperature.

D. Receiver Operating Characteristic (ROC) of Viral Microarrays

Postprint staining and random primer hybridization are both used for quality control of arrays. Viral microarrays can be easily assessed for accuracy and sensitivity by applying a ROC analysis by using random primed viral genomic DNA as the target for hybridization to the microarray probes. The ROC analysis gives the summary statistics of diagnostic accuracy for a particular microarray experiment (area under the curve) and can be used for determining, instead of assuming, a threshold based on experimental constrains for specificity or sensitivity. A ROC curve (or plot for nonparametric analysis) can be calculated by grouping signals into positives (1) and negatives (0). These data are then plotted as the hit-rate 1-β (true positives or sensitivity) versus the false-positive rate α for a series of thresholds covering the entire signal range. The area under the curve is an estimate of the accuracy of the test for signal detection. The test is better the closer this value is to 1 (small alpha *and* beta error). A ROC plot is essentially a graphical representation of the signal-to-noise ratio for a single chip and thus provides a tool for working with single arrays. For small control groups a nonparametric ROC plot can be used, where a parametric analysis (e.g., estimate of sample standard deviation) is not possible. As an example, we show a typical ROC analysis of hybridization of total nick-translated HSV DNA to the HSV microarray. In this experiment, the cellular transcript probes serve as the negative control (Fig. 5). In addition, as a diagnostic tool, ROC plots can be used to compare different microarray manufacturing and hybridization techniques (e.g., coating, length of target DNA).

E. Normalization and Statistical Evaluation of Chip Data

To compare data from repeat and time-varying experiments, the chip hybridization data must be normalized following the following procedures. First the SSC background is calculated as the median of all the SSC spots located throughout the chip. Provided that the overall values do not indicate a defective region of the chip (i.e., one with abnormally high background, or some

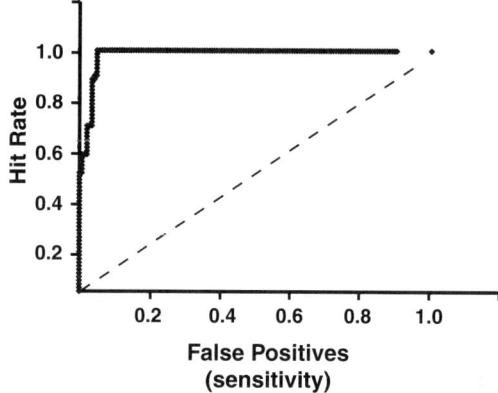

FIG. 5. A typical ROC plot of fluorescent-labeled HSV DNA hybridization to the HSV-1 microarray. In this experiment, a 0.5-μg aliquot of total HSV-1 DNA was nick-translated with Cy3-labeled dCTP, and the resulting labeled material was hybridized to an HSV-1 DNA microarray. Data are plotted as described in the text. The dashed line shows the expected values for random hybridization.

other flaw), the median SSC background is subtracted from the intensities of all probes. Then analysis begins by taking the median signal of the three-, six-, or nine-replicate probe values for each transcript set, and calculating the 75th percentile rank for the total viral hybridization. In this way, chips belonging to each experimental group are scaled accordingly. For publishing and statistical confidence in any findings, the recommended minimum number of statistically independent samples is three per group. This not only helps to indicate data variation representative of a population but also allows rudimentary hypothesis testing by nonparametric statistical methods. Obviously, the reliability is greater with more independent samples.

Hybridization data can be expressed as a ratio between a control and an experimental determination, as a numerical value related to fluorescence signal intensity, or as a measure of relative abundance based upon the ratio of a given probe signal to the total viral signal under any given conditions. Since laser power and photomultiplier settings are variable, ratios are (perhaps) the most popular way of expressing data (62). Still, it has been shown above that signal strength can be reliably related to scanning parameters; therefore, we think it important to include numerical signal strengths as a measure of hybridization no mater what the final form of presentation. Further, knowledge of the background signal seen with either nonspecific probes or SSC-spotted blanks provides an index of general reliability of the hybridization in question.

Ratio analysis between matched pairs as a measure of changes in transcript abundance is intuitively appealing because it eliminates one major problem: the sequence-dependent variation of hybridization strength. It is important to realize, however, that ratios themselves show considerable variations. In particular, ratio variability has a reverse correlation with signal strength. The unspecific nature of label interaction with control sequences or coating material makes negative controls more susceptible to methodical errors. Consequently, negative controls show large, random signal fluctuations (as assessed by internal repeats variability) translating into a larger ratio variability. Thus, dealing with ratios means dealing with a considerable degree of uncertainty about the reproducibility unless the data are appropriately analyzed.

It is often assumed that a twofold difference between control and experimental signals is a reliable "cutoff" value when deciding significance, but it is not *a priori* evident that a 100% increase in fluorescence intensity is always due to a sequence-specific hybridization. It is entirely possible that a twofold difference between two observations is a chance event and not the result of differences in cDNA concentration; while in another instance, a 50% increase or decrease in signal strength vis-à-vis a control can be reliable. Indeed, it can be argued that each printing of a given microarray, and even each gene, will have a unique threshold. For a gene-by-gene analysis on single chips, limited sample size greatly decreases the confidence in signal detection. If the change detected equals the sample standard deviation, the power of the test to detect a difference (the "confidence") for a sample size of $n = 5$ is about 30–40%. The confidence level increases with an increased sample size, $n = 10$, to about 60%. In other words, for an experiment with only five observations (per gene), the confidence level would indeed be low (less than 50%), if we assumed that there was no change in gene expression for signals close to the sample standard deviation. A threshold equaling two sample standard deviations increases the confidence (power) to about 80% (i.e., to be wrong in one out of five cases) (63). Thus, it should become clear that acceptable confidence limits for small sample sizes require a threshold far from the sample mean (i.e., reporting sample mean ± 3 SD). Of course, setting a higher threshold value (e.g., 2 or 3 SD) to increase specificity (reducing false positives) comes at the expense of sensitivity of signal detection, and we are left making statements only about the most obvious changes in our data (e.g., threefold or more).

We use two approaches to deal with such problems. First (and as discussed above), the vast majority of *HSV* genes are represented by at least two nonoverlapping probes. Second, each experimental condition is tested with separate RNA isolations with duplicate hybridizations of at least one of them, for a minimum of three replicates. Repeating the hybridization with three to five independent chips, each with three to six repeats, gives not only sample sizes adequate to assess signal variability within a chip but also chip-to-chip variability for signal detection across multiple experiments. Data from replicate experiments are then analyzed for variance using a nonparametric test (e.g., Wilcoxon Matched-Pairs Signed-Ranks Test) by using Microsoft Excel to calculate the ranking of each value of hybridization obtained followed by calculation of the p value using Student's t test. This is superior to the direct calculation of p values from data sets with less than five observations, since there is no basis to assume normal distribution.

The value of using this approach is illustrated in Table III. Here, HSV RNA was isolated at 1, 3, 6, and 9 h after infection of human foreskin fibroblasts with 5 PFU HSV-1 per cell. Data for three HSV transcripts of different kinetic classes are shown. The *ICP0* gene is immediate-early, U_L50 (dUTPase) is early, and U_L44 (gC) is strict-late. It is well known from kinetic analysis that *ICP0* is expressed very early, and its steady-state level remains more or less constant over the time period tested, although its rate of transcription varies with time. The U_L50 transcript is expressed most abundantly within the first 4 h following infection, and its steady-state levels tend to reach a maximum at intermediate times and then decline. Finally, the gC transcript is expressed abundantly only following a delay during which DNA replication proceeds, and the level of this transcript increases throughout the time window chosen for the present experiment (3). While the average hybridization values seen for each of the probes follows these patterns, the standard deviation of the values are such that no significance can be given to the differences in levels seen.

A paired ranking of values for each probe, however, demonstrates a high degree of significance between characteristic time points. Thus, the difference in ICP0 transcript levels is only significant when the earliest and latest times are compared; in contrast, the increase of U_L50 transcript levels from 1 to 6 h is quite significant, and a limited, but measurable decline in transcript levels between the peak at 6 h postinfection and that at the 9-h point is evident. Finally, the slow initial accumulation of the gC transcript and its continued increase in abundance throughout the time points tested are clearly significant.

From this discussion, it should be clear that p values demonstrate the probability of making an erroneous statement about detecting a real change, as discussed above, and emphasize the importance of replicate experiments. This requirement, of course, is a major criterion that we used in choosing the approach outlined in this review. It should be evident that an adequately large number of replicates are also important in surveying the changes in levels of abundance of the cellular transcripts that are surveyed. This is especially important when attempting to analyze small differences between control and experimental values. At a p value of 0.05, 1 out of 20 observations will be a false-positive, but p values do not imply that a change has not occurred in gene expression patterns where no change could be detected. It may simply be that the change is too small to be detected. Increasing either the sensitivity of hybridization or more sensitive probes can overcome some of these problems.

TABLE III
Variance Tests for HSV Transcript Abundance in Infected HFF Cells

Probe	1 h-Exp 1	1 h-Exp 2	3 h-Exp 1	3 h-Exp 2	6 h-Exp 1	6 h-Exp 2	9 h-Exp 1	9 h-Exp 2
RICP0-A	9,086	8,102	20,219	9,757	24,809	12,777	40,300	16,038
	8,255	4,038	9,820	11,439	28,138	11,122	28,385	8,015
	9,092	18,361	7,293	16,022	24,232	10,952	33,251	14,180
RICP0-B	9,548	18,381	23,373	13,821	32,091	36,823	56,793	31,825
	10,623	18,951	13,739	31,814	39,726	34,807	43,453	12,157
	10,920	17,432	17,250	17,618	47,873	14,612	29,146	11,091
Median—SD	**10,085**	**5,023**	**14,922**	**6,809**	**26,474**	**12,274**	**28,766**	**15,143**
Variance	1 vs 3	0.037362		p 1 vs 9	0.005434		p 6 vs 9	0.935515
(p)	1 vs 6	0.000016		p 3 vs 9	0.069961		p 3 vs 6	0.052000
U50-A	8,769.54	6,865.92	31,244.26	34,530.43	54,938.98	35,722.32	47,086.25	20,428.26
	6,627.78	3,211.29	29,712.80	25,953.56	53,517.74	37,327.18	55,202.73	31,200.50
	7,567.02	3,653.90	24,997.96	33,486.20	39,458.58	41,083.68	45,551.50	15,174.20
U50-B	8,391.89	6,543.89	34,951.06	27,060.15	52,680.05	28,978.67	44,201.25	22,240.47
	5,508.25	5,237.75	29,316.93	25,779.63	45,796.85	42,616.18	51,315.08	29,464.48
	8,257.00	6,013.39	17,002.95	27,383.29	33,595.64	33,070.28	32,104.45	15,942.51
Median—SD	**6,586**	**1,776**	**28,350**	**4,992**	**40,271**	**8,608**	**31,652**	**14,119**
Variance	1 vs 3	0.000002		p 1 vs 9	0.000000		p 6 vs 9	0.023651
(p)	1 vs 6	0.000001		p 3 vs 9	0.461903		p 3 vs 6	0.000038
U44-5′-A	364.11	369.21	6,908.61	3,805.68	28,059.63	22,815.77	32,609.43	24,859.85
	235.23	143.30	8,912.07	5,167.42	36,206.23	36,685.87	37,560.62	28,868.48
	469.65	431.32	7,222.06	5,850.32	31,596.46	29,178.02	38,305.60	28,640.15
U44-5′-B	1,008.95	506.13	3,833.88	5,628.20	25,452.64	21,963.89	30,881.90	13,744.38
	961.30	537.15	4,113.33	5,142.63	27,438.12	24,013.62	32,935.17	21,535.51
	396.14	377.17	3,627.85	10,033.59	25,612.48	31,242.58	39,859.81	28,060.39
Median—SD	**414**	**258**	**5398**	**2,069**	**27,749**	**4841**	**29875**	**7437**
(p)	1 vs 3	0.000022		p 1 vs 9	0.000010		p 6 vs 9	0.000220
(p)	1 vs 6	0.000043		p 3 vs 9	0.000005		p 3 vs 6	0.000038

F. Representation and Standardization of Experimental Data

As discussed above, the results of any given single experiment analyzed on a single chip can readily be presented as a graph of overall net (i.e., less the SSC background) hybridization signal intensity of a given probe along with the standard deviation of the replicate values from the chip versus the relative position of the probe on the HSV-1 genome. An example of such a representation is shown in Fig. 2, where the effect of multiplicity of infection on hybridization intensity is presented. As long as the conditions of the scanning are essentially equivalent, this representation is visually satisfying and readily interpretable. This type of representation is not particularly illuminating, however, when it is important to analyze the difference in relative (or absolute) levels of abundance of a given transcript or group of transcripts expressed under varying conditions of infection.

To control for individual variations between samples, we have included several oligonucleotides, specific for the bacterial β-galactosidase gene in our standard HSV-1 chips, and individual "salt" poly(A)-containing RNA preparations with *in vitro* synthesized β-galactosidase-specific RNA prior to generation of cDNA using hexamer primers. In principle, this approach could also be used for oligo-dT-primed cDNA synthesis if a poly(A) sequence were to be recombined into the β-galactosidase DNA template used to synthesize the control RNA. Other control sequences such as CAT, GFP, or any other handy DNA sequence could also be used—this will be of value when recombinant viruses expressing β-galactosidase as a reporter gene under the control of modified HSV promoters are subjected to chip analysis (5). In general, we find that individual variation between chips in a single experiment is no more than +25%, although, if necessary, corrections by as much as a factor of 2 can be justified.

Examples of the comparison of a complete set of analyses using viral RNA extracted from different types of cultured cells are illustrated in data shown in Fig. 6. Here, experiments with HFF, neuron-derived SKN (64), and HeLa cells were carried out in which individual confluent cultures of 7×10^6 cells were infected at a multiplicity of infection of 5 PFU per cell, and then poly(A) RNA was isolated at various times following a 30-min virus-adsorption period. Following hybridization of fluorescent dye-labeled cDNA to DNA chips, chips were scanned and normalized, and the total net HSV hybridization signal was determined at each time point. All chips for a given experimental group were from the same printing, but these printings were different for the various experiments. It is clear from the data shown that the rate of increase of the total HSV signal varies slightly but significantly in the different infected cells. Since the steady-state level of abundance of any viral transcript is a function of both the rate of synthesis and the rate of degradation, both of which may vary under any

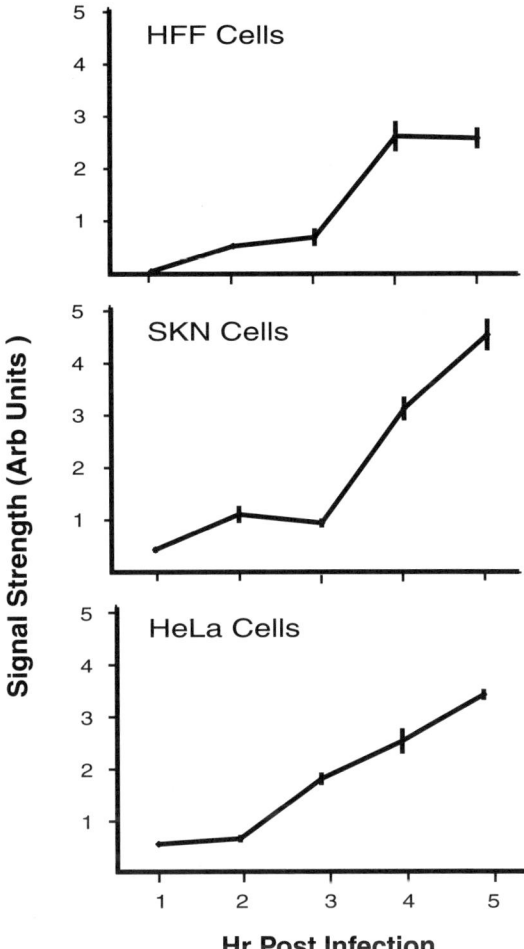

FIG. 6. Normalized hybridization signals obtained with cDNAs synthesized from RNAs isolated at various times following infection of FF, SKN, and HeLa cells. All data are normalized to the 75th percentile intensity for a representative hybridization. The normalization protocols are described in the text and provide data equivalent to those generated by scanning at the same laser power and photomultiplier settings for all hybridizations.

specific set of experimental and physiological parameters, further comparison between cell types requires detailed analysis of individual transcripts.

To directly compare HSV-specific transcript abundance, we feel that expression of each transcript or partially coterminal transcript family group as a proportion of the total viral signal is a reliable and defensible approach. This method is extensively explored in the next section to analyze the time-dependent variation in viral transcript levels following infection. The measured relative proportion of a given transcript to the total value of viral transcripts is independent of chip-to-chip variation, as is any ratio, but can be directly related to total signal strength. Of course a major advantage to this approach is the ability to directly compare the role of cell and/or organ type on the relative proportion of viral transcripts under varying conditions of infection.

Any such normalization procedure is problematical when applied to cellular transcript abundance because the number of transcripts assayed is only a very small proportion of the total number of cellular transcripts, and compared to viral signals, most cellular signals are low. For this reason, cellular transcript abundance is best expressed in terms of numerical signal strength. Despite variability between experiments, the general trend of changes of cellular transcript accumulation is relatively consistent for the cell types we have assayed. Some examples are briefly discussed in the next section.

V. Exemplary Applications

A. Genotype Profiling of HCMV

Naturally occurring mutations in virus-infected patients have important implications for not only therapy but also the outcome of clinical drug studies. Thus, viral genotyping methods for detecting the prevalence of either drug-resistant mutations/polymorphisms or virulent strains/isolates from patients are of high priority. For this reason, one of the first applications of viral DNA microarrays was to rapidly identify sequence variation (genotype) among different virus strains/isolates. In the last few years, a wide variety of different microarray technology platforms have been applied to viral genotyping. These include DNA microarrays for human immunodeficiency virus, hepatitis C virus, poliovirus, and influenza viruses (65–69). Overall, these studies clearly show that viral DNA microarray technology provides a useful supplement, and in the future will provide alternatives to PCR-based diagnostic methods. Moreover, a global diagnostic microarray approach is likely to disclose many key clinical features that may well help tailor drug regimes. For example, in DNA microarray sequence studies of HIV, it was found that sequence changes known to contribute to drug resistance occurred as natural polymorphisms in isolates from some patients naive to inhibitors (69). In such cases, an alternative drug combination may well be recommended as well as the incorporation of resistance testing as a standard of care for treatment. Long-oligonucleotide microarrays can also be efficiently used to examine a more marked sequence variation of a viral genome. In the

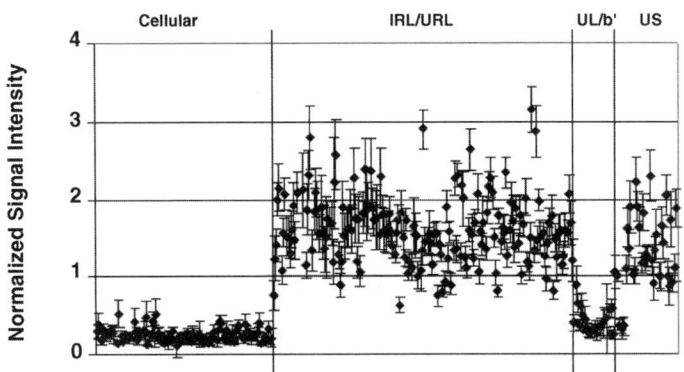

FIG. 7. Normalized signal intensities of HCMV microarrays hybridized to HCMV DNA. A total of 21 independent hybridizations were performed, using AD169 genomic DNA as target. The negative probes, including all human cellular genes and the ORFs belonging to the UL/b' region, show values lower than 0.5°, whereas the rest of the HCMV genome shows signal intensities higher than 0.8.

following section, an example is provided using a HCMV microarray to examine the genetic content of clinical strains.

Human cytomegalovirus (HCMV) is an ancient virus and, like herpes simplex virus (HSV), is closely linked to its natural host. There are many genetically unique strains (probably thousands) distributed among the human population (70). There are a number of attenuated viral strains, commonly used in laboratory studies, which were developed initially as vaccine candidates in clinical trials. It has been hypothesized that part of the genetic information encoded by the viral genome has been lost during long-term passage in cell culture, and the region more susceptible to recombination was mapped to the UL/b' region of HCMV (71). We have used a DNA microarray to genetically map this region in representative laboratory and clinical isolates. A microarray has been fabricated comprising all the ORFs in the HCMV genome and over 100 oligonucleotides for human genes. Microarrays were hybridized to fluorescent-labeled virus genomic DNA, using AD169 (ATCC) DNA. Figure 7 shows the average signal intensities of 21 independent experiments developed with DNA of the HCMV strain AD169. The probes corresponding to the IRL, UL, and US regions show positive signals, and those corresponding to the UL/b' region, known to be absent from the genome of this strain, show low values equivalent to those of the human cellular genes, included in the HCMV array as negative controls. These and other results indicate that the HCMV microarray is a valuable tool for genotyping studies of both clinical and laboratory strains of HCMV. In recent studies, we have determined that all clinical strains examined thus far have most of the

UL/b′ region, which confirms the presence of essential genes for *in vivo* replication (data not shown). Moreover, the analysis of the rest of HCMV genomes revealed other regions that are prone to mutating, expanding the differences between clinical and laboratory strains.

B. Profiling Cell-Specific Temporal Viral Gene Expression

1. CHANGES IN HSV-1 TRANSCRIPT ABUNDANCE AS A FUNCTION OF TIME FOLLOWING INFECTION OF CULTURED HUMAN CELLS

To establish the nominal course of changes in HSV-1 transcript abundance following infection of human foreskin fibroblasts, a number of independent experiments were carried out in which individual culture dishes were seeded at a density of 10^6 cells, allowed to grow to confluence (a density of 7×10^6 cells per culture), and then maintained at confluence for 16–24 h. These were then infected at a multiplicity of infection of 5 PFU per cell, and poly(A) RNA was isolated at various times following a 30-min virus-adsorption period. Fluorescent-labeled cDNA was generated from this RNA using random hexamer primers in a reaction with 1-μg aliquots of poly(A)-containing RNA and then hybridized to HSV-1 DNA microarrays.

Following hybridization, the chips were scanned as described in the previous section, and total net HSV-1-specific fluorescence signals from each chip were used to calculate the normalized relative abundance of specific viral transcripts. These data are summarized in Table IV, and representative resolvable unique transcripts grouped into representative kinetic classes are shown in Fig. 8. The four immediate-early transcripts all achieve a high relative abundance within the first 2 h following infection, but in this experiment it is evident that the ICP4 transcript is unique in achieving its highest relative level of abundance at the earliest time assayed (panel A). The relative abundance of all transcripts declines markedly as infection proceeds, but all transcripts attain nominally steady-state levels by 4 h. This pattern of transcript accumulation is consistent with the known kinetics of this group, where expressions of all are repressed by the binding of ICP4 protein to specific sites on the promoter of each.

We can identify a large group of transcripts characterized by high relative levels at early or intermediate times following infection followed by a marked decline in abundance thereafter. Representative examples are shown graphically in Fig. 8B. Generally, these correspond to the early kinetic class as determined by studies on individual transcripts. Despite shutoff, the individual details of transcript accumulation differ markedly between members of this early group, which includes the U_L23 (thymidine kinase), U_L39 (large subunit ribonucleotide reductase), and U_L50 (dUTPase) transcripts as perhaps the best-characterized representatives. The time of maximum abundance attained by individual transcripts

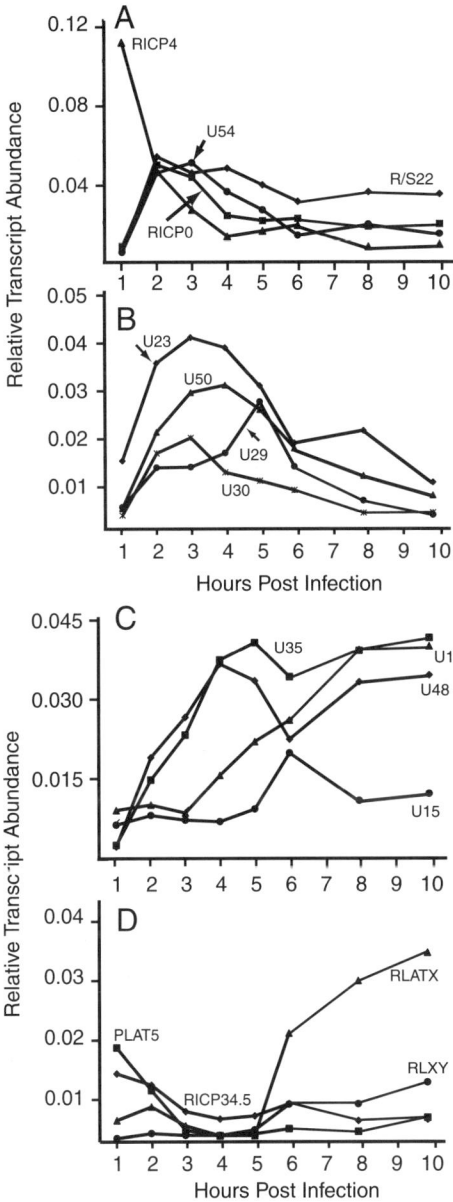

FIG. 8. Relative abundance of different kinetic classes of HSV-1 transcripts isolated at different times following infection of HFF cells with 5 PFU/cell HSV-1. The data are taken from Table V.

TABLE IV
TIME-DEPENDENT CHANGES IN RELATIVE ABUNDANCE OF HSV-1 TRANSCRIPTS IN INFECTED HUMAN FORESKIN FIBROBLASTS

Transcript(s)	1 h $(12,500 \pm 2000)^a$		2 h $(133,000 \pm 11,500)^b$		3 h $(174,000 \pm 42,000)^a$		4 h $(650,000 \pm 71,000)^b$		Var (p) 2 vs 4
	M	SD	M	SD	M	SD	M	SD	
U1	0.0056	0.0015	0.0085	0.0035	0.0086	0.0015	0.0154	0.0047	0.0027
U1X	0.0010	0.0013	0.0018	0.0013	0.0011	0.0012	0.0008	0.0004	0.0503
U3	0.0034	0.0016	0.0020	0.0005	0.0034	0.0015	0.0071	0.0054	0.0093
U4-5'	0.0035	0.0030	0.0082	0.0019	0.0117	0.0030	0.0065	0.0027	0.1361
U4/5	0.0162	0.0026	0.0148	0.0042	0.0206	0.0067	0.0148	0.0051	0.9950
U6/7	0.0162	0.0060	0.0130	0.0047	0.0108	0.0025	0.0127	0.0070	0.9279
U8/9	0.0045	0.0020	0.0079	0.0019	0.0108	0.0051	0.0089	0.0042	0.5174
U8-5'	0.0225	0.0083	0.0146	0.0059	0.0084	0.0018	0.0064	0.0049	0.0060
U10	0.0126	0.0051	0.0094	0.0069	0.0057	0.0017	0.0101	0.0050	0.8318
U11/13	0.0068	0.0010	0.0218	0.0109	0.0458	0.0146	0.0330	0.0085	0.0296
U16/17	0.0070	0.0021	0.0097	0.0019	0.0096	0.0015	0.0159	0.0044	0.0009
U15	0.0059	0.0030	0.0050	0.0026	0.0034	0.0002	0.0045	0.0019	0.6285
U18/20	0.0042	0.0017	0.0124	0.0071	0.0157	0.0026	0.0228	0.0068	0.0062
U19/20	0.0037	0.0012	0.0108	0.0053	0.0206	0.0069	0.0266	0.0063	0.0000
U19-5'	0.0172	0.0017	0.0163	0.0046	0.0142	0.0064	0.0103	0.0079	0.0629
U21	0.0033	0.0025	0.0029	0.0008	0.0021	0.0009	0.0035	0.0020	0.3604
U22	0.0021	0.0012	0.0054	0.0014	0.0063	0.0015	0.0137	0.0046	0.0001
U23	0.0179	0.0053	0.0327	0.0105	0.0407	0.0072	0.0377	0.0090	0.3013
U24	0.0074	0.0057	0.0099	0.0045	0.0096	0.0033	0.0102	0.0036	0.8953
U25/26.5	0.0062	0.0044	0.0105	0.0052	0.0180	0.0105	0.0247	0.0036	0.0000
U27/8	0.0065	0.0007	0.0170	0.0085	0.0257	0.0012	0.0233	0.0049	0.0842
U27-5'	0.0071	0.0014	0.0190	0.0037	0.0236	0.0098	0.0351	0.0090	0.0001
U29	0.0045	0.0024	0.0120	0.0062	0.0120	0.0009	0.0134	0.0061	0.6262
U30	0.0060	0.0007	0.0160	0.0048	0.0221	0.0019	0.0173	0.0071	0.6462
U31/34	0.0055	0.0009	0.0077	0.0023	0.0090	0.0012	0.0112	0.0029	0.0116
U35	0.0042	0.0009	0.0080	0.0015	0.0121	0.0010	0.0237	0.0031	0.0000
U36	0.0040	0.0021	0.0052	0.0017	0.0054	0.0005	0.0101	0.0034	0.0010
U37	0.0038	0.0012	0.0074	0.0023	0.0120	0.0054	0.0105	0.0040	0.0601
U38	0.0171	0.0098	0.0153	0.0082	0.0092	0.0011	0.0125	0.0061	0.4318
U39-5'	0.0102	0.0051	0.0345	0.0159	0.0344	0.0085	0.0257	0.0066	0.1664
U39/40	0.0043	0.0013	0.0241	0.0172	0.0461	0.0133	0.0304	0.0083	0.3596
U41	0.0022	0.0015	0.0012	0.0004	0.0012	0.0001	0.0031	0.0028	0.0690
U42	0.0128	0.0007	0.0129	0.0033	0.0145	0.0018	0.0336	0.0053	0.0000
U43	0.0594	0.0132	0.0217	0.0140	0.0103	0.0030	0.0091	0.0067	0.0341
U43.5-5'	0.0047	0.0052	0.0033	0.0020	0.0018	0.0012	0.0011	0.0014	0.0217
U44-5'	0.0107	0.0019	0.0063	0.0028	0.0054	0.0024	0.0137	0.0042	0.0004
U44/45	0.0154	0.0043	0.0160	0.0056	0.0131	0.0008	0.0250	0.0034	0.0010
U46	0.0603	0.0206	0.0474	0.0192	0.0345	0.0087	0.0310	0.0068	0.0353
U48	0.0030	0.0019	0.0107	0.0059	0.0135	0.0042	0.0230	0.0057	0.0004
U49	0.0128	0.0062	0.0207	0.0063	0.0271	0.0040	0.0279	0.0053	0.0201
U50	0.0078	0.0014	0.0178	0.0083	0.0280	0.0040	0.0311	0.0071	0.0025
U51	0.0165	0.0096	0.0141	0.0042	0.0107	0.0018	0.0156	0.0103	0.6924
U52-5'	0.0043	0.0020	0.0039	0.0013	0.0029	0.0004	0.0020	0.0009	0.0027
U52/53	0.0074	0.0035	0.0118	0.0036	0.0099	0.0007	0.0140	0.0063	0.3608
U54	0.0196	0.0157	0.0451	0.0118	0.0516	0.0069	0.0348	0.0096	0.0625
U55	0.0034	0.0037	0.0022	0.0007	0.0016	0.0004	0.0042	0.0043	0.1753
U56	0.0094	0.0073	0.0069	0.0027	0.0069	0.0019	0.0090	0.0058	0.3312

TABLE IV (Continued)

5 h $(640{,}000 \pm 50{,}000)^a$		6 h $(1{,}275{,}000 \pm 115{,}000)^b$		8 h $(1{,}580{,}000 \pm 129{,}000)^a$		10 h $(1{,}960{,}000 \pm 80{,}000)^b$		Var (p)
M	SD	M	SD	M	SD	M	SD	4 vs 10
0.0131	0.0112	0.0172	0.0079	0.0260	0.0053	0.0288	0.0051	0.0011
0.0002	0.0001	0.0010	0.0002	0.0005	0.0002	0.0006	0.0003	0.4275
0.0043	0.0024	0.0142	0.0085	0.0219	0.0024	0.0229	0.0016	0.0002
0.0023	0.0007	0.0077	0.0006	0.0080	0.0045	0.0067	0.0036	0.0178
0.0117	0.0062	0.0152	0.0008	0.0097	0.0039	0.0070	0.0023	0.7779
0.0069	0.0003	0.0185	0.0065	0.0141	0.0028	0.0133	0.0020	0.8800
0.0036	0.0026	0.0157	0.0055	0.0127	0.0047	0.0069	0.0024	0.4130
0.0023	0.0004	0.0067	0.0040	0.0061	0.0025	0.0044	0.0007	0.4536
0.0069	0.0038	0.0182	0.0028	0.0259	0.0057	0.0273	0.0057	0.0003
0.0278	0.0012	0.0279	0.0100	0.0235	0.0021	0.0229	0.0030	0.0474
0.0151	0.0042	0.0185	0.0025	0.0186	0.0023	0.0206	0.0019	0.0710
0.0021	0.0008	0.0100	0.0064	0.0085	0.0029	0.0083	0.0022	0.0103
0.0343	0.0027	0.0202	0.0042	0.0215	0.0012	0.0221	0.0014	0.8435
0.0347	0.0115	0.0240	0.0059	0.0177	0.0016	0.0207	0.0016	0.0987
0.0083	0.0072	0.0111	0.0105	0.0118	0.0011	0.0119	0.0029	0.7122
0.0056	0.0041	0.0093	0.0077	0.0118	0.0036	0.0097	0.0056	0.0163
0.0168	0.0046	0.0131	0.0020	0.0219	0.0023	0.0215	0.0008	0.0080
0.0259	0.0107	0.0246	0.0076	0.0215	0.0012	0.0147	0.0053	0.0009
0.0154	0.0022	0.0099	0.0028	0.0126	0.0042	0.0131	0.0052	0.2747
0.0364	0.0073	0.0199	0.0043	0.0198	0.0042	0.0195	0.0028	0.0309
0.0338	0.0057	0.0170	0.0059	0.0148	0.0015	0.0174	0.0019	0.0451
0.0269	0.0116	0.0259	0.0123	0.0283	0.0082	0.0288	0.0082	0.2672
0.0267	0.0157	0.0089	0.0048	0.0112	0.0065	0.0081	0.0062	0.1847
0.0161	0.0061	0.0109	0.0038	0.0063	0.0041	0.0051	0.0021	0.0078
0.0095	0.0009	0.0148	0.0042	0.0167	0.0051	0.0176	0.0007	0.0019
0.0238	0.0114	0.0258	0.0030	0.0252	0.0049	0.0274	0.0055	0.1590
0.0069	0.0049	0.0139	0.0085	0.0187	0.0045	0.0215	0.0041	0.0004
0.0054	0.0028	0.0151	0.0065	0.0173	0.0022	0.0171	0.0006	0.0100
0.0064	0.0019	0.0233	0.0001	0.0257	0.0032	0.0285	0.0048	0.0010
0.0258	0.0110	0.0133	0.0034	0.0171	0.0024	0.0151	0.0053	0.0203
0.0326	0.0060	0.0209	0.0052	0.0145	0.0018	0.0161	0.0018	0.0075
0.0017	0.0008	0.0112	0.0105	0.0122	0.0018	0.0119	0.0012	0.0002
0.0299	0.0136	0.0311	0.0041	0.0315	0.0094	0.0324	0.0074	0.7412
0.0026	0.0012	0.0175	0.0111	0.0157	0.0034	0.0151	0.0015	0.1187
0.0002	0.0002	0.0020	0.0018	0.0005	0.0001	0.0006	0.0002	0.4574
0.0244	0.0144	0.0160	0.0041	0.0169	0.0016	0.0180	0.0014	0.0818
0.0289	0.0117	0.0293	0.0038	0.0292	0.0057	0.0284	0.0063	0.2440
0.0239	0.0025	0.0274	0.0054	0.0250	0.0055	0.0302	0.0074	0.8493
0.0271	0.0008	0.0225	0.0062	0.0255	0.0048	0.0252	0.0019	0.4859
0.0347	0.0060	0.0218	0.0082	0.0175	0.0007	0.0197	0.0010	0.0126
0.0339	0.0073	0.0204	0.0036	0.0139	0.0026	0.0122	0.0060	0.0011
0.0184	0.0056	0.0159	0.0013	0.0205	0.0019	0.0185	0.0021	0.5882
0.0017	0.0005	0.0015	0.0010	0.0012	0.0007	0.0009	0.0003	0.0436
0.0092	0.0039	0.0146	0.0045	0.0158	0.0046	0.0112	0.0075	0.5130
0.0260	0.0058	0.0192	0.0068	0.0209	0.0032	0.0211	0.0029	0.0209
0.0046	0.0023	0.0064	0.0058	0.0039	0.0044	0.0020	0.0017	0.3766
0.0050	0.0007	0.0110	0.0065	0.0068	0.0058	0.0075	0.0053	0.6809

(continued)

TABLE IV (Continued)

Transcript(s)	1 h $(12{,}500 \pm 2000)^a$		2 h $(133{,}000 \pm 11{,}500)^b$		3 h $(174{,}000 \pm 42{,}000)^a$		4 h $(650{,}000 \pm 71{,}000)^b$		Var (p) 2 vs 4
	M	SD	M	SD	M	SD	M	SD	
RLXY	0.0049	0.0024	0.0019	0.0009	0.0018	0.0009	0.0017	0.0007	0.7274
RLX	0.0347	0.0068	0.0232	0.0160	0.0132	0.0047	0.0093	0.0121	0.0592
RLAT-5	0.0159	0.0024	0.0077	0.0043	0.0025	0.0010	0.0017	0.0014	0.0016
RHA6	0.0013	0.0009	0.0019	0.0014	0.0010	0.0007	0.0011	0.0011	0.2209
RLAT-I	0.0105	0.0064	0.0081	0.0057	0.0029	0.0011	0.0018	0.0021	0.0094
RICP0	0.0626	0.0589	0.0485	0.0119	0.0347	0.0105	0.0184	0.0031	0.0000
ROP	0.0171	0.0083	0.0088	0.0053	0.0039	0.0004	0.0018	0.0016	0.0025
RICP34.5	0.0167	0.0021	0.0 127	0.0080	0.0056	0.0009	0.0043	0.0018	0.0114
RLATX	0.0078	0.0017	0.0091	0.0058	0.0034	0.0004	0.0038	0.0041	0.0439
RLAT-3'	0.0180	0.0056	0.0136	0.0045	0.0070	0.0006	0.0060	0.0066	0.0099
RICP4	0.1675	0.0182	0.0616	0.0226	0.0388	0.0041	0.0158	0.0026	0.0000
R/S22	0.0389	0.0344	0.0569	0.0187	0.0510	0.0008	0.0432	0.0111	0.0871
US2	0.0115	0.0055	0.0067	0.0040	0.0057	0.0018	0.0103	0.0061	0.1434
US3-5'	0.0245	0.0126	0.0154	0.0074	0.0090	0.0013	0.0064	0.0031	0.0056
US3/4	0.0056	0.0045	0.0056	0.0016	0.0086	0.0044	0.0103	0.0048	0.0096
US5-5'	0.0224	0.0136	0.0132	0.0079	0.0069	0.0008	0.0046	0.0025	0.0100
US5/6/7	0.0086	0.0072	0.0148	0.0073	0.0163	0.0021	0.0187	0.0060	0.2397
US8-5'	0.0175	0.0137	0.0131	0.0033	0.0139	0.0041	0.0078	0.0038	0.0053
US8/9	0.0132	0.0099	0.0247	0.0158	0.0277	0.0142	0.0246	0.0086	0.9875
US10/11/12	0.0229	0.0134	0.0272	0.0106	0.0357	0.0051	0.0316	0.0084	0.3579

[a] Four replicates   [b] Three replicates

varies widely in this group. Surprisingly, in HFF cells, the U_L43, U_L52, U_S3, U_S5, and U_S8 transcripts all attain their maximum levels before U_L39 does, despite the fact that this latter transcript is expressed at extremely early times and even in the presence of high levels of cycloheximide in some cells (see Ref. 51 for pertinent references). The U_L23 and U_L30 transcripts attain maximum levels of abundance measurably earlier than do the U_L27 and U_L50 transcripts, while the U_L29 (DNA polymerase) transcript attains its maximal level measurably later. The U_L15 and U_L21 transcripts, which have recently been classified by a number of laboratories including ours (see Refs. 3 and 57), fit into the present grouping of early transcripts because of the clearly evident decline in abundance at the latest times assayed. The abundance of the U_L24 transcript remained nearly constant from 4 through 8 h following infection and only declined at the latest time. Finally, expression of the U_L43 transcript shows a complex pattern with a maximum at the earliest times measured but with a second peak of abundance at 6 h following infection.

Transcripts that attain a high level of abundance without subsequent marked decline tend to include a number of those classified as late based on transcriptional analysis using other methods. Some examples are shown in Fig. 8C. Four transcripts, U_L22, U_L42, U_L35, and U_L48 attain relatively high levels by 4 h after infection, and the high level is maintained throughout the time period assayed.

TABLE IV (Continued)

5 h (640,000 ± 50,000)[a]		6 h (1,275,000 ± 115,000)[b]		8 h (1,580,000 ± 129,000)[a]		10 h (1,960,000 ± 80,000)[b]		Var (p)
M	SD	M	SD	M	SD	M	SD	4 vs 10
0.0012	0.0003	0.0067	0.0063	0.0111	0.0015	0.0169	0.0036	0.0000
0.0039	0.0032	0.0169	0.0066	0.0084	0.0020	0.0064	0.0013	0.6543
0.0012	0.0005	0.0027	0.0009	0.0022	0.0012	0.0037	0.0014	0.0419
0.0005	0.0000	0.0024	0.0009	0.0029	0.0010	0.0046	0.0015	0.0010
0.0010	0.0005	0.0025	0.0014	0.0016	0.0004	0.0036	0.0015	0.1550
0.0178	0.0016	0.0141	0.0015	0.0110	0.0025	0.0130	0.0017	0.0094
0.0010	0.0003	0.0025	0.0014	0.0016	0.0006	0.0029	0.0007	0.2304
0.0037	0.0015	0.0064	0.0004	0.0040	0.0016	0.0038	0.0006	0.5557
0.0010	0.0003	0.0128	0.0064	0.0216	0.0023	0.0251	0.0052	0.0000
0.0050	0.0003	0.0132	0.0021	0.0120	0.0026	0.0128	0.0020	0.0731
0.0161	0.0021	0.0152	0.0038	0.0108	0.0039	0.0108	0.0033	0.0152
0.0450	0.0052	0.0357	0.0076	0.0302	0.0088	0.0296	0.0056	0.0457
0.0102	0.0055	0.0147	0.0031	0.0109	0.0055	0.0097	0.0046	0.8554
0.0042	0.0018	0.0060	0.0042	0.0033	0.0013	0.0025	0.0009	0.0371
0.0055	0.0041	0.0183	0.0115	0.0206	0.0012	0.0186	0.0031	0.0114
0.0026	0.0008	0.0072	0.0006	0.0046	0.0031	0.0048	0.0010	0.9312
0.0399	0.0051	0.0152	0.0037	0.0237	0.0023	0.0217	0.0018	0.3490
0.0068	0.0043	0.0091	0.0049	0.0141	0.0010	0.0124	0.0013	0.0398
0.0351	0.0037	0.0166	0.0022	0.0209	0.0032	0.0184	0.0006	0.1883
0.0384	0.0095	0.0215	0.0100	0.0179	0.0008	0.0182	0.0020	0.0121

[a]Four replicates [b]Three replicates

This pattern is consistent with the leaky-late ($\beta\gamma$) or intermediate kinetic classification. The $U_L18/20$ complex also fits this pattern. Interestingly, the relative abundance of all these transcripts appears to dip at around 6 h post-infection—a time of maximal rate of viral DNA replication. This dip is not seen in the relative levels of expression of five strict-late transcripts: U_L10, U_L35, U_L38, U_L41, and U_L44. It was also noted that the relative level of U_L38 is significantly higher at the earliest times assayed than that at the 3- and 4-h time points, but the low total signal at these times would amplify nonspecific background effects.

The relative proportion of the U_L37 attains high levels at two discrete times: 3 and 6–8 h following infection. This type of pattern of transcript accumulation cannot be readily placed into any simple kinetic classification. As has been shown elsewhere, the specific variation of abundance with time for this transcript is influenced by the nature of the host cell being infected (51).

Besides encoding the three immediate early transcripts, ICP4, ICP0, and U_S1 (ICP22), the repeat regions of the HSV-1 genome encode a number of low-abundance transcripts. Although the functions of a number of these are unknown, clearly several are involved with aspects of HSV latency. For example, the expression of the 5′ portion of the HSV LAT transcript facilitates reactivation, and the ICP34.5 transcript blocks the interferon response through inhibition of interferon-induced eIF2 phosphorylation. The relative abundance of both LAT

TABLE V
TIME-DEPENDENT CHANGES IN RELATIVE ABUNDANCE OF HSV-1 TRANSCRIPTS IN INFECTED SKN CELLS[a]

Transcript(s)	1 h (269,504 ± 24,112)		2 h (581,285 ± 75,433)		3 h (505,667 ± 40,897)		4 h (1,507,828 ± 104,530)		5 h (2,158,324 ± 139,130)	
	Md	SD	Md	SD	Md	SD	Md	SD	Md	SD
U1	0.0039	0.0014	0.0082	0.0013	0.0182	0.0053	0.0079	0.0043	0.0101	0.0051
U1X	0.0022	0.0012	0.0007	0.0009	0.0004	0.0005	0.0003	0.0005	0.0004	0.0003
U3	0.0025	0.0018	0.0039	0.0019	0.0046	0.0016	0.0057	0.0018	0.0070	0.0016
U4-5'	0.0045	0.0021	0.0078	0.0029	0.0060	0.0019	0.0075	0.0015	0.0058	0.0023
U4/5	0.0273	0.0042	0.0293	0.0015	0.0249	0.0024	0.0270	0.0019	0.0252	0.0059
U6/7	0.0186	0.0029	0.0168	0.0065	0.0130	0.0017	0.0126	0.0011	0.0114	0.0013
U8/9	0.0207	0.0058	0.0120	0.0020	0.0110	0.0024	0.0097	0.0022	0.0084	0.0013
U8-5'	0.0197	0.0030	0.0116	0.0077	0.0104	0.0046	0.0074	0.0016	0.0052	0.0014
U10	0.0226	0.0064	0.0122	0.0016	0.0107	0.0023	0.0085	0.0017	0.0127	0.0031
U11/13	0.0012	0.0005	0.0129	0.0057	0.0216	0.0064	0.0243	0.0055	0.0250	0.0047
U16/17	0.0069	0.0013	0.0067	0.0013	0.0105	0.0018	0.0112	0.0013	0.0149	0.0038
U15	0.0073	0.0025	0.0043	0.0008	0.0034	0.0008	0.0018	0.0007	0.0022	0.0009
U18/20	0.0010	0.0020	0.0080	0.0027	0.0124	0.0021	0.0192	0.0023	0.0222	0.0060
U19/20	0.0044	0.0011	0.0112	0.0018	0.0106	0.0018	0.0158	0.0059	0.0231	0.0051
U19-5'	0.0133	0.0053	0.0091	0.0045	0.0095	0.0018	0.0075	0.0014	0.0078	0.0007
U21	0.0015	0.0020	0.0019	0.0016	0.0027	0.0014	0.0043	0.0010	0.0030	0.0037
U22	0.0013	0.0010	0.0057	0.0018	0.0128	0.0029	0.0136	0.0031	0.0140	0.0074
U23	0.0169	0.0073	0.0218	0.0042	0.0369	0.0035	0.0231	0.0070	0.0217	0.0070
U24	0.0069	0.0016	0.0089	0.0019	0.0092	0.0017	0.0165	0.0070	0.0165	0.0053
U25	0.0033	0.0016	0.0061	0.0017	0.0162	0.0037	0.0312	0.0046	0.0304	0.0109
U27/8	0.0116	0.0006	0.0199	0.0028	0.0264	0.0034	0.0316	0.0035	0.0368	0.0063
U27-5'	0.0073	0.0017	0.0162	0.0036	0.0256	0.0079	0.0226	0.0092	0.0267	0.0115
U29	0.0010	0.0014	0.0079	0.0018	0.0145	0.0028	0.0229	0.0101	0.0151	0.0081
U30	0.0071	0.0012	0.0123	0.0005	0.0202	0.0034	0.0186	0.0047	0.0183	0.0039

U31/34	0.0084	0.0028	0.0084	0.0011	0.0069	0.0005	0.0089	0.0027	0.0092	0.0036
U35	0.0068	0.0030	0.0067	0.0030	0.0139	0.0050	0.0168	0.0063	0.0182	0.0052
U36	0.0020	0.0013	0.0011	0.0005	0.0014	0.0005	0.0009	0.0005	0.0017	0.0008
U37	0.0031	0.0019	0.0037	0.0012	0.0032	0.0010	0.0048	0.0018	0.0045	0.0009
U38	0.0179	0.0025	0.0107	0.0018	0.0068	0.0013	0.0070	0.0018	0.0081	0.0020
U39-5'	0.0043	0.0021	0.0231	0.0061	0.0226	0.0041	0.0267	0.0077	0.0247	0.0051
U39/40	0.0047	0.0014	0.0262	0.0026	0.0390	0.0094	0.0398	0.0062	0.0426	0.0108
U41	0.0004	0.0004	0.0005	0.0005	0.0014	0.0007	0.0021	0.0012	0.0040	0.0021
U42	0.0089	0.0027	0.0097	0.0015	0.0222	0.0088	0.0201	0.0101	0.0285	0.0125
U43	0.0210	0.0057	0.0100	0.0035	0.0078	0.0023	0.0068	0.0016	0.0066	0.0018
U43.5-5'	0.0028	0.0013	0.0019	0.0012	0.0012	0.0008	0.0010	0.0007	0.0009	0.0007
U44-5'	0.0104	0.0023	0.0065	0.0045	0.0055	0.0014	0.0065	0.0024	0.0199	0.0064
U44/45	0.0335	0.0065	0.0180	0.0069	0.0201	0.0020	0.0204	0.0034	0.0239	0.0050
U46	0.1006	0.0257	0.0510	0.0152	0.0318	0.0082	0.0251	0.0071	0.0274	0.0054
U48	0.0022	0.0018	0.0153	0.0062	0.0201	0.0063	0.0256	0.0040	0.0281	0.0045
U49	0.0140	0.0060	0.0225	0.0042	0.0289	0.0040	0.0347	0.0039	0.0312	0.0027
U50	0.0093	0.0022	0.0185	0.0043	0.0377	0.0081	0.0402	0.0107	0.0325	0.0073
U51	0.0243	0.0054	0.0171	0.0028	0.0143	0.0049	0.0148	0.0078	0.0191	0.0046
U52-5'	0.0039	0.0005	0.0022	0.0013	0.0024	0.0007	0.0009	0.0003	0.0008	0.0002
U52/53	0.0052	0.0005	0.0105	0.0020	0.0134	0.0017	0.0138	0.0021	0.0110	0.0070
U54	0.0295	0.0185	0.0645	0.0191	0.0513	0.0118	0.0311	0.0073	0.0267	0.0123
U55	0.0023	0.0018	0.0051	0.0005	0.0082	0.0011	0.0104	0.0056	0.0092	0.0061
U56	0.0132	0.0023	0.0113	0.0045	0.0117	0.0041	0.0150	0.0065	0.0116	0.0046
RLXY	0.0024	0.0019	0.0002	0.0004	0.0010	0.0004	0.0006	0.0006	0.0019	0.0011
RLX	0.0504	0.0181	0.0250	0.0056	0.0112	0.0088	0.0183	0.0124	0.0145	0.0108

(continued)

TABLE V (Continued)

Transcript(s)	1 h (269,504 ± 24,112)		2 h (581,285 ± 75,433)		3 h (505,667 ± 40,897)		4 h (1,507,828 ± 104,530)		5 h (2,158,324 ± 139,130)	
	Md	SD	Md	SD	Md	SD	Md	SD	Md	SD
RLAT-5	0.0132	0.0023	0.0069	0.0021	0.0020	0.0019	0.0018	0.0018	0.0019	0.0015
RHA6	0.0087	0.0049	0.0045	0.0019	0.0021	0.0010	0.0016	0.0011	0.0011	0.0017
RLAT-I	0.0102	0.0026	0.0072	0.0028	0.0034	0.0019	0.0015	0.0021	0.0011	0.0011
RICP0	0.0284	0.0289	0.0303	0.0047	0.0145	0.0013	0.0147	0.0026	0.0160	0.0044
ROP	0.0135	0.0031	0.0085	0.0023	0.0044	0.0017	0.0034	0.0033	0.0020	0.0016
RICP34.5	0.0141	0.0042	0.0093	0.0035	0.0078	0.0021	0.0039	0.0007	0.0028	0.0006
RLATX	0.0192	0.0019	0.0072	0.0019	0.0043	0.0024	0.0032	0.0020	0.0036	0.0016
RLAT-3'	0.0282	0.0123	0.0178	0.0082	0.0093	0.0057	0.0072	0.0095	0.0102	0.0047
RICP4	0.1146	0.0248	0.0805	0.0364	0.0324	0.0168	0.0226	0.0181	0.0153	0.0110
R/S22	0.0370	0.0196	0.0613	0.0124	0.0496	0.0121	0.0356	0.0152	0.0323	0.0144
US2	0.0053	0.0011	0.0064	0.0027	0.0051	0.0023	0.0096	0.0027	0.0146	0.0050
US3-5'	0.0278	0.0054	0.0224	0.0068	0.0141	0.0026	0.0105	0.0011	0.0085	0.0032
US3/4	0.0075	0.0012	0.0083	0.0018	0.0103	0.0023	0.0119	0.0023	0.0106	0.0029
US5-5'	0.0243	0.0110	0.0147	0.0051	0.0076	0.0036	0.0082	0.0030	0.0067	0.0044
US5/6/7	0.0043	0.0029	0.0149	0.0035	0.0182	0.0016	0.0232	0.0066	0.0247	0.0089
US8-5'	0.0206	0.0041	0.0180	0.0046	0.0145	0.0032	0.0166	0.0049	0.0155	0.0044
US8/9	0.0114	0.0036	0.0255	0.0124	0.0406	0.0085	0.0395	0.0129	0.0313	0.0127
US10/11/12	0.0054	0.0068	0.0282	0.0047	0.0295	0.0058	0.0384	0.0142	0.0342	0.0081

[a] Three replicates

and ICP34.5 transcripts is highest at the earliest time points. While the ICP34.5 transcript abundance maintains a low but constant level throughout the periods assayed, LAT abundance increases measurably at the latest times (Fig. 8D).

The time-dependent changes in abundance of several other transcripts expressed as members of coterminal families can also be determined by comparing hybridization levels with unique 5' probes to those of 3' probes specific for both transcripts. Some examples are as follows. The general correspondence between the increase in abundance of the $U_L44/45$ transcripts and that of UL44 as determined by the 5' specific probe is expected because both transcripts are of the strict-late kinetic class. The abundance profile seen with the probes specific for the U_L27/U_L28 transcript group reveals a relative maximum between 3 and 5 h postinfection; this is followed by a marked decline followed by a slow increase in abundance. This is consistent with the earlier classifications of U_L28 (gB) as being expressed at high levels at early to intermediate times following infection along with a later maximal expression of the U_L27 transcript, as suggested by the hybridization profile using the specific probe. The earlier maximal abundance of the U_L52 helicase/primase transcript as compared to that of U_L53 (gK) and that of U_S3 protein kinase as compared to that of U_S4 (gE) are also evident. The differential expression of the U_S9 transcript versus that of the U_S8 (gE) transcript is also evident and is suggestive of the former accumulating with early kinetics, reaching a maximum at 4 to 5 h postinfection. The very early decline in the relative abundance of the U_S5 (gJ) transcript as compared to those of the partially colinear U_S6 (gD) and U_S7 (gI) transcripts is also apparent, although the specifics of expression of the latter two cannot be inferred.

Parallel experiments were carried out with a human neuroblastoma-derived cell line, SKN. Data for the first 5 h following infection are shown in Table V. While the pattern of total increase in transcript abundance was quite similar (see Fig. 6) and there was a good deal of general similarity in the patterns of changes of transcript abundance in this time window, there were subtle but consistent differences. The immediate-early U_L54 transcript attains its maximal levels of abundance earlier in the neuronal cells, and the decline in relative abundance of the ICP4 transcript is somewhat slower. The early U_L23 and U_L50 transcripts begin to decline in relative abundance earlier in the SKN cells, suggesting a more rapid shutoff. Finally, the relative rise in U_L44 abundance begins somewhat later in the neuron-derived cells.

We also did a series of parallel time-course experiments using HeLa cells with an earlier version of the DNA microarray described here. This earlier chip contained no unique 5' transcript probes and 84 cellular transcript probes as compared to the 144 used with the latest one. The relative HSV-1 transcript abundance measured at 2, 4, 6, 8, and 10 h following infection of cultures of 2×10^7 cells was determined as described in preceding sections and is shown in Table VI. With a few notable exceptions shown in Fig. 9, the changes of

TABLE VI
TIME-DEPENDENT CHANGES IN RELATIVE ABUNDANCE OF HSV-1 TRANSCRIPTS IN INFECTED HeLa CELLS[a]

Transcript(s)	1 h (103,850 ± 5381)		2 h (122,420 ± 13,864)		4 h (492,189 ± 47,968)		6 h (666,864 ± 20,741)		8 h (881,256 ± 36,197)		10 h (824,073 ± 29,958)	
	Md	SD	Md	SD	Md	SD	Md	SD	Md	SD	Md	SD
U1	0.0350	0.0241	0.0187	0.0066	0.0199	0.0104	0.0190	0.0063	0.0222	0.0024	0.0316	0.0004
U1X	0.0053	0.0066	0.0027	0.0000	0.0037	0.0010	0.0042	0.0011	0.0049	0.0003	0.0094	0.0011
U3	0.0109	0.0061	0.0052	0.0019	0.0120	0.0008	0.0157	0.0042	0.0158	0.0021	0.0321	0.0036
U4/5	0.0086	0.0109	0.0067	0.0027	0.0052	0.0012	0.0064	0.0018	0.0071	0.0009	0.0066	0.0024
U6/7	0.0114	0.0130	0.0107	0.0076	0.0115	0.0033	0.0134	0.0007	0.0149	0.0002	0.0170	0.0027
U8/9	0.0130	0.0127	0.0049	0.0013	0.0061	0.0021	0.0064	0.0023	0.0112	0.0007	0.0109	0.0035
U10	0.0095	0.0118	0.0056	0.0014	0.0110	0.0058	0.0184	0.0032	0.0285	0.0107	0.0370	0.0097
U11/13	0.0294	0.0336	0.0181	0.0021	0.0315	0.0087	0.0316	0.0065	0.0244	0.0017	0.0340	0.0002
U15	0.0063	0.0074	0.0043	0.0026	0.0025	0.0009	0.0047	0.0008	0.0052	0.0005	0.0040	0.0007
U16/17	0.0080	0.0084	0.0049	0.0004	0.0153	0.0025	0.0156	0.0012	0.0162	0.0020	0.0123	0.0033
U18/20	0.0177	0.0169	0.0075	0.0010	0.0374	0.0548	0.0404	0.0027	0.0333	0.0279	0.0347	0.0129
U19/20	0.0107	0.0055	0.0050	0.0006	0.0012	0.0005	0.0325	0.0008	0.0137	0.0022	0.0214	0.0106
U21	0.0078	0.0060	0.0025	0.0004	0.0073	0.0006	0.0144	0.0022	0.0118	0.0023	0.0126	0.0021
U22	0.0118	0.0120	0.0067	0.0009	0.0192	0.0023	0.0198	0.0001	0.0197	0.0097	0.0141	0.0008
U23	0.0216	0.0118	0.0264	0.0126	0.0258	0.0130	0.0090	0.0021	0.0089	0.0012	0.0040	0.0008
U24	0.0059	0.0039	0.0030	0.0003	0.0015	0.0005	0.0035	0.0013	0.0020	0.0003	0.0028	0.0007
U25/26	0.0105	0.0055	0.0068	0.0018	0.0278	0.0090	0.0324	0.0056	0.0271	0.0063	0.0308	0.0092
U27/8	0.0318	0.0182	0.0109	0.0012	0.0338	0.0333	0.0387	0.0021	0.0302	0.0101	0.0370	0.0162
U29	0.0188	0.0140	0.0058	0.0002	0.0169	0.0030	0.0137	0.0018	0.0071	0.0014	0.0047	0.0001
U30	0.0117	0.0043	0.0107	0.0011	0.0104	0.0116	0.0056	0.0025	0.0047	0.0002	0.0088	0.0011
U31/34	0.0029	0.0027	0.0041	0.0022	0.0124	0.0046	0.0138	0.0047	0.0178	0.0022	0.0222	0.0026
U35	0.0078	0.0060	0.0056	0.0030	0.0414	0.0683	0.0289	0.0045	0.0395	0.0291	0.0345	0.0051
U36	0.0035	0.0027	0.0033	0.0019	0.0200	0.0110	0.0194	0.0016	0.0339	0.0024	0.0314	0.0045
U37	0.0032	0.0151	0.0075	0.0030	0.0175	0.0151	0.0129	0.0009	0.0269	0.0032	0.0262	0.0036
U38	0.0030	0.0020	0.0029	0.0005	0.0099	0.0079	0.0100	0.0003	0.0171	0.0072	0.0138	0.0014
U39/40	0.0508	0.0404	0.0371	0.0292	0.0417	0.0709	0.0399	0.0160	0.0382	0.0294	0.0431	0.0010

U41	0.0035	0.0009	0.0020	0.0007	0.0149	0.0095	0.0095	0.0020	0.0126	0.0014	0.0139	0.0038
U42	0.0103	0.0045	0.0224	0.0065	0.0494	0.1323	0.0367	0.0112	0.0451	0.0660	0.0404	0.0007
U43	0.0066	0.0059	0.0058	0.0006	0.0100	0.0060	0.0099	0.0032	0.0133	0.0006	0.0161	0.0034
U44	0.0057	0.0060	0.0117	0.0044	0.0434	0.0811	0.0472	0.0034	0.0528	0.1223	0.0501	0.0343
U46	0.0171	0.0162	0.0066	0.0032	0.0239	0.0036	0.0281	0.0049	0.0287	0.0140	0.0278	0.0034
U48	0.0236	0.0181	0.0144	0.0089	0.0330	0.0318	0.0313	0.0079	0.0249	0.0104	0.0223	0.0045
U49	0.0310	0.0239	0.0171	0.0056	0.0237	0.0063	0.0299	0.0039	0.0193	0.0063	0.0210	0.0030
U50	0.0291	0.0296	0.0307	0.0150	0.0369	0.0428	0.0356	0.0118	0.0247	0.0092	0.0082	0.0027
U51	0.0053	0.0057	0.0048	0.0016	0.0247	0.0019	0.0254	0.0094	0.0258	0.0161	0.0212	0.0086
U52/3	0.0041	0.0043	0.0062	0.0028	0.0024	0.0012	0.0101	0.0025	0.0063	0.0033	0.0047	0.0014
U54	0.1182	0.2509	0.0599	0.1527	0.0337	0.0303	0.0278	0.0103	0.0191	0.0029	0.0073	0.0027
U55	0.0085	0.0067	0.0035	0.0028	0.0077	0.0000	0.0050	0.0014	0.0030	0.0005	0.0007	0.0006
U56	0.0081	0.0084	0.0085	0.0015	0.0172	0.0027	0.0133	0.0040	0.0107	0.0038	0.0046	0.0001
RLXY	0.0073	0.0070	0.0058	0.0019	0.0058	0.0012	0.0056	0.0010	0.0067	0.0015	0.0080	0.0004
RLX	0.0051	0.0062	0.0062	0.0010	0.0010	0.0005	0.0021	0.0009	0.0034	0.0010	0.0041	0.0030
RHA-6	0.0053	0.0046	0.0042	0.0012	0.0018	0.0003	0.0016	0.0004	0.0025	0.0002	0.0036	0.0008
RICP0	0.0766	0.0380	0.0328	0.0200	0.0198	0.0010	0.0209	0.0076	0.0277	0.0053	0.0238	0.0071
RICP34.5	0.0058	0.0070	0.0076	0.0012	0.0032	0.0011	0.0022	0.0015	0.0033	0.0003	0.0029	0.0008
RLAT-3'	0.0048	0.0038	0.0021	0.0015	0.0015	0.0006	0.0064	0.0012	0.0059	0.0017	0.0035	0.0002
RICP4	0.0674	0.0188	0.0514	0.1022	0.0288	0.0238	0.0262	0.0075	0.0213	0.0124	0.0103	0.0012
R/S22	0.0473	0.0032	0.0529	0.0841	0.0450	0.0911	0.0387	0.0043	0.0385	0.0264	0.0379	0.0093
S2	0.0035	0.0011	0.0058	0.0043	0.0069	0.0052	0.0073	0.0024	0.0066	0.0002	0.0056	0.0012
S3/4	0.0059	0.0024	0.0117	0.0047	0.0138	0.0058	0.0122	0.0011	0.0102	0.0013	0.0076	0.0007
S5/6/7	0.0156	0.0017	0.0175	0.0075	0.0306	0.0022	0.0304	0.0089	0.0326	0.0052	0.0332	0.0034
S8/9	0.0319	0.0063	0.0338	0.0061	0.0404	0.0648	0.0327	0.0069	0.0347	0.0139	0.0396	0.0052
S10/11/12	0.0954	0.0960	0.0228	0.0019	0.0376	0.0477	0.0337	0.0068	0.0381	0.0380	0.0446	0.0046

[a]Three replicates

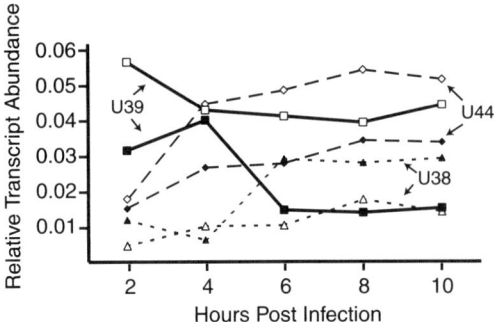

FIG. 9. Changes of relative abundance of selected HSV-1 transcripts as a function of time following infection in HeLa vs HFF cells. The open symbols are HeLa cell values.

relative transcript abundance with time were equivalent to those seen in infected HFF cells. Differences include a significantly earlier maximal level of the U_L39 transcript along with a notably less robust decline in its proportion. The U_L37 transcript attains its highest abundances later in HeLa cells than in HFFs; the U_L38 transcript does not reach as great a relative proportion in HeLa cells at late times, and the opposite is the case for the U_L44 transcript.

VI. Conclusions

Oligonucleotide-based DNA microarrays combine a number of attractive features that make them especially valuable for the analysis of genetic variation and patterns of gene expression by large DNA-containing viruses. Consistent with inferences based on extensive work with individual probes, the patterns of viral transcript abundance differ sufficiently in various cell types, suggesting that certain aspects of viral pathogenesis must take this fact into account. Combined with large cellular microarrays for the assay of cellular gene expression, global descriptions of transcript abundance in specific cells and tissues will provide an important component of a total molecular description of the course of viral infection and pathogenesis, which can be readily applied to the development of strategies for treatment and palliation of virus-induced morbidity.

ACKNOWLDGMENTS

This work was supported in part by Grants CA1186, CA90287, and the Chao Family Cancer Center to E.K.W and BBSRC to P.G. The excellent technical assistance provided by M. K. Rice, Danielle Foster, Douglas Roy, and Alan Ross is greatly appreciated.

References

1. M. Schena, D. Shalon, R. W. Davis, and P. O. Brown, Quantitative monitoring of gene expression patterns with a complementary DNA microarray. *Science* **270,** 467–470 (1995).
2. V. R. Iyer, M. B. Eisen, D. T. Ross, G. Schuler, T. Moore, J. C. F. Lee, J. M. Trent, L. M. Staudt, J. Hudson, Jr., M. S. Boguski, D. Lashkari, D. Shalon, D. Botstein, and P. O. Brown, The transcriptional program in the response of human fibroblasts to serum. *Science* **283,** 83–87 (1999).
3. E. K. Wagner, J. F. Guzowski, and J. Singh, Transcription of the Herpes Simplex Virus Genome during Productive and Latent Infection. in "Progress in Nucleic Acid Research and Molecular Biology" (W. E. Cohen and K. Moldave, eds.), Vol. 51, pp. 123–168. Academic Press, San Diego, 1995.
4. E. K. Wagner and D. C. Bloom, The experimental investigation of herpes simplex virus latency. *Clin. Microbiol. Rev.* **10,** 419–443 (1997).
5. E. K. Wagner, M. D. Petroski, N. T. Pande, P. T. Lieu, and M. K. Rice, Analysis of factors influencing the kinetics of herpes simplex virus transcript expression utilizing recombinant virus. *Methods* **16,** 105–116 (1998).
6. R. M. Sandri-Goldin, M. K. Hibbard, and M. A. Hardwicke, The C-terminal repressor region of herpes simplex virus type 1 ICP27 is required for the redistribution of small nuclear ribonucleoprotein particles and splicing factor SC35. However, these alterations are not sufficient to inhibit host cell splicing. *J. Virol.* **69,** 6063–6076 (1995).
7. T. M. Soliman, R. M. Sandri-Goldin, and S. J. Silverstein, Shuttling of the herpes simplex virus type 1 regulatory protein ICP27 between the nucleus and cytoplasm mediates the expression of late proteins. *J. Virol.* **71,** 9188–9197 (1997).
8. P. Xiao and J. P. Capone, A cellular factor binds to the herpes simplex virus type 1 transactivator Vmw65 and is required for Vmw65-dependent protein-DNA complex assembly with Oct-1. *Mol. Cell. Biol.* **10,** 4974–4977 (1990).
9. D. N. Arnosti, C. M. Preston, M. Hagmann, W. Schaffner, R. G. Hope, G. Laughlan, and B. F. Luisi, Specific transcriptional activation *in vitro* by the herpes simplex virus protein VP16. *Nucleic Acids Res.* **21,** 5570–5576 (1993).
10. A. Peyman, M. Helsberg, G. Kretzschmar, M. Mag, S. Grabley, and E. Uhlmann, Inhibition of viral growth by antisense oligonucleotides directed against the IE110 and the UL30 mRNA of herpes simplex virus type-1. *Biol. Chem. Hoppe Seyler* **376,** 195–198 (1995).
11. N. A. DeLuca and M. J. Carrozza, Interaction of the viral activator protein ICP4 with TFIID and through TAF250. *Mol. Cell. Biol.* **16,** 3085–3093 (1996).
12. J. G. Stevens, in "Overview of Herpesvirus Latency" (E. K. Wagner, ed.), Vol. 5, pp. 191–196. Academic Press, London, 1994.
13. E. K. Wagner, Herpesvirus transcription—General aspects. in "Herpesvirus Transcription and its Regulation" (E. K. Wagner, ed.), pp. 1–15. CRC Press, Boca Raton, FL, 1991.
14. J. G. Stevens, Human herpesviruses: a consideration of the latent state. *Microbiol. Rev.* **53,** 318–332 (1989).
15. R. Ahmed and J. G. Stevens, Viral persistence. in "Virology" (B. N. Fields and D. M. Knipe, eds.), pp. 241–266. Raven Press, New York, 1990.
16. E. K. Wagner, Herpesvirus latency. *Seminars Virol.* **5,** 89–90 (1994).
17. R. T. Javier, J. G. Stevens, V. B. Dissette, and E. K. Wagner, A herpes simplex virus transcript abundant in latently infected neurons is dispensable for establishment of the latent state. *Virology* **166,** 254–257 (1988).
18. J. P. Katz, E. T. Bodin, and D. M. Coen, Quantitative polymerase chain reaction analysis of herpes simplex virus DNA in ganglia of mice infected with replication-incompetent mutants. *J. Virol.* **64,** 4288–4295 (1990).

19. F. Sedarati, T. P. Margolis, and J. G. Stevens, Latent infection can be established with drastically restricted transcription and replication of the HSV-1 genome. *Virology* **192,** 687–691 (1993).
20. J. G. Stevens, E. K. Wagner, G. B. Devi-Rao, M. L. Cook, and L. T. Feldman, RNA complementary to a herpesvirus alpha gene mRNA is prominent in latently infected neurons. *Science* **235,** 1056–1059 (1987).
21. G. A. Lewandowski, D. Lo, and F. E. Bloom, Interference with major histocompatibility complex class II-restricted antigen presentation in the brain by herpes simplex virus type 1: A possible mechanism of evasion of the immune response. *Proc. Nat. Acad. Sci. U.S.A.* **90,** 2005–2009 (1993).
22. A. Simmons, D. Tscharke, and P. Speck, The role of immune mechanisms in control of herpes simplex virus infection of the peripheral nervous system. *Curr. Top. Microbiol. Immunol.* **179,** 31–56 (1992).
23. S. L. Deshmane and N. W. Fraser, During latency, herpes simplex virus type 1 DNA is associated with nucleosomes in a chromatin structure. *J. Virol.* **63,** 943–947 (1989).
24. D. L. Rock and N. W. Fraser, Detection of the HSV-1 Genome in the Central Nervous System of Latently Infected Mice. *Nature* **302,** 523–525 (1983).
25. D. L. Rock and N. W. Fraser, Latent herpes simplex virus type 1 DNA contains two copies of the virion joint region. *J. Virol.* **62,** 3820–3826 (1985).
26. G. B. Devi-Rao, D. C. Bloom, J. G. Stevens, and E. K. Wagner, Herpes simplex virus type 1 DNA replication and gene expression during explant induced reactivation of latently infected murine sensory ganglia. *J. Virol.* **68,** 1271–1282 (1994).
27. D. C. Bloom, G. B. Devi-Rao, J. M. Hill, J. G. Stevens, and E. K. Wagner, Molecular analysis of herpes simplex virus type 1 during epinephrine induced reactivation of latently infected rabbits in vivo. *J. Virol.* **68,** 1283–1292 (1994).
28. S. Silver and B. Roizman, gamma 2-Thymidine kinase chimeras are identically transcribed but regulated a gamma 2 genes in herpes simplex virus genomes and as beta genes in cell genomes. *Mol. Cell. Biol.* **5,** 518–528 (1985).
29. R. H. Costa, K. G. Draper, G. B. Devi-Rao, R. L. Thompson, and E. K. Wagner, Virus-induced modification of the host cell is required for expression of the bacterial chloramphenicol acetyltransferase gene controlled by a late herpes simplex virus promoter (VP5). *J. Virol.* **56,** 19–30 (1985).
30. J. R. Smiley, C. Smibert, and R. D. Everett, Expression of a cellular gene cloned in herpes simplex virus: Rabbit beta-globin is regulated as an early viral gene in infected fibroblasts. *J. Virol.* **61,** 2368–2377 (1987).
31. S. L. McKnight, Constitutive transcriptional control signals of the herpes simplex virus tk gene. *Cold Spring Harbor Symp. Quant. Biol.* **47,** 945–958 (1983).
32. J. R. Smiley, B. Panning, and C. A. Smibert, Regulation of cellular genes by HSV products. in "Herpesvirus Transcription and Its Regulation" (E. K. Wagner, ed.), pp. 151–180. CRC Press, Boca Raton, FL, 1991.
33. S. L. McKnight and R. Tjian, Transcriptional selectivity of viral genes in mammalian cells. *Cell* **46,** 795–805 (1986).
34. S. P. Eisenberg, D. M. Coen, and S. L. McKnight, Promoter domains required for expression of plasmid-borne copies of the herpes simplex virus thymidine kinase gene in virus-infected mouse fibroblasts and microinjected frog oocytes. *Mol. Cell. Biol.* **5,** 1940–1947 (1985).
35. L. R. Stanberry, Animal modles and HSV latency. *Seminars Virol.* 213–220 (1994).
36. L. R. Stanberry, Pathogenesis of herpes simplex virus infection and animal models for its study. *Curr. Top. Microbiol. Immunol.* **179,** 15–30 (1992).
37. J. M. Hill, F. Sedarati, R. T. Javier, E. K. Wagner, and J. G. Stevens, Herpes simplex virus latent phase transcription facilitates in vivo reactivation. *Virology* **174,** 117–125 (1990).

38. P. Ghazal, A. Gonzalez, R. Garcia, S. Kurz, and A. Angulo, Viruses: Hostages to the cell. *Virology* **275**, 233–237 (2000).
39. N. Soderberg, K. N. Fish, and J. A. Nelson, Reactivation of latent human cytomegalovirus by allogeneic stimulation of blood cells from healthy donors. *Cell* **91**, 119–126 (1997).
40. W. Taylor, J. G. Sissons, L. K. Borysiewicz, and J. H. Sinclair, Monocytes are a major site of persistence of human cytomegalovirus in peripheral blood mononuclear cells. *J. Gen. Virol.* 72(Pt9), 2059–2064 (1991).
41. J. A. Nelson, C. Ibañez, R. Schrier, C. A. Wiley, and P. Ghazal, Expression of cytomegalovirus in mononuclear cells. *in* "Viruses That Affect the Immune System" (H. Y. Fan, I. S. Y. Chen, N. Rosenberg, and W. Sugden, eds.), pp. 231–246. American Society for Microbilogy, Washington, D. C., 2001.
42. J. L. Pollock, R. M. Presti, S. Paetzold, and H. W. Virgin, Latent murine cytomegalovirus infection in macrophages. *Virology* **227**, 168–179 (1997).
43. M. J. Reddehase, M. Balthesen, M. Rapp, S. Jonjiç, I. Paviç, and U. H. Koszinowski, The conditions of primary infection define the load of latent viral genome in organs and the risk of recurrent cytomegalovirus disease. *J. Exp. Med.* **179**, 185–193 (1994).
44. J. L. Pollock and H. W. Virgin, Latency, without persistence, of murine cytomegalovirus in the spleen and kidney. *J. Virol.* **69**, 1762–1768 (1995).
45. G. Hahn, R. Jores, and E. S. Mocarski, Cytomegalovirus remains latent in a common precursor of dendritic and myeloid cells. *Proc. Natl. Acad. Sci. U.S.A.* **95**, 3937–3942 (1998).
46. B. Poliç, H. Hengel, A. Krmpotiç, J. Trgovcich, I. Paviç, P. Luccaronin, S. Jonjiç, and U. H. Koszinowski, Hierarchical and redundant lymphocyte subset control precludes cytomegalovirus replication during latent infection. *J. Exp. Med.* **188**, 1047–1054 (1998).
47. R. M. Presti, J. L. Pollock, C. Dal, A. K. Guin, and H. W. Virgin, Interferon gamma regulates acute and latent murine cytomegalovirus infection and chronic disease of the great vessels. *J. Exp. Med.* **188**, 577–588 (1998).
48. S. Jonjiç, I. Paviç, B. Poliç, I. Crnkoviç, P. Lucin, and U. H. Koszinowski, Antibodies are not essential for the resolution of primary cytomegalovirus infection but limit dissemination of recurrent virus. *J. Exp. Med.* **179**, 1713–1717 (1994).
49. S. K. Kurz and M. J. Reddehase, Patchwork pattern of transcriptional reactivation in the lungs indicates sequential checkpoints in the transition from murine cytomegalovirus latency to recurrence. *J. Virol.* **73**, 8612–8622 (1999).
50. J. Chambers, A. Angulo, D. Amarantunga, H. Cuo, Y. Jian, J. S. Wan, A. Bittner, K. Frueh, M. R. Jackson, P. A. Peterson, M. G. Erlander, and P. Ghazal, DNA microarrays of the complex human cytomegalovirus genome: Profiling kinetic class with drug sensitivity of viral gene expression. *J. Virol.* **73**, 5757–5766 (1999).
51. S. W. Stingley, J. J. G. Ramirez, S. A. Aguilar, K. Simmen, R. M. Sandri-Goldin, P. Ghazal, and E. K. Wagner, Global analysis of herpes simplex virus type 1 transcription using an oligonucleotide-based DNA microarray. *J. Virol.* **74**, 9916–9927 (2000).
52. R. G. Jenner, M. M. Alba, C. Boshoff, and P. Kellam, Kaposi's sarcoma-associated herpesvirus latent and lytic gene expression as revealed by DNA arrays. *J. Virol.* **75**, 891–902 (2001).
53. W. A. Bresnahan and T. Shenk, A subset of viral transcripts packaged within human cytomegalovirus particles. *Science* **288**, 2373–2376 (2000).
54. R. G. Jenner, M. M. Alba, C. Boshoff, and P. Kellam, Kaposi's sarcoma-associated herpesvirus latent and lytic gene expression as revealed by DNA arrays. *J. Virol.* **75**, 891–902 (2001).
55. P. T. Lieu and E. K. Wagner, The kinetics of VP5 mRNA expression is not critical for viral replication in cultured cells. *J. Virol.* **74**, 2770–2776 (2000).
56. R. G. Jarman, E. K. Wagner, and D. C. Bloom, LAT expression during an acute HSV infection in the mouse. *Virology* **262**, 384–397 (1999).

57. E. K. Wagner, Herpes simplex virus—Molecular biology. *in* "Encyclopedia of Virology" (R. G. Webster and A. Granoff, eds.), pp. 686–697. Academic Press, London, 1999.
58. G. B. Devi-Rao, J. S. Aguilar, M. K. Rice, H. H. Garza, Jr., D. C. Bloom, J. M. Hill, and E. K. Wagner, Herpes simplex virus genome replication and transcription during induced reactivation in the rabbit eye. *J. Virol.* **71**, 7039–7047 (1997).
59. W. C. Hahn, C. M. Counter, A. S. Lundberg, R. L. Beijersbergen, M. W. Brooks, and R. A. Weinberg, Creation of human tumour cells with defined genetic elements. *Nature* **400**, 464–468 (1999).
60. J. A. Warrington, A. Nair, M. Mahadevappa, and M. Tsyganskaya, Comparison of human adult and fetal expression and identification of 535 housekeeping/maintenance genes. *Physiological Genomics* **2**, 143–147 (2000).
61. A. Angulo, P. Ghazal, and M. Messerle, The major immediate-early gene ie3 of mouse cytomegalovirus is essential for viral growth. *J. Virol.* **74**, 11,129–11,136 (2000).
62. D. J. Duggan, M. Bittner, Y. Chen, P. Meltzer, and J. M. Trent, Expression profiling using cDNA microarrays. *Nat. Genet.* **21**, 10–14 (1999).
63. S. A. Glanz, "Primer of Biostatistics." McGraw-Hill, New York, 1997.
64. L. C. Gilbert and J. T. Wachsman, Characterization and partial purification of the plasminogen activator from human neuroblastoma cell line, SK-N-SH. A comparison with human urokinase. *Biochim. Biophys. Acta* **704**, 450–460 (1982).
65. T. Livache, B. Fouque, A. Roget, J. Marchand, G. Bidan, R. Teoule, and G. Mathis, Polypyrrole DNA chip on a silicon device: example of hepatitis C virus genotyping. *Anal. Biochem.* **255**, 188–194 (1998).
66. P. Bean and J. Wilson, HIV genotyping by chip technology. *Am. Clin. Lab.* **19**, 16–17 (2000).
67. J. W. Wilson, P. Bean, T. Robins, F. Graziano, and D. H. Persing, Comparative evaluation of three human immunodeficiency virus genotyping systems: The HIV-GenotypR method, the HIV PRT GeneChip assay, and the HIV-1 RT line probe assay. *J. Clin. Microbiol.* **38**, 3022–3028 (2000).
68. D. Proudnikov, E. Kirillov, K. Chumakov, J. Donlon, G. Rezapkin, and A. Mirzabekov, Analysis of mutations in oral poliovirus vaccine by hybridization with generic oligonucleotide microchips. *Biologicals* **28**, 57–66 (2000).
69. M. J. Kozal, N. Shah, N. Shen, R. Yang, R. Fucini, T. C. Merigan, D. D. Richman, D. Morris, E. Hubbell, M. Chee, and T. R. Gingeras, Extensive polymorphisms observed in HIV-1 clade B protease gene using high-density oligonucleotide arrays. *Nat. Med.* **2**, 753–759 (1996).
70. C. A. Alford and R. F. Pass, Epidemiology of chronic congenital and perinatal infections of man. *Clin. Perinatol.* **8**, 397–414 (1981).
71. T. A. Cha, E. Tom, G. W. Kemble, G. M. Duke, E. S. Mocarski, and R. R. Spaete, Human cytomegalovirus clinical isolates carry at least 19 genes not found in laboratory strains. *J. Virol.* **70**, 70–83 (1996).
72. J. Chou, E. R. Kern, R. J. Whitley, and B. Roizman, Mapping of herpes simplex virus-1 neurovirulence to gamma$_1$34.5, a gene nonessential for growth in culture. *Science* **250**, 1262–1266 (1990).
73. J. Chou and B. Roizman, Herpes simplex virus 1 gamma$_1$34.5 gene function, which blocks the host response to infection, maps in the homologous domain of the genes expressed during growth arrest and DNA damage. *Proc. Nat. Acad. Sci. U.S.A.* **91**, 5247–5251 (1994).
74. T. Ben-Hur, M. Moyal, A. Rosen-Wolff, G. Darai, and Y. Becker, Characterization of RNA transcripts from herpes simplex virus-1 DNA fragment BamHI-B. *Virology* **169**, 1–8 (1989).
75. A. Rosen-Wolff, J. Scholz, and G. Darai, Organotropism of latent herpes simplex virus type 1 is correlated to the presence of a 1.5 kb RNA transcript mapped within the BamHI DNA fragment B (0.738 to 0.809 map units). *Virus Res.* **12**, 43–51 (1989).

76. L. E. Hann, W. J. Cook, S. L. Uprichard, D. M. Knipe, and D. M. Coen, The role of herpes simplex virus ICP27 in the regulation of *UL24* gene expression by differential polyadenylation. *J. Virol.* **72,** 7709–7714 (1998).
77. G. B. Devi-Rao, S. A. Goodart, L. B. Hecht, R. Rochford, M. K. Rice, and E. K. Wagner, The relationship between polyadenylated and non-polyadenylated herpes simplex virus type 1 latency associated transcripts. *J. Virol.* **65,** 2179–2190 (1991).
78. K. P. Anderson, R. J. Frink, G. B. Devi-Rao, B. H. Gaylord, R. H. Costa, and E. K. Wagner, Detailed characterization of the mRNA mapping in the HindIII fragment K region of the herpes simplex virus type 1 genome. *J. Virol.* **37,** 1011–1027 (1981).
79. W. M. Flanagan, A. G. Papavassiliou, M. Rice, L. B. Hecht, S. J. Silverstein, and E. K. Wagner, Analysis of the herpes simplex virus type 1 promoter controlling the expression of U_L38, a true late gene involved in capsid assembly. *J. Virol.* **65,** 769–786 (1991).
80. K. L. Carter, P. L. Ward, and B. Roizman, Characterization of the products of the U(L)43 gene of herpes simplex virus 1: Potential implications for regulation of gene expression by antisense transcription. *J. Virol.* **70,** 7663–7668 (1996).
81. P. L. Ward, D. E. Barker, and B. Roizman, A novel herpes simplex virus 1 gene, $U_L43.5$, maps antisense to the U_L43 gene and encodes a protein which colocalizes in nuclear structures with capsid proteins. *J. Virol.* **70,** 2684–2690 (1996).

Sphingosine Kinases: A Novel Family of Lipid Kinases

HONG LIU,[*]
DEBYANI CHAKRAVARTY,[*]
MICHAEL MACEYKA,[*]
SHELDON MILSTIEN,[†]
AND SARAH SPIEGEL[*]

[*]*Department of Biochemistry*
Medical College of Virginia Campus
Virginia Commonwealth University
Richmond, Virginia 23298
[†]*Laboratory of Cellular and Molecular Regulation*
National Institute of Mental Health
Bethesda, Maryland 20892

I.	Pleiotropic Functions of Sphingosine-1-Phosphate	494
II.	Sphingosine Kinase and Sphingosine-1-Phosphate in Yeast and Plants	495
III.	Cellular Functions of Sphingosine Kinase in Mammalian Cells	497
IV.	How Is Sphingosine Kinase Activated?	498
V.	Cloning of Mammalian Sphingosine Kinases	500
VI.	Sphingosine Kinase Family	504
VII.	Five Conserved Domains of the SPHK Superfamily	505
VIII.	Phylogenetic Analysis of Sphingosine Kinases	508
IX.	Concluding Remarks	508
	References	509

Sphingosine kinase (SPHK) catalyzes the formation of sphingosine-1-phosphate (S1P). S1P plays an important role in regulation of a variety of biological processes through intracellular and extracellular actions. S1P has recently been shown to be the ligand for the EDG-1 family of G-protein-coupled receptors. To date, seven cloned SPHKs have been reported with confirmed SPHK activity, including human, mouse, yeast, and plant. A computer search of various databases suggests that a new SPHK family is emerging. The cloning and manipulation of SPHK genes will no doubt provide us with important information about the functions of S1P in a wide range of organisms. © 2002, Elsevier Science (USA).

Abbreviations: DAGKc, diacylglycerol kinase catalytic domain; EGF, epidermal growth factor; ERK, extracellular signal regulated kinases; GPCRs, G-protein coupled receptors; HDL, high-density lipoproteins; InsP$_3$, inositol-1,4,5-triphosphate; DMS, N,N-dimethylsphingosine; NGF, nerve growth factor; PDGF, platelet-derived growth factor; S1P, sphingosine-1-phosphate; SPHK, sphingosine kinase; SPHKc, sphingosine kinase catalytic domain; S1PRs, S1P receptors.

I. Pleiotropic Functions of Sphingosine-1-Phosphate

Sphingosine kinase (SPHK) catalyzes the ATP-dependent phosphorylation of sphingosine on its primary hydroxyl group to form sphingosine-1-phosphate (S1P). S1P is a bioactive sphingolipid metabolite that plays an important role in the regulation of a variety of biological processes, including calcium mobilization, cell-growth differentiation, survival, motility, and cytoskeletal reorganization (reviewed in Refs. *1–4*).

S1P, similar to other well-known phospholipid mediators, performs dual actions in cells. S1P acts intracellularly as a second messenger and extracellularly as a ligand for G-protein-coupled receptors (GPCRs), previously named the EDG-1 family and now known as S1P receptors (S1PRs) (*1–4*). To date, five members of the S1PR family have been cloned, $S1P_1$ (EDG-1), $S1P_2$ (EDG-5), $S1P_3$ (EDG-3), $S1P_4$ (EDG-6), and $S1P_5$ (EDG-8). All these receptors bind and are activated specifically by S1P and a structural analog of S1P, dihydro-S1P, which lacks the *trans* 4–5 double bond.

This family of receptors is differentially expressed and coupled to a variety of G proteins. Thus, S1P can elicit and regulate a diverse array of signal transduction pathways depending on both the cell type and the relative expression of G proteins and receptors, resulting in a variety of responses (*1–4*). Despite their diversity, amazingly, all of the S1PRs have been implicated in regulation of cell motility—be it positive or negative (*5–9*). For example, it has been demonstrated that activation of $S1P_1$ or $S1P_3$ by either S1P or dihydro-S1P in several cell types induces chemotaxis and membrane ruffling, whereas activation of $S1P_2$ in the same cell types inhibits chemotaxis and membrane ruffling (*10*). In agreement, stimulation of these receptors either activates or inhibits members of the Rho family of small GTPases, particularly Rho and Rac (*11*). Rho GTPases are downstream of heterotrimeric G proteins and are consequential in cytoskeletal rearrangements: activation of Rho induces stress fiber formation and Rac induces cortical actin formation (*12, 13*). Activation of $S1P_1$ directs Rac-coupled cortical actin formation (*11*), whereas activation of $S1P_2$ inhibits Rac-activated cortical actin formation (*10*) and therefore abolishes Rac-induced chemotaxis and membrane ruffling (*10*). However, binding of S1P to $S1P_3$ or to $S1P_2$ elicits Rho-coupled stress fiber assembly (*10*). In this way, differential expression of the S1P receptors can affect cellular chemotactic responses to extracellular gradients of S1P.

Cell motility is important in many physiological and pathological processes, such as inflammation, wound healing, tumor growth, metastasis, and angiogenesis, and plays a crucial role during development. Several recent reports substantiate the importance of S1PRs in motility and development. Disruption of the zebrafish *miles apart* gene, a homolog of $S1P_2$, causes defective migration of myocardial cells during vertebrate heart development, attributing a unique role to $S1P_2$ in regulation of cell migration in organogenesis of the heart (*9*).

Furthermore, disruption of the murine $s1p_1$ gene was lethal and revealed that $S1P_1$ signaling is essential for vascular maturation (14), as the $S1P_1$ knockout mice died *in utero* due to a severe deficiency of smooth muscle cells and pericytes in the vessel walls, resulting in massive hemorrhage (14). Examination of the $s1p_1$ null embryos revealed physiologically normal blood vessel networks (14). Furthermore, fibroblasts from these embryos showed an attenuation of Rac-mediated chemotactic responses when placed in a S1P gradient, emphasizing that $S1P_1$ is essential in the recruitment of cells during microvasculature maturation.

Subsequently, $S1P_1$ was also found to be essential for cell migration toward a gradient of platelet-derived growth factor (PDGF) (15). It is now well established that PDGF stimulates SPHK and thereby increases the intracellular concentration of S1P in a variety of cell types (16). Recently, we found that PDGF activated $S1P_1$, as measured by its phosphorylation and translocation of β-arrestin, suggesting a new mechanistic concept for cross-communication between a tyrosine kinase receptor, PDGFR, and a GPCR such as $S1P_1$ (15). According to this new paradigm, PDGF activates SPHK resulting in the increased synthesis of S1P, which consequently stimulates $S1P_1$, in either a paracrine or an autocrine fashion, resulting in the cascade of downstream signals critical for cell locomotion (15). In agreement with this concept, it was recently found that PDGFR is tethered to $S1P_1$, thereby providing an integrative signaling platform for these receptors (17). Importantly, PDGF-B and PDGFR-β knockouts share a common phenotype with the $S1P_1$ knockouts, where the embryos also died from incomplete vasculogenesis (18, 19). This not only highlights the pivotal role of S1P in cross-communication between receptors but also emphasizes the importance of pericyte/smooth muscle cell recruitment in vascular maturation (14). The discovery that $S1P_1$ plays a critical role in directed cell motility unveils the underlying mechanism by which S1P acts as a regulator of angiogenesis (7, 11, 20).

There is also ample evidence that S1P acts as an intracellular second messenger. Despite the ability to bind to and activate all of the S1PRs, dihydro-S1P does not mimic all of the biological effects elicited by S1P, especially those related to cell survival, suggesting that S1P has unique intracellular targets (16, 21–24). The next major step in deciphering the intracellular actions of S1P would be to identify and characterize these targets which elicit such diverse cellular responses as the opening/closure of calcium channels (25–28) and inhibition of apoptosis (16, 22–24).

II. Sphingosine Kinase and Sphingosine-1-Phosphate in Yeast and Plants

Sphingolipids are ubiquitous constituents of eukaryotic membranes, and there is growing evidence that sphingolipid signaling is a feature not only of the animal kingdom but also of yeast and plants. The sphingolipids of yeast (29) and plants (30) differ from those of animals in that the major sphingoid base is phytosphingosine, which lacks the 4–5 double bond of sphingosine and has an additional 4-hydroxyl group.

The brewer's yeast, *Saccharomyces cerevisiae*, has two long-chain sphingoid base kinases, Lcb4p and Lcb5p (31), which are highly similar to each other and to mammalian SPHK (see below). These proteins are 53% identical, 65% similar, catalyze the phosphorylation of dihydrosphingosine, phytosphingosine, and sphingosine, and have similar K_M values for these substrates. We will refer collectively to the phosphorylated forms of these sphingoid bases as S1P for purposes of this review. Interestingly, deletion of either or both Lcbps has no effect on yeast growth on fermentable carbon sources (31), even though levels of S1P are undetectable. In contrast, yeast deficient in the two major enzymes responsible for degradation of S1P have 500-fold or greater increases in S1P levels and a marked decrease (32) or abolition (33) of cell growth. These data suggest that excess S1P negatively regulates growth of yeast, perhaps due to aberrant S1P signaling.

Several lines of evidence suggest that there is a role for sphingolipids in mediating stress responses, particularly heat stress, in yeast. First, yeast strains unable to make sphingolipids fail to grow in response to osmotic shock, low pH, or high temperature (34). Subsequently, it was found that heat stress induced a transient increase in sphingoid bases and ceramide (35, 36), as well as in S1P (32, 33). Exogenous addition of these sphingolipid metabolites likewise increased thermotolerance (35, 36). The effect was likely due to activation of a signaling pathway rather than to nonspecific changes in membrane properties, as stress-response-specific transcription was activated (35). These results implicated one or more sphingolipid metabolites in a signaling pathway involved in the heat stress response. A clue to the identity of the metabolite(s) involved came with the identification of two specific S1P phosphatases in yeast (37, 38). Inactivating mutations in these S1P phosphatase genes led to increased thermotolerance, while overexpression of these phosphatases reduced thermotolerance (38), suggesting a role for S1P in this process. Further support for a role of S1P in the yeast stress response emerged from the observation that cells differentially deficient in S1P degradation showed a dose-dependent increase in thermotolerance (32). Recent elegant work by Jenkins and Hannun (39) demonstrated that SPHK is required for exit from heat-induced cell-cycle arrest, likely for removal of sphingosine.

S1P may also be required for a calcineurin-mediated-signaling pathway in yeast, as it has recently been shown that S1P can activate a plasma membrane calcium channel leading to calcimeurin activation (40). Moreover, in recent intriguing studies, it was reported that *de novo* sphingoid base synthesis is required for endocytosis in *Saccharomyces cerevisiae* and for proper actin organization

(*41, 42*). This is the first example of a physiological role for sphingoid base synthesis in eukaryotes, rather than as a precursor for ceramide or phosphorylated sphingoid base synthesis (*41, 42*).

Plants also have SPHK activity (*30*), and recently, a SPHK homolog was cloned from *Arabidopsis thaliana* with 27% identity and 50% similarity to murine SPHK (*43*). While the roles of SPHK and S1P in plants have been less well studied, a recent report provided strong evidence for a role of S1P in plant cell stomata closure (*44*). When plants were grown under drought conditions, the levels of endogenous S1P rose. Exogenously applied S1P stimulated calcium oscillations and stomata closure in a similar manner under drought conditions. Moreover, the effects of the stomata-closing plant hormone abscisic acid could be partially blocked by treatment with the known SPHK inhibitor, D,L-*threo*-dihydrosphingosine. Together, these data suggest that, just as in animal cells, S1P acts as a second messenger in plants to regulate calcium homeostasis and ion channels.

III. Cellular Functions of Sphingosine Kinase in Mammalian Cells

Ceramide (*N*-acyl sphingosine) and its metabolite, sphingosine, have been implicated in programmed cell death, known as apoptosis (*45–47*). Ceramide and sphingosine are also usually associated with cellular growth retardation. In contrast, S1P generally stimulates cellular proliferation, promotes survival (*48, 49*), and protects against ceramide-mediated apoptosis (*22, 50–52*). It has therefore been suggested that the dynamic balance between the intracellular concentration of S1P versus that of ceramide and sphingosine—the so-called "sphingolipid rheostat"—and the consequent activities of opposing signaling pathways are pivotal factors in determining mammalian cell fate (*22, 23, 53, 54*). One of the best examples substantiating the importance of the sphingolipid rheostat in cell-fate decisions emerged from studies by Tilly and colleagues (*23, 53*). They showed that apoptosis of unfertilized mouse oocytes treated with the anticancer drug doxorubicin, a known inducer of ceramide formation leading to apoptosis, was prevented by pretreatment with S1P (*53*). In addition, acidic sphingomyelinase null mouse oocytes or wild-type oocytes treated with S1P were resistant to developmental apoptosis as well as to apoptosis induced by anticancer treatment (*23*). Importantly, in adult wild-type female mice, radiation-induced oocyte loss via ceramide-mediated apoptosis was completely prevented by *in vivo* treatment with S1P (*23*). This experiment is of particular clinical value, considering the side effects of premature ovarian failure and infertility in female cancer patients occur as a consequence of radiation anticancer therapy. Thus, it might be possible to formulate practical therapeutic adjuncts for the prevention of oocyte destruction after cancer therapy based on the principle of increasing intracellular S1P levels, conceivably by activation of SPHK.

Recent data have also connected the sphingolipid rheostat and essentially the regulatory role of SPHK to several other biological processes including calcium homeostasis and allergic responses. Calcium mobilization from internal stores is of primary importance in the initiation of the allergic response and is considered to be mainly mediated by inositol-1,4,5-triphosphate (InsP$_3$). However, Kinet and colleagues found that FcεRI cross-linking caused activation of SPHK and production of S1P which can mobilize intracellular calcium independently of InsP$_3$ (55). Indeed, inhibition of SPHK with dihydrosphingosine eliminated the calcium-mobilization effects of allergic stimulation, despite the presence of functional muscarinic receptors (GPCRs that activate InsP$_3$-mediated calcium mobilization), indicating that FcεRI principally utilizes a SPHK-dependent pathway for initiation of the allergic response.

Using CPII mouse mast cells and bone marrow-derived mast cells, Prieschl *et al.* demonstrated that sphingosine acts as a potent inhibitor of IgE/Ag-mediated leukotriene synthesis and cytokine production (the late stage of the allergic response) via abrogation of the extracellular signal-regulated kinases (ERK1/2) and AP-1-mediated transcription (56). S1P, in further support of the sphingolipid rheostat principle, counteracts these inhibitory effects by activating ERK1/2 and AP-1 (57). In this way, SPHK acts as a "permissive switch" by determining the ratio of sphingosine to S1P and, as a consequence, might regulate the allergic response (56).

SPHK might also play a role within the cardiovascular system. A fundamental event in the initiation of atherosclerosis is the activation of endothelial cells. Stimulation of SPHK and generation of S1P were shown to be critically involved in mediating TNF-α-induced endothelial cell activation (58) and to trigger a signaling pathway that protects endothelial cells against apoptosis (54). High-density lipoproteins (HDL) have the capacity to protect the heart against atherosclerosis and associated heart disease by inhibiting the expression of cytokine-induced adhesion molecules. A recent study has demonstrated that HDL inhibits SPHK and thereby blocks S1P generation while simultaneously enhancing ceramide levels (59). The ability of HDL to inhibit cytokine-induced adhesion molecule expression has been correlated with resetting of the sphingolipid rheostat (59). This has important implications for understanding the mechanism of the protective function of HDL against the development of atherosclerosis and associated coronary heart disease.

IV. How Is Sphingosine Kinase Activated?

The precise mechanism of activation of SPHK, which in turn increases the intracellular concentrations of S1P, remains largely unknown. To resolve the *modus operandi* utilized by SPHK activators is a problem that is daunting due to the scarcity of molecular tools and to the plethora of biological stimuli that can increase the activity of this kinase. These include PDGF (*49*), nerve growth factor (NGF) (*50*), vitamin D3 (*52*), epidermal growth factor (EGF) (*60*), TNF-α (*54, 58*), cross-linking of the immunoglobin receptors FcεR1 and FcγR1 (*56, 61*), LPA (*26, 62*), acetylcholine (muscarinic agonists) (*26*), as well as other GPCR agonists (*26, 63*). In certain cell types, an increase in the intracellular concentration of cAMP or activation of protein kinase C may result in the activation of SPHK (*64–67*). The formyl-peptide receptor activates sphingosine SPHK through a G_i-dependent pathway (*63*), whereas FcγRI-induced stimulation of SPHK is mediated by sequential activation of a tyrosine kinase and phospholipase D (*61*).

To examine in more detail how PDGF stimulates SPHK, Olivera *et al.* studied the effect of PDGF on SPHK activation in TRMP cells expressing wild-type or various mutant PDGF-β receptors (*68*). SPHK was stimulated by PDGF in cells expressing wild-type receptors but not in cells expressing kinase-inactive receptors (R634). Cells were transfected with mutant PDGF receptors with phenylalanine substitutions at five major tyrosine phosphorylation sites, 740/751/771/1009/1021 (F5 mutants). These mutants are unable to associate with PLCγ, phosphatidylinositol 3-kinase, Ras GTPase-activating protein, or protein tyrosine phosphatase SHP-2, respectively. These cells not only failed to increase DNA synthesis in response to PDGF but also were unable to activate SPHK. Moreover, mutation of tyrosine-1021 of the PDGF receptor to phenylalanine, which impairs its association with PLCγ, abrogated PDGF-induced activation of SPHK. In contrast, PDGF was still able to stimulate SPHK in cells expressing the PDGF receptor mutated at tyrosines 740/751 and 1009, responsible for binding of phosphatidylinositol 3-kinase and SHP-2, respectively. In agreement, PDGF did not stimulate SPHK activity in F5-receptor "add-back" mutants in which association with the Ras GTPase-activating protein, phosphatidylinositol 3-kinase, or SHP-2 was individually restored. However, a mutant PDGF receptor that was able to bind PLCγ (tyrosine-1021), but not other signaling proteins, restored SPHK sensitivity to PDGF. These data indicate that the tyrosine residue responsible for binding of PLCγ is required for PDGF-induced activation of SPHK. Moreover, calcium mobilization downstream of PLCγ, but not protein kinase C activation, appears to be required for stimulation of SPHK by PDGF (*68*). Interestingly, stimulation of SPHK by the P2Y(2) receptor in HL-60 cells also required calcium mobilization (*27*). Thus, calcium mobilization might be the common denominator by which tyrosine kinases and GPCRs stimulate SPHK. In this regard, it should be noted that SPHK binds with high affinity to calmodulin in the presence of calcium (*69*), suggesting that SPHK activity might be modulated by Ca^{2+}-calmodulin either by direct allosteric regulation or by affecting its cellular localization.

V. Cloning of Mammalian Sphingosine Kinases

Due to the pivotal role of S1P in regulating many biological responses, in the last few years, a major effort of our lab was aimed at cloning mammalian SPHK, the enzyme responsible for its formation. After 12 steps of purification, rat kidney SPHK was purified 600,000-fold to homogeneity (69). The purified enzyme had an apparent molecular mass of 49 kDa, was active as a monomer, and exhibited high-substrate specificity for D-*erythro*-sphingosine. On the basis of peptide sequences derived from the purified enzyme, we subsequently cloned and characterized the first mammalian SPHK, named mSPHK1 (70). mSPHK1 specifically phosphorylated D-*erythro*-sphingosine and did not catalyze phosphorylation of phosphatidylinositol, diacylglycerol, ceramide, D,L-*threo*-dihydrosphingosine or N,N-dimethylsphingosine. The latter two sphingolipids were competitive inhibitors of recombinant SPHK (70), as was previously found with the purified rat kidney enzyme (69). Hydropathy plots of the predicted amino acid sequence of SPHK1 indicate that it lacks hydrophobic transmembrane sequences. Indeed, when overexpressed, SPHK1 is located predominantly in the cytosol, although a small proportion is also associated with membrane fractions. Northern analysis revealed that expression of SPHK1 was highest in adult mouse lung and spleen, with barely detectable levels in skeletal muscle and liver. Simultaneously, independent studies by Dickson and colleagues, identified two yeast genes *Lcb4* (*YOR171c*) and *Lcb5* (*YLR260w*) which encode long-chain base kinases (31). LCB4 accounts for most (97%) of the *Saccharomyces* LCB kinase activity. Both enzymes can use phytosphingosine, dihydrosphingosine, or sphingosine as substrate (31). The two isoforms of SPHK cloned from *S. cerevisiae* show substantial sequence homology to mammalian SPHK1 (31, 70).

Based on sequence identity to murine mSPHK1, we cloned and characterized the first human SPHK (hSPHK1) (71). The open reading frame of hSPHK1 encodes a 384-amino acid protein with 85% identity and 92% similarity to mSPHK1 at the amino acid level. Similar to mSPHK1, hSPHK1 also specifically phosphorylated D-*erythro*-sphingosine and to a lesser extent sphinganine. Subsequently, Pitson *et al.* also cloned hSPHK1 using a different approach (72). Northern analysis revealed that hSPHK1 was widely expressed with highest levels in adult liver, kidney, heart, and skeletal muscle (70). These results have subsequently been confirmed by Melendez *et al.* (73), who also found the highest expression in adult lung and spleen, followed by peripheral blood leukocyte and thymus. Recently, staining with a specific SPHK antibody revealed that SPHK1 was expressed in the white matter of the cerebrum and cerebellum, the red nucleus and cerebral peduncle in the midbrain, the uriniferous tubules in the kidney, the endothelial cells of blood vessels, and in megakaryocytes and platelets (74).

Enforced expression of SPHK1 increased the proportion of cells in the S phase of the cell cycle, promoted the G_1-S transition, and reduced the doubling time, especially under low-serum conditions, indicating that intracellular S1P is an important regulator of cell growth (75). In addition, it was recently found that cells overexpressing hSPHK1 acquired a transformed phenotype, as determined by focus formation, colony growth in soft agar, and by the ability to form tumors in NOD/SCID mice (76). This is the first demonstration that a wild-type lipid kinase gene acts as an oncogene. Using a SPHK inhibitor or a dominant-negative mutant that inhibits enzyme activation, it was shown that SPHK is involved in oncogenic H-Ras-mediated transformation, suggesting a novel signaling pathway for Ras activation (76). Collectively, these findings point to a potential role of SPHK1 and S1P in cell growth, transformation, and cancer.

Recently, we cloned a second type of mouse and human SPHK (mSPHK2 and hSPHK2). mSPHK2 and hSPHK2 encode proteins of 617 and 618 amino acids, respectively, both much larger than SPHK1 (77). Although both contain five conserved domains previously found in SPHK1 (see below), their sequences diverge considerably in the centers and at the amino termini (Fig. 1). Despite the overall homology between SPHK1 and SPHK2, the sequences of these two proteins diverge sufficiently to suggest that they did not arise from a simple gene-duplication event. Northern blot analysis revealed that the mRNAs corresponding to these two forms of SPHK differ in their tissue distributions. Whereas SPHK1 expression is highest in lung and spleen, SPHK2 is most abundant in liver and heart. Furthermore, there are differences in the temporal patterns of their appearance during development, with SPHK1 transcripts appearing before SPHK2 transcripts (77). The enzymatic properties of these types of SPHK were also found to differ markedly (Table I). Thus, (i) whereas SPHK1 shows a preference for sphingosine over dihydrosphingosine as substrate and does not recognize phytosphingosine, SPHK2 shows a slight preference for *erythro*-dihydrosphingosine and is also able to phosphorylate phytosphingosine, albeit to a lesser extent; (ii) SPHK1 is inhibited by N,N-dimethylsphingosine (DMS) and

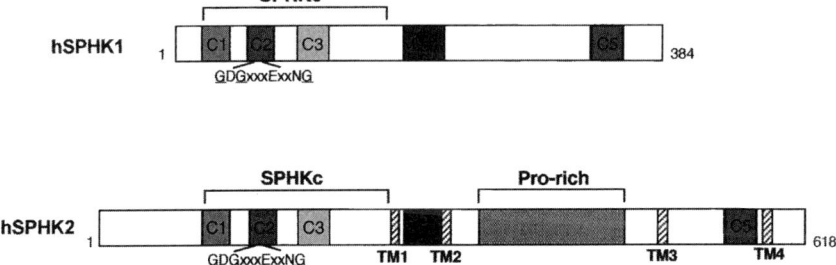

FIG. 1. Schematic representation of domains present in hSPHK1 and hSPHK2. Five conserved domains are labeled C1 to C5. hSPHK2 contains a proline-rich region. All SPHKc domains contain the conserved ATP-binding sequence, GDGxxxExxNG. TM, transmembrane regions.

TABLE I
COMPARISON OF THE TWO ISOFORMS OF SPHK

	SPHK1	SPHK2
MW (kDa)	42	66
Number of AA	384	618
Conserved domains	5	5
Tissue expression (mouse)	Lung, spleen ≫ kidney, liver, brain	Liver, heart ≫ kidney, brain, testes
Substrate specificity	D-*erythro*-Sph (Km 2.2 μM)	D-*erythro*-DH-Sph > D-*erythro*-Sph
Membrane-association	17%	20%
Inhibitors	DMS (competitive) DHS (competitive)	DMS (noncompetitive)
Salt	Inhibits	Stimulates
Detergents	Stimulates	Inhibits
BSA	Slightly stimulates	Inhibits
Acidic phospholipids	Stimulate	Stimulate

by D,L-*threo*-dihydrosphingosine in a competitive manner, whereas SPHK2 is inhibited in a noncompetitive manner by DMS, and D,L-*threo*-dihydrosphingosine is a substrate; (iii) whereas high-salt concentrations markedly inhibit SPHK1, they activate SPHK2; (iv) Triton X-100 activates SPHK1 and inhibits SPHK2; and (v) presentation of the substrate as a complex with BSA gives slightly more activity with SPHK1 and gives lower activity with SPHK2. Collectively, these data suggest that type 1 and type 2 SPHKs most likely have different functions and regulate levels of S1P in different manners.

Although acidic phospholipids increase recombinant SPHK1 and SPHK2 activities, neither contain a consensus binding site for acidic phospholipids, suggesting that activation might occur through either another protein or a factor that mediates interaction of the enzyme with membranes, by either post-translational modification or translocation. Both kinases contain five conserved domains, C1 to C5 (Fig. 2). Interestingly, the invariant GGKGK positively charged motif in the C1 domain of all type 1 SPHKs is modified to GGRGL in SPHK2, suggesting that it may not be part of the ATP-binding site as was originally proposed (70). The motif search also revealed that a region beginning just before the conserved C1 domains of mSPHK2 and hSPHK2 (amino acids 147 to 284) also has homology to the diacylglycerol kinase catalytic (DAGKc) domain. Both types of SPHKs have a putative ATP-binding motif which has the consensus sequence GDGxxxExxNG (see below). Surprisingly, although only a minor portion of SPHK2 activity was detected in plasma membrane fractions, hSPHK2 also contains four putative transmembrane domains, suggesting that SPHK2 may have the potential to associate with or translocate to membranes. Further studies are needed to clarify

SPHINGOSINE KINASES

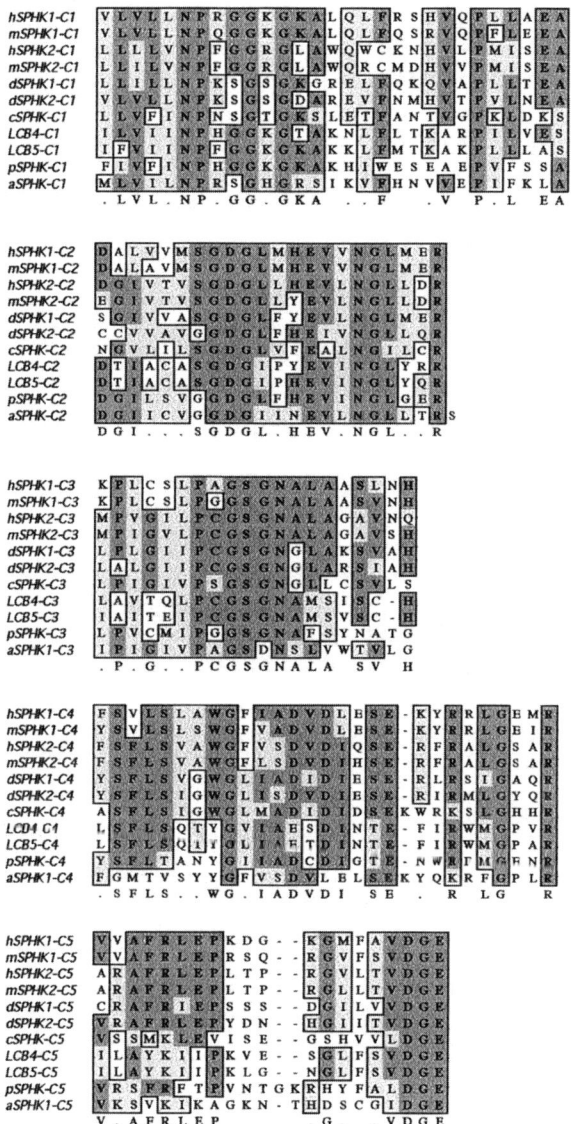

FIG. 2. Amino acid sequence alignments of five conserved domains (C1 to C5) from 11 members of SPHK superfamily. The alignments were made by the CLUSTAL method. The consensus sequences are listed below each alignment. The most conserved amino acids for these domains are: C1, NPxxGxG(x)$_{11}$P; C2, GDGxxxExxNGxxxR (ATP-binding domain); C3, PxGSxN; C4, G(x)$_8$S/TE(x)$_{5-6}$GxxR; C5, DGE. SPHK sequences from the following organism were used: h, *Homo sapiens*; m, *Mus musculus*; d, *Drosophila melanogaster*; c, *Caenorhabditis elegans*; p, *Saccharomyces pombe*; Lcb4, Lcb5, *Saccharomyces cerevisiae*; a, *Arabidopsis thaliana*.

the localization of SPHK2 within the cell. Interestingly, hSPHK2 also contains a proline-rich region, and within this region there is a putative SH3-binding domain, suggesting that this kinase may associate with other SH3-containing proteins. SH3 domains are protein–protein interaction modules which bind to specific proline-rich sequences and are required for the function of proteins that regulate cytoskeletal architecture, signal transduction cascades (78), and endocytosis (79).

VI. Sphingosine Kinase Family

To date, seven cloned SPHKs have been reported with confirmed SPHK activity. These include hSPHK1 and mSPHK1 (70–72), hSPHK2 and mSPHK2 (77), Lcb4 and Lcb5 from *S. cerevisiae* (31), and a recently reported plant SPHK from *Arabidopsis thaliana* (43). BLAST searches of protein database using known SPHK amino acid sequences revealed at least an additional four putative members of the SPHK family based on their high-sequence homologies. As shown in Table II, these proteins include two SPHKs from *Drosophila melanogaster* (dSPHK1 and dSPHK2), one from *Caenorhabditis elegans* (cSPHK), and one from *Saccharomyces pombe* (pSPHK). Designations as either type 1 or type 2 are based on the degree of similarity to the corresponding mammalian SPHK isoform. The predicted sizes of these proteins range from 384 amino acids (hSPHK1) to 907 amino acids (dSPHK2). The SMART (Simple Modular Architecture Research Tool) program searches indicate that all of the SPHKs possess a domain similar to that of the DAGKc which we have named the SPHKc domain (see below). It is also noteworthy that besides

TABLE II
THE SPHK SUPERFAMILY FAMILY

Name (accession no.)	Species	Amino acids	Confirmed activity	TM[a]
hSPHK1 (AAF73423)	*H. sapiens*	384	Yes	1
mSPHK1 (AAC61697)	*M. musculus*	388	Yes	0
hSPHK2 (AAF74124)	*H. sapiens*	618	Yes	4
mSPHK2 (AAF74125)	*M. musculus*	617	Yes	4
dSPHK1 (AAF48045)	*D. melanogaster*	641	No	1
dSPHK2 (AAF47706)	*D. melanogaster*	907	No	3
cSPHK (CAA91259)	*C. elegans*	473	No	1
LCB4 (NP-014814)	*S. cerevisiae*	624	Yes	0
LCB5 (NP-013361)	*S. cerevisiae*	687	Yes	0
pSPHK1 (T38776)	*S. pombe*	458	No	1
aSPHK1 (BAB07787)	*A. thaliana*	763	Yes	2

[a] Transmembrane domains.

the SPHKc domain, dSPHK2 also has a C-terminal SEC14 domain. SEC14 is a lipid-binding domain that is present in a homolog of a *S. cerevisiae* phosphatidylinositol transfer protein and in RhoGAPs, RhoGEFs, RasGEF, and neurofibromin. *Sec14* is an essential gene that codes for the major PC/phosphatidylinositol transfer protein in yeast required for Golgi-mediated protein transport (*80*).

Table II also shows the predicted number of putative transmembrane regions in each sequence determined by the TMpred program. Whereas type 1 SPHKs have either zero or one transmembrane domain, most type 2 SPHKs have three or four transmembrane domains.

VII. Five Conserved Domains of the SPHK Superfamily

There are five domains that are conserved among all members of the mammalian SPHK family (Fig. 2). These domains were originally identified by comparison of the sequence of mSPHK1 to the yeast homologs *Lcb4* and *Lcb5* (*70*). Based on cloning and identification of the SPHK superfamily, these domains have now been slightly modified (Fig. 2). A region containing conserved domains C1, C2, and C3 has high homology to the putative ATP-binding site of the DAGKc (*81*). This region in DAGK displays some resemblance to the glycine-rich loop within the ATP-binding site of many protein kinases (*82*) but is more divergent in both DAGKc and SPHKc. A glycine residue in this region is known to be essential for DAGK catalytic activity, since its mutation to aspartate ablates activity in all DAGKs examined (*81, 83*). All of the different members of the SPHK superfamily have the putative ATP-binding (GDGxxExxNGxxR) sequence present within the conserved C2 domain (Fig. 2). Indeed, site-directed mutagenesis of the second conserved glycine residue to aspartate was recently used to prepare a catalytically inactive, dominant-negative mutant of hSPHK1 (*84*). Moreover, expression of this mutant blocked stimulation of SPHK activity by TNF-α, IL-1β, and phorbol ester in HEK293T cells (*84*), suggesting that dominant-negative SPHKs should be a useful approach for dissecting the role of SPHK in diverse biological responses.

ClustalW alignments of SPHKc from 11 members of the SPHK family and DAGKc from 11 members of the DAGK family are shown in Fig. 3. (See color insert.) All of the DAGK catalytic domains have at least one presumed ATP-binding site with the consensus GxGxxG that is also found in protein kinases (*82*). However, this ATP-binding motif differs from that of the protein kinases, where there are essential lysine 14–23 amino acids downstream of the glycines (*82*). Although many of the DAGKc have a lysine in a similar position, site-directed mutagenesis of this amino acid in DAGKα does not alter activity (*85*), indicating that the ATP-binding pockets of DAGKs have a different conformation from

TABLE III
PAIRWISE COMPARISONS OF SPHINGOSINE KINASE CATALYTIC DOMAINS[a]

	hSPHK1	mSPHK1	hSPHK2	mSPHK2	dSPHK1	dSPHK2	cSPHK	LCB4	LCB5	pSPHK	aSPHK
					Percentage amino acid similarity						
hSPHK1		93	77	76	65	63	56	58	61	57	57
mSPHK1	86		72	71	61	61	56	56	59	56	56
hSPHK2	54	53		93	61	59	52	58	57	57	61
mSPHK2	54	53	86		61	60	52	57	55	56	57
dSPHK1	46	44	41	42		71	62	52	50	50	52
dSPHK2	43	41	39	38	53		55	52	52	49	55
cSPHK	33	34	33	32	39	37		45	46	49	54
LCB4	33	34	35	33	32	29	29		87	58	51
LCB5	36	34	31	31	34	31	31	73		60	50
pSPHK	36	38	36	32	29	28	31	42	41		57
aSPHK	31	30	37	33	31	31	37	32	33	33	
					Percentage amino acid identity						

[a] Values were determined by alignment of the SPHK catalytic domain sequences using ClustalW.

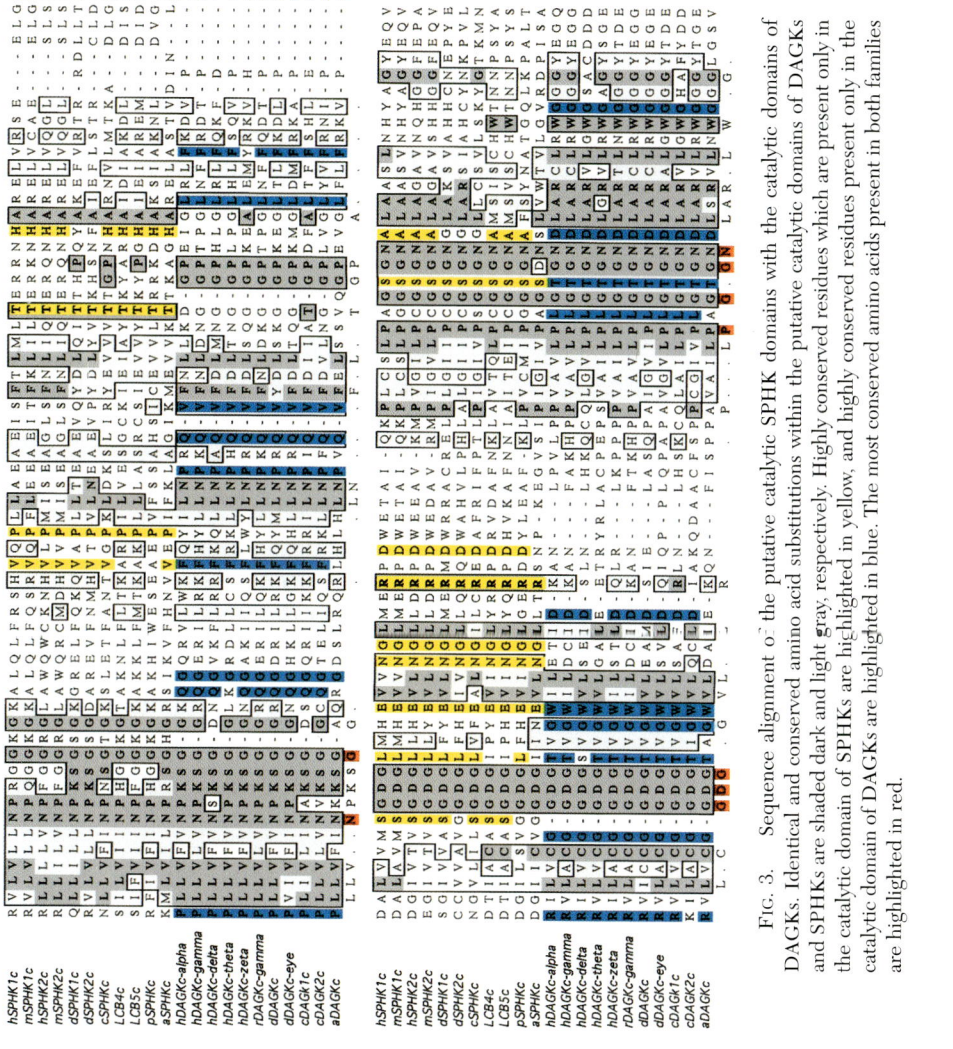

FIG. 3. Sequence alignment of the putative catalytic SPHK domains with the catalytic domains of DAGKs. Identical and conserved amino acid substitutions within the putative catalytic domains of DAGKs and SPHKs are shaded dark and light gray, respectively. Highly conserved residues which are present only in the catalytic domain of SPHKs are highlighted in yellow, and highly conserved residues present only in the catalytic domain of DAGKs are highlighted in blue. The most conserved amino acids present in both families are highlighted in red.

that of the protein kinases. In contrast, the catalytic domains of SPHK have the consensus sequence of the presumed ATP-binding site, $GDGx_7G$. Somewhat similar to protein kinases, SPHKs have a conserved arginine residue, 12 amino acids after the second glycine. Further site-directed mutagenesis studies are necessary to identify critical residues important for substrate binding and catalysis. There are 13 amino acids that are conserved among members of the SPHK family (highlighted as yellow, Fig. 3) and 19 amino acids conserved only in the DAGK family (highlighted in blue). Also, there are 9 amino acids that are conserved in both families (highlighted in red). In sum, although both SPHKc and DAGKc have a high degree of similarity, it is clear that they diverge considerably and belong to different subfamilies of lipid kinases. Hence, we propose that the catalytic domain of SPHK should be named SPHKc and not DAGKc as designated by SMART and Pfam programs.

Because the SPHK family members are very divergent but contain a conserved catalytic domain, we compared amino acid identities and similarities of the SPHKc domains by pairwise alignments (Table III). In general, corresponding human and murine amino acid sequences of the same isotype have >90% similarity to one another, but the similarity is much lower when comparisons are

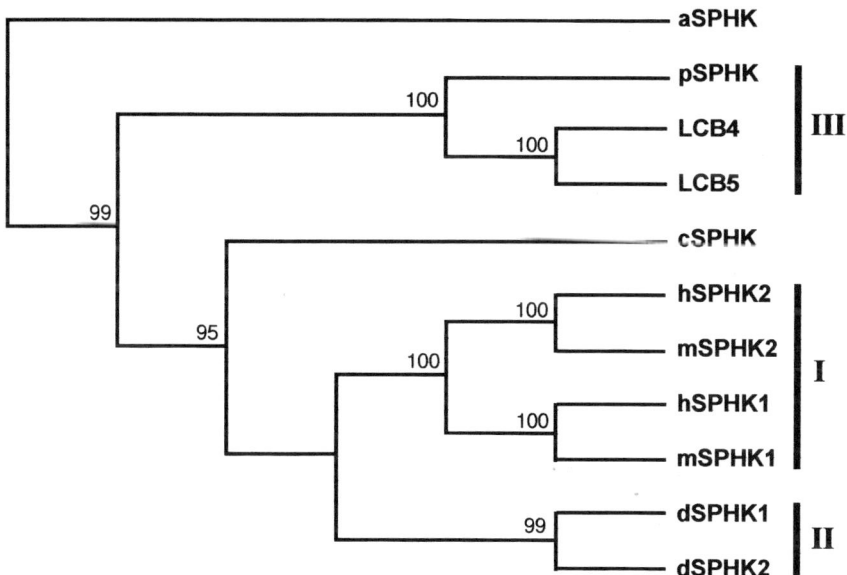

FIG. 4. Phylogenetic tree of the SPHK family. Alignment of known and putative SPHK sequences was done in MacVector by the CLUSTAL method. The phylogenetic tree was generated using the Neighbor Joining method. The numbers at the branches indicate the number of times the best tree was supported out of 100 bootstrap tries.

made to species which are more separated evolutionarily. Different isotype pairs (i.e., mSPHK1/mSPHK2, Lcb4/Lcb5) within the same organism shared >70% overall similarity, which is greater than the similarity to any DAGKc.

VIII. Phylogenetic Analysis of Sphingosine Kinases

Figure 4 shows a phylogenetic tree depicting the evolutionary relationships of the 11 members of the SPHK superfamily. The SPHKs can be divided into three subgroups. Human and mouse SPHK1 and SPHK2 are closely related to each other, form their own subgroup, and are clearly distinct from the two SPHKs from *D. melanogaster* (dSPHK1 and dSPHK2). Subgroup III contains three members from yeast: LCB4, LCB5 from *S. cerevisiae*, and a putative SPHK from *S. pombe*. SPHK from *C. elegans* has higher homology to subgroups I and II. The recently identified *Arabidopsis thaliana* SPHK (43) is the most diverse and is distinct from all other members.

IX. Concluding Remarks

Although it has long been known that S1P has important biological functions, only recently has SPHK, the critical enzyme regulating its formation, been purified and cloned. This has led to the identification of a new superfamily of related SPHKs. Based on motif comparisons, it is clear that SPHKs are a distinct and novel class of lipid kinases that are expressed throughout evolution. The structural diversity and complex pattern of tissue expression of SPHKs are reminiscent of the DAGK family, and it is probably not a coincidence that these two families share a high degree of homology in their catalytic domains. However, it should be pointed out that SPHKs have a unique catalytic domain (SPHKc) which is different from those of other known lipid kinases, and is even distinct from that of its close relative, the DAGKc. This growing class of SPHKs is likely to play an important role in the regulation of many cellular processes, including mitogenesis, apoptosis, neuronal development, chemotaxis, angiogenesis, and inflammatory responses. Manipulations of the genes that encode SPHKs will no doubt provide important insights into the functions of S1P in different organisms.

ACKNOWLEDGMENTS

We thank the many members of the Spiegel lab for their contributions to the studies that were quoted in this review. This work was supported by research grants from the National Institutes of Health (GM43880 and CA61774 to SS).

References

1. S. Pyne and N. J. Pyne, *Biochem. J.* **349**, 385–402 (2000).
2. S. Spiegel and S. Milstien, *Biochim. Biophys. Acta* **1484**, 107–116 (2000).
3. E. J. Goetzl, *Prostaglandins* **64**, 11–20 (2001).
4. T. Hla, *Prostaglandins* **64**, 135–142 (2001).
5. F. Wang, J. R. Van Brocklyn, J. P. Hobson, S. Movafagh, Z. Zukowska-Grojec, S. Milstien, and S. Spiegel, *J. Biol. Chem.* **274**, 35,343–35,350 (1999).
6. D. English, A. T. Kovala, Z. Welch, K. A. Harvey, R. A. Siddiqui, D. N. Brindley, and J. G. Garcia, *J. Hematother. Stem Cell Res.* **8**, 627–634 (1999).
7. O. H. Lee, Y. M. Kim, Y. M. Lee, E. J. Moon, D. J. Lee, J. H. Kim, K. W. Kim, and Y. G. Kwon, *Biochem. Biophys. Res. Commun.* **264**, 743–750 (1999).
8. H. Lee, E. J. Goetzl, and S. An, *Am. J. Physiol. Cell Physiol.* **278**, C612–C618 (2000).
9. E. Kupperman, S. An, N. Osborne, S. Waldron, and D. Y. Stainier, *Nature* **406**, 192–195 (2000).
10. H. Okamoto, N. Takuwa, T. Yokomizo, N. Sugimoto, S. Sakurada, H. Shigematsu, and Y. Takuwa, *Mol. Cell Biol.* **20**, 9247–9261 (2000).
11. M. J. Lee, S. Thangada, K. P. Claffey, N. Ancellin, C. H. Liu, M. Kluk, M. Volpi, R. I. Sha'afi, and T. Hla, *Cell* **99**, 301–312 (1999).
12. A. Hall, *Science* **280**, 2074–2075 (1998).
13. D. Bar-Sagi and A. Hall, *Cell* **103**, 227–238 (2000).
14. Y. Liu, R. Wada, T. Yamashita, Y. Mi, C. X. Deng, J. P. Hobson, H. M. Rosenfeldt, V. E. Nava, S. S. Chae, M. J. Lee, C. H. Liu, T. Hla, S. Spiegel, and R. L. Proia, *J. Clin. Invest.* **106**, 951–961 (2000).
15. J. P. Hobson, H. M. Rosenfeldt, L. S. Barak, A. Olivera, S. Poulton, M. G. Caron, S. Milstien, and S. Spiegel, *Science* **291**, 1800–1803 (2001).
16. S. Spiegel and S. Milstien, *FEBS Lett.* **476**, 55–67 (2000).
17. F. Alderton, S. Rakhit, K. K. Choi, T. Palmer, B. Sambi, S. Pyne, and N. J. Pyne, *J. Biol. Chem.* **276**, 12,452–13,460 (2001).
18. P. Lindahl, B. R. Johansson, P. Leveen, and C. Betsholtz, *Science* **277**, 242–245 (1997).
19. M. Hellstrom, M. Kaln, P. Lindahl, A. Abramsson, and C. Betsholtz, *Development* **126**, 3047–3055 (1999).
20. D. English, Z. Welch, A. T. Kovala, K. Harvey, O. V. Volpert, D. N. Brindley, and J. G. Garcia, *FASEB J.* **14**, 2255–2265 (2000).
21. J. R. Van Brocklyn, M. J. Lee, R. Menzeleev, A. Olivera, L. Edsall, O. Cuvillier, D. M. Thomas, P. J. P. Coopman, S. Thangada, T. Hla, and S. Spiegel, *J. Cell Biol.* **142**, 229–240 (1998).
22. O. Cuvillier, G. Pirianov, B. Kleuser, P. G. Vanek, O. A. Coso, S. Gutkind, and S. Spiegel, *Nature* **381**, 800–803 (1996).
23. Y. Morita, G. I. Perez, F. Paris, S. R. Miranda, D. Ehleiter, A. Haimovitz-Friedman, Z. Fuks, Z. Xie, J. C. Reed, E. H. Schuchman, R. N. Kolesnick, and J. L. Tilly, *Nature Med.* **6**, 1109–1114 (2000).
24. L. C. Edsall, O. Cuvillier, S. Twitty, S. Spiegel, and S. Milstien, *J. Neurochem.* **76**, 1573–1584 (2001).
25. M. Mattie, G. Brooker, and S. Spiegel, *J. Biol. Chem.* **269**, 3181–3188 (1994).
26. D. Meyer zu Heringdorf, H. Lass, R. Alemany, K. T. Laser, E. Neumann, C. Zhang, M. Schmidt, U. Rauen, K. H. Jakobs, and C. J. van Koppen, *EMBO J.* **17**, 2830–2837 (1998).
27. R. Alemany, B. Sichelschmidt, D. M. zu Heringdorf, H. Lass, C. J. van Koppen, and K. H. Jakobs, *Mol. Pharmacol.* **58**, 491–497 (2000).
28. C. J. van Koppen, D. Meyer zu Heringdorf, R. Alemany, and K. H. Jakobs, *Life Sci.* **68**, 2535–2540 (2001).

29. R. C. Dickson and R. L. Lester, *Biochim. Biophys. Acta.* **1438**, 305–321 (1999).
30. R. C. Dickson and R. L. Lester, *Biochim. Biophys. Acta* **1426**, 347–357 (1999).
31. M. M. Nagiec, M. Skrzypek, E. E. Nagiec, R. L. Lester, and R. C. Dickson, *J. Biol. Chem.* **273**, 19,437–19,442 (1998).
32. M. S. Skrzypek, M. M. Nagiec, R. L. Lester, and R. C. Dickson, *J. Bacteriol.* **181**, 1134–1140 (1999).
33. S. Kim, H. Fyrst, and J. Saba, *Genetics* **156**, 1519–1529 (2000).
34. J. L. Patton, B. Srinivasan, R. C. Dickson, and R. L. Lester, *J. Bacteriol.* **174**, 7180–7184 (1992).
35. R. C. Dickson, E. E. Nagiec, M. Skrzypek, P. Tillman, G. B. Wells, and R. L. Lester, *J. Biol. Chem.* **272**, 30,196–30,200 (1997).
36. G. M. Jenkins, A. Richards, T. Wahl, C. Mao, L. Obeid, and Y. Hannun, *J. Biol. Chem.* **272**, 32,566–32,572 (1997).
37. S. M. Mandala, R. Thornton, Z. Tu, M. B. Kurtz, J. Nickels, J. Broach, R. Menzeleev, and S. Spiegel, *Proc. Nat. Acad. Sci. U.S.A.* **95**, 150–155 (1998).
38. C. Mao, J. D. Saba, and L. M. Obeid, *Biochem. J.* **342**, 667–675 (1999).
39. G. M. Jenkins and Y. A. Hannun, *J. Biol. Chem.* **276**, 8574–8581 (2001).
40. C. J. Birchwood, J. D. Saba, R. C. Dickson, and K. W. Cunningham, *J. Biol. Chem.* **276**, 11,712–11,718 (2001).
41. S. Friant, B. Zanolari, and H. Riezman, *EMBO J.* **19**, 2834–2844 (2000).
42. B. Zanolari, S. Friant, K. Funato, C. Sutterlin, B. J. Stevenson, and H. Riezman, *EMBO J.* **19**, 2824–2833 (2000).
43. H. Nishiura, K. Tamura, Y. Morimoto, and H. Imai, *Biochem. Soc. Trans.* **28**, 747–748 (2000).
44. C. K. Ng, K. Carr, M. R. McAinsh, B. Powell, and A. M. Hetherington, *Nature* **410**, 596–599 (2001).
45. Y. Hannun, *Science* **274**, 1855–1859 (1996).
46. R. N. Kolesnick and M. Kronke, *Annu. Rev. Physiol.* **60**, 643–665 (1998).
47. S. Spiegel and A. H. Merrill, Jr., *FASEB J.* **10**, 1388–1397 (1996).
48. H. Zhang, N. N. Desai, A. Olivera, T. Seki, G. Brooker, and S. Spiegel, *J. Cell Biol.* **114**, 155–167 (1991).
49. A. Olivera and S. Spiegel, *Nature* **365**, 557–560 (1993).
50. L. C. Edsall, G. G. Pirianov, and S. Spiegel, *J. Neurosci.* **17**, 6952–6960 (1997).
51. O. Cuvillier, D. S. Rosenthal, M. E. Smulson, and S. Spiegel, *J. Biol. Chem.* **273**, 2910–2916 (1998).
52. B. Kleuser, O. Cuvillier, and S. Spiegel, *Cancer Res.* **58**, 1817–1824 (1998).
53. G. I. Perez, C. M. Knudson, L. Leykin, S. J. Korsmeyer, and J. L. Tilly, *Nature Med.* **3**, 1228–1232 (1997).
54. P. Xia, L. Wang, J. R. Gamble, and M. A. Vadas, *J. Biol. Chem.* **274**, 34,499–34,505 (1999).
55. O. Choi, H., J.-H. Kim, and J.-P. Kinet, *Nature* **380**, 634–636 (1996).
56. E. E. Prieschl, R. Csonga, V. Novotny, G. E. Kikuchi, and T. Baumruker, *J. Exp. Med.* **190**, 1–8 (1999).
57. Y. Su, D. Rosenthal, M. Smulson, and S. Spiegel, *J. Biol. Chem.* **269**, 16,512–16,517 (1994).
58. P. Xia, J. R. Gamble, K. A. Rye, L. Wang, C. S. T. Hii, P. Cockerill, Y. Khew-Goodall, A. G. Bert, P. J. Barter, and M. A. Vadas, *Proc. Natl. Acad. Sci. U.S.A.* **95**, 14,196–14,201 (1998).
59. P. Xia, M. A. Vadas, K. A. Rye, P. J. Barter, and J. R. Gamble, *J. Biol. Chem.* **274**, 33,143–33,147 (1999).
60. D. Meyer zu Heringdorf, H. Lass, I. Kuchar, R. Alemany, Y. Guo, M. Schmidt, and K. H. Jakobs, *FEBS Lett.* **461**, 217–222 (1999).
61. A. Melendez, R. A. Floto, D. J. Gillooly, M. M. Harnett, and J. M. Allen, *J. Biol. Chem.* **273**, 9393–9402 (1998).

62. K. W. Young, R. A. Challiss, S. R. Nahorski, and J. J. MacKrill, *Biochem. J.* **343**, 45–52 (1999).
63. R. Alemany, D. Meyer zu Heringdorf, C. J. van Koppen, and K. H. Jakobs, *J. Biol. Chem.* **274**, 3994–3999 (1999).
64. N. Mazurek, T. Megidish, S.-I. Hakomori, and Y. Igarashi, *Biochem. Biophys. Res. Commun.* **198**, 1–9 (1994).
65. B. M. Buehrer, E. S. Bardes, and R. M. Bell, *Biochim. Biophys. Acta* **1303**, 233–242 (1996).
66. R. A. Rius, L. C. Edsall, and S. Spiegel, *FEBS Lett.* **417**, 173–176 (1997).
67. M. Machwate, S. B. Rodan, G. A. Rodan, and S. I. Harada, *Mol. Pharmacol.* **54**, 70–77 (1998).
68. A. Olivera, L. Edsall, S. Poulton, A. Kazlauskas, and S. Spiegel, *FASEB J.* **13**, 1593–1600 (1999).
69. A. Olivera, T. Kohama, Z. Tu, S. Milstien, and S. Spiegel, *J. Biol. Chem.* **273**, 12,576–12,583 (1998).
70. T. Kohama, A. Olivera, L. Edsall, M. M. Nagiec, R. Dickson, and S. Spiegel, *J. Biol. Chem.* **273**, 23,722–23,728 (1998).
71. V. E. Nava, E. Lacana, S. Poulton, H. Liu, M. Sugiura, K. Kono, S. Milstien, T. Kohama, and S. Spiegel, *FEBS Lett.* **473**, 81–84 (2000).
72. S. M. Pitson, J. D'Andrea, R. L. Vandeleur, P. A. Moretti, P. Xia, J. R. Gamble, M. A. Vadas, and B. W. Wattenberg, *Biochem. J.* **350**, 429–441 (2000).
73. A. J. Melendez, E. Carlos-Dias, M. Gosink, J. M. Allen, and L. Takacs, *Gene* **251**, 19–26 (2000).
74. T. Murate, Y. Banno, T. K. K. K. Watanabe, N. Mori, A. Wada, Y. Igarashi, A. Takagi, T. Kojima, H. Asano, Y. Akao, S. Yoshida, H. Saito, and Y. Nozawa, *J. Histochem. Cytochem.* **49**, 845–855 (2001).
75. A. Olivera, T. Kohama, L. C. Edsall, V. Nava, O. Cuvillier, S. Poulton, and S. Spiegel, *J. Cell Biol.* **147**, 545–558 (1999).
76. P. Xia, J. R. Gamble, L. Wang, S. M. Pitson, P. A. Moretti, B. W. Wattenberg, R. J. D'Andrea, and M. A. Vadas, *Curr. Biol.* **10**, 1527–1530 (2000).
77. H. Liu, M. Sugiura, V. E. Nava, L. C. Edsall, K. Kono, S. Poulton, S. Milstien, T. Kohama, and S. Spiegel, *J. Biol. Chem.* **275**, 19,513–19,520 (2000).
78. B. K. Kay, M. P. Williamson, and M. Sudol, *FASEB J.* **14**, 231–241 (2000).
79. P. S. McPherson, *Cell. Signal* **11**, 229–238 (1999).
80. S. E. Phillips, B. Sha, L. Topalof, Z. Xie, J. G. Alb, V. A. Klenchin, P. Swigart, S. Cockcroft, T. F. Martin, M. Luo, and V. A. Bankaitis, *Mol. Cell* **4**, 187–197 (1999).
81. M. K. Topham and S. M. Prescott, *J. Biol. Chem.* **274**, 11,447–11,450 (1999).
82. S. K. Hanks, A. M. Quinn, and T. Hunter, *Science* **241**, 42–51 (1988).
83. M. K. Topham, M. Bunting, G. A. Zimmerman, T. M. McIntyre, P. J. Blackshear, and S. M. Prescott, *Nature* **394**, 697–700 (1998).
84. S. M. Pitson, P. A. Moretti, J. R. Zebol, P. Xia, J. R. Gamble, M. A. Vadas, R. J. D'Andrea, and B. W. Wattenberg, *J. Biol. Chem.* **275**, 33,945–33,950 (2000).
85. F. Sakane, M. Kai, I. Wada, S. Imai, and H. Kanoh, *Biochem. J.* **318**, 583–590 (1996).

Mechanisms of EF-Tu, a Pioneer GTPase

Ivo M. Krab* and
Andrea Parmeggiani*

Laboratory of Biophysics
Ecole Polytechnique
F-91128 Palaiseau Cedex, France

I. Introduction 514
   A. The GTPase EF-Tu 514
   B. EF-Tu Cycle in Polypeptide Elongation 514
   C. Other Properties and Functions 515
   D. Three-Dimensional Structure 517
   E. Interaction with Macromolecular Ligands 517
II. Structure–Function Relationships 524
   A. Catalytic Domain 1 524
   B. Regulatory Domains 2 and 3 525
   C. Nucleotide-Binding Pocket 525
   D. Magnesium Coordination Network 527
III. EF-Ts as a Steric Chaperone for EF-Tu Folding 529
IV. EF-Tu as Target of Antibiotics 531
   A. Mode of Action 532
   B. Resistance 534
   C. Binding Sites 535
   D. Kirromycin and EF-1α 537
V. Specific Aspects of EF-Tu GTPase Activity 538
   A. Stoichiometry, Cooperation 538
   B. The "Hydrophobic Gate" 540
VI. Conclusions and Perspectives 542
   References 543

This review considers several aspects of the function of EF-Tu, a protein that has greatly contributed to the advancement of our knowledge of both protein biosynthesis and GTP-binding proteins in general. A number of topics are described with emphasis on the function–structure relationships, in particular of EF-Tu's domains, the nucleotide-binding site, and the magnesium-binding network. Aspects related to the interaction with macromolecular ligands and

Abbreviations: EF, elongation factor; GBP, GTP binding protein; GEF, guanine nucleotide exchange factor; GAP, GTPase-activating protein.

*Current affiliation: Centre for Protein Engineering, MRC Centre, Cambridge CB2 2QH, United Kingdom.

antibiotics and to folding and GTPase activity are also presented and discussed. Comments and criticism are offered to draw attention to remaining discrepancies and problems. © 2002, Elsevier Science (USA).

I. Introduction

Elongation factor (EF) Tu is one of the best studied components of the machinery for protein biosynthesis. Moreover, for many years it has been a functional and structural model for GTP-binding proteins (GBP). Despite intensive studies, many aspects of its function remain to be clarified. This review does not intend to be a general review of EF-Tu properties and functions; it reflects personal experience and objectives. Section I is devoted to the context of this work and introduces the more specific problems treated in the following sections. For recent reviews of the general or specific properties of EF-Tu, see Refs. *1–6*.

A. The GTPase EF-Tu

EF-Tu is a ubiquitous GTPase (EC 3.6.1.-) that was identified and isolated from its complex with EF-Ts in *Escherichia coli* more than 3 decades ago (*7*). As a GBP, EF-Tu is a member of a superfamily of proteins that, as a carrier of information or biological components, regulates a myriad of cellular processes. It had become a major object of investigation more than a decade before the observations in the 1980s that GBPs are universal "on" and "off" switches (*8*). Common characteristics of the members of this superfamily are (i) the cycling between an active and an inactive conformation depending on whether either GTP ("on"-state) or GDP ("off"-state) is bound and (ii) a very low intrinsic GTPase and GDP/GTP exchange activity. These two activities, which control the level of the functionally active GTP-bound conformation, are activated by two specific ligands in most GBP systems, generally called GAP (GTPase-activating protein) and GEF (GDP/GTP exchange factor) (*6, 9, 10*). For EF-Tu, these roles are fulfilled by the ribosome and the EF-Ts, respectively.

B. EF-Tu Cycle in Polypeptide Elongation

In the bacterial cell, the key function of EF-Tu · GTP is to transport the aa-tRNA to the A site of the mRNA-programmed ribosome carrying peptidyl-tRNA in the P site. The formation of a stable ternary complex, EF-Tu · GTP · aa-tRNA, is the prerequisite for the elongation cycle of the polypeptide chain on the ribosome. Cognate codon–anticodon interaction triggers the fast hydrolysis of the EF-Tu-bound GTP, resulting in EF-Tu · GDP release due to its low affinity for aa-tRNA and ribosome. This allows a correct positioning of aa-tRNA into the ribosomal A site leading to the synthesis of a peptide bond between the NH_2

group of the amino acid and the C-terminal ester group of peptidyl-tRNA situated in the P site, forming the so-called "pretranslocational" ribosome complex. There is evidence that the hydrolysis of GTP and release of EF-Tu are parts of a proofreading mechanism that discourage misincorporation of near-cognate aa-tRNA (*11*; see also Section I,E,3).

The next step, the translocation of peptidyl-tRNA from the A to the P site promoted by EF-G—another GBP—displaces the discharged tRNA situated in the P site to the E site. The vacant A site that results from the formation of this "post-translocational" ribosome complex allows the binding of a new ternary complex, starting the subsequent elongation cycle.

Various factors contribute to the recycling of the released EF-Tu · GDP: (i) the action of EF-Ts that favors the dissociation of the nucleotide from EF-Tu, (ii) the interaction between EF-Tu and aa-tRNA strongly increasing the affinity for GTP, and (iii) the 5- to 10-fold higher concentration of GTP than that of GDP in the cell. The EF-Tu function in the elongation cycle is summarized in Fig. 1, which reports a traditional scheme that—although disputed in some details—has remained a fundamental reference during past decades.

C. Other Properties and Functions

EF-Tu from *E. coli* (EF-Tu$_{Ec}$) is the most abundant cell protein (5–10% of the protein content); it is approximately equimolar to aa-tRNA and 10-fold more abundant than the ribosome (*12*). Nearly all cellular aa-tRNA is trapped by EF-Tu · GTP · EF-Tu$_{Ec}$ and is encoded by two unlinked genes: *tuf*A (in the *str* region of the chromosome) and *tuf* B (in the *rif* region) (*12*). Their sequences are nearly identical, and the products only differ in the C terminus—Gly in EF-TuA and Ser in EF-TuB. They do not show any functional difference (*13*). Either *tuf* gene is dispensable in EF-Tu without loss of cell viability (*14, 15*), making the reason for the existence of more than one gene, as yet, unclear. Two *tuf* genes can also occur in other bacterial species (*16*).

The abundance of EF-Tu in the cell has suggested possible functions other than those in elongation. It has been proven to form one of the four subunits of RNA-phage replicase, acting in complex with EF-Ts in the initiation step (*17*). Its possible role in the regulation of transcription has received a new impulse with the recent finding that EF-Tu interacts directly with the C-terminal region of the β-subunit of the RNA polymerase (*18*). In this context, its involvement in the stringent response has been suggested, as already proposed in the 1970s (*19*). *In vitro* evidence indicates that EF-Tu can facilitate the refolding of several denatured proteins, such as rhodanese (*20*), citrate-synthase, and α-glucosidase, and possesses a disulfide isomerase activity favoring the correct redox status of nascent proteins (*21*). However, EF-Tu has not been mentioned as one of the proteins associated with the nascent α-galactosidase (*22*). Noteworthy is the property of the eukaryotic counterpart, EF-1α

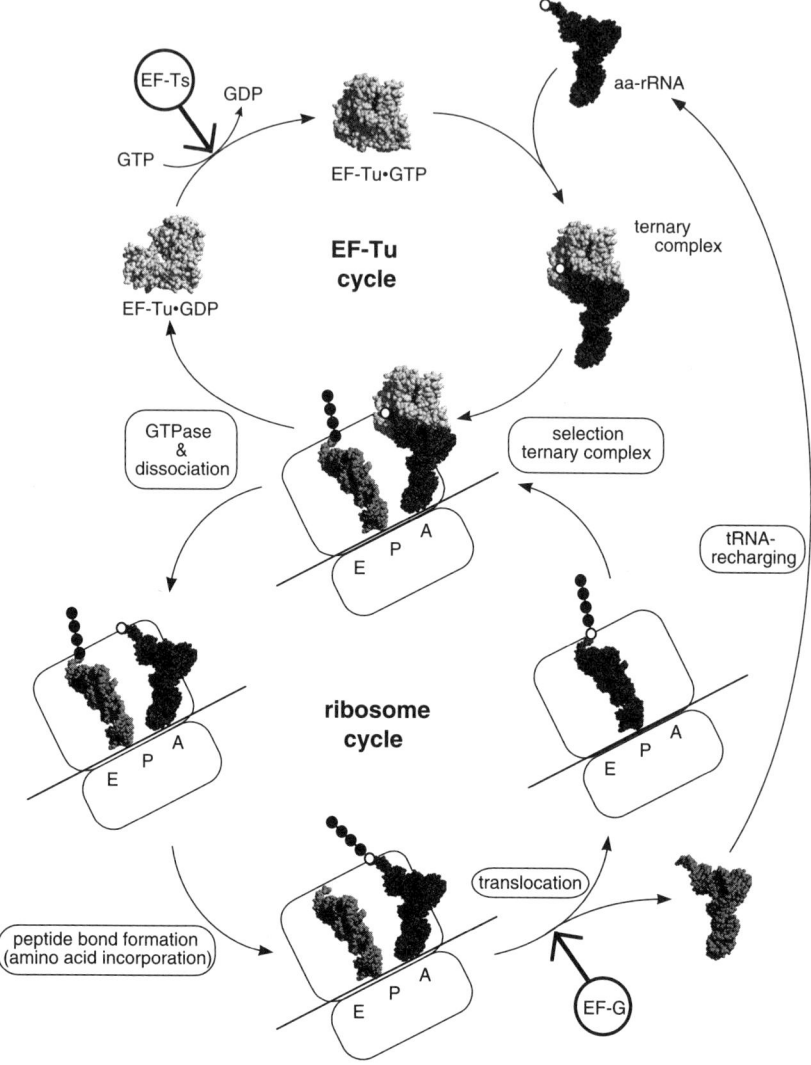

FIG. 1. Scheme of the EF-Tu function in the elongation cycle of protein biosynthesis.

to form stable complexes with cytoskeleton components such as actin and tubulin (23).

EF-Tu is, after the ribosome, the most important target for antibiotics inhibiting protein biosynthesis (see Section IV). These agents have contributed to providing insight into several of the basic properties of the factor.

D. Three-Dimensional Structure

EF-Tu is a monomeric protein with a molecular mass ranging from 40 to 45 kDa depending on the bacterial species (EF-Tu$_{Ec}$: 393 aa; 43 kDa). EF-1α is larger ($\sim$52 kDa) due to a few insertions. Three-dimensional (3D) models of EF-Tu from various bacterial sources (EF-Tu$_{Ec}$, EF-Tu$_{Tt}$ from *Thermus thermophilus*; EF-Tu$_{Ta}$ from *Thermus aquaticus*) exist for GTP- and GDP-induced conformations (*24–29*). In this review, EF-Tu without specification refers in general to EF-Tu$_{Ec}$. Moreover, if not otherwise stated, residue numbers refer to EF-Tu$_{Ec}$; even in the case of EF-Tu from other bacterial species.

The EF-Tu molecule consists of three domains, whose relative positions change dramatically upon binding of GTP and GDP (Fig. 2). EF-Tu · GDP displays a hole and limited contacts between the three domains, while the EF-Tu in complex with GTP is compact without the hole and with extensive contacts between the three domains (Figs. 2A and 2B). The N-terminal Domain 1 consists of a core formed by six β-strands (a–f) surrounded on both sides by six major α-helices (A–F) in a fold shared by all GBPs. The nucleotide-binding pocket is delimited by structural elements containing three GBP consensus motifs (GBP I–III) and two conserved motifs for prokaryotic elongation factors (EF I–II, see Figs. 2 and 3). The structures modified by the type of bound nucleotide, the Switch 1 (in the effector region, so called by homology with the Ha-Ras p21 region interacting with ligands) and Switch 2 regions delimit the guanine nucleotide-binding pocket. Consensus motifs I and II and the switch 1 region each contain a residue involved in the coordination ring of the essential Mg^{2+} bound to the β- and γ-phosphates of the nucleotide (Fig. 2C and D). Domains 2 and 3 of EF-Tu$_{Ec}$ contain only β-strands. The transition from the GDP- to the GTP-bound state is associated with overall conformational changes in which Domains 2 and 3 together behave as a rigid unit, leading to a displacement of certain residues by as much as $\sim$40 Å (*25 28*) (Figs. 2E and 2F).

E. Interaction with Macromolecular Ligands

Under specific conditions, it is possible to isolate stable complexes between EF-Tu and three of the macromolecular ligands involved in protein synthesis: EF-Ts, aa-tRNA, and the ribosome. EF-Tu · EF-Ts represents a true physiological intermediate of the elongation process, whereas with the other two ligands the formation of stable complexes suitable for 3D studies requires the use of nonhydrolysable analogs and, in the case of ribosomes, also antibiotics. The recent progress of structural analysis has made 3D models of these complexes available, enabling the identification of either the interacting sites or, in the case of the ribosome, of the EF-Tu domains and ribosomal proteins involved in the binding.

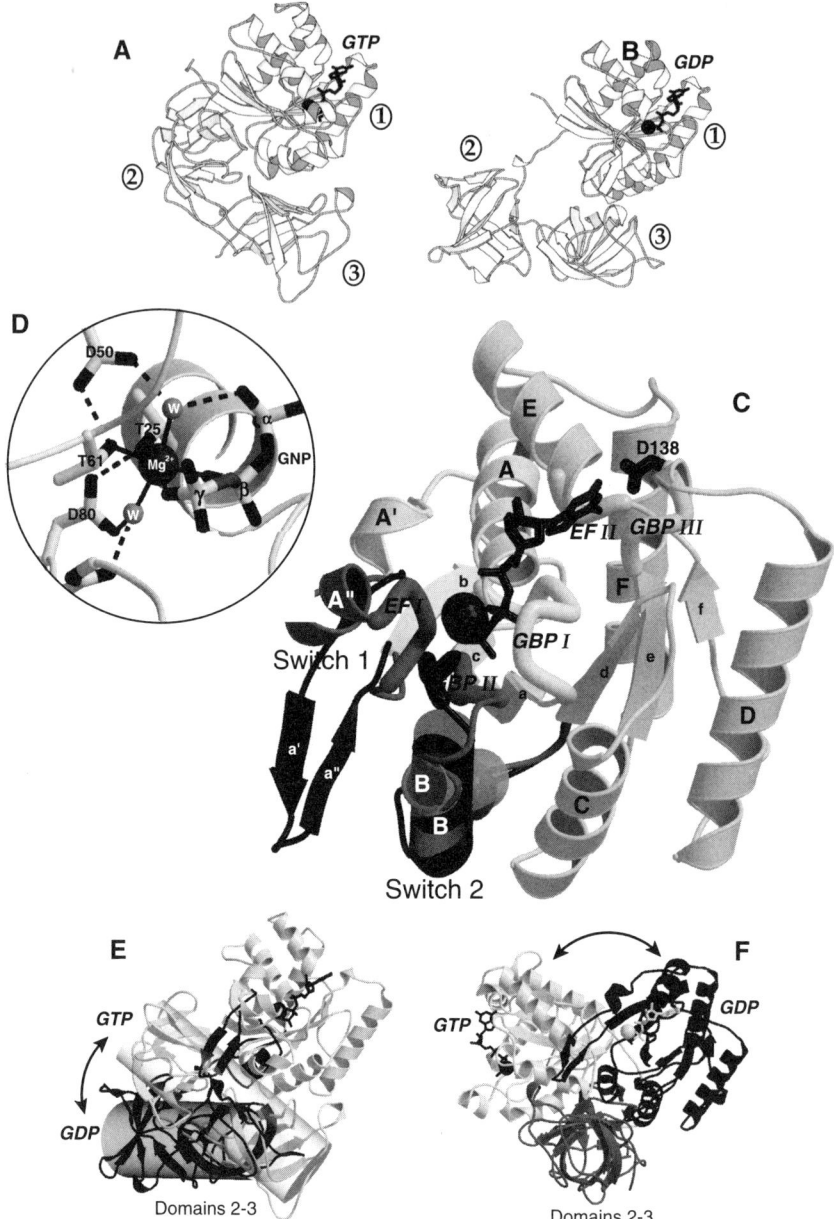

FIG. 2. 3D structure of EF-Tu. Top: Overview of the basic conformations. (A) The compact GTP-bound form and (B) the GDP-bound form, showing the opening between the three domains (designated by numbers). (C) Detail of Domain 1 showing both conformations of the switch regions

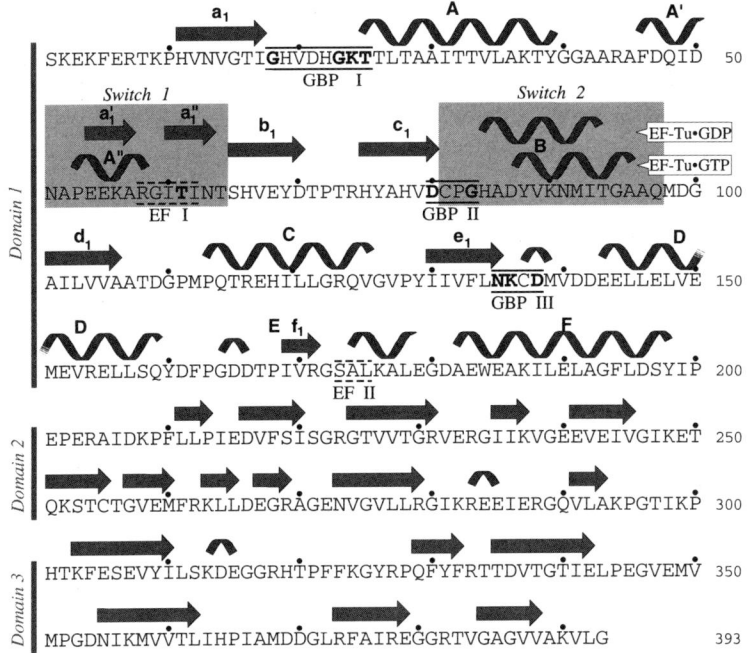

FIG. 3. Summary of the primary and secondary structures of EF-Tu$_{Ec}$, showing the differences between GDP- and GTP-bound conformation in the switch regions (gray background) and the consensus motifs GBP I–III and EF I–II mentioned in the text.

(GDP conformation in black, GTP conformation medium gray) with labels locating secondary structure elements and consensus motifs GBP I–III and EF I–II (represented by thicker backbone tubes, see also Fig. 3). The side chain of base-discriminating residue D138 is also shown. Note the unwinding of α-helix A″ present only in the GTP conformation to a β-hairpin bridging toward Domain 3 in the GDP conformation and the rearrangement of helix B (enveloped in a transparent cylinder to better distinguish the two conformations). (D) A close-up of the magnesium ion and its octahedrally arranged ligands in the GTP conformation: to the right and front, oxygens from the β- resp. γ-phosphate; to the left and back, the Thr61 and Thr25 OH groups; at the top and bottom, water molecules (W) hydrogen-bonded to Asp50/α-phosphate and Asp80, respectively. Panels (E) and (F): Representations of the overall conformational change of EF-Tu. In (E) the G domains are superimposed. Both conformations of Domains 2 and 3 and the switch regions in Domain 1 are shown. The GDP conformation is dark, and the GTP conformation is bright. Transparent cylinders enveloping Domains 2 and 3 show how together these domains move as a unit. (F) shows the converse: Domains 2 and 3 of both conformations are aligned (shown in medium gray for contrast—looking from the right down the axis of the cylinders in Panel (E). In this view, Domain 1 can be seen to "roll over" the Domain 2–3 unit. All 3D structures in this and subsequent figures were rendered using Molscript (153) and in some cases Raster3D (154).

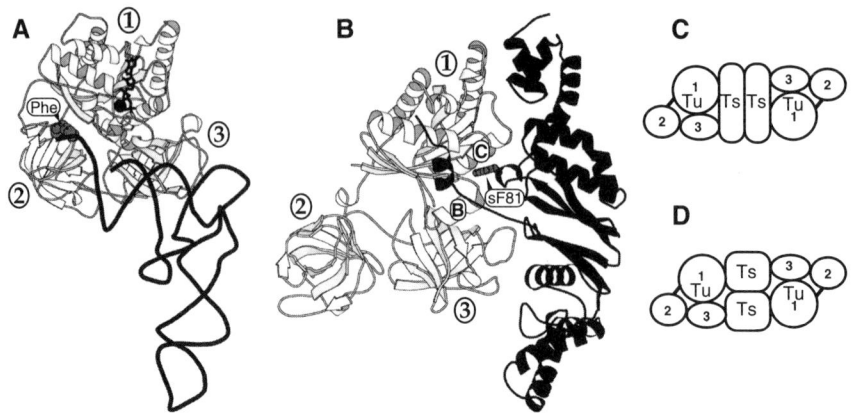

FIG. 4. (A) 3D model of the ternary complex EF-Tu$_{Ta}$·GMPPNP·Phe-tRNA^{Phe}. (B) View of one EF-Tu·EF-Ts dimer of the *E. coli* (EF-Tu·EF-Ts)$_2$ complex. EF-Tu is in white and EF-Ts is in gray, with EF-Ts residue F81 intruding between EF-Tu helices B and C. On the right: a schematic depiction of the difference in interaction between the two EF-Tu and EF-Ts in the tetramers of *E. coli* (C) and *T. thermophilus* (D).

1. TERNARY COMPLEX

The 3D model of the EF-Tu$_{Ta}$·GMPPNP·Phe-tRNA^{Phe} complex (Fig. 4A) shows that the tRNA-bound amino acid is accommodated in the cleft between Domains 1 and 2 (*30*). This followed cross-linking and protection studies, which had suggested that the aminoacylated CCA 3' end of tRNA interacts with Domains 1 and 2 (*31*). Major contacts take place between Domain 3 and one side of the T stem, while the 5' phosphate is located in the three-domain cross point. The anticodon arm points away from EF-Tu. EF-Tu·GMPPNP as a component of the ternary complex essentially conserves the same conformation as that of free EF-Tu·GMPPNP. A second ternary complex structure, that of EF-Tu$_{Ta}$·GMPPNP·Cys-tRNA^{Cys} (*32*), has confirmed the general principles of the EF-Tu·tRNA interaction derived from the Phe-tRNA^{Phe} ternary complex, while specific differences in the way the 3' ends are bound suggest how all the 20 different aminoacylated 3' ends may be accommodated. The 3D model of the ternary complex confirms earlier functional results, indicating that the interaction of aa-tRNA with EF-Tu takes place on one side of the aminoacylated 3' terminal end and stem, T stem, and extra loop (*33–35*). It is noteworthy that the action of the aminoacylated CCA 3' end can be influenced *via* EF-Tu by the remaining regions of tRNA (*36*).

An interesting contribution is that of bovine mitochondrial EF-Tu, whose 3D structure has been resolved at 1.94 Å resolution (*37*) and displays a different orientation of Domain 1 to the other two domains as compared to bacterial

EF-Tu. A C-terminal extension shows structural similarity to zinc finger proteins, suggesting that it may be involved in the recognition of RNA. The lower affinity for nucleotides of mitochondrial EF-Tu compared to that for the bacterial one is ascribed to a possibly increased flexibility of elements surrounding the nucleotide-binding site.

2. EF-Tu · EF-Ts

In the crystal of EF-Tu · EF-Ts$_{Ec}$, the first 3D structure of a GBP · GEF complex, two EF-Tu · EF-Ts form a heterotetramer in which the two EF-Ts interact extensively, each being associated with only one EF-Tu. In this complex, EF-Tu shows a conformation resembling that of EF-Tu · GDP (38) (Figs. 4B and 4C). This quaternary structure is not a general feature of this complex. EF-Tu · EF-Ts$_{Tt}$ forms a different heterotetramer (Fig. 4D) in which each EF-Ts interacts with both EF-Tu's (39), though the interaction surface conserves the same essential features of the E. coli complex. The major contacts with EF-Ts are in Domain 1, while, of the other two EF-Tu domains, only the tip of a Domain 3 hairpin interacts directly with EF-Ts. In Domain 1 structures, such as the "P loop," the amino terminus of α-helix B—tightly associated with Domain 3—and those of α-helices C and D are involved. As a remarkable feature, the residue F81 of EF-Ts (sF81) intrudes into EF-Tu between α-helices B and C. Intrusion of a ligand's residue into a GBP has thereafter also been observed in other complexes, such as those formed by Ha-Ras p21 (40) and ARF-1 (41) with GAP and GEF, respectively. The amino acid residue involved has been proposed to represent an important functional element—dubbed "finger"—that either participates in catalysis or shifts structural elements of the host GBP.

As for the mechanism by which EF-Ts destabilizes the EF-Tu · guanine nucleotide complex, sF81 was proposed, with some variation in the details (38, 39), to (i) destabilize residues from the magnesium coordination ring by shifting α-helices B and C, thus favoring the release of the nucleotide; and/or (ii) to generate a cascade of interactions leading to a peptide flip in the "P loop." This would break hydrogen bonds fixing the β-phosphate of the bound nucleotide, thus weakening the binding of the nucleotide.

3. Ribosome

In recent years, exciting progress has been made in elucidating the 3D structure of the Tt 30S ribosomal subunit at 3.0 Å (42), the Tt 50S subunit at 2.4 Å (43), and the Tt 70S ribosome with RNAs bound at 5.5 Å (44). The direct visualization of ribosomal RNA (rRNA) and most proteins showed that the intersubunit areas consist primarily of rRNA, while proteins mostly form an outer shell on the rRNA skeleton.

Concerning the factors, older observations (45–47) indicated that the 50S ribosome is responsible for enhancing the EF-Tu GTPase activity, whereas the 30S ribosome exerts a modulatory effect. More recently, cross-linking, rRNA protection, and functional experiments found that the most probable region of the 50S subunit binding to EF-Tu includes the stalk formed by the (L7–L12)$_2$ dimers, the nearby proteins L10 and L11, and the α-sarcin loop (also called the ribotoxin loop) from domain V of 23S rRNA (48). In the 16S rRNA of the 30S subunit, the 530-loop is also implicated in the interaction with EF-Tu (49).

Some of the ribosome structures interacting with EF-Tu and EF-G have been available at high resolution before the whole subunit structure, like the isolated α-sarcin loop (50), the complex of protein L11, and the underlying region of 23S rRNA (51). Both structures are implicated in conformational switches. It is noteworthy that the concerning parts of 23S rRNA bind to EF-G free in solution (52). Although this has not been shown for EF-Tu, several synthetic RNAs selected for binding to EF-Tu show a consensus motif found in the α-sarcin region (53).

Even if without L7/L12 some factor-dependent amino acid incorporation was described (54), fast protein synthesis requires an intact stalk (55). A single-headed dimer of L7/L12 was recently found to be sufficient for protein synthesis (56). Whereas the EF-Tu GTPase is not stimulated by free L7/L12 (57), a recent report finds stimulation of EF-G GTPase by high concentrations of L7/L12 ($K_M = 43\mu M$); but mutagenesis shows that the single arginine in the head of L7/L12 is not important for either this stimulation or translation (58, 59). Thus, it is likely not an "arginine finger" in a GAP-like action of L7/L12 on elongation factors. The role of L7/L12 in binding EF-Tu, but not in GTP hydrolysis, is supported by the observation that methanol allows ribosomal cores deprived of L7/L12 to enhance the EF-Tu GTPase activity (46).

Direct visualization of an elongation factor on the ribosome was first achieved with a kirromycin-stalled EF-Tu · GDP · aa-tRNA complex at 18 Å resolution (60). Domain 1 is in contact with both the C-terminal region of the (L7–L12)$_2$ stalk and the underneath region of the 50S subunit, and Domain 2 is close to the 30S subunit on the side opposite to the platform where proteins S4, S5, and S12 are located.

For EF-G, several stable complexes with the ribosome were visualized in both "pre-" and "post-translocational" states. For either of these, depending on whether fusidic acid, thiostrepton, or a noncleavable GTP analog was used to prepare the complex, considerable differences in the relative orientation of EF-G domains and their contacts with the ribosome were observed (61–63, 93). These discrepancies are probably due to these complexes being stalled in non-physiological states. For example, thiostrepton binds to and interferes with just the region at the base of the stalk (see also Ref. 43) found not to interact with EF-G in the thiostrepton "post-translocational" complex (63). Consequently, the structure of the EF-Tu · GDP · kirromycin · aa-tRNA complex on the ribosome ought to be interpreted with caution, and comparison to a ternary complex containing a nonhydrolyzable GTP analog would be informative, because binding of a GMPPNP-containing ternary complex, but not of a GDP · kirromycin ternary complex, increased cross-linking from specific positions of L7/L12 to several ribosomal proteins (64).

The overall similarity between the EF-Tu ternary complex and EF-G · GDP, termed "macromolecular mimicry" (30), suggests common ribosomal-binding sites, in line with competition phenomena between EF-Tu and EF-G for binding to ribosomes (65). This is also supported by the protection of the α-sarcin loop by either EF-Tu or EF-G (48) and by the common functional role of proteins L7/L12 in the activity of both factors (45, 46, 66). Further studies are needed for improving the visualization of the EF-G-binding site on the ribosome and for reexamining the competition between EF-Tu and EF-G. It is noteworthy that EF-1α from mouse ascite cells was reported not to dissociate from the ribosome during elongation, thus allowing the simultaneous binding of EF-2 (67).

Some experimental evidence has been interpreted to suggest that the EF-Tu effector region interacts with the ribosome (68, 69); however, this is far from proven, even if GAP has been shown to bind to the homologous region in Ha-Ras (70).

Kinetic parameters of partial reactions of the elongation cycle have been measured by stop–flow techniques and some parameters derived by global fitting (11). The initial steps in the EF-Tu part of the cycle are all relatively fast, but the dissociation of EF-Tu · GDP from the ribosome is the slowest ($3\ \text{s}^{-1}$), thus rate-limiting, while aa-tRNA accommodation is only slightly faster. The low dissociation rate does not account for the 10–20 aa/s elongation speed that can be achieved *in vivo* and therefore reflects *in vitro* conditions, suggesting caution in extrapolating these observations to the *in vivo* situation.

The parameters also show how fidelity of selection is achieved: Compared to binding of a cognate aa-tRNA, a near-cognate aa-tRNA shows an 85-fold higher reverse rate of the codon recognition step, a 10-fold lower GTPase activation rate (10^4-fold lower for a noncognate aa-tRNA), and a 70-fold decreased rate of aa-tRNA accommodation after EF-Tu release resulting in at least 20-fold increase in aa-tRNA rejection. Thus, at each of several steps, a near-cognate is "discouraged" from proceeding. In the proposed "induced-fit" mechanism of recognition, a cognate aa-tRNA more efficiently induces a conformation of the ribosome that triggers the EF-Tu GTPase and accommodates the aa-tRNA in the A site. The coupling of the codon recognition signal to GTPase activation was also examined by aa-tRNA fragments, an A-site-binding anticodon/D loop, and an EF-Tu-bound acceptor stem/T hairpin. This "split ternary complex" is not able to trigger EF-Tu GTPase hydrolysis on A-site binding like an intact aa-tRNA · EF-Tu · GTP complex would (71). This shows that a cognate codon–anticodon interaction of the anticodon stem-loop by itself does not generate the signal for activating the EF-Tu GTPase trigger in the ribosome, but unfortunately it does not allow one to decide whether the trigger signal passes through the aa-tRNA, which therefore must be intact, or whether the entire aa-tRNA on codon–anticodon interaction induces a conformation in the ribosome that causes the trigger on EF-Tu.

II. Structure–Function Relationships

The multidomain and multifunctional nature of EF-Tu makes it a model protein for the study of structure–function relationships. The identification of kirromycin as the first antibiotic acting specifically on EF-Tu (72) created the possibility of carrying out a selective screening for mutants (see Section IV,B). In the 1980s, the rapid development of genetic engineering opened further horizons to the study of structure–functional aspects. The investigation of the function of its domains, in particular that of Domain 1 containing the GBP consensus motifs, became a priority.

A. Catalytic Domain 1

To investigate whether the G domain could express some of the functions of the full-length molecule, it was overexpressed from an engineered gene and purified (73). This construct proved active in GTP and GDP binding and in GTP hydrolysis. However, a characteristic property of EF-Tu, its differential affinity for GDP and GTP—a consequence of the very slow dissociation of EF-Tu · GDP—is lost. The affinity for both nucleotides lies in the micromolar range, which, compared to that of EF-Tu · GDP, represents a dramatic decrease (1000 times) but versus that if EF-Tu · GTP is only 10 times less. Also the catalytic activity is modified. The one-round course of the intrinsic GTP hydrolysis of full-length EF-Tu (74) has changed into a linear multiround turnover with a velocity close to the initial velocity of the one-round hydrolysis. This modification—a consequence of the much faster dissociation rate of GDP that is no longer the rate-limiting step in the regeneration of the GTP complex—demonstrates that the intrinsic catalytic activity is fully preserved. In contrast, the interaction with macromolecular ligands is either absent (aa-tRNA) or very weak (ribosome) and anomalous (EF-Ts) (73, 75). Accordingly, the G domain was found to interact with 23S rRNA (J. M. Robertson, A. Parmeggiani, and H. F. Noller, quoted in Ref. 48). These results show that the G domain can sustain only the basic activities of EF-Tu (nucleotide-binding and GTPase activity). Circular dichroism (76) and NMR studies (77) emphasize that the nucleotide-binding region of the G domain retains an organized structure. Thus, the behavior of the isolated Domain 1 underlines the prominent role of Domains 2 and 3 in the regulation of the EF-Tu functions.

B. Regulatory Domains 2 and 3

The structure–function relationships between Domain 1 and the other two domains were explored with constructs in which either Domain 2 (EF-Tu[1,3]) or Domain 3 (EF-Tu[1,2]) was deleted (78). The existence of functionally active conformations and selective properties proved the conservation of a physiological or near-physiological folding in both constructs. Circular dichroism spectra and thermostability curves indicated a structural organization of the domains not markedly deviating from that in full-length EF-Tu. Model building concerning orientation and folding shows that the interaction forces between Domains 1 and 3 are sufficiently strong to conserve the interface, avoiding significant changes of the relative orientations and distances.

Both EF-Tu[1,2] and EF-Tu[1,3] bind GTP and GDP with an affinity in the micromolar range, resembling in this regard more the G domain than the full-length factor. The affinity for GDP is decreased ~1000 times; that for GTP, only 10 times. Thus, the differential behavior of these two nucleotides is a selective property of only the intact molecule, resulting from cooperative effects involving all three domains. On the other hand, different from the G domain, EF-Tu[1,2] and [1,3] can only in part express intrinsic GTP hydrolysis, very likely due to long-range effects altering the catalytic site.

A major finding was the ability of EF-Ts to efficiently stimulate the release of GDP from both constructs (78). Since only Domain 3 directly interacts with EF-Ts (38), it is remarkable that each of the noncatalytic domains is sufficient to induce a Domain 1 conformation capable of perceiving the EF-Ts signal, even if both domains are required for optimum activity. In contrast, no stimulation by ribosome on the intrinsic GTPase activity of either EF-Tu[1,2] or EF-Tu[1,3] was observed, showing that the transmission of the ribosome signal to the Domain 1 catalytic center requires a more integrated action of the two noncatalytic domains.

Both regulatory domains of EF-Tu are required for poly(Phe) synthesis *in vitro,* as expected from a process involving the multistep formation of highly selective complexes. Either one of the mutated constructs can form a stable complex with aa-tRNA but only to a low extent (<10% that of EF-Tu). This result is nevertheless interesting, since no binding of aa-tRNA was observed with the G domain (73, 75). As suggested by the 3D model (30), an efficient binding of aa-tRNA requires the combination of all three domains.

C. Nucleotide-Binding Pocket

The nucleotide-binding site of EF-Tu is formed by the GBP consensus motifs I ("P loop"), II, and III. The Mg^{2+} coordinates the phosphate groups of the nucleotide with the first two motifs and the Switch 1 region (Fig. 2D). Residues from motif III interact with the base (24, 25). In both this and the next section, the function of several residues of both the guanine nucleotide-binding site and the Mg^{2+} coordination ring is considered. V20 and H84 from consensus motifs I and II, respectively, are considered in Section V,B.

C81 from motif II (D80 CPG83) is highly conserved in elongation factors and is covalently modified by the inhibitor of aa-tRNA-binding N-tosyl-L-phenylalanine chloromethane (79). However, substitution C81($\rightarrow$G) only partially inhibits poly(Phe) synthesis and aa-tRNA interaction (80). The lower activity of this EF-Tu mutant is probably due to long-range effects mediated by an overall destabilization of the molecule, more pronounced for the GDP-bound state. Remarkably, with mutation P82T in the G domain (81), the disappearance of GTPase activity is accompanied by autophosphorylation of the substituting threonine, an effect also found with the homologous residue in either human Ha-Ras p21 (82) or yeast Ras2p (83). The autokinase activity points to the close proximity of T82 to the γ-phosphate of GTP without direct contact, since compared to EF-Tu[wt] the affinity for GDP and GTP is not affected.

Concerning the nucleotide base, N135 and D138 are two interesting residues. The 3D model shows that the amino group N(2) of the base forms a strong hydrogen bond with D138 (Fig. 5). Accordingly, EF-Tu[D138N] recognizes XTP (84–86) (see also Section V,A). The situation concerning the exocyclic 0(6) of the base was less clear, for it was first thought that either this oxygen could form a hydrogen bond with N135 (87) or at least the existence of a weak

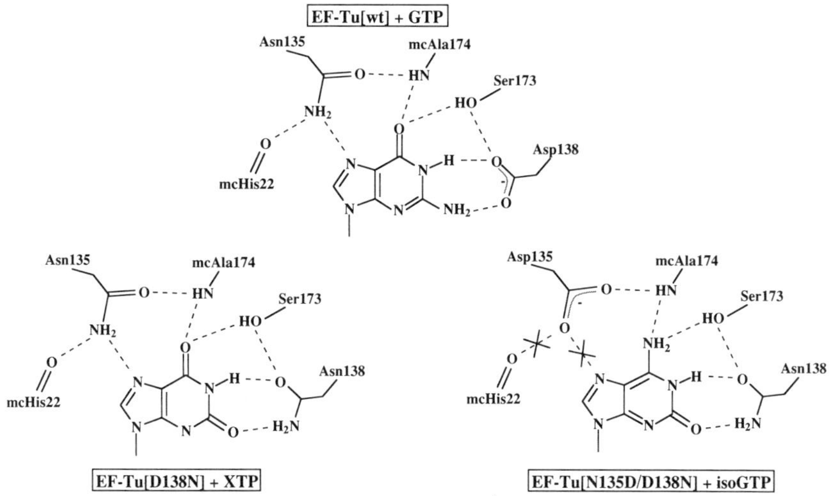

FIG. 5. Diagram representing the hydrogen bond interactions of guanine in EF-Tu[wt], xanthine in EF-Tu[D138N], and isoguanine in EF-Tu[N135D/D138N] according to the refined model of (24). Adapted from Ref. 89.

interaction between this residue and the base was not excluded (88). Later on, refined models of EF-Tu · GDP (24) indicated that the side chain of N135 either hydrogen-bonds with O(6) and N(7) or, as a more probable alternative, hydrogen-bonds with the main chain carbonyl group of H22 on consensus motif I ("P loop") involved in phosphate binding. The lack of a productive interaction between double-mutant EF-Tu[N135D/D138N] and *iso*-GTP (Fig. 5) demonstrates that N135 is unable to recognize the exocyclic keto group of the guanine base (89). However, since substitution N135D in EF-Tu[D138N] abolishes any response to XTP, N135 must be essential for the correct folding of the nucleotide-binding pocket.

D. Magnesium Coordination Network

In EF-Tu, as in all GBPs, magnesium plays a central role in the interaction with the nucleotide and fulfills several functions. Chelation of Mg^{2+} dramatically decreases (1000 times) the affinity for GDP (90), an effect used to obtain nucleotide-free-EF-Tu (91). Evidence was obtained assigning a low-affinity role to Mg^{2+} influencing nucleotide binding and GTPase activity via changes of the overall conformation and a high-affinity role important to the catalysis (92, 93). Relatively high concentrations of EDTA completely eliminate the kirromycin-dependent GTPase activity (92). On the other hand, a recent study on *Thermus thermophilus* EF-Tu reports intrinsic and kirromycin-stimulated GTPase activity to be substantially identical in either the absence or the presence of Mg^{2+} (94).

In EF-Tu$_{Ec}$, the residues involved in nucleotide–Mg^{2+} coordination are T25 and T61, which bind to the ion directly *via* the OH–oxygen, and D80 and D50, each of which hydrogen-bonds a water molecule of the coordination sphere of Mg^{2+} (Fig. 2D). T25 is a strictly conserved residue on α-helix A flanking at the C terminal side the "P loop" (loop L1). Substitution of Thr by Ser (95), the amino acid most similar to Thr, causes a slight decrease in the affinity for GDP, an effect probably related to a loss of hydrophobic interactions of the T25 side chain methyl group stabilizing the effector region. Quite dramatic is the effect of substitution T25A that suppresses the link to Mg^{2+} and causes a 1000 times decrease in affinity for GDP. Surprisingly, EF-Tu[T25A] activity in protein synthesis is still preserved to some extent even in the absence of this bond, emphasizing that other interactions are also important. Accordingly, EF-Tu[T25S] can efficiently bind aa-tRNA, and even EF-Tu[T25A] is partially active. However, the lack of protection against spontaneous hydrolysis of the aa-tRNA ester bond by the latter mutant reveals an anomalous interaction with the aa-end of tRNA.

An interesting effect is obtained with the introduction of mutation H22Y in EF-Tu[T25S] that reduces the affinity for GDP as markedly as that for T25A (95), probably by a mechanism similar to the nucleotide release induced by EF-Ts (H118→ Q114→"P loop") (39). Both EF-Tu[T25A] and EF-Tu[D80N] show a dominant-negative phenotype caused by an increased stability of the complex with EF-Ts, concomitant with a markedly decreased nucleotide affinity. EF-Tu[H22Y/T25S] has a similarly decreased nucleotide affinity yet is not dominant-negative, demonstrating that the increased stability of the complex with EF-Ts in the dominant-negative phenotype is related to the ability of EF-Tu to complex Mg^{2+} and not simply to the affinity for the nucleotide. These results show that the Mg^{2+}–T25 bond plays a central role in the mechanism of dissociation of EF-Tu · GDP, in agreement with the 3D model (38) in which the phosphate cavity is not completely open due to a peptide flip in the "P loop." If a stable Mg^{2+}–T25[OH] interaction is hindered, the insertion of the nucleotide into the binding site and the release of EF-Ts must be impaired. The same could occur in the reverse process, the dissociation of the EF-Tu · nucleotide by EF-Ts, in which breaking the Mg^{2+}–T25[OH] bond could constitute the major energetic barrier.

In the GTP conformation, T61 lies at the C-terminal side of the short α-helix A″ of the Switch 1 region (27, 28), while D80 is strictly conserved in the D80CPG83 motif preceding the Switch 2 region (Figs. 2 and 3). Both residues were implicated in the GTPase reaction: the former as a trigger of the ribosomal signal activating the EF-Tu GTPase (68) and the latter for the property of G domain[D80N] to display a higher intrinsic GTPase activity, an effect associated with a destabilization of the protein (96). Surprisingly, the T61A mutation does not influence the affinity for GDP and only very modestly that for GTP (95). Ribosomes slightly stimulate the intrinsic GTPase, an effect not found for the homologous mutation in EF-Tu$_{Tt}$ (68). The weakness of the T61 interaction despite the prominence of the T61–Mg^{2+} link in the crystal structure contrasts to the drastic change caused by mutation of the oppositely located T25($\rightarrow$A) (95), but this was also found for the homologous threonine in Ha-Ras (97, 98).

Substitution D80N in both EF-Tu and its G domain causes a dramatic change of the nucleotide interaction (99); the GDP affinity lies in the micromolar range, close to that for GTP. A rationale is that Asp (pK_a: 4.4) is mostly ionized at pH 7, making it an exclusive hydrogen bond acceptor, whereas the Asn side chain is poorly ionizable and can act as both a hydrogen bond acceptor and a donor. The lack of a negative charge attracting Mg^{2+} and the changed H-bonding pattern make the bonding by Asn weaker, decreasing nucleotide affinity and at the same time inducing dominant-negative properties in EF-Tu (Fig. 6) (see above). The finding that the GTPase activity of G domain[D80N] is strongly enhanced up to 60–80 times was striking (96). This increase is not found in the full-length EF-Tu[D80N] (99), making a physiological role of this effect improbable. However, an increase of the GTPase of the ternary complex points to an involvement in the tight regulation of the GTPase activity. Poly(Phe) synthesis of EF-Tu[D80N] is impaired, whereas the enzymatic binding at 10 mM Mg^{2+} and its (empty) ribosome-stimulated GTPase activity are not. Thus, it seems that the decreased activity of EF-Tu[D80N] in the elongation cycle originates from

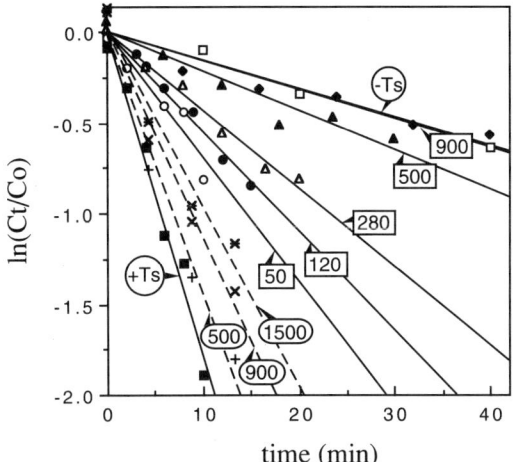

FIG. 6. Inhibition of EF-Ts stimulation of EF-Tu[wt] · [^{3}H]GDP dissociation by increasing concentrations of EF-Tu[wt] · GDP (dashed lines and encircled labels) or EF-Tu[D80N] · GDP (solid lines and square labels). The numbers in the labels indicate the concentrations in μM of competing EF-Tu. The curves labeled +Ts and −Ts represent the control reactions without competitor. Adapted and reproduced with permission from Ref. 99.

a reduced (re)formation of the ternary complex due to the lower GTP affinity, whereas the slowest step for EF-Tu[T61A] may be the GTPase activation.

In conclusion, we can say that the characterization of mutants of all these conserved residues has yielded a fairly broad picture of the involvement of the Mg^{2+} network in both GTPase activity and protein synthesis.

Last minute note: an interesting very recent report (100) describing the 3D structure of the *Sulfolobus solfataricus* EF-1α · GDP complex at 1.8 Å resolution shows that the protein–nucleotide interaction is not mediated by Mg^{2+}, the conserved residues usually coordinating Mg^{2+} in GBPs being located quite differently. This surprising result needs to be carefully evaluated, keeping in mind that this hyperthermophilic protein was crystallized far below its physiological temperature. It may, however, serve as an illustration of how nature may take different approaches to regulating a similar activity. It is noteworthy that the intrinsic GTPase activity was dependent on the presence of Mg^{2+}.

III. EF-Ts as a Steric Chaperone for EF-Tu Folding

EF-Tu$_{Ec}$ is a rather unstable protein; 50% inactivation takes place after 8 min at ∼52°C for EF-Tu · GDP and at ∼47°C for EF-Tu · GTP (80). Nucleotide-free EF-Tu is even less stable; at 4°C it loses its activity within a few hours.

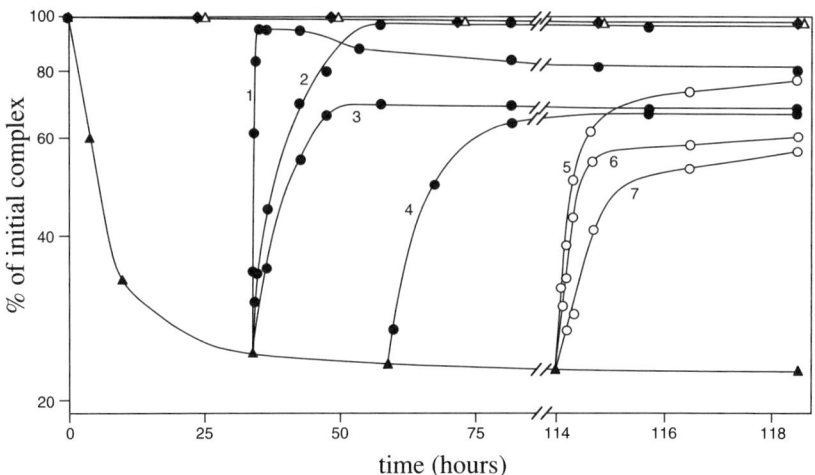

FIG. 7. Time course of the inactivation of nucleotide-free EF-Tu at 0°C and its reactivation in the presence of GDP and kirromycin. Inactivation control profiles: (▲), without additions; (△), plus kirromycin; (♦), plus GDP. Reactivation profiles: aliquots of the solution containing nucleotide-free EF-Tu without GDP or kirromycin were incubated with GDP (curves 1, 3, 4, and 6) or kirromycin (7) and GDP plus kirromycin (2 and 5) at 0 (2–4) and 23 (1, 5–7) °C for 15 min. After addition of [^{3}H]GDP, the activities were determined kinetically as [^{3}H]GDP-bound EF-Tu retained on nitrocellulose filters. Adapted from and for more details see Ref. 91.

Inactivation can be reversed totally or in part—depending on the length of the period of inactivation—by GDP, an effect accelerated by kirromycin (91) (Fig. 7). In contrast, the nucleotide-free complex with EF-Ts is stable (101), underlining the need for either nucleotide and bound magnesium or, in their absence, EF-Ts for maintaining a correct fold.

More information about this was obtained using a vector that coexpresses glutathione-S-transferase (GST)-fused EF-Tu and EF-Ts (102). Low-affinity EF-Tu mutants, such as EF-Tu[T25A], [H22Y/T25S], and [D80N], were used, as well as EF-Tu[D138N]-binding xanthine nucleotides only (see Section II,C). Coexpression with EF-Ts turned out to be very effective in increasing the soluble portion of the former three mutants. A different behavior was observed with EF-Tu[D138N], for which there are no stabilizing nucleotides in the cell. In this case, formation of a soluble product occurs only during the stationary phase, whereas EF-Ts is immediately soluble (Fig. 8A). This suggests a slow conformational rearrangement of initially misfolded and insoluble protein, forming a stable, soluble complex upon binding with EF-Ts. The existence of a slow refolding process is clearly indicated by the course of reactivation depicted in Fig. 8B, in which protein synthesis is stopped after 8 h and initially insoluble protein is found in the supernatant after 25 h. Thus, EF-Ts appears to act as "a steric molecular chaperone" for the folding of EF-Tu. Ellis (103) distinguishes between

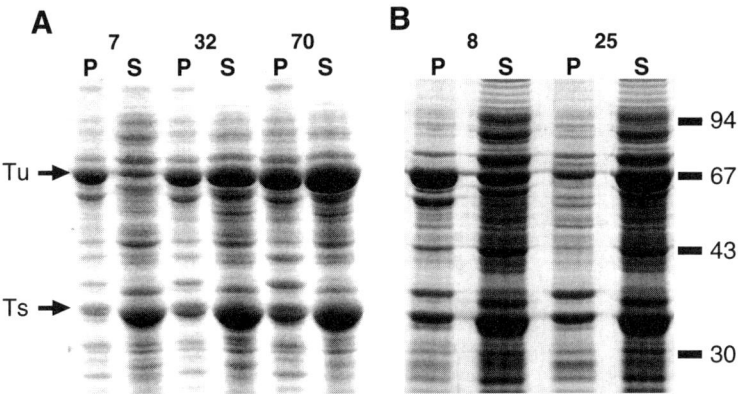

FIG. 8. Long induction of pTu[D138N]Ts. (A) SDS–PAGE showing S30 supernatant (S) and pellet (P) fractions after 7, 32, and 70 h induction time at 24°C. (B) SDS–PAGE showing the same at 8-h induction at 25°C, after which protein synthesis was arrested with chloramphenicol and the cells were again analyzed after 25 h. Adapted and reproduced with permission from Ref. *102*.

"classical chaperones" that assist refolding of other proteins in a nonspecific way by protecting exposed hydrophobic residues and "steric chaperones" that use information contained in their structure, which they communicate through shape complementarity. Since the EF-Ts effect is derived from a very specific interaction, we employ the restricted term of "steric chaperone."

For the other three mutants with reduced nucleotide affinity, the residual binding of the nucleotide is sufficient for it to act as a folding template at least in part, even if the lower affinity either limits its effectiveness and/or reduces the *in vivo* stability of the resulting protein. Consequently, the presence of EF-Ts in a 1:1 ratio to EF-Tu is necessary for maximum solubility and long-term stability. T25A and especially D80N induce a dominant-negative phenotype *in vivo* caused by the sequestration of EF-Ts through stable complex formation, as also shown by their interfering effect *in vitro* on the EF-Ts-dependent stimulation of the EF-Tu[wt] · GDP dissociation (95, 99). Growth inhibition takes place *in vivo* shortly after their expression, indicating a relatively fast folding assisted by the residual guanine nucleotide-binding activity. This differs from D138N, for which the folding lags so far behind expression that no significant inhibition takes place. Coexpression of EF-Ts relieves the growth inhibition by the former two mutants.

IV. EF-Tu as Target of Antibiotics

EF-Tu is the specific and only target of four (families of) antibiotics inhibiting protein biosynthesis: kirromycin, pulvomycin, GE2270 A (= MDL 62,879)

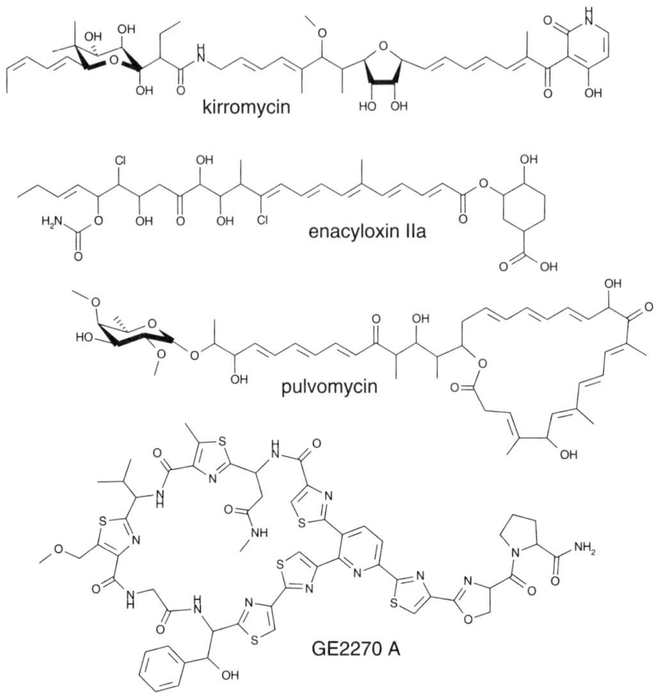

FIG. 9. Structures of the four EF-Tu-specific antibiotics.

(6, 104–106), and enacyloxin IIa (107, 108). The four groups share no structural similarities (Fig. 9). Kirromycin, the first identified (72) and best studied of these antibiotics, is the prototype of a family comprising many analogs, among them aurodox (methyl-kirromycin). The GE2270 A family comprises several analogs, among them amythiamicin. The last few years have registered a substantial progress concerning the mode of action, function–structure relationships, and resistance mechanisms of these antibiotics.

A. Mode of Action

These inhibitors of protein biosynthesis were important for defining the functions of the *tuf* genes and EF-Tu and its ligands (6). A general property of these antibiotics is to induce a very stable EF-Tu · GTP complex, characterized by a slow dissociation rate. They also bind efficiently to EF-Tu · GDP, affecting its kinetics of association and dissociation to various extents depending on the type of antibiotic. Independent of the bound nucleotide, kirromycin and enacyloxin IIa "freeze" EF-Tu in a GTP-like "on state," while pulvomycin and GE2270 A induce a GDP-like "off-state." The former two antibiotics inhibit protein biosynthesis by hindering peptide bond formation after hydrolysis of the bound GTP. Since EF-Tu · GDP · kirromycin has a conformation similar enough to that of EF-Tu · GTP, it remains bound to aa-tRNA. This is demonstrated by both the comigration of EF-Tu · GDP · kirromycin with Phe-tRNA · poly(U) · ribosomes after GTP hydrolysis and the observation of EF-Tu · GDP · kirromycin-stimulated binding of aa-tRNA to the ribosome (*109, 110*). The failure of EF-Tu to dissociate hinders the incorporation of a new amino acid into the polypeptide chain. A selective effect of kirromycin is the marked enhancement of the intrinsic EF-Tu GTPase activity that can be stimulated to various extents (up to 100-fold) depending on both the concentration and the nature of the monovalent cation (*92*). The kirromycin-induced EF-Tu GTPase activity historically provided the first proof that EF-Tu contains an intrinsic catalytic center, even if the velocity of this reaction in the presence of kirromycin remains thousands of times smaller than that following codon–anticodon interaction on the ribosome. Also in the absence of the antibiotic, the intrinsic GTPase activity of EF-Tu is dependent on M^+ (*74*).

A retarded release of EF-Tu · GDP from the ribosome can also be inferred for enacyloxin IIa (*108*) although unlike that for kirromycin it is not directly observed (*111*). A number of differences distinguish the action of this antibiotic from that of kirromycin (*111*). Enacyloxin IIa makes EF-Tu less sensitive than kirromycin, to the action of trypsin, and the binding of aa-tRNA to the ribosome by EF-Tu · GDP · enacyloxin IIa is less efficient than that by EF-Tu · GDP · kirromycin. This shows that the conformation induced on EF-Tu · GDP by enacyloxin IIa is less "GTP-like" than that induced by kirromycin. Moreover enacyloxin IIa decreases, less than kirromycin, the protection by EF-Tu · GTP of the spontaneous deacylation of aa-tRNA. Enacyloxin IIa binding to EF-Tu is uncharacteristically weak: it is released from EF-Tu · GTP on gel filtration, an effect even more pronounced with aa-tRNA. In contrast, kirromycin binding to EF-Tu is very tight and can only be overcome competitively by EF-Ts (*109, 112*). Also enacyloxin IIa binding is outcompeted by EF-Ts (*107*). Finally, another difference from kirromycin is the inability of enacyloxin IIa to enhance the intrinsic GTPase activity of EF-Tu.

Both pulvomycin (*113*) and GE2270 A (*106*) prevent the interaction between EF-Tu and aa-tRNA. Their binding to EF-Tu is extremely tight and once it occurs no conditions but denaturation can release the antibiotic (see Section IV,C). Specific differences characterize the effects of pulvomycin and GE2270 A, such as a pronounced stimulation by the former of the dissociation and to a lesser extent of the association rate of EF-Tu · GDP, whereas GE2270 A virtually does not influence the kinetics of this interaction. Nonetheless, the binding to EF-Tu · GDP of the latter antibiotic is dominant over that of pulvomycin (*106*).

B. Resistance

In 1977, it was possible to isolate the first EF-Tu mutants in *E. coli* strains containing a *tufA* resistant to kirromycin mutated in position 375 [A375V (*114*) and A375T (*115*)]. EF-Tu A375V was the only *tuf* product found in the cell, *tufB* likely carrying a mutation knocking out the product, whereas EF-Tu A375T was associated with a mutated *tufB* product (EF-TuB[G222D] called also EF-TuBo) with recessive kirromycin sensitivity that was later shown to depend on a deficient interaction with the ribosome (*116*). Since the mode of action of kirromycin, blocking EF-Tu · GDP on the ribosome, implies dominance of kirromycin sensitivity over resistance, the existence in *E. coli* of two genes coding for EF-Tu makes the phenotypic expression of kirromycin resistance *in vivo* only possible if the cell lacks a dominant-sensitive second *tuf* gene. The dominance becomes amplified in a polysome system, in which a single EF-Tu · kirromycin-blocked ribosome causes the following ribosomes to also be blocked, even if they interact with a kirromycin-resistant EF-Tu, inducing a "traffic jam" effect. Dominance is also evident in poly(Phe) synthesis *in vitro* as shown by the failure of a mixture of kirromycin-sensitive and -resistant populations to express resistance (*114*).

Several other residues were identified, whose mutation induces kirromycin resistance: L120, Q124, Y160, G316, Q329, K357, and E378 (*117, 118*). In EF-Tu · GTP, all of these residues are clustered on the interface between Domains 1 and 3. In EF-Tu · GDP, this interface is much less extensive and virtually only formed by Q124. Substitution of Q124 and A375, buried in the EF-Tu · GTP interface, induces a stronger resistance than that of the peripheral G316 (*117*). Binding of kirromycin to this interface could inhibit the transition between EF-Tu · GTP and EF-Tu · GDP in agreement with the property of this agent to deregulate all of the EF-Tu-supported reactions (see 3D model of EF-Tu · GDP · aurodox in the next section).

The characterization of both kirromycin action and EF-Tu kirromycin-resistant mutants has suggested several mechanisms for kirromycin resistance (*117, 119*): (i) a decreased binding affinity of mutant EF-Tu for the antibiotic (*120, 121*) related to a stabilization of the interface 1,3, (ii) a reduced affinity for kirromycin of EF-Tu · GDP (*117*), and iii) the impaired affinity for aa-tRNA of several kirromycin-resistant EF-Tu (*122*), facilitating a quick release of the EF-Tu · GDP · kirromycin complex after delivery of aa-tRNA to the ribosome. These mechanisms can act in concert to yield high resistance to kirromycin.

Recently, the kirromycin resistance-inducing mutations Q124K, G316D, Q329H, and A375T—all located in the interface—have also been shown to cause resistance to enacyloxin IIa (*108*). However, the degree of resistance to the two antibiotics is different. As measured on native zonal PAGE electrophoresis (*108, 117*), the resistance to kirromycin of EF-Tu[G316D], [A375T], and [Q124K] is 4, 20, and more than 200 times, respectively, whereas that to enacyloxin IIa is

10, 7, and 70 times. Remarkably, EF-Tu[A375V], which in poly(Phe) synthesis is 80–100 times more resistant to kirromycin compared to EF-Tu[wt] (*108, 114*), was found to be 3 times more sensitive to enacyloxin than EF-Tu[wt] (*108*). Hopefully, a future elucidation of the 3D structure of EF-Tu · enacyloxin IIa will explain these differences.

Mutations inducing resistance to pulvomycin [R230C, the double mutation R230V/R233F, R233C/S and T334A (*123, 124*)] are located in Domains 2 and 3 in the three-domain junction, an area crucial for binding aa-tRNA. Considerations from the 3D model suggest that they should destabilize the three-domain interactions. Mutation V228A in Domain 2 of *Bacillus subtilis* EF-Tu (residue V226 in EF-Tu$_{Ec}$) induces resistance to amythiamicin, a GE2270 A analog (*125*), and substitution G278A in *B. subtilis* EF-Tu (G275 in EF-Tu$_{Ec}$) to GE2270 A (*126*). More recently (*127*) two EF-Tu$_{Ec}$ mutations, G257S and G275A, that induce resistance to this antibiotic have been identified, which nevertheless in the EF-Tu · GTP state bind the antibiotic as efficiently as EF-Tu[wt]. The drop in affinity for aa-tRNA caused by the antibiotic, which for EF-Tu[wt] · GTP is 4 orders of magnitude, is reduced in EF-Tu[G257S] and EF-Tu[G275A] to 100 and 10 times, respectively. Interestingly, the *Streptomyces* strain producing GE2270 A shows amino acids inducing resistance in the corresponding positions.

C. Binding Sites

The analysis of EF-Tu[1,2] and EF-Tu[1,3] (*78*) shows that the former is consistently reactive to all four antibiotics, more weakly than full-length EF-Tu but in a similar manner. In contrast, EF-Tu[1,3] · GTP response is very small; that of the G domain is virtually abolished. The observation that after removal of Domain 3 kirromycin can still productively interact with EF-Tu shows that (i) Domain 1 is essential for the binding of the antibiotic and (ii) that the contribution of Domain 2 to induce an active kirromycin-binding site on Domain 1 is more important than that of Domain 3. In Section II,B it has been mentioned that the dissociating effect of EF-Ts on the GDP complexes of EF-Tu[1,2] and EF-Tu[1,3] is conserved. The same concerns the ability of kirromycin to compete with the action of EF-Ts (Fig. 10). This proves that both noncatalytic domains participate either directly (Domain 3) or indirectly (Domain 2) to the interaction of kirromycin with Domain 1. Moreover, these findings further confirm that the removal of either noncatalytic domain does not cause a general conformational disorder.

Functional studies have also indicated different binding sites on EF-Tu for kirromycin and pulvomycin (*128*) and for kirromycin and GE2270 (*129*). Interestingly, in the case of aurodox and GE2270 A, the specific binding sites as predicted from residues inducing resistance (*107*) turned out to be not far from the actual location recently found by crystallographic analysis. In the 3D

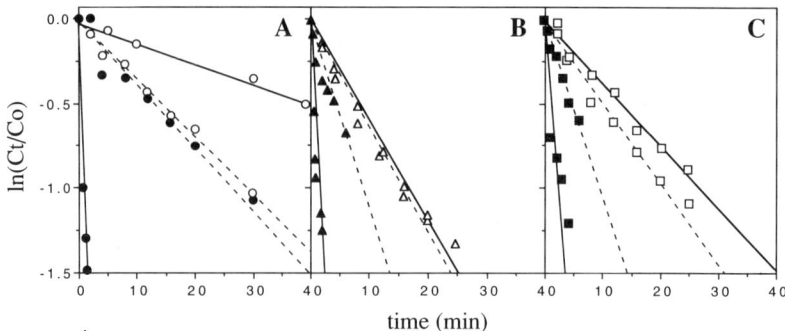

FIG. 10. Kirromycin competes with EF-Ts for binding to EF-Tu[wt], EF-Tu[1,2], and EF-Tu[1,3] as determined by the dissociation rate of the EF-Tu · [^{3}H]GDP complex stimulated by EF-Ts in either the absence or the presence of the antibiotic. Solid lines: no antibiotic; dashed lines: with antibiotic; (●, ▲, ■), plus EF-Ts; (○, △, □), controls minus EF-Ts. (A) EF-Tu[wt]; (B) EF-Tu[1,2]; (C) EF-Tu[1,3]. The concentration of EF-Tu and EF-Ts was 0.1 and 0.01 μM, respectively, that of kirromycin, 50 μM. Adapted and reproduced with permission from Ref. 78.

model of EF-Tu$_{Ec}$·GDP · GE2270 A, elucidated at 2.35 Å resolution (130), the antibiotic binds to Domain 2 (Fig. 11A) and contacts three amino acid stretches (aa: 215–230, 256–264, and 273–277), where mutations (V226A, G257S, and G275A) induce resistance to either this drug or (R230C/V) pulvomycin. Extrapolation of this structure to the EF-Tu · GTP conformation shows that the bound GE2270 A would sterically collide with Domain 1, explaining why this complex blocks the transition to the GTP-specific conformation. Moreover, since a region of the antibiotic binds to the same site occupied by aa-tRNA, the interaction with the latter is prevented. Remarkably, R223 and G259 form a salt bridge spanning over the side chain of GE2270 A. This could explain why the bound antibiotic is locked on EF-Tu. The similarity in the action of GE2270 A and that of pulvomycin and the overlapping resistance mutants favor a location of the pulvomycin-binding site on Domain 2 not far from that of GE2270 A, even though the 3D structure of EF-Tu · GTP · pulvomycin has yet to be elucidated.

A 3D model of the EF-Tu$_{Tt}$·GDP · aurodox(methyl kirromycin) complex at 2.0 Å resolution has also been reported (131). This complex displays an overall similarity to EF-Tu · GMPPNP, further confirming that kirromycin induces an "EF-Tu · GTP-like" conformation, independently of whether GDP or GTP is bound. This also explains the anomalous ability of EF-Tu · GDP · kirromycin to interact productively with aa-tRNA. In the 3D model, aurodox—that on EF-Tu exerts exactly the same effects as does kirromycin (105)—binds to the interface between Domains 1 and 3 (Fig. 11B), and its accommodation is favored by a moderate rotation of Domains 2 and 3 relative to Domain 1 and each other as compared to the GMPPNP-bound state. The analysis of the relationships

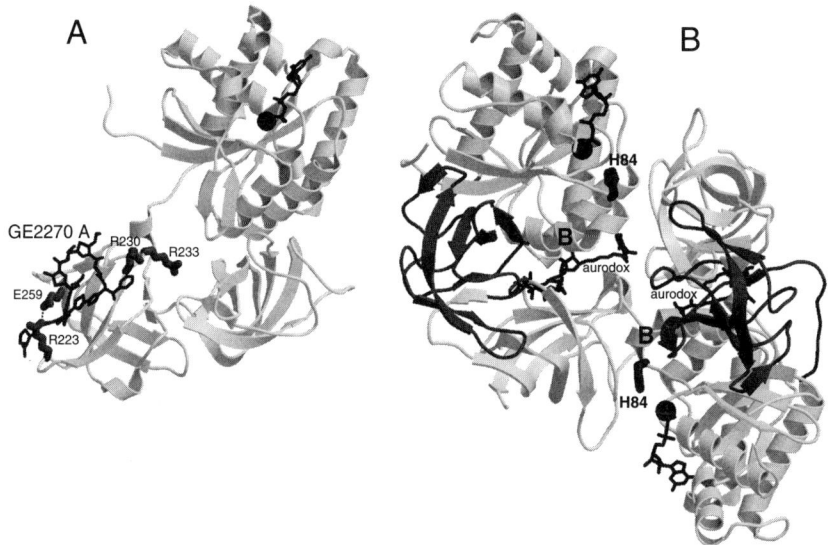

FIG. 11. Crystal structures of EF-Tu complexes with antibiotics. (A) GE2270 A bound to Domain 2 of EF-Tu · GDP, with the salt bridge on its left (R223 to E259) locking the antibiotic in place, and residues R230 and R233 on the right side, whose mutation causes pulvomycin resistance, and of which R230 is in contact with the GE2270 A. (B) The two aurodox-bound EF-Tu · GDP molecules in the unit cell. For contrast, the EF-Tu Domains 2 are darker gray. Histidine-84 (actually His85 in *T. thermophilus*) is highlighted pointing towards the nucleotide. It is situated in the loop L4-tip of helix B region (label B) that is in contact with Domain 3 of the second EF-Tu molecule (see text).

between surface 1,3 and antibiotic suggests that the binding is preceded by both opening and reorganization of the interface following a ligand-induced fit mechanism. The authors also describe a reorientation of the H84 side chain that now points to the nucleotide-binding site (see also Section V,B). The common properties of the action of kirromycin and enacyloxin IIa on EF-Tu also suggest that the latter antibiotic binds to the interface between Domains 1 and 3, but, even though very probable, this remains to be demonstrated.

D. Kirromycin and EF-1α

Kirromycin does not act on protein biosynthesis from eukaryotic or archeal organisms (*105*). Despite this, a study (*132, 133*) carried out on calf brain EF-1α has shown that kirromycin can bind to EF-1α and influence its interaction with ligands, even though its effect differs in part from that on EF-Tu. In fact, it stimulates the association and dissociation rates of both GDP and GTP complexes of EF-1α (Fig. 12), thus enhancing the exchange of the EF-1α-bound nucleotide with the free one reminiscent of the effect of EF-Ts and EF-1β. In contrast,

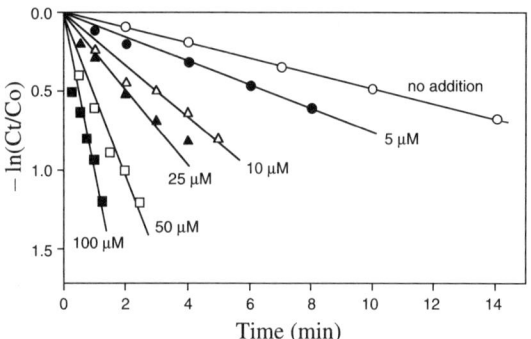

FIG. 12. Time course of the dissociation of GDP from calf brain EF-1α in the presence of kirromycin at the indicated concentrations. Adapted from Ref. 133.

kirromycin inhibits the dissociation rate of bacterial EF-Tu · GTP, favoring GTP binding but not true exchange. The concentration of the antibiotic inducing a 50% effect (ca. 25 μM) on EF-1α is 10- to 100-fold higher than that for the bacterial counterpart. Moreover, the actions of kirromycin and EF-1β are not mutually exclusive, as in the case of bacterial EF-Ts, but additive. Different from EF-Tu, the intrinsic GTPase activity of EF-1α is not stimulated by the antibiotic but, as a consequence of a faster regeneration of the EF-1α · GTP complex, is enhanced by aa-tRNA. In the presence of the ribosome, the kirromycin action on EF-1α is relieved, in agreement with its lack of effect on eukaryotic protein synthesis. This is in line with older observations (67) (see Section I,E,3) that, different from EF-Tu, EF-1α · GDP does not leave the ribosome each elongation cycle. The information obtained with EF-1α demonstrates how useful antibiotics can be as tools for investigating the mechanisms underlying the interactions with specific ligands.

V. Specific Aspects of EF-Tu GTPase Activity

The importance of GTP hydrolysis for the EF-Tu physiological activities was already recognized in the 1960s (134), but despite several studies, the mechanism and function of this reaction in EF-Tu still remain in large part unclear. We will discuss some specific aspects such as its stoichiometry versus amino acid incorporation, the cooperative action of EF-Tu mutants, and some recent structure–function developments.

A. Stoichiometry, Cooperation

The initial studies in the 1960s (134) received new impulses in the 1980/1990s by the observation that, in a poly(Phe) system, the minimal stoichiometry

(EF-Tu- plus EF-G activity) was found to be 2.5–3 GTP-hydrolyzed for each amino acid incorporated, the excess over 2 being attributed to uncoupled and unspecific activity (135). More recently, a 2:1 stoichiometry for the GTP consumption of EF-Tu alone has been obtained by synthesizing not only long poly(Phe) chains (85, 86, 136) but also short heteropolypeptides (137). Our group (85) utilized the xanthine nucleotide-specific EF-Tu[D138N] (see Section II,C) to eliminate the problem of unspecific GTPases that to some extent are present in any purified system for protein synthesis *in vitro*. In fact, this mutant EF-Tu is the only XTPase present in the system. In contrast to the 2:1 stoichiometry in poly(Phe) synthesis, step-by-step analysis of the synthesis of tripeptides dependent on hetero-mRNA-programmed ribosomes (138) resulted in a 1:1 stoichiometry of GTP hydrolyzed versus amino acid incorporation, except for the sequence UUUUUC yielding a 2:1 ratio restricted to the first Phe incorporated. The latter result was attributed to frameshifting. The possibility, that in systems synthesizing long poly(Phe) stretches the precise 2:1 ratio is due to frameshifting taking place at each interaction with the ternary complex, is interesting but does not sound convincing and should at least be substantiated by comparing the poly(Phe) stoichiometries directed by long UUC and UUU polymers.

As a general criticism to these kinds of studies, the tendency of elongation factors to express GTPase activity in excess over amino acid incorporation, the so-called "uncoupled GTPase," could depend on mechanisms of relevant importance. Although restriction to a minimum of the energy consumption in experiments may be useful in shedding light on some basic aspects, it could hide the expression of other processes. Thus, doubts about the physiological meaning of results from minimal systems are justified. In conclusion, despite some recent claims (139), the question of how many GTPs per amino acid incorporated are hydrolyzed remains open and should be reexamined.

Related to this aspect is the genetic and biochemical evidence for the involvement of EF-Tu in cooperative interactions. It is known that the combination of kirromycin-resistant EF-Tu[A375T] with EF-Tu[G222D], in which kirromycin sensitivity is recessive to resistance (see Section IV,B), suppresses nonsense codons (140), and $-1/+1$ frameshifting (141). Substitution of either one of these mutants with EF-Tu[wt] abolishes the cooperative effect. Even though the importance of EF-Tu[G222D] for the suppression of nonsense codons by EF-Tu [A375T] has been played down (142), evidence *in vitro* confirms the cooperation of these two EF-Tu species (Fig. 13) (13). However, the evidence for the involvement of EF-Tu[wt] in cooperative interactions is still lacking. Observations that could have supported this evidence, such as the formation of a quinary complex by two EF-Tu · GTPs and one aa-tRNA (136, 138), have not been confirmed by observations using X-ray scattering and neutron analysis (143). Therefore, the meaning of the cooperative effects of EF-Tu, since they were thus far only observed with two mutated EF-Tu species, still remains unclear.

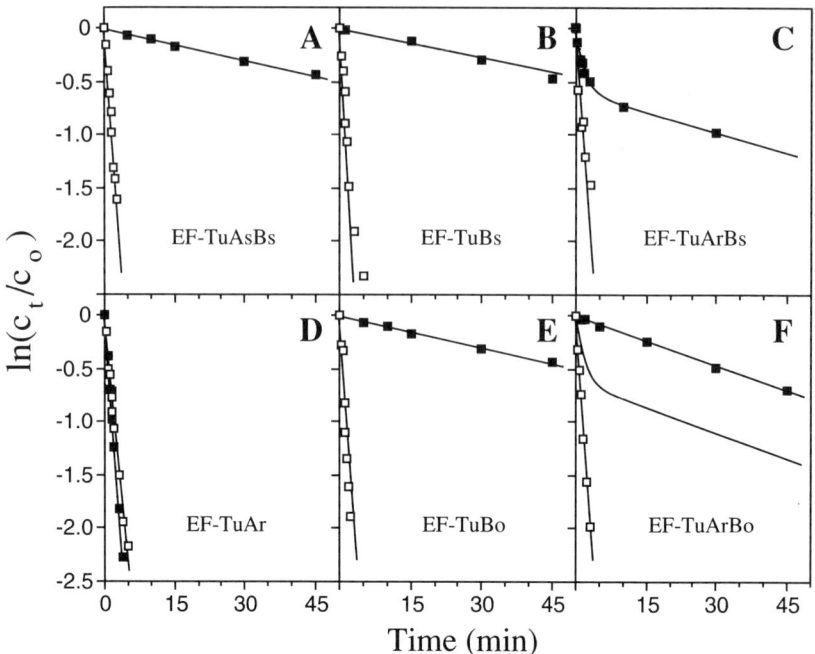

FIG. 13. Effect of kirromycin on the GTP dissociation rates of kirromycin-resistant and -sensitive EF-Tu species alone or in combination as shown in the Panels A–F. Nucleotide-free EF-Tu (1 μM) and [γ^{32}P]GTP (2.5 μM) were incubated in buffer at 0°C for 15 min in either the absence ($\square$) or the presence ($\blacksquare$) of kirromycin, after which the dissociation reaction was started by adding a 1000-fold excess of unlabeled GTP. In Panel F, the broken line represents the theoretical dissociation of an equimolar mixture of EF-Tu[A375T] · GTP (half-life: 70 s) and EF-Tu[G222D] · GTP · kirromycin (half-life: 75 min) as obtained in Panel C. As/Bs: kirromycin-sensitive EF-Tu[wt] products from *tufA* and *tufB*; Ar: EF-Tu[A375T] resistant to kirromycin; Bo: EF-Tu[G222D] sensitive to kirromycin. Adapted from Ref. 13.

B. The "Hydrophobic Gate"

Histidine-84, a residue conserved only in initiation and elongation factors, was proposed to participate in the activation of a water molecule acting as a nucleophile on the γ-phosphate of GTP (*144*) at a time in which the only available 3D model of G domain · GDP was showing that the H84 side chain points away from the γ-phosphate of the nucleotide. This possibility, implying a reorientation of H84 in the GTP · bound state, was augmented by the report that the side chain of Q61 in Ha-Ras p21 · GTP, the residue homologous to EF-Tu H84, was properly positioned in one of its possible orientations for the activation of a H$_2$O molecule in line with the γ-phosphate (*145*). Doubts on a direct participation of H84 in the catalytic reaction came from site-directed mutagenesis.

H84(→G) in the G domain strongly reduces (to 5–10%) but does not abolish the GTPase activity. This led to the proposal of a transition-state stabilization mechanism (146). In full-length EF-Tu, substitution H84(→Q) decreases the GTPase activity and poly(Phe) synthesis by one-third, whereas H84(→A) abolishes the activity in poly(Phe) synthesis but still leaves a 10% residual intrinsic GTPase activity (147). Ribosomes can enhance the GTPase of EF-Tu[H84Q] but not that of EF-Tu[H84A], underlining the importance of H84 for the physiological GTPase and polypeptide synthesis.

An interesting hypothesis was inspired by the 3D model of EF-Tu · GMPPNP (25), in which a "hydrophobic gate" shields the γ-phosphate of the nucleotide from both the bulk solvent and the H84 side chain. The side chains of V20 from the P loop and those of I60 from the effector (Switch 1) region form the wings of this "gate" (Fig. 14). It was proposed that opening of one or both wings of this "hydrophobic gate" could allow the H84 side chain to approach the γ-phosphate, thus allowing its participation in GTP hydrolysis. However, neither substitution V20G (148, 149) nor I60A (99) increases the EF-Tu GTP hydrolysis, in contrast to the prediction that reducing the hydrophobic barrier should increase the intrinsic catalytic rate. A more convincing hypothesis about the mechanism

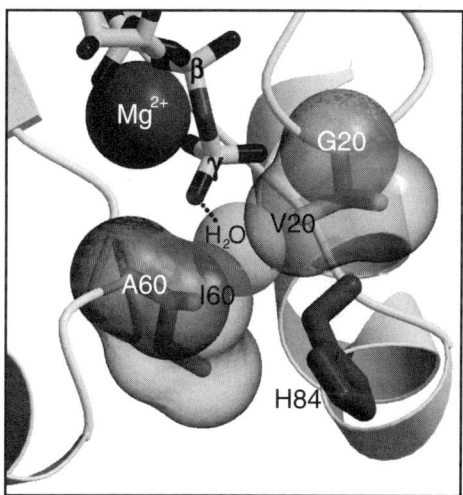

FIG. 14. Close-up of the hydrophobic barrier in EF-Tu · GMPPNP. H84 is prevented from approaching the water molecule (center) in line with the γ-phosphate by residues V20 and I60, while either alanine in position 60 or glycine in position 20 leaves more space for the barrier to open through thermal motion. Transparent surfaces indicate the Van der Waals envelopes of the side chains of both V20 and I60 (bright) as well as those of the α-methylene of a glycine in position 20 and the side chain of an alanine in position 60 (darker and smaller), assuming no dramatic changes of the local conformation.

of the intrinsic EF-Tu (150) and H-Ras p21 (151) GTPase was derived from the so-called "substrate-assisted" GTP hydrolysis (152).

Interestingly, in the recent 3D model of the EF-Tu$_{Tt}$·GDP·aurodox complex, the side chain of H84 is oriented toward the nucleotide, while I60 and all the Switch 1 segment are disordered, suggesting a loss of the hydrophobic barrier and the possibility that with kirromycin and ribosomes H84 could play a role in the EF-Tu GTPase. However, one "caveat" is that the tip of the α-helix is in contact with, Domain 3 of the second EF-Tu in the unit cell, probably influences the structure of the region (Fig. 14). Thus, the relevance for the physiological mechanism of the situation found in the EF-Tu$_{Tt}$·GDP·aurodox complex is still questionable. The implied assumption that ribosomes induce dramatic conformational changes on EF-Tu is likely true; however, the fact that kirromycin enhances the EF-Tu GTPase thousands of times less than the codon–anticodon interaction suggests large differences between their mechanisms of action. This is also supported by the observation that mutations of V20, T25, and T61 induce a different response of the EF-Tu GTPase to kirromycin and to ribosome (99, 148, 149). Even if this is, thus far, purely speculative, an outside GAP "arginine finger"-like residue furnished by an external component such as a ribosomal protein (other than L7/L12, see Section I,E,3) or rRNA remains possible for the fast codon–anticodon-induced GTP hydrolysis. A more detailed characterization of the atomic details of the 3D structure of ribosome·EF-Tu complexes may provide some clues for this possibility.

VI. Conclusions and Perspectives

The impressive progress of our knowledge of the 3D structure of biological components is now allowing, in many cases, detailed studies of the structure–function relationships and of the huge macromolecular complexes, as in the case of EF-Tu·ribosome, in time period unimaginable a few years ago. The recent visualization of EF-Tu on the ribosome now allows a more targeted approach for determining the interacting structures and thus their functional relationships. This permits the reexamination of fundamental as yet unsolved aspects. However, the unexpected findings from functional studies have shown time and again in the past that structural studies alone do not provide all the answers; functional results complement them and often indicate which questions to address. The mechanisms of the rapid GTP catalysis induced by the interaction with programmed ribosomes, the proofreading process, the problems correlated with the "uncoupled" GTPase activities, and the stoichiometry between GTP hydrolysis versus polypeptide synthesis continue to be fundamental aspects that should be reexamined.

The role of EF-Tu in the cell also merits further study with respect to gene expression, cell growth, and multifunctional activities. In particular, the involvement of EF-Tu in viral replicase, which has been neglected for a long period of time, requires renewed attention. As a general comment, the existence in EF-Tu$_{Ec}$ and from other organisms of more than one gene may have hindered the detection of specific phenotypes of EF-Tu, thus making the identification of other functions not associated with protein synthesis difficult. It is moreover probable that other drugs exist that can influence bacterial (eukaryotic) protein biosynthesis by interacting with EF-Tu (EF-1α), therefore making them candidates for therapeutic applications. EF-Tu pioneered the field of the GBPs and still represents a valuable model for cell physiology, structure function–relationships, and drug action.

References

1. B. F. C. Clark, M. Kjeldgaard, J. Barciszewski, and M. Sprinzl, Recognition of aminoacyl-tRNA: Structure, biosynthesis and function by protein elongation factors. in "tRNA: Structure, Biosynthesis and Function" (S. Söll and U. Rajbhandary, Eds.), pp. 423–442. American Society for Microbiology Press, Washington DC, 1995.
2. J. Czworkowski and P. B. Moore, The elongation phase of protein synthesis. *Prog. Nucleic Acid Res. Mol. Biol.* **54**, 293–332 (1996).
3. M. Kjeldgaard, J. Nyborg, and B. F. C. Clark, The GTP binding motif: Variations on a theme. *FASEB J.* **10**, 1347–1368 (1996).
4. K. Abel and F. Jurnak, A complex profile of protein elongation: Translating chemical energy into molecular movement. *Structure* **4**, 229–238 (1996).
5. J. Nyborg, Possible evolution of factors involved in protein biosynthesis. *Acta Biochim. Polonica* **45**, 883–894 (1998).
6. I. M. Krab and A. Parmeggiani, EF-Tu, a GTPase odyssey. *Biochim. Biophys. Acta* **1443**, 1–22 (1998).
7. J. Lucas-Lenard and F. Lipmann, Protein biosynthesis. *Annu. Rev. Biochem.* **40**, 409–448 (1971).
8. H. R. Bourne, D. A. Sanders, and F. McCormick, The GTPase superfamily: A conserved switch for diverse cell functions. *Nature* **348**, 125–132 (1990).
9. M. S. Boguski and F. McCormick, Proteins regulating Ras and its relatives. *Nature* **366**, 643–654 (1993).
10. S. R. Sprang, G protein mechanisms: Insights from structural analysis. *Annu. Rev. Biochem.* **66**, 639–678 (1997).
11. M. V. Rodnina and A. Wintermeyer, Ribosome fidelity: tRNA discrimination, proofreading and induced fit. *Trends Biochem. Sci.* **26**, 124–130 (2001).
12. L. Bosch, B. Kraal, P. H. Van der Meide, F. J. Duisterwinkel, and J. M. van Noort, The elongation factor Tu and its two encoding genes. *Prog. Nucleic Acids Res. Mol. Biol.* **30**, 91–126 (1983).
13. P. H. Anborgh, G. W. M. Swart, and A. Parmeggiani, Kirromycin-induced modifications facilitate the separation of EF-Tu species and reveal intermolecular interactions. *FEBS Lett.* **292**, 232–236 (1991). Erratum in *FEBS Lett.* **295**, 232 (1991).
14. D. Hughes, Both genes for EF-Tu in *Salmonella typhimurium* are individually dispensable for growth. *J. Mol. Biol.* **215**, 41–51 (1990).

15. A. M. Zuurmond, A. K. Rundlöf, and B. Kraal, Either of the chromosomal *tuf* genes of *E. coli* K-12 can be deleted without loss of cell viability. *Mol. Gen. Genet.* **260**, 603–607 (1999).
16. E. Vijgenboom, L. P. Woudt, P. W. Heinstra, K. Rietveld, J. van Haarlem, G. P. van Wezel, S. Shochat, and L. Bosch, Three *tuf*-like genes in the kirromycin producer *Streptomyces ramocissimus*. *Microbiology* **140**, 993–998 (1994).
17. T. A. Landers, T. Blumenthal, and K. Weber, Function and structure in ribonucleic acid phage Qβ ribonucleic acid replicase. The roles of the different subunits in transcription of synthetic templates. *J. Biol. Chem.* **249**, 5801–5808 (1974).
18. A. Trigwell and R. E. Glass, Function *in vivo* of separate segment of the β-subunit of *Escherichia coli* RNA polymerase. *Genes Cells* **3**, 635–647 (1998).
19. A. Travers, Control of ribosomal RNA synthesis *in vitro*. *Nature* **244**, 15–18 (1973).
20. W. Kudlicki, A. Coffman, G. Kramer, and B. Hardesty, Renaturation of rhodanese by translational elongation factor (EF) Tu. *J. Biol. Chem.* **272**, 32,206–32,210 (1997).
21. G. Richarme, Protein-disulfide isomerase activity of elongation factors EF-Tu. *Biochem. Biophys. Res. Commun.* **252**, 156–161 (1998).
22. T. Hesterkamp, S. Hauser, H. Lütcke, and B. Bukau, *Escherichia coli* trigger factor is a prolyl isomerase that associates with nascent polypeptide chains. *Proc. Natl. Acad. Sci. U.S.A.* **93**, 4437–4441 (1996).
23. J. Condeelis, Elongation factor 1α translation and the cytoskeleton. *Trends Biochem. Sci.* **20**, 169–170 (1995).
24. M. Kjeldgaard and J. Nyborg, Refined structure of elongation factor EF-Tu from *Escherichia coli*. *J. Mol. Biol.* **223**, 721–742 (1992).
25. H. Berchtold, L. Reshetnikova, C. O. Reiser, N. K. Schirmer, M. Sprinzl, and R. Hilgenfeld, Crystal structure of active elongation factor Tu reveals major domain rearrangements. *Nature* **365**, 126–132 (1993). Erratum *Nature* **365**, 368 (1993).
26. M. Kjeldgaard, P. Nissen, S. Thirup, and J. Nyborg, The crystal structure of elongation factor EF-Tu from *Thermus aquaticus* in the GTP conformation. *Structure* **1**, 35–50 (1993).
27. K. Abel, M. D. Yoder, R. Hilgenfeld, and F. Jurnak, An alpha to beta conformational switch in EF-Tu. *Structure* **4**, 1153–1159 (1996).
28. G. Polekhina, S. Thirup, M. Kjeldgaard, P. Nissen, C. Lippmann, and J. Nyborg, Helix unwinding in the effector region of elongation factor EF-Tu-GDP. *Structure* **4**, 1141–1151 (1996).
29. H. Song, M. R. Parsons, S. Rowsell, G. Leonard, and S. E. Phillip, Crystal structure of intact elongation factor EF-Tu from *Escherichia coli* in GDP conformation at 2.05 Å resolution. *J. Mol. Biol.* **285**, 1245–1256 (1999).
30. P. Nissen, M. Kjeldgaard, S. Thirup, G. Polekhina, L. Reshetnikova, B. F. C. Clark, and J. Nyborg, Crystal structure of the ternary complex of Phe-tRNA[Phe], EF-Tu, and a GTP analog. *Science* **270**, 1464–1472 (1995).
31. M. H. Metz-Boutigue, J. Reinbolt, J. P. Ebel, C. Ehresmann, and B. Ehresmann, Crosslinking of elongation factor Tu to tRNA(Phe) by trans-diamminedichloroplatinum (II). Characterization of two crosslinking sites on EF-Tu. *FEBS Lett.* **245**, 194–200 (1989).
32. P. Nissen, S. Thirup, M. Kjeldgaard, and J. Nyborg, The crystal structure of Cys-tRNA[Cys] · GDPNP reveals general and specific features in the ternary complex and in tRNA. *Structure Fold. Des.* **7**, 143–156 (1999).
33. G. Parlato, J. Guesnet, J. B. Créchet, and A. Parmeggiani, The GTPase activity of elongation factor Tu and the 3'-terminal end of aminoacyl-tRNA. *FEBS Lett.* **125**, 257–260 (1981).
34. F. P. Wikman, G. E. Siboska, H. U. Petersen, and B. F. C. Clark, The site of interaction of aminoacyl-tRNA with elongation factor Tu. *EMBO J.* **1**, 1095–1100 (1982).
35. G. Parlato, R. Pizzano, D. Picone, J. Guesnet, O. Fasano, and A. Parmeggiani, Different regions of aminoacyl-tRNA regulate the function of elongation factor Tu. *J. Biol. Chem.* **258**, 995–1000 (1983).

36. D. Picone and A. Parmeggiani, Transfer ribonucleic acid deprived of the C-C-A 3'-extremity can interact with elongation factor Tu. *Biochemistry* **22**, 4400–4405 (1983).
37. G. R. Andersen, S. Thirup, L. L. Spremulli, and J. Nyborg, High resolution crystal structure of bovine mitochondrial EF-Tu in complex with GDP. *J. Mol. Biol.* **297**, 421–436 (2000).
38. T. Kawashima, C. Berthet-Colominas, M. Wulff, S. Cusack, and R. Leberman, The structure of the *Escherichia coli* EF-Tu·EF-Ts complex at 2.5 Å resolution. *Nature* **379**, 511–518 (1996).
39. Y. Wang, Y. Jiang, M. Meyering-Voss, M. Sprinzl, and P. B. Sigler, Crystal structure of the EF-Tu·EF-Ts complex from *Thermus thermophilus*. *Nature Struct. Biol.* **4**, 650–656 (1997).
40. M. R. Ahmadian, P. Stege, K. Sheffzek, and A. Wittinghofer, Confirmation of the arginine-finger hypothesis for the GAP-stimulated GTP hydrolysis reaction of Ras. *Nat. Struct. Biol.* **4**, 686–689 (1997).
41. S. Baraud-Dufour, P. Robineau, P. Chardin, S. Paris, M. Chabre, J. Cherfils, and B. Antonny, A glutamic finger in the guanine nucleotide exchange factor ARNO displaces Mg^{2+} and the β-phosphate to destabilize GDP on ARF1. *EMBO J.* **17**, 3651–3659 (1998).
42. B. T. Wimberly, D. E. Brodersen, W. M. J. Clemons, R. J. Morgan-Warren, A. P. Carter, C. Vonrhein, T. Hartsch, and V. Ramakrishnan, Structure of the 30S ribosomal subunit. *Nature* **407**, 327–339 (2000).
43. N. Ban, P. Nissen, J. Hansen, P. B. Moore, and T. A. Steitz, The complete atomic structure of the large ribosomal subunit at 2.4 Å resolution. *Science* **289**, 905–920 (2000).
44. M. M. Yusupov, G. Z. Yusupova, A. Baucom, K. Lieberman, T. N. Earnest, J. H. D. Cate, and H. F. Noller, Crystal structure of the ribosome at 5.5 Å. *Science* **292**, 883–996 (2001).
45. E. Hamel, M. Koka, and T. Nakamoto, Requirement of an *Escherichia coli* 50S ribosomal protein component for effective interaction of the ribosome with T and G factors and with guanosine triphosphate. *J. Biol. Chem.* **247**, 805–814 (1972).
46. G. Sander, R. C. Marsh, J. Voigt, and A. Parmeggiani, A comparative study of the 50S ribosomal subunit and several 50S subparticles in EF-T-and EF-G-dependent activities. *Biochemistry* **14**, 1805–1814 (1975).
47. G. Sander, R. C. Marsh, and A. Parmeggiani, Activity of the 30-S CsCl core in elongation-factor-dependent GTP hydrolysis. *Eur. J. Biochem.* **61**, 317–323 (1976).
48. H. F. Noller, Ribosomal RNA and Translation. *Annu. Rev. Biochem.* **60**, 191–227 (1991).
49. T. Powers and H. F. Noller, Evidence for functional interaction between elongation factor Tu and 16S ribosomal RNA. *Proc. Natl. Acad. Sci. U.S.A.* **90**, 1364–1368 (1993).
50. C. C. Correll, A. Munishkin, Y. L. Chan, Z. Ren, I. G. Wool, and T. A. Steitz, Crystal structure of the ribosomal RNA domain essential for binding elongation factors. *Proc. Natl. Acad. Sci. U.S.A.* **95**, 13,436–13,441 (1998).
51. B. T. Wimberly, R. Guyomon, J. P. McCutcheon, S. W. White, and V. Ramakrishnan, A detailed view of a ribosomal active site: The structure of the L11-RNA complex. *Cell* **97**, 491–502 (1999).
52. A. Munishkin and I. G. Wool, The ribosome-in-pieces: Binding of elongation factor EF-G to oligoribonucleotides that mimic the sarcin/ricin and thiostrepton domains of 23S ribosomal RNA. *Proc. Natl. Acad. Sci. U.S.A.* **94**, 12,280–12,284 (1997).
53. V. Hornung, H. P. Hofmann, and M. Sprinzl, *In vitro* selected RNA molecules that bind to elongation factor Tu. *Biochemistry* **37**, 7260–7267 (1998).
54. V. E. Koteliansky, S. P. Domogatsky, A. T. Gudkov, and A. S. Spirin, Elongation factor-dependent reactions of ribosomes deprived of proteins L7 and L12. *FEBS Lett.* **73**, 6–11 (1977).
55. I. Pettersson and C. G. Kurland, Ribosomal protein L7/L12 is required for optimal translation. *Proc. Natl. Acad. Sci. U.S.A.* **77**, 4007–4010 (1980).
56. A. V. Oleinikov, G. G. Jokhadze, and R. R. Traut, A single-headed dimer of *Escherichia coli* ribosomal protein L7/L12 supports protein synthesis. *Proc. Natl. Acad. Sci. U.S.A.* **95**, 4215–4218 (1998).

57. G. Sander, R. Ivell, J. B. Créchet, and A. Parmeggiani, Interaction of elongation factor Tu with the ribosome. A study using the antibiotic kirromycin. *Biochemistry* **19,** 865–870 (1980).
58. I. Kusser, C. Lowing, C. Rathlef, A. K. Kopke, and A. T. Matheson, Structure-function relationships in the ribosomal protein L12 in the archeon *Sulfolobus acidocaldarius*. *Arch. Biochem. Biophys.* **365,** 254–261 (1999).
59. A. Savelsbergh, D. Mohr, B. Wilden, W. Wintermeyer, and M. V. Rodnina, Stimulation of the GTPase activity of translation elongation factor G by ribosomal protein L7/L12. *J. Biol. Chem.* **275,** 890–894 (2000).
60. H. Stark, M. V. Rodnina, J. Rinke-Appel, R. Brimacombe, W. Wintermeyer, and M. van Heel, Visualization of elongation factor Tu on the *Escherichia coli* ribosome. *Nature* **389,** 403–406 (1997).
61. R. K. Agrawal, P. Penczek, R. A. Grassucci, and J. Frank, Visualization of elongation factor G on the *Escherichia coli* 70S ribosome: The mechanism of translocation. *Proc. Natl. Acad. Sci. U.S.A.* **95,** 6134–6138 (1998).
62. R. K. Agrawal, A. B. Heagle, P. Penczek, R. A. Grassucci, and J. Frank, EF-G-dependent GTP hydrolysis induces translocation accompanied by large conformational changes in the 70S ribosome. *Nat. Struct. Biol.* **6,** 643–647 (1999).
63. H. Stark, M. V. Rodnina, H. J. Wieden, M. Van Heel, and W. Wintermeyer, Large-scale movement of elongation factor G and extensive conformational change of the ribosome during translocation. *Cell* **100,** 301–309 (2000).
64. D. Dey, D. E. Bochkariov, G. G. Jokhadze, and R. R. Traut, Cross-linking of selected residues in the N- and C-terminal domains of *E. coli* protein L7/L12 to other ribosomal proteins and the effect of elongation factor Tu. *J. Biol. Chem.* **273,** 1670–1676 (1998).
65. J. Lucas-Lenard and L. Beres, Protein synthesis-peptide chain elongation. *in* "Enzymes" (P. D. Boyer ed.), Vol. 10, pp. 53–86. Academic press, New York/San Francisco/London, 1974.
66. G. Sander, R. C. Marsh, and A. Parmeggiani, Role of split proteins from 30 S subunits in the ribosome-EF-T GTPase reaction. *FEBS Lett.* **33,** 132–134 (1973).
67. H. Grasmuk, R. D. Nolan, and J. Drews, Further evidence that elongation factor 1 remains bound to ribosomes during peptide chain elongation. *Eur. J. Biochem.* **79,** 93–102 (1977).
68. M. R. Ahmadian, R. Kreutzer, B. Blechschmidt, and M. Sprinzl, Site-directed mutagenesis of *Thermus thermophilus* EF-Tu: The substitution of threonine-62 by serine or alanine. *FEBS Lett.* **377,** 253–257 (1995).
69. C. R. Knudsen and B. F. C. Clark, Site-directed mutagenesis of Arg58 and Asp86 of elongation factor Tu from *Escherichia coli*: effects on the GTPase reaction and aminoacyl-tRNA binding. *Protein Eng.* **12,** 1267–1273 (1995).
70. K. Scheffzek, M. R. Ahmadian, W. Kabsch, L. Wiesmüller, A. Lautwein, F. Schmitz, and A. Wittinghofer, The Ras-RasGAP complex: Structural basis for GTPase activation and its loss in oncogenic Ras mutants. *Science* **18,** 333–338 (1997).
71. O. Piepenburg, T. Pape, J. A. Pleiss, A. Wintermeyer, O. C. Uhlenbeck, and M. V. Rodnina, Intact aminoacyl-tRNA is required to trigger GTP hydrolysis by elongation factor Tu on the ribosome. *Biochemistry* **39,** 1734–1738 (2000).
72. H. Wolf, G. Chinali, and A. Parmeggiani, Kirromycin, an inhibitor of protein biosynthesis that acts on elongation factor Tu. *Proc. Natl. Acad. Sci. U.S.A.* **71,** 4910–4914 (1974).
73. A. Parmeggiani, G. W. M. Swart, K. K. Mortensen, M. Jensen, B. F. C. Clark, L. Dente, and R. Cortese, Properties of a genetically engineered G domain of elongation factor Tu. *Proc. Natl. Acad. Sci. U.S.A.* **84,** 3141–3145 (1987).
74. O. Fasano, E. De Vendittis, and A. Parmeggiani, Hydrolysis of GTP by elongation factor Tu can be induced by monovalent cations in the absence of other effectors. *J. Biol. Chem.* **257,** 3145–3150 (1982).

75. M. Jensen, R. H. Cool, K. K. Mortensen, B. F. C. Clark, and A. Parmeggiani, Structure-function relationships of elongation factor Tu. Isolation and activity of the guanine-nucleotide-binding domain. *Eur. J. Biochem.* **182,** 247–255 (1989).
76. A. Pingoud, M. Wehrmann, U. Pieper, F.-U. Gast, C. Urbanke, J. Alves, J. Feuerstein, and A. Wittinghofer, Spectroscopic and hydrodynamic studies reveal structural differences in normal and transforming H-ras gene products. *Biochemistry* **27,** 4735–4740 (1988).
77. D. F. Lowry, R. H. Cool, A. G. Redfield, and A. Parmeggiani, NMR study of the phosphate-binding elements of *Escherichia coli* elongation factor Tu catalytic domain. *Biochemistry* **30,** 10,872–10,877 (1991).
78. R. Cetin, P. H. Anborgh, R. H. Cool, and A. Parmeggiani, Functional role of the noncatalytic domains of elongation factor Tu in the interaction with ligands. *Biochemistry* **37,** 486–495 (1998).
79. J. Jonák, T. E. Petersen, B. F. C. Clark, and I. Rychlík, N-Tosyl-L-phenylalanylchloromethane reacts with cysteine 81 in the molecule of elongation factor Tu from *Escherichia coli*. *FEBS Lett.* **150,** 485–488 (1982).
80. P. H. Anborgh, A. Parmeggiani, and J. Jonák, Site-directed mutagenesis of elongation factor Tu. The functional and structural role of residue Cys81. *Eur. J. Biochem.* **208,** 251–257 (1992).
81. R. H. Cool, M. Jensen, J. Jonák, B. F. C. Clark, and A. Parmeggiani, Substitution of proline 82 by threonine induces autophosphorylating activity in GTP-binding domain of elongation factor Tu. *J. Biol. Chem.* **265,** 6744–6749 (1990).
82. T. Y. Shih, A. G. Papageorge, P. E. Stokes, M. O. Weeks, and E. M. Scolnick, Guanine nucleotide-binding and autophosphorylating activities associated with the p21src protein of Harvey murine sarcoma virus. *Nature* **287,** 686–691 (1980).
83. P. Wagner, C. M. Molenaar, A. J. Rauh, R. Brokel, H. D. Schmitt, and D. Gallwitz, Biochemical properties of the ras-related YPT protein in yeast: A mutational analysis. *EMBO J.* **6,** 2373–2379 (1987).
84. Y. W. Hwang and D. L. Miller, A mutation that alters the nucleotide specificity of elongation factor Tu, a GTP regulatory protein. *J. Biol. Chem.* **262,** 13,081–13,085 (1987).
85. A. Weijland and A. Parmeggiani, Toward a model for the interaction between elongation factor Tu and the ribosome. *Science* **259,** 1311–1314 (1993).
86. A. Weijland, G. Parlato, and A. Parmeggiani, Elongation factor Tu D138N, a mutant with modified substrate specificity, as a tool to study energy consumption in protein biosynthesis. *Biochemistry* **33,** 10,711–10,717 (1994).
87. T. F. La Cour, J. Nyborg, S. Thirup, and B. F. C. Clark, Structural details of the binding of guanosine diphosphate to elongation factor Tu from *E. coli* as studied by X-ray crystallography. *EMBO J.* **4,** 2385–2388 (1985).
88. F. Jurnak, Structure of the GDP domain of EF-Tu and location of the amino acids homologous to ras oncogene proteins. *Science* **230,** 32–36 (1985).
89. A. Weijland, R. Sarfati, O. Barzu, and A. Parmeggiani, Asparagine-135 of elongation factor Tu is a crucial residue for the folding of the guanine nucleotide binding pocket. *FEBS Lett.* **330,** 334–338 (1993).
90. D. L. Miller and H. Weissbach, Factors involved in the transfer of aminoacyl-tRNA to the ribosome. *in* "Molecular Mechanisms in Protein Biosynthesis" (H. Weissbach and S. Pestka, eds.), pp. 323–373. Academic Press, New York, 1977.
91. O. Fasano, J. B. Créchet, and A. Parmeggiani, Preparation of nucleotide-free elongation factor Tu and its stabilization by the antibiotic kirromycin. *Anal. Biochem.* **124,** 53–58 (1982).
92. R. Ivell, G. Sander, and A. Parmeggiani, Modulation by monovalent and divalent cations of the guanosine-5′-triphosphatase activity dependent on elongation factor Tu. *Biochemistry* **20,** 6852–6859 (1981).

93. M. Y. Mistou, R. H. Cool, and A. Parmeggiani, Effects of ions on the intrinsic activities of c-H-ras protein p21. A comparison with elongation factor Tu. *Eur. J. Biochem.* **204,** 179–185 (1992).
94. H. Rütthard, A. Banerjee, and W. Makinen, Mg^{2+} is not catalytically required in the intrinsic and kirromycin stimulated GTPase action of *Thermus thermophilus* EF-Tu. *J. Biol. Chem.* **276,** 18,728–18,733 (2001).
95. I. M. Krab and A. Parmeggiani, Functional-structural analysis of threonine 25, a residue coordinating the nucleotide-bound magnesium in Elongation Factor Tu. *J. Biol. Chem.* **274,** 11,132–11,138 (1999).
96. K. Harmark, P. H. Anborgh, M. Merola, B. F. C. Clark, and A. Parmeggiani, Substitution of aspartic acid-80, a residue involved in coordination of magnesium, weakens the GTP binding and strongly enhances the GTPase of the G domain of elongation factor Tu. *Biochemistry* **31,** 7367–7372 (1992).
97. C. T. Farrar, C. J. Halkides, and D. J. Singel, The frozen solution structure of p21 ras determined by ESEEM spectroscopy reveals weak coordination of Thr35 to the active site metal ion. *Structure* **5,** 1055–1066 (1997).
98. C. J. Halkides, C. T. Farrar, R. G. Larsen, A. G. Redfield, and D. J. Singel, Characterization of the active site of p21 ras by electron spin-echo envelope modulation spectroscopy with selective labeling: Comparisons between GDP and GTP forms. *Biochemistry* **33,** 4019–4035 (1994). Erratum. *Biochemistry* **34,** 14,270 (1995).
99. I. M. Krab and A. Parmeggiani, Mutagenesis of three residues, Ile-60, Thr-61 and Asp-80, implicated in the GTPase activity of *Escherichia coli* elongation factor Tu. *Biochemistry* **38,** 13,035–13,041 (1999).
100. L. Vitagliano, M. Masullo, F. Sica, A. Zagari, and V. Bocchini, The crystal structure of *Sulfolobus Solfataricus* elongation factor 1α in complex with GDP reveals novel features in nucleotide binding and exchange. *EMBO J.* **20,** 5305–5311 (2001).
101. A. Parmeggiani and G. Sander, Properties and regulation of the GTPase activities of elongation factors Tu and G, and of initiation factor 2. *Mol. Cell. Biochem.* **35,** 129–158 (1981).
102. I. M. Krab, R. te Biesebeke, A. Bernardi, and A. Parmeggiani, Elongation factor Ts can act as a steric chaperone by increasing the solubility of nucleotide binding-impaired EF-Tu. *Biochemistry* **40,** 8531–8535 (2001).
103. R. J. Ellis, Steric chaperones. *Trends Biochem. Sci.* **23,** 43–45 (1998).
104. A. Parmeggiani and G. Sander, Properties and action of kirromycin (mocimycin) and related antibiotics. *in* "Topics in Antibiotic Chemistry" (P. G. Sammes ed.), Vol. 5, pp. 161–221. Ellis Horwood Ltd., Chichester, 1980.
105. A. Parmeggiani and G. W. M. Swart, Mechanism of action of kirromycin-like antibiotics. *Annu. Rev. Microbiol.* **39,** 557–577 (1985).
106. P. H. Anborgh and A. Parmeggiani, New antibiotic that acts specifically on the GTP-bound form of elongation factor Tu. *EMBO J.* **10,** 779–784 (1991).
107. R. Cetin, P. H. Anborgh, R. H. Cool, T. Watanabe, T. Sugiyama, K. Izaki, and A. Parmeggiani, Enacyloxin IIa, an inhibitor of protein biosynthesis that acts on elongation factor Tu and the ribosome. *EMBO J.* **15,** 2604–2611 (1996).
108. A. M. Zuurmond, L. N. Olsthoorn-Tieleman, J. Martien de Graaf, A. Parmeggiani, and B. Kraal, Mutant EF-Tu species reveal novel features of the enacyloxin IIa inhibition on the ribosome. *J. Mol. Biol.* **294,** 627–637 (1999).
109. G. Chinali, H. Wolf, and A. Parmeggiani, Effect of kirromycin on elongation factor Tu. Location of the catalytic center for ribosome-elongation-factor-Tu GTPase activity on the elongation factor. *Eur. J. Biochem.* **75,** 55–65 (1977).
110. H. Wolf, G. Chinali, and A. Parmeggiani, Mechanism of the inhibition of protein synthesis by kirromycin. Role of elongation factor Tu and ribosomes. *Eur. J. Biochem.* **75,** 67–75 (1977).

111. I. M. Krab, Functional Studies of Elongation Factor Tu from *Escherichia coli*. Ph.D. Thesis. Leiden University, The Netherlands, 2001.
112. G. W. M. Swart, A. Parmeggiani, B. Kraal, and L. Bosch, Effects of the mutation glycine-222—Aspartic acid on the functions of elongation factor Tu. *Biochemistry* **26,** 2047–2054 (1987).
113. H. Wolf, D. Assmann, and E. Fischer, Pulvomycin, an inhibitor of protein biosynthesis preventing ternary complex formation between elongation factor Tu, GTP, and aminoacyl-tRNA. *Proc. Natl. Acad. Sci. U.S.A.* **75,** 5324–5328 (1978).
114. E. Fischer, H. Wolf, K. Hantke, and A. Parmeggiani, Elongation factor Tu resistant to kirromycin in an *Escherichia coli* mutant altered in both *tuf* genes. *Proc. Natl. Acad. Sci. U.S.A.* **74,** 4341–4345 (1977).
115. J. A. van de Klundert, P. H. van der Meide, P. van de Putte, and L. Bosch, Mutants of *Escherichia coli* altered in both genes coding for the elongation factor Tu. *Proc. Natl. Acad. Sci. U.S.A.* **75,** 4470–4473 (1978).
116. G. W. M. Swart and A. Parmeggiani, tRNA and the guanosinetriphosphatase activity of elongation factor Tu. *Biochemistry* **28,** 327–332 (1989).
117. J. R. Mesters, L. A. H. Zeef, R. Hilgenfeld, J. M. de Graaf, B. Kraal, and L. Bosch, The structural and functional basis for the kirromycin resistance of mutant EF-Tu species in *Escherichia coli*. *EMBO J.* **20,** 4877–4885 (1994).
118. F. Abdulkarim, L. Liljas, and D. Hughes, Mutations to kirromycin resistance occur in the interface of domains I and III of EF-Tu · GTP. *FEBS Lett.* **352,** 118–122 (1994).
119. B. Kraal, L. A. H. Zeef, J. R. Mesters, K. Boon, E. L. Vorstenbosch, L. Bosch, P. H. Anborgh, A. Parmeggiani, and R. Hilgenfeld, Antibiotic resistance mechanisms of mutant EF-Tu species in *Escherichia coli*. *Biochem. Cell Biol.* **73,** 1167–1177 (1995).
120. O. Fasano and A. Parmeggiani, Altered regulation of the guanosine 5′-triphosphate activity in a kirromycin-resistant elongation factor Tu. *Biochemistry* **20,** 1361–1366 (1981).
121. P. H. Van der Meide, F. J. Duisterwinkel, J. M. de Graaf, B. Kraal, L. Bosch, J. Douglass, and T. Blumenthal, Molecular properties of two mutant species of the elongation factor Tu. *Eur. J. Biochem.* **117,** 1–6 (1981).
122. F. Abdulkarim, M. Ehrenberg, and D. Hughes, Mutants of EF-Tu defective in binding aminoacyl-tRNA. *FEBS Lett.* **382,** 297–303 (1996).
123. L. A. H. Zeef, L. Bosch, P. H. Anborgh, R. Cetin, A. Parmeggiani, and R. Hilgenfeld, Pulvomycin-resistant mutants of *E. coli* elongation factor Tu. *EMBO J.* **13,** 5113–5120 (1994).
124. K. Boon, I. Krab, A. Parmeggiani, L. Bosch, and B. Kraal, Substitution of Arg230 and Arg233 in *Escherichia coli* elongation factor Tu strongly enhances its pulvomycin resistance. *Eur. J. Biochem.* **227,** 816–822 (1995).
125. K. Shimanaka, H. Iinuma, M. Hamada, S. Ikeno, K. S. Tsuchiya, M. Arita, and M. Hori, Novel antibiotics, amythiamicins. IV. A mutation in the elongation factor Tu gene in a resistant mutant of B. subtilis. *J. Antibiot.* **48,** 182–184 (1995).
126. M. Sosio, G. Amati, C. Cappellano, E. Sarubbi, F. Monti, and S. Donadio, An elongation factor Tu (EF-Tu) resistant to the EF-Tu inhibitor GE2270 in the producing organism *Planobispora rosea*. *Mol. Microbiol.* **22,** 43–51 (1996).
127. A. M. Zuurmond, J. Martien de Graaf, L. N. Olsthoorn-Tieleman, B. van Duyl, V. G. Möhrle, F. Jurnak, J. R. Mesters, R. Hilgenfeld, and B. Kraal, GE2270A-resistant mutations in elongation factor Tu allow productive aminoacyl-tRNA binding to EF-Tu · GTP · GE2270A. *J. Mol. Biol.* **304,** 995–1005 (2000).
128. A. Pingoud, W. Block, C. Urbanke, and H. Wolf, The antibiotics kirromycin and pulvomycin bind to different sites on the elongation factor Tu from *Escherichia coli*. *Eur. J. Biochem.* **123,** 261–265 (1982).

129. P. Landini, A. Soffientini, F. Monti, S. Lociuro, E. Marzorati, and K. Islam, Antibiotics MDL 62,879 and kirromycin bind to distinct and independent sites of elongation factor Tu (EF-Tu). *Biochemistry* **35**, 15,288–15,294 (1996).
130. S. E. Heffron and F. Jurnak, Structure of an EF-Tu complex with a thiazolyl peptide antibiotic determined at 2.35 Å resolution: Atomic basis for GE2270A inhibition of EF-Tu. *Biochemistry* **39**, 37–45 (2000).
131. L. Vogeley, G. J. Palm, J. R. Mesters, and R. Hilgenfeld, Conformational change of elongation factor Tu (EF-Tu) induced by antibiotic binding. Crystal structure of the complex between EF-Tu · GDP and aurodox. *J. Biol. Chem.* **276**, 17,149–17,155 (2001).
132. J. B. Créchet and A. Parmeggiani, Characterization of the elongation factors from calf brain. 3. Properties of the GTPase activity of EF-1α and mode of action of kirromycin. *Eur. J. Biochem.* **161**, 655–660 (1986).
133. J. B. Créchet and A. Parmeggiani, Characterization of the elongation factors from calf brain. 2. Functional properties of EF-1α, the action of physiological ligands and kirromycin. *Eur. J. Biochem.* **161**, 647–653 (1986).
134. Y. Nishizuka and F. Lipmann, Comparison of guanosine triphosphate split and polypeptide synthesis with a purified *E. coli* system. *Proc. Natl. Acad. Sci. U.S.A.* **55**, 212–219 (1966).
135. G. Chinali and A. Parmeggiani, The coupling with polypeptide synthesis of the GTPase activity dependent on elongation factor G. *J. Biol. Chem.* **255**, 7455–7459 (1980).
136. M. Ehrenberg, A. M. Rojas, J. Weiser, and C. G. Kurland, How many EF-Tu molecules participate in aminoacyl-tRNA binding and peptide bond formation in *Escherichia coli* translation? *J. Mol. Biol.* **211**, 739–749 (1990).
137. M. Ehrenberg, N. Bilgin, V. Dinçbas, R. Karimi, D. Hughes, and F. Abdulkarim, tRNA-ribosome interaction. *Biochem. Cell Biol.* **73**, 1049–1054 (1995).
138. M. V. Rodnina and W. Wintermeyer, GTP consumption of elongation factor Tu during translation of heteropolymeric mRNAs. *Proc. Natl. Acad. Sci. U.S.A.* **92**, 1945–1949 (1995).
139. D. M. Freymann and P. Walter, GTPases in protein translocation and protein elongation. "GTPases" (H. Hall ed.), Oxford University Press, Oxford/New York, 2000.
140. E. Vijgenboom, T. Vink, B. Kraal, and L. Bosch, Mutants of the elongation factor EF-Tu, a new class of nonsense suppressors. *EMBO J.* **4**, 1049–1052 (1985).
141. D. Hughes, J. F. Atkins, and S. Thompson, Mutants of elongation factor Tu promote ribosomal frameshifting and nonsense readthrough. *EMBO J.* **6**, 4235–4239 (1987).
142. S. Tapio and C. G. Kurland, Mutant EF-Tu increases missense error *in vitro*. *Mol. Gen. Genet.* **205**, 186–188 (1986).
143. N. Bilgin, M. Ehrenberg, C. Ebel, N. Bilgin, M. Ehrenberg, C. Ebel, G. Zaccai, Z. Sayers, M. H. Koch, D. I. Svergun, C. Barberato, V. Volkov, P. Nissen, and J. Nyborg, Solution structure of the ternary complex between aminoacyl-tRNA, elongation factor Tu and guanosine triphosphate. *Biochemistry* **37**, 8163–8172 (1998).
144. G. W. M. Swart, The Polypeptide Chain Elongation Factor Tu from *Escherichia coli*. Ph.D. Thesis. Leiden University, The Netherlands, 1987.
145. M. Frech, T. A. Darden, L. G. Pedersen, C. K. Foley, P. S. Charifson, M. W. Anderson, and A. Wittinghofer, Role of glutamine-61 in the hydrolysis of GTP by p21H-ras: An experimental and theoretical study. *Biochemistry* **33**, 3237–3244 (1994).
146. R. H. Cool and A. Parmeggiani, Substitution of histidine-84 and the GTPase mechanism of elongation factor Tu. *Biochemistry* **30**, 362–366 (1991).
147. G. Scarano, I. M. Krab, V. Bocchini, and A. Parmeggiani, Relevance of histidine-84 in the elongation factor Tu GTPase activity and in poly(Phe) synthesis: its substitution by glutamine and alanine. *FEBS Lett.* **365**, 214–218 (1995).
148. E. Jacquet and A. Parmeggiani, Structure-function relationships in the GTP binding domain of EF-Tu: Mutation of Val20, the residue homologous to position 12 in p21. *EMBO J.* **7**, 2861–2867 (1988).

149. E. Jacquet and A. Parmeggiani, Substitution of Val20 by Gly in elongation factor Tu. Effects on the interaction with elongation factors Ts, aminoacyl-tRNA and ribosomes. *Eur. J. Biochem.* **185,** 341–346 (1989).
150. R. Hilgenfeld, How do the GTPases really work? *Nat. Struct. Biol.* **2,** 3–6 (1995).
151. T. Schweins, M. Geyer, K. Scheffzek, A. Warshel, H. R. Kalbitzer, and A. Wittinghofer, Substrate-assisted catalysis as a mechanism for GTP hydrolysis of p21ras and other GTP-binding proteins. *Nat. Struct.Biol.* **2,** 36–44 (1995).
152. R. Langen, T. Schweins, and A. Warshel, On the mechanism of guanosine triphosphate hydrolysis in ras p21proteins. *Biochemistry* **31,** 8691–8696 (1992).
153. P. J. Kraulis, MOLSCRIPT: A program to produce both detailed and schematic plots of protein structures. *J. Appl. Cryst.* **24,** 946–950 (1991).
154. E. A. Merritt and D. J. Bacon, Raster3D-photorealistic molecular graphics. *Methods Enzymol.* **277,** 505–524 (1997).

Index

A

A375 cells, Muc4/SMC, 174
AAUAAA signal
  hybridization, 302–303
  polyadenylation, 291–292
  poly A site UV crosslinking, 308
Accessory proteins, reverse transcription
  fidelity, 110–112
Acetate, methanogens
  conversion to CO_2 and methane, 238
  proton translocation, 256–258
bis-Acetatoamminedichloro(cyclohexylamine)
  platinum(IV)–DNA interactions, 33–34
Adenocarcinoma cells, Muc4/SMC, 173
Adenovirus, primary transcript processing, 290
AICAR, see 5-Aminoimidazole-4-carboxamide
  riboside
Airway, mucociliary transport, Muc4/SMC,
  163–164
Aliphatic amines, DNA–transplatin analogs, 42
Alphaviruses
  characteristics, 187–188
  nonstructural polyprotein P1234, 197
  nsP1
    guanylyltransferase activity, 198–201
    membrane association, 201–203
    methyltransferase activity, 198–201
    minus-strand RNA synthesis, 203–204
  nsP2
    neuropathogenicity, 206–208
    NTPase activity, 204–205
    nuclear transport, 206–208
    protease activity, 206
    RNA helicase activity, 204–205
    RNA triphosphatase activity, 205–206
  nsP3
    features, 210
    phosphorylation, 209–210
    sequence conservation, 208–209
  nsP4, 210–211
  replication complex, 211–214
  replication cycle, 188–190
  RNA replication
    genome complement, 192–194
    26S mRNA, 196–197
    plus-strand RNA synthesis, 194–196
Alphavirus-like superfamily, 190–191
trans-Amine(cyclohexylamine)
  dichlorodihydroxoplatinum(IV), 42
Amine ligand, DNA–transplatin analogs, 38–40
Amino acid residues, HIV-1 RT
  dNTP-binding site, 114–121
  mutational analysis overview, 112–113
  other residues, 127–128
  primer strand, 123–126
  template strand, 121–123
5-Aminoimidazole-4-carboxamide riboside,
  AMPK activation, 72–73
Aminophosphine platinum(II)
  compounds–DNA interactions, 31–32
AMP-activated protein kinase
  characteristics and function, 70
  gene transcription downstream targets,
    77–78
  liver gene expression, 71–73
  muscle tissue gene expression, 77
  pancreatic β-cell gene expression, 74–77
AMPK, see AMP-activated protein kinase
AMV, see Avian myeloblastosis virus
Antiadhesion agent, Muc4/SMC, 155
Antibiotic resistance, EF-Tu target, 534–535
Antibiotics, EF-Tu target
  binding sites, 535–537
  kirromycin and EF-1α, 537–538
  mode of action, 532–533
  overview, 531–532
  resistance, 534–535
Antibodies, ASGP-1 and ASGP-2, 154–155
Antirecognition agent, Muc4/SMC, 155
Antitumor activity, cisplatin–DNA adducts,
  24–25
Antitumor ruthenium compounds–DNA
  interactions
  chloropolypyridyl ruthenium compounds, 53
  dimethyl sulfoxide complexes, 49–52

Antitumor ruthenium compounds (*cont.*)
  heterocyclic complexes, 52
  heterodinuclear (Ru,Pt) compounds, 53–54
Apoptosis, repression by Muc4/SMC, 161–163
Arabidopsis thaliana, SPHK and S1P, 497
Archaea
  characteristics, 224–225
  methanogenesis, cofactors, 232
Archaeoglobus fulgidus, $F_{420}H_2$ dehydrogenase, 249–253
Arginine, HIV-1 RT
  Arg-72, 120–122
  Arg-78, 122–123
Ascites sialoglycoprotein 1
  antibody production, 154–155
  cloning, 153–154
  isolation, 153
Ascites sialoglycoprotein 2
  antibody production, 154–155
  isolation, 153
  Muc4/SMC, 175–176
  soluble Muc4/SMC production, 156–157
ASGP-1, *see* Ascites sialoglycoprotein 1
ASGP-2, *see* Ascites sialoglycoprotein 2
Aspartic acid residue 76, HIV-1 RT, 122–123
Assays
  GEF, 404–405
  retroviral RT fidelity, 99–108
Asymmetric aliphatic amines, DNA–transplatin analogs, 42
ATP, methanogenic archaea, 240–242
ATP synthases, *Methanosarcina*, 265–270
Avian myeloblastosis virus, misinsertion fidelity assay, 102–103

B

Bacteria
  characteristics, 224–225
  methanogenesis, cofactors, 232
BBR3464–DNA interactions
  overview, 42–44
  trinuclear compounds, 47–49
BBR3535–DNA interactions, 47
BBR3537–DNA interactions, 47
BBR3571–DNA interactions, 47
BCAR3, *see* Breast cancer anti-estrogen-resistance gene 3
$\beta 8$-αE loop, Gln-151, HIV-1 RT mutations, 119–120
Biological assays, GEF, 404–405
Blastocyst implantation, Muc4/SMC, 164–167
BMV, *see* Brome mosaic virus
Breast cancer, Muc4/SMC, 170–175
Breast cancer anti-estrogen-resistance gene 3, 422–423
Brome mosaic virus
  nsP1, 199–201
  nsP2, 205

C

C3G, *see* Crk SH3 domain-binding guanine nucleotide exchange factor
Calcitonin/CGRP, pre-mRNA, poly A sites, 337–341
Carbon compounds, methylated
  methanogenesis, 235–238
  proton translocation, 256
Carbon dioxide
  acetate conversion, 238
  methanogens, $H_2 + CO_2$
  methanogenesis, 232–234
  proton translocation, 256
Carboplatin–DNA interactions, 25–26
CDC25^{Mm}, *see* Guanine nucleotide releasing factor 1
CDC25p
  early identification, 393–395
  homology domains, 396–401
Cell motility, sphingosine-1-phosphate receptor, 494–495
Cellular proteins, DNA adduct recognition
  cisplatin-damaged DNA-binding proteins, 23
  DNA photolyase, 20–21
  DNA-repair proteins, 19
  histone H1, 18–19
  HMG-domain proteins, 15–18
  overview, 14–15
  p53 protein, 21
  T4 endonuclease VII, 21
  TATA-binding protein, 18
  Y-box binding protein 1, 18
Cellular resistance, cisplatin-DNA adducts
  mechanisms, 11
  mismatch repair, 14
  nucleotide excision repair, 13

INDEX

Cervix, Muc4/SMC, 177
$CF1_m$ cleavage factors, 331–333
CF1 complex, 354
$CF11_m$ cleavage factors, 331–333
CF11 complex, 355–356
Chaperones, EF-Ts in EF-Tu folding, 529–531
Chips, HSV-1 DNA microarray
    data evaluation, 465–468
    experimental data, 470–472
    fabrication, 464–465
Chlorodiethylenetriamineplatinum(II) chloride–DNA interactions, 7–8
Chloropolypyridyl ruthenium compounds–DNA interactions, 53
Cisplatin, DNA adducts
    adducts *in vitro*, 3–5
    adducts *in vivo*, 6–7
    adduct stability, 5–6
    antitumor activity, 24–25
    cellular resistance
        mismatch repair, 14
        nucleotide excision repair, 13
        resistance mechanisms, 11
    crosslinks, 8
    damaged DNA-binding proteins recognition, 23
    monodentate platinum(II) compounds, 34–38
    monofunctional adducts, 7–8
    replication effects, 10
    targeted analogs, 29–32
    tetravalent analogs, 32–34
    transcription effects, 11
Cleavage/polyadenylation
    cleavage stimulation factor, 312–313
    CPSF, 313–316
    CPSF–RNA binding, 315–316
    CPSF subunit interactions, 314
    mechanism, 299–301
    poly A-binding protein, 317–318
    poly A polymerase, 316–317
    SnRNP in CPSF, 314–315
Cleavage/polyadenylation proteins
    complex assembly, 309–311
    component separation, 304–305
    nuclear extract fractionation, 305–308
    poly A sites, crosslinking to proteins, 309
    UV crosslinking, 308
Cleavage and polyadenylation specificity factor
    cDNA clones, 328

mRNA 3' end processing, 347–349
poly A-binding protein, 317–318
RNA binding, 315–316
SnRNP, 314–315
subunit interactions, 314
Cleavage stimulation factor
    50-kDa subunit, 326–327
    64-kDa subunit, 326, 343–347
    77-kDa subunit, 327–328
    cDNA clones, 325–326
    characterization, 312–313
Cloning
    mammalian sphingosine kinases, 500–503
    poly A polymerase cDNA, 320–322
CLs, *see* Cross-links
CMV, *see* Cytomegaloviruses
Coding exons, poly A sites
    calcitonin/CGRP pre-mRNA, 337–341
    immunoglobulin heavy-chain pre-mRNA, 341–347
Coenzyme B, methanogenesis, 231
Coenzyme F_{420}, methanogenesis, 231, 273
Coenzyme M, methanogenesis, 231
Complementary DNA
    CPSF, cloning, 328
    CstF, cloning, 325–326
    poly A polymerases, 320–322
Conjunctiva, Muc4/SMC, 177–179
Cooperative interactions, EF-Tu, 539
Cornea, Muc4/SMC, 177–179
CPSF, *see* Cleavage and polyadenylation specificity factor
CPVs, *see* Cytoplasmic vacuoles
Crk SH3 domain-binding guanine nucleotide exchange factor, 412–414
Cross-links
    cisplatin–DNA adducts
        adducts *in vitro*, 3–5
        adducts *in vivo*, 6–7
        adduct stability, 5–6
        monofunctional adducts, 7–8
        nucleotide excision repair, 13
    DNA–BBR3464, 48
    DNA–dinuclear platinum compounds, 45–47
    HMG-domain protein DNA adduct recognition, 15–18
    oxaliplatin–DNA interactions, 26–29
    [PtCl(dien)]Cl–DNA, 7–8
    transplatin–DNA, 4, 8–9
    UV, proteins to poly A sites, 308

CstF, *see* Cleavage stimulation factor
Cyclohexylamine ligands, DNA–transplatin analogs, 42
Cytomegaloviruses, infection and latency, 450–451
Cytoplasmic vacuoles
  alphavirus plus-strand RNA, 195–196
  alphavirus replication complex, 212, 214

D

DACH, *see* Oxaliplatin
DAGKc, *see* Diacylglycerol kinase catalytic domain
Defective-interfering RNAs, alphavirus, 192–194
Degradosome, poly A sequence, 372
Deoxynucleoside triphosphate, RT fidelity
  misinsertion fidelity assays, 100–102
  mutational analysis
    Arg-72, 120–121
    Gln-151, 119–120
    Lys-65, 120–121
    Met-184, 117–119
    overview, 112–113
    Tyr-115, 114–117
  pre-steady-state kinetic assays, 105
  reverse transcription initiation, 108–109
Diacylglycerol kinase catalytic domain, SPHK domains, 504–508
1,2-Diaminocyclohexane, *see* Oxaliplatin
Dihydrofolate reductase, poly A site, 336
Dimethyl sulfoxide complexes–DNA interactions, 49–52
Dinuclear platinum compounds–DNA interactions, 44–47
DI-RNAs, *see* Defective-interfering RNAs
Disease, GEF relationship, 425–426
Distamycin–DNA interactions, 31
DNA
  minor groove, HIV-1 RT, 126
  plasmids, PAP1 regulation, 366–367
  templates, fidelity assay, 103
DNA adducts
  cellular protein recognition
    cisplatin-damaged DNA-binding proteins, 23
    DNA photolyase, 20–21
    DNA-repair proteins, 19
    histone H1, 18–19
    HMG-domain proteins, 15–18
    overview, 14–15
    p53 protein, 21
    T4 endonuclease VII, 21
    TATA-binding protein, 18
    Y-box binding protein 1, 18
  cisplatin
    adducts *in vitro*, 3–5
    adducts *in vivo*, 6–7
    adduct stability, 5–6
    antitumor activity, proposed mechanism, 24–25
    cellular resistance, 11, 13–14
    crosslinks, 8
    monofunctional adducts, 7–8
    replication effects, 10
    transcription effects, 10–11
    [PtCl(dien)]Cl, 7–8
    telomerase effects, 23
    topoisomerase effects, 23–24
    transplatin, crosslinks, 8–9
DNA-binding proteins, cisplatin-damaged, 23
DNA interactions
  carboplatin, 25–26
  chloropolypyridyl Ru compounds, 53
  dimethyl sulfoxide complexes, 49–52
  heterodinuclear (Ru,Pt) compounds, 53–54
  monodentate Pt(II) compounds, 34–38
  oxaliplatin, 26–29
  polynuclear platinum antitumor drugs
    dinuclear compounds, 44–47
    overview, 42–44
    trinuclear compounds, 47–49
  ruthenium heterocyclic complexes, 52
  targeted cisplatin analogs, 29–32
  tetravalent cisplatin analogs, 32–34
  transplatin analogs
    asymmetric aliphatic amines, 42
    cyclohexylamine ligands, 42
    imino ether groups, 40–41
    planar amine ligand, 38–40
DNA microarrays, herpesvirus
  design, 454–455
  development rationale, 446–447
  glass slide-based, HSV-1
    chip data, 465–468
    chip fabrication and scanning, 464–465
    experimental data, 470–472
    hybridization, 455

INDEX

receiver operating characteristic, 465
scanning protocols, 456–459
specific oligonucleotides, 461–464
transcript labeling, 455–456
oligonucleotide-based approach, 453–454
PCR fragment-based approach, 453
transcription, 452–453
DNA photolyase, DNA adduct recognition, 20–21
DNA-repair proteins, DNA adduct recognition, 19
DNA synthesis
proviral DNA, 97
reverse transcriptase
amino acid role, 127–128
Arg-72, 120–121
Gln-151, 119–120
Lys-65, 120–121
Met-184, 117–119
minor groove binding track residues, 126
overview, 112–113
primer grip residues, 124–126
template strand residues, 121–123
Tyr-115, 114–117
Tyr-183, 123–124
dNTPs, *see* Deoxynucleoside triphosphate
Dominant inhibitory Ras proteins, GEF targeting, 402–404
Drug-resistant viruses, enhanced polymerase fidelity, 129
Drugs, RT mutation rates, 130–131

E

Ech hydrogenase, methanogens, 248–249
Electron transport
methanogenesis, 231–232
Methanosarcina mazei, 255
Elongation factor 1α, kirromycin, 537–538
Elongation factor Ts
EF-Tu · EF-Ts, 521
EF-Tu folding, 529–531
Elongation factor Tu
antibiotic target
binding sites, 535–537
kirromycin and EF-1α, 537–538
mode of action, 532–533
overview, 531–532
resistance, 534–535
catalytic domain 1, 524
cooperative interactions, 539
EF-Tu · EF-Ts, 521
folding, EF-Ts role, 529–531
hydrophobic gate, 540–542
identification and isolation, 514
magnesium coordination network, 527–529
nucleotide-binding pocket, 525–527
overview, 517
polypeptide elongation, 514–515
properties and functions, 515–516
regulatory domains 2 and 3, 525
ribosome, 521–524
stoichiometry, 538–539
ternary complex, 520–521
3D structure, 517
Energy conservation, methanogens
ATP synthases, 265–270
Ech hydrogenase, 248–249
$F_{420}H_2$ dehydrogenase, 249–253
F_{420}-nonreducing hydrogenase, 246–248
formyl-methanofuran dehydrogenase system, 261–265
F_{420}-reducing hydrogenase, 245–246
growth, proton translocation, 256–258
heterodisulfide reductase, 253–255
membrane-bound electron transport systems, 255
membrane-bound methyltransferases, 259–261
obligate hydrogenotrophic methanogens, 270–274
proton-translocating pyrophosphatases, 258–259
redox-driven proton dislocation, 242–245
sodium ion pump, 260
Epac/cyclic AMP-guanine nucleotide exchange factors, 414–416
Epithelium, Muc4/SMC
cornea and conjunctiva, 177–179
mucociliary transport, 163–164
oviduct, 167–170
uterus, 164–167
vagina and cervix, 177
ErbB2, Muc4/SMC
lacrimal gland, 176
localized forms, 167–170
mammary acinar cells, 170–171
ocular protection, 178–179
vagina and cervix, 177

ErbB2/HER2/Neu, Muc4/SMC
    ligand quality, 157–160
    tumor progression, 162
ERKs, see Extracellular regulated kinases
Error catastrophe, mutator RT, 129–130
Escherichia coli
    EF-Tu isolation, 514
    EF-Tu properties and functions, 515
    polyadenylation overview, 362–363
    poly A sequences
        degradosome, 372
        mRNA decay, 370
        mRNA-degrading RNases, 370
        overview, 363–366
        PAP1, 366–367
        PAP1-stimulated mRNA decay, 370–372
        poly A polymerase gene, 366
        second gene search, 372–374
        target RNA decay, 367–369
Eukarya
    characteristics, 224–225
    methanogenesis, 232
Exons, poly A sites
    calcitonin/CGRP pre-mRNA, 337–341
    immunoglobulin heavy-chain pre-mRNA, 341–347
    3' noncoding exons, 336–337
Exon skipping model, calcitonin/CGRP pre-mRNA, 338–339
Extracellular regulated kinases, ERK2, 79

F

$F_{420}H_2$ dehydrogenase, methanogens, 249–253
$F_{420}H_2$:heterodisulfide oxidoreductase
    Methanosarcina mazei, 242–243
    proton translocation, 256
Fidelity, retroviral transcriptase
    accessory proteins, 110–112
    amino acid role, 127–128
    Arg-72, 120–121
    assays *in vitro*, 99–100
    enhanced fidelity, 129
    genetic assays, 105–108
    Gln-151, 119–120
    Lys-65, 120–121
    Met-184, 117–119
    minor groove binding track residues, 126
    misinsertion fidelity assay, 100–103
    mispair extension fidelity assay, 103–104
    mutational analysis overview, 112–113
    mutation rate variations, 130–131
    mutator, error catastrophe, 129–130
    pre-steady-state kinetic assays, 104–105
    primer grip residues, 124–126
    strand transfer, 109–110
    template strand residues, 121–123
    transcription initiation, 108–109
    Tyr-115, 114–117
    Tyr-183, 123–124
FIP1 gene, yeast mRNA 3' end-processing factor, 360–361
F_{420}-nonreducing hydrogenase
    methanogens, 246–248
    Methanothermobacter strains, 271–273
Formate, methanogenesis, 234–235
Formyl-methanofuran dehydrogenase system, *Methanosarcina*, 261–265
Fractionation, cleavage/polyadenylation protein nuclear extract, 305–308
F_{420}-reducing hydrogenase, methanogens, 245–246

G

Gag–Pol, retroviral RT, 94
GAP, see GTPase-activating proteins
GE2270 A
    EF-Tu, 532–533
    EF-Tu binding sites, 535–536
    EF-Tu resistance, 535
GEF, see Guanine nucleotide exchange factors
Genes
    BCAR3, 422–423
    poly A polymerase, 372–374
    yeast mRNA 3' end-processing factor
        conditional mutants, 358–359
        FIP1, 360–361
        genetic approaches, 359
        mammalian homologs, 361–362
        PCF11, 361
        synergistic lethality, 359
        two-hybrid system, 360–361
Genetic assays, retroviral RT, 105–108
Gene transcription
    AMPK regulation
        downstream targets, 77–78
        liver, 71–73

INDEX

muscle tissue gene expression, 77
pancreatic β-cell, 74–77
MAPK and SAPK regulation, 81–82
Genomes
  alphavirus, complementary RNA, 192–194
  cloning, poly A polymerase cDNA, 322
  yeast mRNA 3′ end-processing factor genes, 361–362
Genotype profiling
  cell-specific temporal viral gene expression, 474, 478–479, 483, 486
  HCMV, 472–474
GFR, 416
GK, see Glucokinase
Glandular secretory epithelia, Muc4/SMC
  lacrimal gland, 176
  mammary acinar cells, 170–175
  salivary glands, 175
Glass slides, HSV-1 DNA microarray
  chip data, 465–468
  chip fabrication and scanning, 464–465
  experimental data, 470–472
  hybridization, 455
  receiver operating characteristic, 465
  scanning protocols, 456–459
  specific oligonucleotides, 461–464
  transcript labeling, 455–456
Glucokinase, glucose in gene expression, 72
Glucose
  liver gene expression, 71–73
  preproinsulin gene expression, 81–82
Glucose repression, yeast SNF1, 71
Glutamate receptor-interacting protein, 412
Glutamate receptor-interacting protein-associated protein, 412
Glutamic acid-89, HIV-1 RT, 123
Glutamine, residue 151, HIV-1 RT, 119–120, 122
GMP, nsP1, 198, 201
GRASP-1, Glutamate receptor–interacting protein–associated protein
GRIP, see Glutamate receptor-interacting protein
Growth
  methanogens, 256–258
  primary tumor, Muc4/SMC, 161–163
GRP/CalDAG-GEF family
  characteristics, 409–410
  expression, 411–412
  founding member isolation, 410

Ras activation, 410–411
Ras and Rap specificity, 411
GTP, nsP1, 198–201
GTPase, EF-Tu, see Elongation factor Tu
GTPase-activating proteins, Ras-GTP *in vivo* regulation, 393
GTP binding protein, EF-Tu, see Elongation factor Tu
Guanine nucleotide exchange factors
  BCAR3, 422–423
  biological assays, 404–405
  C3G, 412–414
  disease relationship, 425–426
  dominant inhibitory Ras protein targeting, 402–404
  Epac/cyclic AMP-GEFs, 414–416
  GRASP-1, 412–414
  GRF1 and 2, 407–409
  GRP/CalDAG-GEF family, 409–412
  MR-GEF, 416
  nuclear exchange reaction, 396, 401
  PDZ-GEFs, 416–418
  phospholipase C_ϵ, 418–420
  RalGDS family, 420–421
  RalGPS, 421–422
  Rap1 GEFs, 420
  Ras-family, early identification, 393–396
  SmgGDS, 423–425
  Sos1 and 2, 405–407
  structure, 396, 401
Guanine nucleotide releasing factor 1, 407–409
Guanine nucleotide releasing factor 2, 407–409
Guanylyltransferase, nsP1 activity, 198–201

H

β12–β13 Hairpin, HIV-1 RT, 124–126
HB[*trans*-Ru(III)B$_2$Cl$_4$]–DNA interactions, 52
HCMV, see Human cytomegalovirus
Herpesviruses, DNA microarrays
  design, 454–455
  development rationale, 446–447
  oligonucleotide-based approach, 453–454
  PCR fragment-based approach, 453
  transcription, 452–453
Heterodinuclear (Ru,Pt) compounds–DNA interactions, 53–54

Heterodisulfide reductase
  methanogens, 253–255
  Methanothermobacter strains, 271–273
Histidine, residue 84, EF-Tu, 540–542
Histone H1, DNA adduct recognition, 18–19
HIV-1, *see* Human immunodeficiency virus type 1
HMGB1, DNA adduct recognition, 15–18
HMGB2, DNA adduct recognition, 15, 18
HMG-domain proteins
  DNA adduct recognition, 15–18
  DNA interactions, 45–46
H$_4$MPT, *see* Tetrahyromethanopterin
HSV, *see* Human herpesvirus
HSV-1, *see* Human herpesvirus type 1
HTLV-I, *see* Human T-cell leukemia virus I
Human cells, HSV-1 transcript abundance, 474, 478–479, 483, 486
Human cytomegalovirus
  genotype profiling, 472–474
  infection and latency, 450–451
  microarray analysis, 452–453
Human herpesvirus
  ease of study, 447–448
  genome maintenance, 449
  productive infection, 448–449
  reactivation, 449–450
  replication and pathogenesis, 448
Human herpesvirus type 1
  glass slide-based DNA microarrays
    chip data, 465–468
    chip fabrication and scanning, 464–465
    experimental data, 470–472
    hybridization, 455
    receiver operating characteristic, 465
    scanning protocols, 456–459
    specific oligonucleotides, 461–464
    transcript labeling, 455–456
    transcript abundance changes, 474, 478–479, 483, 486
Human immunodeficiency virus type 1, RT
  accessory proteins, 110–112
  amino acid role, 127–128
  Arg-72, 120–121
  enhanced fidelity, 129
  genetic assays, 106–108
  Gln-151, 119–120
  Lys-65, 120–121
  Met-184, 117–119
  minor groove binding track residues, 126
  misinsertion fidelity assay, 102–103
  mispair extension fidelity assay, 103–104
  mutational analysis overview, 112–113
  mutation rate variations, 130–131
  mutator, error catastrophe, 129–130
  pre-steady-state kinetic assays, 105
  primer grip residues, 124–126
  strand transfer fidelity, 109–110
  structure, 94–95
  template strand residues, 121–123
  Tyr-115, 114–117
  Tyr-183, 123–124
Human sphingosine kinases, 500–502
Human T-cell leukemia virus I, mutation rates, 98–99
Hybridization
  AAUAAA sequence, 302–303
  HSV-1 microarrays, 455
Hydrogen
  methanogens, H$_2$ + CO$_2$, 232–234, 256
  Methanosarcina mazei, 242–243
Hydrogenases, methanogens, 245–249
Hydrophobic gate, EF-Tu, 540–542

I

Imino ether groups, DNA–transplatin analogs, 40–41
Immunoglobulin heavy-chain, pre-mRNA, poly A sites
  CstF64 levels, 343–347
  poly A site competition, 341–342
  poly A site complex stability, 342–343
Infection
  cytomegaloviruses, 450–451
  human cytomegalovirus, 450–451
  human herpesvirus, 448–449
  murine cytomegalovirus, 450–451
  viral mutation rates, 98–99
Intron enhancer, poly A site choice, 339–341

J

JM216, *see bis*-Acetatoamminedichloro(cyclohexylamine) platinum(IV)
JM335, *see trans*-Amine(cyclohexylamine)dichlorodihydroxoplatinum(IV)
JNK, isoforms, 79

INDEX

K

Kaposi's sarcoma-associated herpesvirus, microarray analysis, 452
KIAA0277, 416
Kirromycin
  EF-1α, 537–538
  EF-Tu binding sites, 535–537
  EF-Tu resistance, 534–535
KSHV, see Kaposi's sarcoma-associated herpesvirus

L

Labeling, transcripts, HSV-1 microarrays, 455–456
Lacrimal gland, Muc4/SMC, 176
Liver, gene expression, 71–73
Long-oligonucleotides, herpesvirus DNA microarrays, 454
Lysine-65, HIV-1 RT, 120–121

M

Macromolecular ligands, EF-Tu interaction
  EF-Tu · EF-Ts, 521
  overview, 517
  ribosome, 521–524
  ternary complex, 520–521
Magnesium coordination network, EF-Tu, 527–529
Mammalian cells
  gene transcription, AMPK
    liver gene expression, 71–73
    muscle tissue gene expression, 77
    pancreatic β-cell, 74–77
  sphingosine kinase function, 497–498
Mammalian sphingosine kinases, cloning, 500–503
Mammary acinar cells, Muc4/SMC, 170–175
MAPKs, see Mitogen-activated protein kinases
Mast cells, mouse, CPII, SPHK, 498
Matrigel, Muc4/SMC, 172–173
MCF-7 cells, Muc4/SMC, 174
MCMV, see Murine cytomegalovirus
MCR, see Methyl-S-CoM reductase
Membrane anchoring, nsP1, 201–203
Membrane mucins
  characteristics, 150–151
  forms, 151–152
  members, 152–153
  transmembrane domain, 151
Messenger RNA
  26S mRNA, synthesis, 196–197
  polyadenylation
    cleavage mechanism, 299–301
    complexes, 302–304
    poly A(+), polymerases, 288–290
    3′ end formation, 298–299
    3′ end processing, 347–349
  polyadenylation, pre-mRNA
    calcitonin/CGRP, 337–341
    immunoglobulin heavy-chain, 341–347
    nuclei, 295–297
  poly A sequences, *Escherichia coli*
    decay, 370
    overview, 363–366
    PAP1, 366–367
    PAP1-stimulated mRNA decay, 370–372
    RNases degrading, 370
    target RNA decay, 367–369
  3′ end-processing factor, yeast genes
    CF1 complex, 354
    CF11 complex, 355–356
    conditional mutants, 358–359
    genetic approaches, 359
    mammalian homologs, 361–362
    overview, 353–354
    PF1 fraction, 356–357
    synergistic lethality, 359
    two-hybrid system, 360–361
  Vaccinia virus, poly A signals
    mechanism, 377–378
    VP39, 379–380
  yeast poly A signals, 350–353
Metastasis, Muc4/SMC, 160–161
Methane
  acetate conversion, 238
  methylated C_1 compounds, 235–238
Methanococcoides, energy conservation, 273–274
Methanofuran, methanogenesis, 228–233
Methanogenesis
  acetate conversion, 238
  cofactors, 228–232
  ecological role, 225–227
  formate, 234–235
  $H_2 + CO_2$, 232–234

Methanogenesis (cont.)
  methylated C_1 compounds, 235–238
  pathways, 238–240
Methanogens
  ATP synthesis, 240–242
  cellular characteristics, 227–228
  coenzyme F_{420}, 273
  formate, 234–235
  growth, proton translocation
    acetate, 256–258
    $H_2 + CO_2$, 256
    methylated C_1 compounds, 256
  membrane-bound methyltransferases, 259–261
  proton-translocating pyrophosphatases, 258–259
  taxony, 227–228
Methanosarcina
  ATP synthases, 265–270
  Ech hydrogenase, 248–249
  F_{420}-nonreducing hydrogenase, 246–248
  formyl-methanofuran dehydrogenase system, 261–265
  F_{420}-reducing hydrogenase, 245–246
Methanosarcina barkeri, heterodisulfide reductase, 253–255
Methanosarcinaceae
  acetate conversion, 238
  methane from methylated C_1 compounds, 235–238
Methanosarcina mazei
  $F_{420}H_2$ dehydrogenase, 249–253
  membrane-bound electron transport systems, 255
  proton-translocating pyrophosphatases, 258–259
  redox-driven proton dislocation, 242–245
  sodium ion pump, 260
Methanosarcina thermophila, heterodisulfide reductase, 253–255
Methanothermobacter strains
  F_{420}-nonreducing hydrogenase, 271–273
  heterodisulfide reductase, 271–273
Methionine
  residue 184, HIV-1 RT, 117–119, 123–124
  residue 230, Mo-MLV RT, 125–126
Methylase, 5′ cap, *Vaccinia virus* mRNA poly A signals, 379–380
Methyl-S-CoM reductase, methanogenesis, 238–240

Methyltransferases
  membrane-bound, methanogenic archaea, 259–261
  nsP1 activity, 198–201
MFR, *see* Methanofuran
Milk, Muc4/SMC, 170–175
Minus-strand RNA, nsP1 role, 203–204
Misinsertion fidelity assay, retroviral RT, 100–103
Mismatch repair, cisplatin-DNA adducts, 14
Mispair extension fidelity assay, retroviral RT, 103–104
Mitogen-activated protein kinase 2, 80
Mitogen-activated protein kinases
  overview, 78–81
  preproinsulin gene expression, 81–82
MMR, *see* Mismatch repair
Moloney murine leukemia virus, RT
  mutational analysis, Tyr-222, 124
  strand transfer fidelity, 109–110
Mo-MLV, *see* Moloney murine leukemia virus
Monodentate platinum(II) compounds–DNA interactions, 34–38
Mouse sphingosine kinases, cloning, 501–502
MR-guanine nucleotide exchange factor, 416
MUC1 forms, 151–152
Muc4/SMC
  antiadhesion agent, 155
  antirecognition agent, 155
  breast cancer, 170–175
  forms, 151–152
  isolation, 153
  ligand for ErbB2/HER2/Neu, 157–160
  mammary acinar cells, 170–175
  metastasis, 160–161
  milk, 170–175
  molecular cloning, 153–154
  mucociliary transport, 163–164
  ocular protection, 177
  oviduct, 167–170
  primary tumor growth, 161–163
  soluble form, 155–156
  soluble form production, 156–157
  uterus, 164–167
  vagina and cervix, 177
Mucociliary transport, Muc4/SMC, 163–164
Murine cytomegalovirus, infection and latency, 450–451

INDEX

Mutagenesis, retroviral RT, 93–94
Mutants, yeast mRNA 3′ end-processing factor genes, 358–359
Mutational analysis, HIV-1 RT fidelity
   amino acid role, 127–128
   Arg-72, 120–121
   Gln-151, 119–120
   Lys-65, 120–121
   Met-184, 117–119
   minor groove binding track residues, 126
   overview, 112–113
   primer grip residues, 124–126
   template strand residues, 121–123
   Tyr-115, 114–117
   Tyr-183, 123–124
Mutation rates, reverse transcriptase, 98–99, 130–131
Mutator reverse transcriptase, error catastrophe, 129–130

N

Na[trans-Ru(III)((CH$_3$)$_2$SO)Cl$_4$Im]–DNA interactions, 52
Neuropathogenicity, nsP2, 206–208
NLS, see Nuclear localization signal
Nonstructural proteins
   NsP1
      guanylyltransferase activity, 198–201
      membrane association, 201–203
      methyltransferase activity, 198–201
      minus-strand RNA synthesis, 203–204
   NsP2
      neuropathogenicity, 206–208
      NTPase activity, 204–205
      nuclear transport, 206–208
      protease activity, 206
      RNA helicase activity, 204–205
      RNA triphosphatase activity, 205–206
   NsP3
      features, 210
      phosphorylation, 209–210
      sequence conservation, 208–209
   NsP4, 210–211
   polyprotein P1234 processing, 197
NsP, see Nonstructural proteins
NTPase, see Nucleosidetriphosphatase
Nuclear exchange reaction, GEFs, 396, 401

Nuclear localization signal
   nsP2, 207
   poly A polymerase, 323
Nuclear transport, nsP2, 206–208
Nucleosidetriphosphatase, nsP2 activity, 204–205
Nucleotide-binding pocket, EF-Tu, 525–527
Nucleotide excision repair, cisplatin-DNA adducts, 13
Nucleotide selectivity, pre-steady-state kinetic assays, 104–105
Nucleus, pre-mRNA processing, 295–297

O

Obligate hydrogenotrophic methanogens, 270–274
Octahedral platinum(IV) complexes–DNA interactions, 32–33
Ocular protection, Muc4/SMC, 177–179
Oligonucleotides, HSV, DNA microarrays
   basic approach, 453–454
   chip data, 465–468
   chip fabrication and scanning, 464–465
   experimental data, 470–472
   hybridization, 455
   receiver operating characteristic, 465
   scanning protocols, 456–459
   specific oligonucleotides, 461–464
   transcript labeling, 455–456
Oviduct, Muc4/SMC, 167–170
Oxaliplatin–DNA interactions, 26–29
Oxoplatin–DNA interactions, 33

P

p38/MAPK
   identification, 79–80
   isoforms, 80–81
   preproinsulin gene regulation, 81–82
p53 protein, DNA adduct recognition, 21
PABII, see Poly A-binding protein
Palmitoylation, nsP1, 201–203
Pancreatic islet β-cells, MIN6
   AMPK-activated gene expression, 74–77
   preproinsulin gene expression, 81–82
PAP, see Poly A polymerase

PCF11 gene, yeast mRNA 3′ end-processing factor, 361
PCR, see Polymerase chain reaction
PDGF, see Platelet-derived growth factor
PDZ-guanine nucleotide exchange factors, 416–418
PEPCK, see Phospho *enol* pyruvate carboxykinase
Phenylalanine-160, HIV-1 RT, 114–117
Phospho *enol* pyruvate carboxykinase, 72–73
Phospholipase C
  PLC$_\epsilon$, 418–420
  PLCγ, SPHK activation, 499
Phospholipids, nsP1 activity, 203
Phosphorylation
  nsP3, 209–210
  poly A polymerase, 334–335
Phylogenetic analysis, sphingosine kinases, 508
Plants, SPHK and S1P, 497
Plasmids, DNA replication, PAP1 regulation, 366–367
Platelet-derived growth factor, SPHK activation, 499
trans-Platinum complexes–DNA interactions, 38–40
Platinum compounds–DNA interactions, 29–32
Platinum drugs–DNA interactions, 31
PLC, see Phospholipase C
Plus-strand RNA, alphavirus, synthesis, 194–196
Poly A-binding protein
  CF1$_m$ and CF11$_m$ cleavage factors, 331–333
  cleavage/polyadenylation, 317–318
  clones, 330–331
  mRNA 3′ end processing, 347–349
  sequence motifs and homologs, 331
Polyadenylation
  AAUAAA signal, 291–292
  adenovirus primary transcripts, 290
  cleavage, see Cleavage/polyadenylation
  complexes, 302–304
  core proteins, 319–320
  Escherichia coli, overview, 362–363
  Escherichia coli, poly A sequences
    degradosome, 372
    mRNA decay, 370
    mRNA-degrading RNases, 370
    overview, 363–366
    PAP1, 366–367
    PAP1-stimulated mRNA decay, 370–372
  second gene search, 372–374
  target RNA decay, 367–369
  mRNA 3′ end formation, 298–299
  mRNA 3′ end processing, 347–349
  poly A polymerase phosphorylation, 334–335
  poly A site choice, 335–336
  poly A site downstream sequences, 292–295
  poly A site processing
    calcitonin/CGRP pre-mRNA, 337–341
    immunoglobulin heavy-chain pre-mRNA, 341–347
    3′ noncoding exons, 336–337
  pre-mRNA processing, 295–297
  snRNP role, 304
  transcription
    coupling to 3′ end processing, 297–298
    initiation, 297
  Vaccinia virus
    mRNA, poly A signals, 377–380
    overview, 374–375
    poly A polymerase, 375–377
  yeast
    overview, 349–350
    poly A polymerase, 357–358
    poly A signals, 350–353
  yeast, 3′ end-processing factors
    CF1 complex, 354
    CF11 complex, 355–356
    conditional mutants, 358–359
    genetic approaches, 359
    mammalian homologs, 361–362
    overview, 353–354
    PF1 fraction, 356–357
    synergistic lethality, 359
    two-hybrid system, 360–361
Polyadenylation complexes, assembly from nuclear fractions, 309–311
Poly A polymerase
  cDNA clones, 320–322
  characterization, 316–317
  CPSF complex, 328
  CstF complex cDNA clones, 325–326
  CstF complex 50-kDa subunit, 326–327
  CstF complex 64-kDa subunit, 326
  CstF complex 77-kDa subunit, 327–328
  genomic cloning, 322
  phosphorylation, 334–335
  plasmid DNA replication, 366–367
  poly A-binding proteins, 330–333
  poly A(+) mRNA, 288–290

INDEX

second gene, 372–374
sequence motifs
  catalytic site, 323–324
  characteristics, 322–323
  NLS signal, 323
  RNA-binding domain, 324
  serine/threonine-rich domain, 324–325
  sequence motifs and homologs, 328–330
  stimulated mRNA decay, 370–372
  Vaccinia virus, 375–377
  yeast polyadenylation, 357–358
Poly A sequences
  Escherichia coli polyadenylation
    degradosome, 372
    mRNA decay, 370
    mRNA-degrading RNases, 370
    overview, 363–366
    PAP1, 366–367
    PAP1-stimulated mRNA decay, 370–372
    second gene search, 372–374
    target RNA decay, 367–369
  mRNA, polymerases, 288–290
Poly A sites
  choice regulation, 335–336
  coding exons
    calcitonin/CGRP pre-mRNA, 337–341
    immunoglobulin heavy-chain pre-mRNA, 341–347
  crosslinking to proteins, 309
  downstream sequences, 292–295
  3′ noncoding exons, 336–337
  UV crosslinking of proteins, 308
  Vaccinia virus mRNA, 377–380
  yeast, 350–353
Polymerase, drug-resistant viruses, 129
Polymerase chain reaction, HSV DNA microarrays, 453
Polynuclear platinum antitumor drugs–DNA interactions
  dinuclear compounds, 44–47
  overview, 42–44
  trinuclear compounds, 47–49
Polypeptides, EF-Tu, 514–515
Polyprotein P123, alphavirus replication complex, 211–212
Polyprotein P1234
  alphavirus replication complex, 211–212
  processing, 197
Polypyrimidine tract protein, intron enhancer, 339–340

Positive-strand RNA viruses, 190–191
PPI, *see* Preproinsulin
Pregnancy, virgin rat, Muc4/SMC, 172
Preproinsulin, pancreatic β-cell expression
  AMPK regulation, 74–75
  glucose regulation, 81–82
Pre-steady-state kinetic assays, retroviral RT, 104–105
Primer grip, HIV-1 RT, 124–126
Primer strand, HIV-1 RT
  minor groove binding track residues, 126
  primer grip residues, 124–126
  Tyr-183, 123–124
Protease, nsP2 activity, 206–208
Protein folding, EF-Tu, EF-Ts role, 529–531
Proton dislocation, redox-driven, *Methanosarcina mazei*, 242–245
Proton translocation, methanogens
  Ech hydrogenase, 248–249
  $F_{420}H_2$ dehydrogenase, 249–253
  F_{420}-nonreducing hydrogenase, 246–248
  F_{420}-reducing hydrogenase, 245–246
  growth, 256–258
  heterodisulfide reductase, 253–255
  pyrophosphatases, 258–259
PTB, *see* Polypyrimidine tract protein
[PtCl(dien)]Cl, *see* Chlorodiethylenetriamineplatinum(II) chloride
cis-[PtCl(NH$_3$)$_2$(A)]$^+$ compounds–DNA interactions, 34–37
trans-[PtCl$_2$(NH$_3$)(quinoline)], DNA modification, 38–40
trans-[PtCl$_2$(NH$_3$)(thiazole)], DNA modification, 38–40
Pulvomycin
  EF-Tu, 532–533
  EF-Tu binding sites, 535
  EF-Tu resistance, 535
Pyrophosphatases, proton-translocating type, 258–259

R

RalGDS, 420–421
RalGPS, 421–422
Rap1 guanine nucleotide factors, 420
Rap proteins, GRP/CalDAG-GEF family specificity, 411–412

Ras family-guanine nucleotide exchange factors
  BCAR3, 422–423
  C3G, 412–414
  disease relationship, 425–426
  dominant inhibitory Ras protein targeting, 402–404
  Epac/cyclic AMP-GEFs, 414–416
  GRASP-1, 412–414
  GRF1 and 2, 407–409
  GRP/CalDAG-GEF family, 409–412
  MR-GEF, 416
  nuclear exchange reaction, 396, 401
  PDZ-GEFs, 416–418
  phospholipase C_ϵ, 418–420
  RalGDS family, 420–421
  RalGPS, 421–422
  Rap1 GEFs, 420
  Ras-family, early identification, 393–396
  SmgGDS, 423–425
  Sos1 and 2, 405–407
  structure, 396, 401
Ras-GTP, regulation *in vivo*, 393
Ras proteins
  definition, 390–391
  dominant inhibitory, GEF targeting, 402–404
  subfamily activators, 391–393
Rat, Muc4/SMC, Muc4/SMC, 170–175
Receiver operating characteristic, viral microarrays, 465
Repac, 416
Replication
  alphavirus RNA
    genome complement, 192–194
    26S mRNA, 196–197
    plus-strand RNA synthesis, 194–196
  DNA, cisplatin adduct effects, 10
Replication complex, alphavirus, 211–214
Replication cycle, alphaviruses, 188–190
Resistance, antibiotic, EF-Tu, 534–535
Retroviral mutagenesis, RT, 93–94
Retroviral reverse transcriptase, HIV-1
  amino acid role, 127–128
  Arg-72, 120–121
  enhanced fidelity, 129
  fidelity, accessory proteins, 110–112
  fidelity at initiation, 108–109
  fidelity *in vitro* assays, 99–100
  genetic assays, 105–108
  Gln-151, 119–120
  Lys-65, 120–121
  Met-184, 117–119
  minor groove binding track residues, 126
  misinsertion fidelity assay, 100–103
  mispair extension fidelity assay, 103–104
  mutagenesis, 93–94
  mutational analysis overview, 112–113
  mutation rates, 98–99
  mutation rate variations, 130–131
  mutator RT, error catastrophe, 129–130
  pre-steady-state kinetic assays, 104–105
  primer grip residues, 124–126
  proviral DNA, 97
  strand transfer fidelity, 109–110
  structure, 94–97
  template strand residues, 121–123
  Tyr-115, 114–117
  Tyr-183, 123–124
Reverse transcriptase, *see* Retroviral reverse transcriptase
Ribonucleoside triphosphate, RT fidelity, Tyr-115, 114–117
Ribosomes–EF-Tu interactions, 521–524
RNA
  alphavirus, replication
    genome complement, 192–194
    26S mRNA, 196–197
    plus-strand RNA synthesis, 194–196
  CPSF binding, 315–316
  defective-interfering, alphavirus, 192–194
  messenger, *see* Messenger RNA
  minus-strand, nsP1 role, 203–204
  plus-strand, alphavirus, 194–196
  Semliki Forest virus, 188–190
RNA1, poly A sequences, 367–369
RNA-binding domain, poly A polymerase, 324
RNA-capping pathway, nsP1, 198–201
RNA helicase, nsP2 activity, 204–205
RNA polymerase II, retroviral mutagenesis, 93–94
RNases, mRNA-degrading, *Escherichia coli*, 370
RNA templates, misinsertion fidelity assay, 103
RNA triphosphatase, nsP2 activity, 205–206
RNA viruses, positive-strand, members, 190–191
ROC, *see* Receiver operating characteristic

INDEX

RT, see Retroviral reverse transcriptase
cis-[Ru(II)(bpy)$_2$Cl$_2$]–DNA interactions, 53
cis-[Ru(II)((CH$_3$)$_2$SO)$_4$Cl$_2$]–DNA interactions, 49–52
trans-[Ru(II)((CH$_3$)$_2$SO)$_4$Cl$_2$]–DNA interactions, 49–52
{cis-Ru(II)((CH$_3$)$_2$SO)$_3$Cl$_2$}NH$_2$(CH$_2$)$_4$NH$_2${cis-PtCl$_2$(NH$_3$)}–DNA interactions, 53–54
[Ru(II)Cl(bpy)(terpy)]Cl–DNA interactions, 53
mer-[Ru(II)(terpy)Cl$_3$]–DNA interactions, 53
Ruthenium(III) heterocyclic complexes–DNA interactions, 52

S

S1P, see Sphingosine-1-phosphate
S1PR, see Sphingosine-1-phosphate receptor
Saccharomyces cerevisiae, SPHK and S1P, 496–497
Salivary glands, Muc4/SMC, 175
SAPKs, see Stress-activated protein kinases
Scanning, HSV-1 microarrays
  chips, 464–465
  protocols, 456–459
Semliki Forest virus
  26S mRNA synthesis, 196
  nsP1, minus-strand RNA synthesis, 203
  nsP2, neuropathogenicity, 207–208
  nsP2, RNA helicase, 204–205
  nsP3, 209–210
  replication cycle, 188–190
  RNA complementary to genome, 192–194
Serine, nsP3 phosphorylation, 209–210
Serine/threonine-rich domain, poly A polymerase, 324–325
SFV, see Semliki Forest virus
SHEP1, 422–423
SIN, see Sindbis virus
Sindbis virus
  26S mRNA synthesis, 196
  nsP3, 209–210
  nsP4, 211
  RNA complementary to genome, 192–194
Skeletal muscle, AMPK-activated gene expression, 77
Small-molecular-weight G protein guanine nucleotide-dissociation stimulator, 423–425

SmgGDS, see Small-molecular-weight G protein guanine nucleotide-dissociation stimulator
SNF1, see Sucrose nonfermenting-1 complex
SnRNP
  CPSF, 314–315
  3′ end-processing complexes, 304
Sodium ion pump, methanogenic archaea, 259–261
Sos1, 405–407
Sos2, 405–407
Sphingosine kinase
  activation, 498–499
  cloning, 500–503
  mammalian cells, 497–498
  phylogenetic analysis, 508
  plants, 497
  SPHK family, 503–504
  SPHK superfamily domains, 504–508
  yeast, 495–497
Sphingosine-1-phosphate
  function in mammalian cells, 497–498
  plants, 497
  pleiotropic functions, 494–495
  yeast, 495–497
Sphingosine-1-phosphate receptor, 494–495
SPHK, see Sphingosine kinase
Strand transfer, RT fidelity, 109–110
Stress-activated protein kinases
  overview, 78–81
  preproinsulin gene expression, 81–82
Structural proteins, 26S mRNA, 196–197
Structure–function relationships, EF-Tu
  catalytic domain 1, 524
  magnesium coordination network, 527–529
  nucleotide-binding pocket, 525–527
  regulatory domains 2 and 3, 525
Sucrose nonfermenting-1 complex, glucose repression in yeast, 71
Synergistic lethality, yeast mRNA 3′ end-processing factor genes, 359

T

T4 endonuclease VII, DNA adduct recognition, 21
TATA-binding protein, DNA adduct recognition, 18
TBP, see TATA-binding protein

Telomerase, DNA adduct effects, 23
Template strand, HIV-1 RT, 121–123
Tetrahyromethanopterin, methanogenesis, 228–232, 234–235
TGFβ, see Transforming growth factor β
3′ end processing
  mRNA, polyadenylation, 298–299, 347–349
  polyadenylation, coupled transcription, 297–298
  snRNP role, 304
  yeast mRNA
    CF1 complex, 354
    CF11 complex, 355–356
    conditional mutants, 358–359
    genetic approaches, 359
    mammalian homologs, 361–362
    overview, 353–354
    PF1 fraction, 356–357
    synergistic lethality, 359
    two-hybrid system, 360–361
3′ noncoding exons, poly A site, 336–337
Threonine, nsP3 phosphorylation, 209–210
Topoisomerase, DNA adduct effects, 23–24
Transcription
  DNA, adduct effects, 10–11
  genes, see Gene transcription
  herpesvirus, microarray analysis, 452
  polyadenylation
    coupling to 3′ end processing, 297–298
    initiation, 297
    pre-mRNAs in nuclei, 295–297
  RT fidelity, accessory proteins, 110–112
  RT fidelity at initiation, 108–109
Transforming growth factor β, Muc4/SMC, 172–173
Translation initiation factor, eIF2α, poly A site, 336–337
Transplatin–DNA
  adducts, 4, 11
  crosslinks, 8–9
  imino ether groups, 40–41
  planar amine ligand, 38–40
Tumors, Muc4/SMC
  metastasis, 160–161
  primary growth, 161–163
Two-hybrid system, yeast mRNA 3′ end-processing factor genes, 360–361

Tyrosine
  residue 115, HIV-1 RT, 114–117
  residue 183, HIV-1 RT, 123–124
  residue 222, Mo-MLV RT, 124
Tyrosine receptor kinase, ErbB2/HER2/Neu, Muc4/SMC ligand, 157–160

U

Ultraviolet irradiation, protein crosslinking to poly A sites, 308
Uterus, Muc4/SMC, 164–167

V

Vaccinia virus, polyadenylation
  mRNA, poly A signals, 377–380
  overview, 374–375
  poly A polymerase, 375–377
Vacuoles, cytoplasmic
  alphavirus plus-strand RNA, 195–196
  alphavirus replication complex, 212, 214
Vagina, Muc4/SMC, 177
Viral protein, mRNA 3′ end processing, 347–349
Viral transcripts
  HSV-1, abundance changes, 474, 478–479, 483, 486
  microarray analysis, 455–456, 461–464
Viruses, drug-resistant, enhanced polymerase fidelity, 129
VP39, *Vaccinia virus*
  mRNA poly A signals, 379–380
  poly A polymerase, 376
VP55, *Vaccinia virus* mRNA poly A signals, 377–378

Y

Y-box binding protein 1, 18
Yeast
  glucose repression, SNF1, 71
  polyadenylation
    overview, 349–350
    poly A polymerase, 357–358

poly A signals, 350–353
polyadenylation, 3' end-processing
  CF1 complex, 354
  CF11 complex, 355–356
  conditional mutants, 358–359
  genetic approaches, 359
  mammalian homologs, 361–362
  overview, 353–354
  PF1 fraction, 356–357

synergistic lethality, 359
two-hybrid system, 360–361
SPHK and S1P, 495–497

Z

Zidovudine, RT mutation rates, 130–1311

ISBN 0-12-540071-3